Graduate Texts in Physics

Graduate Texts in Physics publishes core learning/teaching material for graduate- and advanced-level undergraduate courses on topics of current and emerging fields within physics, both pure and applied. These textbooks serve students at the MS- or PhD-level and their instructors as comprehensive sources of principles, definitions, derivations, experiments and applications (as relevant) for their mastery and teaching, respectively. International in scope and relevance, the textbooks correspond to course syllabi sufficiently to serve as required reading. Their didactic style, comprehensiveness and coverage of fundamental material also make them suitable as introductions or references for scientists entering, or requiring timely knowledge of, a research field.

Stephen P. Martin

Quantum Mechanics

A Comprehensive Graduate Text

Springer

Stephen P. Martin
Physics Department
Northern Illinois University
DeKalb, IL, USA

ISSN 1868-4513 ISSN 1868-4521 (electronic)
Graduate Texts in Physics
ISBN 978-3-032-21118-7 ISBN 978-3-032-21119-4 (eBook)
https://doi.org/10.1007/978-3-032-21119-4

This Springer imprint is published by the registered company Springer Nature Switzerland AG
The registered company address is: Gewerbestrasse 11, 6330 Cham, Switzerland

Preface

This book is intended as a text for a core graduate course on quantum physics. The modern form of quantum mechanics emerged in a remarkable burst of activity from 1924 to 1930. In the subsequent century, there have been many excellent textbooks written on the topic, and one might well wonder why someone would decide to create another. The short answer is this: I've attempted to imagine, and then write, the book that would have made me happiest to read, if I were a graduate student today. Whether my choices of topics, explanations, and derivations work well for someone else is an entirely different and personal question, of course!

Although this book aims to be a self-contained discussion, starting from the basics, it will probably be more readily accessible for those who have already taken undergraduate courses in classical mechanics, electromagnetism, and quantum mechanics. It is also assumed that the reader is already familiar with matrices and linear algebra, multivariable calculus, simple differential equations, the algebra of complex numbers, and basic concepts of probability.

Quantum mechanics is a big subject, and it is not possible to cover everything. So I didn't. Even so, there is more material in this book than can be reasonably covered in a typical academic year of lectures. This is intentional, because every course and every student will have different priorities. I did consider it essential to include certain modern subjects that are not always covered in older books at this level, notably the hidden-variables alternative, Bell inequalities, entangled subsystems and open systems, generalized measurements, decoherence, and quantum information. The instructor will likely have to make some hard choices of what to leave out, informed by the interests and preparation of the students.

It would be impossible to draw up a universal plan for a course based on this book, appropriate for everyone. But at least I can offer the instructor and student a guide to those chapters and sections that are *not* prerequisites for later parts of the text. These can therefore be covered as topics of special interest, or not at all. In order of appearance:

- Sections 5.7 and 8.8 on motion reversal (the transformation) and time reversal (the corresponding symmetry), which require antilinear and antiunitary operators
- Section 9.3 on Landau levels for a charged particle in a magnetic field
- Section 11.2 on the unbound states of Coulomb potentials
- Section 15.4 on Brillouin–Wigner perturbation theory

- Section 15.8 on almost-degenerate perturbation theory
- Section 18.4 on multi-electron atoms
- Chapter 19 on periodic potentials, Bloch waves, the Kronig–Penney model, and electronic band structure
- Chapter 20 on the Born–Oppenheimer approximation, simple molecules, and quantization of rigid bodies
- Chapter 23 on absorption and emission of light
- Chapters 25 and 26 on the related topics of entanglement, subsystems and open systems, dynamical maps, the Lindblad equation, decoherence, and the emergence of classical behavior
- Chapter 27, which is an invitation to the ideas of quantum information
- Chapter 28 on relativistic quantum mechanics and the Dirac equation
- Chapter 29 on the Feynman path integral approach, including the Ehrenberg–Siday–Aharonov–Bohm effect and Dirac magnetic monopoles

The exercises at the end of each chapter vary in length and difficulty. For this I make no apology, since at the wild frontiers of research, problems do not present themselves with convenient labels saying "I am easy, you can solve me in one line" or "I am challenging, good luck." With that said, none of the exercises given here are remotely close to research level, and each has been chosen to reinforce some concept or method.

A few brief words about conventions and notations are in order. Vectors in three-dimensional space are denoted by boldface, except unit vectors, which carry hats. Thus, $\boldsymbol{r} = \hat{x}x + \hat{y}y + \hat{z}z$ is a position vector. Operators do not carry hats, but are usually distinguished by capitalization, except when long-standing tradition says otherwise. Thus, $\boldsymbol{R} = \hat{x}X + \hat{y}Y + \hat{z}Z$ is the vector position operator for a particle in three dimensions, with individual component operators X, Y, and Z, but a and $a^\dagger$ are the ladder operators for the harmonic oscillator. When electromagnetic quantities play a role, I use Gaussian cgs units, avoiding the execrable ϵ_0 and μ_0 of the SI units system. However, translations between the two systems do appear, for example in the very first equation of the text.

I am grateful to the many students who provided feedback, with special thanks to those who pointed out corrections to earlier drafts: Sameneh Ahmadinejad, Jared Coles, Matthew Dudak, William Emark, David Iglesias Tinoco, Abdulrahman Kauther, Spencer Kelham, Mark Mekosh, Dillon Merenich, Olabisi Olayinka, Zhenghao Pan, Cassandra Phillips, Deeksha Sinha, Vladimir Tsitrin, and Tasfia Yeashna.

Finally, I thank my wife Jeanette for her constant support, incredible patience, and endless encouragement.

DeKalb, IL, USA
January 2026

Stephen Martin

Contents

Introduction: The Quantum Revolution 1

1.1 Classical Instability of Charged Matter

Quantum mechanics is often portrayed as nonintuitive, weird, or even paradoxical. We begin with a retort: it is actually classical physics that is conspicuously incompatible with basic features of the world we live in. Most strikingly, if classical mechanics governed the universe, then matter made up of charged particles would necessarily be unstable against rapid and catastrophic collapse.

Consider, for example, a classical model of the hydrogen atom, consisting of a point-like[1] electron and a much heavier proton, separated by a distance r. The electric potential energy is

$$V(r) = -\frac{Q_e^2}{4\pi\epsilon_0 r} = -\frac{e^2}{r}, \tag{1.1}$$

where in SI units, $Q_e = -1.60218 \times 10^{-19}\,\mathrm{C}$ is the electronic charge and $\epsilon_0 = 8.85419 \times 10^{-12}\,\mathrm{C}^2/\mathrm{N}\cdot\mathrm{m}^2$ is the permittivity of free space. The positive quantity e, which will appear often in this book, is equal to the proton's charge in the Gaussian (cgs) system of units. Its square is given numerically by

$$e^2 = 1.43996 \times 10^{-9}\,\mathrm{eV}\cdot\mathrm{m} = 2.30708 \times 10^{-28}\,\mathrm{J}\cdot\mathrm{m}. \tag{1.2}$$

[1] All experiments to the present day are consistent with the electron having no substructure. Protons and nuclei are certainly not point-like, as they are now known to be composed of quarks and gluons. However, the important thing for the following discussion is just that they are tiny ($\sim 10^{-15}$ m) compared to atoms ($\sim 10^{-10}$ m). This had become apparent by 1911, before the development of modern quantum theory, from the results of Ernest Rutherford's experiments with Hans Geiger and Ernest Marsden in which alpha particles were observed scattering at large angles off of gold nuclei.

S. P. Martin, *Quantum Mechanics*, Graduate Texts in Physics,
https://doi.org/10.1007/978-3-032-21119-4_1

Clearly, classical physics has a serious problem: the potential $V(r)$ is unbounded from below as r approaches 0, implying that a classical atom should release an arbitrarily large amount of energy as it shrinks to zero size.

While this may seem dangerous (or possibly useful, depending on your imagination!), it is certainly not what is observed. One might suppose that safety could be achieved by somehow forcing the electron to travel in a fixed orbit about the much heavier proton. However, this cannot work in a classical theory, because of energy conservation. Maxwell's equations imply that the classical electron will continuously lose energy in the form of electromagnetic radiation, due to its centripetal acceleration. For a circular orbit, the acceleration is

$$a = \frac{v^2}{r} = \frac{F}{m_e} = \frac{e^2}{m_e r^2}, \tag{1.3}$$

where $m_e = 9.109384 \times 10^{-31}$ kg $= 0.510999$ MeV$/c^2$ is the electron's mass, with $c = 2.99792458 \times 10^8$ m/sec, the speed of light in vacuum. The Larmor formula for the radiated power of an accelerating charge,

$$P = -\frac{dE}{dt} = \frac{Q_e^2 a^2}{6\pi\epsilon_0 c^3} = \frac{2e^2a^2}{3c^3}, \tag{1.4}$$

therefore tells us that the classical electron must lose energy at a rate proportional to $1/r^4$, causing its orbit size to shrink at an ever-increasing rate.

Just for fun, let us estimate the tragic fate of the electronic orbit in this classical model, making some simplifying assumptions. If the orbit stays nearly circular, and nonrelativistic, as it decays, then the energy will be

$$E = \frac{1}{2}m_e v^2 - \frac{e^2}{r} = -\frac{e^2}{2r}. \tag{1.5}$$

Combining (1.4) and (1.5) gives

$$3r^2\frac{dr}{dt} = -K, \tag{1.6}$$

where $K \equiv 4e^4/m_e^2c^3$ is a constant. If $r = r_0$ at $t = 0$, this integrates to $r^3 - r_0^3 = -Kt$, so

$$r = r_0\left(1 - Kt/r_0^3\right)^{1/3}. \tag{1.7}$$

This shows that the decay of the classical electron's orbit is even worse than asymptotic; it collapses all the way to $r = 0$ in a finite time r_0^3/K, which turns out to be very short (see Exercise 1.1 to find out just how short). Heavier atoms, molecules, and larger structures of electrons and nuclei would have similar instabilities if classical physics governed them.

Quantum mechanics resolves this catastrophe, and allows matter composed of charged-particle constituents to be stable, by changing the rules. In the quantum theory, there is no counterpart to the decaying classical orbit with unbounded negative energy. In particular, there are no physical quantum states of the hydrogen atom with arbitrarily low energy, corresponding to a classical electron localized arbitrarily near $r = 0$. Instead, as we will see, there is a single state with the lowest possible energy (about 13.6 eV below a state of ionization), which therefore is stable since there is no lower energy state into which it could decay by emitting electromagnetic radiation. In this way, quantum mechanics saves the universe. More generally, in quantum mechanics the energies of bound states turn out to be quantized (discrete). There are also unbound (ionized) states with a continuum of allowed energies, but those energies are bounded from below by zero.

1.2 Black-Body Radiation and Planck's Constant

In the late 19th century, it became apparent that classical physics theory makes a similarly catastrophic prediction for electromagnetic radiation. Consider a cavity whose walls are as close as possible to idealized black (a perfect absorber and emitter of electromagnetic radiation), heated as an oven so that it is kept in thermal equilibrium at temperature T with the electromagnetic radiation inside it. The spectrum of the electromagnetic radiation is characterized by the energy density ρ_E, per unit volume V and per unit frequency ν, in terms of which the total energy inside the cavity is

$$\text{Total energy} = V \int_0^\infty d\nu\, \rho_E(\nu, T). \tag{1.8}$$

The function $\rho_E(\nu, T)$ can be determined experimentally by making a small hole in the walls of the cavity and measuring the radiation that escapes, analyzed for different frequencies using diffraction gratings, for example.

To obtain a theoretical prediction for $\rho_E(\nu, T)$, we first need to quantify the density of electromagnetic modes per unit frequency. For simplicity, assume that the cavity is a cubic box of side L, and that the electromagnetic radiation modes satisfy periodic boundary conditions, with allowed wavevectors $\boldsymbol{k} = (2\pi/L)\boldsymbol{n}$, where $\boldsymbol{n} = \hat{x}n_x + \hat{y}n_y + \hat{z}n_z$ for integers n_x, n_y, n_z. The corresponding wavelengths and frequencies are

$$\lambda = L/n, \qquad \nu = cn/L, \tag{1.9}$$

where $n = |\boldsymbol{n}| = \sqrt{n_x^2 + n_y^2 + n_z^2}$. Because the allowed n_x, n_y, and n_z are integers, the number of modes with frequency between ν and $\nu + d\nu$ can be enumerated by integrating the volume in $\boldsymbol{n}$ space, using radial coordinate n, with

$$d^3\boldsymbol{n} \to 4\pi n^2 dn = \frac{4\pi L^3}{c^3}\nu^2 d\nu. \tag{1.10}$$

So, if $\overline{E}$ is the average energy of a mode with frequency ν when in thermal equilibrium with the walls at temperature T, then the total energy for frequencies between ν and $\nu + d\nu$ is

$$2\overline{E}\,\frac{4\pi L^3}{c^3}\nu^2 d\nu, \tag{1.11}$$

where the first factor of 2 accounts for the fact that each electromagnetic mode can have two transverse polarizations. Setting (1.11) equal to $L^3 \rho_E\, d\nu$ in accord with (1.8), we find

$$\rho_E(\nu, T) = \frac{8\pi}{c^3}\nu^2\,\overline{E}. \tag{1.12}$$

We next need to evaluate $\overline{E}$ as a function of temperature.

A fundamental prediction of statistical mechanics is that for an ensemble of identical systems in thermal equilibrium, the probability for one of the systems to have energy E is proportional to the **Boltzmann factor**, named after Ludwig Boltzmann,

$$P(E) \propto e^{-E/k_B T}, \tag{1.13}$$

where T is the temperature in Kelvin, and Boltzmann's constant

$$k_B = 8.617333 \times 10^{-5}\,\text{eV/K} = 1.380649 \times 10^{-23}\,\text{J/K} \tag{1.14}$$

is the conversion factor between units of temperature and energy. If the allowed energies of electromagnetic modes are continuous and unrestricted, we therefore have the simple classical prediction for the average energy,

$$\overline{E} = \left(\int_0^\infty dE\, e^{-E/k_B T}\, E\right) \Big/ \int_0^\infty dE\, e^{-E/k_B T} = k_B T, \tag{1.15}$$

independent of ν. Plugging this into (1.12) gives the result

$$\rho_E(\nu, T) = 8\pi k_B T \nu^2/c^3. \tag{1.16}$$

This is the Rayleigh–Jeans prediction, developed by John W. Strutt, 3rd Baron Rayleigh, and James Jeans. Even without consulting experimental data, this formula is clearly problematic, since it claims that the energy density grows quadratically with frequency. After integrating $\int_0^\infty d\nu\, \rho_E(\nu, T)$, we would find an infinite total energy density per unit volume in black-body radiation. This impossible prediction of classical physics is called the **ultraviolet catastrophe**.

In order to explain the existing data and avoid the ultraviolet catastrophe, Max Planck proposed in 1900 that the black-body walls can only emit modes with frequency ν if the energy is equal to an integer multiple of $h\nu$, where, using modern data,

$$h = 4.1356677 \times 10^{-15}\,\text{eV}\cdot\text{s} = 6.62607015 \times 10^{-34}\,\text{J}\cdot\text{s} \tag{1.17}$$

is known as **Planck's constant**. In that case, the integrals in (1.15) are replaced by sums over only the discrete allowed energies,

$$\overline{E} = \Big(\sum_{n=0}^{\infty} nh\nu\, e^{-nh\nu/k_BT}\Big)\Big/\sum_{n=0}^{\infty} e^{-nh\nu/k_BT} = \frac{h\nu}{e^{h\nu/k_BT}-1}. \tag{1.18}$$

[To get the last expression, note that the sum in the denominator of the middle expression is a geometric series $f(x) = \sum_{n=0}^{\infty} e^{-nx} = 1/(1-e^{-x})$, and the numerator is proportional to $-f'(x) = e^{-x}/(1-e^{-x})^2$.] For very small $h\nu/k_BT$, this expression for $\overline{E}$ agrees with (1.15). However, unlike that formula, it depends on the frequency, and is exponentially suppressed in the ultraviolet limit of large $h\nu/k_BT$. Putting (1.18) into (1.12) yields[2]

$$\rho_E(\nu, T) = \frac{8\pi h}{c^3}\frac{\nu^3}{e^{h\nu/k_BT}-1}. \tag{1.19}$$

Figure 1.1 compares Planck's formula (1.19) with the Rayleigh–Jeans prediction (1.16), for black-body radiation with $T = 1000$ K. In the far infrared, the

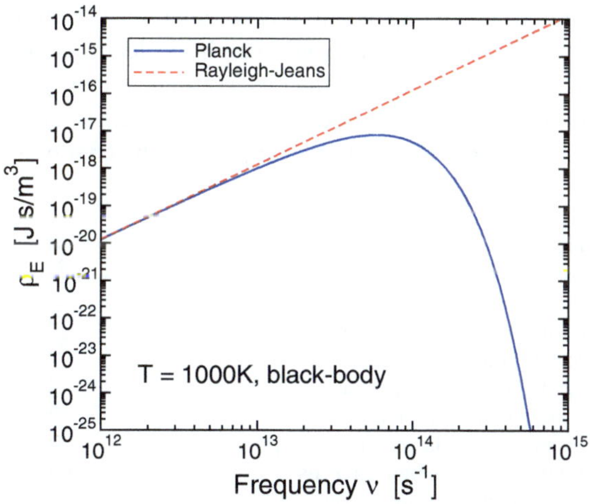

Fig. 1.1 Comparison of the classical Rayleigh–Jeans (dashed line) and Planck (solid curve) predictions for the electromagnetic radiation energy per unit volume and unit frequency, ρ_E, in a black-body cavity, as a function of the frequency ν, for temperature $T = 1000$ K

[2] Historically, Planck's original proposal of this formula was simply an inspired guess to fit the data, and his subsequent derivation of it was not particularly compelling by modern standards, but the result was correct. The origin of (1.18) from deeper principles is derived in Sect. 3.5, see (3.103).

two results agree, but they differ significantly in the near infrared, and the Planck prediction for ρ_E decreases with frequency above $\nu = 5.9 \times 10^{13}$ Hz. In the visible range of ν between about 4×10^{14} and 8×10^{14} Hz, the Planck formula is many orders of magnitude smaller, in accord with observation. Furthermore, the Planck prediction for the total energy density per unit volume is finite,

$$\int_0^\infty d\nu\, \rho_E(\nu, T) = \frac{8\pi^5 (k_B T)^4}{15 h^3 c^3}, \tag{1.20}$$

resolving the ultraviolet catastrophe. The agreement of this prediction with observation provided evidence that the electromagnetic radiation modes inside the cavity are quantized in energy units equal to Planck's constant multiplied by the frequency.

1.3 Photoelectric Effect and Particle-Like Features of Light

The photoelectric effect is the ejection of electrons from the surface of a material when light shines on it. This is observed to occur if the angular frequency ω of the light is sufficiently high. For ω less than a certain threshold value characteristic of the materials used in the experiment, almost no electrons are ejected, even as the intensity of the light is increased. The threshold angular frequency $\omega_{\min}$ typically corresponds to visible or ultraviolet light. For $\omega > \omega_{\min}$, the maximum kinetic energy of ejected electrons is observed to rise linearly, as sketched in Fig. 1.2.

This behavior was unexpected when it was first found, because in classical electrodynamics, the frequencies and energies of electromagnetic waves are continuous and independent of each other. One might have supposed that increasing the intensity of the light would result in electrons being ejected for any ω, no matter how small. To explain the observations, Albert Einstein proposed in 1905 that light of a given angular frequency always occurs in discrete chunks, or quanta, which are now called photons. For each photon, the energy is related to the angular frequency

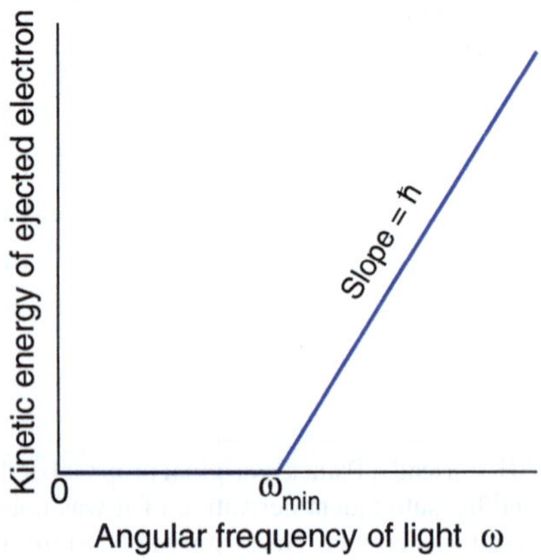

Fig. 1.2 Sketch of results for the photoelectric effect. Shining electromagnetic radiation on the surface of a metal produces a current of ejected electrons, but to a good approximation this occurs only if the angular frequency of the light exceeds a minimum value $\omega_{\min}$. The maximum kinetic energy of the ejected electrons then rises like $\hbar(\omega - \omega_{\min})$

by the same formula proposed by Planck, which can be rewritten as

$$E = \hbar\omega, \tag{1.21}$$

where $\omega = 2\pi\nu$, and the **reduced Planck's constant** (or just "h bar", when speaking) is defined to be related to the ordinary Planck's constant by

$$\hbar = \frac{h}{2\pi} = 6.58212 \times 10^{-16}\,\text{eV}\cdot\text{s} = 1.05457 \times 10^{-34}\,\text{J}\cdot\text{s}. \tag{1.22}$$

Einstein's quantization condition conceptually generalized Planck's proposal, which was only intended to apply to electromagnetic modes absorbed and emitted by the black-body cavity.

It follows from special relativity and wave kinematics that the momentum p and the wavelength λ of each photon are related by

$$p = E/c = \hbar\omega/c = 2\pi\hbar/\lambda. \tag{1.23}$$

In the photoelectric effect, the discrete particle-like nature of light explains the existence of ω_{min}, because $\hbar\omega_{\text{min}}$ is the minimum energy jump needed for the electron to escape the metal and be collected on another plate when it is struck by a single photon.[3] Experiments reported by Robert A. Millikan in 1916 later verified, using sodium and lithium as the targets, that the maximum kinetic energy of the ejected electron indeed behaves like

$$E_e = \hbar(\omega - \omega_{\text{min}}). \tag{1.24}$$

In practice, the measured value of ω_{min} depends on the collector plate material as well as the material being illuminated, but the most important result of the experiment is the slope of the line in Fig. 1.2, which is $\hbar$. Millikan's measured value for $\hbar$ was consistent with Planck's result from black-body radiation, but was significantly more accurate and precise.

Although light behaves like a particle in the photoelectric effect, it also has interference properties governed by the wavelength. These effects appear in interference and diffraction experiments like the double-slit experiment illustrated in Fig. 1.3. Light from a coherent source passing through holes or slits in a screen yields a pattern of intensity maxima and minima, where the interference of amplitudes is constructive and destructive, respectively. However, the interpretation of this effect is slightly different in quantum mechanics than in the corresponding classical theory. Classically, the intensity is proportional to the magnitude of the time-averaged Poynting vector,

[3] Actually, two photons can team up to eject a single electron, as has been observed in experiments with high-power lasers, but the rate for this is very small except when the intensity is extremely large.

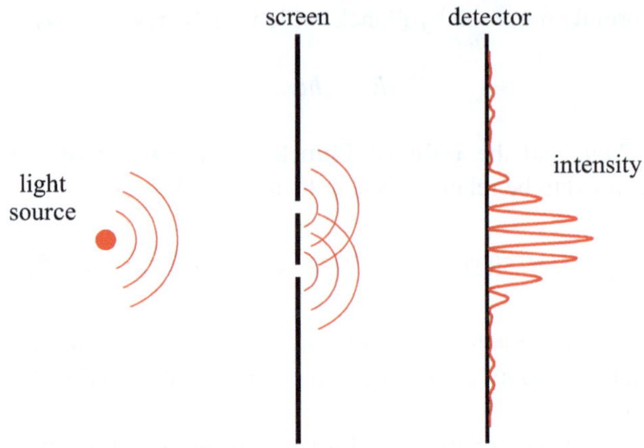

Fig. 1.3 The wave nature of electromagnetic radiation causes interference and diffraction phenomena when light from a coherent source passes through holes or slits in a screen

$$\text{Classical intensity} \propto |\boldsymbol{E} \times \boldsymbol{B}|, \tag{1.25}$$

where $\boldsymbol{E}$ and $\boldsymbol{B}$ are the electric and magnetic fields, which each obey superposition and interference due to the linearity of Maxwell's equations. In the quantum theory, the *real* quantity $\boldsymbol{E} \times \boldsymbol{B}$ is replaced by the squared magnitude of a *complex* quantity ψ,

$$\text{Quantum intensity} \propto |\psi|^2. \tag{1.26}$$

Here $\psi(x, y, z, t)$ is called a **wavefunction**, and it is a probability density amplitude. This means that $|\psi(x, y, z, t)|^2$ is proportional to the probability to detect a photon near position (x, y, z) at time t. It is a fundamental feature of quantum mechanics that we must deal with probabilities, rather than definite outcomes for individual events.

1.4 Electron Diffraction and Wave-Like Features of Matter

Inspired by the dual wave-particle nature of light exhibited in the photoelectric effect, Louis de Broglie in his 1924 PhD thesis suggested that matter particles, such as electrons, also have wavefunctions $\psi(x, y, z, t)$ that are subject to interference. He proposed that what is now called the **de Broglie wavelength** λ (and the corresponding wavenumber k) of the wavefunction should be related to the momentum of a particle in the same way as for light, (1.23). That is, for a particle with mass m and energy $E = \sqrt{p^2c^2 + m^2c^4}$,

$$\lambda \equiv 2\pi/k = 2\pi\hbar/p. \tag{1.27}$$

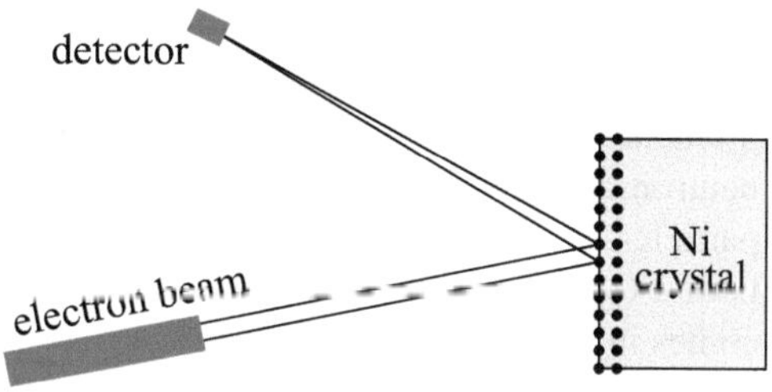

Fig. 1.4 Schematic setup for the Davisson–Germer electron diffraction experiments

De Broglie's wave hypothesis for matter was verified by the discovery of electron diffraction in a series of experiments by Clinton Davisson and Lester Germer from 1923–1928.

In the Davisson–Germer experiments, electrons with controlled energies (tens of eV) were made to hit a nickel crystal[4] target, and the intensity of electrons scattered at various fixed angles was measured, as shown schematically in Fig. 1.4. Their data clearly showed maxima and minima of constructive and destructive interference, depending on the differences in path lengths of the electrons scattering from the regularly spaced face-centered cubic crystal sites, with a lattice spacing of about 3.52×10^{-10} m, and a nearest-neighbor distance a factor of $\sqrt{2}$ smaller. This interference effect was similar to the previously known phenomenon of Bragg peaks in X-ray scattering. The electrons' wavelength corresponded to the prediction of the de Broglie relation to momentum in (1.27). This provided the first direct evidence that matter particles are also described by a wavefunction that can be interpreted as a probability amplitude, with the crucial feature that this amplitude obeys superposition and interference. It also provided another example of the central role of Planck's constant in the quantum theory, through the universal connection between momentum, wavelength, and $\hbar$ in (1.27).

1.5 Spin and the Stern–Gerlach Experiment

In classical mechanics, angular momentum takes on continuous values, but in quantum mechanics it always occurs in multiples of a fundamental unit. For angular momentum associated with the motion of particles, called **orbital angular momentum**, the fundamental quantum unit is $\hbar$, as we will show in Sect. 5.3. However, quantum mechanics also allows for **intrinsic angular momentum**, or **spin**, which has no classical counterpart. The projection of a spin measured along any direction is always an integer multiple of $\hbar/2$, as we will prove in Sect. 8.1, and its maximum possible magnitude depends only on the identity of the particle. Electrons, protons, neutrons, muons, tau leptons, neutrinos, and quarks all carry

[4] Their original motivation was to study the surface of nickel, not to check de Broglie's idea, which they learned of later.

this type of intrinsic angular momentum. The magnitude of the projection of their spin angular momentum is always half of the quantized unit for orbital angular momentum, so they are called spin-1/2 particles. Of these, all but the proton and neutron are currently modeled as fundamental (not composite collections of other particles). But even composite particles like atoms, atomic nuclei, and hadrons bound together by the strong nuclear force always have total spin projection that comes in integer multiples of $\hbar/2$.

The first experimental hints of the existence of spin came from the otherwise mysterious doubling of certain spectral lines from atomic transitions. Wolfgang Pauli suggested in 1924 that this was due to the presence of some extra quantum number that takes on exactly two values. The following year, Samuel Goudsmit and George Uhlenbeck proposed the intrinsic angular momentum interpretation that we now know as spin. This idea took some time to gain acceptance, and the many doubters famously included Pauli. Part of the problem was that, at first, the spin was erroneously thought to be due to some kind of internal rotational motion of the structure of the electron, but nobody could get this idea to work in detail, and it turns out to be incorrect. Spins of fundamental particles like the electron are now understood to be a form of angular momentum that is distinct from the mechanical type.

In 1922, Otto Stern and Walther Gerlach reported an experiment that eventually provided profound insights into spin and the emerging quantum theory. They heated silver atoms in a furnace to vaporize them. The atoms escaped through a narrow collimating structure to form a beam, which then moved through a region where the atoms were deflected by an inhomogeneous magnetic field, and finally were collected on a measurement screen, as shown in Fig. 1.5.

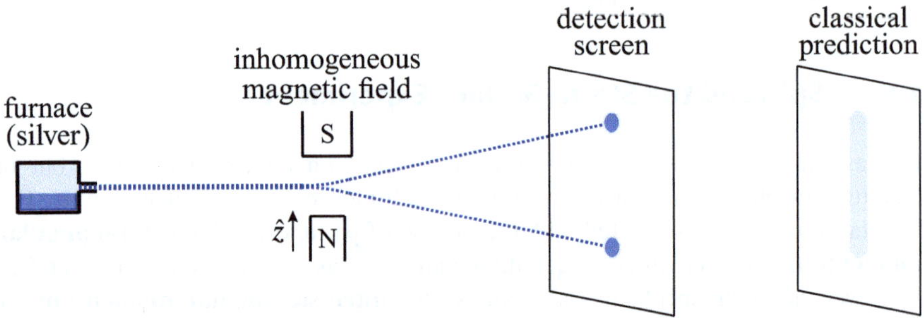

Fig. 1.5 The Stern–Gerlach experiment. Silver atoms are heated in a furnace and passed through an inhomogeneous magnetic field, which exerts a force on them proportional to the $\hat{z}$ component of the atomic magnetic dipole moment. The observed result is that the silver beam splits into two "spatially quantized" components on the detection screen. In contrast, the (incorrect) classical prediction is a continuum of deflection magnitudes

The total magnetic dipole moment for the silver atom is a vector $\boldsymbol{\mu}$ whose magnitude μ is a constant, very nearly the same as that of a single electron,[5] which in turn is proportional to the electron's spin, although Stern and Gerlach did not know this at the time. The furnace thoroughly randomizes the directions of the magnetic moments. As a result of the inhomogeneous magnetic field $\boldsymbol{B}$, there is a classical force on the atoms,

$$\boldsymbol{F} = \nabla(\boldsymbol{\mu} \cdot \boldsymbol{B}), \tag{1.28}$$

causing them to be deflected. For simplicity, suppose the magnetic field in the deflection region has a form invariant under rotations about the z axis,[6]

$$\boldsymbol{B}(x, y, z) = \left[\hat{x}\frac{x}{2a} + \hat{y}\frac{y}{2a} + \hat{z}\left(1 - \frac{z}{a}\right)\right] B_0, \tag{1.29}$$

where the coordinate system has unit vector $\hat{z}$ (up in Fig. 1.5) perpendicular to the beam, and a is a length that is large compared to the size of the region where the magnetic field acts on the atoms in the beam. The atomic dipole moment experiences a torque, and since $\boldsymbol{\mu}$ is proportional to the angular momentum, it obeys an equation of motion of the form

$$\frac{d\boldsymbol{\mu}}{dt} \propto \boldsymbol{\mu} \times \boldsymbol{B}. \tag{1.30}$$

The dominant magnetic field component $\hat{z}B_0$ therefore causes $\boldsymbol{\mu}$ to rotate rapidly about the $\hat{z}$ direction, keeping μ_z nearly constant but causing the oscillating μ_x and μ_y components to average to 0 over the time scale in which the atom is moving through the macroscopic magnetic field region. This implies that μ_x and μ_y can be neglected when computing the deflection of the atom. Since only μ_z contributes, (1.28) becomes

$$\boldsymbol{F} = \mu_z \nabla B_z = \hat{z}\mu_z \frac{\partial B_z}{\partial z} = -\hat{z}(B_0/a)\mu_z. \tag{1.31}$$

[5] The explanation for this is as follows. Silver atoms have 47 electrons, each of them carrying a magnetic dipole moment along that electron's spin direction. However, 46 of the electrons pair up in such a way that their spins are opposite, and cancel. Furthermore, there is no net orbital angular momentum of the electrons, and the contribution of the heavy nucleus to the atomic magnetic moment is relatively small.

[6] The $\hat{x}$ and $\hat{y}$ components of $\boldsymbol{B}$ play no essential role here, except being necessary for consistency in order to satisfy the magnetostatic field equations $\nabla \cdot \boldsymbol{B} = 0$ and $\nabla \times \boldsymbol{B} = 0$. In reality, the field need not be symmetric under rotations about the z axis, but at least one of B_x or B_y must be nonzero. The $\boldsymbol{B}$ given here has the form that would result from a dipole magnet fixed at a distance $3a$ below the beam, to linear order in $1/a$.

Because the gradient of the vertical magnetic field $\partial B_z/\partial z = -B_0/a$ is known and fixed, measuring the deflection of the atom is equivalent to measuring the $\hat{z}$ component of its magnetic dipole moment.

Note that in the limit of a homogeneous field ($a \to \infty$) there would be no deflection at all. Thus, the role of the large homogeneous part of the magnetic field $\hat{z}B_0$ is to determine which component of $\boldsymbol{\mu}$ will be measured, by washing out the effects of the other components, while the smaller inhomogeneous part $-\hat{z}zB_0/a$ provides the force needed to produce the deflection and actually make the measurement.

Since the magnetic dipole moments of silver atoms emerging from the furnace are random in direction, classical physics reasoning suggests that the measured values of μ_z should have equal likelihood to be anything between $-|\mu|$ and $|\mu|$. This in turn would imply that their deflections as observed on the screen should form a continuum between two extremes. Instead, they form two spots with equal numbers of atoms deposited, with a gap in between. The important and surprising conclusion is that the result of measuring μ_z for silver atoms can only give two discrete, quantized values, with equal probabilities.

The amount of deflection can be related to the spin carried by each atom, which in the case of silver comes mostly from a single unpaired electron. The proportionality between magnetic moment and spin for an electron is very close to

$$\boldsymbol{\mu} = -\frac{e}{m_e c}\boldsymbol{S}. \tag{1.32}$$

The numerical magnitudes of the deflections observed in the Stern–Gerlach experiment imply that the measurement of S_z for an electron can only give the values

$$\text{measured } S_z = \pm\hbar/2. \tag{1.33}$$

There is nothing special about the $\hat{z}$ direction, so this also applies to $\hat{n}\cdot\boldsymbol{S}$ for any unit vector $\hat{n}$.

The Stern–Gerlach experiment has been performed for other types of atoms and nuclei,[7] in which the angular momentum (and its relationship to the magnetic moment) can be different, resulting in more than two spots on the screen. This can be used to reveal the possible angular momentum properties ("quantum numbers") of the atom in question. A Stern–Gerlach apparatus can even be used to isolate a sample with particular desired angular momentum quantum numbers. The results of such experiments are always consistent with quantization of the components of any

[7] However, the Stern–Gerlach setup does not work for free electrons, because the necessarily nonzero B_x and/or B_y cause a Lorentz force $-e\boldsymbol{v}\times\boldsymbol{B}$ on the electron. This produces a large deflection due to the electron's small mass, washing out the magnetic moment effect. The Stern–Gerlach setup relies on the deflected particles being electrically neutral, or heavy, or both as in the case of silver atoms.

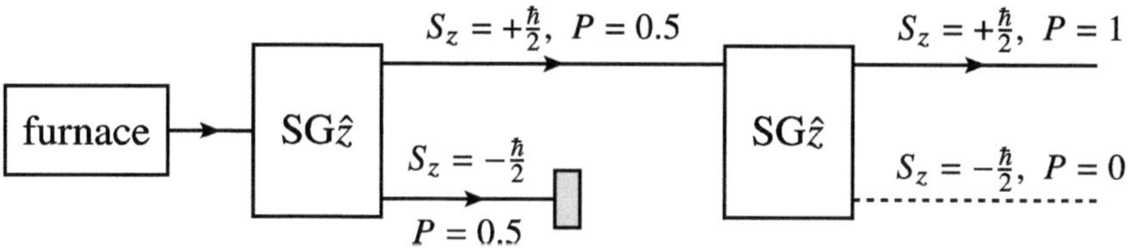

Fig. 1.6 An experiment with two sequential Stern–Gerlach analyzers. The first analyzer prepares a sample of atoms with $S_z = +\hbar/2$, which are then fed into the second analyzer, which again measures $S_z = +\hbar/2$, with probability $P = 1$ in the idealized case

angular momentum vector in integer multiples of $\hbar/2$. The quantum theory must account for this property, and we will see how in Chap. 8.

One can have more general Stern–Gerlach analyzers (called SG$\hat{n}$ in the following) with the inhomogeneous magnetic field element oriented in any chosen unit vector $\hat{n}$ direction. The result of analyzing silver atoms fresh from the furnace with SG$\hat{n}$ is that half of them will be found to have $\hat{n} \cdot \boldsymbol{S} = +\hbar/2$ and the other half will have $-\hbar/2$, for any $\hat{n}$. Stern–Gerlach analyzers play a dual role in further efforts to understand quantum mechanics. First, they are measuring devices, if the output beams are sent directly to a detection screen. Second, because the outgoing beams are separated (sometimes called "spatial quantization"), SG$\hat{n}$ provides a way of preparing a sample of atoms in which the spin component along $\hat{n}$ is known to be either $+\hbar/2$ or $-\hbar/2$. Instead of impacting a detection screen, one or both of the output beams can be sent off to some other component of the experiment. Idealized versions of the Stern–Gerlach deflection analyzers are commonly used as modular components in experiments, real or imagined, to probe the implications of quantum theory.

For example, consider the sequence of two ideal Stern–Gerlach analyzers shown in Fig. 1.6. The atoms in the experiment start in a furnace, which is assumed to produce completely randomized spins. After passing through a first analyzer SG$\hat{z}$, the atoms that had the result $S_z = +\hbar/2$ are sent into a second analyzer SG$\hat{z}$, while those that had $S_z = -\hbar/2$ are thrown away. In this case, the prediction of quantum mechanics for the output of the second analyzer is unlikely to surprise anyone. All of the output atoms on the far right again have $S_z = +\hbar/2$; the second analyzer simply confirms the measurement made by the first.

A more interesting setup is shown in Fig. 1.7, which differs only by placing a SG$\hat{x}$ analyzer between the two SG$\hat{z}$ analyzers. As before, the experiment uses the first analyzer to select a pure sample of atoms with $S_z = +\hbar/2$. However, now the second analyzer separates the sample by measuring S_x. Since the $\hat{z}$ direction has no way of preferring one of $\pm\hat{x}$ over the other, it is no surprise that the output of SG$\hat{x}$ is 50% for each of $S_x = +\hbar/2$ and $-\hbar/2$.

The experiment in Fig. 1.7 then throws away the atoms with $S_x = -\hbar/2$, and feeds those with $S_x = +\hbar/2$ into a third analyzer SG$\hat{z}$. The final results for S_z can then be determined with a detection screen (not shown). One might perhaps suppose that we should find that the final atoms will all have $S_z = +\hbar/2$, since the

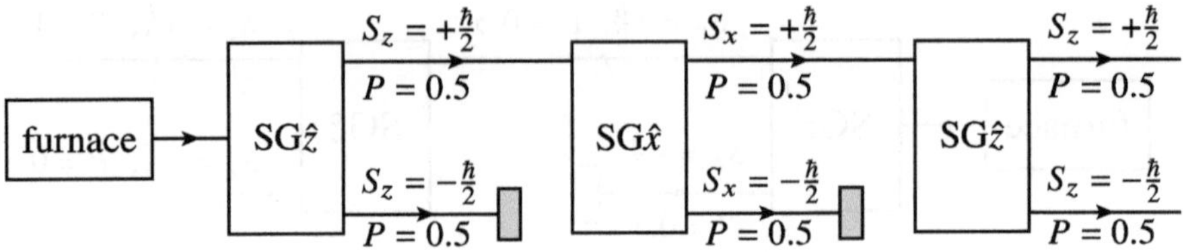

Fig. 1.7 An experiment with three sequential Stern–Gerlach analyzers. The first analyzer prepares a sample of atoms with $S_z = +\hbar/2$, which are then fed into the second analyzer, which measures $S_x = +\hbar/2$ and $-\hbar/2$, each with probability 0.5. The atoms with $S_x = +\hbar/2$ are then sent to the third analyzer, which measures $S_z = +\hbar/2$ and $-\hbar/2$, each with probability 0.5. The act of measuring S_x restores the possibility of measuring $S_z = -\hbar/2$ at the end, even though the first analyzer had been used to select only atoms with $S_z = +\hbar/2$

first analyzer already selected only atoms with that property. However, this is wrong. In reality, equal numbers are measured to have $S_z = +\hbar/2$ and $-\hbar/2$. Inserting the SG$\hat{x}$ analyzer in the middle of the chain affects the atoms in such a way as to restore the possibility of obtaining $S_z = -\hbar/2$.

Another way of thinking about the experiment shows that the restoration of the $S_z = -\hbar/2$ outcomes is a logical necessity, assuming that the spin is the only thing that makes a difference. (In particular, this assumes that the velocity direction of the atoms has no impact on the spin measurement, and furthermore that there are no "hidden variables" associated with the atom that are involved in the measurements in some mysterious way that we do not know how to take into account.) To see this, cover up everything in Fig. 1.7 except the last analyzer. Feeding directly into it are atoms with spin known to be aligned in the $+\hat{x}$ direction. By assumption, the $+\hat{x}$ direction has no reason to prefer $+\hat{z}$ over $-\hat{z}$, or vice versa. So, no matter what may have occurred earlier, the only possibility is that the final probabilities for $S_z = +\hbar/2$ and $-\hbar/2$ are equal.

It is important that this result does *not* have anything to do with the fact that we threw away the atoms that were measured to have $S_x = -\hbar/2$ coming out of the second analyzer. We could feed those atoms into the final analyzer as well, and they would also be measured to have probability $P = 0.5$ for each of $S_z = +\hbar/2$ and $-\hbar/2$, by the same argument. Evidently, it is the mere act of measuring S_x that causes the restoration of the $S_z = -\hbar/2$ outcomes. When we give the postulates of quantum mechanics in Sect. 3.1, the act of measurement will play a special role in one of them (Postulate 5), consistent with the preceding discussion. This also carries with it the implication that quantum mechanics, as defined by these postulates, is an inherently probabilistic, rather than deterministic, theory.

If you are familiar with the behavior of polarizing filters for light, you may recognize that the preceding example is closely analogous to a similar experiment that is common and easy to do in optics. Two linear polarization filters arranged with axes of polarization at right angles will not allow any light to pass through, but a third polarizer inserted between the two, with axis at a 45° angle with respect to the axes of each of the others, will restore the transmission of a fraction of the light. This analogy carries over into the quantum regime. Many of the most sensitive experiments testing quantum mechanics use the polarization of photons to replace the role of the spin in Stern–Gerlach type experiments. One famous example will be discussed in detail in Sect. 14.4.

Exercises

1.1. Estimate the time in seconds needed for a classical hydrogen atom to reach zero size and infinite binding energy, under the assumptions used to obtain (1.7), and taking $r_0 = 5 \times 10^{-11}$ m as the initial condition.

1.2. The cosmic background radiation is nearly black-body with a present temperature of 2.73 K. Compute the numerical energy per cubic meter in this radiation:

(a) within the visible frequency range, defined here as 4×10^{14} Hz $< \nu <$ 8×10^{14} Hz, assuming (1.16), the classical Rayleigh–Jeans formula. (For comparison, the energy density in starlight is roughly 10^{-15} J/m^3.)
(b) integrated over all frequencies, as given by (1.19), the quantum Planck formula. (Most of this is in the microwave spectrum; convince yourself that the visible range contribution is tiny.)

1.3. (a) What velocity should an electron have so that its de Broglie wavelength is 1 m? What if its wavelength is 4.5×10^{-7} m (the same as for blue light)?
(b) Electrons are accelerated from rest by a potential difference of 1 volt, and then pass through a screen with two very narrow long parallel slits. What is the de Broglie wavelength of the electrons? How far apart must the slits be in order for the first minimum of the interference pattern at a distant detector to be at an angle $\theta = 0.2°$ away from the central maximum?

1.4. A coin is a thin disk of 30 g of pure silver, with radius 2 cm. Each silver atom has one unpaired electron spin with magnitude $\hbar/2$ and a nucleus that also has spin magnitude $\hbar/2$. In the (extraordinarily unlikely) case that all of these spins were lined up in the same direction, what would be the total intrinsic angular momentum of the (nonrotating) coin? How does this compare to the ordinary angular momentum that the coin would have if it rotated once per second about the symmetry axis perpendicular to the disk?

Math Tools: States, Operators, and Representations

2

2.1 Complex Linear Vector Spaces

From the results of many experiments, physicists have arrived at a set of postulates that govern quantum theory. These will be given in the next chapter, but first we introduce the mathematical language necessary to frame these ideas. This chapter therefore contains many definitions of key concepts and some useful theorems. In doing so, we employ Paul A.M. Dirac's **bra-ket** notation, which is the modern standard in quantum mechanics.

Quantum mechanics is based on the algebra of complex numbers. For any complex number

$$c = a + ib, \tag{2.1}$$

where a and b are real and $i = \sqrt{-1}$, the real and imaginary parts are denoted in this book by

$$\text{Re}[c] = a, \qquad \text{Im}[c] = b, \tag{2.2}$$

and the **complex conjugate** is denoted using an asterisk,

$$c^* = (a + ib)^* = a - ib. \tag{2.3}$$

The **magnitude** (also known as the **modulus**) is defined by

$$|c| = |a + ib| \equiv \sqrt{cc^*} = \sqrt{a^2 + b^2}. \tag{2.4}$$

It follows that for every nonzero complex number c, the multiplicative inverse is

$$1/c = c^*/|c|^2. \tag{2.5}$$

S. P. Martin, *Quantum Mechanics*, Graduate Texts in Physics,
https://doi.org/10.1007/978-3-032-21119-4_2

This formula is useful for simplifying expressions.

Any complex number can be written as a product of its magnitude and a phase factor,

$$c = a + ib = |c|e^{i\phi}, \tag{2.6}$$

and Leonhard Euler's formula says that the unit-magnitude phase factor is

$$e^{i\phi} = \cos\phi + i\sin\phi, \tag{2.7}$$

so that $\cos\phi = a/|c|$ and $\sin\phi = b/|c|$ and $\tan\phi = b/a$. We write $\phi = \arg(c)$, and say that ϕ is the phase of the complex number c, and $e^{i\phi}$ is the phase factor. (However, one often just says "phase" to refer to the phase factor.)

A **complex linear vector space** is a set of vectors, also known as **kets**, denoted for example as $|v\rangle$, $|w\rangle$, . . ., such that the following properties hold:

- Additive closure: if $|v\rangle$ and $|w\rangle$ are kets, then so is their sum, or **superposition**, $|v\rangle + |w\rangle$.
- Multiplicative closure: if $|v\rangle$ is a ket, then so is the product $c\,|v\rangle$, where c is any complex number. In particular, $-\,|v\rangle$ is the additive inverse of $|v\rangle$.
- There exists a **null ket**, $|\text{Null}\rangle$, with the properties that $|v\rangle + |\text{Null}\rangle = |v\rangle$ for every ket $|v\rangle$, and $c\,|\text{Null}\rangle = |\text{Null}\rangle$ for every complex number c. Multiplying any ket by the complex number 0 results in the null ket: $0\,|v\rangle = |\text{Null}\rangle$.
- Addition of kets and multiplication by complex numbers satisfy the usual commutative, associative, and distributive properties. This means that

$$|v\rangle + |w\rangle = |w\rangle + |v\rangle\,, \tag{2.8}$$

$$(|v\rangle + |w\rangle) + |x\rangle = |v\rangle + (|w\rangle + |x\rangle), \tag{2.9}$$

$$c_1(c_2\,|v\rangle) = (c_1c_2)\,|v\rangle\,, \tag{2.10}$$

$$(c_1 + c_2)\,|v\rangle = c_1\,|v\rangle + c_2\,|v\rangle\,, \tag{2.11}$$

$$c(|v\rangle + |w\rangle) = c\,|v\rangle + c\,|w\rangle\,. \tag{2.12}$$

The null ket, although it is part of the vector space, cannot describe any physical state. There is also a convenient and obvious notational shortcut: if we want to write down that some expression is equal to the null ket, we will just write " $= 0$" instead of " $= |\text{Null}\rangle$". Therefore, the null ket will not appear explicitly any more, but one should always consider the possibility that a ket arising in some calculation might actually be the null ket. A warning: it is common to use the notation $|0\rangle$ for some specific ket that does represent a physical state and must not be confused with the null ket. For example, $|0\rangle$ might represent the lowest energy state of a system, also known as the **ground state**.

In quantum mechanics, the physical state of a system is completely described by a non-null ket, known as the **state ket** or **state vector**, which we will often denote by $|\psi\rangle$. However, one of the rules is that the ket $c\,|\psi\rangle$ represents the same physical state as $|\psi\rangle$, provided that c is a nonzero complex number. This is true even though $c\,|\psi\rangle$ and $|\psi\rangle$ are mathematically distinct members of the vector space; they are identified with each other physically.

The **global phase** of a single state ket is not an observable quantity, as $|\psi\rangle$ and $e^{i\theta}\,|\psi\rangle$ represent exactly the same physical state. However, the relative phases between different kets can be physically meaningful if one considers their sum, also known as their superposition. This is because if we multiply two non-null kets by different phase factors,

$$|v\rangle \rightarrow e^{i\theta_v}\,|v\rangle\,, \qquad\qquad |w\rangle \rightarrow e^{i\theta_w}\,|w\rangle\,, \tag{2.13}$$

then their sum is replaced by

$$|v\rangle + |w\rangle \rightarrow e^{i\theta_v}\left(|v\rangle + e^{i(\theta_w-\theta_v)}\,|w\rangle\right), \tag{2.14}$$

which, by removing the overall global phase factor $e^{i\theta_v}$, is physically equivalent to

$$|v\rangle + e^{i(\theta_w-\theta_v)}\,|w\rangle\,. \tag{2.15}$$

In general, this is not proportional to, and therefore not physically equivalent to, $|v\rangle + |w\rangle$, unless $|v\rangle$ and $|w\rangle$ are proportional or θ_w and θ_v differ by an integer multiple of 2π.

Let us list some examples of complex linear vector spaces, which you can check satisfy the properties in the definition.

Example 2.1 The set of complex numbers z.

Example 2.2 The set of all ordered triples (z_1, z_2, z_3), where z_1, z_2, and z_3 are complex numbers.

Example 2.3 The set of all complex linear combinations $c_1\,|\!\uparrow\rangle + c_2\,|\!\downarrow\rangle$ of two basic kets $|\!\uparrow\rangle$ and $|\!\downarrow\rangle$. (This turns out to be the state space for a single spin-1/2 quantum system.)

Example 2.4 The set of all complex linear combinations of an infinite number of basic kets $|0\rangle$, $|1\rangle$, ..., $|n\rangle$, ..., in one-to-one correspondence with the non-negative integers. (This turns out to be a standard notation for the states of fixed energies of a harmonic oscillator in one dimension.)

Example 2.5 The set of all complex functions of a real variable, $f(x)$, defined on the domain $-\infty < x < \infty$. One can choose to add extra conditions on the functions in a variety of ways, for example requiring that they satisfy the L^2 integrability condition that $\int_{-\infty}^{\infty} dx\,|f(x)|^2$ is finite, and/or that they vanish at certain points or on specified intervals.

Intuitively, these vector spaces have different sizes. To make this precise, we define the notions of linear independence, dimension, and basis. A set of kets $|\psi_j\rangle$ is **linearly independent** if, for any finite N, the equation $\sum_{j=1}^{N} c_j|\psi_j\rangle = 0$ can only be satisfied by taking all $c_j = 0$. In other words, the kets $|\psi_j\rangle$ are linearly independent if we cannot write any of them as a complex linear combination of a finite number of the others. Otherwise, the kets are **linearly dependent**.

A vector space is said to have **dimension** d if one can choose a set of d, but not more, linearly independent vectors. The d linearly independent vectors are then said to form a **basis** for the vector space. The choice of basis is certainly not unique for $d > 1$; it is a common problem that one wants to change the choice of basis, either to make some calculation easier or to make some result simpler to interpret.

You can now check that for our five examples, the dimensions are, respectively, 1, 3, 2, ∞, and ∞. Regarding the last two, the cardinality (countable vs. uncountable) of the basis for infinite-dimensional vector spaces involves subtle mathematical issues that we will find it convenient to mostly ignore, because they have no impact on practical calculations. In physics, we are interested in kets that are members of a vector space with the additional structure of a Hilbert space with an inner product, as discussed in the next section. In almost all cases, the Hilbert space is **separable**, which means that it has a countable basis of kets that are orthonormal with respect to the inner product.[1] Despite this, we will see that it is often a very convenient fiction to include continuous and uncountable sets of vectors that are not members of the physical Hilbert state space but are nevertheless extremely useful, both as physical idealizations and in practical calculations. The most common examples are kets that describe idealized states in which either a particle's position or its momentum (but not both!) is perfectly known. These continuous sets of kets can be chosen to obey other useful properties of a basis, so we will often simply call them basis vectors, as a slight abuse of terminology.

Given a specific basis set $\{|\beta_j\rangle\}$, any vector $|v\rangle$ can be expressed as

$$|v\rangle = \sum_{j=1}^{d} v_j|\beta_j\rangle, \tag{2.16}$$

where the d complex numbers v_j are called the **components** of $|v\rangle$ in that basis. Using the preceding definitions, one can show that, for a given $|v\rangle$ and a given

[1] Technically, for an infinite-dimensional separable Hilbert space, the **Hilbert dimension** corresponding to the number of orthonormal basis states is countable, but the **algebraic dimension** as defined above in terms of the vector-space structure is uncountable.

choice of basis $\{|\beta_j\rangle\}$, the components v_j are unique. In writing (2.16) as a sum, we have implicitly assumed that the basis vectors are discrete and countable. If the basis kets are instead continuous,[2] then the sum must be replaced by an integral. For example, if the basis set is $\{|\beta_q\rangle\}$ where q is a continuous real variable with domain $a < q < b$, then we can write any ket $|v\rangle$ as

$$|v\rangle = \int_a^b dq\; v(q)|\beta_q\rangle. \tag{2.17}$$

Here $v(q)$ are the components, which in this case form a function of q. The symbol q might represent a coordinate on ordinary space, but it could also be a momentum, or an energy, or some other continuous quantity of interest.

A **subspace** of a vector space is a subset that forms a vector space by itself. Below, we will often work with subspaces that consist of states with some feature in common, such as a fixed energy or fixed angular momentum magnitude.

2.2 Inner Products, Hilbert Spaces, and Orthonormal Bases

In quantum mechanics, the vector space of states has the additional structure of a **Hilbert space**, which requires it to be endowed with an inner product. Given a ket $|v\rangle$ and a second ket $|w\rangle$, the **inner product** (also known as a **scalar product**) returns a complex number, denoted $\langle w|v\rangle$, which must satisfy the following rules:

- The order matters, in such a way that exchanging the two kets gives the complex conjugate,

$$\langle v|w\rangle = (\langle w|v\rangle)^*\,. \tag{2.18}$$

  It immediately follows that $\langle v|v\rangle$ is always real. But also...
- If $|v\rangle$ is not the null ket, then $\langle v|v\rangle$ is positive.[3]
- If either $|v\rangle$ or $|w\rangle$ is the null ket, then $\langle v|w\rangle = \langle w|v\rangle = 0$.
- Linearity holds. The inner product of $c_1\,|v\rangle + c_2\,|w\rangle$ and $|x\rangle$, in that order, is

$$\langle x|\big(c_1\,|v\rangle + c_2\,|w\rangle\big) = c_1\,\langle x|v\rangle + c_2\,\langle x|w\rangle\,. \tag{2.19}$$

[2] Notice that we are already engaging in the slight abuse of terminology mentioned in the previous paragraph.

[3] In quantum field theories with gauge invariance, it is sometimes useful, as a book-keeping trick, to modify the rules by allowing some kets to satisfy $\langle v|v\rangle \le 0$. However, these kets represent fictitious (unphysical) states, which must decouple from the true physical states. We will not encounter that issue in this book.

It follows from (2.18) and (2.19) that the inner product of $|x\rangle$ and $c_1|v\rangle + c_2|w\rangle$, in that order, must be

$$c_1^* \langle v|x\rangle + c_2^* \langle w|x\rangle . \tag{2.20}$$

The inner product is analogous to the dot product in the familiar three-dimensional real vector space. However, because quantum mechanics uses complex linear vector spaces, the inner product treats the two input vectors asymmetrically, and interchanging them is the same as taking the complex conjugate.

The **norm** of a ket $|v\rangle$ is defined by $\sqrt{\langle v|v\rangle}$. It follows from the preceding that the norm of a non-null ket is real and positive, and it is 0 if $|v\rangle$ is the null ket. It is sometimes written as $\||v\rangle\| \equiv \sqrt{\langle v|v\rangle}$.

Two useful inequalities that govern the inner product follow.

Theorem 2.1 (Cauchy–Schwarz Inequality) *For any two kets $|v\rangle$ and $|w\rangle$,*

$$|\langle v|w\rangle|^2 \leq \langle v|v\rangle \langle w|w\rangle , \tag{2.21}$$

Also, equality holds if and only if $|w\rangle$ and $|v\rangle$ are proportional to each other or one of them is the null ket.

Proof If either $|v\rangle$ or $|w\rangle$ is the null ket, then (2.21) is trivially satisfied with equality. Therefore, we can assume for the remainder of the proof that neither of them is null. Consider the ket $|z\rangle = c_1|v\rangle - c_2|w\rangle$, where c_1 and c_2 are complex numbers. Since $|z\rangle$ is a ket by the additive closure property, $\langle z|z\rangle$ must be non-negative, which gives

$$\langle z|z\rangle = |c_1|^2 \langle v|v\rangle + |c_2|^2 \langle w|w\rangle - c_1 c_2^* \langle w|v\rangle - c_1^* c_2 \langle v|w\rangle \geq 0. \tag{2.22}$$

Choosing $c_1 = \langle w|w\rangle$ and $c_2 = \langle w|v\rangle$, and using $\langle w|v\rangle = \langle v|w\rangle^*$, (2.22) becomes

$$(\langle w|w\rangle)^2 \langle v|v\rangle - \langle w|w\rangle\, |\langle v|w\rangle|^2 \geq 0. \tag{2.23}$$

Now, since $|w\rangle$ is not null, we can divide by $\langle w|w\rangle$ to get (2.21). If the equality condition holds, then it follows that $|z\rangle$ is the null ket, which implies that $|v\rangle$ and $|w\rangle$ are proportional. □

Theorem 2.2 (Triangle Inequality) *The norms of the kets $|v\rangle$ and $|w\rangle$ and their superposition $|v\rangle + |w\rangle$ must obey*

$$\|(|v\rangle + |w\rangle)\| \leq \||v\rangle\| + \||w\rangle\|, \tag{2.24}$$

with equality if and only if $|w\rangle = n|v\rangle$ for some real number $n > 0$ or one of them is the null ket.

The proof can be obtained from the Cauchy–Schwarz inequality, and is left to Exercise 2.1. The triangle inequality is similar to the statement in ordinary plane geometry that the sum of the lengths of two sides of a triangle must exceed that of the third side.

Given a non-null ket $|v\rangle$, one can define a new ket by dividing it by its norm. The result

$$|v\rangle / \sqrt{\langle v|v\rangle} \tag{2.25}$$

then has norm 1, and in quantum mechanics it is physically equivalent to the original ket, in the sense that it represents the same physical state. We say that the ket has been normalized to unity. This still leaves the freedom to multiply the ket by a complex phase factor, because

$$e^{i\theta}\, |v\rangle \tag{2.26}$$

has the same norm as $|v\rangle$, for any real number θ. This freedom will appear very often as an ambiguity in the determination of a ket that has to satisfy some other specified properties. The resolution of such ambiguities is arbitrary, and equivalent to a choice of convention.

Two kets $|v\rangle$ and $|w\rangle$ are **orthogonal** (also known as **perpendicular**) if $\langle w|v\rangle = 0$. An **orthonormal basis**, which we will call an **orthobasis** for short, is a basis of kets $\{|\varphi_j\rangle\}$ with $j = 1, 2, \ldots$ that satisfies the additional property

$$\langle \varphi_j|\varphi_k\rangle = \delta_{jk}. \tag{2.27}$$

Here δ_{jk} is the **Kronecker delta symbol,**

$$\delta_{jk} = \begin{cases} 1 & \text{for } j = k, \\ 0 & \text{for } j \neq k, \end{cases} \tag{2.28}$$

and is defined only when j and k are labels that take on discrete values.

As noted in the previous section, we will also often want to deal with basis kets labeled by one or more continuous parameters. In that case, we need a different orthonormality condition, called **Dirac orthonormality**, in which the Kronecker delta symbol is replaced by a **Dirac delta function**. The Dirac orthonormality condition for kets $|\varphi_q\rangle$ labeled by a single continuous real variable q can be taken to be

$$\langle \varphi_q|\varphi_{q'}\rangle = \delta(q - q'). \tag{2.29}$$

Here q could be, for example, a position coordinate of a particle. The delta function $\delta(x)$ has the properties that

$$\delta(x) = \begin{cases} 0 & \text{for } x \neq 0, \\ \infty & \text{for } x = 0, \end{cases} \tag{2.30}$$

in such a way that

$$\int_{-\infty}^{\infty} dx\, \delta(x) = 1. \tag{2.31}$$

More generally, for sufficiently well-behaved functions $f(x)$,

$$\int_{-\infty}^{\infty} dx\, \delta(x-c) f(x) = f(c). \tag{2.32}$$

For most purposes, (2.32) can be taken as the definition of the delta function. Indeed, technically, the Dirac delta function is not a function at all, but a distribution in mathematical language. This means that it is only well-defined when appearing in expressions in which the argument of the delta function is integrated over.

An important technical note: as we have already warned in the previous section, kets satisfying the Dirac orthonormality condition (2.29) cannot be physical states, strictly speaking. This is because they do not have finite norm, due to $\delta(0) = \infty$. Such kets with continuous labels are still very useful in practical calculations, where they often represent idealizations with perfectly known position, or momentum, or some other continuous quantity. These are sometimes known as **generalized kets** or **non-normalizable kets**, and they act as a basis for the Hilbert space of physical states, even though they are not themselves part of the Hilbert space. They are so useful as idealizations that mere non-normalizability is not a good enough reason for us to eliminate them from our toolbox.

It is often useful to think of $\delta(x)$ as the "limit" (in a sense that we will not bother to try to make mathematically rigorous) of a sequence of increasingly narrow and sharply peaked functions that have unit area, in various different ways. For example, we can take a sequence of rectangular functions with width Δ and height $1/\Delta$,

$$\delta(x) = \lim_{\Delta \to 0} \begin{cases} 1/\Delta & \text{for } |x| \leq \Delta/2, \\ 0 & \text{for } |x| > \Delta/2. \end{cases} \tag{2.33}$$

Another useful representation of $\delta(x)$ is

$$\delta(x) = \frac{1}{2\pi} \int_{-\infty}^{\infty} dk\, e^{ikx}. \tag{2.34}$$

To make better sense of this technically ill-defined expression, we can "cut off" the integration over k. One way to do this is to insert a convergence factor $e^{-k^2\Delta^2/2}$ into the integrand, and then take $\Delta \to 0$,

$$\delta(x) = \lim_{\Delta \to 0} \frac{1}{2\pi} \int_{-\infty}^{\infty} dk\, e^{ikx} e^{-k^2\Delta^2/2} = \lim_{\Delta \to 0} \frac{\exp(-x^2/2\Delta^2)}{\sqrt{2\pi}\,\Delta}. \tag{2.35}$$

This interprets $\delta(x)$ as the $\Delta \to 0$ limit of Gaussian functions with height $1/\sqrt{2\pi}\,\Delta$ and full width at half maximum (FWHM) equal to $2\sqrt{2\ln 2}\,\Delta \approx 2.3548\Delta$. A different way to insert a convergence factor is

$$\delta(x) = \lim_{\Delta \to 0} \frac{1}{2\pi} \int_{-\infty}^{\infty} dk\, e^{ikx} e^{-\Delta|k|} = \lim_{\Delta \to 0} \frac{\Delta}{\pi(x^2 + \Delta^2)}, \tag{2.36}$$

which represents $\delta(x)$ as a limit of increasingly narrow Breit–Wigner lineshapes. Yet another way to make sense of (2.34) is to limit the k integration to a large but finite range $-1/\Delta < k < 1/\Delta$,

$$\delta(x) = \lim_{\Delta \to 0} \frac{1}{2\pi} \int_{-1/\Delta}^{1/\Delta} dk\, e^{ikx} = \lim_{\Delta \to 0} \frac{\sin(x/\Delta)}{\pi x}, \tag{2.37}$$

or to do so multiplied by a triangular integrand function to maintain continuity in k,

$$\delta(x) = \lim_{\Delta \to 0} \frac{1}{2\pi} \int_{-1/\Delta}^{1/\Delta} dk\, e^{ikx} (1 - \Delta|k|) = \lim_{\Delta \to 0} \frac{2\Delta}{\pi x^2} \sin^2(x/2\Delta). \tag{2.38}$$

Each of the expressions in (2.33) and (2.35)–(2.38) has the crucial property of unit integrated area for all Δ, even before taking the limit $\Delta \to 0$. This justifies the normalization factor of $1/2\pi$ in (2.34). They are illustrated in Fig. 2.1. This variety is not just for entertainment purposes; we will have occasion to use all of them, as well as the formal integral representation of (2.34), in the following chapters, depending on the situation.

The representation of the delta function in (2.34) is related to the theory of Fourier transforms. To see how this works, we can use it to write, for any sufficiently well-behaved function $f(x)$,

$$f(x) = \int_{-\infty}^{\infty} dx'\, f(x')\, \delta(x - x') = \int_{-\infty}^{\infty} dx'\, f(x') \left(\frac{1}{2\pi} \int_{-\infty}^{\infty} dk\, e^{ik(x-x')} \right). \tag{2.39}$$

By rearranging the order of integrations, this becomes

$$f(x) = \frac{1}{2\pi} \int_{-\infty}^{\infty} dk\, e^{ikx} \int_{-\infty}^{\infty} dx'\, e^{-ikx'} f(x'). \tag{2.40}$$

Now, we define the Fourier transform of $f(x)$ by the second integral,

$$F(k) \equiv \frac{1}{\sqrt{2\pi}} \int_{-\infty}^{\infty} dx'\, e^{-ikx'} f(x'), \tag{2.41}$$

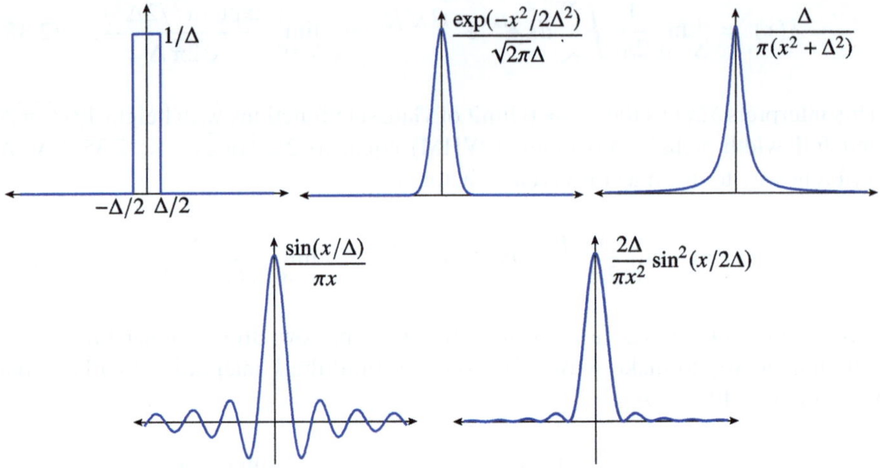

Fig. 2.1 The Dirac delta function $\delta(x)$ can be thought of as the $\Delta \to 0$ "limit" of functions peaked near $x = 0$ with characteristic width proportional to Δ and unit integrated area, as in the rectangular functions in (2.33) [top left], the Gaussian functions in (2.35) [top center], the Breit–Wigner lineshape in (2.36) [top right], the normalized sine functions in (2.37) [bottom left], or the normalized sine-squared functions in (2.38) [bottom right]. Each of the last four will prove useful to us later, in different circumstances

for $-\infty < k < \infty$. Then (2.40) becomes

$$f(x) = \frac{1}{\sqrt{2\pi}} \int_{-\infty}^{\infty} dk\, e^{ikx} F(k), \tag{2.42}$$

which is the inverse Fourier transform relation.

For a finite-dimensional[4] Hilbert space, there is a systematic way to find an orthobasis:

Theorem 2.3 *Given a finite-dimensional Hilbert space with an inner product and a known basis $\{|\beta_j\rangle\}$ that is not necessarily orthonormal, one can always construct an orthobasis $\{|\varphi_j\rangle\}$ by a systematic procedure known as the* ***Gram–Schmidt process****.*

Proof We prove this constructively, by giving the steps of the process.

Step 1: Choose any ket in the original basis; call it $|\beta_1\rangle$. The first member of the orthobasis will be the rescaled version of this ket with unit norm, $|\varphi_1\rangle = |\beta_1\rangle / \sqrt{\langle\beta_1|\beta_1\rangle}$.

[4] In the infinite-dimensional case, the Gram–Schmidt algorithm cannot end in a finite number of steps, but in cases relevant to quantum mechanics a suitable orthobasis is often easy to identify anyway. For example, the Hilbert space may naturally split into an infinite number of mutually orthogonal finite-dimensional subspaces.

Step 2: Choose a second ket $|\beta_2\rangle$ from the original basis set. From it, define a new vector $|\tilde{\varphi}_2\rangle = |\beta_2\rangle - |\varphi_1\rangle \langle\varphi_1|\beta_2\rangle$, which we can describe in words as subtracting off the projection along the vector $|\varphi_1\rangle$. Thus $|\tilde{\varphi}_2\rangle$ is orthogonal to $|\varphi_1\rangle$ by construction. Now take $|\varphi_2\rangle = |\tilde{\varphi}_2\rangle / \sqrt{\langle\tilde{\varphi}_2|\tilde{\varphi}_2\rangle}$, which has norm 1. This is the second member of the orthobasis we are constructing.

Step 3: Choose a third ket $|\beta_3\rangle$ from the original basis set. From it, define a new vector $|\tilde{\varphi}_3\rangle = |\beta_3\rangle - |\varphi_1\rangle \langle\varphi_1|\beta_3\rangle - |\varphi_2\rangle \langle\varphi_2|\beta_3\rangle$. In words, we are subtracting off the projections along both of the vectors $|\varphi_1\rangle$ and $|\varphi_2\rangle$. Now $|\tilde{\varphi}_3\rangle$ is orthogonal to both $|\varphi_1\rangle$ and $|\varphi_2\rangle$ by construction, so again we just need to re-scale it to have unit norm. Therefore, define $|\varphi_3\rangle = |\tilde{\varphi}_3\rangle / \sqrt{\langle\tilde{\varphi}_3|\tilde{\varphi}_3\rangle}$ as the third member of the orthobasis.

...

Step n: Take the nth ket $|\beta_n\rangle$ from the original basis set, and subtract off the projections along all of the previously found orthobasis kets, by defining

$$|\tilde{\varphi}_n\rangle = |\beta_n\rangle - \sum_{j=1}^{n-1} |\varphi_j\rangle\langle\varphi_j|\beta_n\rangle. \tag{2.43}$$

By construction, this is orthogonal to all of the previously found kets $|\varphi_j\rangle$ with $j = 1, \ldots, n-1$. It is not null, because of the linear independence of the original basis. So, to include it in the orthobasis, we only need to re-scale it to also have unit norm,

$$|\varphi_n\rangle = |\tilde{\varphi}_n\rangle / \sqrt{\langle\tilde{\varphi}_n|\tilde{\varphi}_n\rangle}. \tag{2.44}$$

Continuing in this way, when we are finished with Step d, where d is the dimension of the Hilbert space, we will have constructed the full orthobasis $\{|\varphi_n\rangle\}$. □

Suppose that we have chosen an orthobasis $\{|\varphi_j\rangle\}$. To obtain the components v_j of an arbitrary ket

$$|v\rangle = \sum_j v_j|\varphi_j\rangle, \tag{2.45}$$

we can take the inner product with $|\varphi_k\rangle$, resulting in

$$\langle\varphi_k|v\rangle = \sum_j v_j\langle\varphi_k|\varphi_j\rangle = \sum_j v_j\delta_{jk} = v_k, \tag{2.46}$$

where the second equality relies on the orthonormality property (2.27). Thus, the components of the ket are simply given by its inner products with the orthobasis vectors.

2.3 Dual Vector Spaces

Given a Hilbert space of kets with an inner product, it is useful to construct another complex linear vector space, called the **dual vector space**. Dual vectors are defined mathematically as linear maps from the space of kets to the complex numbers. For each ket $|v\rangle$, there is a corresponding dual vector denoted $\langle v|$, and we write the association as

$$\begin{array}{ccc} \text{vector} & & \text{dual vector} \\ |v\rangle & \rightarrow & \langle v| \,. \end{array} \tag{2.47}$$

Specifically, the dual vector $\langle v|$ is defined to map each ket $|w\rangle$ to the complex number given by the inner product of the ket $|w\rangle$ with the corresponding ket $|v\rangle$,

$$\langle v| \big(|w\rangle \big) \equiv \langle v|w\rangle \,. \tag{2.48}$$

A dual vector defined in this way is also called a **bra**, a whimsical bit of terminology evoking that the inner product is a "bra-ket", or bracket. From the properties of the inner product, specifically by comparing (2.19) and (2.20), one finds that the bra associated with a linear combination of kets is

$$\begin{array}{ccc} \text{vector} & & \text{dual vector} \\ c_1 |v\rangle + c_2 |w\rangle & \rightarrow & c_1^* \langle v| + c_2^* \langle w| \,. \end{array} \tag{2.49}$$

One must remember to take the complex conjugates of the coefficients.

The bra $\langle v|$ is also known as the **Hermitian adjoint** of the corresponding ket $|v\rangle$, and vice versa. To understand the relation better, it is useful to consider the following linear algebra analogy: kets are like complex d-dimensional column vectors, and bras are like complex d-dimensional row vectors, with

$$\boldsymbol{w} = \begin{pmatrix} w_1 \\ w_2 \\ \vdots \\ w_d \end{pmatrix}, \qquad \boldsymbol{v}^\dagger = \begin{pmatrix} v_1^* \; v_2^* \; \cdots \; v_d^* \end{pmatrix}. \tag{2.50}$$

The fact that a dual vector maps vectors to complex numbers is just expressed as

$$\boldsymbol{v}^\dagger \boldsymbol{w} = \sum_k v_k^* w_k \,. \tag{2.51}$$

As we will discuss more fully in Sect. 2.5, this is not just an analogy; if one has chosen an orthobasis $\{|\varphi_k\rangle\}$, then $w_k = \langle \varphi_k|w\rangle$ are the components of the ket $|w\rangle$, while $v_k^* = \langle v|\varphi_k\rangle$ are the components of the bra $\langle v|$. However, the bra-ket notation has the great virtue of being independent of any particular choice of orthobasis.

According to the dual vector definition, for each ket $|v\rangle$, there is always a unique corresponding bra $\langle v|$. If the Hilbert space has a finite dimension, then there is also a unique ket for every bra, and the correspondence is one-to-one. However, in the infinite-dimensional case it is possible to construct bras that have no associated ket within the Hilbert space, although this fact is of little practical consequence. For example, consider the generalized kets like those that satisfy the Dirac orthonormality condition (2.29); as we have noted, these have infinite norm and are therefore not part of the physical Hilbert space. The corresponding bras are nevertheless well-defined maps from the physical Hilbert space to the complex numbers, and so are perfectly respectable members of the dual vector space.

You may find it useful to think of kets (vectors) as representing possible states of a system, while bras (dual vectors) represent possible questions that one may ask about the state. For example, we can interrogate the actual state of a system $|\psi\rangle$ about its overlap with another possible state $|\chi\rangle$. Then the question that we are asking is associated with the dual vector $\langle\chi|$. As we will discuss further near the end of Sect. 3.2, the answer that one receives is that the probability of finding the system in the state $|\chi\rangle$ is nothing other than $|\langle\chi|\psi\rangle|^2$, assuming that both $|\psi\rangle$ and $|\chi\rangle$ are normalized to 1.

2.4 Operators

An **operator** A is a map from the space of kets to itself. This means that the result of acting with A on any ket $|v\rangle$ must be another ket in the Hilbert space, which we can call either $|Av\rangle$, or equivalently, $A\,|v\rangle$. The second notation is almost always used. As a notational convention, we will usually use capital letters to represent operators, except when established tradition says otherwise.

In quantum mechanics, we are almost always[5] interested in **linear operators**, which obey

$$A\big(c_1\,|v\rangle + c_2\,|w\rangle\big) = c_1 A\,|v\rangle + c_2 A\,|w\rangle\,. \tag{2.52}$$

Because there is a bra associated with each ket, a linear operator also maps the dual space to itself. Specifically, for each bra $\langle w|$, the bra $\langle w|\,A$ resulting from the operation of A is defined by the relation

$$\big(\langle w|\,A\big)\,|v\rangle = \langle w|\,\big(A\,|v\rangle\big). \tag{2.53}$$

Adopting this definition, it follows that the operation of A on the dual space also obeys linearity,

[5] The most notable exception, and the only one in this book, is the case of the antilinear operator associated with time-reversal (or motion-reversal). It is confined to Sects. 5.7 and 8.8.

$$\big(c_1 \langle v| + c_2 \langle w|\big)A \;=\; c_1 \langle v| A + c_2 \langle w| A. \tag{2.54}$$

Because of the equality of the two sides of (2.53), we can define the **matrix element** of the operator A between $\langle w|$ and $|v\rangle$ as their common value $\langle w|A|v\rangle$, without parentheses.

Addition and subtraction of linear operators and multiplication by complex numbers are defined in the obvious ways, such that

$$\big(c_1 A + c_2 B\big) |v\rangle \;=\; c_1 A |v\rangle + c_2 B |v\rangle . \tag{2.55}$$

The product of two operators A and B is defined by

$$(AB) |v\rangle \;\equiv\; A\big(B |v\rangle\big) \;=\; AB |v\rangle . \tag{2.56}$$

As before, the parentheses make no difference and can be omitted, as indicated in the last equality. Operators also obey associativity; for any three operators A, B, and C,

$$(AB)C \;=\; A(BC). \tag{2.57}$$

The order of operators matters, so that BA and AB are different, in general. One therefore defines the **commutator** of A and B as

$$[A, B] \;=\; AB - BA, \tag{2.58}$$

and the **anticommutator** by

$$\{A, B\} \;=\; AB + BA. \tag{2.59}$$

The simplest example of a linear operator is the **identity operator** I, defined by

$$I |v\rangle \;=\; |v\rangle , \qquad \langle v| I \;=\; \langle v| \tag{2.60}$$

for every $|v\rangle$. The **inverse of an operator** A, if it exists, is denoted A^{-1}, and is defined by the requirements[6]

$$A^{-1}A \;=\; I \;=\; AA^{-1}. \tag{2.61}$$

However, it is important to recognize that not all operators have an inverse.

---

[6] For a finite-dimensional vector space, the second equality in (2.61) is redundant, because $BA = I$ can be shown to imply $AB = I$ as well. However, in an infinite-dimensional vector space this is not true. For example, consider the vector space with general element $v = (v_1, v_2, v_3, \ldots)$, and define the right-shift and left-shift operators by $Rv = (0, v_1, v_2, v_3, \ldots)$ and $Lv = (v_2, v_3, v_4, \ldots)$. Then $LR = I$, but $RL \neq I$.

There is an enormously useful way of writing the identity operator, given an orthobasis. Equations (2.45) and (2.46) can be combined to write

$$|v\rangle = \sum_j |\varphi_j\rangle\langle\varphi_j|v\rangle. \tag{2.62}$$

Since this is true for all $|v\rangle$, one has simply

$$I = \sum_j |\varphi_j\rangle\langle\varphi_j|\,. \tag{2.63}$$

This extraordinarily important identity is known as the **completeness relation**, or the **closure relation**. We will use it very often.

Given any two kets $|v\rangle$ and $|w\rangle$, one can construct the linear operator

$$A = |v\rangle\langle w|\,, \tag{2.64}$$

called the **outer product** of $|v\rangle$ and $\langle w|$. It is defined by its action on any other ket $|x\rangle$,

$$A\,|x\rangle = \big(|v\rangle\langle w|\big)\,|x\rangle = |v\rangle\,\big(\langle w|x\rangle\big). \tag{2.65}$$

As a useful special case, the **projection operator** P_v onto a ket $|v\rangle$ is defined by

$$P_v = \frac{|v\rangle\langle v|}{\langle v|v\rangle}. \tag{2.66}$$

If $|v\rangle$ has norm 1, then one can simply write

$$P_v = |v\rangle\langle v|\,. \tag{2.67}$$

Projection operators have the property

$$P_v^2 = P_v. \tag{2.68}$$

Intuitively, the projection operator P_v acts on a ket by throwing away the part orthogonal to $|v\rangle$, and keeping the part proportional to $|v\rangle$ intact; doing this twice has the same effect as doing it once. As an aside, this is a good example of an operator that has no inverse (with the trivial exception of the case that the state space is one-dimensional). The reason is that when P_v acts on any ket $|w\rangle$ that is orthogonal to $|v\rangle$, it yields 0 (the null ket), and then there is no way to resurrect $|w\rangle$ by acting with another linear operator, the purported inverse.

The completeness relation (2.63) can now be equivalently expressed as the statement that the identity operator is equal to

$$I = \sum_j P_{\varphi_j}, \tag{2.69}$$

which is the sum over the projection operators for all of the members of an orthobasis, spanning the whole Hilbert space. Similarly, one can project onto a subspace of the Hilbert space, by summing over projection operators for only a subset of the orthobasis vectors.

It is often important to be able to recognize when an operator defined in some other way must be proportional to the identity operator. An intuitive and useful tool in this regard is the following:

Theorem 2.4 (Schur's Lemma) *For a linear operator A on a Hilbert space, the following statements are equivalent:*

(i) A commutes with every operator,
(ii) A is diagonal in every orthobasis,
(iii) $A = aI$ is a constant multiple of the identity operator.

The proof is left to Exercise 2.7.

For any linear operator A, we define the Hermitian adjoint (or just adjoint) $A^\dagger$ by its action on dual vectors,

$$\langle v|\, A^\dagger \;=\; \langle Av|\,, \tag{2.70}$$

where $\langle Av|$ is the bra corresponding to the ket $|Av\rangle = A\,|v\rangle$. Consider a generic matrix element involving $A^\dagger$,

$$\langle w|A^\dagger|v\rangle \;=\; \langle Aw|v\rangle \;=\; (\langle v|Aw\rangle)^*\,, \tag{2.71}$$

where the second equality made use of (2.18). This can be restated as the useful relation

$$\langle w|A^\dagger|v\rangle \;=\; \big(\langle v|A|w\rangle\big)^*. \tag{2.72}$$

It is not hard to use the definition to show the following facts. The adjoint of the operator $A = |w\rangle\langle v|$ is

$$\big(|w\rangle\langle v|\big)^\dagger \;=\; |v\rangle\langle w|\,. \tag{2.73}$$

The adjoint of the operator cI, where c is a complex number, is $(cI)^\dagger = c^*I$. The adjoint of a product of operators is

$$(AB)^\dagger \;=\; B^\dagger A^\dagger, \tag{2.74}$$

where the order matters, in general.

It is useful to generalize the concept of taking the adjoint to whole expressions and equalities involving kets, bras, and operators. Given any expression, define the adjoint of it according to the following rules:

- Substitute $A \to A^\dagger$ for all operators.
- Substitute $c \to c^*$ for all complex numbers.
- Substitute $|v\rangle \leftrightarrow \langle v|$ for all kets and bras.
- Reverse the order of kets, bras, and operators within each term.

(Of course, the complex-number factors of any term can be written in any desired order.) Then one can show that the adjoint of any valid equality will also be a valid equality. Equations (2.18), (2.72), (2.73) and (2.74) are examples of this. As a more random example, if one has

$$ABC\,|v\rangle\,\langle v|w\rangle \;+\; B\,|x\rangle \;=\; c\,|z\rangle\,, \tag{2.75}$$

where A, B, and C are operators and c is a complex number, then one also has

$$\langle w|v\rangle\,\langle v|\,C^\dagger B^\dagger A^\dagger \;+\; \langle x|\,B^\dagger \;=\; c^*\,\langle z|\,. \tag{2.76}$$

An operator A is called **Hermitian** (or **self-adjoint**) if it is the same as its adjoint, $A^\dagger = A$. Hermitian operators are particularly important in quantum mechanics, where they are associated with physically measurable quantities. An **anti-Hermitian** operator is one that satisfies $A^\dagger = -A$. Note that any operator can be written as the sum of a Hermitian part and an anti-Hermitian part,

$$A = \left(\frac{A + A^\dagger}{2}\right) + \left(\frac{A - A^\dagger}{2}\right). \tag{2.77}$$

The product of two Hermitian operators is Hermitian if, and only if, they commute. It is also not hard to show that for two Hermitian operators A and B, the commutator is anti-Hermitian, so that $i[A, B]$ is Hermitian.

An operator U is **unitary** if its adjoint is equal to its inverse, so $U^\dagger = U^{-1}$. The product of two unitary operators is always unitary. In quantum mechanics, unitary operators are associated with a change of orthobasis, as we will discuss in the next section. They often appear in the context of defining or exploiting the symmetries of the physical system. The time evolution of a system will also be associated with a unitary operator.

It is possible to define operators that are functions of other operators. As a simple example, polynomials in operators are also operators. More generally, one can define an operator $f(A)$ as a power series in A. The most common example of this that we will encounter in quantum mechanics is the exponentiation of an operator A to get the operator $\exp(A)$, which can be defined as

$$e^A = \sum_{n=0}^{\infty} \frac{A^n}{n!}, \tag{2.78}$$

where of course $A^0 = I$. This definition treats A just as if it were an ordinary number, exploiting the facts that it obviously commutes with itself, and no other operators are involved. Another way to define the exponential of an operator uses the usual limit,

$$e^A = \lim_{N\to\infty} \left(I + \frac{A}{N}\right)^N . \tag{2.79}$$

We will unveil a more general and powerful way of defining functions of operators in Sect. 2.7; see (2.186).

The operator e^A is always invertible, and the inverse is e^{-A}. Taking $A = iB$, where B is Hermitian, we then find:

Theorem 2.5 *If B is a Hermitian operator, then the operator* $\exp(iB)$ *is unitary.*

We now give some other useful results involving functions of operators.

Theorem 2.6 *For any two operators A and B,*

$$e^A B e^{-A} = B + [A, B] + \frac{1}{2!}[A, [A, B]] + \frac{1}{3!}[A, [A, [A, B]]] + \cdots . \tag{2.80}$$

Proof Define the operator $F(\lambda) = e^{\lambda A} B e^{-\lambda A}$, where λ is a real variable. Now

$$\frac{dF}{d\lambda} = AF - FA = [A, F(\lambda)], \tag{2.81}$$

where we have used the fact that $\frac{d}{d\lambda} e^{\lambda A} = Ae^{\lambda A} = e^{\lambda A} A$. Repeating this gives

$$\frac{d^2 F}{d\lambda^2} = [A, [A, F(\lambda)]], \qquad \frac{d^3 F}{d\lambda^3} = [A, [A, [A, F(\lambda)]]], \tag{2.82}$$

etc. Using these to compute the Taylor series expansion for $F(\lambda)$ about the point $\lambda = 0$, and using $F(0) = B$, we have

$$F(\lambda) = B + \lambda[A, B] + \frac{\lambda^2}{2!}[A, [A, B]] + \frac{\lambda^3}{3!}[A, [A, [A, B]]] + \cdots , \tag{2.83}$$

and (2.80) now follows by taking $\lambda = 1$. □

As a special case application of Theorem 2.6, we have

Theorem 2.7 *If A and B are operators such that* $[A, B] = cB$ *where c is a number, then*

$$e^A B e^{-A} = e^c B. \tag{2.84}$$

Another useful result, established in Exercise 2.5, is:

Theorem 2.8 *For any two operators A and B such that* $[A, B]$ *commutes with B, and the function* $f(B)$ *can be expressed as a power series in B,*

$$[A, f(B)] = [A, B]f'(B). \tag{2.85}$$

The following formula is also useful:

Theorem 2.9 (Baker–Campbell–Hausdorff, Special Case) *For any two operators A and B such that* $[A, B]$ *commutes with both A and B,*

$$e^A e^B = e^{A+B} e^{\frac{1}{2}[A,B]}. \tag{2.86}$$

The proof is left as Exercise 2.6. The previous two theorems hold in particular if the commutator of A and B is proportional to the identity operator.

2.5 Matrix Representations of Operators

In this section, we specialize to the case of a Hilbert space with finite dimension d. Suppose we have selected an orthobasis $\{|\varphi_j\rangle\}$, and consider two kets $|v\rangle$, $|w\rangle$ given as

$$|v\rangle = \sum_j v_j |\varphi_j\rangle, \qquad |w\rangle = \sum_j w_j |\varphi_j\rangle, \tag{2.87}$$

where v_j and w_j are their components in the chosen orthobasis. The inner product can be written as

$$\langle w|v\rangle = \sum_j \sum_k w_j^* v_k \langle\varphi_j|\varphi_k\rangle = \sum_j w_j^* v_j = \begin{pmatrix} w_1^* \cdots w_d^* \end{pmatrix} \begin{pmatrix} v_1 \\ \vdots \\ v_d \end{pmatrix}, \tag{2.88}$$

where the second equality was obtained using the orthonormality property (2.27). We can therefore consider the bra $\langle w|$ as associated with a row vector of complex numbers, and the ket $|v\rangle$ with a column vector,

$$\langle w| \leftrightarrow \begin{pmatrix} w_1^* \cdots w_d^* \end{pmatrix}, \qquad |v\rangle \leftrightarrow \begin{pmatrix} v_1 \\ \vdots \\ v_d \end{pmatrix}. \tag{2.89}$$

The double-arrow notation is used to indicate that the bra and ket are represented by the corresponding row and column vectors. However, it is important to remember that while this representation depends on the choice of basis, the bra and ket themselves exist as abstract objects independent of the choice of basis.

Given an operator A, one can construct a matrix representation for it corresponding to the orthobasis $\{|\varphi_j\rangle\}$. We start with

$$A|v\rangle = A\sum_k v_k |\varphi_k\rangle = \sum_k v_k A|\varphi_k\rangle . \tag{2.90}$$

Now taking the inner product with an orthobasis ket $|\varphi_j\rangle$ gives

$$\langle\varphi_j|A|v\rangle = \sum_k \langle\varphi_j|A|\varphi_k\rangle\, v_k. \tag{2.91}$$

We therefore define the $d \times d$ complex matrix with elements

$$A_{jk} = \langle\varphi_j|A|\varphi_k\rangle. \tag{2.92}$$

It then follows that the components of

$$|v'\rangle = A|v\rangle \tag{2.93}$$

are

$$v'_j = \langle\varphi_j|v'\rangle = \sum_k A_{jk} v_k, \tag{2.94}$$

or, in matrix form,

$$\begin{pmatrix} v'_1 \\ \vdots \\ v'_d \end{pmatrix} = \begin{pmatrix} A_{11} & \cdots & A_{1d} \\ \vdots & \ddots & \vdots \\ A_{d1} & \cdots & A_{dd} \end{pmatrix} \begin{pmatrix} v_1 \\ \vdots \\ v_d \end{pmatrix}. \tag{2.95}$$

Thus, we complete the correspondence of (2.89) by writing, for any operator A, the matrix representation

$$A \leftrightarrow \begin{pmatrix} A_{11} & \cdots & A_{1d} \\ \vdots & \ddots & \vdots \\ A_{d1} & \cdots & A_{dd} \end{pmatrix}, \tag{2.96}$$

with the orthobasis matrix elements given by (2.92).

It is often convenient to dispense with formalities by replacing the $\leftrightarrow$ symbol in equations like (2.89) and (2.96) with an $=$ symbol. However, again we emphasize that the kets $|v\rangle$, bras $\langle w|$, and operators A are objects that are conceptually independent of the choice of orthobasis, while their components v_j, w_j^*, and A_{jk} do depend on the choice of orthobasis, and will be different if a different orthobasis is used.

The matrix representation for a product of operators AB is given by

$$(AB)_{jk} = \sum_n A_{jn} B_{nk}. \tag{2.97}$$

The proof of this is an almost trivial consequence of the completeness relation (2.63), because it can also be read as $\langle \varphi_j | AB | \varphi_k \rangle = \sum_n \langle \varphi_j | A | \varphi_n \rangle \langle \varphi_n | B | \varphi_k \rangle$. It is also easy to show that the orthobasis matrix elements of an operator A and its adjoint $A^\dagger$ are related by

$$(A^\dagger)_{jk} = (A_{kj})^*. \tag{2.98}$$

Thus, the matrix representation of $A^\dagger$ is the conjugate transpose (or Hermitian conjugate) of the matrix representation of A.

Many operator relations and properties are easiest to see by using the matrix representation in an appropriately chosen orthobasis, using the tools of linear algebra. For instance, a Hermitian (or unitary) operator A is represented by a Hermitian (or unitary) matrix with elements A_{jk}. The matrix representation of A^{-1}, if it exists, is equal to the inverse of the matrix representation of A. It is a standard result in linear algebra that A^{-1} exists if and only if the determinant of the matrix A is nonzero,

$$\det(A) \neq 0. \tag{2.99}$$

In that case, the inverse matrix for A is

$$A^{-1} = \frac{1}{\det(A)} \mathrm{cof}[A]^T \tag{2.100}$$

where cof$[A]$ is the cofactor matrix of A, and T denotes the transpose. For a general $d \times d$ matrix A, the cofactor matrix is defined by

$$(\mathrm{cof}[A])_{jk} = (-1)^{j+k} M_{jk} \tag{2.101}$$

where the minor M_{jk} is in turn defined as the determinant of the $(d-1) \times (d-1)$ submatrix obtained from A by deleting row j and column k.

A unitary operator U, as defined in the previous section, satisfies

$$U^\dagger U = UU^\dagger = I. \tag{2.102}$$

The components of the matrix representation of U in an orthobasis $\{|\varphi_k\rangle\}$,

$$U = \begin{pmatrix} U_{11} & U_{12} & \cdots & U_{1d} \\ U_{21} & U_{22} & \cdots & U_{2d} \\ \vdots & \vdots & \ddots & \vdots \\ U_{d1} & U_{d2} & \cdots & U_{dd} \end{pmatrix}, \tag{2.103}$$

obey the rule

$$U_{jk} = \langle\varphi_j|U|\varphi_k\rangle = j\text{th component of } |\varphi'_k\rangle, \tag{2.104}$$

where

$$|\varphi'_k\rangle = U\,|\varphi_k\rangle\,. \tag{2.105}$$

Now, using (2.102), it is not difficult to show that the set $\{|\varphi'_k\rangle\}$ defined in this way is *another* orthobasis. From (2.103) and (2.104), we see that the components of the new orthobasis element $|\varphi'_k\rangle$ are the entries in the kth column of the matrix representation of U in the original orthobasis. In this sense, a unitary operator U performs a change of orthobasis, which can be thought of as a complex rotation in the Hilbert space. The operator $U^\dagger$ performs the inverse change of basis,

$$|\varphi_k\rangle = U^\dagger|\varphi'_k\rangle, \tag{2.106}$$

and it follows that the bras for the two orthobases are related by

$$\langle\varphi'_k| = \langle\varphi_k|\,U^\dagger, \qquad \langle\varphi_k| = \langle\varphi'_k|\,U. \tag{2.107}$$

Conversely, given any two orthonormal bases $\{|\varphi_k\rangle\}$ and $\{|\varphi'_k\rangle\}$, the inner products are the elements of a unitary matrix,

$$\langle\varphi_j|\varphi'_k\rangle = \langle\varphi_j|U|\varphi_k\rangle = U_{jk}. \tag{2.108}$$

To verify directly that U as defined by (2.108) is indeed unitary, one can use the completeness relation (2.63) to show

$$\sum_n U^*_{nj} U_{nk} = \delta_{jk}, \tag{2.109}$$

or, rewriting,

$$\sum_n \left(U^\dagger\right)_{jn} U_{nk} = \delta_{jk}. \tag{2.110}$$

This expresses the unitarity of the matrix representation of the operator U, and is equivalent to the operator equation (2.102). Also,

$$|\det(U)| = 1, \tag{2.111}$$

follows from the general linear algebra facts that $\det(AB) = (\det A)(\det B)$ and $\det(A^\dagger) = (\det A)^*$, for matrices A, B.

An important feature of such a unitary change of basis is that it preserves inner products. To see this, consider the transformation[7]

$$|v\rangle \rightarrow |v'\rangle = U\,|v\rangle\,, \tag{2.112}$$

$$|w\rangle \rightarrow |w'\rangle = U\,|w\rangle\,, \tag{2.113}$$

so that we have performed the same complex rotation on both $|v\rangle$ and $|w\rangle$. Then

$$\langle w|v\rangle \rightarrow \langle w'|v'\rangle = \langle w|U^\dagger U|v\rangle = \langle w|v\rangle\,, \tag{2.114}$$

so that inner products are unaffected by a change of orthobasis.

We can also see how the unitary change of basis is realized on the matrix elements of any operator A. For any two kets $|v\rangle$ and $|w\rangle$,

$$\langle w'|A|v'\rangle = \langle w|U^\dagger AU|v\rangle. \tag{2.115}$$

If we now define a transformation of the operator A according to

$$A \rightarrow A' = UAU^\dagger, \tag{2.116}$$

to go along with (2.112) and (2.113), then

$$\begin{aligned}\langle w'|A'|v'\rangle &= (\langle w|\,U^\dagger)(UAU^\dagger)(U\,|v\rangle) = \langle w|\,(U^\dagger U)A(U^\dagger U)\,|v\rangle \\ &= \langle w|A|v\rangle\,.\end{aligned} \tag{2.117}$$

In words, the matrix element for the transformed states of the transformed operator A' is the same as the original matrix element of A.

Functions of operators are often easiest to deal with using their matrix representations. If the operator A has a matrix representation that is diagonal in some

[7] Here, the arrow "$\rightarrow$" can be read as "transforms to".

appropriate orthobasis,

$$A \leftrightarrow \mathrm{diag}(a_1, a_2, \ldots, a_d), \tag{2.118}$$

then immediately from the series definition of (2.78), with A replaced by iA, one finds

$$\exp(iA) \leftrightarrow \mathrm{diag}(e^{ia_1}, e^{ia_2}, \ldots, e^{ia_d}). \tag{2.119}$$

Also, if the powers of A obey some recurrence relation, then the series for $\exp(iA)$ can often be resummed in the matrix representation, even if it is not diagonal. As an example, suppose A has matrix representation

$$A \leftrightarrow a \begin{pmatrix} 0 & 1 \\ 1 & 0 \end{pmatrix}, \tag{2.120}$$

where a is a real number. Then we recognize that $A^2 = a^2 I$, and so the unitary operator e^{iA} has matrix representation, obtained by separating the even powers of A from the odd powers of A,

$$\sum_{n=0}^{\infty} \frac{(ia)^{2n}}{(2n)!} \begin{pmatrix} 1 & 0 \\ 0 & 1 \end{pmatrix} + \sum_{n=0}^{\infty} \frac{(ia)^{2n+1}}{(2n+1)!} \begin{pmatrix} 0 & 1 \\ 1 & 0 \end{pmatrix}, \tag{2.121}$$

or, summing each series,

$$\exp(iA) \leftrightarrow \begin{pmatrix} \cos(a) & i\sin(a) \\ i\sin(a) & \cos(a) \end{pmatrix}. \tag{2.122}$$

2.6 Eigenvalues and Eigenvectors

In quantum mechanics, we will often need to solve eigenvalue problems. If, for some operator A, we can find a complex number α and a non-null ket $|v\rangle$ such that

$$A\,|v\rangle = \alpha\,|v\rangle\,, \tag{2.123}$$

then α is called the **eigenvalue** and $|v\rangle$ is the corresponding **eigenvector** or **eigenket**, associated with the **eigenstate** of A. When (2.123) holds, it follows immediately that

$$\langle v|\,A^\dagger = \alpha^*\,\langle v|\,, \tag{2.124}$$

or, in words, $\langle v|$ is an eigenbra of $A^\dagger$, with eigenvalue α^*.

The **eigenvalue problem** is to find all solutions for both α and $|v\rangle$ as a pair, given A. Rewriting (2.123) as

$$(A - \alpha I)\,|v\rangle = 0, \tag{2.125}$$

we see that for a solution with a particular α to exist, the operator $(A - \alpha I)^{-1}$ must not exist. Otherwise, we could act with it on both sides to discover that $|v\rangle$ could only be the null ket, in contradiction of the assumption.

If the Hilbert space is finite-dimensional, the non-existence of the inverse of the matrix representation of $(A - \alpha I)$ implies that a solution for α must satisfy

$$\det(A - \alpha I) = 0. \tag{2.126}$$

This is called the characteristic equation, and the left side is a polynomial of degree d, the same as the dimension of the Hilbert space. The fundamental theorem of algebra states that this always has exactly d complex solutions, $\alpha_1, \ldots, \alpha_d$, allowing for the possibility that some of them may be repeated. The integer number of times a particular eigenvalue α_n is repeated in the list is called its **degeneracy**, and we will denote it g_{α_n}.

After the eigenvalues α_n have been found, for each of them we can then solve

$$(A - \alpha_n I)\,|v_n\rangle = 0 \tag{2.127}$$

for the corresponding eigenkets $|v_n\rangle$. If a particular eigenvalue α_n has degeneracy g_{α_n}, then for Hermitian or unitary A there will be a set of g_{α_n} linearly independent solutions $|v_n\rangle$ with that eigenvalue. Sometimes the solution of (2.127) can be found by inspection, sometimes it will require linear algebra, and in still other cases it will involve solving some differential equation(s). It is often not feasible to solve it exactly, which makes life interesting and necessitates approximation methods. In any case, note that if $|v_n\rangle$ satisfies the equation, then so does $c\,|v_n\rangle$ for every nonzero complex constant c. That constant can always be chosen, for each eigenket, to normalize it as desired. Even after doing so, the complex phase of each eigenket remains an arbitrary choice.

The most trivial example is the case that $A = I$, the identity operator. Since $I\,|v\rangle = |v\rangle$, every non-null ket $|v\rangle$ is an eigenket, and the corresponding eigenvalue is 1.

Another simple case is the projection operator for a unit-norm ket $|w\rangle$, as introduced in (2.67),

$$P_w = |w\rangle\langle w|\,. \tag{2.128}$$

Then the equation $P_w\,|v\rangle = \alpha\,|v\rangle$ becomes

$$|w\rangle\,\langle w|v\rangle = \alpha\,|v\rangle\,. \tag{2.129}$$

$$\begin{aligned} U_{jk} &= \langle\varphi_j|U|\varphi_k\rangle = \langle\varphi_j|\varphi_k'\rangle \\ &= j\text{th component of } \big(k\text{th orthonormal eigenvector of } A\big). \end{aligned} \tag{2.148}$$

The eigenvalue equation (2.145) can be rewritten, by taking the inner product with $\langle\varphi_n|$,

$$\sum_j A_{nj} U_{jk} = \alpha_k U_{nk} = \sum_l U_{nl}(\delta_{lk}\alpha_l). \tag{2.149}$$

In index-free matrix notation, this reads

$$AU = U A_{\text{diag}}. \tag{2.150}$$

Now, multiplying from the left by $U^\dagger$, we obtain

Theorem 2.16 (Diagonalization of Hermitian Operators) *The matrix representation of a Hermitian operator A in an arbitrary orthobasis $\{|\varphi_k\rangle\}$ is diagonalized by the unitary matrix U specified in (2.148), according to*

$$U^\dagger A U = A_{diag}, \tag{2.151}$$

where A_{diag} is the matrix representation of the operator A in the new orthobasis $\{|\varphi_k'\rangle\}$ consisting of its eigenvectors.

Let us do an extended example to illustrate some of the preceding results. Consider a Hilbert space of dimension 3 with an operator A defined by its action on an orthobasis $\{|\varphi_1\rangle, |\varphi_2\rangle, |\varphi_3\rangle\}$,

$$A|\varphi_1\rangle = 3|\varphi_1\rangle, \tag{2.152}$$

$$A|\varphi_2\rangle = 4|\varphi_2\rangle - i|\varphi_3\rangle, \tag{2.153}$$

$$A|\varphi_3\rangle = i|\varphi_2\rangle + 4|\varphi_3\rangle. \tag{2.154}$$

The matrix representation of A in this original orthobasis is, using (2.92) and (2.96),

$$A \leftrightarrow \begin{pmatrix} 3 & 0 & 0 \\ 0 & 4 & i \\ 0 & -i & 4 \end{pmatrix}. \tag{2.155}$$

Since the matrix representation is Hermitian, A is a Hermitian operator. The characteristic equation for its eigenvalues is $\det(A-\alpha I) = (3-\alpha)(15-8\alpha+\alpha^2) = 0$, which yields

$$\alpha = 3, 3, 5. \tag{2.156}$$

These eigenvalues are real, as guaranteed by Theorem 2.12. For the non-degenerate eigenvalue $\alpha = 5$, the solution of the matrix equation for the corresponding eigenvector gives, after normalization and choice of an arbitrary phase, and conversion to ket form,

$$|5\rangle = \frac{1}{\sqrt{2}} \left(|\varphi_2\rangle - i\, |\varphi_3\rangle \right) . \tag{2.157}$$

Note that we have named the eigenket after its eigenvalue. For the eigenvalue $\alpha = 3$, the degeneracy is 2, so there are two linearly independent eigenkets, which can be chosen to be

$$|3, 1\rangle = \frac{1}{\sqrt{2}} \left(|\varphi_2\rangle + i\, |\varphi_3\rangle \right) , \tag{2.158}$$

$$|3, 2\rangle = |\varphi_1\rangle . \tag{2.159}$$

The second entry (1 or 2) on each $\alpha = 3$ ket is a degeneracy label. These kets are automatically orthogonal to $|5\rangle$, as promised by Theorem 2.14, but we had to make a choice of linear combination to make them orthonormal, since arbitrary linear combinations of $|3, 1\rangle$ and $|3, 2\rangle$ would not be orthogonal to each other and would not have norm 1. In this example, the linear combination that makes an orthobasis is not hard to find, but given any other (non-orthonormal) basis for the degenerate eigenvalue subspace, the Gram–Schmidt procedure of Theorem 2.3 can be used to construct an orthobasis.

Our orthobasis of eigenkets of A is related to the original orthobasis by a unitary transformation. Let us define an operator U by

$$U\, |\varphi_1\rangle = |5\rangle , \tag{2.160}$$

$$U\, |\varphi_2\rangle = |3, 1\rangle , \tag{2.161}$$

$$U\, |\varphi_3\rangle = |3, 2\rangle . \tag{2.162}$$

Now using (2.157)–(2.159), and applying the operator $\leftrightarrow$ matrix correspondence rules of (2.92) and (2.96) again, one finds the matrix representation of U in the $\{|\varphi_1\rangle , |\varphi_2\rangle , |\varphi_3\rangle\}$ basis,

$$U \leftrightarrow \begin{pmatrix} 0 & 0 & 1 \\ 1/\sqrt{2} & 1/\sqrt{2} & 0 \\ -i/\sqrt{2} & i/\sqrt{2} & 0 \end{pmatrix} , \tag{2.163}$$

which is indeed a unitary matrix.

Acting on (2.160)–(2.162) with $U^\dagger$, and using $U^\dagger U = I$, we get

$$U^\dagger\, |5\rangle = |\varphi_1\rangle = |3, 2\rangle , \tag{2.164}$$

$$U^\dagger |3,1\rangle = |\varphi_2\rangle = (|5\rangle + |3,1\rangle)/\sqrt{2}, \tag{2.165}$$

$$U^\dagger |3,2\rangle = |\varphi_3\rangle = i(|5\rangle - |3,1\rangle)/\sqrt{2}. \tag{2.166}$$

The second equality in each of these equations follows by inverting (2.157)–(2.159) to solve for $|\varphi_1\rangle$, $|\varphi_2\rangle$, and $|\varphi_3\rangle$. Equations (2.164)–(2.166) provide the matrix representation of $U^\dagger$ in the orthobasis $\{|5\rangle, |3,1\rangle, |3,2\rangle\}$,

$$U^\dagger \leftrightarrow \begin{pmatrix} 0 & 1/\sqrt{2} & i/\sqrt{2} \\ 0 & 1/\sqrt{2} & -i/\sqrt{2} \\ 1 & 0 & 0 \end{pmatrix}. \tag{2.167}$$

As a check, this matrix is indeed equal to the conjugate transpose of (2.163). As promised by Theorem 2.16 on diagonalization of Hermitian operators,

$$U^\dagger A U = A_{\text{diag}} = \begin{pmatrix} 5 & 0 & 0 \\ 0 & 3 & 0 \\ 0 & 0 & 3 \end{pmatrix}, \tag{2.168}$$

where A, U, and $U^\dagger$ on the left side are taken to be the matrices given by (2.155), (2.163), and (2.167), respectively. Note that A_{diag} on the right side is the matrix representation of the operator A in the orthobasis $\{|5\rangle, |3,1\rangle, |3,2\rangle\}$.

The choice of orthobasis made in (2.157)–(2.159) is not the unique one composed of eigenkets of A, due to the superposition principle of Theorem 2.10. The most general orthonormal pair of basis kets with A eigenvalue $\alpha = 3$ is obtained using the freedom to do unitary transformations within the degenerate eigenvalue subspace, and so is parameterized by two complex numbers c, s, subject to the constraint $|c|^2 + |s|^2 = 1$,

$$|3,1'\rangle = c|3,1\rangle + s|3,2\rangle, \tag{2.169}$$

$$|3,2'\rangle = -s^*|3,1\rangle + c^*|3,2\rangle. \tag{2.170}$$

These kets have new degeneracy labels $(1', 2')$ to distinguish this orthobasis from the unprimed one with $c = 1$, $s = 0$. The change in orthobasis is brought about by another unitary operator V, defined by

$$V|5\rangle = |5\rangle, \tag{2.171}$$

$$V|3,1\rangle = |3,1'\rangle, \tag{2.172}$$

$$V|3,2\rangle = |3,2'\rangle, \tag{2.173}$$

or, in a matrix representation in the basis $\{|5\rangle, |3,1\rangle, |3,2\rangle\}$,

$$V \leftrightarrow \begin{pmatrix} 1 & 0 & 0 \\ 0 & c & -s^* \\ 0 & s & c^* \end{pmatrix}, \qquad (|5\rangle\,,\ |3,1\rangle\,,\ |3,2\rangle \text{ basis}). \tag{2.174}$$

By combining the information in (2.157)–(2.159) and (2.169)–(2.173), one also finds

$$V\,|\varphi_1\rangle = c^*\,|\varphi_1\rangle - \frac{s^*}{\sqrt{2}}\,|\varphi_2\rangle - i\frac{s^*}{\sqrt{2}}\,|\varphi_3\rangle\,, \tag{2.175}$$

$$V\,|\varphi_2\rangle = \frac{s}{\sqrt{2}}\,|\varphi_1\rangle + \frac{1}{2}(1+c)\,|\varphi_2\rangle + \frac{i}{2}(c-1)\,|\varphi_3\rangle\,, \tag{2.176}$$

$$V\,|\varphi_3\rangle = -i\frac{s}{\sqrt{2}}\,|\varphi_1\rangle + \frac{i}{2}(1-c)\,|\varphi_2\rangle + \frac{1}{2}(1+c)\,|\varphi_3\rangle\,, \tag{2.177}$$

so that the matrix representation of V in the original orthobasis is

$$V \leftrightarrow \begin{pmatrix} c^* & s/\sqrt{2} & -is/\sqrt{2} \\ -s^*/\sqrt{2} & (1+c)/2 & i(1-c)/2 \\ -is^*/\sqrt{2} & i(c-1)/2 & (1+c)/2 \end{pmatrix}, \qquad (|\varphi_1\rangle\,,\ |\varphi_2\rangle\,,\ |\varphi_3\rangle \text{ basis}). \tag{2.178}$$

Comparison of (2.174) and (2.178) illustrates the point that even though V is unambiguous as a Hilbert space operator, its matrix representation looks very different depending on the orthobasis used.

The **trace of an operator** in a finite-dimensional Hilbert space can be defined to be equal to the trace of its matrix representation. For an operator A and any orthobasis $\{|\varphi_k\rangle\}$,

$$\mathrm{Tr}(A) = \sum_k \langle\varphi_k|A|\varphi_k\rangle = \sum_k A_{kk}. \tag{2.179}$$

It is left as Exercise 2.4 to show that the trace of an operator is independent of the choice of orthobasis, and in particular is equal to the sum of its eigenvalues, and that

$$\mathrm{Tr}(AB) = \mathrm{Tr}(BA), \tag{2.180}$$

$$\mathrm{Tr}\big(\,|v\rangle\,\langle w|\,\big) = \langle w|v\rangle\,, \tag{2.181}$$

which can be proved very quickly using completeness of the orthobasis.

We conclude this section by stating two useful theorems using unitary operators, omitting the proofs (which are not difficult).

Theorem 2.17 *The eigenvalues of a unitary operator U have magnitude 1.*

Note that (2.137) illustrates this.

Theorem 2.18 *If an operator A has eigenkets* $|\alpha_n\rangle$ *with eigenvalues* α_n*, and U is a unitary operator, then* $UAU^\dagger$ *has eigenvectors* $U|\alpha_n\rangle$ *with the same eigenvalues* α_n.

This implies the important result that the eigenvalues of an operator do not depend on the choice of basis we use to calculate them.

2.7 Observables

An operator A is an **observable** if it is Hermitian and its eigenkets form an orthobasis $\{|\alpha, u\rangle\}$, which satisfies the orthonormality relation (2.146) and the completeness relation

$$\sum_{\alpha}\sum_{u=1}^{g_\alpha} |\alpha, u\rangle\langle\alpha, u| = I. \tag{2.182}$$

Here, α are the eigenvalues with degeneracies g_α, and u are the degeneracy labels. For a Hilbert space of finite dimension, the completeness part of this definition is redundant, following automatically from Theorem 2.15. However, it is not automatic for a Hermitian operator in an infinite-dimensional Hilbert space; that is one reason why it must be included in the definition here. The other reason is that we will later find it convenient to slightly generalize our definition of an observable to include ordered sets of Hermitian operators, so that their eigenvalues can be not just numbers, but ordered sets of numbers. (An example is the vector position operator in three-dimensional space, which is really an ordered triple of three Hermitian operators X, Y, and Z.) In such cases, (2.182) is again certainly not automatic. The crucial requirement we are ensuring with (2.182) is that if A is really supposed to be an observable, then every state in the Hilbert space can always be expressed as a linear combination of its eigenstates.

The way that (2.182) is written assumes that the eigenvalues of A are countable and discrete. If instead they are continuous (and therefore uncountable), then the orthonormality and completeness conditions are of the Dirac type [compare to (2.29)],

$$\langle\alpha', u'|\alpha, u\rangle = \delta(\alpha - \alpha')\,\delta_{uu'}, \tag{2.183}$$

$$\int d\alpha \sum_{u=1}^{g_\alpha} |\alpha, u\rangle\,\langle\alpha, u| = I, \tag{2.184}$$

where the integral is over the range or ranges of α that occur as eigenvalues. Similarly, it is possible for the degeneracy labels u to be continuous, in which case the Kronecker $\delta_{uu'}$ is replaced by a Dirac delta function $\delta(u-u')$ and the sum over u is replaced by $\int du$.

It is even possible for the eigenvalues α or the degeneracy labels u to have a spectrum that includes both discrete (countable) and continuous (uncountable) components. In that case, the orthonormality relations will include both Kronecker and Dirac deltas, and the completeness relation will include both a sum and an integral. We will see an example of this when we study the bound and unbound states of the hydrogen atom, in (11.90)–(11.92) and (11.101).

A particularly useful consequence of completeness follows from acting with A on both sides of (2.182). Then, evaluating $A = \alpha$ when acting on its eigenkets, we obtain the **spectral decomposition** of an observable operator A,

$$A = \sum_{\alpha} \sum_{u=1}^{g_\alpha} \alpha \, |\alpha, u\rangle \langle \alpha, u| \,, \tag{2.185}$$

with an obvious counterpart for continuous eigenvalues by replacing summation with integration. Solving problems in quantum mechanics is often the art of turning expressions involving operators into numbers. The spectral decomposition idea allows us to do this in a systematic way. When we see an observable operator A appearing in a matrix element, a standard trick is to insert a completeness sum directly before or after it, so as to replace A by a sum (or integral) over its eigenvalues, as in (2.185). Of course, a prerequisite for using this trick is to solve the eigenvalue problem for A.

At the end of Sect. 2.4, we mentioned that it is possible to define functions of operators, and gave the most common example, that of an exponential of an operator. If we have solved the eigenvalue problem for an observable A, completeness now gives us yet another way to define a completely general function of it, $F(A)$. For the case of discrete eigenvalues α, the spectral decomposition of the operator $F(A)$ is

$$F(A) = \sum_{\alpha} \sum_{u=1}^{g_\alpha} F(\alpha) \, |\alpha, u\rangle \langle \alpha, u| \,, \tag{2.186}$$

again with an obvious counterpart for the continuous case involving integration rather than summation.

We next state a very useful theorem about commuting observables, which we will prove only in the case of a Hilbert space of finite dimension.

Theorem 2.19 (Compatible Observables) *If A and B are observables, and $[A, B] = 0$, then there exists an orthobasis of kets that are eigenkets of both A and B. In other words, for commuting observables A and B, there is an orthobasis in which their matrix representations are both diagonal.*

Proof From the definition of an observable, we already know that we can find an orthobasis of kets $|\alpha, u\rangle$ that are eigenkets of A. Furthermore, since $[A, B] = 0$, we know from Theorem 2.13 that whenever $\alpha \neq \alpha'$,

$$\langle \alpha', u' | B | \alpha, u \rangle = 0. \tag{2.187}$$

Therefore, the matrix representation of B has a block-diagonal form,

$$\begin{pmatrix} B_{\alpha_1} & 0 & 0 & \cdots & 0 \\ 0 & B_{\alpha_2} & 0 & \cdots & 0 \\ 0 & 0 & B_{\alpha_3} & \cdots & 0 \\ \vdots & \vdots & \vdots & \ddots & \vdots \\ 0 & 0 & 0 & 0 & B_{\alpha_n} \end{pmatrix}, \tag{2.188}$$

where $n \leq d$ is the number of distinct eigenvalues α_j of A, and the blocks B_{α_j} along the diagonal represent the possibly nonzero entries, which are confined to the subspaces of the Hilbert space that have the same eigenvalues α_j. Each B_{α_j} is a Hermitian $g_{\alpha_j} \times g_{\alpha_j}$ matrix, where g_{α_j} is the degeneracy[9] of α_j, and has entries

$$\langle \alpha_j, u' | B | \alpha_j, u \rangle, \qquad (u, u' = 1, \ldots, g_{\alpha_j}). \tag{2.189}$$

Therefore, according to Theorem 2.16, each of the matrices B_{α_j} can be diagonalized by a change of basis corresponding to a unitary operator. Each of these unitary operators acts nontrivially only within the corresponding B_{α_j} block, and is the identity operator outside of that block. After doing these basis transformations, we will be left in an orthobasis in which the operators A and B are represented by diagonal matrices of the form

$$A = \text{diag}(\underbrace{\alpha_1, \ldots, \alpha_1}_{g_{\alpha_1}\text{ times}}, \underbrace{\alpha_2, \ldots, \alpha_2}_{g_{\alpha_2}\text{ times}}, \cdots, \underbrace{\alpha_n, \ldots, \alpha_n}_{g_{\alpha_n}\text{ times}}) \tag{2.190}$$

and

$$B = \text{diag}(\beta_{\alpha_1,1}, \ldots, \beta_{\alpha_1,g_{\alpha_1}}, \beta_{\alpha_2,1}, \ldots, \beta_{\alpha_2,g_{\alpha_2}}, \cdots, \beta_{\alpha_n,1}, \ldots, \beta_{\alpha_n,g_{\alpha_n}}), \tag{2.191}$$

where the $\beta_{\alpha_j,k}$ with $k = 1, \ldots, g_{\alpha_j}$ are the eigenvalues of the B_{α_j} submatrix. This is the desired orthobasis that achieves the requirements of the theorem. □

[9] Note that $j = 1, \ldots, n$, and $\sum_{j=1}^{n} g_{\alpha_j} = d$ is the dimension of the Hilbert space.

Even with two observables A and B, there can be unresolved degeneracies, by which we mean that there can be more than one orthobasis ket with the same eigenvalues α and β. This raises the question of how to tell apart the corresponding degenerate states. From a physics perspective, if two states are genuinely distinct, then there must be some observable that distinguishes them. This leads to the important concept of a **Complete Set of Commuting Observables**, called a **CSCO** from here on. In this book, a CSCO is a finite set of observables $A, B, \ldots, C$ that all commute with each other, and whose common eigenkets form an orthobasis with no degeneracies. Thus, we can write

$$A\,|\alpha, \beta, \ldots, \gamma\rangle = \alpha\,|\alpha, \beta, \ldots, \gamma\rangle\,, \tag{2.192}$$

$$B\,|\alpha, \beta, \ldots, \gamma\rangle = \beta\,|\alpha, \beta, \ldots, \gamma\rangle\,, \tag{2.193}$$

$$\ldots$$

$$C\,|\alpha, \beta, \ldots, \gamma\rangle = \gamma\,|\alpha, \beta, \ldots, \gamma\rangle\,, \tag{2.194}$$

where each of the eigenkets $|\alpha, \beta, \ldots, \gamma\rangle$ is *uniquely* determined by specifying its CSCO eigenvalues. For each one of the eigenvalue labels α, β, γ, etc., the others can be viewed as degeneracy labels.

If we have a set of commuting observables, but the degeneracy in the eigenkets is not fully eliminated, it means that we have not actually found a CSCO yet, and at least one more observable (commuting with all of the others) can be found to add to the list. Conversely, if the degeneracy has been fully eliminated by a list of commuting observables, then we have a CSCO; there is no need to add more observables to the list, even if it is easy to find more observables that commute with all of the others. As we will see in many examples, the number of observables needed to form a CSCO is always finite (and generally small) for quantum systems with a fixed number of particles, as considered in this book. This is true even if the Hilbert space is infinite-dimensional.

For a given quantum system, the choice of which operators to include in a CSCO is not unique. For example, in the case of a spinless particle moving in three dimensions, one CSCO choice is just the three spatial coordinates X, Y, Z that specify the position of the particle, but another CSCO choice consists of the three momentum components P_x, P_y, P_z of the particle. In cases with rotational symmetry, still another CSCO choice is the Hamiltonian (total energy) operator of the particle together with two operators associated with its angular momentum with respect to the origin. In general, the best choice of CSCO depends on what physics questions we would like to answer. If a CSCO includes (or commutes with) the Hamiltonian, the corresponding eigenvalues are known as **good quantum numbers**, because this is a convenient basis in which to describe time evolution. It is often useful to be able to translate between the orthobases defined by different CSCOs.

Note that the word "complete" in CSCO does not have the same meaning as it does in the completeness relation. For a CSCO it is the operators that are complete, while for the completeness relation it is the orthobasis that is complete. Also, the second C in CSCO can equivalently be interpreted as standing for the word "compatible". Observables are called **compatible** if they commute with each other, and are called **incompatible** if they have a nonzero commutator.

2.8 Wavefunctions

In the preceding, we have been mainly concerned with matrix representations in a finite-dimensional Hilbert space. Let us now consider in more detail the case of an observable with a continuous spectrum of eigenvalues. A quintessential example is the position coordinate operator X for a particle moving in one dimension. The eigenvalues and eigenkets of X will be labeled as x and $|x\rangle$, respectively, so that the eigenvalue equation is

$$X\,|x\rangle = x\,|x\rangle\,. \tag{2.195}$$

The physical interpretation is that if the state is $|x\rangle$, then the position of the particle is known to be x, possibly because it has just been measured to be there. The allowed values of x form an uncountable, continuous set, perhaps $-\infty < x < \infty$, or $a < x < b$ if the particle is confined to that range. In the following, we will assume the latter, in order to be more general, and reserve the option to take $a \to -\infty$ and $b \to \infty$.

Since a particle measured to be at position x is not at any other position x' at that precise moment, the states $|x\rangle$ and $|x'\rangle$ are orthogonal. Furthermore, a measurement of the particle's position must return one of the allowed values. These two statements are made precise as a special case of (2.183) and (2.184),

$$\langle x|x'\rangle = \delta(x-x'), \tag{2.196}$$

$$\int_a^b dx\;|x\rangle\langle x| = I. \tag{2.197}$$

To check that this is consistent, and that $\delta(x-x')$ in (2.196) is really the Dirac delta function, consider an arbitrary ket $|\psi\rangle$, and define the **position wavefunction**

$$\psi(x) \equiv \langle x|\psi\rangle\,. \tag{2.198}$$

The wavefunction $\psi(x)$ can also be viewed as comprising the components of the ket $|\psi\rangle$ in the **position representation**. Using completeness, we have

$$\psi(x) = \int_a^b dx'\,\langle x|x'\rangle\langle x'|\psi\rangle = \int_a^b dx'\,\delta(x-x')\,\psi(x'). \tag{2.199}$$

The fact that this should hold for every function $\psi(x)$ is just the definition of the Dirac delta function $\delta(x-x')$; see (2.32).

The kets $|x\rangle$ do not have finite norm, since $\langle x|x\rangle = \delta(0)$ is not finite. This means that they are really generalized kets in the sense defined earlier [see the paragraph after (2.32)], and are not associated with physical states in the Hilbert space, strictly speaking. However, (2.197)–(2.199) show that they do satisfy the requirements of orthonormality and completeness in the Dirac sense. Therefore, we will refer to the uncountable continuous set $\{|x\rangle\}$ as the **position eigenstate orthobasis**, despite the fact that they are only idealized limits of physical states.

The inner product of two kets can be written in terms of an integral of the wavefunctions, using the completeness relation, as

$$\langle\phi|\psi\rangle = \int_a^b dx\ \langle\phi|x\rangle\langle x|\psi\rangle = \int_a^b dx\, \phi^*(x)\, \psi(x). \tag{2.200}$$

In particular, the squared norm of a state can be written in the manifestly non-negative form

$$\langle\psi|\psi\rangle = \int_a^b dx\, |\psi(x)|^2. \tag{2.201}$$

For valid physical states, both (2.200) and (2.201) should be finite, and the latter must be nonzero. In other words, physical states correspond to wavefunctions that are **normalizable**. This set of functions forms a Hilbert space. However, for physics purposes, one imposes the further constraint that the wavefunctions are also continuous. The reason for this is as follows. The norm of a ket is not physically significant, but it is often most convenient to choose its normalization so that it has norm 1. In particular, if the wavefunction for a single particle is chosen to have unit norm, $\int_a^b dx\, |\psi(x)|^2 = 1$, then $|\psi(x)|^2$ can be interpreted as the probability density, in the sense that the probability to find the particle between x and $x + dx$ is

$$d\mathcal{P}(x) = |\psi(x)|^2 dx. \tag{2.202}$$

This is known as the **Born rule** for the probabilistic interpretation of the wavefunction, after its original proponent Max Born. On physical grounds, the probability density $|\psi(x)|^2$ can in principle be measured, and therefore should be unambiguous; this is ensured if $\psi(x)$ (and hence $|\psi(x)|^2$) is continuous.

There is a one-to-one correspondence between kets $|\psi\rangle$ and their wavefunctions $\psi(x) = \langle x|\psi\rangle$. Now consider the ket corresponding to the derivative of the wavefunction, $d\psi/dx$. Let us write

$$|d\psi/dx\rangle = D\,|\psi\rangle\,, \tag{2.203}$$

which defines a linear operator D on the space of kets, so that

$$\langle x|D|\psi\rangle = \langle x|d\psi/dx\rangle = d\psi/dx. \tag{2.204}$$

Using completeness gives a condition satisfied by the position-eigenket matrix elements of D,

$$d\psi/dx = \int_a^b dx' \langle x|D|x'\rangle\langle x'|\psi\rangle = \int_a^b dx' \langle x|D|x'\rangle\psi(x'). \tag{2.205}$$

We also have

$$d\psi/dx = \frac{d}{dx}\left(\int_a^b dx' \psi(x')\delta(x-x')\right) = \int_a^b dx' \psi(x')\frac{d}{dx}\delta(x-x'). \tag{2.206}$$

A necessary and sufficient condition for the consistency of (2.205) and (2.206) is

$$\langle x|D|x'\rangle = \frac{d}{dx}\delta(x-x'). \tag{2.207}$$

To see the necessity, plug in $\psi(x') = \delta(x'-x'')$, do the integrations, and then relabel $x'' \to x'$.

We now see that D cannot be an observable, because it is certainly not Hermitian, since

$$\langle x'|D^\dagger|x\rangle = \left(\langle x|D|x'\rangle\right)^* = \frac{d}{dx}\delta(x-x') = -\frac{d}{dx'}\delta(x'-x) = -\langle x'|D|x\rangle. \tag{2.208}$$

Although D is not Hermitian, the minus sign at the end of (2.208) suggests that it is anti-Hermitian, and we can simply multiply by a factor of i or $-i$ to make an observable. We therefore define[10] the **momentum operator**,

$$P = -i\hbar D, \tag{2.209}$$

which could be Hermitian and thus a candidate to be an observable. However, since this is an infinite-dimensional Hilbert space, and (2.208) was derived using the technically ill-defined derivatives of the Dirac delta function, it is prudent to be suspicious about the Hermiticity of P. So, let us check.

From (2.203),

$$P\,|\psi\rangle = -i\hbar\,|d\psi/dx\rangle\,. \tag{2.210}$$

[10] The sign and the $\hbar$ normalization in the definition of P are mysteriously arbitrary at this point, but will be justified in what follows; see the discussions following (2.233) and surrounding (3.40) and (3.41). For now, note that the presence of $\hbar$ at least has the correct units to make P a momentum.

Therefore, given any two kets $|f\rangle$ and $|g\rangle$, with wavefunctions $\langle x|f\rangle = f(x)$ and $\langle x|g\rangle = g(x)$,

$$\langle g|P|f\rangle = \int_a^b dx\, \langle g|x\rangle\langle x|P|f\rangle = -i\hbar \int_a^b dx\, g(x)^* \frac{d}{dx} f(x), \quad (2.211)$$

where the first equality used completeness. Interchanging the roles of f and g, we also have

$$\langle f|P|g\rangle = -i\hbar \int_a^b dx\, f(x)^* \frac{d}{dx} g(x). \quad (2.212)$$

Therefore, the difference between $P^\dagger$ and P, in an arbitrary matrix element, is

$$\langle f|P^\dagger|g\rangle - \langle f|P|g\rangle = (\langle g|P|f\rangle)^* - \langle f|P|g\rangle \quad (2.213)$$

$$= i\hbar \int_a^b dx\, g(x) \frac{d}{dx} f(x)^* + i\hbar \int_a^b dx\, f(x)^* \frac{d}{dx} g(x) \quad (2.214)$$

$$= i\hbar \int_a^b dx\, \frac{d}{dx} \left[f(x)^* g(x)\right] \quad (2.215)$$

$$= i\hbar [f(b)^* g(b) - f(a)^* g(a)]. \quad (2.216)$$

So, for P to be a Hermitian operator, we must have

$$f(b)^* g(b) = f(a)^* g(a). \quad (2.217)$$

This is a condition on allowed wavefunctions (and their corresponding kets) in the physical Hilbert space of states if P is to be an observable.

For example, for a particle confined to a box with impenetrable walls at $x = a$ and $x = b$, (2.217) is satisfied because all wavefunctions are required to obey $\psi(a) = \psi(b) = 0$. It is also satisfied if all wavefunctions satisfy the weaker condition $\psi(b) = e^{i\varphi}\psi(a)$, where φ is any fixed constant. This can occur if x is interpreted to be some kind of periodic or angular coordinate, defined for example on the interval $0 \leq x \leq 2\pi$; then $\varphi = 0$ implies that wavefunctions are single-valued at the point $x = 0$, which is identified with $x = 2\pi$.

If the domain of x is infinite, with $a = -\infty$ and $b = \infty$, then things are trickier. In practical solutions for wavefunctions, one sometimes finds that the condition $\psi(x) \to 0$ at large spatial distances is naturally satisfied. States with this property are called **bound states**. If either $|f\rangle$ or $|g\rangle$ is one of these, then (2.217) is satisfied. However, there may also be other solutions that do not obey this property, which are examples of **unbound states**, also known as **scattering states**. States without finite norm are very useful idealized solutions to simple problems, but strictly speaking they are generalized ket states rather than physical ones. In fact, finiteness of the norms for true physical states $|f\rangle$ and $|g\rangle$ with finite energy ensures that $|f(x)|^2$ and $|g(x)|^2$ must approach 0 for large $|x|$. One resolution is that in a more precise formulation that maintains Hermiticity of P, one should only consider unbound states with finite norm, which can be realized as wave packet superpositions of the infinite norm unbound states. Another possibility is to confine the system to a very large box with sides of length L = several light-years, to be safe, so that (2.217) holds with $a = -L/2$ and $b = L/2$. This can be done either by imposing vanishing boundary conditions $\psi(-L/2) = \psi(L/2) = 0$ or just periodicity $\psi(-L/2) = \psi(L/2)$. These modifications will have a negligible effect on physical questions about phenomena localized far from the edges of the box, and in practice one can use the idealized unbound states with impunity in most cases.

Next, consider the eigenvalue problem for the momentum operator. The eigenkets with definite and constant momentum p, corresponding to a **free particle**, satisfy

$$P\,|p\rangle = p\,|p\rangle\,. \tag{2.218}$$

Similarly to the case of position eigenstates, these are taken to obey Dirac orthonormality and completeness conditions, again as a special case of (2.183) and (2.184),

$$\langle p|p'\rangle = \delta(p - p'), \tag{2.219}$$

$$\int_{-\infty}^{\infty} dp\;|p\rangle\langle p| = I. \tag{2.220}$$

Here the allowed range of continuous p is taken to be from $-\infty$ to ∞. As should now be familiar, the generalized kets $|p\rangle$ are, strictly speaking, not associated with physical states, because they do not have finite norm, but are still very useful as idealizations, because they satisfy the properties of an orthobasis in the Dirac sense. For any ket $|\psi\rangle$, we define the **momentum wavefunction**

$$\widetilde{\psi}(p) \equiv \langle p|\psi\rangle\,, \tag{2.221}$$

which can be viewed as the components of the ket $|\psi\rangle$ in the **momentum representation**. Using the completeness relation (2.220), the inner product of any two states can be written

$$\langle\phi|\psi\rangle = \int_{-\infty}^{\infty} dp\;\widetilde{\phi}^*(p)\,\widetilde{\psi}(p). \tag{2.222}$$

If one has chosen the ket $|\psi\rangle$ to have norm 1, then

$$\int_{-\infty}^{\infty} dp\;|\widetilde{\psi}(p)|^2 = 1, \tag{2.223}$$

and we can interpret

$$d\mathcal{P}(p) = |\widetilde{\psi}(p)|^2 \, dp \tag{2.224}$$

as the probability for the particle in the state $|\psi\rangle$ to have momentum between p and $p + dp$.

Consider the position wavefunctions of the free-particle momentum eigenstates, defined by

$$\phi_p(x) \equiv \langle x|p\rangle \,. \tag{2.225}$$

Now we have

$$\langle x|P|p\rangle = p\,\langle x|p\rangle = p\phi_p(x), \tag{2.226}$$

but also, using the completeness relation for position eigenkets,

$$\begin{aligned}\langle x|P|p\rangle &= \int_{-\infty}^{\infty} dx'\,\langle x|P|x'\rangle\langle x'|p\rangle = \int_{-\infty}^{\infty} dx' \Big(-i\hbar\frac{d}{dx}\delta(x-x')\Big)\phi_p(x') \\ &= -i\hbar\frac{d}{dx}\bigg(\int_{-\infty}^{\infty} dx'\,\delta(x-x')\phi_p(x')\bigg) = -i\hbar\frac{d}{dx}\phi_p(x).\end{aligned} \tag{2.227}$$

Comparing this to (2.226), we see that the momentum eigenvalue equation in the position representation is a linear first-order differential equation,

$$\frac{d}{dx}\phi_p(x) = i\frac{p}{\hbar}\phi_p(x), \tag{2.228}$$

with the solution

$$\phi_p(x) = c_p e^{ipx/\hbar}, \tag{2.229}$$

where c_p is a nonzero complex number. To fix the magnitude of c_p, we use completeness in x,

$$\langle p'|p\rangle = \int_{-\infty}^{\infty} dx\,\langle p'|x\rangle\langle x|p\rangle = \int_{-\infty}^{\infty} dx\,\phi_{p'}(x)^*\phi_p(x) \tag{2.230}$$

$$= c_{p'}^* c_p \int_{-\infty}^{\infty} dx\, e^{ix(p-p')/\hbar} \tag{2.231}$$

$$= |c_p|^2\, 2\pi\hbar\, \delta(p-p'), \tag{2.232}$$

where the last equality used the delta function expression (2.34). So, in order to enforce the Dirac orthonormality condition of (2.219), we choose $c_p = 1/\sqrt{2\pi\hbar}$, and so

$$\langle x|p\rangle = \phi_p(x) = \frac{1}{\sqrt{2\pi\hbar}} e^{ipx/\hbar}, \tag{2.233}$$

up to an unavoidably arbitrary choice of constant phase. From the physical consideration that the wavefunctions $\phi_p(x)$ should not blow up at large $|x|$, p must be real, as appropriate for the eigenvalue of a Hermitian operator P. The position dependence of the momentum eigenstate is seen to have the form of a plane wave, with wavenumber $p/\hbar$, and therefore wavelength $2\pi\hbar/p$. Comparison with de Broglie's (1.27) confirms that we made the correct choice of magnitude of normalization of the momentum operator P in (2.209).

As a further check of the completeness condition for momentum eigenkets in (2.220), take the matrix element of it between two arbitrary position eigenkets,

$$\langle x'|x\rangle = \int_{-\infty}^{\infty} dp\, \langle x'|p\rangle\langle p|x\rangle = \frac{1}{2\pi\hbar}\int_{-\infty}^{\infty} dp\, e^{ip(x'-x)/\hbar} = \delta(x-x'). \tag{2.234}$$

In particular, this confirms that we must integrate over the whole range $-\infty < p < \infty$ in the momentum completeness relation (2.220).

We can also use completeness to find formulas that convert between the momentum wavefunction and the position wavefunction. Using completeness in x,

$$\widetilde{\psi}(p) = \langle p|\psi\rangle = \int_{-\infty}^{\infty} dx\, \langle p|x\rangle\langle x|\psi\rangle = \frac{1}{\sqrt{2\pi\hbar}}\int_{-\infty}^{\infty} dx\, e^{-ipx/\hbar}\psi(x). \tag{2.235}$$

Conversely, given $\widetilde{\psi}(p)$, we can reconstruct $\psi(x)$ using completeness in p,

$$\psi(x) = \langle x|\psi\rangle = \int_{-\infty}^{\infty} dp\, \langle x|p\rangle\langle p|\psi\rangle = \frac{1}{\sqrt{2\pi\hbar}}\int_{-\infty}^{\infty} dp\, e^{ipx/\hbar}\widetilde{\psi}(p). \tag{2.236}$$

In words, the position and momentum wavefunctions are Fourier transforms of each other, with some $\hbar$'s included. It is also easy to check that

$$\langle x|X|\psi\rangle = x\psi(x), \qquad \langle x|P|\psi\rangle = -i\hbar\frac{d}{dx}\psi(x), \tag{2.237}$$

$$\langle p|P|\psi\rangle = p\widetilde{\psi}(p), \qquad \langle p|X|\psi\rangle = i\hbar\frac{d}{dp}\widetilde{\psi}(p). \tag{2.238}$$

These are analogous to matrix representations of observables in a finite-dimensional Hilbert space. If one works in the position representation, with wavefunctions of x, then the observable X is represented by x and the observable P is represented by $-i\hbar d/dx$. Inner products are accomplished by integration over all x, with complex conjugation for the wavefunction of the bra vector. If, instead, one works in the

Table 2.1 Summary of the position and momentum representations for a particle moving in one dimension. The position and momentum wavefunctions are related to each other as in (2.235) and (2.236)

Representation	$\|\psi\rangle$	$\langle\psi\|$	X	P	$\langle\psi_1\|\psi_2\rangle$
Position	$\psi(x)$	$\psi(x)^*$	x	$-i\hbar\dfrac{d}{dx}$	$\int dx\ \psi_1(x)^*\psi_2(x)$
Momentum	$\widetilde{\psi}(p)$	$\widetilde{\psi}(p)^*$	$i\hbar\dfrac{d}{dp}$	p	$\int dp\ \widetilde{\psi}_1(p)^*\widetilde{\psi}_2(p)$

momentum representation with wavefunctions $\widetilde{\psi}(p)$, then P is represented by p and X is represented by $i\hbar d/dp$, and inner products are done by integration over all p. For example,

$$\langle\psi_1|X|\psi_2\rangle = \int_{-\infty}^{\infty} dx\ \psi_1(x)^*\, x\, \psi_2(x) = \int_{-\infty}^{\infty} dp\ \widetilde{\psi}_1(p)^*\Big(i\hbar\frac{d}{dp}\Big)\widetilde{\psi}_2(p), \tag{2.239}$$

$$\langle\psi_1|P|\psi_2\rangle = \int_{-\infty}^{\infty} dx\ \psi_1(x)^*\Big(-i\hbar\frac{d}{dx}\Big)\psi_2(x) = \int_{-\infty}^{\infty} dp\ \widetilde{\psi}_1(p)^*\, p\, \widetilde{\psi}_2(p). \tag{2.240}$$

The position and momentum representation rules are summarized in Table 2.1.

The position representation provides a convenient way to find the commutator of X and P. First we evaluate

$$\langle x|XP|\psi\rangle = x\Big(-i\hbar\frac{d}{dx}\Big)\psi(x) = -i\hbar x\frac{d\psi}{dx}, \tag{2.241}$$

$$\langle x|PX|\psi\rangle = -i\hbar\frac{d}{dx}\big(x\psi(x)\big) = -i\hbar x\frac{d\psi}{dx} - i\hbar\psi(x). \tag{2.242}$$

Therefore, for every ket $|\psi\rangle$,

$$\langle x|[X,P]|\psi\rangle = i\hbar\,\langle x|\psi\rangle\,, \tag{2.243}$$

so we can conclude that

$$[X,P] = i\hbar, \tag{2.244}$$

where the identity operator on the right side is understood. We derived this position-momentum commutation relation after defining the operator P by its action on the orthobasis of position eigenkets, but one could just as easily have worked in the other direction, taking (2.244) to be the fundamental definition, and then deriving the operation of P on the kets.

All of the preceding generalizes in a straightforward way to three position and three momentum observables. We define position and momentum operators that are vectors in real space (not the Hilbert vector space, in which they are Hermitian operators), by

$$\boldsymbol{R} = \hat{x}X + \hat{y}Y + \hat{z}Z, \tag{2.245}$$

$$\boldsymbol{P} = \hat{x}P_x + \hat{y}P_y + \hat{z}P_z. \tag{2.246}$$

They satisfy commutation relations:

$$[X, P_x] = i\hbar, \qquad [Y, P_y] = i\hbar, \qquad [Z, P_z] = i\hbar, \tag{2.247}$$

with all other relevant commutators vanishing. It is convenient to adopt a notation in which the components of $\boldsymbol{R}$ and $\boldsymbol{P}$ are given an index $a = 1, 2, 3$ corresponding to x, y, z respectively, so that $R_1 = X$, $R_2 = Y$, $R_3 = Z$, and $P_1 = P_x$, $P_2 = P_y$, $P_3 = P_z$. Then the commutation relations take the form

$$[R_a, P_b] = i\hbar\delta_{ab}, \qquad [R_a, R_b] = 0, \qquad [P_a, P_b] = 0. \tag{2.248}$$

The operators $\boldsymbol{R}$ and $\boldsymbol{P}$ have eigenkets $|\boldsymbol{r}\rangle$ and $|\boldsymbol{p}\rangle$ with eigenvalues[11] $\boldsymbol{r} = \hat{x}x + \hat{y}y + \hat{z}z$ and $\boldsymbol{p} = \hat{x}p_x + \hat{y}p_y + \hat{z}p_z$, so that

$$\boldsymbol{R}\,|\boldsymbol{r}\rangle = \boldsymbol{r}\,|\boldsymbol{r}\rangle\,, \qquad \boldsymbol{P}\,|\boldsymbol{p}\rangle = \boldsymbol{p}\,|\boldsymbol{p}\rangle\,. \tag{2.249}$$

These eigenkets satisfy completeness relations

$$I = \int d^3\boldsymbol{r}\ \ |\boldsymbol{r}\rangle\langle\boldsymbol{r}|\,, \tag{2.250}$$

$$I = \int d^3\boldsymbol{p}\ \ |\boldsymbol{p}\rangle\langle\boldsymbol{p}|\,, \tag{2.251}$$

and have inner products realizing Dirac orthonormality,

$$\langle\boldsymbol{r}'|\boldsymbol{r}\rangle = \delta^{(3)}(\boldsymbol{r} - \boldsymbol{r}'), \qquad \langle\boldsymbol{p}'|\boldsymbol{p}\rangle = \delta^{(3)}(\boldsymbol{p} - \boldsymbol{p}'). \tag{2.252}$$

Here the three-dimensional delta function is given in rectangular coordinates by

$$\delta^{(3)}(\boldsymbol{r} - \boldsymbol{r}') = \delta(x - x')\delta(y - y')\delta(z - z'), \tag{2.253}$$

---

[11] Here we have stretched the previous notion of eigenvalue slightly, as foreshadowed after (2.182), because our eigenvalues here are actually not just numbers, but vectors in real space, or equivalently ordered triples of numbers (x, y, z) or (p_x, p_y, p_z). This was a sneaky thing to do, but it is convenient, and is perfectly valid *if* the components are compatible observables, as here. This will be discussed further in Sect. 3.2.

and in spherical coordinates by

$$\delta^{(3)}(\boldsymbol{r}-\boldsymbol{r}') = \frac{1}{r^2}\,\delta(r-r')\,\delta(\phi-\phi')\,\delta(\cos\theta-\cos\theta'), \tag{2.254}$$

with a practical definition that, when integrating over a volume V,

$$\int_V d^3\boldsymbol{r}'\, f(\boldsymbol{r}')\,\delta^{(3)}(\boldsymbol{r}-\boldsymbol{r}') = \begin{cases} f(\boldsymbol{r}) & \text{if } \boldsymbol{r} \text{ is inside the volume } V, \\ 0 & \text{if } \boldsymbol{r} \text{ is outside the volume } V. \end{cases} \tag{2.255}$$

The wavefunction for a free particle with momentum eigenvalue $\boldsymbol{p}$ is a plane wave,

$$\langle \boldsymbol{r}|\boldsymbol{p}\rangle = \frac{1}{(2\pi\hbar)^{3/2}} e^{i\boldsymbol{p}\cdot\boldsymbol{r}/\hbar}. \tag{2.256}$$

The position and momentum wavefunctions for a general state $|\psi\rangle$ are

$$\psi(\boldsymbol{r}) = \langle \boldsymbol{r}|\psi\rangle\,, \qquad \widetilde{\psi}(\boldsymbol{p}) = \langle \boldsymbol{p}|\psi\rangle\,, \tag{2.257}$$

and they are related by Fourier transforms,

$$\widetilde{\psi}(\boldsymbol{p}) = \frac{1}{(2\pi\hbar)^{3/2}} \int d^3\boldsymbol{r}\; e^{-i\boldsymbol{p}\cdot\boldsymbol{r}/\hbar}\,\psi(\boldsymbol{r}), \tag{2.258}$$

$$\psi(\boldsymbol{r}) = \frac{1}{(2\pi\hbar)^{3/2}} \int d^3\boldsymbol{p}\; e^{i\boldsymbol{p}\cdot\boldsymbol{r}/\hbar}\,\widetilde{\psi}(\boldsymbol{p}), \tag{2.259}$$

which follows from (2.256) and the completeness relations (2.250) and (2.251). The squared norm of a state can be written in terms of either the position or momentum wavefunction, as

$$\langle\psi|\psi\rangle = \int d^3\boldsymbol{r}\; |\psi(\boldsymbol{r})|^2 = \int d^3\boldsymbol{p}\; |\widetilde{\psi}(\boldsymbol{p})|^2. \tag{2.260}$$

The interpretation of the integrands is that the probability for the particle described by the state $|\psi\rangle$ to be found in an infinitesimal volume $d^3\boldsymbol{r}$ near $\boldsymbol{r}$ is given by the Born rule,

$$d\mathcal{P}(\boldsymbol{r}) = d^3\boldsymbol{r}\; |\psi(\boldsymbol{r})|^2, \tag{2.261}$$

while the probability for it to have momentum in a volume $d^3\boldsymbol{p}$ in momentum space near $\boldsymbol{p}$ is

$$d\mathcal{P}(\boldsymbol{p}) = d^3\boldsymbol{p}\; |\widetilde{\psi}(\boldsymbol{p})|^2. \tag{2.262}$$

The last claims are based on postulates that will be stated more generally in Sect. 3.1.

Table 2.2 Summary of the position and momentum representations for a particle moving in three dimensions. The position and momentum wavefunctions are related as in (2.258) and (2.259). In rectangular coordinates, $\nabla_p = \hat{x}\partial/\partial p_x + \hat{y}\partial/\partial p_y + \hat{z}\partial/\partial p_z$

Representation	$\|\psi\rangle$	$\langle\psi\|$	$\boldsymbol{R}$	$\boldsymbol{P}$	P^2	$\langle\psi_1\|\psi_2\rangle$
Position	$\psi(\boldsymbol{r})$	$\psi(\boldsymbol{r})^*$	$\boldsymbol{r}$	$-i\hbar\nabla$	$-\hbar^2\nabla^2$	$\int d^3\boldsymbol{r}\ \psi_1(\boldsymbol{r})^*\psi_2(\boldsymbol{r})$
Momentum	$\widetilde{\psi}(\boldsymbol{p})$	$\widetilde{\psi}(\boldsymbol{p})^*$	$i\hbar\nabla_p$	$\boldsymbol{p}$	p^2	$\int d^3\boldsymbol{p}\ \widetilde{\psi}_1(\boldsymbol{p})^*\widetilde{\psi}_2(\boldsymbol{p})$

We also note, for future reference, that in the position representation, $\boldsymbol{P}$ is represented by $-i\hbar\nabla$, where ∇ is the gradient, and the momentum squared operator $P^2 = \boldsymbol{P}\cdot\boldsymbol{P}$ is represented by $-\hbar^2\nabla^2$, where ∇^2 is the Laplacian. The rules for the position and momentum representations for a particle in three dimensions are summarized in Table 2.2.

2.9 Tensor-Product Hilbert Spaces

We will often want to work with state spaces that are combinations of simpler state spaces. For example, consider a particle moving in three dimensions. In the previous section, we worked with a Hilbert space spanned by an orthobasis of kets that are eigenkets of all three position coordinates. An equivalent way to proceed is to first define three separate Hilbert spaces that only describe the x, y, and z degrees of freedom, and then combine them to form the Hilbert space that describes all three. As another example, a Hilbert space for a system of two particles can be constructed from the Hilbert spaces of the two particles separately.

The formal construction of a Hilbert space from simpler ones is called a **tensor product**. Consider two distinct Hilbert spaces $\mathcal{H}_1$ and $\mathcal{H}_2$ with dimensions d_1 and d_2, spanned respectively by orthobases $\{|\alpha_j\rangle\}$ and $\{|\beta_k\rangle\}$. Then the tensor product Hilbert space $\mathcal{H} = \mathcal{H}_1 \otimes \mathcal{H}_2$ is defined to be a complex linear vector space with an inner product obeying the following properties:

- There is an orthobasis for $\mathcal{H}$ denoted $|\alpha_j\rangle \otimes |\beta_k\rangle$ for $j = 1, \ldots, d_1$ and $k = 1, \ldots, d_2$. Thus, $\mathcal{H}$ has dimension $d = d_1 d_2$, and any ket in $\mathcal{H}$ can be written in a unique way as a linear combination of the d kets $|\alpha_j\rangle \otimes |\beta_k\rangle$.
- Tensor product kets satisfy linearity and distributive properties. This means that if $|\psi\rangle$ and $|\phi\rangle$ are any kets in $\mathcal{H}_1$, and $|v\rangle$ and $|w\rangle$ are any kets in $\mathcal{H}_2$, and a, b, c, d are complex numbers, then

$$(a\,|\psi\rangle + b\,|\phi\rangle) \otimes (c\,|v\rangle + d\,|w\rangle) = ac\,|\psi\rangle \otimes |v\rangle + ad\,|\psi\rangle \otimes |w\rangle + bc\,|\phi\rangle \otimes |v\rangle + bd\,|\phi\rangle \otimes |w\rangle\,. \quad (2.263)$$

- The dual tensor product Hilbert space has an orthobasis $\langle\alpha_j| \otimes \langle\beta_k|$.
- The inner product of orthobasis kets in $\mathcal{H}$ is inherited from the inner products of the $\mathcal{H}_1$ and $\mathcal{H}_2$ orthobasis kets, according to

$$\big(\langle\alpha_j|\otimes\langle\beta_k|\big)\big(|\alpha_l\rangle\otimes|\beta_m\rangle\big) = \delta_{jl}\,\delta_{km}. \tag{2.264}$$

- Given an operator A that acts on $\mathcal{H}_1$ and an operator B that acts on $\mathcal{H}_2$, the tensor product operator $A\otimes B$ is defined to act on $\mathcal{H}$ according to

$$(A\otimes B)(|\alpha_j\rangle\otimes|\beta_k\rangle) = (A|\alpha_j\rangle)\otimes(B\,|\beta_k\rangle). \tag{2.265}$$

In the very common case that B is the identity operator, we may simply write A instead of $A\otimes I$. Thus, if it is understood that the operator A acts nontrivially only on $\mathcal{H}_1$, then we write $A(|\alpha_j\rangle\otimes|\beta_k\rangle) = (A|\alpha_j\rangle)\otimes|\beta_k\rangle$. Similarly, if A is the identity operator, then we simply write B instead of $I\otimes B$.

Even though an orthobasis of kets for $\mathcal{H}$ can be chosen so that its elements are tensor products $|\alpha_j\rangle\otimes|\beta_k\rangle$, not all kets in $\mathcal{H}$ are tensor products. Only special kets, called **separable**, can be written as a tensor product of kets,

$$|\psi\rangle\otimes|v\rangle\,. \tag{2.266}$$

A ket in a tensor product Hilbert space that cannot be written in this form, for any choice of $|\psi\rangle$ and $|v\rangle$, is said to be an **entangled state**, a concept first highlighted by Erwin Schrödinger in 1935. Because $\mathcal{H}$ is defined to be a vector space, all linear combinations of its elements must also be included in it. Therefore, for example,

$$|\psi_1\rangle\otimes|v_1\rangle + |\psi_2\rangle\otimes|v_2\rangle \tag{2.267}$$

is certainly an element of $\mathcal{H}$. However, provided that neither term is the null ket, it cannot be written as a tensor product of kets $|\psi\rangle\otimes|v\rangle$ unless either $|\psi_1\rangle = c\,|\psi_2\rangle$ or $|v_1\rangle = c\,|v_2\rangle$ for some complex constant c. We will have much more to say about the properties of entangled states in Chaps. 14, 25, 26, and 27.

One of the common uses of the tensor product formalism is as a divide-and-conquer strategy for solving problems. This includes situations where we can make use of an already-solved problem for one or more components of the tensor product. As a simple example, the Hilbert space $\mathcal{H}$ for a free particle moving in three dimensions can be written as a tensor product of Hilbert spaces that would describe a particle moving in the three rectangular coordinates separately,

$$\mathcal{H} = \mathcal{H}_x\otimes\mathcal{H}_y\otimes\mathcal{H}_z. \tag{2.268}$$

An orthobasis for this Hilbert space is

$$|\boldsymbol{r}\rangle = |x\rangle\otimes|y\rangle\otimes|z\rangle\,, \tag{2.269}$$

where $|x\rangle$ describes a state in which a particle is known to have X eigenvalue x. In this example, as in many other cases, the tensor product notation $|x\rangle\otimes|y\rangle\otimes|z\rangle$ has no real advantage over just writing the typographically cleaner

$$|\boldsymbol{r}\rangle = |x, y, z\rangle \,, \tag{2.270}$$

so that is what we will do from now on. Then, $X\,|x, y, z\rangle = x\,|x, y, z\rangle$ and $Y\,|x, y, z\rangle = y\,|x, y, z\rangle$, and $Z\,|x, y, z\rangle = z\,|x, y, z\rangle$. This is similar to the separation-of-variables strategy for solving partial differential equations with several independent variables.

Later, we will study the orbital angular momentum eigenvalue problem by isolating only the relevant angular (θ, ϕ) component of the three-dimensional particle Hilbert space, and then apply the results more generally to the whole Hilbert space. Schematically, this relies on using an orthobasis that is a tensor product of the form

$$|\text{radial part}\rangle \otimes |\text{angular momentum part}\rangle \,, \tag{2.271}$$

although it is often not notationally convenient to write it that way explicitly. This will be made more precise in Sect. 8.6 after we develop the theory of angular momentum.

Another important use of tensor product spaces is to describe systems of more than one particle. For example, we can construct a Hilbert space that describes two distinct spinless particles labeled 1 and 2, with an orthobasis of kets

$$|\boldsymbol{r}_1, \boldsymbol{r}_2\rangle = |\boldsymbol{r}_1\rangle \otimes |\boldsymbol{r}_2\rangle \,. \tag{2.272}$$

Here, each $|\boldsymbol{r}_n\rangle$ is an element of the Hilbert space for just one particle moving in three dimensions. This construction can naturally be generalized to a tensor product Hilbert space for N spinless particles with orthobasis kets

$$|\boldsymbol{r}_1, \ldots, \boldsymbol{r}_N\rangle = |\boldsymbol{r}_1\rangle \otimes \cdots \otimes |\boldsymbol{r}_N\rangle \,. \tag{2.273}$$

However, when the particles have spin (intrinsic angular momentum), or are identical, things are more complicated, as we will discuss in Chap. 18.

Exercises

2.1 Use the Cauchy–Schwarz inequality (2.21) to prove the triangle inequality (2.24), and show that equality holds if and only if $|w\rangle = n\,|v\rangle$ where n is a non-negative real number, or one of them is the null ket. (Hint: start with $\|(|v\rangle + |w\rangle)\|^2 = (\langle v| + \langle w|)(|v\rangle + |w\rangle)$, and use the property of complex numbers $\text{Re}(z) \le |z|$.)

2.2 Use the definition of the adjoint to show that $(|w\rangle\,\langle v|)^\dagger = |v\rangle\,\langle w|$ and that $(AB)^\dagger = B^\dagger A^\dagger$.

2.3 Suppose that U is a unitary operator.

(a) Show that if A is a Hermitian operator, then $U^\dagger AU$ is also Hermitian.
(b) Show that if V is another unitary operator, then VU and $U^\dagger VU$ are each also unitary.

2.4 Consider the trace of an operator A defined in terms of an orthobasis $|\varphi_k\rangle$ by (2.179).

(a) Show that $\mathrm{Tr}(A)$ does not depend on the choice of orthobasis. To do this, consider any other orthobasis of kets $|\chi_q\rangle$, and show that $\mathrm{Tr}(A)$ is also equal to $\sum_q \langle\chi_q|A|\chi_q\rangle$.
(b) Show that $\mathrm{Tr}(AB) = \mathrm{Tr}(BA)$.
(c) Show that the trace of the outer product is the inner product, $\mathrm{Tr}\big(|v\rangle\langle w|\big) = \langle w|v\rangle$.

2.5 Suppose A, B, and C are operators.

(a) Show that $[A, BC] = [A, B]C + B[A, C]$.
(b) Prove the **Jacobi identity**, $[A, [B, C]] + [B, [C, A]] + [C, [A, B]] = 0$.
(c) If $[[A, B], B] = 0$, show that $[A, B^n] = n[A, B]B^{n-1}$. (Hint: do this by mathematical induction, by showing that if it is true for some n, then it is also true for $n + 1$.)
(d) Again assuming $[[A, B], B] = 0$, show that for any function $f(x)$ that has a power series expansion,

$$[A, f(B)] = [A, B]f'(B). \tag{2.274}$$

This is Theorem 2.8.

2.6 Prove Theorem 2.9. (Hints: define $F(\lambda) = e^{\lambda(A+B)}e^{-\lambda B}e^{-\lambda A}$. Then show that $dF/d\lambda = c\lambda[A, B]F(\lambda)$ where c is a number that you will find. You may use the fact that $\frac{d}{d\lambda}e^{\lambda X} = Xe^{\lambda X} = e^{\lambda X}X$ for any operator X, and the power series expansion for the exponential function, and the result of Exercise 2.5(c). Then solve the differential equation for $F(\lambda)$ using the boundary condition at $\lambda = 0$, and plug in the value $\lambda = 1$ to get the desired result.)

2.7 Prove Theorem 2.4 (Schur's lemma) in the case of a finite-dimensional Hilbert space. [Hints: To prove (ii) from (i), consider $[A, P_j] = 0$, where $P_j = |\varphi_j\rangle\langle\varphi_j|$ is the projection operator for an arbitrary orthobasis ket $|\varphi_j\rangle$. To prove (iii) from (ii), for any two distinct elements $|\varphi_j\rangle$ and $|\varphi_k\rangle$ of an arbitrary orthobasis, consider a new orthobasis which includes elements $|\varphi'_j\rangle = (|\varphi_j\rangle + |\varphi_k\rangle)/\sqrt{2}$ and $|\varphi'_k\rangle = (|\varphi_j\rangle - |\varphi_k\rangle)/\sqrt{2}$. Then compute $\langle\varphi'_j|A|\varphi'_k\rangle$ in terms of matrix elements in the original orthobasis.]

2.8 Consider a vector space spanned by three orthobasis kets $|1\rangle$, $|2\rangle$, and $|3\rangle$, and a Hermitian operator A defined by

$$A\,|1\rangle = 5\,|1\rangle + i\,|2\rangle\,, \quad A\,|2\rangle = -i\,|1\rangle + 5\,|2\rangle\,, \quad A\,|3\rangle = 4\,|3\rangle\,. \tag{2.275}$$

(a) Write down the 3×3 matrix representation of A in this basis.
(b) Find the eigenvalues of A. Two of them should be equal (degenerate).
(c) For the non-degenerate eigenvalue, show that the most general form of the eigenvector is, in ket form, $c(|1\rangle + i\,|2\rangle)$, where c is an arbitrary complex number.
(d) For the degenerate eigenvalue, find the most general form for a single eigenvector, writing it in terms of two arbitrary complex numbers. Give your answer both in three-component column vector and ket forms. Check that this vector is orthogonal to the result in part (c).

2.9 Consider a state space spanned by an orthobasis consisting of three kets $|1\rangle$, $|2\rangle$, and $|3\rangle$. Let two Hermitian operators A and B be defined by

$$A\,|1\rangle = 2a\,|1\rangle + a\,|3\rangle\,, \qquad A\,|2\rangle = a\,|2\rangle\,, \qquad A\,|3\rangle = a\,|1\rangle + 2a\,|3\rangle\,, \tag{2.276}$$

$$B\,|1\rangle = b\,|1\rangle + 2b\,|3\rangle\,, \qquad B\,|2\rangle = 0, \qquad B\,|3\rangle = 2b\,|1\rangle + b\,|3\rangle\,, \tag{2.277}$$

where a and b are real constants. The orthobasis kets can be represented by vectors

$$|1\rangle \leftrightarrow v_1 = \begin{pmatrix}1\\0\\0\end{pmatrix}, \qquad |2\rangle \leftrightarrow v_2 = \begin{pmatrix}0\\1\\0\end{pmatrix}, \qquad |3\rangle \leftrightarrow v_3 = \begin{pmatrix}0\\0\\1\end{pmatrix}. \tag{2.278}$$

(a) What arc the 3×3 matrix representations of the operators A and B in this basis?
(b) Show that $[A, B] = 0$, which implies that A and B can be simultaneously diagonalized.
(c) Find the eigenvalues of A and the eigenvalues of B. (One of them has non-degenerate eigenvalues; you would be wise to use this to plan your strategy for the next part.)
(d) Find an orthonormal basis v_1', v_2', and v_3' of eigenvectors of both A and B. For each of them, give their eigenvalues with respect to each of A and B.
(e) Write the matrix U that transforms the original orthobasis into the one you found in the previous part. This means that $v_1' = Uv_1$ and $v_2' = Uv_2$ and $v_3' = Uv_3$. Check by direct computation that U is unitary.
(f) Check that U diagonalizes the matrices A and B, by computing $U^\dagger AU$ and $U^\dagger BU$. Note that this is an illustration of Theorem 2.16.

2.10 Find expressions for the following matrix elements, where X and P are the position and momentum operators for a particle moving in one dimension, with eigenkets $|x\rangle$ and $|p\rangle$ respectively, with the Dirac normalizations $\langle x'|x\rangle = \delta(x - x')$ and $\langle p'|p\rangle = \delta(p - p')$.
(a) $\langle x|X|p\rangle$, (b) $\langle p|P|x\rangle$, (c) $\langle x|PX|p\rangle$, (d) $\langle p|PX|x\rangle$, (e) $\langle x|P^2X^2|p\rangle$.

The Core Principles of Quantum Mechanics

3

3.1 The Basic Postulates

We are now ready to present the six basic postulates that define quantum mechanics. Although they are not postulates in the traditional formal and rigorous mathematical sense, they constitute the key principles that should govern all self-contained quantum systems and can be used to make physical predictions. Some sources give versions that differ from the ones presented here, by combining two or all three of Postulates 2, 3, and 4, or by leaving out the controversial Postulate 5, or by including a requirement that the state space of a composite system is always the tensor product of its subsystems. Others include a postulate governing identical particles, which we do not include for reasons discussed in Chap. 18.

Postulate 1: States The state of a quantum system is specified by a non-null ket $|\psi\rangle$, a vector in a complex Hilbert space (a complex linear vector space with an inner product). For a physical state, $\langle\psi|\psi\rangle$ is real and positive, but its value is not otherwise physically significant; two state kets $|\psi\rangle$ and $c|\psi\rangle$ are physically equivalent if c is any nonzero complex number.

Postulate 2: Observables Physically measurable quantities, called observables, correspond to Hermitian operators whose eigenstates can be chosen to be a complete orthonormal basis (orthobasis) of the state space. This means that for an observable A there is a basis $|\alpha, u\rangle$ satisfying $A|\alpha, u\rangle = \alpha|\alpha, u\rangle$, where the α are the eigenvalues, and the u are the degeneracy labels for each α, with

$$\langle\alpha', u'|\alpha, u\rangle = \delta_{\alpha\alpha'}\,\delta_{uu'}, \tag{3.1}$$

and the completeness relation

S. P. Martin, *Quantum Mechanics*, Graduate Texts in Physics,
https://doi.org/10.1007/978-3-032-21119-4_3

$$I = \sum_{\alpha} \sum_{u} |\alpha, u\rangle\langle\alpha, u|. \tag{3.2}$$

The preceding assumes that α and u have discrete values. If instead α has continuous values, then Dirac orthonormality and completeness are used: $\delta_{\alpha\alpha'}$ is replaced by $\delta(\alpha - \alpha')$ and $\sum_{\alpha}$ is replaced by a definite integral $\int d\alpha$ over the range of allowed α. Similarly, if u is continuous, then $\delta_{uu'}$ is replaced by $\delta(u - u')$, and $\sum_{u}$ is replaced by a definite integral $\int du$. In some cases α and/or u can have both some discrete and some continuous values, which are then summed and integrated over the possible values.

Postulate 3: Allowed Results of Measurements The result of the measurement of an observable A is always one of its eigenvalues, α. This rule is sensible and consistent because the eigenvalues of a Hermitian operator are always real, and do not depend on the arbitrary choice of orthobasis used to calculate them, as we observed following Theorem 2.18.

Postulate 4: Probabilities of Results of Measurements If a system is in a state given by a ket $|\psi\rangle$ which has been normalized so that $\langle\psi|\psi\rangle = 1$, and an observable A is measured, then the probability of getting a particular discrete result α is

$$\mathcal{P}(\alpha) = \sum_{u} |\langle\alpha, u|\psi\rangle|^2 = \langle\psi|P_{\alpha}|\psi\rangle, \tag{3.3}$$

where $|\alpha, u\rangle$ are normalized as in (3.1), and

$$P_{\alpha} = \sum_{u} |\alpha, u\rangle\langle\alpha, u| \tag{3.4}$$

is the projection operator to the subspace of states with eigenvalue α for A. If instead α is continuous, then the probability of getting a result between α and $\alpha + d\alpha$ is

$$d\mathcal{P}(\alpha) = d\alpha \sum_{u} |\langle\alpha, u|\psi\rangle|^2 = d\alpha\, \langle\psi|P_{\alpha}|\psi\rangle. \tag{3.5}$$

This postulate generalizes the Born rule for the probabilistic interpretation of the position wavefunction, (2.261), and the corresponding rule for momentum, (2.262). Note that (3.3) ensures that the probabilities are always non-negative,

$$\mathcal{P}(\alpha) \geq 0, \tag{3.6}$$

and the completeness relation (3.2) ensures that the sum (or integral) of the probabilities for all possible outcomes for a measurement must be unity,

$$\sum_{\alpha} \mathcal{P}(\alpha) = 1. \tag{3.7}$$

Thus, these important consistency requirements demanded by any sensible theory of probability are built into the postulates of quantum mechanics.

Postulate 5: Collapse of the State due to Measurement If a system is in a state $|\psi\rangle$ and an observable A is measured and found to have the value α, then the state immediately after the measurement will be an eigenstate of A with eigenvalue α, specifically,

$$P_\alpha|\psi\rangle, \tag{3.8}$$

where P_α is the projection operator given by (3.4). If α is discrete, this ket has a norm that is typically less than 1; in fact, from (3.3) and the projection operator rules $P_\alpha^\dagger = P_\alpha$ and $P_\alpha^2 = P_\alpha$, we see that the squared norm is simply equal to the probability $\mathcal{P}(\alpha)$. Therefore, one can divide the ket (3.8) by its norm, to obtain the normalized state after the measurement,

$$\frac{P_\alpha|\psi\rangle}{\sqrt{\langle\psi|P_\alpha|\psi\rangle}} = \frac{P_\alpha|\psi\rangle}{\sqrt{\mathcal{P}(\alpha)}}. \tag{3.9}$$

There is a nice shortcut in the special case that the eigenvalue α is not degenerate, because then this post-measurement state is simply the corresponding eigenstate $|\alpha\rangle$, up to an unobservable phase. More generally, it is always a linear combination of the states $|\alpha, u\rangle$.

Postulate 6: Time Evolution of the Quantum State Between measurements described by Postulate 5, the time dependence of a state obeys the Schrödinger differential equation,

$$i\hbar\frac{d}{dt}|\psi(t)\rangle = H(t)|\psi(t)\rangle, \tag{3.10}$$

where $H(t)$ is a Hermitian operator, called the **Hamiltonian**. When $H(t)$ does not depend explicitly on time, its eigenvalues are the allowed energies of the system.

3.2 Valid and Invalid Questions

Scientific advances often reveal not just how to answer certain difficult questions, but that other seemingly sensible questions do not need to be answered or even considered. From the theory of special relativity, we learn that it makes no sense to ask questions about collisions of particles with relative speed larger than the speed of light in vacuum c, because the structure of the theory forbids such collisions entirely. One also learns not to ask questions concerning spatially separated events that are supposedly simultaneous in two different inertial reference frames that are moving with respect to each other, because simultaneity is frame-dependent in special relativity. Even though our experience with nonrelativistic systems might make such questions seem legitimate, they are in fact fundamentally invalid.

Similarly, in quantum mechanics, there are questions that have no good answer because they are not valid to begin with. A prominent example is "what are the position and momentum of this particle at time t?". In classical mechanics this question makes perfect sense, and we learn to calculate the answer from some initial conditions and the equations of motion. However, in quantum mechanics, even in the most idealized case, we can only ask for the probability that the measurement of an observable has a specific result from among the allowed list of eigenvalues. In any particular measurement, that observable could be the position of a particle, or it could be the momentum, but it cannot be both simultaneously.

To see why, suppose we attempt to define a clever new multi-component operator

$$\Omega = (\boldsymbol{R},\ \boldsymbol{P}), \tag{3.11}$$

an ordered pair whose components are the position and momentum vectors of a particle. If one could measure Ω, the result would give the simultaneous position and momentum of a particle. Since $\boldsymbol{R}$ and $\boldsymbol{P}$ are each Hermitian, Ω may indeed be defined as a Hermitian operator. However, it is not an observable, because it fails the part of the definition that calls for the eigenstates to form an orthobasis that satisfies the completeness relation. In fact, Ω has no eigenstates at all; this is directly related to the fact that the components of $\boldsymbol{R}$ and $\boldsymbol{P}$ do not all commute. Notice that if $\boldsymbol{R}$ and $\boldsymbol{P}$ had a vanishing commutator, then Theorem 2.19 would guarantee a whole orthobasis of simultaneous eigenstates for them, and thus for Ω, which could therefore have been an acceptable observable. But, as things are, Ω is not an observable, and the postulates of quantum mechanics do not allow for its measurement.

More generally, we can ask whether a set of observables can be measured simultaneously (as a multi-component operator), and whether it makes sense to ask what the results of such a joint measurement are. This depends on the commutation relations of the observable operators.

First, suppose that the operators $A, B, C, \ldots$ all commute with each other (are compatible). Then the **simultaneous measurement** can be performed, and can be defined as a sequence of consecutive measurements, performed immediately after one another so that there is no time for the system to evolve between the measurements. It is left to Exercise 3.1 to show, using Postulates 4 and 5, that for compatible observables the final results for the probabilities of the different outcomes $(\alpha, \beta, \gamma, \ldots)$ and the corresponding final state do not depend on which order one performs the measurements, as long as they are all performed with no intervening time delay, so that Postulate 6 does not come into play.

Next, suppose that two of the operators are incompatible, with a commutator $[A, B]$ that has no zero eigenvalues. This includes the case that $[A, B]$ is a nonzero multiple of the identity operator, notably if A and B are a position operator and the corresponding momentum. In this case, Postulate 5 tells us that after measuring B the system will be left in some state $|\beta\rangle$ that is an eigenstate of B, but it is definitely not an eigenstate of A. (Otherwise, it would be an eigenstate of $[A, B]$ with eigenvalue 0, which we assumed does not exist.) Similarly, if A is measured, the system will be left in a state that is definitely not an eigenstate of B. Therefore, the order of making the measurements certainly makes a difference, and one cannot define their simultaneous measurement.

A third possibility is that $[A, B]$ is an operator that has both vanishing and non-vanishing eigenvalues. In this case, one might measure A and find a result that leaves the state in an eigenstate of B, but this will not always happen. The same is true if B is measured first. An evaluation using Postulates 4 and 5 will be necessary on a case-by-case basis to decide under what circumstances A and B might be simultaneously determined in the final state.

Consider a classical observable $f(a, b, c, \ldots)$ where $a, b, c, \ldots$ are quantities that have quantum observable counterparts $A, B, C, \ldots$. Then, there is always at least one quantum operator $F(A, B, C, \ldots)$ which is also an observable. However, one must be careful in defining it if $A, B, C, \ldots$ do not all commute, due to quantum ordering ambiguity. For example, if $f(x, p) = xp$ where x and p are the position and momentum of a particle moving in one dimension, we could try quantum operator versions $F(X, P) = XP$ or $F(X, P) = PX$ or the symmetrized form $F(X, P) = (XP + PX)/2$. The first two of these are not Hermitian. However, the last is Hermitian, and is an observable. More generally, by completely symmetrizing each term of $F(A, B, C, \ldots)$, we can always construct a Hermitian version of it, as can easily be proved using (2.74).

For the converse, some quantum observables have no classical counterpart; the quintessential example is spin, or intrinsic angular momentum. The maximum magnitude of the spin component of a particle is a fixed multiple of $\hbar$. Crucially, unlike ordinary angular momentum, there are no states in which it can take on classical values arbitrarily large compared to $\hbar$.

A perhaps unexpected—but very useful—example of a valid question in quantum mechanics is: "Given a system in a state $|\psi\rangle$, what is the probability of finding it in another state $|\chi\rangle$?". (Such a question has a valid counterpart in classical physics, but there it is somewhat trivial since all information about the state of a classical system is, in principle, more directly accessible.) The observable corresponding to this question is the projection operator

$$P_\chi = |\chi\rangle\langle\chi|, \tag{3.12}$$

where $|\chi\rangle$ is assumed to have unit norm. The operator P_χ is Hermitian, and has eigenvalues 1 (with eigenstate $|\chi\rangle$) and 0 (with eigenstates consisting of all states orthogonal to $|\chi\rangle$). Thus, the result of the measurement of P_χ will always be either 1 ("yes, the system is in the state $|\chi\rangle$") or 0 ("no, it is not in the state $|\chi\rangle$"), even though the state $|\psi\rangle$ before the measurement need not have had either of these definite attributes. The probability of the outcome 1 is $|\langle\chi|\psi\rangle|^2$, assuming both kets are normalized. If we do measure P_χ and obtain the result 1, then the state after the measurement will be $P_\chi|\psi\rangle$, which is simply the same as $|\chi\rangle$, up to normalization. If we obtain the result 0, then the state after the measurement will be $(I - P_\chi)|\psi\rangle$, which is orthogonal to $|\chi\rangle$. A crucial feature of quantum mechanics, as embodied in the collapse Postulate 5, is that making the measurement changes the state; it is not the same after the measurement as it was before, unless the system was already in an eigenstate of P_χ.

One can also construct an observable that generalizes (3.12) in a natural way to a weighted sum of projection operators for any orthobasis $\{|\varphi_n\rangle\}$ with $n = 1, \ldots, d$, where d is the dimension of the Hilbert space. Such an observable is associated with the valid question "Given a state $|\psi\rangle$, what are the probabilities of finding it in each of these orthobasis states?" For example, consider the Hermitian operator

$$Q = \sum_{n=1}^{d} n P_n, \tag{3.13}$$

where $P_n = |\varphi_n\rangle\langle\varphi_n|$ are the orthobasis state projection operators. The eigenstates of Q are the orthobasis states $|\varphi_n\rangle$, and the corresponding eigenvalues are the coefficients of the orthogonal projection operators, namely the integers $n = 1, \ldots, d$. Therefore, the possible outcomes of measuring Q are $n = 1, \ldots, d$, as visualized in Fig. 3.1. The probabilities for these outcomes are $\mathcal{P}(n) = |\langle\varphi_n|\psi\rangle|^2$, assuming that the ket $|\psi\rangle$ has norm 1. After the measurement outcome n, the system will be left in the state $|\varphi_n\rangle$. Thus, measuring the observable Q changes the state by resolving it into one of the given orthobasis states, and the outcome n tells you which one.

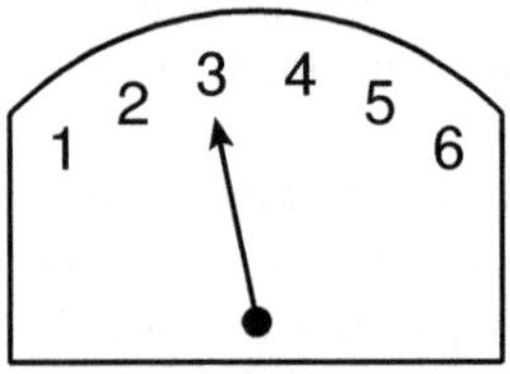

Fig. 3.1 A meter for the Q observable of (3.13), for a system with $d = 6$, showing the result after measurement of Q has resolved the quantum system with initial state $|\psi\rangle$ into the orthobasis state $|\varphi_3\rangle$ from among the possibilities $|\varphi_n\rangle$ with $n = 1, 2, 3, 4, 5, 6$. The probability of this outcome is $|\langle\varphi_3|\psi\rangle|^2$

3.3 Expectation Values and Uncertainties

In addition to the quantities directly associated to single measurements of an observable, there are statistical quantities that result from making many measurements. Consider an idealized situation in which we have access to an arbitrarily large number N of copies of a system, all in the same state $|\psi\rangle$. This mythical group of identical and independent quantum systems is called a **pure ensemble**. The **expectation value** of an operator A in the state $|\psi\rangle$ is defined to be the average value obtained by measuring A in these independent experiments, as $N \to \infty$. According to the frequentist interpretation of probabilities, this is equal to the sum of the possible outcomes for each experiment multiplied by their respective probabilities, which can be evaluated using Postulate 4 as

$$\begin{aligned}\sum_\alpha \alpha \mathcal{P}(\alpha) &= \sum_\alpha \sum_u \alpha \langle\psi|\alpha, u\rangle\langle\alpha, u|\psi\rangle = \sum_\alpha \sum_u \langle\psi|A|\alpha, u\rangle\langle\alpha, u|\psi\rangle \\ &= \langle\psi|A|\psi\rangle. \end{aligned} \tag{3.14}$$

Here, the first equality used (3.3) and assumed that $|\psi\rangle$ is normalized to unity, the second equality used the fact that $|\alpha, u\rangle$ are eigenstates of A with eigenvalue α, and the completeness relation was used to get the last equality. In cases where the state $|\psi\rangle$ is understood by context, it is customary to denote the expectation value by

$$\langle A\rangle \equiv \langle\psi|A|\psi\rangle, \tag{3.15}$$

still assuming that $|\psi\rangle$ is normalized to unity. If that is not convenient for some reason, one has the more general relation

$$\langle A\rangle = \frac{\langle\psi|A|\psi\rangle}{\langle\psi|\psi\rangle}. \tag{3.16}$$

The expectation value of an observable is easily shown to be a real number.

Another quantity defined through measurements on a large pure ensemble of identical states $|\psi\rangle$ is the **uncertainty** of an observable A, which we will denote by ΔA. Here again the notation assumes that $|\psi\rangle$ is understood by context. It is defined by

$$(\Delta A)^2 = \langle (A - \langle A\rangle I)^2\rangle = \langle\psi|(A - \langle\psi|A|\psi\rangle I)^2|\psi\rangle, \tag{3.17}$$

where we have taken $\langle\psi|\psi\rangle = 1$ to get the final expression. Intuitively, the uncertainty tells us how much the measurement of the observable is expected to fluctuate about its expectation value, over the course of many independent measurements, each performed starting in the same state $|\psi\rangle$. It is the same as the concept of standard deviation in statistics. By expanding (3.17), one obtains the equivalent form that is most commonly used in practical calculations,

$$(\Delta A)^2 = \langle\psi|A^2|\psi\rangle - (\langle\psi|A|\psi\rangle)^2\,, \tag{3.18}$$

again assuming $\langle\psi|\psi\rangle = 1$.

There is a fundamental obstacle to the existence of states with arbitrarily small uncertainties for incompatible observables, imposed by the following result due to Howard P. Robertson and Schrödinger:

Theorem 3.1 (Uncertainty Relation) *In any state $|\psi\rangle$, the uncertainties of two observables A and B obey*

$$(\Delta A)(\Delta B) \geq \frac{1}{2}\big|\langle[A, B]\rangle\big|. \tag{3.19}$$

Proof Define observables $\widetilde{A} = A - \langle A\rangle I$ and $\widetilde{B} = B - \langle B\rangle I$. These are Hermitian, because $\langle A\rangle$ and $\langle B\rangle$ are real numbers. It follows from the definition of uncertainty that

$$\begin{aligned}(\Delta A)^2\,(\Delta B)^2 &= \langle\psi|\widetilde{A}^2|\psi\rangle\langle\psi|\widetilde{B}^2|\psi\rangle = \langle\widetilde{A}\psi|\widetilde{A}\psi\rangle\langle\widetilde{B}\psi|\widetilde{B}\psi\rangle \\ &\geq \big|\langle\widetilde{A}\psi|\widetilde{B}\psi\rangle\big|^2,\end{aligned} \tag{3.20}$$

where the Cauchy–Schwarz inequality (2.21) was used at the end. Therefore, we have

$$(\Delta A)^2\,(\Delta B)^2 \geq \big|\langle\psi|\widetilde{A}\widetilde{B}|\psi\rangle\big|^2 = \left|\frac{1}{2}\langle\psi|[\widetilde{A}, \widetilde{B}]|\psi\rangle + \frac{1}{2}\langle\psi|\{\widetilde{A}, \widetilde{B}\}|\psi\rangle\right|^2. \tag{3.21}$$

Using the Hermiticity of $\widetilde{A}$ and $\widetilde{B}$ yet again, $\langle\psi|[\widetilde{A}, \widetilde{B}]|\psi\rangle$ is a purely imaginary number and $\langle\psi|\{\widetilde{A}, \widetilde{B}\}|\psi\rangle$ is a purely real number. Therefore, the squared magnitude of the sum is equal to the sum of the squared magnitudes, so

$$(\Delta A)^2\,(\Delta B)^2 \geq \frac{1}{4}\big|\langle\psi|[\widetilde{A}, \widetilde{B}]|\psi\rangle\big|^2 + \frac{1}{4}\big|\langle\psi|\{\widetilde{A}, \widetilde{B}\}|\psi\rangle\big|^2. \tag{3.22}$$

Now, because the last term is certainly non-negative, we can drop it without affecting the validity of the inequality. Furthermore, $[\widetilde{A}, \widetilde{B}]$ is just equal to $[A, B]$. So, (3.19) follows from taking the square root of both sides of (3.22).

A famous special case is obtained by taking $A = X$ and $B = P$ for a particle moving in one dimension. Because $[X, P] = i\hbar$, the uncertainties must satisfy the **Heisenberg position-momentum uncertainty relation**, named after Werner Heisenberg (who originally proposed a weaker version of it) but derived rigorously first by Earle H. Kennard,

$$(\Delta X)\,(\Delta P) \geq \hbar/2, \tag{3.23}$$

in any state. It might seem that an even stronger version is possible, since we simply discarded the non-negative last term in (3.22). However, for the case of position and momentum, we will show later, in Sect. 6.1, that there do exist states (those with Gaussian wavefunctions), in which (3.23) is saturated, in other words, equality holds. The same wavefunctions will reappear in Sect. 7.4. So, (3.23) is the strongest possible general version of the position-momentum uncertainty relation.

For a particle moving in three dimensions, one finds in the same way that each of $(\Delta X)(\Delta P_x)$ and $(\Delta Y)(\Delta P_y)$ and $(\Delta Z)(\Delta P_z)$ cannot be less than $\hbar/2$. However, since X and P_y commute, there is no uncertainty inequality for the product $(\Delta X)(\Delta P_y)$. This means that, in principle, one could simultaneously specify the exact values of a particle's coordinate along some direction and the momentum component in an orthogonal direction.

3.4 How States Change

According to Postulate 6, the time evolution of a quantum state obeys a linear first-order differential equation, the Schrödinger equation. Let the initial condition for the state at time $t = t_0$ be $|\psi(t_0)\rangle$. We then define the **time-evolution operator** $U(t, t_0)$ such that the state ket at time t is

$$|\psi(t)\rangle = U(t, t_0)|\psi(t_0)\rangle. \tag{3.24}$$

It follows from this definition, and Schrödinger's equation (3.10), that the operator $U(t, t_0)$ satisfies

$$i\hbar\frac{d}{dt}U(t, t_0) = H(t)U(t, t_0). \tag{3.25}$$

Our goal is to solve this differential equation for $U(t, t_0)$ as a function of t, subject to the boundary condition $U(t_0, t_0) = I$. As we will see, it is a unitary operator if $H(t)$ is Hermitian.

First, consider the case that the Hamiltonian does not depend on t. In that case, the solution is simply

$$U(t - t_0) \equiv U(t, t_0) = e^{-i(t-t_0)H/\hbar}, \tag{3.26}$$

which only depends on the time difference $t - t_0$, not the individual times. To check this claimed solution, note that it clearly satisfies the initial condition at $t = t_0$, and that

$$\begin{aligned} i\hbar\frac{d}{dt}|\psi(t)\rangle &= i\hbar\frac{d}{dt}\left[e^{-i(t-t_0)H/\hbar}|\psi(t_0)\rangle\right] = i\hbar\left[(-iH/\hbar)e^{-i(t-t_0)H/\hbar}\right]|\psi(t_0)\rangle \\ &= H|\psi(t)\rangle \end{aligned} \tag{3.27}$$

recovers the Schrödinger equation, as required. It is crucial in the preceding derivation that the operator H does not depend on time, so that in the exponential it can be treated just like a number, as it obviously commutes with itself.

The unitarity of $U(t - t_0)$ in (3.26) is simple to prove, given that H is Hermitian. From the rules for taking adjoints,

$$\begin{aligned} U(t - t_0)^\dagger &= \left[e^{-i(t-t_0)H/\hbar}\right]^\dagger = e^{i(t-t_0)H^\dagger/\hbar} = e^{i(t-t_0)H/\hbar} \\ &= U(t - t_0)^{-1} = U(t_0 - t). \end{aligned} \tag{3.28}$$

The last equality is a bonus, which shows that evolving a state backward in time is the inverse operation of evolving it forward in time by the same amount, as one might expect. Since $U(t - t_0)$ is a unitary operator, time evolution can be regarded as equivalent to a change of orthobasis.

While (3.26) is a neat formal solution of the Schrödinger equation, in practice it leaves more to do, because the infinite series form of an exponential of an operator can be nontrivial to evaluate in matrix elements. To make further progress, we can apply the spectral decomposition trick of (2.186) to the operator $U(t - t_0)$ as given by (3.26). To accomplish this, first consider the eigenvalue problem for the Hamiltonian, which we are still assuming is Hermitian and does not depend on time. This eigenvalue equation is called the **time-independent Schrödinger equation**, and is written as

$$H|\psi_E\rangle = E|\psi_E\rangle, \tag{3.29}$$

where $|\psi_E\rangle$ does not depend on t. Suppose that this equation has been solved completely for all energy eigenvalues E and all corresponding orthobasis eigenstates $|\psi_E\rangle = |E, u\rangle$, where u is a degeneracy label. Then, using completeness of the energy orthobasis,

$$U(t - t_0) = \sum_E \sum_u e^{-i(t-t_0)E/\hbar}|E, u\rangle\langle E, u|, \tag{3.30}$$

where we have turned the operator H into the number E when acting on each of its eigenstates. Applying this to (3.24) gives

$$|\psi(t)\rangle = \sum_E \sum_u e^{-i(t-t_0)E/\hbar}|E, u\rangle\langle E, u|\psi(t_0)\rangle, \tag{3.31}$$

the spectral decomposition of the state with respect to energy.

Finding solutions of the time-independent Schrödinger equation is one of the main problems of quantum mechanics. This often amounts to solving a matrix eigenvalue equation and/or a differential equation for the wavefunction. As an example of the latter, suppose that the Hamiltonian is that of a particle moving in a potential V in one dimension, with

$$H = \frac{1}{2m}P^2 + V(X), \tag{3.32}$$

where P and X are the momentum and position operators. Taking the inner product of (3.29) with the position eigenstate $|x\rangle$, and using the toolbox of Table 2.1 so that $X \to x$ and $P \to -i\hbar d/dx$ and $|\psi_E\rangle \to \psi_E(x)$, we obtain

$$\left(-\frac{\hbar^2}{2m}\frac{d^2}{dx^2} + V(x) - E\right)\psi_E(x) = 0 \tag{3.33}$$

in the position representation. Similarly, for a spinless particle moving in a potential $V(\boldsymbol{r})$ in three dimensions, the time-independent Schrödinger differential equation is

$$\left(-\frac{\hbar^2}{2m}\nabla^2 + V(\boldsymbol{r}) - E\right)\psi_E(\boldsymbol{r}) = 0, \tag{3.34}$$

to be solved simultaneously for E and $\psi_E(\boldsymbol{r})$. The cases of multiple particles, and particles coupled to an electromagnetic field, will be discussed in Sects. 4.2 and 4.3.

Note that the unitary time-evolution operator in (3.30) is nontrivial only because the phases are different for the eigenstates with different energies. As a special case, acting on an initial-state ket $|\psi_E(t_0)\rangle$ that happens to be an eigenstate of energy, time evolution just multiplies by a global phase (that is, a single phase factor that multiplies the whole state ket), namely $e^{-i(t-t_0)E/\hbar}$. Because such a global phase is not physically significant, the state has not really changed. For this reason, a Hamiltonian eigenstate is also known as a **stationary state**.

To illustrate this, consider the time evolution of a stationary state, starting from $t = 0$,

$$|\psi_E(t)\rangle = e^{-itE/\hbar}|\psi_E(0)\rangle, \tag{3.35}$$

and suppose that at time t we measure some observable A, which is assumed to have no explicit time dependence built into it.[1] Recall from Postulate 4 that the probability of getting a particular measurement result α is

$$\begin{aligned}\mathcal{P}(\alpha, t) &= \sum_u |\langle\alpha, u|\psi_E(t)\rangle|^2 = \sum_u |e^{-itE/\hbar}\langle\alpha, u|\psi_E(0)\rangle|^2 \\ &= \sum_u |\langle\alpha, u|\psi_E(0)\rangle|^2 \\ &= \mathcal{P}(\alpha, 0).\end{aligned} \tag{3.36}$$

[1] By "explicit time dependence", we just mean an explicit appearance of t in the definition of the operator. For example, the position operator X has no explicit time dependence, but the operator $A = X\sin(\omega t)$ does depend explicitly on time, which we express as $\partial A/\partial t = \omega X\cos(\omega t)$.

In a stationary state, the probability of measuring each particular possible result α stays the same for all time, as long as the time evolution is not interrupted by a measurement.

Returning to the case of a general state $|\psi(t)\rangle$ and a general observable A, the time dependence of the expectation value obeys

$$\frac{d}{dt}\langle\psi|A|\psi\rangle = \langle\psi|A\left(\frac{d}{dt}|\psi\rangle\right) + \left(\frac{d}{dt}\langle\psi|\right)A|\psi\rangle + \langle\psi|\frac{\partial A}{\partial t}|\psi\rangle, \tag{3.37}$$

where $\partial A/\partial t$ is the derivative of the explicit time dependence of A. Applying Schrödinger's equation and its Hermitian conjugate to the first two terms on the right, respectively, gives

$$\frac{d}{dt}\langle\psi|A|\psi\rangle = -\frac{i}{\hbar}\langle\psi|[A, H]|\psi\rangle + \langle\psi|\frac{\partial A}{\partial t}|\psi\rangle, \tag{3.38}$$

or, in the more compact notation of expectation values, just

$$\frac{d}{dt}\langle A\rangle = -\frac{i}{\hbar}\langle[A, H]\rangle + \left\langle\frac{\partial A}{\partial t}\right\rangle. \tag{3.39}$$

This general result is known as **Ehrenfest's Theorem**, after Paul Ehrenfest.

In the special case that $|\psi\rangle$ is a stationary state, $H|\psi\rangle = E|\psi\rangle$ and $\langle\psi|H = E\langle\psi|$, so $\langle[A, H]\rangle$ evaluates to $(E - E)\langle A\rangle = 0$. In that case, $d\langle A\rangle/dt = \langle\partial A/\partial t\rangle$. The change in an expectation value in a stationary state is due only to the explicit time dependence of the observable operator.

Suppose that the Hamiltonian is that of a particle moving in a time-independent potential $V(X)$ in one dimension, (3.32), and consider $A = P$, the momentum operator. The operator P has no explicit time dependence, so $\partial P/\partial t = 0$. Also, $[P, H] = [P, V(X)] = -i\hbar dV/dX$, where we used Theorem 2.8 to evaluate $[P, V(X)]$. Therefore, the expectation values in a general state obey

$$\frac{d}{dt}\langle P\rangle = -\left\langle\frac{dV}{dX}\right\rangle, \tag{3.40}$$

which is the quantum-mechanical version of Isaac Newton's second law. Similarly, for $A = X$, using $\partial X/\partial t = 0$ and $[X, H] = [X, P^2]/2m = i\hbar P/m$, we get

$$\frac{d}{dt}\langle X\rangle = \langle P\rangle/m. \tag{3.41}$$

This generalizes in a straightforward way to a particle moving in three dimensions. Ehrenfest's Theorem says that the momentum and position expectation values in quantum mechanics obey the same equations of motion as the corresponding quantities in classical mechanics. Equations (3.40) and (3.41) therefore confirm the choices of sign and magnitude in our definition of the momentum operator P in (2.209).

As another application of Ehrenfest's Theorem, apply (3.39) to the time-independent observable $A = \frac{1}{2}(XP + PX)$. Evaluating $[A, H]$, one finds that for expectation values in a general time-dependent state, the **virial theorem** holds:

$$\frac{d}{dt}\langle A\rangle = \frac{1}{m}\langle P^2\rangle - \langle X V'(X)\rangle. \tag{3.42}$$

For the special case of a stationary state, since we found that $\langle A\rangle$ is independent of time when $\partial A/\partial t = 0$, the left side vanishes and the virial theorem becomes

$$\frac{1}{m}\langle P^2\rangle = \langle X V'(X)\rangle \qquad \text{(in a stationary state).} \tag{3.43}$$

Specializing further to the case of a power-law potential $V(X) = kX^n$, this becomes

$$2\langle T\rangle = n\langle V\rangle, \qquad \text{(in a stationary state, if } V \propto X^n\text{),} \tag{3.44}$$

where $T = P^2/2m$ is the kinetic energy operator. The generalization of the virial theorem to a particle moving in three dimensions is straightforward, and is left to Exercise 3.6.

We now turn to the more difficult case that the Hamiltonian operator depends on time. First, consider time evolution over an infinitesimal interval from time t_0 to time $t_0 + \Delta t$. Then, the content of the Schrödinger equation can be expressed as

$$|\psi(t_0 + \Delta t)\rangle = \left[1 - \frac{i}{\hbar}\Delta t\, H(t_0)\right]|\psi(t_0)\rangle. \tag{3.45}$$

Note that because Δt is infinitesimal, it does not matter here whether we use $H(t_0)$, or $H(t_0 + \Delta t)$, or H evaluated at some intermediate time, because the difference will be higher order in Δt. Up to terms of order $(\Delta t)^2$, we can rewrite this as an exponential,

$$|\psi(t_0 + \Delta t)\rangle = \exp\left[-\frac{i}{\hbar}\Delta t\, H(t_0)\right]|\psi(t_0)\rangle. \tag{3.46}$$

An advantage of writing it this way is that the exponential is a unitary operator if $H(t_0)$ is Hermitian, so that $|\psi(t_0 + \Delta t)\rangle$ has the same norm as $|\psi(t_0)\rangle$. Now, if we evolve the state further from time $t_0 + \Delta t$ to $t_0 + 2\Delta t$ in the same way, we have

$$|\psi(t_0 + 2\Delta t)\rangle = \exp\left[-\frac{i}{\hbar}\Delta t\, H(t_0 + \Delta t)\right]\exp\left[-\frac{i}{\hbar}\Delta t\, H(t_0)\right]|\psi(t_0)\rangle. \tag{3.47}$$

Here, the exponentials cannot easily be combined into a single exponential, because $H(t_0 + \Delta t)$ and $H(t_0)$ are different operators, and need not commute. Continuing in this way,

$$|\psi(t_0 + N\Delta t)\rangle = \left(\prod_{n=0}^{N-1} \exp\left[-\frac{i}{\hbar}\Delta t\, H(t_0 + n\Delta t)\right]\right)|\psi(t_0)\rangle, \tag{3.48}$$

where it is important that the terms in the product are understood to be arranged from higher to lower n, reading from left to right. Since each of the terms in the product is a unitary operator, the whole product is a unitary operator as well. Now we can take

$$N = (t - t_0)/\Delta t \to \infty, \tag{3.49}$$

to obtain

$$|\psi(t)\rangle = U(t, t_0)|\psi(t_0)\rangle, \tag{3.50}$$

where the unitary time-evolution operator is

$$U(t, t_0) = \lim_{N\to\infty} \prod_{n=0}^{N-1} \exp\Big[-\frac{i}{\hbar}\Delta t\, H(t_0 + n\Delta t)\Big]. \tag{3.51}$$

From its construction, it satisfies

$$[U(t_2, t_1)]^\dagger = [U(t_2, t_1)]^{-1} = U(t_1, t_2), \tag{3.52}$$

$$U(t_3, t_2)U(t_2, t_1) = U(t_3, t_1). \tag{3.53}$$

This time-evolution operator can depend on both arguments separately; in general $U(t, t_0) \neq U(t - t_0)$ if the Hamiltonian depends on time.

Another useful way of expressing the unitary time evolution is the **Dyson series**, named after Freeman J. Dyson:

$$U(t, t_0) = I + \sum_{N=1}^{\infty} \left(-\frac{i}{\hbar}\right)^N \int_{t_0}^{t} dt_N \int_{t_0}^{t_N} dt_{N-1} \cdots$$
$$\int_{t_0}^{t_2} dt_1\, H(t_N)H(t_{N-1}) \cdots H(t_1), \tag{3.54}$$

where each t_k integration has lower limit t_0 and upper limit t_{k+1}, for $k = 1, \ldots, N$, with t_{N+1} interpreted as t. You can check directly that (3.54) satisfies the differential equation (3.25), by plugging it in. Formally, one can also write for (3.54)

$$U(t, t_0) = T \exp\Big[-\frac{i}{\hbar}\int_{t_0}^{t} dt'\, H(t')\Big], \tag{3.55}$$

where the T is a time-ordering symbol, an instruction to rearrange the Hamiltonians in the expansion of the exponential so that each $H(t_j)$ appears to the right of $H(t_k)$ whenever $t_j < t_k$ (that is, later times appear to the left). The nested upper limits of integration in (3.54) have neatly removed the need for the $1/N!$ usually present in the series expansion of the exponential function. Another Dyson series, based on the interaction picture of quantum mechanics and useful for time dependent perturbation theory, will be discussed in Sect. 21.2.

In (3.54) the ordering of the Hamiltonians is important because in general they do not commute at different times. In the special case that they do all commute,

$$U(t, t_0) = \exp\left[-\frac{i}{\hbar}\int_{t_0}^{t} dt'\, H(t')\right], \tag{3.56}$$

which agrees with the result (3.26) when H does not depend on time at all.

We have seen that there are two very different ways that a state can change in quantum mechanics: Schrödinger time evolution governed by the Hamiltonian, and collapse of the state ket due to measurement. It is natural to ask whether the latter might be a special case of the former. At least within the standard formulation of quantum mechanics, as given by the postulates listed in Sect. 3.1, the answer is "No!". Hamiltonian time evolution is accomplished by multiplying the state by a *unitary* operator, while collapse of the state due to measurement is associated with multiplying by a projection operator, which is instead *Hermitian*. Thus, there is a fundamental difference between Hamiltonian time evolution and measurement collapse.

The time evolution due to the Hamiltonian is perfectly causal and deterministic; the state at a given time is uniquely determined by the state at earlier times, just as in classical physics, provided that a measurement of the type described in Postulates 4 and 5 has not taken place in the interim. In contrast, the collapse of the wavefunction associated with measurement is inherently probabilistic, rather than deterministic. This means that, unlike in classical physics, it is not possible, even in principle, to predict the future. But the situation is actually worse than that: we cannot even predict the past. Given complete knowledge of the present state of a quantum system, the state before the most recent measurement cannot be known, because Postulate 5 says that the act of measurement changes the state in a way that destroys information, in an irreversible way. If someone or something measured an observable A and got a result α, leaving the system in a state $|\alpha\rangle$, then this tells you that the state of the system $|\psi\rangle$ before the measurement must have had a nonzero matrix element $\langle\alpha|\psi\rangle$, but that is all. This is clearly very incomplete information.

The insistence on a fundamental distinction between the measuring agent and the quantum system being measured, as required by Postulate 5, was developed and promoted by Niels Bohr and collaborators, and is often referred to as the **Copenhagen interpretation** of quantum mechanics. According to the Copenhagen interpretation, the measuring agent apparently can be thought of in classical terms, or at least we do not ask questions about its quantum behavior. This seems troubling, as one can always imagine treating any particular measuring apparatus as itself a quantum system undergoing unitary time evolution. For this reason, many people have advocated changing the postulates of quantum mechanics to modify or completely eliminate Postulate 5 dealing with the collapse of the state due to measurement.

In particular, Hugh Everett, in his 1957 PhD thesis, proposed that the state ket always undergoes unitary time evolution, so that the apparent collapse of the state due to measurement is an illusion of limited human powers of discernment. In this **Everett interpretation** of quantum mechanics, all possible alternatives for every measurement outcome are actually realized, through superposition, in some parts of the great state ket of the universe. This can be imagined as a continuous branching of possibilities, which has inspired the alternative name **many-worlds interpretation** and many science fiction story ideas. It is an observed fact that parts of the Everett superposition corresponding to non-classical-like outcomes for macroscopic systems seem to have negligible amplitudes. This can be explained by a phenomenon called decoherence, discussed in Chap. 26, which involves entanglement correlations between subsystems.

It is difficult not to be sympathetic to the no-collapse view. Indeed, we could draw a big sphere of radius several hundred light years around the Earth, and think of the contents (including us, all other observers who might be able to communicate with us, and all of our measuring devices) as one big quantum system evolving strictly according to unitary time evolution. In any case, there seems to be no reasonable scientific principle that could tell us exactly where we should put the boundary separating the measuring apparatus from the quantum system that supposedly undergoes collapse.

However, from a practical point of view, Postulate 5 is indispensable, because it provides a straightforward, consistent, and reliable way of making predictions for the actual experiments that we do in the real world. No matter how philosophically compelling it might be to discard the measurement collapse of the wavefunction, it is not scientifically necessary, with the exception of some interesting and ambitious proposals to treat the quantum dynamics of the universe as a whole. So far, the postulates of quantum mechanics as given in Sect. 3.1 have stood the test of time, successfully providing accurate predictions of every experimental phenomenon with which they have been confronted.

3.5 Mixed Ensembles and the Density Matrix Operator

The expectation value and uncertainty for an operator were defined for a single quantum state in Sect. 3.3, using the concept of a pure ensemble, consisting of many identical systems in the same state. However, it is often more realistic to suppose that in a large ensemble of quantum systems of the same type, some fraction p_1 of them will be in a state $|\psi_1\rangle$, a fraction p_2 will be in a different state $|\psi_2\rangle$, etc. Such a large collection of systems of the same type, but in different states, is called a **mixed ensemble**. If we choose one of the systems at random from a mixed ensemble, there is a probability p_I that it will be in the state $|\psi_I\rangle$, with $p_I \geq 0$ for each I, and

$$\sum_I p_I = 1. \tag{3.57}$$

The kets $|\psi_I\rangle$ are assumed to have unit norm in the following, but there is no reason why they must be linearly independent or orthogonal to each other. There is not even any constraint on the number of distinct states $|\psi_I\rangle$ that are found in the ensemble, and it could be larger than the dimension of the state space. The mixed ensemble generalizes the concept of a pure ensemble, for which one of the p_I is equal to 1 and all others are 0.

Suppose we choose one system at random from the mixed ensemble, and measure an observable A with eigenvalues α and orthonormal eigenstates $|\alpha, u\rangle$. Then, the probability of getting a particular result α is equal to the sum over $|\psi_I\rangle$ of the product of the probability of choosing a system in that state and the probability from Postulate 4 that a measurement in that state will give α,

$$\mathcal{P}(\alpha)_{\text{mixed}} = \sum_I p_I \sum_u |\langle\alpha, u|\psi_I\rangle|^2. \tag{3.58}$$

We can similarly compute the average result obtained by measuring the observable A many times on systems chosen at random from the mixed ensemble,

$$\overline{A} = \sum_I p_I \langle\psi_I|A|\psi_I\rangle. \tag{3.59}$$

We use an overline notation to denote this **mixed ensemble average**, to distinguish it from the expectation value associated with measuring A in a single state in a pure ensemble.

There are two very different types of probabilities at work in (3.58) and (3.59). First, there are the **ensemble probabilities** p_I, which simply reflect the fact that the mixed ensemble is populated by different states. These ensemble probabilities would exist even if our systems were classical. Second, we have the probabilities associated with the inherently non-deterministic nature of measurement in quantum systems, which are manifested in $\sum_u |\langle\alpha, u|\psi_I\rangle|^2$ and in the expectation value $\langle\psi_I|A|\psi_I\rangle$. The results for $\mathcal{P}(\alpha)_{\text{mixed}}$ and $\overline{A}$ incorporate both types of probabilities.

One should not confuse the concepts of a *superposition* of different quantum states in a pure ensemble and a *population* of the same quantum states in a mixed ensemble. A simple example will illustrate the distinction. Consider a state space with two orthobasis kets $|1\rangle$ and $|2\rangle$. Suppose that initially we have a pure ensemble, with all systems in the superposition state

$$|\psi\rangle = c_1|1\rangle + c_2|2\rangle, \tag{3.60}$$

where c_1 and c_2 are complex numbers subject to the normalization condition $|c_1|^2 + |c_2|^2 = 1$. Now we can conduct a measurement to ask if a system is in the state $|1\rangle$. The probability of finding the result 1 (yes) is $|c_1|^2$, and the probability of finding the result 0 (no) is $|c_2|^2$. If we do this measurement on each and every system in the pure ensemble, we will afterwards have a mixed ensemble, with $p_1 = |c_1|^2$ for the state $|1\rangle$, and $p_2 = |c_2|^2$ for the state $|2\rangle$. This illustrates that one way to prepare a mixed ensemble is to conduct measurements on the members of a pure ensemble.

Continuing with this example, if we now make the same measurement again on the mixed ensemble, we will get the same results as for the pure ensemble; the probability is still $|c_1|^2$ to find the state $|1\rangle$. But now consider instead the probability that measurement of some other observable A will yield the result α, with corresponding eigenket $|\alpha\rangle$. For the pure ensemble with state $|\psi\rangle$, this is

$$\mathcal{P}(\alpha)_{\text{pure}} = |\langle\alpha|\psi\rangle|^2 = \left|c_1\langle\alpha|1\rangle + c_2\langle\alpha|2\rangle\right|^2, \tag{3.61}$$

but for the mixed ensemble, we find instead, from (3.58),

$$\mathcal{P}(\alpha)_{\text{mixed}} = |c_1|^2|\langle\alpha|1\rangle|^2 + |c_2|^2|\langle\alpha|2\rangle|^2. \tag{3.62}$$

The key difference is that in the pure ensemble result there are interference terms that do not appear in the mixed ensemble result. There is relative phase information present in the pure state that is absent in the mixed ensemble.

A standard realization of this same example is found in the double-slit diffraction experiment. Suppose that we have a source of particles that can impact on a plane detection screen, after having passed through one of two very narrow slits in a diffraction screen, as shown in Fig. 1.3. The source of particles is said to be **coherent** if we can describe the state of a given particle as a superposition like (3.60), where $|1\rangle$ represents the state in which the particle travels through slit 1, and $|2\rangle$ represents the state in which it travels through slit 2. If we let the operator A in the above discussion be the position operator X on the detection screen perpendicular to the slit directions, then the probability to find the particle between positions x and $x+dx$ is given by the continuous version of (3.61),

$$d\mathcal{P}(x) = \left|c_1\psi_1(x) + c_2\psi_2(x)\right|^2 dx. \tag{3.63}$$

The individual wavefunctions for the different slit contributions interfere destructively or constructively, depending on x, to produce an interference pattern as shown in Fig. 1.3. But now suppose that there are detectors that measure whether the particle goes through slit 1 or slit 2. This measurement results in a mixed ensemble of particles at the detection screen, and we will have instead the continuous version of (3.62),

$$d\mathcal{P}(x) = \left(|c_1|^2|\psi_1(x)|^2 + |c_2|^2|\psi_2(x)|^2\right)dx, \tag{3.64}$$

in which the interference terms (also known as coherences) are absent. The act of measuring which slit each particle goes through destroys the interference pattern. This "which-slit" measurement might effectively be done by the environment in which the experiment takes place, rather than an official detector monitored by a certified experimentalist. This reduction of a pure ensemble of coherent superpositions to a mixed ensemble of populations is an example of decoherence, which will be discussed further in Chap. 26.

All of the physical information about a mixed ensemble is encoded, in an elegant way, in the **density matrix operator**, or just the **density operator**, due to John von Neumann. It is defined in terms of the ensemble states and probabilities by

$$\rho = \sum_I p_I |\psi_I\rangle\langle\psi_I|. \tag{3.65}$$

In the special case of a pure ensemble, ρ is simply the projection operator for the state $|\psi\rangle$. More generally, it is a sum of projection operators weighted by the frequencies of occurrence of states within the ensemble. In terms of the density operator, the result of (3.58) for the probability of getting the result α for a single measurement of A can be written as

$$\mathcal{P}(\alpha)_{\text{mixed}} = \sum_u \langle\alpha, u|\,\rho\,|\alpha, u\rangle. \tag{3.66}$$

We can rewrite this by choosing an arbitrary orthobasis of kets $\{|\phi_k\rangle\}$, and then using completeness followed by a rearrangement,

$$\begin{aligned}\mathcal{P}(\alpha)_{\text{mixed}} &= \sum_u \langle\alpha, u|\left(\sum_{k=1}^{d} |\phi_k\rangle\langle\phi_k|\right)\rho|\alpha, u\rangle \\ &= \sum_{k=1}^{d} \langle\phi_k|\rho P_\alpha|\phi_k\rangle = \mathrm{Tr}[\rho P_\alpha],\end{aligned} \tag{3.67}$$

where P_α is the projection operator for the result α, defined in (3.4), and at the end we used the definition of the trace of an operator in (2.179). Recall that the trace is independent of the choice of orthobasis.

The average of the results of many measurements of A in a mixed ensemble, as computed in (3.59), can also be re-expressed in terms of the density operator. Again using completeness followed by a rearrangement,

$$\overline{A} = \sum_I p_I \langle\psi_I|A\left(\sum_{k=1}^{d} |\phi_k\rangle\langle\phi_k|\right)|\psi_I\rangle = \sum_{k=1}^{d} \langle\phi_k|\rho A|\phi_k\rangle = \mathrm{Tr}[\rho A]. \tag{3.68}$$

As a special case,

$$\mathrm{Tr}[\rho] = 1, \tag{3.69}$$

which simply re-expresses the conservation of probability from (3.57). It is also straightforward to show $\mathrm{Tr}[\rho^2] \leq 1$, with equality only in the special case that the ensemble is a pure one, in which case one also has $\rho^2 = \rho$.

From its definition in (3.65), ρ is clearly a Hermitian operator. So, according to Theorem 2.15, one can find a special orthobasis consisting of its eigenkets, call them $|\varphi_k\rangle$, with eigenvalues p_k, which are real because of Theorem 2.12. Therefore, no matter what states $|\psi_I\rangle$ were involved in the original preparation of the mixed ensemble, or how many such states there were, we can always rewrite the density operator as

$$\rho = \sum_{k=1}^{d} p_k |\varphi_k\rangle\langle\varphi_k|. \tag{3.70}$$

This is very similar in appearance to (3.65), but has the important difference that the index k now takes on a limited number of values up to the dimension d of the state space. The p_k can be interpreted as the ensemble probabilities for the orthobasis. This illustrates the more general fact that the density operator is not tied to any specific set of states $|\psi_I\rangle$, even if we used one when preparing the mixed ensemble. The density operator can also be used to summarize our (incomplete) information about a single system chosen at random from the mixed ensemble, called a **mixed state**.

A pure ensemble is one extreme special case of a mixed ensemble, in which all systems are in the same state. The opposite extreme is the **completely random ensemble**, which we can define by choosing *any* orthobasis $|\varphi_k\rangle$, and writing

$$\rho = \frac{1}{d}\sum_{k=1}^{d} |\varphi_k\rangle\langle\varphi_k|. \tag{3.71}$$

Here $1/d$ is the ensemble probability for each of the orthobasis states. This density operator is proportional to the identity operator, so it is actually independent of the choice of orthobasis, and is the unique one associated with maximum randomness of states in the ensemble.

The extent to which an ensemble of quantum systems is randomized can more generally be quantified by the **von Neumann entropy**,[2]

$$\sigma = -\mathrm{Tr}[\rho \ln \rho] = -\sum_{k=1}^{d} p_k \ln p_k, \tag{3.72}$$

[2] Many sources replace the natural logarithm in (3.72) by the base-2 logarithm, which is especially convenient for quantum information applications where the systems of interest have 2-state orthobases. This simply changes the normalization of the entropy by a multiplicative factor $1/\ln(2)$, since $\log_2(x) = \ln(x)/\ln(2)$.

where the final result is in terms of the orthobasis ensemble probabilities in (3.70), which are defined to be the eigenvalues of ρ. When $p_k = 0$, one should interpret $p_k \ln p_k$ as 0. The von Neumann entropy is the quantum mechanical analog of the **Shannon entropy** introduced (two decades later!) by Claude E. Shannon in the study of classical information and communication theory, but the Shannon entropy has different properties when combining classical subsystems. It also has the same form, up to a multiplicative factor, as the **Gibbs entropy** defined by Josiah Willard Gibbs in classical statistical mechanics and thermodynamics,

$$S = -k_B \sum_{k=1}^{d} p_k \ln p_k, \tag{3.73}$$

where k_B is Boltzmann's constant, and in this case the p_k are interpreted as the probabilities for microstates $|\varphi_k\rangle$ to occur in the fluctuations of a system.

In our two extreme cases,

$$\sigma = 0 \qquad \text{(pure ensemble)}, \tag{3.74}$$

$$\sigma = -n\left[\frac{1}{n}\ln(1/n)\right] = \ln(n) \qquad \text{(completely random ensemble)}, \tag{3.75}$$

where n is the number of orthobasis states available to the systems in the ensemble, usually the same as the dimension d of the state space. In the case of a mixed state, the entropy is a measure of our ignorance, and it is always between 0 and $\ln(n)$. The result $S = k_B \ln(n)$ obtained for the special case of the completely random ensemble is the **Boltzmann entropy**.

In general, the density operator for a mixed state or a mixed ensemble depends on time. From the general form in (3.65), we have

$$\frac{d\rho}{dt} = \sum_I p_I \left[\left(\frac{d}{dt}|\psi_I\rangle\right)\langle\psi_I| + |\psi_I\rangle\left(\frac{d}{dt}\langle\psi_I|\right)\right], \tag{3.76}$$

and evaluating the time derivatives using the Schrödinger equation, we obtain

$$\frac{d\rho}{dt} = -\frac{i}{\hbar}[H, \rho]. \tag{3.77}$$

Note that this vanishes in the special case of a completely random ensemble: random ensembles stay random. It also vanishes in the case of a pure ensemble if the state is an energy eigenstate, but not if it is a superposition of states with different energies.

The historical reason for the name "density operator" is that this equation is analogous to Liouville's Theorem in classical mechanics, which says that for an ensemble of classical systems, the phase-space density $\rho_{\text{classical}}$ (the number of classical ensemble members per unit position and momentum) obeys

$$\frac{\partial \rho_{\text{classical}}}{\partial t} = \{H, \rho_{\text{classical}}\}_{\text{PB}}, \tag{3.78}$$

in which the commutator is replaced by the classical Poisson bracket without the factor of $-i/\hbar$. This is an example of the classical–quantum correspondence principle to be discussed further in Sect. 4.1.

Equation (3.77) gives the instantaneous change in the density operator as the ensemble of systems undergoes time evolution. More generally, in terms of the unitary time evolution operator that we defined in (3.24),

$$\rho(t) = U(t, t_0)\,\rho(t_0)\,U(t, t_0)^\dagger, \tag{3.79}$$

with a short proof left as an exercise. Since $\rho(t)$ and $\rho(t_0)$ are related by a unitary transformation, they have the same eigenvalues, according to Theorem 2.18. It follows that under time evolution, a pure state stays pure, and a mixed state remains mixed. Furthermore, from the expression for the entropy in (3.72) in terms of the eigenvalues p_k,

$$\sigma(t) = \sigma(t_0) \qquad \text{(for unitary time evolution)}, \tag{3.80}$$

in the absence of external measurements or other influences. So, we have found that the entropy of a closed system does not change with time.

Let us now work out what the density operator will be after a measurement on a mixed state. Suppose we have a density operator ρ as given in (3.70) in terms of orthobasis states $|\varphi_k\rangle$ with ensemble probabilities p_k, and we make a measurement on a randomly chosen ensemble state of an observable A and obtain the result α. For each of the ensemble states $|\varphi_k\rangle$, Postulate 5 tells us that the state after the measurement will be, as in (3.9),

$$\frac{P_\alpha|\varphi_k\rangle}{\sqrt{\langle\varphi_k|P_\alpha|\varphi_k\rangle}}, \tag{3.81}$$

where P_α is the projection operator for the result α. Therefore, we can write the post-measurement density operator as

$$\rho_\alpha = \sum_{k=1}^{d} \mathcal{P}(k|\alpha)\left(\frac{P_\alpha|\varphi_k\rangle}{\sqrt{\langle\varphi_k|P_\alpha|\varphi_k\rangle}}\right)\left(\frac{\langle\varphi_k|P_\alpha}{\sqrt{\langle\varphi_k|P_\alpha|\varphi_k\rangle}}\right), \tag{3.82}$$

where $\mathcal{P}(k|\alpha)$ is the conditional probability that the state selected from the ensemble was $|\varphi_k\rangle$, given that the result α was obtained for A. To evaluate this, we use **Bayes' Theorem**, the fundamental result in the theory of conditional probabilities, which says

$$\mathcal{P}(k|\alpha) = \frac{\mathcal{P}(\alpha|k)\, p_k}{\mathcal{P}(\alpha)}. \tag{3.83}$$

Here, as given by Postulate 4,

$$\mathcal{P}(\alpha|k) = \langle\varphi_k|\,P_\alpha\,|\varphi_k\rangle \tag{3.84}$$

is the conditional probability that the result of a measurement of A is α, assuming that the state was $|\varphi_k\rangle$, while $\mathcal{P}(\alpha) = \mathrm{Tr}[\rho\, P_\alpha]$ is the probability, obtained in (3.67), that a measurement of A in the ensemble resulted in α. Putting these results together in (3.82), the density operator after the measurement becomes

$$\rho_\alpha = \frac{P_\alpha\,\rho\, P_\alpha}{\mathcal{P}(\alpha)} = \frac{P_\alpha\,\rho\, P_\alpha}{\mathrm{Tr}[\rho\, P_\alpha]}. \tag{3.85}$$

This is the density operator version of Postulate 5. Note that it obeys $\rho_\alpha^\dagger = \rho_\alpha$ and $\mathrm{Tr}[\rho_\alpha] = 1$, so that the post-measurement density operator is still Hermitian and has trace 1.

We have now succeeded in finding the rules for measurement probabilities and averages, post-measurement collapse, and time evolution for a mixed state in terms of the density matrix operator ρ, in (3.67), (3.68), (3.77), (3.79), and (3.85). Since each of these results only depends on ρ, and not on the individual ensemble states or probabilities, we have justified the assertion that all of the physical properties of the mixed ensemble or mixed state are contained in the density operator.

A common situation is that a measurement of A has been made, but we do not know the result. Perhaps we were not looking, or we lost the data, or the measurement was made by something or someone else who is unwilling or unable to communicate with us. In that case, the density operator describing the mixed state after the measurement will be

$$\rho' = \sum_\alpha \mathcal{P}(\alpha)\,\rho_\alpha = \sum_\alpha P_\alpha \rho P_\alpha. \tag{3.86}$$

We will now show that the entropy always increases when this occurs, except in the trivial special case $\rho' = \rho$. To do so, we will use the following fact from linear algebra about traces of functions of matrices, due to Oskar Klein, the proof of which is omitted.

Theorem 3.2 (Klein's Inequality, General) *Suppose that the function $f(x)$ is differentiable and strictly convex ($f''(x) > 0$) for all $0 < x < \infty$, and that A and B are Hermitian operators with non-negative eigenvalues. Then*

$$Tr[f(A) - f(B) + (B - A)f'(B)] \geq 0, \tag{3.87}$$

with equality if and only if $A = B$.

We now apply this to the case $f(x) = x \ln x$, and let $A = \rho$ and $B = \tilde{\rho}$ be any two density operators. Then, using $\mathrm{Tr}[\rho] = \mathrm{Tr}[\tilde{\rho}] = 1$, we obtain:

Theorem 3.3 (Klein's Inequality for Density Operators) *Suppose that ρ and $\tilde{\rho}$ are any two density operators on a common state space. Then*

$$Tr[\rho(\ln \rho - \ln \tilde{\rho})] \geq 0, \tag{3.88}$$

with equality if and only if $\rho = \tilde{\rho}$.

Now consider the case of ρ' given by (3.86), which arose from having made a measurement of A on a mixed state described by ρ, with unknown result. The entropy after the measurement is

$$\sigma' = -\mathrm{Tr}[\rho' \ln \rho'] = -\sum_{\alpha} \mathrm{Tr}[P_\alpha \, \rho \, P_\alpha \ln \rho'] = -\sum_{\alpha} \mathrm{Tr}[\rho \, P_\alpha \ln \rho' P_\alpha], \tag{3.89}$$

where the cyclic property of the trace was used at the end. Since $P_\alpha^2 = P_\alpha$, (3.86) shows that P_α commutes with ρ', which implies that P_α also commutes with $\ln \rho'$. Therefore, $P_\alpha \ln \rho' P_\alpha = P_\alpha^2 \ln \rho' = P_\alpha \ln \rho'$. Then, using completeness in the form $\sum_\alpha P_\alpha = I$, we obtain

$$\sigma' = -\mathrm{Tr}[\rho \ln \rho']. \tag{3.90}$$

Using $\sigma = -\mathrm{Tr}[\rho \ln \rho]$ and Klein's inequality (3.88) with $\tilde{\rho} = \rho'$, we finally obtain the claimed result,

$$\sigma' \geq \sigma. \tag{3.91}$$

The entropy increases whenever a nontrivial measurement is made but the result is unknown.

Note that this increase of the von Neumann entropy does not apply to a situation in which we made a measurement on a single mixed state and the result is known. In fact, if the known result of the measurement α is a non-degenerate eigenvalue, then the resulting density operator describing the system will be that of a pure state, with vanishing entropy.

As an important example, consider an ensemble consisting of a bottle of, say, $\sim 10^{22}$ gas molecules, each of which can be in states characterized by energy eigenvalues E and degeneracy labels u. The molecules interact with each other and with the bottle walls, but weakly enough that they can be considered an ensemble of independent quantum states of the same type. Intuitively, each interaction can be thought of as a sort of external measurement on a gas molecule, but the results of these measurements remain unknown. Therefore, (3.91) applies so as to make the entropy as large as it can be, subject to the constraint of energy conservation for the gas as a whole. Thus, when the molecules reach thermal equilibrium, the ensemble probability to find one of them in a particular orthobasis state $|E, u\rangle$ can be determined by the principle that the entropy should be maximized, subject to the constraint, due to energy conservation, that the ensemble average energy has a fixed value $\overline{E}$.

To see the implications of this, we write the density operator in the form

$$\rho = \sum_E \sum_u p_E |E, u\rangle\langle E, u|. \tag{3.92}$$

Here, we have already implemented the idea that maximizing the entropy will require the density operator to correspond to complete randomization within each subspace of fixed energy eigenvalue E, but the relative probabilities p_E remain to be found. Equation (3.92) gives

$$\sigma = -\sum_E g_E\, p_E \ln p_E, \tag{3.93}$$

where g_E is the degeneracy of the energy eigenvalue E. Now, to maximize σ subject to the constraints $1 = \mathrm{Tr}[\rho] = \sum_E g_E\, p_E$ and fixed $\overline{E} = \sum_E g_E\, p_E E$, it is simplest to use the method of Lagrange multipliers. The function to be extremized is

$$f(p_E, \alpha, \beta) = -\sum_E g_E\, p_E \ln p_E + \alpha\Big(1 - \sum_E g_E\, p_E\Big) + \beta\Big(\overline{E} - \sum_E g_E\, p_E E\Big), \tag{3.94}$$

where α and β are the Lagrange multipliers for the trace constraint and the energy constraint, respectively, and $\overline{E}$ is fixed. We then obtain, for each E,

$$0 = \frac{\partial f}{\partial p_E} = -g_E(\ln p_E + 1) - \alpha g_E - \beta g_E E, \tag{3.95}$$

which has the solution

$$p_E = e^{-(\beta E + \alpha + 1)}. \tag{3.96}$$

The $e^{-(\alpha+1)}$ factor is independent of E, and so can be absorbed into a common normalization factor; the important point is that we have derived that the canonical ensemble probabilities that maximize the entropy must be proportional to the **Boltzmann factor,**

$$p_E \propto e^{-\beta E}. \tag{3.97}$$

The Lagrange multiplier β is related to the temperature by the definition

$$\beta = \frac{1}{k_B T}. \tag{3.98}$$

Since β has units of 1/energy, this definition shows that Boltzmann's constant is really just a conversion factor between energy and temperature. If we agreed to measure temperature in units of energy, then Boltzmann's constant would be 1.

The result of (3.92) with $p_E \propto e^{-\beta E}$ is called the **canonical ensemble**, with density operator

$$\rho = \frac{1}{Z} \sum_E \sum_u e^{-\beta E} |E, u\rangle\langle E, u|, \tag{3.99}$$

where the normalization factor Z is called the **partition function**. Thus, the canonical ensemble is completely randomized at each fixed energy level E, but with relative probabilities between energy levels determined by the Boltzmann factor. The partition function can be computed using the requirement $\mathrm{Tr}[\rho] = 1$, which gives

$$Z = \sum_E \sum_u e^{-\beta E} = \sum_E g_E \, e^{-\beta E}. \tag{3.100}$$

Equation (3.99) can be recognized as the spectral decomposition form [see (2.186)] of

$$\rho = \frac{1}{Z} e^{-\beta H}. \tag{3.101}$$

This obviously commutes with H, so according to (3.77), ρ is constant in time. For any observable A defined for each molecule, the canonical ensemble average is, from (3.68),

$$\overline{A} = \frac{1}{Z} \mathrm{Tr}\left[e^{-\beta H} A\right] = \frac{1}{Z} \sum_E \sum_u e^{-\beta E} \langle E, u|A|E, u\rangle. \tag{3.102}$$

In particular, the average energy for states in the canonical ensemble is

$$\overline{E} = \frac{1}{Z} \sum_E g_E E e^{-\beta E} = -\frac{\partial}{\partial \beta} \ln Z. \tag{3.103}$$

It is left as an exercise to check that, with the entropy definition $S = k_B \sigma$,

$$\overline{E} - TS = -\frac{1}{\beta} \ln Z = F, \tag{3.104}$$

where F is called the **Helmholtz free energy** after Hermann von Helmholtz. Let us stop our discussion of the canonical ensemble here, before this book accidentally turns into one on statistical mechanics.

In all of the preceding, we have assumed for notational simplicity that the states $|\psi_I\rangle$ appearing in the ensemble are discrete and countable. As usual, one can also consider a continuum of states, which entails turning summations into integrals. If the ensemble states $|\psi_u\rangle$ are labeled by some continuous parameter u (instead of the discrete label I), then the density operator is

$$\rho = \int du\; p(u)\, |\psi_u\rangle\langle\psi_u|, \tag{3.105}$$

where the ensemble probability density $p(u)$ must satisfy the constraint $\int du\; p(u) = 1$.

Exercises

3.1 Consider compatible observables A and B, whose (possibly degenerate) eigenvalues include α and β, respectively. Suppose that they are both measured, with a negligible time delay so that Schrödinger time evolution does not come into play. Show that the probability of obtaining the joint outcome (α, β) for (A, B) does not depend on the order in which they are measured.

3.2 Show that if the position-space wavefunction $\psi(\boldsymbol{r}) = \langle \boldsymbol{r}|\psi\rangle$ is real, then the expectation value of the momentum operator $\boldsymbol{P}$ must vanish: $\langle \boldsymbol{P}\rangle = 0$. This helps to explain why the states for quantum mechanics must form a *complex* vector space.

3.3 Consider a state $|\psi\rangle$ with position wavefunction $\langle \boldsymbol{r}|\psi\rangle = \psi(\boldsymbol{r})$ and momentum expectation value $\langle\psi|\boldsymbol{P}|\psi\rangle = \boldsymbol{p}$. Show that the state with position-space wavefunction $e^{i\boldsymbol{k}\cdot\boldsymbol{r}}\psi(\boldsymbol{r})$ must have momentum expectation value $\boldsymbol{p} + \hbar\boldsymbol{k}$.

3.4 Consider observables A and B with the following matrix representations on a state space with dimension 3,

$$A = \begin{pmatrix} 1 & 0 & 0 \\ 0 & 0 & 0 \\ 0 & 0 & -1 \end{pmatrix}, \qquad B = \frac{1}{\sqrt{2}} \begin{pmatrix} 0 & -i & 0 \\ i & 0 & -i \\ 0 & i & 0 \end{pmatrix}. \tag{3.106}$$

These matrices are representations of the operators in an orthonormal basis $|1\rangle$, $|0\rangle$, $|-1\rangle$, which are labeled by the non-degenerate eigenvalues of A, which we can therefore call "the A basis".

(a) Find the eigenvalues and the corresponding normalized eigenkets of B (in the A basis).
(b) In the state $|-1\rangle$, calculate the expectation value $\langle B\rangle$ and the uncertainty ΔB.
(c) If the particle is in the state with $A = 1$, and B is measured, what are the possible outcomes and their respective probabilities?
(d) If the particle is in the state with $B = 0$, and A is measured, what are the possible outcomes and their respective probabilities?
(e) Consider the state $|\psi\rangle = \frac{1}{\sqrt{6}}|1\rangle + \frac{1}{\sqrt{6}}|0\rangle + \frac{2}{\sqrt{6}}|-1\rangle$. If the operator A^2 is measured and a result $+1$ is obtained, what is the normalized state ket immediately after the measurement? What was the probability of this result? If A is then immediately measured, what are the possible outcomes and their respective probabilities?

3.5 Consider a particle free to move throughout all space in three dimensions. Suppose that its wavefunction in spherical coordinates is $\psi(\boldsymbol{r}) = Ce^{-r/a}$, where C and a are constants.

(a) If the wavefunction is normalized to unity, what is the magnitude of the constant C?
(b) What is the probability that the particle will be found to be farther from the origin than a?
(c) What are the expectation values and uncertainties of the radial coordinate operator R and of the momentum squared operator P^2?

3.6 Consider a particle moving in three dimensions, governed by the Hamiltonian

$$H = \frac{1}{2m}P^2 + V(\boldsymbol{R}), \tag{3.107}$$

where $P^2 = \boldsymbol{P}\cdot\boldsymbol{P} = P_x^2 + P_y^2 + P_z^2$. Apply Ehrenfest's Theorem (3.39) to the observable $A = \frac{1}{2}(\boldsymbol{R}\cdot\boldsymbol{P} + \boldsymbol{P}\cdot\boldsymbol{R})$ to obtain the **virial theorem in three dimensions**. You should find that for a general state,

$$\frac{d}{dt}\langle A\rangle = \frac{1}{m}\langle P^2\rangle - \langle \boldsymbol{R}\cdot\nabla V\rangle, \tag{3.108}$$

so that for a stationary state,

$$\frac{1}{m}\langle P^2\rangle = \langle \boldsymbol{R}\cdot\nabla V\rangle, \qquad \text{(in a stationary state)}, \tag{3.109}$$

and for the special case of a spherically symmetric power-law potential

$$2\langle T\rangle = n\langle V\rangle, \qquad \text{(in a stationary state, if } V \propto R^n\text{)}, \tag{3.110}$$

where $T = P^2/2m$ is the kinetic energy operator.

3.7 Consider a particle of mass m moving in a one-dimensional potential, with Hamiltonian $H = P^2/2m + V(X)$. Suppose that the eigenstates and energy eigenvalues of H are denoted by $|\psi_n\rangle$ and E_n, where n is a discrete label, so that $H|\psi_n\rangle = E_n|\psi_n\rangle$.

(a) Show that $\langle\psi_n|P|\psi_k\rangle = \alpha_{nk}\langle\psi_n|X|\psi_k\rangle$, where α_{nk} is a quantity that you will determine, which depends on the difference between E_n and E_k. (Hint: consider the commutator $[X, H]$.)

(b) From the result of the previous part, show that

$$\langle\psi_n|P^2|\psi_n\rangle = \beta\sum_k (E_n - E_k)^2|\langle\psi_n|X|\psi_k\rangle|^2, \tag{3.111}$$

where β is a constant quantity that you will find. (Hint: completeness is your friend.)

(c) How does this rule generalize to cases with some continuous energy eigenvalues?

(d) Derive the corresponding results for a particle moving in three dimensions.

3.8 Prove (3.79) for the unitary time evolution of the density operator.

3.9 Consider a quantum system with a state space of dimension d.

(a) Show that to specify a general pure state of the system requires $2d - 2$ real parameters.

(b) Show that to specify the density matrix for a general mixed state of the system requires $d^2 - 1$ real parameters.

3.10 A spin-1/2 system with Hamiltonian $H = \omega S_z$ has energy eigenstates $|\uparrow\rangle$ and $|\downarrow\rangle$ with S_z eigenvalues $\hbar/2$ and $-\hbar/2$ respectively. For the canonical ensemble of a large number of such spins at temperature T, what are the density operator ρ and the partition function Z? What is the Gibbs entropy? What is the average result for a measurement of S_z?

Canonical Variables and the Hamiltonian 4

4.1 Quantum Observables from Classical Dynamics

The postulates of quantum mechanics refer to observable operators with time evolution governed by a Hamiltonian, but do not specifically identify these objects. This is intentional, in order to be general. In many physical situations, one can determine the appropriate Hamiltonian and observables by considering a classical version, but this is not always possible. Indeed, one should think of classical physics as an approximate limit obtained from quantum mechanics, and not the reverse. Still, in favorable circumstances the classical limit of a system can be used to infer the observable operators, including the Hamiltonian, and their algebraic commutator properties in the quantum theory.

In the Lagrangian formulation of classical mechanics, one starts with some dynamical variables q_n, often called generalized coordinates, which we will label by an index n. The Lagrangian is a function of the q_n and their first time derivatives $\dot{q}_n = dq_n/dt$,

$$L(q_n, \dot{q}_n, t). \tag{4.1}$$

The classical equations of motion for the system are then

$$\frac{d}{dt}\frac{\partial L}{\partial \dot{q}_n} = \frac{\partial L}{\partial q_n}, \tag{4.2}$$

for each n. A short calculation, found in any good classical physics textbook, shows that this follows from a variational principle involving the action obtained by integrating the Lagrangian with respect to time. However, we will postpone our own discussion of that until Sect. 29.6. The reason for doing so is that rather than accept the variational principle as a postulate of classical mechanics, we will be able to *derive* it as a consequence of the path integral formulation of quantum mechanics.

S. P. Martin, *Quantum Mechanics*, Graduate Texts in Physics,
https://doi.org/10.1007/978-3-032-21119-4_4

The Hamiltonian formulation of classical mechanics recasts the Lagrangian formulation by defining a **canonical momentum** conjugate to each generalized coordinate,

$$p_n = \frac{\partial L}{\partial \dot{q}_n}. \tag{4.3}$$

Now one defines the Hamiltonian as

$$H(q_n, p_n, t) = \sum_n p_n \dot{q}_n - L(q_n, \dot{q}_n, t), \tag{4.4}$$

where it is important that the $\dot{q}_n$ are to be completely eliminated in favor of the p_n using (4.3). Thus H is a function only of the generalized coordinates and their canonical conjugate momenta, and not their time derivatives. The q_n and p_n are collectively called the **phase-space coordinates**. As shown in the same good classical mechanics textbook, the Lagrangian equations of motion (4.2) are equivalent to the Hamiltonian equations of motion,

$$\dot{q}_n = \frac{\partial H}{\partial p_n}, \qquad \dot{p}_n = -\frac{\partial H}{\partial q_n}. \tag{4.5}$$

The generalized coordinates q_n need not be the rectangular coordinates of a particle, but rather could be any quantities that fully specify the classical state of the system at a given time. Likewise, the canonical momenta need not coincide with mechanical momenta (equal to the product of mass and velocity for particles).

Consider any quantity $a(q_n, p_n, t)$, built out of the generalized coordinates and their conjugate momenta. According to the same good classical physics textbook, the time evolution of a is

$$\frac{da}{dt} = \{a, H\}_{\text{PB}} + \frac{\partial a}{\partial t}, \tag{4.6}$$

where the **Poisson bracket** for any two functions on phase space a and b is defined as

$$\{a, b\}_{\text{PB}} \equiv \sum_n \left(\frac{\partial a}{\partial q_n}\frac{\partial b}{\partial p_n} - \frac{\partial b}{\partial q_n}\frac{\partial a}{\partial p_n} \right), \tag{4.7}$$

and $\partial a/\partial t$ is the explicit time dependence of a. Dirac noted that the Poisson brackets $\{a, b\}_{\text{PB}}$ of classical mechanics are closely analogous to the commutators $[A, B]$ for the corresponding observables in the quantum theory. Both are antisymmetric under interchange of the observables, and at least for the position and momentum observables, one has the correspondence

$$\underline{\text{classical}} \qquad\qquad\qquad \underline{\text{quantum}}$$

$$\{q_n, p_k\}_{\text{PB}} = \delta_{nk} \quad \longleftrightarrow \quad [Q_n, P_k] = i\hbar\delta_{nk}. \tag{4.8}$$

Commutators of this form are called **canonical commutation relations**. Furthermore, (4.6) has a striking resemblance to Ehrenfest's Theorem in quantum mechanics, see (3.39). Indeed, one finds (using Theorem 2.8 to compute the commutators) that

$$\frac{d}{dt}\langle Q_n\rangle = \left\langle\frac{\partial H}{\partial P_n}\right\rangle, \qquad \frac{d}{dt}\langle P_n\rangle = -\left\langle\frac{\partial H}{\partial Q_n}\right\rangle, \tag{4.9}$$

directly analogous to the Hamiltonian equations of motion (4.5).

For a single particle of mass m moving in three dimensions in a potential V, it is natural to choose the Q_n to be the usual rectangular coordinate operators $X = R_x$, $Y = R_y$, and $Z = R_z$. They and their conjugate **canonical momenta** P_x, P_y, and P_z satisfy the commutation relations already given in (2.248),

$$[R_a, P_b] = i\hbar\delta_{ab}, \qquad [R_a, R_b] = 0, \qquad [P_a, P_b] = 0, \tag{4.10}$$

for $a, b = x, y, z$. Writing $P^2 \equiv \boldsymbol{P}\cdot\boldsymbol{P}$, the Hamiltonian operator is $H = \frac{P^2}{2m} + V(\boldsymbol{R})$.

The connection between classical and quantum physics just described is often called the **correspondence principle**, but it has some weaknesses. First, there are some quantum observables that do not have a classical counterpart at all, for example spin, also known as intrinsic angular momentum. Second, there is no guarantee that every generalized coordinate and its momentum will obey a canonical commutation relation. Although it is true for the rectangular coordinates of a particle, or a collection of particles, in other cases one might encounter corrections higher order in $\hbar$, or ambiguities in connecting the classical observables to the quantum ones. This is why we preferred to derive the position-momentum commutation relations by the method given in Sect. 2.8. More generally, the most logical (but perhaps not the simplest) way to make the connection is to derive the classical theory as an approximation to the quantum theory, as we will do using the Feynman sum-over-paths approach in Sect. 29.6.

4.2 The Two-Body Problem

An important special case is that of two particles that interact only through a potential that depends only on their separation. This occurs, for example, in the hydrogen atom to be treated in Chap. 11, where the two particles are the electron and the (much heavier) proton. Another example is neutron-proton scattering, to be studied in Sect. 24.8, where the two masses are almost the same.

In general, the two-body Hamiltonian has the form

$$H = \frac{P_1^2}{2m_1} + \frac{P_2^2}{2m_2} + V(\boldsymbol{R}_1 - \boldsymbol{R}_2), \tag{4.11}$$

where we allow for the possibility that the potential energy depends on the vector displacement (not just its magnitude), and the two particles have masses m_1 and m_2 and canonical position and momentum operators $(\boldsymbol{R}_1, \boldsymbol{P}_1)$ and $(\boldsymbol{R}_2, \boldsymbol{P}_2)$. The components of these observables satisfy commutation relations $[R_{1a}, P_{1b}] = i\hbar\delta_{ab}$ and $[R_{2a}, P_{2b}] = i\hbar\delta_{ab}$, for $a, b = x, y, z$, with all other commutators vanishing. In particular, each of the observables for particle 1 commutes with those of particle 2. As an orthobasis, one can choose the tensor product of the eigenkets of $\boldsymbol{R}_1$ and $\boldsymbol{R}_2$,

$$|\boldsymbol{r}_1, \boldsymbol{r}_2\rangle = |\boldsymbol{r}_1\rangle \otimes |\boldsymbol{r}_2\rangle, \tag{4.12}$$

defined to satisfy the eigenvalue equations

$$\boldsymbol{R}_1|\boldsymbol{r}_1, \boldsymbol{r}_2\rangle = \boldsymbol{r}_1|\boldsymbol{r}_1, \boldsymbol{r}_2\rangle, \qquad \boldsymbol{R}_2|\boldsymbol{r}_1, \boldsymbol{r}_2\rangle = \boldsymbol{r}_2|\boldsymbol{r}_1, \boldsymbol{r}_2\rangle. \tag{4.13}$$

However, the solution of the Hamiltonian eigenvalue problem is complicated by the fact that the potential couples the two particle degrees of freedom together.

Fortunately, as in classical mechanics, such cases can be reduced to a simpler problem that is effectively the same as for two decoupled particles, by making a change of variables to separate the relative motion from that of the center of mass. The relative motion is described by

$$\boldsymbol{R} = \boldsymbol{R}_1 - \boldsymbol{R}_2, \qquad \boldsymbol{P} = \frac{m_2\boldsymbol{P}_1 - m_1\boldsymbol{P}_2}{m_1 + m_2}, \tag{4.14}$$

and the motion of the center of mass by

$$\boldsymbol{R}_{\text{cm}} = \frac{m_1\boldsymbol{R}_1 + m_2\boldsymbol{R}_2}{m_1 + m_2}, \qquad \boldsymbol{P}_{\text{tot}} = \boldsymbol{P}_1 + \boldsymbol{P}_2. \tag{4.15}$$

It is a short exercise to show that the pairs $(\boldsymbol{R}, \boldsymbol{P})$ and $(\boldsymbol{R}_{\text{cm}}, \boldsymbol{P}_{\text{tot}})$ each satisfy canonical commutation relations, and do not interfere with each other. Furthermore, if one defines the total mass M and the **reduced mass** μ according to

$$M = m_1 + m_2, \qquad \mu = \frac{m_1 m_2}{m_1 + m_2}, \tag{4.16}$$

then the Hamiltonian (4.11) can be rewritten in the decoupled form

$$H = H_{\text{cm}} + H_{\text{rel}}, \qquad H_{\text{cm}} = \frac{P_{\text{tot}}^2}{2M}, \qquad H_{\text{rel}} = \frac{P^2}{2\mu} + V(\boldsymbol{R}). \tag{4.17}$$

The center-of-mass degrees of freedom have the same Hamiltonian as that of a free particle with mass M, with an eigenvalue problem that is easy to solve (plane waves). The dynamics of H_{rel} are the same as for a single particle with mass equal to μ, moving in the potential $V(\boldsymbol{R})$. One can now choose a new orthobasis as the tensor product of eigenkets of $\boldsymbol{R}$ and $\boldsymbol{R}_{\text{cm}}$,

$$|\boldsymbol{r}, \boldsymbol{r}_{\text{cm}}\rangle = |\boldsymbol{r}\rangle \otimes |\boldsymbol{r}_{\text{cm}}\rangle, \tag{4.18}$$

and look for stationary-state wavefunction solutions of the form

$$\Psi(\boldsymbol{r}, \boldsymbol{r}_{\text{cm}}) = \langle \boldsymbol{r}, \boldsymbol{r}_{\text{cm}}|\Psi\rangle = \frac{1}{(2\pi\hbar)^{3/2}} e^{i\boldsymbol{k}_{\text{tot}}\cdot\boldsymbol{r}_{\text{cm}}}\, \psi(\boldsymbol{r}). \tag{4.19}$$

Here $\boldsymbol{p}_{\text{tot}} = \hbar\boldsymbol{k}_{\text{tot}}$ is the eigenvalue of $\boldsymbol{P}_{\text{tot}}$, and the relative-coordinate wavefunction satisfies

$$\left(-\frac{\hbar^2\nabla^2}{2\mu} + V(\boldsymbol{r}) - E\right)\psi(\boldsymbol{r}) = 0, \tag{4.20}$$

where E is the relative-motion energy, and the total energy eigenvalue is $E + \hbar^2 k_{\text{tot}}^2/2M$. The eigenvalue problem in (4.20) can now be solved for E and $\psi(\boldsymbol{r})$ as if it were a single particle. If one of the particles is much heavier than the other, as in the case of electrons compared to atomic nuclei, then μ is nearly equal to the mass of the lighter particle.

If the (fixed) number of particles N is three or more, then one can construct appropriate **Jacobi coordinates**, and their canonical momenta, by iteration. First, choose two of the particles and define their relative and center-of-mass coordinates and momenta, just as in the preceding. The two-particle center-of-mass coordinates and momenta are then combined in the same way with those of a third particle, to give another relative coordinate/momentum pair and a three-particle center-of-mass coordinate and momentum. The three-particle center-of-mass coordinate and momentum are combined with those of a fourth particle, etc. In the end, there will be a single center-of-mass coordinate $\boldsymbol{R}_{\text{cm}} = \frac{1}{M}\sum_{i=1}^{N} m_i \boldsymbol{R}_i$ and momentum $\boldsymbol{P}_{\text{tot}} = \sum_{i=1}^{N} \boldsymbol{P}_i$ for the whole system with total mass $M = \sum_{i=1}^{N} m_i$, and $N-1$ translation-invariant coordinate/momentum pairs, each satisfying canonical commutation relations. The kinetic energy terms for these momenta are all decoupled from each other, but with the complication that they have different effective masses even if the particle masses m_i are all the same. If there are no external forces, so that the potential energy is invariant under translations, then $\boldsymbol{R}_{\text{cm}}$ will not appear in the Hamiltonian at all. The energy eigenstates will therefore have the form of a tensor product, of plane-wave free particle eigenstates of $\boldsymbol{P}_{\text{tot}}$ and $P_{\text{tot}}^2/2M$, and eigenstates of the remaining, more complicated, part of the Hamiltonian.

In the case of multi-electron atoms, where one of the particles is a nucleus that is much more massive than the electrons, it is much more common to make the simple and good approximation that the nucleus is infinitely massive and fixed at the origin in the center-of-mass frame, and the remaining coordinates and momenta are (very nearly) those of the individual electrons.

4.3 Charged Particle in External Electromagnetic Fields

We now turn our attention to the dynamics of charged particles in external electromagnetic fields. We will follow the example of most quantum mechanics books by using Gaussian cgs units for electrodynamic quantities, rather than the SI units that you may be more familiar with. The convention in this book is that the electric charge for a particle is given by $q = Qe$, where Q is a dimensionless number and e is the proton's charge, numerically given by (1.2), and positive. (Some other sources define e to be negative, referring to the electron.) Thus, for the electron, $Q = -1$ and $q = -e$, and for the proton, $Q = 1$ and $q = e$. All known particles have Q equal to integer multiples[1] of $1/3$, and the ones unconfined by the strong nuclear force have integer Q. For example, $Q = 2/3$ for up, charm, and top quarks, and $Q = -1/3$ for down, strange, and bottom quarks, and $Q = -1$ for the electron, muon, and tau lepton.

In the cgs system, Maxwell's equations for the electric and magnetic fields are (with the SI versions indicated parenthetically, for comparison)

$$\nabla \cdot \boldsymbol{E} = 4\pi\rho \qquad (\text{SI: } \rho/\epsilon_0), \tag{4.21}$$

$$\nabla \cdot \boldsymbol{B} = 0 \qquad (\text{SI: } 0), \tag{4.22}$$

$$\nabla \times \boldsymbol{E} = -\frac{1}{c}\frac{\partial \boldsymbol{B}}{\partial t} \qquad (\text{SI: } -\frac{\partial \boldsymbol{B}}{\partial t}), \tag{4.23}$$

$$\nabla \times \boldsymbol{B} = \frac{1}{c}\frac{\partial \boldsymbol{E}}{\partial t} + \frac{4\pi}{c}\boldsymbol{j} \qquad (\text{SI: } \frac{1}{c^2}\frac{\partial \boldsymbol{E}}{\partial t} + \mu_0\boldsymbol{j}). \tag{4.24}$$

It follows that the charge density ρ and the current density $\boldsymbol{j}$ obey local charge conservation,

$$\nabla \cdot \boldsymbol{j} + \frac{\partial \rho}{\partial t} = 0. \tag{4.25}$$

The electromagnetic energy density and Poynting vector (energy flux, or power per unit area) are

[1] Within the assumed structure of the Standard Model of particle physics, this remarkable charge quantization can be understood as a requirement of anomaly cancellation, a consistency constraint on quantum field theories with gauge interactions. Grand Unified Theories (based on non-Abelian gauge groups like $SU(5)$, $SO(10)$, or E_6) go further, elegantly explaining why all particles, known or unknown, must have integer values of $3Q$, but it is not presently known if these theories are correct.

$$u_{\text{EM}} = \frac{1}{8\pi}(E^2 + B^2) \qquad \left(\text{SI: } \frac{\epsilon_0}{2}E^2 + \frac{1}{2\mu_0}B^2\right), \tag{4.26}$$

$$\boldsymbol{S} = \frac{c}{4\pi}\boldsymbol{E} \times \boldsymbol{B} \qquad \left(\text{SI: } \frac{1}{\mu_0}\boldsymbol{E} \times \boldsymbol{B}\right). \tag{4.27}$$

The electromagnetic fields are obtained as derivatives of the scalar and vector potentials,

$$\boldsymbol{E} = -\nabla\Phi - \frac{1}{c}\frac{\partial \boldsymbol{A}}{\partial t}, \tag{4.28}$$

$$\boldsymbol{B} = \nabla \times \boldsymbol{A}. \tag{4.29}$$

The fields $\boldsymbol{E}$ and $\boldsymbol{B}$ remain unchanged if one makes a simultaneous change in Φ and $\boldsymbol{A}$, called a **gauge transformation**,

$$\Phi \to \Phi - \frac{1}{c}\frac{\partial \Lambda}{\partial t}, \qquad \boldsymbol{A} \to \boldsymbol{A} + \nabla\Lambda, \tag{4.30}$$

where $\Lambda(\boldsymbol{r}, t)$ is an arbitrary function of position and time.

In classical electrodynamics, the Lagrangian for a nonrelativistic particle with mass m and charge q and position $\boldsymbol{r}(t)$, moving in the potentials Φ and $\boldsymbol{A}$, is

$$L = \frac{1}{2}m\left(\frac{d\boldsymbol{r}}{dt}\right)^2 + \frac{q}{c}\frac{d\boldsymbol{r}}{dt} \cdot \boldsymbol{A}(\boldsymbol{r}, t) - q\Phi(\boldsymbol{r}, t). \tag{4.31}$$

The equation of motion resulting from applying (4.2) is the Lorentz force law,

$$m\frac{d^2\boldsymbol{r}}{dt^2} = q\left(\boldsymbol{E} + \frac{1}{c}\frac{d\boldsymbol{r}}{dt} \times \boldsymbol{B}\right). \tag{4.32}$$

The motion of a classical[2] charged particle is thus determined only by the local values of $\boldsymbol{E}$ and $\boldsymbol{B}$. Although the Lagrangian is written in terms of the potentials Φ and $\boldsymbol{A}$, they are not physical observables, because the physics is equally well described if they are modified by any gauge transformation of the form (4.30).

Applying the procedure in Sect. 4.1 to (4.31), the classical canonical momentum conjugate to $\boldsymbol{r}$ is

$$\boldsymbol{p} = m\frac{d\boldsymbol{r}}{dt} + \frac{q}{c}\boldsymbol{A}(\boldsymbol{r}, t), \tag{4.33}$$

[2] In quantum mechanics, the potential $\boldsymbol{A}$ affects charged particles in ways that are *not* encoded locally in the fields $\boldsymbol{E}$ and $\boldsymbol{B}$. This is demonstrated by the Ehrenberg–Siday–Aharonov–Bohm effect, discussed in Sect. 29.4.

and the classical Hamiltonian is

$$H = \frac{1}{2m}\left(\boldsymbol{p} - \frac{q}{c}\boldsymbol{A}(\boldsymbol{r}, t)\right)^2 + q\Phi(\boldsymbol{r}, t). \tag{4.34}$$

Like the Lagrangian, the Hamiltonian and canonical momentum are not written directly in terms of the fields $\boldsymbol{E}$ and $\boldsymbol{B}$, but rather in terms of the potentials Φ and $\boldsymbol{A}$, even though the latter are gauge-dependent. One must be careful to distinguish the canonical momentum $\boldsymbol{p}$ of the particle from the **kinetic momentum** (also known as **mechanical momentum**), defined as the product of mass and velocity,

$$\boldsymbol{\pi} \equiv m\frac{d\boldsymbol{r}}{dt} = \boldsymbol{p} - \frac{q}{c}\boldsymbol{A}(\boldsymbol{r}, t). \tag{4.35}$$

The kinetic momentum $\boldsymbol{\pi}$ is a gauge-invariant observable, since $d\boldsymbol{r}/dt$ is determined by the observed trajectory of the particle, and so cannot depend on the choice of gauge. In contrast, the canonical momentum $\boldsymbol{p}$ is not a gauge-invariant observable, due to the appearance of $\boldsymbol{A}$ in (4.33).

In quantum mechanics, we promote the classical position $\boldsymbol{r}$ and canonical momentum $\boldsymbol{p}$ to operators, and thus the electromagnetic potentials become operators $\Phi(\boldsymbol{R}, t)$ and $\boldsymbol{A}(\boldsymbol{R}, t)$ that are functions of $\boldsymbol{R}$. So, naively, the quantum Hamiltonian should be

$$H = \frac{1}{2m}\left(\boldsymbol{P} - \frac{q}{c}\boldsymbol{A}(\boldsymbol{R}, t)\right)^2 + q\Phi(\boldsymbol{R}, t), \tag{4.36}$$

where $\boldsymbol{R}$ and $\boldsymbol{P}$ satisfy the usual canonical commutation relations of (2.248). This implies that in the position representation, these canonical operators are represented by

$$\boldsymbol{R} \leftrightarrow \boldsymbol{r}, \qquad \boldsymbol{P} \leftrightarrow -i\hbar\nabla, \tag{4.37}$$

as in Table 2.2. One must be careful with the ordering of $\boldsymbol{P}$ and $\boldsymbol{A}$, since they do not commute, owing to the dependence of $\boldsymbol{A}$ on $\boldsymbol{R}$. Thus, $(\boldsymbol{P} - \frac{q}{c}\boldsymbol{A}(\boldsymbol{R}, t))^2$ should be interpreted as the symmetrized form $P^2 - \frac{q}{c}(\boldsymbol{P}\cdot\boldsymbol{A} + \boldsymbol{A}\cdot\boldsymbol{P}) + \frac{q^2}{c^2}A^2$, so that H is Hermitian.

However, (4.36) is still not complete, because it does not include the (purely quantum) effect of intrinsic angular momentum, or spin. We will discuss spin more thoroughly in Sect. 8.2, but for the present discussion we only need to know that the spin for a particle is a vector operator $\boldsymbol{S}$. The intrinsic magnetic moment of a fundamental particle is proportional to its spin (because, in the rest frame of the particle, there is no other special direction in which it could point). We therefore write

$$\boldsymbol{\mu} = \gamma\boldsymbol{S}. \tag{4.38}$$

The constant of proportionality γ is a property of the particle type, called the **gyromagnetic ratio**. For the electron, the gyromagnetic ratio is often written as

$$\gamma_e = -\frac{g_e e}{2m_e c}, \tag{4.39}$$

where the dimensionless quantity g_e is called the g-factor of the electron. The Dirac equation of relativistic quantum mechanics predicts $g_e = 2$, as shown in Sect. 28.4, but there are small corrections to this coming from the quantum field theory of relativistic quantum electrodynamics (QED). It has been predicted very precisely by calculations at fifth order in perturbation theory in QED, and measured experimentally with comparable accuracy, with the results

$$g_e = 2.00231930436321(46) \quad \text{(QED prediction)}, \tag{4.40}$$

$$g_e = 2.00231930436118(26) \quad \text{(experiment)}, \tag{4.41}$$

a famous agreement between theory and experiment to within about 1 part in 10^{12}. The quantity $(g_e - 2)/2$ is called the **anomalous magnetic moment** of the electron. In this book, we will usually simply use the approximation $g_e = 2$.

For the proton and the neutron, the gyromagnetic ratios are often written as

$$\gamma_p = \frac{g_p e}{2m_p c}, \tag{4.42}$$

$$\gamma_n = \frac{g_n e}{2m_p c}, \tag{4.43}$$

which again define dimensionless g-factors. Note that the neutron has a magnetic moment, even though it has no net charge, because it is a composite particle with charged constituents (quarks). The conventional definition of g_n for the neutron in (4.43) uses the proton's charge and mass. The nucleon g-factors are also very accurately known experimentally,

$$g_p = 5.5856946893(16), \tag{4.44}$$

$$g_n = -3.82608552(90), \tag{4.45}$$

but the theoretical predictions of these quantities are not nearly as accurate. The reason is that, unlike the electron, the proton and neutron are complicated composite particles made up of quarks and gluons (and virtual quark–antiquark pairs) held together by the strong nuclear force (quantum chromodynamics, or QCD). Perturbation theory fails to converge in this context, and non-perturbative methods are highly advanced but limited by finite computing power.

For electrons, protons, and neutrons, the magnitude of $\boldsymbol{S}$ is always the same, $\hbar/2$. The magnitude of the electron's intrinsic magnetic moment is within about 0.1% of the **Bohr magneton**,

$$\mu_B = \frac{e\hbar}{2m_e c} = 5.78838 \times 10^{-5} \frac{\text{eV}}{\text{Tesla}} = 9.27401 \times 10^{-21} \frac{\text{erg}}{\text{gauss}}, \tag{4.46}$$

and the proton and neutron have magnetic moment magnitudes equal to about 2.79 and 1.91 times the **nuclear magneton**,

$$\mu_N = \frac{e\hbar}{2m_p c} = 3.15245 \times 10^{-8} \frac{\text{eV}}{\text{Tesla}} = 5.05078 \times 10^{-24} \frac{\text{erg}}{\text{gauss}}. \tag{4.47}$$

Because $\mu_B/\mu_N = m_p/m_e \approx 1836$, the magnetic moments of atoms with unpaired electrons are typically 3 orders of magnitude larger than nuclear magnetic moments.

The classical energy of interaction of a magnetic moment $\boldsymbol{\mu}$ with an external magnetic field is $-\boldsymbol{\mu} \cdot \boldsymbol{B}$. So, adding this to the Hamiltonian for a nonrelativistic particle, we get

$$H = \frac{1}{2m}\left(\boldsymbol{P} - \frac{q}{c}\boldsymbol{A}(\boldsymbol{R}, t)\right)^2 + q\Phi(\boldsymbol{R}, t) - \gamma \boldsymbol{S} \cdot \boldsymbol{B}(\boldsymbol{R}, t), \tag{4.48}$$

where q is the electric charge and γ is the gyromagnetic ratio for the particle. Even this Hamiltonian is not complete, for it does not include relativistic effects suppressed by further powers of $1/c$. These will be discussed when needed for the fine and hyperfine contributions to the hydrogen atom, in Sects. 17.1 and 17.2, and derived in a more fundamental way from the Dirac equation in Sects. 28.4 and 28.5.

If the Hamiltonian (4.48) is written in terms of the kinetic momentum operator

$$\boldsymbol{\Pi} = \boldsymbol{P} - \frac{q}{c}\boldsymbol{A}(\boldsymbol{R}, t), \tag{4.49}$$

it will appear simpler, since it then does not depend explicitly on the vector potential $\boldsymbol{A}$,

$$H = \frac{1}{2m}\Pi^2 + q\Phi - \gamma \boldsymbol{S} \cdot \boldsymbol{B}. \tag{4.50}$$

However, the kinetic momentum does *not* obey canonical commutation relations. In the position representation, $\boldsymbol{P} \leftrightarrow -i\hbar\nabla$, so that $\boldsymbol{\Pi} \leftrightarrow -i\hbar\nabla - \frac{q}{c}\boldsymbol{A}(\boldsymbol{r}, t)$, and

$$[R_a, \Pi_b] = i\hbar\delta_{ab}, \qquad (a, b = x, y, z), \tag{4.51}$$

just as for the canonical momentum, but

$$[\Pi_a, \Pi_b] = i\frac{q\hbar}{c}(\nabla_a A_b - \nabla_b A_a). \tag{4.52}$$

Rewriting this directly in terms of the magnetic field,

$$[\Pi_x, \Pi_y] = i\frac{q\hbar}{c}B_z, \qquad [\Pi_y, \Pi_z] = i\frac{q\hbar}{c}B_x, \qquad [\Pi_z, \Pi_x] = i\frac{q\hbar}{c}B_y. \tag{4.53}$$

This should be contrasted with the canonical commutator $[P_a, P_b] = 0$. If one chooses to write the Hamiltonian in terms of the kinetic momentum, the simplicity comes with a cost; the information about the vector potential and the magnetic field is hidden in the commutation relations for the components of $\boldsymbol{\Pi}$ with each other. Since $\boldsymbol{\Pi}$ is the product of mass and velocity for the particle, we see that in the presence of a magnetic field one cannot simultaneously specify the three components of the velocity of a charged particle, because they are not compatible observables.

The freedom to make gauge transformations as in (4.30) always allows us to select **Coulomb gauge** (also known as **transverse gauge**), defined by

$$\nabla \cdot \boldsymbol{A} = 0. \tag{4.54}$$

Since $\boldsymbol{P}$ is given in the position representation by $-i\hbar\nabla$ acting on everything to its right, the Coulomb gauge condition implies $\boldsymbol{P} \cdot \boldsymbol{A} = \boldsymbol{A} \cdot \boldsymbol{P}$, with the consequent advantage that (unlike other gauge choices) there is no operator-ordering issue associated with the cross-terms in the Hamiltonian, of the type mentioned after (4.37). So, in Coulomb gauge, one is free to write

$$H = \frac{P^2}{2m} - \frac{q}{mc}\boldsymbol{A} \cdot \boldsymbol{P} + \frac{q^2}{2mc^2}A^2 + q\Phi - \gamma \boldsymbol{S} \cdot \boldsymbol{B}. \tag{4.55}$$

This form will be useful for discussing absorption and emission of light, in Chap. 23.

Exercises

4.1 The purpose of this problem is to illustrate the possible ambiguities in connecting classical observables to their quantum counterparts due to operator ordering. Consider a particle moving in one dimension with position x and momentum p.

(a) Consider the classical quantity x^2p^2. The quantum operator X^2P^2 is not Hermitian, but two versions of it that are symmetrized in different ways to obtain Hermitian operators are $A = (X^2P^2+P^2X^2)/2$ and $B = (XP+PX)^2/4$. Show that A and B differ by a certain rational multiple of $\hbar^2 I$ that you will find.

(b) Similarly, consider the classical quantity x^3p^3 and two candidate quantum Hermitian operator versions of it, $C = (X^3P^3+P^3X^3)/2$ and $D = (XP+PX)^3/8$. Show that these differ by an observable operator that is not proportional to the identity.

4.2 Show that the pairs $(\boldsymbol{R}, \boldsymbol{P})$ and $(\boldsymbol{R}_{\text{cm}}, \boldsymbol{P}_{\text{tot}})$ defined in (4.14) and (4.15) satisfy canonical commutation relations. Then check that the Hamiltonian of (4.11) decomposes into H_{cm} and H_{rel} as claimed in (4.17).

Transformations, Symmetries, and Conservation Laws

5

5.1 Continuous Unitary Transformations, Symmetries, and Noether's Principle

As we saw in Sect. 2.5, a unitary operator can be viewed as implementing a change of orthobasis for the space of states. Certain unitary operators also have the interpretation of physical changes, or **transformations**, on a system. Notable examples include translations, rotations, inversion of the coordinate system, and displacements in time.

If a transformation leaves the Hamiltonian unchanged, then we say, as a definition, that the transformation is a **symmetry** of the quantum system. In this section we will show, in a general way, that there is always a conserved quantity corresponding to each continuous symmetry. As special cases, we will see that the conserved quantities associated with time translation, spatial translation, and rotation symmetries are, respectively, energy, momentum, and angular momentum.

Consider the unitary transformation operator

$$U(\alpha) = \exp(-i\alpha_a G_a), \tag{5.1}$$

where the G_a form a set of N Hermitian operators, typically observables without explicit time dependence, called the **generators** of the transformations, and the α_a are N real numbers that parameterize the transformations. Here, and in the following, repeated indices $a = 1, \ldots, N$ are implicitly summed over. Theorem 2.5 confirms that since $\alpha_a G_a$ is Hermitian, $U(\alpha)$ is a unitary operator. Following the discussion surrounding (2.112)–(2.117), the transformations for the state ket and all observables A are defined by

$$|\psi\rangle \rightarrow |\psi'\rangle = U|\psi\rangle, \tag{5.2}$$

$$A \rightarrow A' = UAU^\dagger, \tag{5.3}$$

S. P. Martin, *Quantum Mechanics*, Graduate Texts in Physics,
https://doi.org/10.1007/978-3-032-21119-4_5

so that matrix elements are invariant under the transformation, because $U^\dagger U = I$. The inverse of the transformation parameterized by α_a is parameterized by $-\alpha_a$,

$$U(\alpha)^\dagger = U(\alpha)^{-1} = U(-\alpha). \tag{5.4}$$

The closure property says that the combination of two transformations α_a and β_a must be another transformation, parameterized by some set of real numbers γ_a,

$$U(\beta)U(\alpha) = U(\gamma). \tag{5.5}$$

A set of transformations obeying these properties has the mathematical structure of a group, and if the parameters α_a are continuous it is a **Lie group**, named after the mathematician Marius Sophus Lie.

Consider the following sequence of unitary transformations that are infinitesimally close to the identity,

$$\begin{aligned} U(\epsilon)U(\delta)U(\epsilon)^\dagger U(\delta)^\dagger &= (I - i\epsilon_a G_a)(I - i\delta_b G_b)(I + i\epsilon_c G_c)(I + i\delta_d G_d) \\ &= I - \epsilon_a \delta_b [G_a, G_b] + \cdots , \end{aligned} \tag{5.6}$$

where terms of higher order in either ϵ_a or δ_b have been dropped. The closure property implies that this must also be a unitary transformation infinitesimally close to the identity, so $[G_a, G_b]$ must be a linear combination of generators. Therefore, for continuous transformations we have

$$[G_a, G_b] = i f_{abc} G_c \tag{5.7}$$

for some set of numbers f_{abc}. Equation (5.7) is called the **Lie algebra** of the Lie group, and f_{abc} are called its **structure constants**. Since the commutator of any two Hermitian operators $[G_a, G_b]$ is anti-Hermitian, and the G_c are Hermitian, the quantities f_{abc} must all be real. This is the reason for the conventional factor of i in (5.7). If the structure constants f_{abc} are all 0, so that the generators all commute, then the group is said to be Abelian; otherwise, it is non-Abelian. The mathematical study and classification of Lie groups is a rich and beautiful subject that we will not address here.

For the remainder of this section, let us consider unitary transformations that are symmetries of a quantum system. According to our definition, this means that H does not change, so (5.3) gives

$$U(\alpha) H U(\alpha)^\dagger = H. \tag{5.8}$$

Taking the special case that the parameters $\alpha_a = \epsilon_a$ are infinitesimal, we have

$$(I - i\epsilon_a G_a + \cdots)\, H\, (I + i\epsilon_b G_b + \cdots) = H, \tag{5.9}$$

which implies $\epsilon_a[H, G_a] = 0$. Since this is supposed to hold for any ϵ_a, we learn that

$$[H, G_a] = 0. \tag{5.10}$$

It follows from Theorem 2.19 that an orthobasis of energy eigenstates can also be chosen to be eigenstates of any subset of the symmetry generators G_a that are mutually commuting. (If this subset is maximal, it is called a Cartan subalgebra, after mathematician Élie Cartan, and the number of linearly independent generators of it is called the rank of the Lie algebra).

Consider an eigenstate $|g_a\rangle$ of one of the symmetry generators G_a, labeled by its eigenvalue g_a. Since the symmetry generators G_a commute with the Hamiltonian, $|g_a\rangle$ will remain an eigenstate of G_a at later times, because

$$G_a \left(e^{-itH/\hbar}\,|g_a\rangle\right) = e^{-itH/\hbar} G_a\,|g_a\rangle = g_a \left(e^{-itH/\hbar}\,|g_a\rangle\right). \tag{5.11}$$

This holds even if the Hamiltonian depends on time, provided that $[H(t), G_a] = 0$ for all t, as one can see by replacing the factors $e^{-itH/\hbar}$ by the time-evolution operator $U(t, t_0)$ from (3.54).

Furthermore, the symmetry generators G_a are **conserved quantities**, also known as **constants of the motion**. The precise meaning of this is given by the following result.

Theorem 5.1 (Noether's Principle) *Suppose that an observable G_a has no explicit time dependence, and that it commutes with the (possibly time-dependent) Hamiltonian $H(t)$ for all t. Then, for an arbitrary state, the probability of each possible outcome g_a for a measurement of G_a is independent of time.*

Proof The key point is that $[H(t), G_a] = 0$ implies that, for any particular time $t = t_0$, one can find an orthobasis of common eigenkets $|g_a, E, u\rangle$, defined such that $G_a\,|g_a, E, u\rangle = g_a\,|g_a, E, u\rangle$ and $H(t_0)\,|g_a, E, u\rangle = E\,|g_a, E, u\rangle$. Here u labels possible degeneracies for the pair (g_a, E). These orthobasis kets are fixed, determined by the Hamiltonian at the time t_0. Now, Postulate 4 tells us that if the system is in a state $|\psi(t)\rangle$, then the probability of measuring G_a at time t and getting the result g_a is

$$\mathcal{P}(g_a, t) = \sum_{E,u} \langle g_a, E, u|\psi(t)\rangle\,\langle\psi(t)|g_a, E, u\rangle\,. \tag{5.12}$$

Here E and u are summed over, as the degeneracy labels for g_a. Taking the time derivative, and then using the time-dependent Schrödinger equation $\frac{d}{dt}\,|\psi\rangle = -\frac{i}{\hbar}H(t)\,|\psi\rangle$ and its adjoint $\frac{d}{dt}\,\langle\psi| = \frac{i}{\hbar}\,\langle\psi|\,H(t)$, we get

$$\frac{d}{dt}\mathcal{P}(g_a, t) = -\frac{i}{\hbar}\sum_{E,u}\Big(\langle g_a, E, u|H(t)|\psi(t)\rangle\,\langle\psi(t)|g_a, E, u\rangle$$
$$-\langle g_a, E, u|\psi(t)\rangle\,\langle\psi(t)|H(t)|g_a, E, u\rangle\Big). \tag{5.13}$$

Evaluating (5.13) at $t = t_0$, the two terms on the right-hand side cancel after using $H(t_0)\,|g_a, E, u\rangle = E\,|g_a, E, u\rangle$ and $\langle g_a, E, u|\,H(t_0) = E\,\langle g_a, E, u|$. Thus, we obtain

$$\frac{d}{dt}\mathcal{P}(g_a, t)\Big|_{t=t_0} = 0. \tag{5.14}$$

Since t_0 is arbitrary, this shows that, for an arbitrary state, $\mathcal{P}(g_a, t)$ has a vanishing time derivative at any given $t = t_0$, so it must be constant in time, as claimed. □

The expectation value of an operator in a state is the sum of its measurement outcomes weighted by the probabilities; see (3.14). Therefore, Theorem 5.1 immediately implies a weaker but still interesting and important result, that the expectation value of a symmetry generator G_a in an arbitrary state does not depend on time:

$$\frac{d}{dt}\langle G_a\rangle = 0. \tag{5.15}$$

This also follows directly from Ehrenfest's Theorem (3.39) with G_a playing the role of A. Theorem 5.1 and (5.15) are quantum-mechanical versions of Emmy Noether's celebrated 1918 theorem in classical mechanics, that every continuous symmetry implies a conservation law. The observables G_a that generate the symmetries of the Hamiltonian are conserved quantities.

The most straightforward special case is that of **time translation** symmetry, for which the generator is the Hamiltonian itself, as expressed by the Schrödinger equation (3.10) in Postulate 6. If the Hamiltonian does not depend explicitly on time, then the energy is a conserved quantity, since $[H, H] = 0$ is trivially satisfied. In that case, for any state, the probability of measuring the energy to be any particular value is independent of time, and the expectation value of the energy does not change.

5.2 Translations

For a particle moving in one dimension with position operator X and momentum operator P, consider the operator

$$T(a) = e^{-iaP/\hbar}, \tag{5.16}$$

where a is a constant length. Because P is Hermitian, Theorem 2.5 says that $T(a)$ is unitary,

$$T(a)^{\dagger} = T(a)^{-1} = T(-a). \tag{5.17}$$

Since the momentum operator commutes with itself, in the general language of Sect. 5.1 where $T(a)$ plays the role of $U(\alpha)$ and P plays the role of G_a, the Lie algebra for translations is simply

$$[P, P] = 0. \tag{5.18}$$

This is an Abelian algebra; the antisymmetric structure constants defined in general by (5.7) vanish trivially in this case, because there is only one generator, P.

Using $[X, P] = i\hbar$, Theorem 2.8 gives the commutator

$$[X, T(a)] = aT(a). \tag{5.19}$$

From this we get, acting on a position eigenstate $|x\rangle$,

$$X(T(a)|x\rangle) = T(a)(X+a)|x\rangle = (x+a)(T(a)|x\rangle). \tag{5.20}$$

This shows that $T(a)|x\rangle$ is an eigenstate of X with eigenvalue $x+a$, which means that it must be equal to $|x+a\rangle$, up to a multiplicative constant. If $|x\rangle$ is normalized according to the Dirac condition, then so will be $T(a)|x\rangle$, since $T(a)$ is unitary. Therefore, the constant is just a phase factor, and $T(a)|x\rangle = e^{i\theta}|x+a\rangle$ for some θ. The only way to resolve the ambiguity represented by θ is to arbitrarily choose a value for it, and $\theta = 0$ is a particularly nice choice, so we define

$$|x+a\rangle = T(a)|x\rangle. \tag{5.21}$$

In words, $T(a)$ operates by changing a state in which the particle is known to be at x into a state where it is known to be at $x+a$. We therefore call $T(a)$ a **translation operator**, and we say that the momentum operator is the generator of translations.

Taking the adjoint of (5.21), and using (5.17), gives

$$\langle x|T(a) = \langle x-a|. \tag{5.22}$$

This implies that for any state $|\psi\rangle$, the wavefunction obeys

$$\psi(x-a) = \langle x|T(a)|\psi\rangle. \tag{5.23}$$

Thus the action of $T(a)$ can be interpreted as shifting the particle's position to the right by a (the "active view"), or equivalently as leaving the particle's position unchanged but shifting the coordinate system to the left by a (the "passive view").

If we do a translation on any normalized state $|\psi\rangle$,

$$|\psi\rangle \to T(a)|\psi\rangle, \tag{5.24}$$

then the expectation value of position is changed according to

$$\langle\psi|X|\psi\rangle \rightarrow \langle\psi|T(a)^{\dagger}XT(a)|\psi\rangle = \langle\psi|(X+a)|\psi\rangle = \langle\psi|X|\psi\rangle + a. \quad (5.25)$$

The effect of $T(a)$ on momentum eigenstates is just to multiply by a phase factor,

$$T(a)\,|p\rangle = e^{-iap/\hbar}\,|p\rangle\,, \quad (5.26)$$

and the expectation value of momentum is unaffected by the transformation,

$$\langle\psi|P|\psi\rangle \rightarrow \langle\psi|T(a)^{\dagger}PT(a)|\psi\rangle = \langle\psi|P|\psi\rangle\,. \quad (5.27)$$

Following the general example of a unitary transformation of an operator in (5.3), we can also define for an arbitrary observable A the translated version $A' = T(a)AT(a)^{\dagger}$. Now if we do a simultaneous transformation of both states and operators,

$$|\psi\rangle \rightarrow |\psi'\rangle = T(a)\,|\psi\rangle\,, \quad (5.28)$$

$$A \rightarrow A' = T(a)AT(a)^{\dagger}, \quad (5.29)$$

then matrix elements will be unchanged. Using $[X, T(a)] = aT(a)$, the translated position operator is

$$X' = T(a)XT(a)^{\dagger} = X - a, \quad (5.30)$$

while $[P, T(a)] = 0$ implies that the momentum operator does not change,

$$P' = T(a)PT(a)^{\dagger} = P. \quad (5.31)$$

From the rule $T(a)\,|x\rangle = |x+a\rangle$ found in (5.21), it follows that translations simply add,

$$T(a)T(b) = T(a+b). \quad (5.32)$$

One consequence of this is that any finite translation can be viewed as the combination of many infinitesimal translations. For an infinitesimal translation, the exponential can be truncated to linear order,

$$T(\epsilon) = 1 - i\frac{\epsilon}{\hbar}P. \quad (5.33)$$

Now suppose that a quantum-mechanical system has translations as a symmetry, by which we mean that the Hamiltonian is invariant. In particular, for infinitesimal translations, we require $T(\epsilon)HT(\epsilon)^{\dagger} = H$, from which it follows that

$$\left(1 - i\frac{\epsilon}{\hbar}P\right) H \left(1 + i\frac{\epsilon}{\hbar}P\right) - H = i\frac{\epsilon}{\hbar}[H, P] + O(\epsilon^2) \tag{5.34}$$

must vanish, so

$$[H, P] = 0. \tag{5.35}$$

In general, this requires the potential V to have no dependence on X. The Hamiltonian could be that of a free particle with $H = P^2/2m$, but it could also have some extra terms that may involve other functions of P or other degrees of freedom (for example, spin), but not X. In any case, if (5.35) holds, there must be an orthobasis of common eigenstates of P and H. Then an eigenstate of P with momentum p remains so at later times, since

$$P\left(e^{-itH/\hbar}|p\rangle\right) = p\left(e^{-itH/\hbar}|p\rangle\right). \tag{5.36}$$

Also, Ehrenfest's Theorem (3.39) says

$$\frac{d}{dt}\langle P\rangle = 0, \tag{5.37}$$

and Noether's Principle (Theorem 5.1) says even more, that the probability to measure the momentum within any given range is constant in time. Note that these statements are true for any state, including an arbitrary superposition of free particle eigenstates with different momenta, provided that $[H, P] = 0$. A nontrivial illustration will be provided at the end of Sect. 6.2.

For a particle moving in three dimensions, the preceding discussion can be adapted in a straightforward way. The translation operator for a displacement vector $\boldsymbol{a}$ is

$$T(\boldsymbol{a}) = e^{-i\boldsymbol{a}\cdot\boldsymbol{P}/\hbar}, \tag{5.38}$$

and it satisfies

$$T(\boldsymbol{a})T(\boldsymbol{b}) = T(\boldsymbol{a}+\boldsymbol{b}), \tag{5.39}$$

$$T(\boldsymbol{a})^{-1} = T(\boldsymbol{a})^\dagger = T(-\boldsymbol{a}). \tag{5.40}$$

The operation on position and momentum eigenkets is

$$T(\boldsymbol{a})|\boldsymbol{r}\rangle = |\boldsymbol{r}+\boldsymbol{a}\rangle, \qquad \langle\boldsymbol{r}|T(\boldsymbol{a}) = \langle\boldsymbol{r}-\boldsymbol{a}|, \tag{5.41}$$

$$T(\boldsymbol{a})|\boldsymbol{p}\rangle = e^{-i\boldsymbol{a}\cdot\boldsymbol{p}/\hbar}|\boldsymbol{p}\rangle, \qquad \langle\boldsymbol{p}|T(\boldsymbol{a}) = e^{-i\boldsymbol{a}\cdot\boldsymbol{p}/\hbar}\langle\boldsymbol{p}|, \tag{5.42}$$

and the remaining discussion for the one-dimensional case likewise follows through for the three-dimensional case with P replaced by $\boldsymbol{P}$. Although there are now three generators P_x, P_y, and P_z, they all commute with each other, so the structure constants are all 0, and the Lie group is Abelian. Note that in the case of a charged particle moving in an electromagnetic field as discussed in Sect. 4.3, it is the canonical momentum $\boldsymbol{P}$ that generates translations, not the kinetic momentum $\boldsymbol{\Pi}$.

Consider a quantum system describing two distinguishable particles labeled 1 and 2. Then one can have translation invariance even with a nonzero potential, provided that the Hamiltonian has the form considered in our discussion of the two-body problem in Sect. 4.2,

$$H = \frac{P_1^2}{2m_1} + \frac{P_2^2}{2m_2} + V(\boldsymbol{R}_1 - \boldsymbol{R}_2). \tag{5.43}$$

The individual translation operators for particles 1 and 2 are

$$T_1(\boldsymbol{a}) = e^{-i\boldsymbol{a}\cdot\boldsymbol{P}_1/\hbar}, \qquad T_2(\boldsymbol{a}) = e^{-i\boldsymbol{a}\cdot\boldsymbol{P}_2/\hbar}. \tag{5.44}$$

These are not symmetries of the Hamiltonian unless the potential $V(\boldsymbol{R}_1 - \boldsymbol{R}_2)$ is neglected. However, defining the total momentum operator $\boldsymbol{P}_{\text{tot}} = \boldsymbol{P}_1 + \boldsymbol{P}_2$ as in Sect. 4.2, one can check that each component of $\boldsymbol{P}_{\text{tot}}$ commutes with each component of $\boldsymbol{R}_1 - \boldsymbol{R}_2$, so

$$\left[H, \boldsymbol{P}_{\text{tot}}\right] = 0. \tag{5.45}$$

Therefore, the total translation operator

$$T(\boldsymbol{a}) = T_1(\boldsymbol{a})T_2(\boldsymbol{a}) = \exp\left[-i\boldsymbol{a}\cdot\boldsymbol{P}_{\text{tot}}/\hbar\right] \tag{5.46}$$

leaves the full Hamiltonian (5.43) invariant,

$$T(\boldsymbol{a})HT(\boldsymbol{a})^\dagger = H, \tag{5.47}$$

and is therefore a symmetry of the system. [Note that this would not be true if one added individual potentials $V_1(\boldsymbol{R}_1)$ or $V_2(\boldsymbol{R}_2)$ to the Hamiltonian.] Equation (5.47) simply reflects the feature that the total translation moves both particles 1 and 2 by the same amount, and so does not change the separation between them, nor their individual momenta. It follows that one can find an orthonormal basis of simultaneous eigenstates of H and $\boldsymbol{P}_{\text{tot}} = \boldsymbol{P}_1 + \boldsymbol{P}_2$, that an eigenstate of $\boldsymbol{P}_{\text{tot}}$ remains so at later times, and that in an arbitrary state the probability of measuring a given result for $\boldsymbol{P}_{\text{tot}}$ does not change in time, and its expectation value is conserved.

5.3 Rotations

In this section, we explore the connection between rotations and the angular momentum operators that generate them. Classically, the angular momentum of a particle about the point chosen as the coordinate system origin is defined by

$$\boldsymbol{l} = \boldsymbol{r} \times \boldsymbol{p}. \tag{5.48}$$

In quantum mechanics, we promote this to a vector operator, and define the **orbital angular momentum operator** for a particle as

$$\boldsymbol{L} = \hat{x} L_x + \hat{y} L_y + \hat{z} L_z = \boldsymbol{R} \times \boldsymbol{P}, \tag{5.49}$$

where each of the components

$$L_x = Y P_z - Z P_y, \qquad L_y = Z P_x - X P_z, \qquad L_z = X P_y - Y P_x \tag{5.50}$$

is an observable. There is no operator-ordering problem to worry about here, because $[Y, P_z] = [Z, P_y] = [Z, P_x] = [X, P_z] = [X, P_y] = [Y, P_x] = 0$.

The commutator algebra for the angular momentum components L_x, L_y, and L_z can be computed using the commutators of the position and momentum operators that they are built out of. For example,

$$[L_x, L_y] = [Y P_z, Z P_x] + [Z P_y, X P_z] = -i\hbar Y P_x + i\hbar X P_y = i\hbar L_z. \tag{5.51}$$

Similarly, $[L_y, L_z] = i\hbar L_x$ and $[L_z, L_x] = i\hbar L_y$. These results can be summarized as[1]

$$L_a = \epsilon_{abc} R_b P_c, \tag{5.52}$$

$$[L_a, L_b] = i\hbar \epsilon_{abc} L_c, \tag{5.53}$$

with a, b, c each equal to x, y, or z. Here, ϵ_{abc} is the totally antisymmetric **Levi-Civita symbol**, with

$$\epsilon_{xyz} = \epsilon_{yzx} = \epsilon_{zxy} = 1, \qquad \epsilon_{xzy} = \epsilon_{yxz} = \epsilon_{zyx} = -1, \tag{5.54}$$

and all other components $\epsilon_{abc} = 0$. It obeys the identities

$$\epsilon_{abc}\epsilon_{dec} = \delta_{ad}\delta_{be} - \delta_{ae}\delta_{bd}, \tag{5.55}$$

[1] Here, and from now on, we use the **repeated index summation convention**, which says that repeated indices in a given term are implicitly summed over. Thus, in (5.52), b and c are summed over, but a is not. In (5.53), c is summed over, but a and b are not.

$$\epsilon_{abc}\epsilon_{dbc} = 2\delta_{ad}, \tag{5.56}$$

$$\epsilon_{abc}\epsilon_{abc} = 6. \tag{5.57}$$

In the general language of Sect. 5.1, the role of the generators G_a is taken by $L_a/\hbar$ for rotations, as we are about to show. Comparing (5.53) to (5.7), we see that the Lie algebra of angular momentum operators is non-Abelian, with structure constants $f_{abc} = \epsilon_{abc}$. Note that there is always an arbitrary normalization in the definition of the generators; it was convenient to include the factor of $1/\hbar$ here so that the structure constants are dimensionless.

Rotations have the property that they leave invariant the distance of each point from the origin. The composition of two rotations is another rotation; in mathematical language, infinitesimal rotations correspond to the $SO(3)$ Lie algebra. In quantum mechanics, we can think of rotations as a unitary change of basis corresponding to a change in coordinates; a rotation by an angle α about the axis defined by a unit vector $\hat{n}$ is implemented by a unitary operator $U(\boldsymbol{\alpha})$, where $\boldsymbol{\alpha} = \hat{n}\alpha$. The fact that rotations form a group means that, for any $\boldsymbol{\alpha}$ and $\boldsymbol{\beta}$,

$$U(\boldsymbol{\beta})U(\boldsymbol{\alpha}) = U(\boldsymbol{\gamma}), \tag{5.58}$$

for some vector $\boldsymbol{\gamma}$. In particular, arbitrary finite rotations can be constructed from the limit of a large number of infinitesimal rotations.

Let us now show that (for the case of a single spinless particle) L_z is the generator of rotations about the z-axis (analogously to how P_z is the generator of translations along the z direction). For such a rotation by an angle α, the new coordinates of the rotated point are

$$\begin{pmatrix} x' \\ y' \\ z' \end{pmatrix} = \begin{pmatrix} \cos\alpha & -\sin\alpha & 0 \\ \sin\alpha & \cos\alpha & 0 \\ 0 & 0 & 1 \end{pmatrix} \begin{pmatrix} x \\ y \\ z \end{pmatrix}. \tag{5.59}$$

If $\alpha = \epsilon$ is infinitesimal, then this becomes $x' = x - \epsilon y$ and $y' = y + \epsilon x$, and of course $z' = z$. We therefore define an infinitesimal unitary rotation operator by its action on the position eigenkets,

$$U(\epsilon\hat{z})\,|x,\, y,\, z\rangle = |x - \epsilon y,\; y + \epsilon x,\; z\rangle\,, \tag{5.60}$$

which implies

$$\langle x, y, z|\, U(\epsilon\hat{z})^{\dagger} = \langle x - \epsilon y,\; y + \epsilon x,\; z|\,, \tag{5.61}$$

with $U(\epsilon\hat{z})^{\dagger} = U(\epsilon\hat{z})^{-1} = U(-\epsilon\hat{z})$. It follows that, for any state $|\psi\rangle$,

$$\langle x, y, z|U(\epsilon\hat{z})|\psi\rangle = \langle x + \epsilon y,\; y - \epsilon x,\; z|\psi\rangle = \psi(x + \epsilon y,\; y - \epsilon x,\; z)$$

$$= \psi(x, y, z) + \epsilon y \frac{\partial}{\partial x}\psi(x, y, z) - \epsilon x \frac{\partial}{\partial y}\psi(x, y, z), \quad (5.62)$$

where the last equality is the result of the Taylor series expansion to linear order in ϵ. Now, since $\partial/\partial x \leftrightarrow i P_x/\hbar$ and $\partial/\partial y \leftrightarrow i P_y/\hbar$ in the position representation, we have

$$\langle x, y, z|U(\epsilon\hat{z})|\psi\rangle = \langle x, y, z| \left[I - \frac{i}{\hbar}\epsilon(XP_y - YP_x) \right] |\psi\rangle , \quad (5.63)$$

for every state $|\psi\rangle$, so comparing with the definition of L_z in (5.50), we find

$$U(\epsilon\hat{z}) = I - \frac{i}{\hbar}\epsilon L_z. \quad (5.64)$$

This establishes that L_z generates rotations about the z axis, and that

$$L_z \leftrightarrow -i\hbar \left(x\frac{\partial}{\partial y} - y\frac{\partial}{\partial x} \right) \quad (5.65)$$

in the position wavefunction representation.

There is nothing special about the z-axis in the preceding discussion, so repeating the process for infinitesimal rotations about the x and y axes, one obtains

$$U(\epsilon\hat{x}) = I - \frac{i}{\hbar}\epsilon L_x, \qquad U(\epsilon\hat{y}) = I - \frac{i}{\hbar}\epsilon L_y, \quad (5.66)$$

with L_x and L_y given by (5.50), and position representations

$$L_x \leftrightarrow -i\hbar \left(y\frac{\partial}{\partial z} - z\frac{\partial}{\partial y} \right), \qquad L_y \leftrightarrow -i\hbar \left(z\frac{\partial}{\partial x} - x\frac{\partial}{\partial z} \right). \quad (5.67)$$

Thus, L_x, L_y, and L_z are generators for rotations about the x, y, and z axes, respectively (assuming that there is no intrinsic angular momentum).

The unitary rotation operator for a non-infinitesimal angle α can be built from the limit of a large number N of infinitesimal operators with $\epsilon = \alpha/N$, acting sequentially, so

$$U(\alpha\hat{z}) = \lim_{N\to\infty} \left(I - \frac{i}{\hbar}\frac{\alpha}{N} L_z \right)^N = \exp\left(-\frac{i}{\hbar}\alpha L_z \right), \quad (5.68)$$

where we have used the definition of the exponential of an operator in (2.79). For a rotation by an angle α about an arbitrary axis $\hat{n}$, this generalizes to

$$U(\boldsymbol{\alpha}) = \exp\left(-\frac{i}{\hbar}\boldsymbol{\alpha} \cdot \mathbf{L} \right), \quad (5.69)$$

where $\boldsymbol{\alpha} = \hat{n}\alpha$. For infinitesimal $\boldsymbol{\alpha} = \boldsymbol{\epsilon}$, the generalization of (5.60) is

$$U(\boldsymbol{\epsilon})\,|\boldsymbol{r}\rangle = |\boldsymbol{r} + \boldsymbol{\epsilon} \times \boldsymbol{r}\rangle\,, \tag{5.70}$$

and, as you will show in Exercise 5.1,

$$U(\boldsymbol{\epsilon})\boldsymbol{V}U(\boldsymbol{\epsilon})^{\dagger} = \boldsymbol{V} - \boldsymbol{\epsilon} \times \boldsymbol{V}, \tag{5.71}$$

for each of $\boldsymbol{V} = \boldsymbol{R}$, $\boldsymbol{P}$, and $\boldsymbol{L}$. An operator that behaves in this way is said to transform under rotations as a vector.[2] For the more general case of non-infinitesimal rotations, the transformations of states and operators are given in Exercise 5.3.

For the position representation, it is often convenient to use cylindrical or spherical coordinates instead of rectangular coordinates, so that the position eigenkets are $|r, \phi, z\rangle$ or $|r, \theta, \phi\rangle$, respectively. For spherical coordinates in particular, this has the advantage that one of the coordinates, r, is not involved in the differential operators that represent the angular momenta. In spherical coordinates, (5.62) becomes

$$\langle r, \theta, \phi|U(\epsilon\hat{z})|\psi\rangle = \psi(r, \theta, \phi) - \epsilon\frac{\partial}{\partial\phi}\psi(r, \theta, \phi), \tag{5.72}$$

so that as an equivalent to (5.65),

$$L_z \leftrightarrow -i\hbar\frac{\partial}{\partial\phi}. \tag{5.73}$$

For the special case of rotations about the z axis, this also holds in cylindrical coordinates.

Let us now solve the eigenvalue problem for the operator L_z, using the position representation. Because (5.73) only involves the coordinate ϕ, the following derivation works equally well in spherical or cylindrical coordinates. We start with

$$L_z\,|l_z\rangle = l_z\,|l_z\rangle\,, \tag{5.74}$$

where l_z is the eigenvalue, also used as a label for the eigenstate, and we have suppressed any degeneracy labels. In the position representation, this becomes

$$-i\hbar\frac{\partial}{\partial\phi}\psi(\phi) = l_z\psi(\phi), \tag{5.75}$$

[2] Beware: not all operators that have three components transform under rotations as vectors in this sense. For example, the operator $\boldsymbol{C} = \boldsymbol{c}I$, where $\boldsymbol{c} = c_x\hat{x} + c_y\hat{y} + c_z\hat{z}$ is a constant vector in three-dimensional space, does not transform under rotations as a vector. Instead, it obeys $U(\boldsymbol{\epsilon})\boldsymbol{C}U(\boldsymbol{\epsilon})^{\dagger} = \boldsymbol{C}$.

also suppressing the dependence on other coordinates. The solutions are

$$\psi(\phi) = ce^{il_z\phi/\hbar}, \tag{5.76}$$

where c is a nonzero normalization constant. The range of ϕ is the continuous interval $0 \leq \phi \leq 2\pi$. Therefore, for L_z to be a Hermitian operator, by following exactly the same derivation that led to (2.217) with P replaced by L_z, we find that for any wavefunctions ψ_1 and ψ_2

$$[\psi_1(2\pi)]^*\psi_2(2\pi) = [\psi_1(0)]^*\psi_2(0) \tag{5.77}$$

must hold. A sufficient condition for this to be satisfied is that all wavefunctions obey $\psi(2\pi) = \psi(0)$, and more generally $\psi(\phi + 2\pi) = \psi(\phi)$. It is necessary to impose this periodic boundary condition anyway, in order for the wavefunction to have a unique single-valued expansion in terms of position eigenstate kets. It follows that $e^{il_z 2\pi/\hbar} = 1$, which requires that

$$l_z = \hbar m, \tag{5.78}$$

where m is an integer. The use of the letter m is traditional in this role, and it is sometimes called the **magnetic quantum number**. We therefore use it as a label for the corresponding eigenstates of L_z, and write, from now on,

$$L_z |m\rangle = \hbar m |m\rangle . \tag{5.79}$$

All of the preceding was derived with the assumption that the angular momentum was associated with the motion of a single particle. More generally, a Hilbert space can describe more than one particle. Furthermore, each of the particles may also have an intrinsic angular momentum, or spin, which has no classical counterpart, and is associated with rotations but is not associated at all with the position wavefunction of the particle (and therefore need not have integer m). These distinct types of angular momenta can also be combined to form new angular momenta.

In general, an angular momentum operator

$$\boldsymbol{J} = \hat{x}J_x + \hat{y}J_y + \hat{z}J_z \tag{5.80}$$

by definition must satisfy the same commutator algebra as the orbital angular momentum,

$$[J_a,\, J_b] = i\hbar\epsilon_{abc}J_c. \tag{5.81}$$

Because the components of an angular momentum operator $\boldsymbol{J}$ do not commute with each other, they are not compatible, and one cannot find a complete orthobasis of $\boldsymbol{J}$ eigenstates. The only solution to the eigenvalue equation $\boldsymbol{J}\,|\boldsymbol{j}\rangle = \boldsymbol{j}\,|\boldsymbol{j}\rangle$ has $\boldsymbol{j} = 0$.

If we choose eigenstates of J_z, they cannot also be eigenstates of J_x or J_y, except in the very special case that all of the eigenvalues are 0. Thus, an angular momentum vector $\boldsymbol{J}$ is not an observable, although each of its components is.

However, you can check that the **angular momentum squared operator**

$$J^2 = J_x^2 + J_y^2 + J_z^2 \tag{5.82}$$

does commute with J_z. This means that J^2 and J_z have common eigenkets and can be part of a CSCO. The simultaneous eigenvalue problem for J^2 and J_z will be worked out in Chap. 8.

For each particle, the intrinsic angular momentum, or spin $\boldsymbol{S}$, is a special case of $\boldsymbol{J}$ defined by the property that it has fixed magnitude, in the sense that all particles of a given type have the same eigenvalue of S^2, which can never change. For any single particle, the total angular momentum operator is the sum of the orbital and spin parts, $\boldsymbol{J} = \boldsymbol{L} + \boldsymbol{S}$. (Note that we use the same symbol $\boldsymbol{J}$ as for a generic angular momentum defined in the previous paragraph.)

For the two-body problem of Sect. 4.2, you can check that the total orbital angular momentum operator (not including spins) can be written in two equivalent ways as

$$\boldsymbol{L}_{\text{tot}} = \boldsymbol{L}_1 + \boldsymbol{L}_2 = \boldsymbol{L} + \boldsymbol{L}_{\text{cm}}, \tag{5.83}$$

where $\boldsymbol{L}_1 = \boldsymbol{R}_1 \times \boldsymbol{P}_1$ and $\boldsymbol{L}_2 = \boldsymbol{R}_2 \times \boldsymbol{P}_2$ for the individual particles, and

$$\boldsymbol{L} = \boldsymbol{R} \times \boldsymbol{P}, \qquad \boldsymbol{L}_{\text{cm}} = \boldsymbol{R}_{\text{cm}} \times \boldsymbol{P}_{\text{tot}}, \tag{5.84}$$

define the relative and center-of-mass contributions to the angular momenta. It is often sensible to restrict to the subspace of states with vanishing total momentum in the center-of-mass frame, consisting of eigenstates of $\boldsymbol{P}_{\text{tot}}$ with eigenvalue 0. If we do so, then $\boldsymbol{L}_{\text{cm}}$ vanishes on that subspace, and the relative angular momentum $\boldsymbol{L}$ is equal to the total angular momentum $\boldsymbol{L}_{\text{tot}}$.

More generally, for a given choice of origin, every three-dimensional physical system has a **total angular momentum** operator $\boldsymbol{J}$, which includes both orbital and spin angular momentum contributions for all of the particles that are present. By definition, $\boldsymbol{J}$ is the operator that acts on the full space of states to generate rotations about any axis $\hat{n}$ by any angle α, with

$$U(\boldsymbol{\alpha}) = \exp\left(-i\alpha\hat{n} \cdot \boldsymbol{J}/\hbar\right), \tag{5.85}$$

where $\boldsymbol{\alpha} = \hat{n}\alpha$. To rotate a state $|\psi\rangle$, the unitary transformation is

$$|\psi\rangle \rightarrow |\psi'\rangle = U(\boldsymbol{\alpha})\,|\psi\rangle\,. \tag{5.86}$$

We also define rotated operators

$$A \rightarrow A' = U(\boldsymbol{\alpha}) A U(\boldsymbol{\alpha})^{\dagger}, \tag{5.87}$$

so that, due to the unitarity of $U(\boldsymbol{\alpha})$, matrix elements of the rotated operators between rotated states are the same as the original matrix elements, $\langle \chi' | A' | \psi' \rangle = \langle \chi | A | \psi \rangle$.

Suppose that the Hamiltonian for a system has the symmetry of invariance under rotations generated by an angular momentum component $\hat{n} \cdot \boldsymbol{J}$, so that rotations about the unit vector axis $\hat{n}$ are a symmetry of the system, and

$$[H, \hat{n} \cdot \boldsymbol{J}] = 0. \tag{5.88}$$

(It is traditional to choose the coordinate system so that $\hat{n} = \hat{z}$, unless there is a good reason not to, but let us be more general.) One can then choose an orthobasis of simultaneous eigenstates of H and $\hat{n} \cdot \boldsymbol{J}$, and as shown on general grounds in Sect. 5.1, $\hat{n} \cdot \boldsymbol{J}$ is a conserved quantity. This means that eigenstates of $\hat{n} \cdot \boldsymbol{J}$ will remain so under time evolution, and for an arbitrary state the probability to obtain a given outcome for a measurement of $\hat{n} \cdot \boldsymbol{J}$ is constant, and

$$\frac{d}{dt} \langle \hat{n} \cdot \boldsymbol{J} \rangle = 0. \tag{5.89}$$

If the Hamiltonian is invariant under all rotations, it is convenient to choose a CSCO to include H, J^2, J_z, since these observables commute with each other.

5.4 Parity

Another kind of coordinate transformation is **parity** (also known as **space inversion**), defined as the replacement of each rectangular position coordinate by minus itself. Unlike translations and rotations, parity is a discrete transformation corresponding to the group known as Z_2; acting twice with parity gives back the identity operation, and there are no infinitesimal parity transformations.

Let us start by defining the **parity operator** Π for a particle moving in one dimension by giving its action on the position-eigenstate orthobasis kets,

$$\Pi \, |x\rangle = |-x\rangle \,. \tag{5.90}$$

(Note that $|-x\rangle$ is the ket that describes a particle known to be at the point $-x$, and is not at all the same thing as $-|x\rangle$, which still describes a particle known to be at the point x.) Since $\Pi \, (\Pi \, |x\rangle) = |x\rangle$, we have $\Pi \, \Pi = I$, so that the parity operator is its own inverse,

$$\Pi^{-1} = \Pi. \tag{5.91}$$

Also, taking the Hermitian adjoint of (5.90) gives $\langle x | \, \Pi^{\dagger} = \langle -x |$, so

$$\langle x'|\Pi^\dagger|x\rangle \;=\; \langle -x'|x\rangle \;=\; \delta(x+x') \;=\; \langle x'|-x\rangle \;=\; \langle x'|\Pi|x\rangle\,. \tag{5.92}$$

Since this is true for every $|x\rangle$ and $|x'\rangle$ in the position orthobasis, it must be that

$$\Pi^\dagger \;=\; \Pi. \tag{5.93}$$

Comparing (5.91) and (5.93), we see that the parity operator is both unitary (an invertible map taking an orthobasis to another orthobasis) and Hermitian (an observable). Since all Hermitian operators have real eigenvalues, and all unitary operators have eigenvalues with magnitude 1, the only possible eigenvalues of Π are 1 and -1.

Given a wavefunction $\langle x|\psi\rangle = \psi(x)$ for an arbitrary state $|\psi\rangle$, we have

$$\langle x|\Pi|\psi\rangle \;=\; \langle -x|\psi\rangle \;=\; \psi(-x). \tag{5.94}$$

Applying this to the case of a momentum eigenstate,

$$\langle x|\Pi|p\rangle \;=\; \langle -x|p\rangle \;=\; \frac{1}{\sqrt{2\pi\hbar}}e^{-ipx/\hbar} \;=\; \langle x|-p\rangle\,, \tag{5.95}$$

from which we learn that

$$\Pi\,|p\rangle \;=\; |-p\rangle\,. \tag{5.96}$$

Similarly, it is easy to show that

$$\Pi X\Pi \;=\; -X, \qquad\qquad \Pi P\Pi \;=\; -P, \tag{5.97}$$

so that the position and momentum operators are both said to be odd under parity. More generally, one defines the **parity of an operator** A as

$$\pi_A \;=\; \pm 1, \quad \text{if} \quad \Pi A\Pi \;=\; \pm A. \tag{5.98}$$

However, not all operators have definite parity in this sense.

If A is even under parity, $\pi_A = +1$, it follows that

$$\Pi A - A\Pi \;=\; 0, \tag{5.99}$$

so that Π and A are compatible operators, and can have common eigenstates. For example, the Hamiltonian $H = P^2/2m + V(X)$ is parity-even if, and only if, $V(X)$ is an even function of X. In the special case of a free particle with $V = 0$, the simultaneous eigenstates of H and Π are

$$|E,+1\rangle = \frac{1}{\sqrt{2}}\left(|p\rangle + |-p\rangle\right), \quad |E,-1\rangle = \frac{1}{\sqrt{2}}\left(|p\rangle - |-p\rangle\right), \tag{5.100}$$

where $p = \sqrt{2mE}$. However, since P and Π do not commute, they cannot have simultaneous eigenstates (except for the trivial case $p = 0$ with even parity), and indeed the parity eigenstates $|E, \pm 1\rangle$ with $E \neq 0$ are not eigenstates of momentum.

Just as for translations, parity generalizes straightforwardly to three dimensions. Define

$$\Pi \, |\boldsymbol{r}\rangle = |-\boldsymbol{r}\rangle \,, \tag{5.101}$$

from which it follows that

$$\Pi \, |\boldsymbol{p}\rangle = |-\boldsymbol{p}\rangle \,, \tag{5.102}$$

and position and momentum operators carry odd parity,

$$\Pi \boldsymbol{R} \Pi = -\boldsymbol{R}, \qquad \Pi \boldsymbol{P} \Pi = -\boldsymbol{P}. \tag{5.103}$$

Angular momentum operators, as defined in the previous section, always have even parity,

$$\Pi \boldsymbol{J} \Pi = \boldsymbol{J}. \tag{5.104}$$

In the case of orbital angular momentum, $\Pi \boldsymbol{L} \Pi = \boldsymbol{L}$ follows directly from the definition of the components of $\boldsymbol{L}$ in (5.50). For more general angular momentum operators (including spin), the even parity assignment is consistent with the general commutator algebra structure in (5.81). If the potential energy depends only on the radial coordinate, so that the Hamiltonian has the form $H = P^2/2m + V(R)$, then this H commutes not only with L^2 and L_z, but also with Π. Such a system therefore admits simultaneous eigenstates of all four operators. In that case, parity is a symmetry of the system, and eigenstates of parity remain so under time evolution.

A powerful application of parity is to the identification of selection rules for matrix elements. Suppose that an operator A has definite parity π_A as defined by (5.98), and that the states $|\psi\rangle$ and $|\phi\rangle$ are parity eigenstates with eigenvalues π_ψ and π_ϕ, respectively. Now we note that

$$\langle\phi|A|\psi\rangle = \pi_A \langle\phi|\Pi A \Pi|\psi\rangle = \pi_A \pi_\phi \pi_\psi \langle\phi|A|\psi\rangle \,. \tag{5.105}$$

The product $\pi_A \pi_\phi \pi_\psi$ is either $+1$ or -1. In the former case, (5.105) tells us nothing, but in the latter case, we obtain:

Theorem 5.2 (Parity Selection Rule) *If states $|\psi\rangle$ and $|\phi\rangle$ and an operator A all have definite parities, and $\pi_A \pi_\phi \pi_\psi = -1$, then the matrix element $\langle\phi|A|\psi\rangle$ must vanish.*

As we will see, this result has many practical applications, including neatly explaining the absence of certain atomic transitions. Before investing your valuable time in calculating a quantity, it is always a good idea to consider first whether it must vanish due to a matrix element selection rule.

5.5 Gauge Transformations

Physical systems that involve electromagnetic interactions with charged particles can be formulated in terms of potentials, subject to gauge transformations

$$\Phi \to \Phi' = \Phi - \frac{1}{c}\frac{\partial \Lambda}{\partial t}, \qquad \boldsymbol{A} \to \boldsymbol{A}' = \boldsymbol{A} + \nabla\Lambda, \tag{5.106}$$

where $\Lambda(\boldsymbol{r}, t)$ is an arbitrary function. As discussed in Sect. 4.3, the physical situation described by a pair of potentials $(\Phi, \boldsymbol{A})$ is equally well described by any pair $(\Phi', \boldsymbol{A}')$ related to them by a gauge transformation.

Before proceeding, we pause to emphasize that gauge transformations have a completely different character from the other transformations discussed in the preceding sections. If we do a translation or rotation on a system, we have changed the system into a physically distinct one. A particle that has been translated by 1 cm is in a different place, clearly distinguishable by experiment. To say that a system has translation symmetry means that we can physically move the entire system in an experimentally measurable way without changing the Hamiltonian that describes its time evolution. In contrast, the whole point of gauge transformations is that although they *do* change the potentials and thus the expression for the Hamiltonian, they do so *without* changing the actual physical situation at all. There is no way an experiment can tell whether or not we have chosen to write the Hamiltonian in Coulomb gauge! Although it is common to refer to "gauge symmetries", gauge transformations really are not symmetries in the same way that translations or rotations can be. Instead, they are a manifestation of the fact that the potentials have an arbitrariness, in the form of redundancies in our description that can be eliminated by fixing the gauge.

Consider a classical charged particle in an electromagnetic field. Clearly, if we do a gauge transformation, the position and the velocity of the charged particle are unaffected, since they can be measured experimentally. However, the classical canonical momentum $\boldsymbol{p}$ defined in (4.33) does change with a gauge transformation, because it involves not just the velocity but also the gauge-dependent potential $\boldsymbol{A}$.

In the quantum description, the state ket used to describe the system is similarly gauge-dependent. Naively, this might seem like a problem, but it is not, because the state ket by itself is not a physically measurable observable. For example, we have already pointed out that multiplying a ket by a *constant* complex phase does not change probabilities. In the following, we will show that the gauge transformation of the ket describing a single particle of charge q is realized as a unitary transformation

$$U_\Lambda = \exp\left[i\frac{q}{\hbar c}\Lambda(\boldsymbol{R},t)\right], \tag{5.107}$$

which imparts a *position-dependent* complex phase to the state ket,

$$|\psi\rangle \to |\psi'\rangle = U_\Lambda |\psi\rangle, \tag{5.108}$$

and must be accompanied by changes in the electromagnetic potentials,

$$\Phi(\boldsymbol{R},t) \to \Phi(\boldsymbol{R},t)' = \Phi(\boldsymbol{R},t) - \frac{1}{c}\frac{\partial}{\partial t}\Lambda(\boldsymbol{R},t), \tag{5.109}$$

$$\boldsymbol{A}(\boldsymbol{R},t) \to \boldsymbol{A}(\boldsymbol{R},t)' = \boldsymbol{A}(\boldsymbol{R},t) + \nabla\Lambda(\boldsymbol{R},t). \tag{5.110}$$

A gauge transformation in quantum mechanics is defined as the simultaneous changes in (5.108)–(5.110) with the same Λ. (Note that Λ is an operator because $\boldsymbol{R}$ is.) Primes are used to indicate the state ket and operators after the gauge transformation. For any operator Ω, the gauge transformation

$$\Omega \to \Omega' \tag{5.111}$$

is defined by applying (5.109) and (5.110) to the potentials appearing explicitly in the expression for Ω. We now show that physical predictions are unaffected by such a transformation.

We start by noting that a prerequisite for an operator Ω to be a **gauge-invariant observable** is that its expectation values should be the same before and after the gauge transformation,

$$\langle\psi|\Omega|\psi\rangle = \langle\psi'|\Omega'|\psi'\rangle. \tag{5.112}$$

Requiring this to be true for all state kets $|\psi\rangle$, and using (5.108), we obtain the defining property of a gauge-invariant observable,

$$\Omega' = U_\Lambda \Omega U_\Lambda^\dagger. \tag{5.113}$$

Not all Hermitian operators that would otherwise satisfy the requirements of being an observable will have this property.

For example, consider the canonical operators $\boldsymbol{R}$ and $\boldsymbol{P}$. Since they do not depend explicitly on the potentials, we have

$$\boldsymbol{R}' = \boldsymbol{R}, \qquad \boldsymbol{P}' = \boldsymbol{P}. \tag{5.114}$$

Meanwhile, using the definition of U_Λ in (5.107), we find

$$U_\Lambda \boldsymbol{R} U_\Lambda^\dagger = \boldsymbol{R}, \tag{5.115}$$

$$U_\Lambda \boldsymbol{P} U_\Lambda^\dagger = \boldsymbol{P} - \frac{q}{c}\nabla\Lambda(\boldsymbol{R}, t). \tag{5.116}$$

The first equation is an immediate consequence of the fact that U_Λ involves only $\boldsymbol{R}$ and not $\boldsymbol{P}$, and the second follows from applying Theorem 2.8. Equation (5.115) says that $\boldsymbol{R}$ is a gauge-invariant observable, but (5.116) shows that the canonical momentum operator $\boldsymbol{P}$ is not. In contrast, it follows from (5.110) that the gauge transformation of the kinetic momentum

$$\boldsymbol{\Pi} = \boldsymbol{P} - \frac{q}{c}\boldsymbol{A}(\boldsymbol{R}, t) \tag{5.117}$$

is $\boldsymbol{\Pi}' = \boldsymbol{P} - \frac{q}{c}(\boldsymbol{A} + \nabla\Lambda) = \boldsymbol{\Pi} - \frac{q}{c}\nabla\Lambda$. Using (5.116), this can be rewritten as

$$\boldsymbol{\Pi}' = U_\Lambda \boldsymbol{\Pi} U_\Lambda^\dagger. \tag{5.118}$$

This establishes that the kinetic momentum $\boldsymbol{\Pi}$ is a gauge-invariant observable, according to our defining requirement of (5.113).

Since this was perhaps a bit tricky and unexpected, we reiterate. Even though the canonical momentum operator $\boldsymbol{P}$ does not change when we do a gauge transformation, its expectation values do change, so it is not a gauge-invariant observable. And, although the kinetic momentum operator $\boldsymbol{\Pi}$ changes when we do a gauge transformation, its expectation values do not, and it is a gauge-invariant observable. These are the quantum versions of the statements made about the gauge dependences of the classical quantities $\boldsymbol{p}$ and $\boldsymbol{\pi}$ following (4.35).

The Hamiltonian operator in (4.48) certainly changes when we do a gauge transformation, since it depends explicitly on the potentials Φ and $\boldsymbol{A}$. Using (5.109) and (5.110) gives the gauge transformation of H,

$$H \to H' = \frac{1}{2m}\left(\boldsymbol{P} - \frac{q}{c}\boldsymbol{A} - \frac{q}{c}\nabla\Lambda\right)^2 + q\Phi - \frac{q}{c}\frac{\partial\Lambda}{\partial t} - \gamma \boldsymbol{S}\cdot\boldsymbol{B}. \tag{5.119}$$

Equations (5.115) and (5.116) allow us to rewrite this in the convenient form

$$H' = U_\Lambda H U_\Lambda^\dagger + i\hbar\left(\frac{\partial U_\Lambda}{\partial t}\right)U_\Lambda^\dagger. \tag{5.120}$$

Now, given the Schrödinger equation

$$i\hbar\frac{d}{dt}|\psi(t)\rangle = H\,|\psi(t)\rangle, \tag{5.121}$$

it is straightforward to use $|\psi(t)'\rangle = U_\Lambda\,|\psi(t)\rangle$ and (5.120) to obtain

$$i\hbar\frac{d}{dt}|\psi(t)'\rangle = H'\,|\psi(t)'\rangle. \tag{5.122}$$

This shows the key result that the Schrödinger equation for time evolution is also satisfied if H and $|\psi(t)\rangle$ are replaced by their gauge-transformed counterparts.

Another consequence of (5.120) is that the Hamiltonian is not, in general, a gauge-invariant observable, due to the presence of the last term. This reflects the primary and defining role of H as the generator of time evolution, rather than a measurable observable. If we limit the gauge transformations to those with no explicit time dependence, $\partial\Lambda/\partial t = 0$, then the requirement $H' = U_\Lambda H U_\Lambda^\dagger$ is satisfied, and H is a gauge-invariant observable in that restricted sense. Also, if the potentials Φ and $\boldsymbol{A}$ do not depend on time, then according to the results at the end of Sect. 5.1, H is a conserved quantity, the total energy of the particle. Note that a gauge transformation using a Λ that is linear in time and independent of position just adds a constant term to Φ and thus to H. This corresponds to the classical freedom to add an arbitrary constant to the energy without affecting the equations of motion at all.

The defining requirement for gauge-invariant observables, (5.113), was obtained by demanding that expectation values do not depend on the choice of gauge. Now we will prove the stronger result that all measurement probabilities for such observables are gauge independent. For any gauge-invariant observable Ω, choose an orthobasis $|\omega, u\rangle$ of its eigenkets, so that

$$\Omega\,|\omega, u\rangle = \omega\,|\omega, u\rangle\,, \tag{5.123}$$

where u are possible degeneracy labels. According to Postulates 3 and 4, the probabilities for allowed results of the measurement of Ω in a state $|\psi\rangle$ are

$$\mathcal{P}(\omega) = \sum_u |\langle\omega, u|\psi\rangle|^2. \tag{5.124}$$

Now, we observe that

$$\Omega'\,U_\Lambda\,|\omega, u\rangle = U_\Lambda \Omega U_\Lambda^\dagger\,U_\Lambda\,|\omega, u\rangle = U_\Lambda\Omega\,|\omega, u\rangle = \omega U_\Lambda\,|\omega, u\rangle\,, \tag{5.125}$$

which shows that the states $U_\Lambda\,|\omega, u\rangle$ form an orthobasis of eigenkets of Ω', with the same eigenvalues ω and the same degeneracies. Therefore, Postulates 3 and 4 tell us that the probability to obtain the result ω from a measurement of Ω' in the gauge-transformed description is

$$\mathcal{P}(\omega)' = \sum_u |\langle\omega, u|U_\Lambda^\dagger|\psi'\rangle|^2. \tag{5.126}$$

Using $|\psi'\rangle = U_\Lambda\,|\psi\rangle$, we obtain

$$\mathcal{P}(\omega)' = \sum_u |\langle\omega, u|U_\Lambda^\dagger U_\Lambda|\psi\rangle|^2 = \sum_u |\langle\omega, u|\psi\rangle|^2 = \mathcal{P}(\omega). \tag{5.127}$$

Thus, we have succeeded in our goal of showing that the predictions for measurements of gauge-invariant observables are not changed by the gauge transformation. This is in accord with the general principle that gauge transformations may affect our equations in intermediate steps of calculations, but do not change the final results that reflect physical reality.

For simplicity, in the preceding we have treated the case of a single particle with charge q. In the case of N particles, with charges q_n and masses m_n, the same discussion goes through, with the Hamiltonian

$$H = \sum_n \left(\frac{1}{2m_n} \left[\boldsymbol{P}_n - \frac{q_n}{c} \boldsymbol{A}(\boldsymbol{R}_n, t) \right]^2 + q_n \Phi(\boldsymbol{R}_n, t) - \gamma_n \boldsymbol{S}_n \cdot \boldsymbol{B}(\boldsymbol{R}_n, t) \right) \quad (5.128)$$

subject to gauge transformations as given by (5.108)–(5.110), but with

$$U_\Lambda = \exp\left[\frac{i}{\hbar c} \sum_n q_n \Lambda(\boldsymbol{R}_n, t) \right]. \quad (5.129)$$

The gauge transformation of the wavefunction $\psi(\boldsymbol{r}_1, \ldots, \boldsymbol{r}_N, t) = \langle \boldsymbol{r}_1, \ldots, \boldsymbol{r}_N | \psi(t) \rangle$ is therefore

$$\psi(\boldsymbol{r}_1, \ldots, \boldsymbol{r}_N, t) \to \exp\left[\frac{i}{\hbar c} \sum_n q_n \Lambda(\boldsymbol{r}_n, t) \right] \psi(\boldsymbol{r}_1, \ldots, \boldsymbol{r}_N, t). \quad (5.130)$$

This consists of multiplication by a phase that may depend on time and the particle positions. Thus, gauge invariance can be viewed as the statement that a *local* change of phase factor of this special form can be compensated by a redefinition of the electromagnetic potentials. In practice, one may avoid this redundancy in the description of the physical system by fixing the gauge. This just means that one chooses a specific form of the potentials, or at least imposes some condition on them that will not be satisfied if one does an arbitrary gauge transformation.

It is also possible to turn things around, and view the gauge transformations as the logical starting point that determines the interactions. In quantum mechanics, the physics is always independent of the conventional choice of global phase of the wavefunction. If we generalize this to a local change in phase, as in (5.130), then we are led to introduce the electromagnetic potentials to provide a compensating change, and this leads in turn to the necessity of interactions of charged particles with electromagnetic fields. Although it is beyond our scope here, this approach allows the interactions of the strong and weak nuclear forces to be constructed from their non-Abelian gauge transformation groups, $SU(3)$ and $SU(2) \times U(1)$, respectively.

5.6 Currents and Local Conservation of Probability

From Postulate 4, and the completeness of position eigenstates, we know that the probability to find a particle within an infinitesimal volume $d^3\boldsymbol{r}$ is given by the Born rule,

$$d\mathcal{P} = \rho(\boldsymbol{r}, t)\, d^3\boldsymbol{r}, \tag{5.131}$$

where the probability density per unit volume is

$$\rho(\boldsymbol{r}, t) = |\langle \boldsymbol{r}|\psi(t)\rangle|^2 = |\psi(\boldsymbol{r}, t)|^2. \tag{5.132}$$

Assuming the total probability of finding the particle somewhere is fixed and equal to 1, there must be a law of conservation of probability. The local form of this law is a differential equation

$$\frac{\partial \rho}{\partial t} = -\nabla \cdot \boldsymbol{J}, \tag{5.133}$$

where $\boldsymbol{J}$ is a probability current density. The left side of this equation is the rate at which probability density is accumulating at a point, which the right side tells us is the negative of a source for the vector field $\boldsymbol{J}$.

To prove (5.133), and identify the current density, start with the Schrödinger equation in the position representation with wavefunction ψ, and multiply by $-i\psi^*/\hbar$,

$$\psi^* \frac{\partial \psi}{\partial t} = \frac{i\hbar}{2m}\psi^* \nabla^2 \psi - \frac{i}{\hbar} V \psi^* \psi. \tag{5.134}$$

Adding this to its complex conjugate, we get

$$\frac{\partial \rho}{\partial t} = \frac{i\hbar}{2m}\left(\psi^* \nabla^2 \psi - \psi \nabla^2 \psi^*\right), \tag{5.135}$$

in which the contribution proportional to $V(\boldsymbol{r}, t)$ has canceled due to the fact that the potential is real-valued. Now if we define the probability current density by

$$\boldsymbol{J} \equiv \frac{i\hbar}{2m}\left(\psi \nabla \psi^* - \psi^* \nabla \psi\right), \tag{5.136}$$

then we have

$$\nabla \cdot \boldsymbol{J} = \frac{i\hbar}{2m}\left(\psi \nabla^2 \psi^* - \psi^* \nabla^2 \psi\right), \tag{5.137}$$

where two terms proportional to $\nabla\psi^* \cdot \nabla\psi$ have canceled. Comparison of (5.135) and (5.137) establishes the local conservation of probability, (5.133).

In the case of a charged particle, an important modification is needed if the vector potential $\boldsymbol{A}$ is nonzero. Starting with the Hamiltonian in (4.48), one can show by steps similar to those followed in (5.134)–(5.137) that the current density satisfying $\nabla \cdot \boldsymbol{J} = -\partial\rho/\partial t$ is

$$\boldsymbol{J} = \frac{1}{m}\mathrm{Re}\left[\psi^*\left(-i\hbar\nabla - \frac{q}{c}\boldsymbol{A}\right)\psi\right], \tag{5.138}$$

generalizing (5.136). Here, $-i\hbar\nabla - \frac{q}{c}\boldsymbol{A}$ is the position representation of the kinetic momentum operator, $\boldsymbol{\Pi}$, which was introduced in (4.49) of Sect. 4.3. This expression for $\boldsymbol{J}$ is invariant under gauge transformations, and is related to the electric current density $\boldsymbol{j}$ by $\boldsymbol{j} = q\boldsymbol{J}$, where q is the electric charge of the particle.

As an example, consider a free particle in a plane-wave simultaneous eigenstate of momentum and energy with eigenvalues $\boldsymbol{p} = \hbar\boldsymbol{k}$ and $E = \hbar^2k^2/2m$,

$$\psi(\boldsymbol{r},t) = Ce^{i(\boldsymbol{k}\cdot\boldsymbol{r}-Et/\hbar)}. \tag{5.139}$$

In terms of the complex normalization constant C, the probability and current densities are just constants in both time and position,

$$\rho = |C|^2, \qquad \boldsymbol{J} = \frac{i\hbar}{2m}(-i\boldsymbol{k} - i\boldsymbol{k})|C|^2 = \frac{\boldsymbol{p}}{m}\rho. \tag{5.140}$$

If the domain of the particle is all space, then the wavefunction is not normalizable to unity for any finite C, but at least the ratio of the current density to the probability density is well-defined and equal to the velocity eigenvalue of the particle. Although the probability density is constant and uniform, it is constantly flowing in the direction of $\boldsymbol{v} = \boldsymbol{p}/m$. So, in the case of a free particle,

$$\frac{\partial\rho}{\partial t} = 0, \qquad -\nabla \cdot \boldsymbol{J} = 0, \tag{5.141}$$

satisfying the local conservation of probability in the most trivial possible way.

In the case of one-dimensional problems, the probability density and current are

$$\rho = |\psi(x,t)|^2, \qquad J = \frac{\hbar}{m}\mathrm{Im}\left(\psi^*\frac{\partial\psi}{\partial x}\right), \tag{5.142}$$

and the statement of local conservation of probability is

$$\frac{\partial\rho}{\partial t} = -\frac{\partial J}{\partial x}. \tag{5.143}$$

The current density J is just a number (not a vector) in the one-dimensional case, and is positive for probability density flowing to positive x.

A key ingredient in establishing conservation of probability was that the potential V is a real function in the position representation, or more generally the Hermiticity of the Hamiltonian operator. This is also directly related to the unitarity of the time evolution operator $U(t, t_0)$ that we studied in Sect. 3.1. However, in some situations, one can think of probability as *not* being conserved. For example, suppose that we are using quantum mechanics to model the behavior of an unstable particle. If our Hilbert space only describes the unstable particle, and not the other particles that it decays into, then we should expect that the total integrated probability to find it somewhere should decrease with time. To describe situations like this, we can break the law for a good cause, by taking the Hamiltonian to be non-Hermitian.

To see how this works, suppose that we defy the authority of Postulate 6 by taking H to have complex eigenvalues, with corresponding normalized eigenkets $|\phi_n\rangle$, so that

$$H\,|\phi_n\rangle = (E_n - i\Gamma_n/2)\,|\phi_n\rangle\,, \tag{5.144}$$

where E_n and Γ_n are real numbers. Of course, this is only possible if H is not Hermitian, due to Theorem 2.12. If the state of the system at time $t = 0$ is one of them, $|\psi(0)\rangle = |\phi_n\rangle$, with unit norm, then the time evolution predicted by the Schrödinger equation will be non-unitary,

$$|\psi(t)\rangle = e^{-i(E_n - i\Gamma_n/2)t/\hbar}\,|\phi_n\rangle\,. \tag{5.145}$$

The norm of this ket as a function of time is therefore

$$\langle\psi(t)|\psi(t)\rangle = e^{-\Gamma_n t/\hbar}. \tag{5.146}$$

This can be interpreted as the survival probability that the particle still exists at time $t > 0$, given that it existed at time $t = 0$. [Compare to (3.3) with $P_\alpha = I$.] The mean lifetime of the unstable state $|\phi_n\rangle$ is thus $\tau = \hbar/\Gamma_n$.

The unstable states could correspond to particles that undergo spontaneous decay, such as the neutron, the muon, or many atomic nuclei. The non-Hermitian-Hamiltonian formalism could also be applied to a set of atomic states, if our quantum treatment does not include the complete set of states to which they could decay, including photons released in the process. It could even apply in a description of a single absolutely stable particle like the electron, if it can be removed from the system under discussion by a process such as electron capture, which in some nuclei occurs as $e^- p \to \nu_e n$. This would correspond to an effective potential $V(\boldsymbol{r})$ with negative imaginary part localized at the nuclei. Repeating the steps of (5.134)–(5.137), one finds that for a non-Hermitian potential the equation governing the local probability density and current is

$$\frac{\partial \rho}{\partial t} = -\nabla \cdot \boldsymbol{J} + \frac{2}{\hbar}\,\mathrm{Im}[V(\boldsymbol{r})]\,\rho. \tag{5.147}$$

The last term describes a local probability sink if $\mathrm{Im}[V(\boldsymbol{r})] < 0$.

In all such cases of particles decaying, or changing their numbers through interaction, the use of a non-Hermitian Hamiltonian is really a sign of an incomplete description. There is always a more fundamental description in which the *complete* Hamiltonian is Hermitian. Quantum field theories provide the natural way to incorporate processes that change the numbers of particles, in such a way that the postulates of quantum mechanics hold, including the unitary time evolution of Postulate 6.

5.7 The Curious Case of Time Reversal

Throughout most of this book, all of the operators implicitly obey the linearity rule of (2.52), which governs how they act on linear superpositions of kets. Linearity is a natural and intuitive property for operators to have, because it coincides with the arithmetic rules for ordinary complex matrices acting on ordinary complex vectors. The subject of this section is a quite exceptional case that does not obey these intuitive rules: the **time-reversal operator**, which we will call Θ. It is also known as the **motion-reversal operator**, a terminology favored by its originator, Eugene Wigner, for a good reason that will become apparent shortly.

As we will show, Θ is an antilinear operator. By definition, an operator Θ is **antilinear** if, for any complex numbers c_1 and c_2 and any kets $|\psi_1\rangle$ and $|\psi_2\rangle$,

$$\Theta\big(c_1\,|\psi_1\rangle + c_2\,|\psi_2\rangle\big) = c_1^*\Theta\,|\psi_1\rangle + c_2^*\Theta\,|\psi_2\rangle\,. \tag{5.148}$$

This is almost the same as the rule for linear operators, but with the crucial difference that the coefficients are complex conjugated when they are moved through Θ. Put another way, antilinear operators commute with real numbers, but anticommute with purely imaginary numbers. A warning: the special properties of antilinear operators may cause initial confusion for those who have internalized the rules for linear operators. You may find it reassuring that time reversal is the only antilinear operator that we will encounter, and it appears *only* in this section and in Sect. 8.8. Skipping those two sections will cause no further harm.

Let us start with time-reversal invariance in a classical setting. Consider a ball thrown from the top of a tower, with initial position $\boldsymbol{r}_0$ and momentum $\boldsymbol{p}_0$ at time $t = 0$. The ball accelerates downward due to a uniform gravitational acceleration, following a parabolic path. At time t, the ball has reached position $\boldsymbol{r}$ with momentum $\boldsymbol{p}$. Suppose we then intervene by instantaneously flipping the momentum to $-\boldsymbol{p}$ while keeping the position the same. Neglecting air resistance, the classical equation of motion tells us that the ball will then simply retrace its parabolic trajectory in the opposite direction. So, after another time interval t, it will have returned to its original position $\boldsymbol{r}_0$, but with the opposite momentum $-\boldsymbol{p}_0$. If one made a movie of the reversed trajectory, and played it backwards, it would look exactly the same as the movie of the trajectory before the motion was reversed. This explains the "time reversal" terminology.

In this example, we can describe our intervention by defining the classical operation of motion reversal, which at the instant it is applied keeps the position of the particle the same, but reverses its momentum:

$$\boldsymbol{r} \to \boldsymbol{r}, \qquad \boldsymbol{p} \to -\boldsymbol{p}, \qquad \text{(classical motion reversal).} \tag{5.149}$$

To summarize our classical time-reversal thought experiment, if we

- evolve the system forward in time by t,
- apply the motion-reversal operation,
- evolve the system forward in time by t again,

then the system will be in its original position, but with the momentum reversed. This is the same as if we had just applied the motion reversal operation with no time evolution.

However, not all systems have time-reversal invariance. For example, if we take into account the rotation of the Earth as well as its gravitational field, then the ball will experience a Coriolis force proportional to $\boldsymbol{v} \times \boldsymbol{\Omega}$ where $\boldsymbol{v}$ is the ball's velocity and $\boldsymbol{\Omega}$ is the angular velocity vector of the Earth. This causes it to be deflected in such a way that it does not quite return to its original position. In such cases, we can still define the classical motion-reversal operation as the transformation (5.149), but it is not a symmetry of the system.

The quantum mechanical counterpart of the classical motion-reversal operation is Θ. In the case of a spinless particle, we are inspired by (5.149) to define it as the transformation

$$\Theta\,|\boldsymbol{r}\rangle = |\boldsymbol{r}\rangle\,, \qquad \Theta\,|\boldsymbol{p}\rangle = |-\boldsymbol{p}\rangle\,, \tag{5.150}$$

up to phase factors that can be set to unity. Defining it in this way shows why Θ might better be called motion reversal, rather than time reversal, which is not necessarily assumed to be a symmetry. Now, unlike the classical mechanical analog, in quantum mechanics (5.150) is a logically overloaded definition, since the states $|\boldsymbol{r}\rangle$ by themselves form a basis, but the states $|\boldsymbol{p}\rangle$ form another basis. Therefore, we must check that our definition is self-consistent.

From (5.150), we can act with the inverse operator Θ^{-1} to get $\Theta^{-1}\,|\boldsymbol{r}\rangle = |\boldsymbol{r}\rangle$ and $\Theta^{-1}\,|\boldsymbol{p}\rangle = |-\boldsymbol{p}\rangle$ as well.[3] Acting on these with $\Theta\boldsymbol{R}$ and $\Theta\boldsymbol{P}$, respectively, we

[3] From the form of (5.150), one might conclude that $\Theta^{-1} = \Theta$ and there is no point in distinguishing them. This is indeed true for systems with integer spin. However, as we will see in Sect. 8.8, $\Theta^{-1} = -\Theta$ for a particle with half-integer spin. So we will keep the distinction as a matter of principle.

find

$$\Theta \boldsymbol{R} \Theta^{-1} |\boldsymbol{r}\rangle = \Theta \boldsymbol{R} |\boldsymbol{r}\rangle = \Theta \boldsymbol{r} |\boldsymbol{r}\rangle = \boldsymbol{r} \Theta |\boldsymbol{r}\rangle = \boldsymbol{r} |\boldsymbol{r}\rangle , \tag{5.151}$$

$$\Theta \boldsymbol{P} \Theta^{-1} |\boldsymbol{p}\rangle = \Theta \boldsymbol{P} |-\boldsymbol{p}\rangle = -\Theta \boldsymbol{p} |-\boldsymbol{p}\rangle = -\boldsymbol{p} \Theta |-\boldsymbol{p}\rangle = -\boldsymbol{p} |\boldsymbol{p}\rangle , \tag{5.152}$$

where we have used the fact that regardless of whether Θ is linear or antilinear, it commutes with the real numbers $\boldsymbol{r}$ and $-\boldsymbol{p}$. So, we have operator relations

$$\Theta \boldsymbol{R} \Theta^{-1} = \boldsymbol{R}, \qquad \Theta \boldsymbol{P} \Theta^{-1} = -\boldsymbol{P}. \tag{5.153}$$

Consider how this fits in with the commutation relations $[R_a, P_b] = i\hbar\delta_{ab}$. We have

$$\Theta [R_a, P_b] \Theta^{-1} = \Theta i\hbar\delta_{ab} \Theta^{-1}. \tag{5.154}$$

However, this can also be rewritten as

$$\begin{aligned} \Theta R_a \Theta^{-1} \Theta P_b \Theta^{-1} - \Theta P_b \Theta^{-1} \Theta R_a \Theta^{-1} &= -R_a P_b + P_b R_a \\ &= -i\hbar\delta_{ab}. \end{aligned} \tag{5.155}$$

Comparing the last two equations, we learn that $\Theta\, i\, \Theta^{-1} = -i$, or

$$\Theta i = -i\, \Theta. \tag{5.156}$$

This establishes that Θ must be antilinear. It commutes with real numbers, but anticommutes with all purely imaginary numbers.

Now we can check that the second part of (5.150) is indeed consistent with the first:

$$\begin{aligned} \Theta |\boldsymbol{p}\rangle &= \Theta \int d^3\boldsymbol{r}\, |\boldsymbol{r}\rangle \langle \boldsymbol{r}|\boldsymbol{p}\rangle = \Theta \int d^3\boldsymbol{r} \frac{e^{i\boldsymbol{p}\cdot\boldsymbol{r}/\hbar}}{(2\pi\hbar)^{3/2}} |\boldsymbol{r}\rangle = \int d^3\boldsymbol{r} \frac{e^{-i\boldsymbol{p}\cdot\boldsymbol{r}/\hbar}}{(2\pi\hbar)^{3/2}} \Theta |\boldsymbol{r}\rangle \\ &= \int d^3\boldsymbol{r} \frac{e^{-i\boldsymbol{p}\cdot\boldsymbol{r}/\hbar}}{(2\pi\hbar)^{3/2}} |\boldsymbol{r}\rangle = \int d^3\boldsymbol{r}\, |\boldsymbol{r}\rangle \langle \boldsymbol{r}|-\boldsymbol{p}\rangle \\ &= |-\boldsymbol{p}\rangle . \end{aligned} \tag{5.157}$$

The first and last equalities used completeness of the position eigenstates, and the second and fifth used the explicit form of the momentum eigenstate wavefunction from (2.256). The third equality made crucial use of the antilinearity of Θ, and the fourth is where we used the first part of (5.150).

Equation (5.153) tells us that position and momentum are even and odd, respectively, under a motion-reversal transformation. Because the orbital angular momentum operator is defined in terms of them by $\boldsymbol{L} = \boldsymbol{R} \times \boldsymbol{P}$, it follows immediately that

$$\Theta \boldsymbol{L} \Theta^{-1} = \Theta \boldsymbol{R} \Theta^{-1} \times \Theta \boldsymbol{P} \Theta^{-1} = \boldsymbol{R} \times (-\boldsymbol{P}) = -\boldsymbol{L}. \tag{5.158}$$

More generally, due to the antilinearity of Θ, for any angular momentum operator the form of the commutation relation $[J_a, J_b] = i\hbar\epsilon_{abc}J_c$, with its factor of i, is only consistent with $\boldsymbol{J}$ being odd under motion reversal,

$$\Theta \boldsymbol{J} \Theta^{-1} = -\boldsymbol{J}. \tag{5.159}$$

We will use this relation in Sect. 8.8 to learn about the time-reversal properties of angular momentum eigenstates.

Given any state $|\psi\rangle$, the motion-reversed[4] state is

$$|\Theta\psi\rangle \equiv \Theta\,|\psi\rangle\,. \tag{5.160}$$

Suppose that $|\psi\rangle$ has position wavefunction $\psi(\boldsymbol{r}) = \langle \boldsymbol{r}|\psi\rangle$. We now compute the position wavefunction of $|\Theta\psi\rangle$, using completeness:

$$\begin{aligned}
\langle \boldsymbol{r}|\Theta\psi\rangle &= \int d^3\boldsymbol{r}'\,\langle \boldsymbol{r}|\Theta|\boldsymbol{r}'\rangle\,\langle \boldsymbol{r}'|\psi\rangle = \int d^3\boldsymbol{r}'(\langle \boldsymbol{r}'|\psi\rangle)^*\,\langle \boldsymbol{r}|\Theta|\boldsymbol{r}'\rangle \\
&= \int d^3\boldsymbol{r}'(\langle \boldsymbol{r}'|\psi\rangle)^*\,\langle \boldsymbol{r}|\boldsymbol{r}'\rangle = \int d^3\boldsymbol{r}'(\langle \boldsymbol{r}'|\psi\rangle)^*\delta^{(3)}(\boldsymbol{r}'-\boldsymbol{r}) \\
&= \psi(\boldsymbol{r})^*.
\end{aligned} \tag{5.161}$$

Thus, motion reversal has the effect of complex conjugating the wavefunction.

In the previous derivation, there was a somewhat tricky step. The factor $\langle \boldsymbol{r}'|\psi\rangle$ was a complex number coefficient appearing to the right of the Θ. Therefore, in the second equality it had to be complex conjugated to move it to the left before we could safely use $\Theta\,|\boldsymbol{r}'\rangle = |\boldsymbol{r}'\rangle$ in the third equality. This illustrates a potential source of confusion for antilinear operators: what if we had instead tried to think of Θ as acting to the left, on the bra $\langle \boldsymbol{r}|$? A good way to avoid such issues in a consistent way is to always define antilinear operators like Θ only by their action to the right. Although it is possible to define left multiplication $\langle\psi|\,\Theta$, it is not necessary, and doing so for antilinear operators leads to more potential confusion than it is worth. Therefore, we always interpret

$$\langle\psi_1|\Theta|\psi_2\rangle = \langle\psi_1|\left(\Theta\,|\psi_2\rangle\right), \tag{5.162}$$

and do not bother to define $\left(\langle\psi_1|\,\Theta\right)$ at all. Instead, we make use of the definition

$$\langle\Theta\psi| \equiv \text{bra dual to } |\Theta\psi\rangle\,. \tag{5.163}$$

[4] The state $|\Theta\psi\rangle$ can be (and often is) called the time-reversed state, but this is less precise, because it holds as a definition at fixed time regardless of the actual time evolution of the system.

Likewise, we always interpret Θ^{-1} as acting to the right, and choose to never write, or even define, $\Theta^\dagger$ for an antilinear operator Θ.

As a transformation and a possible symmetry, the operator Θ should preserve probabilities. In other words, for any two kets $|\psi_1\rangle$ and $|\psi_2\rangle$, we must have

$$|\langle\Theta\psi_1|\Theta\psi_2\rangle|^2 = |\langle\psi_1|\psi_2\rangle|^2, \tag{5.164}$$

so that the probability of the transformed states overlapping is equal to the probability of the original states overlapping. A theorem due to Wigner (with a nontrivial proof, omitted here) asserts that the only operators that can obey this relation are either unitary or antiunitary. By definition, an operator Θ is **antiunitary** if it is antilinear and also obeys

$$\langle\Theta\psi_1|\Theta\psi_2\rangle = \langle\psi_2|\psi_1\rangle = \left(\langle\psi_1|\psi_2\rangle\right)^*. \tag{5.165}$$

It is a short exercise to prove that antilinear operators cannot be unitary. Since we have already established that the time-reversal operator Θ is antilinear, it must be antiunitary as well.

Given an antiunitary operator Θ and a choice of orthobasis $\{|\varphi_n\rangle\}$, which by definition satisfies $\langle\varphi_n|\varphi_{n'}\rangle = \delta_{nn'}$, the set of states $\{|\Theta\varphi_n\rangle\}$ can easily be shown to form another orthobasis. So,

$$U = \sum_n |\Theta\varphi_n\rangle\langle\varphi_n| \tag{5.166}$$

is a unitary operator that maps the original basis to the motion-reversed basis. Let us also define the antiunitary operator K that acts as the identity on the orthobasis states, so that $K|\varphi_n\rangle = |\varphi_n\rangle$ for each n, and therefore, for any complex number c,

$$Kc\,|\varphi_n\rangle = c^*\,|\varphi_n\rangle\,. \tag{5.167}$$

Thus K is the complex conjugation operator for the basis. Then Θ can be expressed as

$$\Theta = UK. \tag{5.168}$$

To prove it, consider an arbitrary state

$$|\psi\rangle = \sum_n c_n\,|\varphi_n\rangle\,. \tag{5.169}$$

Then, directly from the antilinearity of Θ, we have

$$\Theta\,|\psi\rangle = \sum_n c_n^*\Theta\,|\varphi_n\rangle\,. \tag{5.170}$$

To match it, we also compute

$$\begin{aligned}
UK\,|\psi\rangle &= U\sum_n Kc_n\,|\varphi_n\rangle = U\sum_n c_n^* K\,|\varphi_n\rangle = U\sum_n c_n^*\,|\varphi_n\rangle \\
&= \sum_n\sum_{n'} c_n^*\,|\Theta\varphi_{n'}\rangle\,\langle\varphi_{n'}|\varphi_n\rangle = \sum_n\sum_{n'} c_n^*\,|\Theta\varphi_{n'}\rangle\,\delta_{nn'} \\
&= \sum_n c_n^*\Theta\,|\varphi_n\rangle\,,
\end{aligned} \tag{5.171}$$

which indeed equals (5.170). Although Θ is independent of the choice of basis, its decomposition into U and K clearly does depend on that choice.

In our discussion so far, we have been treating Θ as the motion-reversal transformation on states, without any assumption about whether it is a symmetry of the Hamiltonian. Let us now consider quantum systems that have time-reversal invariance as a symmetry. Then we can follow the same logic as in our classical thrown-ball example, summarized just after (5.149). In words: if we evolve the system forward in time by t, then apply Θ, and evolve it forward in time again by t, then time-reversal invariance holds if the result is the same as if we had just applied Θ. Translating these words into an equation,

$$e^{-itH/\hbar}\,\Theta\,e^{-itH/\hbar}\,|\psi\rangle = \Theta\,|\psi\rangle \tag{5.172}$$

for an initial state $|\psi\rangle$. (Here we have assumed that H has no explicit time dependence, as a prerequisite for time-reversal invariance.) Assuming this is true for any initial state, we have an operator equation that can be rewritten as

$$e^{-itH/\hbar}\Theta = \Theta e^{itH/\hbar}. \tag{5.173}$$

Expanding this to first order in small t, we find

$$-iH\Theta = \Theta iH, \tag{5.174}$$

where we have cancelled the real quantity $t/\hbar$.

Although we have already established that Θ is antilinear, it is instructive to see what would go wrong here if it were linear. In that case, (5.174) would simplify to $H\Theta = -\Theta H$. Then, for every energy eigenstate $|E\rangle$ with $H\,|E\rangle = E\,|E\rangle$, we would find that $H\Theta\,|E\rangle = -\Theta H\,|E\rangle = -E\Theta\,|E\rangle$. Thus, $\Theta\,|E\rangle$ would be an eigenstate of H with energy $-E$. This would be a disaster, because it would mean that all states with arbitrarily high energies would have corresponding states with negative energies, and the energies of the system would be unbounded below. So this confirms, in a different way, that Θ must be antilinear.

Using the antilinearity of Θ in the form $\Theta i = -i\Theta$, (5.174) becomes

$$H\Theta = \Theta H, \tag{5.175}$$

so that H and Θ are compatible operators, or equivalently,

$$\Theta H \Theta^{-1} = H. \tag{5.176}$$

This is the condition for a system with Hamiltonian H to have the symmetry of time reversal.

An obvious example of a time-reversal invariant system is that of a particle moving in any time-independent potential,

$$H = \frac{P^2}{2m} + V(\boldsymbol{R}), \tag{5.177}$$

which is invariant due to (5.153).

For a system starting at time $t = 0$ in a general state $|\psi(0)\rangle$, the time evolution of the wavefunction is $\psi(\boldsymbol{r}, t) = \langle \boldsymbol{r}|\psi(t)\rangle = \langle \boldsymbol{r}|e^{-itH/\hbar}|\psi(0)\rangle$. Now consider the motion-reversed initial state $\Theta\,|\psi(0)\rangle$. The time evolution of its wavefunction is

$$\begin{aligned}\psi_{\text{rev}}(\boldsymbol{r}, t) &= \langle \boldsymbol{r}|e^{-itH/\hbar}\Theta|\psi(0)\rangle = \langle \boldsymbol{r}|\Theta e^{itH/\hbar}|\psi(0)\rangle = \langle \boldsymbol{r}|\Theta|\psi(-t)\rangle \\ &= [\psi(\boldsymbol{r}, -t)]^*,\end{aligned} \tag{5.178}$$

where we used both antilinearity and $[H, \Theta] = 0$ in the second equality. Thus, the position wavefunction of the time-reversed state is obtained from the original wavefunction by replacing $t \to -t$ and taking the complex conjugate. Another way to see this is to start from the Schrödinger equation in the position representation,

$$\left(-i\hbar\frac{\partial}{\partial t} - \frac{\hbar^2\nabla^2}{2m} + V(\boldsymbol{r})\right)\psi(\boldsymbol{r}, t) = 0. \tag{5.179}$$

Now take the complex conjugate to get

$$\left(i\hbar\frac{\partial}{\partial t} - \frac{\hbar^2\nabla^2}{2m} + V(\boldsymbol{r})\right)[\psi(\boldsymbol{r}, t)]^* = 0, \tag{5.180}$$

and then replace $t \to -t$ to get

$$\left(-i\hbar\frac{\partial}{\partial t} - \frac{\hbar^2\nabla^2}{2m} + V(\boldsymbol{r})\right)[\psi(\boldsymbol{r}, -t)]^* = 0. \tag{5.181}$$

This shows that $\psi_{\text{rev}}(\boldsymbol{r}, t) = [\psi(\boldsymbol{r}, -t)]^*$ satisfies the Schrödinger equation for a system with time-reversal invariance.

A simple consequence of assuming time-reversal symmetry for a spinless system is that the wavefunction $\psi_E(\boldsymbol{r})$ of a non-degenerate energy eigenstate must be real, up to a global phase factor. To prove it, recall that (5.161) tells us that $|\Theta\psi_E\rangle$ has wavefunction $[\psi_E(\boldsymbol{r})]^*$. But we also have $H\Theta\,|\psi_E\rangle = \Theta H\,|\psi_E\rangle = E\,|\Theta\psi_E\rangle$, implying that $|\Theta\psi_E\rangle$ also has energy E. Given our non-degeneracy assumption for the eigenvalue E, it must be that $|\Theta\psi_E\rangle$ is the same state as $|\psi_E\rangle$, and therefore must have the same wavefunction up to a global phase factor $e^{i\phi}$ for some constant ϕ. Therefore, $[\psi_E(\boldsymbol{r})]^* = e^{i\phi}\psi_E(\boldsymbol{r})$, as claimed.

In the case of a charged particle in the presence of static electromagnetic potentials $\Phi(\boldsymbol{R})$ and $\boldsymbol{A}(\boldsymbol{R})$, we determine the status of time-reversal invariance by examining the Hamiltonian given in (4.48), rewritten for convenience here:

$$H = \frac{1}{2m}\left[P^2 - \frac{q}{c}\left(\boldsymbol{A}\cdot\boldsymbol{P} + \boldsymbol{P}\cdot\boldsymbol{A}\right) + \frac{q^2}{c^2}A^2\right] + q\Phi - \gamma\boldsymbol{S}\cdot\boldsymbol{B}. \tag{5.182}$$

Treated as fixed functions of $\boldsymbol{R}$ only, both Φ and $\boldsymbol{A}$ are even under time reversal. It follows that (5.176) holds, and time reversal is a symmetry, if there is a static external electric field, but it is violated if there is a static external magnetic field. The culprits are the terms proportional to $\boldsymbol{A}\cdot\boldsymbol{P} + \boldsymbol{P}\cdot\boldsymbol{A}$ and $\boldsymbol{S}\cdot\boldsymbol{B}$. This corresponds to the classical fact that the Lorentz force on a charged particle in a magnetic field, $\boldsymbol{F} = q\boldsymbol{v}\times\boldsymbol{B}/c$, is linear in the velocity, and so changes sign under reversal of the motion. Here it is important that we are treating the static vector potential as an external source, not as part of the quantum system under consideration. If, instead, we enlarge our system to include the charged particles that are responsible for producing $\boldsymbol{A}$ and $\boldsymbol{B} = \nabla\times\boldsymbol{A}$, then motion reversal for them would amount to reversing their currents, and we would find $\boldsymbol{A}\to-\boldsymbol{A}$ and $\boldsymbol{B}\to-\boldsymbol{B}$, and the whole system would be invariant under time reversal.

The Hamiltonian for the Standard Model of particle physics has interesting properties with respect to time reversal. The electromagnetic forces are invariant, but the weak nuclear interactions are not, although the effects of this are highly suppressed in most processes. For the strong nuclear interactions, things are more mysterious. Experimentally, the strong nuclear Hamiltonian appears to be time-reversal invariant to a very good approximation, despite the fact that there is a plausible term that could be included in it that would violate time-reversal symmetry. The most significant consequence would be an electric dipole moment for the neutron. The effective Hamiltonian governing the electric dipole moment d_n of the neutron in a static external electric field $\boldsymbol{E}$ is

$$H = -\frac{2}{\hbar}d_n\boldsymbol{S}_n\cdot\boldsymbol{E}, \tag{5.183}$$

where $\boldsymbol{S}_n$ is the spin operator for the neutron. This clearly violates time-reversal symmetry, because the external electric field is even, but the spin is odd. The experimental measurement of $|d_n|$ is consistent with 0 within uncertainties, with an upper bound about 10 orders of magnitude smaller than its naive estimate from the strong interactions assuming generic parameters. Understanding theoretically why a large term of this type is not generated by the strong nuclear interactions is an outstanding problem, although there are many conjectured explanations. One of the most promising explanations introduces a new (approximate) symmetry, which in turn predicts the existence of a very light, very weakly coupled particle, the axion. At this writing, experimental searches for the axion are entering interesting levels of sensitivity.

Exercises

5.1 Use the canonical commutation relations (2.248) to show that the components of the position, momentum, and orbital angular momentum operators for a particle satisfy

$$[L_a, R_b] = i\hbar\epsilon_{abc}R_c, \tag{5.184}$$

$$[L_a, P_b] = i\hbar\epsilon_{abc}P_c, \tag{5.185}$$

$$[L_a, L_b] = i\hbar\epsilon_{abc}L_c. \tag{5.186}$$

Use these results to obtain the transformed position and momentum operators $\boldsymbol{R}' = U\boldsymbol{R}U^\dagger$ and $\boldsymbol{P}' = U\boldsymbol{P}U^\dagger$ and $\boldsymbol{L}' = U\boldsymbol{L}U^\dagger$ resulting from a rotation $U = e^{-i\boldsymbol{\epsilon}\cdot\boldsymbol{L}/\hbar}$, to first order in $\boldsymbol{\epsilon}$. Your answers should be the ones given in (5.71).

5.2 An operator Ω is said to transform under rotations as a scalar if

$$U(\boldsymbol{\alpha})\Omega U(\boldsymbol{\alpha})^{-1} = \Omega \tag{5.187}$$

for every rotation angle $\boldsymbol{\alpha}$.

(a) Consider an operator $\Omega = \boldsymbol{V}\cdot\boldsymbol{W}$, which consists of the dot product of any two operators $\boldsymbol{V}$ and $\boldsymbol{W}$ that transform as vectors in the sense of (5.71). Show that Ω obeys (5.187) for infinitesimal $\boldsymbol{\alpha}$. Since a non-infinitesimal rotation can be built out of infinitesimal rotations, it follows that Ω obeys (5.187) for any $\boldsymbol{\alpha}$.
(b) Consider the operator $\Omega = \boldsymbol{B}\cdot\boldsymbol{L}$, where $\boldsymbol{L}$ is the orbital angular momentum operator and $\boldsymbol{B} = B\hat{z}$ is a constant uniform magnetic field. Does Ω transform as a scalar operator under rotations?

5.3 Consider the unitary rotation operator $U(\boldsymbol{\theta}) = e^{-i\boldsymbol{\theta}\cdot\boldsymbol{L}/\hbar}$ for a general non-infinitesimal vector $\boldsymbol{\theta}$. Since this can be built out of infinitesimal rotations, (5.70) implies that

$$U(\boldsymbol{\theta})\,|\boldsymbol{r}\rangle = |\boldsymbol{r}'\rangle\,, \tag{5.188}$$

for some position vector $\boldsymbol{r}'$ that depends linearly on $\boldsymbol{r}$. We can therefore use this to *define* a 3×3 matrix (not an operator) $\mathcal{R}(\boldsymbol{\theta})$, according to

$$r'_a \equiv \mathcal{R}(\boldsymbol{\theta})_{ab}r_b. \tag{5.189}$$

(a) Prove, using the last two equations and without resorting to the infinitesimal rotation case, that

$$U(\boldsymbol{\theta})R_a U(\boldsymbol{\theta})^\dagger = [\mathcal{R}(\boldsymbol{\theta})^{-1}]_{ab} R_b, \tag{5.190}$$

where $[\mathcal{R}(\boldsymbol{\theta})^{-1}]_{ab} = \mathcal{R}(-\boldsymbol{\theta})_{ab}$. It then follows from comparison with the infinitesimal case in (5.71) that, more generally for any operator $\boldsymbol{V}$ transforming as a vector under rotations,

$$U(\boldsymbol{\theta})V_a U(\boldsymbol{\theta})^\dagger = [\mathcal{R}(\boldsymbol{\theta})^{-1}]_{ab} V_b. \tag{5.191}$$

(This includes $\boldsymbol{V} = \boldsymbol{R}$, $\boldsymbol{P}$, and $\boldsymbol{L}$ as notable special cases.)

(b) From the fact that $\boldsymbol{R} \cdot \boldsymbol{R} = R_a R_a$ transforms as a scalar in the sense of Exercise 5.2, and therefore is equal to $U R_a R_a U^\dagger = (U R_a U^\dagger)(U R_a U^\dagger)$, show that $\mathcal{R}(\boldsymbol{\theta})$ must be an orthogonal matrix, obeying

$$[\mathcal{R}(\boldsymbol{\theta})^T]_{ab} \mathcal{R}(\boldsymbol{\theta})_{bc} = \mathcal{R}(\boldsymbol{\theta})_{ba} \mathcal{R}(\boldsymbol{\theta})_{bc} = \delta_{ac}, \tag{5.192}$$

or equivalently

$$[\mathcal{R}(\boldsymbol{\theta})^{-1}]_{ab} = \mathcal{R}(\boldsymbol{\theta})_{ba}. \tag{5.193}$$

(c) Consider the unitary translation operator $T(\boldsymbol{a}) = e^{-i\boldsymbol{a}\cdot\boldsymbol{P}/\hbar}$. Show that

$$U(\boldsymbol{\theta})T(\boldsymbol{a})U(\boldsymbol{\theta})^\dagger = T(\boldsymbol{a}'), \tag{5.194}$$

where

$$a'_b = \mathcal{R}(\boldsymbol{\theta})_{bc} a_c. \tag{5.195}$$

(d) What is the 3×3 matrix $\mathcal{R}(\boldsymbol{\theta})$ explicitly, in the special case of a rotation about the z axis, $\boldsymbol{\theta} = \theta\hat{z}$?

5.4 For a system of two particles labeled 1 and 2 as discussed in Sect. 4.2, show that the total orbital angular momentum can be written in the way claimed in (5.83) in terms of $\boldsymbol{L}$ and $\boldsymbol{L}_{\text{cm}}$ defined in (5.84). Then use the results of Exercise 4.2 to show that $\boldsymbol{L}$, $\boldsymbol{L}_{\text{cm}}$, and $\boldsymbol{L}_{\text{tot}}$ each obey the defining requirement of an angular momentum operator in (5.81).

5.5 Two angular momentum operators $\boldsymbol{L}$ and $\boldsymbol{S}$ each satisfy the commutator algebra of the form in (5.81), and further obey $[L_a, S_b] = 0$. Find the commutator of the scalar operator $\boldsymbol{L} \cdot \boldsymbol{S}$ with each of the vector operators $\boldsymbol{L}$, $\boldsymbol{S}$, and $\boldsymbol{J} = \boldsymbol{L} + \boldsymbol{S}$, and with each of the scalar operators L^2, S^2, and J^2.

5.6 Derive (5.138), the probability current density $\boldsymbol{J}$ in the presence of electromagnetic potentials Φ and $\boldsymbol{A}$. Show that it is invariant under gauge transformations.

5.7 Consider a linear operator A, and an antiunitary operator Θ such as the time-reversal operator. Show that

$$\langle\psi_1|A|\psi_2\rangle = \langle\Theta\psi_2|\Theta A^{\dagger}\Theta^{-1}|\Theta\psi_1\rangle\,. \tag{5.196}$$

(Recall that we defined antilinear and antiunitary operators like Θ and Θ^{-1} only by their actions to the right, on kets multiplied by complex numbers, so you should not use $\langle\Theta\psi_2| = \langle\psi_2|\,\Theta^{-1}$.)

Particle Moving in One Dimension

6

6.1 Gaussian Wavefunctions

Consider a particle moving in one dimension in the domain $-\infty < x < \infty$. We will not need to assume anything in particular about the Hamiltonian of the system in this section. Suppose that at some moment in time, the position wavefunction has a Gaussian form,

$$\langle x|\psi\rangle = \psi(x) = N \exp\left[-(x-a)^2/4\sigma^2\right]. \tag{6.1}$$

Here a and σ are constants with units of length, corresponding respectively to the center and the width of the probability density $|\psi(x)|^2$. More precisely, the full width at half maximum (FWHM) of the Gaussian peak of the probability density is $2\sqrt{2\ln 2}\sigma \approx 2.3548\sigma$. The factor N is a normalization. If we require the ket $|\psi\rangle$ to have unit norm, we need

$$1 = \int_{-\infty}^{\infty} dx\, |N|^2 \exp\left[-(x-a)^2/2\sigma^2\right] = |N|^2\sqrt{2\pi}\sigma. \tag{6.2}$$

Here we have assumed that $|N|$ is a constant, so we can choose $N = e^{i\theta}/(2\pi\sigma^2)^{1/4}$, where θ is any phase. If θ is a constant, then it is a global phase and has no physical significance at all. So, let us choose the next simplest possibility, that θ depends linearly on x, and denote the phase factor of the wavefunction by $e^{i\theta} = e^{ikx}$. The Gaussian wavefunction that we choose to study is thus

$$\psi(x) = \frac{1}{(2\pi\sigma^2)^{1/4}}\, e^{ikx} \exp\left[-(x-a)^2/4\sigma^2\right] \tag{6.3}$$

for some constant k, whose interpretation will become clear soon.

S. P. Martin, *Quantum Mechanics*, Graduate Texts in Physics,
https://doi.org/10.1007/978-3-032-21119-4_6

According to Postulate 4, the probability of finding the particle between x and $x + dx$ is

$$d\mathcal{P}(x) = |\langle x|\psi\rangle|^2 dx = \frac{1}{\sqrt{2\pi}\sigma} \exp\left[-(x-a)^2/2\sigma^2\right] dx. \tag{6.4}$$

Therefore, the probability of finding the particle in a range $b < x < c$ is

$$\mathcal{P}(b < x < c) = \frac{1}{\sqrt{2\pi}\sigma} \int_b^c dx \, \exp\left[-(x-a)^2/2\sigma^2\right]. \tag{6.5}$$

We can also find the expectation value of X in the state $|\psi\rangle$,

$$\begin{aligned}
\langle X\rangle &= \langle\psi|X|\psi\rangle = \int_{-\infty}^{\infty} dx \, \langle\psi|X|x\rangle\langle x|\psi\rangle = \int_{-\infty}^{\infty} dx \, x|\psi(x)|^2 \\
&= \frac{1}{\sqrt{2\pi}\sigma} \int_{-\infty}^{\infty} dx \, x \exp\left[-(x-a)^2/2\sigma^2\right] \\
&= \frac{1}{\sqrt{2\pi}\sigma} \int_{-\infty}^{\infty} du \, (u+a) \exp\left[-u^2/2\sigma^2\right] \\
&= a.
\end{aligned} \tag{6.6}$$

The second equality uses the completeness relation, the third uses $X|x\rangle = x|x\rangle$, and the fourth uses the common trick of "completing the square", which means that we define a new integration variable u so that the exponential in the integrand depends only on u^2. In this case, the change of variable was $x = u + a$. Similarly,

$$\langle X^2\rangle = \frac{1}{\sqrt{2\pi}\sigma} \int_{-\infty}^{\infty} du \, (u+a)^2 \exp\left[-u^2/2\sigma^2\right] = a^2 + \sigma^2. \tag{6.7}$$

Therefore, the uncertainty of X for this state is

$$\Delta X = \sqrt{\langle X^2\rangle - \langle X\rangle^2} = \sigma. \tag{6.8}$$

So far, the constant k has not made any difference at all.

The momentum wavefunction for the same state is

$$\begin{aligned}
\widetilde{\psi}(p) &= \int_{-\infty}^{\infty} dx \, \langle p|x\rangle\langle x|\psi\rangle \\
&= \frac{1}{\sqrt{2\pi\hbar}(2\pi\sigma^2)^{1/4}} \int_{-\infty}^{\infty} dx \, e^{-ipx/\hbar} e^{ikx} e^{-(x-a)^2/4\sigma^2},
\end{aligned} \tag{6.9}$$

where we have used $\langle p|x\rangle$ from the complex conjugate of (2.233). We now use the completing-the-square trick again; this time the appropriate change of variables is

$x = u + b$, where we choose $b = a + i2\sigma^2(k - p/\hbar)$, so

$$\int_{-\infty}^{\infty} dx\, e^{-ipx/\hbar} e^{ikx} e^{-(x-a)^2/4\sigma^2} = e^{(b^2-a^2)/4\sigma^2} \int_{-\infty}^{\infty} du\, e^{-u^2/4\sigma^2}$$
$$= 2\sqrt{\pi}\sigma\, e^{(b^2-a^2)/4\sigma^2}. \tag{6.10}$$

It follows that

$$\widetilde{\psi}(p) = \left(\frac{2\sigma^2}{\pi\hbar^2}\right)^{1/4} e^{ia(k-p/\hbar)} e^{-\sigma^2(p/\hbar-k)^2}. \tag{6.11}$$

The last factor shows that a state with a Gaussian position wavefunction also has a Gaussian momentum wavefunction, centered at $p = \hbar k$. The momentum wavefunction also contains a phase factor that depends on p; this factor encodes the information about a, the center of the position wavefunction Gaussian peak.

It is now clear that our state depends on three physically significant parameters: the average momentum $\hbar k$, the position center a, and the position width σ, which is also the uncertainty in X. There is a duality between the position and momentum wavefunctions, for if one of them has a Gaussian magnitude with a linear phase, then so does the other, with parameters that are related by comparing (6.3) and (6.11). The widths of the position and momentum Gaussian wavefunctions are inversely proportional.

Using (6.11), one now obtains

$$\langle P\rangle = \langle\psi|P|\psi\rangle = \int_{-\infty}^{\infty} dp\ \langle\psi|P|p\rangle\langle p|\psi\rangle$$
$$= \int_{-\infty}^{\infty} dp\ p|\widetilde{\psi}(p)|^2 = \hbar k, \tag{6.12}$$

and similarly,

$$\langle P^2\rangle = \langle\psi|P^2|\psi\rangle = \int_{-\infty}^{\infty} dp\ p^2|\widetilde{\psi}(p)|^2 = \hbar^2(k^2 + 1/4\sigma^2). \tag{6.13}$$

It follows that the momentum uncertainty in the state with a Gaussian position wavefunction is

$$\Delta P = \sqrt{\langle P^2\rangle - \langle P\rangle^2} = \hbar/2\sigma, \tag{6.14}$$

and so the product of the position and momentum uncertainties is

$$(\Delta X)(\Delta P) = \hbar/2. \tag{6.15}$$

This result does not depend on the center of the Gaussian a, or its width σ, or its average momentum $\hbar k$. Thus, Gaussian wavefunctions always realize the minimum possible product of uncertainties consistent with the general position-momentum uncertainty relation, (3.23).

6.2 Time Evolution of Free-Particle State in One Dimension

In this section, we will study the kinematics and time evolution for an arbitrary state of a free particle with mass m moving in one dimension. The Schrödinger equation is

$$i\hbar\frac{d}{dt}|\psi\rangle = H|\psi\rangle = \frac{P^2}{2m}|\psi\rangle, \tag{6.16}$$

where P is the momentum operator. Since $[H, P] = 0$, we know that there is an orthobasis of stationary states that are also eigenstates of P with eigenvalue p. Acting on such states, $E = P^2/2m = p^2/2m$. Therefore, for a given E there are exactly two solutions, $p = p_R = \sqrt{2mE}$ and $p = p_L = -\sqrt{2mE}$, and the plane-wave stationary states can be labeled

$$|E, R\rangle = |p_R\rangle, \qquad |E, L\rangle = |p_L\rangle, \tag{6.17}$$

where R, L in the kets on the left are degeneracy labels that tell us whether the particle is moving right or left.

It is convenient to define the wavenumber $k = p/\hbar$ and the angular frequency $\omega = E/\hbar$. The time-dependent wavefunction for a stationary state with momentum p, combining (2.233) with the time-evolution phase factor $e^{-iEt/\hbar}$, is

$$\psi_p(x, t) = \frac{1}{\sqrt{2\pi\hbar}}e^{i(kx-\omega t)}. \tag{6.18}$$

The position of constant phase is $x = \omega t/k = Et/p = pt/2m$, so the **phase velocity** of one of these waves is

$$v_{\text{phase}} = \omega/k = p/2m. \tag{6.19}$$

This is half of the classical velocity p/m.

To understand the classical speed of propagation, one must consider the **group velocity** for wavepacket superpositions of states with a continuous distribution of k. It is a general feature of wave kinematics that, in the presence of dispersion (that is, ω depending nonlinearly on k), the velocity for a wavepacket is not ω/k, but instead

$$v_{\text{group}} = \partial\omega/\partial k. \tag{6.20}$$

Let us pause from our discussion of the quantum free particle to see how this arises in a more general context. Consider a packet of waves described by the wavefunction

$$\psi(x,t) = \int dk\, a(k)\, e^{i(kx-\omega(k)t)}, \tag{6.21}$$

where, in general, the angular frequency $\omega(k)$ and the amplitude function $a(k)$ both depend on the wavenumber in some arbitrary way. Now suppose that $a(k)$ is such that only a narrow range of k near k_0 contributes appreciably. In that range, we write $k = k_0 + \Delta k$ and

$$\omega(k) = \omega_0 + \Delta k \frac{\partial \omega}{\partial k}\bigg|_{k=k_0} + O(\Delta k)^2, \tag{6.22}$$

where $\omega_0 = \omega(k_0)$. Then (6.21) can be rewritten in the form

$$\psi(x,t) = e^{i(k_0 x - \omega_0 t)} A(x,t), \tag{6.23}$$

where the factor describing the shape of the magnitude of the wavepacket is

$$A(x,t) = \int d(\Delta k)\, a(k_0+\Delta k) \exp\left[i\left(x - \frac{\partial \omega}{\partial k}t\right)\Delta k + \cdots\right]. \tag{6.24}$$

Here the partial derivative is understood to be evaluated at $k = k_0$. In this approximation, the magnitude $|\psi(x,t)| = |A(x,t)|$ depends on position and time only through the combination

$$x - \frac{\partial \omega}{\partial k}t, \tag{6.25}$$

which immediately implies that the velocity of the wavepacket is indeed $v_{\text{group}} = \partial\omega/\partial k$.

Applying this to the special case of a free particle, with $\omega = E/\hbar$, we have

$$v_{\text{group}} = \frac{\partial}{\partial k}\left(\frac{\hbar k^2}{2m}\right) = \hbar k/m = p/m, \tag{6.26}$$

which is the classical value. For wave packets in quantum mechanics with momentum sharply peaked near p, this group velocity corresponds to the motion of the expectation value of the position, as can be seen in general by comparing to $d\langle X\rangle/dt = \langle P\rangle/m$ as found in (3.41). Before the end of this section, we will verify this for the particular case of a Gaussian superposition of plane waves.

Using (3.30), the unitary time evolution operator is

$$U(t) = \int_{-\infty}^{\infty} dp\, e^{-itp^2/2m\hbar}\, |p\rangle\langle p|. \tag{6.27}$$

The matrix element of this operator between different position eigenstates is therefore

$$\begin{aligned}\langle x|U(t)|x'\rangle &= \int_{-\infty}^{\infty} dp\, e^{-itp^2/2m\hbar}\, \langle x|p\rangle\langle p|x'\rangle \\ &= \frac{1}{2\pi\hbar}\int_{-\infty}^{\infty} dp\, e^{-itp^2/2m\hbar} e^{ip(x-x')/\hbar}. \end{aligned} \tag{6.28}$$

This integral is again done by the completing-the-square trick, this time by changing the integration variable to $p' = p - m(x-x')/t$, with the result

$$\langle x|U(t)|x'\rangle = \left(\frac{m}{2\pi\hbar i t}\right)^{1/2} e^{im(x-x')^2/2\hbar t}. \tag{6.29}$$

Applying the completeness relation, we can now evaluate the time dependence of the wavefunction for a free particle, which reduces to an integral involving the wavefunction at time $t = 0$:

$$\begin{aligned}\psi(x,t) = \langle x|\psi(t)\rangle &= \langle x|U(t)|\psi(0)\rangle = \int_{-\infty}^{\infty} dx'\, \langle x|U(t)|x'\rangle\langle x'|\psi(0)\rangle \\ &= \left(\frac{m}{2\pi\hbar i t}\right)^{1/2}\int_{-\infty}^{\infty} dx'\, e^{im(x-x')^2/2\hbar t}\,\psi(x',0). \end{aligned} \tag{6.30}$$

In principle, the time evolution of an arbitrary free-particle state has therefore been solved.

As a good example for which the integral can be done analytically, consider the motion and spreading with time of the wavefunction of a Gaussian wavepacket state. We start at time $t = 0$ with a state $|\psi(0)\rangle$ that has wavefunction

$$\psi(x',0) = \frac{1}{(2\pi\sigma^2)^{1/4}} e^{ip_0x'/\hbar} e^{-x'^2/4\sigma^2}. \tag{6.31}$$

As we saw in the previous section, this wavefunction can be interpreted as having center at $x' = 0$, width σ, and average momentum p_0. At time t, (6.30) gives

$$\psi(x,t) = \left(\frac{m}{2\pi\hbar i t}\right)^{1/2}\frac{1}{(2\pi\sigma^2)^{1/4}}\int_{-\infty}^{\infty} dx'\, \exp\left[-\frac{x'^2}{4\sigma^2} + i\frac{m(x-x')^2}{2\hbar t} + i\frac{p_0x'}{\hbar}\right]. \tag{6.32}$$

By the usual completing-the-square integration variable change trick, this becomes, after some algebraic manipulation,

$$\psi(x,t) = \frac{1}{\left[\sqrt{2\pi}(\sigma + i\hbar t/2m\sigma)\right]^{1/2}}\exp\left[-\frac{(x-p_0t/m)^2}{4\sigma^2 + 2i\hbar t/m}\right]\exp\left[i\frac{p_0}{\hbar}\left(x - \frac{p_0t}{2m}\right)\right], \tag{6.33}$$

or, after further rearrangement to isolate the phase in the exponent,

$$\psi(x,t) = \frac{e^{i\theta(x,t)}}{\left[\sqrt{2\pi}(\sigma + i\hbar t/2m\sigma)\right]^{1/2}} \exp\left[-\frac{(x - p_0 t/m)^2}{4\sigma^2 + \hbar^2 t^2/m^2\sigma^2}\right], \tag{6.34}$$

where

$$\theta(x,t) = \frac{p_0}{\hbar}\left(x - \frac{p_0 t}{2m}\right) + \frac{\hbar t (x - p_0 t/m)^2}{2m(4\sigma^4 + \hbar^2 t^2/m^2)}. \tag{6.35}$$

The probability density at time t is therefore

$$|\psi(x,t)|^2 = \frac{1}{\sqrt{2\pi(\sigma^2 + \hbar^2 t^2/4m^2\sigma^2)}} \exp\left[-\frac{(x - p_0 t/m)^2}{2(\sigma^2 + \hbar^2 t^2/4m^2\sigma^2)}\right]. \tag{6.36}$$

This is a Gaussian function of x, whose peak moves to the right at exactly the constant speed p_0/m of naive classical expectation, but whose width grows with time t. Using the results of the previous section, we obtain

$$\langle X \rangle = p_0 t/m, \tag{6.37}$$

$$\Delta X = \sqrt{\sigma^2 + \hbar^2 t^2/4m^2\sigma^2}. \tag{6.38}$$

The speed at which the expectation value $\langle X \rangle$ moves is also the group velocity $v_{\text{group}} = \partial\omega/\partial k$. At large t, $\Delta X \approx \hbar t/2m\sigma$ grows linearly with time, and ironically is larger for smaller σ; the more we try to confine the wavefunction initially, the more spread out the particle's probability density will be at late times. The explanation is that, due to the uncertainty relation (3.23), a highly constrained particle position has larger amplitudes for momenta that deviate from the central value.

One can also compute the momentum wavefunction as a function of time, by applying the Fourier transform (2.235) to (6.33). After another integration and more algebraic juggling, one finds

$$\widetilde{\psi}(p,t) = \left(\frac{2\sigma^2}{\pi\hbar^2}\right)^{1/4} \exp\left[-\sigma^2(p - p_0)^2/\hbar^2\right] e^{-ip^2 t/2\hbar m}. \tag{6.39}$$

It is now an easy check that the free-particle Schrödinger equation holds: $i\hbar\partial\widetilde{\psi}/\partial t = H\widetilde{\psi} = (p^2/2m)\widetilde{\psi}$. The momentum-space wavefunction is remarkably simple; the time dependence is entirely in the complex phase factor, so

$$|\widetilde{\psi}(p,t)|^2 = \sqrt{\frac{2}{\pi}\frac{\sigma}{\hbar}} \exp\left[-2\sigma^2(p - p_0)^2/\hbar^2\right]. \tag{6.40}$$

This does not depend on time at all, despite the fact that the characteristic width of the position wavefunction grows with time. This is an illustration of Noether's Principle (Theorem 5.1); because the momentum is a conserved quantity for a free particle Hamiltonian, the probability density to measure the momentum between p and $p + dp$ is a constant in time.

6.3 Properties of Stationary States in One-Dimensional Potentials

Now consider a particle moving in one dimension in a potential that is given classically by $V(x)$, so that the Hamiltonian operator is

$$H = \frac{P^2}{2m} + V(X). \tag{6.41}$$

In general, we want to find the stationary states,

$$H|\psi_E\rangle = E|\psi_E\rangle. \tag{6.42}$$

To accomplish this, we multiply on the left by $\langle x|$ to work in the position representation, in which $X \to x$ and $P \to -i\hbar d/dx$. Then the wavefunction $\psi_E(x) = \langle x|\psi_E\rangle$ obeys the eigenvalue differential equation

$$\frac{d^2}{dx^2}\psi_E(x) = \frac{2m}{\hbar^2}\left[V(x) - E\right]\psi_E(x). \tag{6.43}$$

This is the time-independent Schrödinger equation for a particle in one dimension.

Before exploring some special cases, it is helpful to make some general statements about this problem. First, consider a range of positions x over which the potential $V(x)$ is less than the energy eigenvalue E. This is called a classically allowed range, because in classical mechanics the kinetic energy contribution is always positive, so that $E > V(x)$. In a classically allowed range, the wavefunction is oscillatory provided that $V(x)$ varies sufficiently smoothly with x. To illustrate this, consider the case of a nearly constant V, and define the wavenumber $k = \sqrt{2m(E - V)}/\hbar$. Then, within that range, the Schrödinger equation becomes (dropping the subscript E on the wavefunction from here on) $\psi'' \approx -k^2\psi$, which has the general solution

$$\psi \approx ae^{ikx}+be^{-ikx}, \quad \text{(for nearly constant } k^2{=}2m(E{-}V)/\hbar^2 > 0), \tag{6.44}$$

where a and b are constants.

Conversely, classically forbidden ranges of x are those in which $E < V(x)$. If V is nearly constant, we can define a real quantity $\kappa = \sqrt{2m(V - E)}/\hbar$, and the Schrödinger equation becomes $\psi'' \approx \kappa^2\psi$. The corresponding general solution is a sum of real exponentials,

$$\psi \approx ae^{-\kappa x}+be^{\kappa x}, \quad \text{(for nearly constant } \kappa^2=2m(V-E)/\hbar^2 > 0). \quad (6.45)$$

If the classically forbidden domain being considered includes $x = \infty$, then one must have $b = 0$ in order to have a sensible wavefunction without exponential growth at large distances. For the same reason, if the considered domain includes $x = -\infty$, then one must have $a = 0$.

It is often the case that the potential approaches a constant at large distances. Unbound states are those that have $E > V(x)$, and therefore oscillatory behavior, at either $x = \infty$, or $x = -\infty$, or both. Bound states are those that have $E < V(x)$ at both $x = \infty$ and $-\infty$, implying exponentially falling wavefunctions at large distances.

A stationary state with energy E may have both classically allowed and classically forbidden ranges of x, depending on the potential. The points x with $V(x) = E$ that separate them are called classical turning points, because the corresponding classical trajectory for a particle must turn back at those points.

It is also possible that $V(x)$ has some special points where it may not be smooth, or may even diverge. We would like to know what can be said about the behavior of the wavefunction $\psi(x)$ at such special points. Consider a particular special point x_0. Integrating (6.43) with respect to x over a small neighborhood of that point gives

$$\begin{aligned} \frac{2m}{\hbar^2}\int_{x_0-\epsilon}^{x_0+\epsilon} dx\,\big[V(x)-E\big]\psi(x) &= \int_{x_0-\epsilon}^{x_0+\epsilon} dx\,\frac{d}{dx}\left(\frac{d\psi}{dx}\right) \\ &= \psi'(x_0+\epsilon)-\psi'(x_0-\epsilon), \end{aligned} \qquad (6.46)$$

where the last equality follows from the fundamental theorem of calculus.

Now suppose that $V(x)$ is bounded near $x = x_0$. Then the left side of (6.46) tends to 0 as $\epsilon \to 0$, so we can conclude that the first derivative of $\psi(x)$ must be continuous at x_0. Note that this holds even if $V(x)$ is discontinuous, as long as it stays finite in the neighborhood of $x = x_0$. It follows that $\psi(x)$ itself is also continuous at $x = x_0$.

Another possibility is that $V(x) = C\delta(x - x_0) + \cdots$, where C is a constant, and the ellipsis represents a contribution that is possibly discontinuous but bounded near $x = x_0$. In that case, (6.46) reveals that

$$\lim_{\epsilon\to 0}\big[\psi'(x_0+\epsilon)-\psi'(x_0-\epsilon)\big] = \frac{2m}{\hbar^2}C\psi(x_0), \qquad (6.47)$$

so that the derivative of the wavefunction at x_0 is discontinuous by a known amount proportional to the coefficient of the delta function in the potential. Equation (6.47) is consistent with $\psi'(x)$ being bounded in a neighborhood of $x = x_0$, even if it is not continuous. Therefore, $\int_{x_0-\epsilon}^{x_0+\epsilon} dx\,\psi'(x)$ must approach 0 as $\epsilon \to 0$, implying

$$\lim_{\epsilon\to 0}\big[\psi(x_0+\epsilon)-\psi(x_0-\epsilon)\big] = 0. \qquad (6.48)$$

More generally, **continuity of the wavefunction** is a requirement that we always impose on physically sensible states. This is consistent with the fact that $|\psi(x)|^2$ represents the probability density, so that its value at $x = x_0$ has a unique physical meaning and must not depend on whether x approaches x_0 from above or below.

Yet another possibility is that $V(x) = \infty$ for a whole range $x < x_0$ but it is finite for $x > x_0$. In that case, the particle is forbidden to enter the region of infinite potential, and so $\psi(x) = 0$ for $x \leq x_0$. Now, continuity of the wavefunction tells us that $\psi(x_0) = 0$, which acts as a boundary condition for $\psi(x)$. Of course, the same holds if the inequalities are reversed: if $V(x) = \infty$ for $x > x_0$ but it is finite for $x < x_0$, then $\psi(x) = 0$ for $x \geq x_0$.

Bound states in one-dimensional potentials in quantum mechanics never have degenerate energy eigenvalues. To prove it, suppose there are two states $|\psi_1\rangle$ and $|\psi_2\rangle$ with the same energy eigenvalue E. In the position representation, the wavefunctions satisfy

$$-\frac{\hbar^2}{2m}\psi_1'' + V\psi_1 = E\psi_1, \qquad -\frac{\hbar^2}{2m}\psi_2'' + V\psi_2 = E\psi_2. \tag{6.49}$$

Multiplying the first equation by ψ_2 and the second by ψ_1 and taking the difference gives $\psi_2\psi_1'' - \psi_1\psi_2'' = 0$, or

$$\frac{d}{dx}(\psi_2\psi_1' - \psi_1\psi_2') = 0, \tag{6.50}$$

so that integrating with respect to x gives

$$\psi_2\psi_1' - \psi_1\psi_2' = c, \tag{6.51}$$

where c is a constant of integration. Now, by definition the bound states must approach $\psi_1 = \psi_2 = 0$ when $x \to \pm\infty$. So, evaluating (6.51) at $x \to \infty$ in particular, we learn that c must be 0. Therefore $\psi_1'/\psi_1 = \psi_2'/\psi_2$, or

$$\frac{d}{dx}(\ln\psi_1) = \frac{d}{dx}(\ln\psi_2), \tag{6.52}$$

which integrates to

$$\ln(\psi_1) = \ln(\psi_2) + C, \tag{6.53}$$

where C is another constant of integration. So, $\psi_1 = e^C\psi_2$, and since e^C is a nonzero constant, ψ_1 and ψ_2 are proportional and therefore represent the same state. Note that in this proof it is crucial that we are working in one dimension. In two or three dimensions, bound states certainly can have degenerate energies, as we will see by finding examples in Chap. 10.

6.4 Particle in a One-Dimensional Box

Consider a particle of mass m confined to a box of length L, so that its potential energy as a function of x is

$$V(x) = \begin{cases} 0 & \text{for } |x| < L/2, \\ \infty & \text{for } |x| \geq L/2. \end{cases} \tag{6.54}$$

Since this is an even function of x, we know from the discussion in Sect. 5.4 that there must be simultaneous eigenstates of energy and parity. Let us find them.

In the region $|x| < L/2$, (6.43) becomes

$$\psi'' = -\frac{2mE}{\hbar^2}\psi. \tag{6.55}$$

The general solution to this differential equation is

$$\psi(x) = Ae^{ikx} + Be^{-ikx}, \tag{6.56}$$

where A and B are constants and

$$k = \sqrt{2mE}/\hbar. \tag{6.57}$$

Because the particle is completely confined to the box region, one must have $\psi(x) = 0$ for $|x| \geq L/2$, and so by continuity of the wavefunction, $\psi(L/2) = 0$ and $\psi(-L/2) = 0$, or

$$Ae^{ikL/2} + Be^{-ikL/2} = 0, \tag{6.58}$$

$$Ae^{-ikL/2} + Be^{ikL/2} = 0. \tag{6.59}$$

This system of equations will have a nontrivial solution for A and B if and only if the matrix

$$M = \begin{pmatrix} e^{ikL/2} & e^{-ikL/2} \\ e^{-ikL/2} & e^{ikL/2} \end{pmatrix} \tag{6.60}$$

is not invertible, which means $\det(M) = e^{ikL} - e^{-ikL} = 2i\sin(kL) = 0$. The solutions to the particle-in-a-box eigenvalue problem therefore must have $k = n\pi/L$, where n is an integer. It follows that $B = -e^{ikL}A = (-1)^{n+1}A$.

For odd n, we have $B = A$, and so for $-L/2 \leq x \leq L/2$,

$$\psi_n(x) = 2A\cos(n\pi x/L). \tag{6.61}$$

To normalize the states to unity, we require

$$1 = \int_{-\infty}^{\infty} dx\, |\psi_n(x)|^2 = 4|A|^2 \int_{-L/2}^{L/2} dx\, \cos^2(n\pi x/L) = 2|A|^2 L, \quad (6.62)$$

so we can choose $A = 1/\sqrt{2L}$, yielding the even-parity solutions

$$\psi_n(x) = \sqrt{\frac{2}{L}} \cos(n\pi x/L) \qquad (n = 1, 3, 5, \ldots). \quad (6.63)$$

Negative values of n need not be considered, because those just have the same wavefunction as for the corresponding positive n, and are therefore the same states. Similarly, for even n, one finds the odd-parity solutions

$$\psi_n(x) = \sqrt{\frac{2}{L}} \sin(n\pi x/L) \qquad (n = 2, 4, 6, \ldots). \quad (6.64)$$

Here, not only do we not need negative n, but also the case $n = 0$ is excluded, because it would result in $\psi_0(x) = 0$, which would correspond to the null ket, which is not a physical state. For both odd and even n, (6.57) gives the allowed energy levels:

$$E_n = \frac{\hbar^2\pi^2 n^2}{2mL^2} \qquad (n = 1, 2, 3, \ldots). \quad (6.65)$$

These are non-degenerate and discrete, as is true for any bound-state solutions in one dimension.

The ground-state energy for the particle in a box, $E_1 = \hbar^2\pi^2/2mL^2$, becomes very large as the confining box is taken smaller ($L \to 0$). This can be seen to be in accord with the uncertainty relation (3.23), as follows. First, note that $\langle P \rangle = 0$ for each of the stationary states; this can be obtained either by direct evaluation in terms of the wavefunction, or seen as a consequence of the parity selection rule discussed at the end of Sect. 5.4. It follows that

$$\langle H \rangle = \frac{1}{2m}\langle P^2 \rangle = \frac{1}{2m}(\Delta P)^2. \quad (6.66)$$

Now the uncertainty relation (3.23) says that $(\Delta P)^2 \geq (\hbar/2\Delta X)^2$, and it is also clear that $\langle X \rangle = 0$, again either by direct computation or as a consequence of the parity selection rule. Therefore, $(\Delta X)^2 = \langle X^2 \rangle$, and we have a bound

$$\langle H \rangle \geq \frac{\hbar^2}{8m\langle X^2 \rangle}. \quad (6.67)$$

Since the particle is confined to a box of length L, a crude and conservative estimate valid for any state is that $\langle X^2\rangle \leq (L/2)^2$. Using this to compare our estimate of the lower bound on $\langle H\rangle$ from the uncertainty relation to the exact result of (6.65) with $n = 1$, we see that the latter is a factor of π^2 larger. For the ground state, a much better estimate of $\langle H\rangle$ comes from actually computing

$$\langle X^2\rangle = L^2\left(\frac{1}{12} - \frac{1}{2\pi^2}\right), \tag{6.68}$$

which gives the result

$$\langle H\rangle \geq \frac{\hbar^2\pi^2}{2mL^2}\left(\frac{3}{\pi^2 - 6}\right). \tag{6.69}$$

Comparing again with (6.65) with $n = 1$, we see that the ground-state energy is a factor of $\pi^2/3 - 2$, or about 1.29, larger than the estimate (6.69) of its lower bound following from the uncertainty relation.

6.5 Bound States for the One-Dimensional Square Well

As a generalization of the previous section, consider a particle of mass m in a finite square-well potential with width L and depth V_0:

$$V(x) = \begin{cases} V_0 & \text{for } x \leq -L/2 \quad \text{(region I)}, \\ 0 & \text{for } -L/2 < x < L/2 \quad \text{(region II)}, \\ V_0 & \text{for } x \geq L/2 \quad \text{(region III)}. \end{cases} \tag{6.70}$$

Before proceeding, we note that the special case $V_0 \to \infty$ should give the results of the previous section. Again, we expect to find energy eigenstate solutions with definite parity, because the potential is invariant under $x \to -x$.

The strategy for finding the stationary states is to first solve the differential equation (6.43) separately in each of the three regions I, II, and III, and then stitch these solutions together, using (6.47) and (6.48) as boundary conditions at the points $x = \pm L/2$.

In region II, the differential equation is exactly the same as for the particle in a box,

$$\psi'' = -\frac{2mE}{\hbar^2}\psi, \tag{6.71}$$

and so the general solution is an arbitrary linear combination of e^{ikx} and e^{-ikx}, where $k = \sqrt{2mE}/\hbar$, as before. So, we can write

$$\psi_{\text{II}}(x) = A\cos(kx) + B\sin(kx). \tag{6.72}$$

In regions I and III, the differential equation is

$$\psi'' = \frac{2m(V_0 - E)}{\hbar^2}\psi. \tag{6.73}$$

Assume that $E < V_0$, so that we will have a bound state. Then, defining

$$\kappa = \sqrt{2m(V_0 - E)}/\hbar, \tag{6.74}$$

the general solutions in regions I and III are

$$\psi_{\rm I}(x) = Ce^{\kappa x} + De^{-\kappa x}, \tag{6.75}$$

$$\psi_{\rm III}(x) = Fe^{\kappa x} + Ge^{-\kappa x}. \tag{6.76}$$

The coefficient D must be 0, because otherwise $\psi_{\rm I}(x)$ would blow up for $x \to -\infty$, giving a non-normalizable unphysical solution. Similarly, from requiring $\psi_{\rm III}(x)$ to be well-behaved at $x \to \infty$, we get $F = 0$. It remains to solve for A, B, C, and G, using the boundary conditions at the points $x = -L/2$ and $L/2$.

Let us first look for wavefunction solutions that are even under the parity transformation $x \to -x$. It follows that $B = 0$ and $G = C$, and the wavefunction must have the form

$$\psi(x) = \begin{cases} Ce^{\kappa x} & \text{for } x \le -L/2, \\ A\cos(kx) & \text{for } -L/2 \le x \le L/2, \\ Ce^{-\kappa x} & \text{for } x \ge L/2. \end{cases} \tag{6.77}$$

Now we can apply the requirements that the wavefunction and its first derivative are both continuous at $x = L/2$, since the potential is finite. These conditions amount to

$$A\cos(kL/2) = Ce^{-\kappa L/2}, \tag{6.78}$$

$$-Ak\sin(kL/2) = -C\kappa e^{-\kappa L/2}. \tag{6.79}$$

Taking the ratio of these equations gives $k\tan(kL/2) = \kappa$. It is now convenient to define dimensionless quantities

$$X = \frac{kL}{2} = \sqrt{\frac{mE}{2}}\frac{L}{\hbar}, \tag{6.80}$$

$$Y = \frac{\kappa L}{2} = \sqrt{\frac{m(V_0 - E)}{2}}\frac{L}{\hbar}, \tag{6.81}$$

$$R = \sqrt{\frac{mV_0}{2}}\frac{L}{\hbar} \tag{6.82}$$

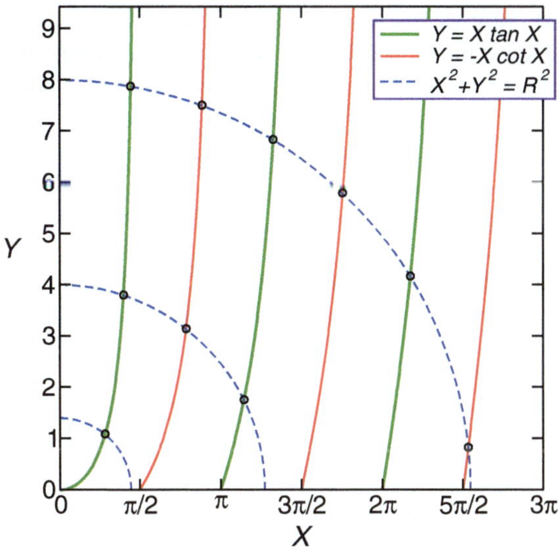

Fig. 6.1 Graphical solutions for the bound energy eigenstates of a particle of mass m in a one-dimensional square well potential with depth V_0 and width L. The dashed circles are $X^2 + Y^2 = R^2 = mV_0L^2/2\hbar^2$, for sample values $R = 1.4$, 4, and 8. Even-parity solutions are the intersections of the appropriate circle with $Y = X \tan X$ (darker, green curves) and odd-parity solutions are the intersections with $Y = -X \cot X$ (lighter, red, curves). The energy eigenvalues for each solution are $E = 2\hbar^2 X^2/mL^2$. The cases $R = 1.4$, 4, and 8 are seen to have 1, 3, and 6 bound-state solutions, respectively.

so that

$$X \tan X = Y, \qquad \text{(even parity)}, \tag{6.83}$$

$$X^2 + Y^2 = R^2. \tag{6.84}$$

It is not possible to solve the simultaneous transcendental equations (6.83) and (6.84) analytically, but one can use graphical methods to understand the solutions and then obtain numerical results. In Fig. 6.1 we graph in the X, Y plane the curves $Y = X \tan X$ (darker solid curves) and the circle (6.84) (dashed), for some sample values (1.4, 4, and 8) of the dimensionless radius R. For a given value of R, the intersections provide the numerical eigenvalue solutions for X and Y, and thus for k and κ. We only need to consider positive X and Y, because κ is positive, and k can be taken positive without loss of generality.

The graphical approach makes clear that there is always at least one bound-state solution, no matter how small V_0 (and thus R) is. The ground state, which we will call $|\psi_1\rangle$ with energy eigenvalue E_1, is the solution with smallest X. It always has even parity and occurs for $0 < X < \pi/2$, which means $0 < k < \pi/L$, and so

$$0 < E_1 < \hbar^2\pi^2/2mL^2. \tag{6.85}$$

The lower bound corresponds to small R, which means the limit of small $V_0 \ll \hbar^2/mL^2$. Conversely, the upper bound is relevant for large R and thus $V_0 \gg \hbar^2/mL^2$, the limit of the particle completely confined to a box as treated in the previous section. If we imagine increasing V_0 (for fixed m, L), the radius of the dashed circle grows, so the number of solutions increases, but for any finite V_0 the number of bound-state solutions is always finite. We number the states by an integer n that increases with energy, so that $n = 1$ is the ground state, $n = 2$ is the first excited state (if it exists), etc.

Consider the lone bound-state solution in the limit of small V_0, which implies X and Y are also very small. By expanding $X \tan X$ for small X, and solving (6.83) and (6.84) in that approximation, we obtain

$$E_1 = \frac{\hbar^2 k^2}{2m} \approx V_0 \left(1 - \frac{m V_0 L^2}{2\hbar^2} + \cdots\right). \tag{6.86}$$

Because this energy is only slightly lower than V_0, the state is very weakly bound.

Now consider the stationary states that are odd under parity ($x \to -x$), with wavefunctions that are therefore of the form

$$\psi(x) = \begin{cases} -Ce^{\kappa x} & \text{for } x \leq -L/2, \\ B\sin(kx) & \text{for } -L/2 \leq x \leq L/2, \\ Ce^{-\kappa x} & \text{for } x \geq L/2. \end{cases} \tag{6.87}$$

The continuity of the wavefunction and its derivative give

$$B\sin(kL/2) = Ce^{-\kappa L/2}, \tag{6.88}$$

$$Bk\cos(kL/2) = -C\kappa e^{-\kappa L/2}. \tag{6.89}$$

Defining X and Y in exactly the same way as before, we now have instead of (6.83),

$$-X\cot X = Y, \qquad \text{(odd parity)}, \tag{6.90}$$

with the same equation for $X^2 + Y^2 = R^2$. The graph of $Y = -X\cot X$ is shown in Fig. 6.1 as the lighter solid lines. This time, we see that if V_0 is small enough, there will not be any odd-parity bound-state solution. In order for there to exist at least one bound-state solution with an odd wavefunction, the dashed circle must have a large enough radius R to intersect with the curve $Y = -X\cot X$ for positive X and Y, specifically, $\sqrt{\frac{mV_0L^2}{2\hbar^2}} > \pi/2$, so that $V_0 > \pi^2\hbar^2/2mL^2$. In this case, the first excited state will have $\pi/2 < X < \pi$, so $\hbar^2\pi^2/2mL^2 < E_2 < 2\hbar^2\pi^2/mL^2$.

From Fig. 6.1 we can see that the energy levels alternate between even and odd parity, so that even parity states (including the ground state) have odd n, and odd parity states have even n. The nth bound state $|\psi_n\rangle$ exists if and only if the potential well is deep enough, which requires that R^2 is sufficiently large,

$$\frac{mV_0L^2}{2\hbar^2} > \frac{\pi^2}{4}(n-1)^2. \tag{6.91}$$

If $|\psi_n\rangle$ does exist as a bound state, then $\pi(n-1)/2 < X_n < \pi n/2$, which implies

$$\frac{\hbar^2\pi^2(n-1)^2}{2mL^2} < E_n < \frac{\hbar^2\pi^2 n^2}{2mL^2} \qquad (n = 1, 2, 3, \ldots). \tag{6.92}$$

Here E_n will approach its upper bound in the limit of large V_0, in agreement with the result found in Sect. 6.4.

The square well potential also has a continuum of unbound states with energies $E \geq V_0$, with wavefunctions whose magnitudes approach a constant for large distances $|x|$. These are known as scattering states. We will discuss the problem of scattering from the one-dimensional square well in Sect. 6.6.

6.6 Scattering Problems in One Dimension

Scattering theory deals with unbound quantum states with continuous energy eigenvalues. The particles described by these states can be thought of as originating far away, in a measured or controlled configuration, typically a superposition of free plane waves moving in a common direction. In a localized region of space, the particles then interact with a potential, or with another group of particles, after which they escape to large distances again where they can be measured. In practice, one can learn about the potential, or interactions between particles, by studying the asymptotic forms of the unbound states. In this section we will consider potential scattering problems in the simplified realm of one dimension, and turn to the problem of scattering in three dimensions in Chap. 24.

Consider two asymptotic regions I and II with constant potentials, and an intermediate region III where the potential can be arbitrary, so

$$V(x) = \begin{cases} 0 & \text{region I } (x < a), \\ U(x) & \text{region III } (a < x < b), \\ V_0 & \text{region II } (x > b), \end{cases} \tag{6.93}$$

as illustrated in Fig. 6.2. The constant potential in region I, which will contain the incident particles, is taken to be 0. This entails no loss of generality, because as usual the effect of a constant contribution to the energy can be absorbed into a global phase that is the same for all states. The potential V_0 in region II can be either positive or negative. We will be interested only in unbound states with energy $E > 0$, although there may also be bound states if the potential $U(x)$ in region III goes negative for some x.

The stationary states for this potential have wavefunctions that might be complicated in region III, but they are definitely simple in regions I and II. Assuming that $E > V_0$, one class of such stationary states is

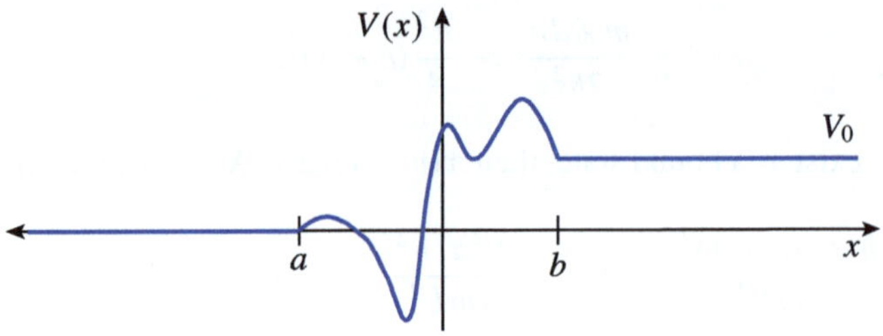

Fig. 6.2 A generic one-dimensional scattering potential of the type in (6.93). The potential V_0 in region II on the far right can be either positive (as shown) or negative

$$\psi(x,t) = \begin{cases} \left(Ae^{ikx} + Be^{-ikx}\right) e^{-iEt/\hbar}, & \text{(region I)}, \\ Ce^{ik_T x} e^{-iEt/\hbar}, & \text{(region II)}. \end{cases} \tag{6.94}$$

Here the component A corresponds to an initial beam of particles moving to the right in region I from $x = -\infty$, and B corresponds to a reflected component going back to $x = -\infty$, with the same wavenumber k. In region II, there is only a transmitted component moving to the right, with wavenumber k_T. As a boundary condition we are imposing that there is no left-moving component that would correspond to incident particles arriving from $x = +\infty$. Note that in one-dimensional scattering problems, we are not interested in eigenstates of parity; the symmetry is broken because the incident particles are coming from the left.

Now, for a given energy E and normalization A treated as known inputs characteristic of the incident beam of particles, we can solve for k, k_T and then, in principle, for B and C. The easy part is that from the Schrödinger equation,

$$\frac{\hbar^2 k^2}{2m} = E, \qquad \frac{\hbar^2 k_T^2}{2m} = E - V_0. \tag{6.95}$$

To solve for B and C requires the scattering potential $U(x)$ to be specified. Once B and C are known, (5.142) can be used to find the probability and current densities in each region. In region I, the probability density is

$$\rho = |\psi|^2 = |A|^2 + |B|^2 + 2\text{Re}[A^* B]\cos(2kx) + 2\text{Im}[A^* B]\sin(2kx). \tag{6.96}$$

The last two terms will each average to 0 over a range of x that is large compared to $1/k$. The current density [see (5.142)] in region I is

$$J = \frac{\hbar k}{m}(|A|^2 - |B|^2), \tag{6.97}$$

where the cross-terms between A and B canceled completely, even without doing any averaging. Meanwhile, in region II, the probability density and current are

$$\rho = |C|^2, \tag{6.98}$$

$$J = \frac{\hbar k_T}{m}|C|^2. \tag{6.99}$$

The interpretation of the probability densities (after spatial averaging) of (6.96) and (6.98) is

$$\text{density of particles in} \begin{pmatrix} \text{incident} \\ \text{reflected} \\ \text{transmitted} \end{pmatrix} \text{beam} = \begin{pmatrix} |A|^2 \\ |B|^2 \\ |C|^2 \end{pmatrix}, \tag{6.100}$$

while the interpretation of the current densities of (6.97) and (6.99) is

$$\text{particles/time in} \begin{pmatrix} \text{incident} \\ \text{reflected} \\ \text{transmitted} \end{pmatrix} \text{beam} = \begin{pmatrix} \hbar k|A|^2/m \\ -\hbar k|B|^2/m \\ \hbar k_T|C|^2/m \end{pmatrix} = \begin{pmatrix} J_A \\ J_B \\ J_C \end{pmatrix}. \tag{6.101}$$

The effect of the scattering potential on the incident monoenergetic beam can therefore be given in terms of reflection and transmission ratios, reminiscent of quantities in classical optics,

$$R = \frac{|J_B|}{|J_A|} = \frac{|B|^2}{|A|^2}, \qquad T = \frac{|J_C|}{|J_A|} = \frac{k_T|C|^2}{k|A|^2}. \tag{6.102}$$

In specific problems, we can always set $A = 1$ from the beginning, since only the ratios B/A and C/A are needed, and B and C will always be proportional to A due to the linearity of the wavefunction in the Schrödinger equation.

To find R and T, we must specify the scattering potential $V(x)$. However, in general, probability conservation implies that they are related by

$$R + T = 1. \tag{6.103}$$

To prove this, we note that for a stationary state,

$$\frac{\partial J}{\partial x} = -\frac{\partial \rho}{\partial t} = -\frac{\partial}{\partial t}|\psi(x,t)|^2 = -\frac{\partial}{\partial t}|\psi(x,t_0)e^{-i(t-t_0)E/\hbar}|^2 = 0, \tag{6.104}$$

so, by the fundamental theorem of calculus,

$$0 = \int_{-\infty}^{\infty} dx\, \frac{\partial J}{\partial x} = J(\infty) - J(-\infty). \tag{6.105}$$

This can be rewritten as

$$J_C = |J_A| - |J_B|, \tag{6.106}$$

from which (6.103) follows immediately. Since R and T are manifestly non-negative, they must also be between 0 and 1, inclusive, with the extremes reached only in very special cases. This probabilistic situation is in contrast to classical mechanics, where T is always exactly either 1 or 0, depending on whether or not the particle has enough energy to overcome the potential barrier set by the maximum value of the potential.

To find R and T in specific examples, it is necessary to solve the Schrödinger equation in region III and to apply boundary conditions in the form of continuity of the wavefunction, and its derivative if the potential is finite, at the boundaries between adjacent regions. If the potential at a boundary has a delta function, then (6.47) can be used to obtain the boundary condition for the first derivative of the wavefunction.

As a first example, consider a step-function potential

$$V(x) = \begin{cases} 0 & \text{region I } (x < 0), \\ V_0 & \text{region II } (x > 0), \end{cases} \tag{6.107}$$

for which region III does not exist, and regions I and II meet at $x = 0$. Continuity of the wavefunction (6.94) and its first derivative at $x = 0$ give

$$1 + B = C, \tag{6.108}$$

$$ik + (-ik)B = ik_T C, \tag{6.109}$$

where we have set $A = 1$, and

$$k_T/k = \sqrt{1 - V_0/E} \tag{6.110}$$

from (6.95). The solutions for B and C are

$$B = \frac{k - k_T}{k + k_T}, \qquad C = \frac{2k}{k + k_T}. \tag{6.111}$$

The reflection and transmission coefficients are therefore

$$R = \left|\frac{1 - k_T/k}{1 + k_T/k}\right|^2, \qquad T = \frac{4k_T/k}{(1 + k_T/k)^2}. \tag{6.112}$$

Note that we need $E > V_0$ in order for k_T to be real. Otherwise, region II is a classically forbidden region, and the wavefunction for $x > 0$ is a decaying exponential of the form $\psi_{\text{II}}(x) = Ce^{-\kappa x}$ with $\hbar^2\kappa^2/2m = V_0 - E$. In this case the transmitted flux vanishes, since (5.142) tells us that $J = (\hbar/m)\text{Im}(\psi^* d\psi/dx) = 0$ due to the fact that the wavefunction is real in region II. Thus $R = 1$ and $T = 0$

in that case, and the particles are all reflected, although they still have a nonzero probability to be found at any point $x > 0$, proportional to $e^{-2\kappa x}$.

As a second example, consider the symmetric rectangular barrier potential

$$V(x) = \begin{cases} 0 & \text{region I } (x < -a/2), \\ V & \text{region III } (|x| < a/2), \\ 0 & \text{region II } (x > a/2), \end{cases} \tag{6.113}$$

as illustrated in Fig. 6.3.

Let us first consider the case that $E > V > 0$, so that transmission past the barrier is classically allowed. Then we can write the wavefunctions in the three regions as

$$\psi_{\text{I}}(x) = e^{ikx} + Be^{-ikx}, \tag{6.114}$$

$$\psi_{\text{III}}(x) = De^{ik'x} + Fe^{-ik'x}, \tag{6.115}$$

$$\psi_{\text{II}}(x) = Ce^{ikx}, \tag{6.116}$$

where $k' = \sqrt{2m(E-V)}/\hbar$ and $k = k_T = \sqrt{2mE}/\hbar$ are real numbers, and again we choose $A = 1$. There are four remaining unknowns, B, C, D, F, but only the first two of these enter into the measurable quantities $R = |B|^2$ and $T = |C|^2$. Continuity of the wavefunction and its first derivative at $x = -a/2$, using ψ_{I} and ψ_{III}, give us

$$e^{-ika/2} + Be^{ika/2} = De^{-ik'a/2} + Fe^{ik'a/2}, \tag{6.117}$$

$$ike^{-ika/2} - ikBe^{ika/2} = ik'De^{-ik'a/2} - ik'Fe^{ik'a/2}. \tag{6.118}$$

Also, at $x = a/2$ we find from ψ_{II} and ψ_{III},

$$Ce^{ika/2} = De^{ik'a/2} + Fe^{-ik'a/2}, \tag{6.119}$$

$$ikCe^{ika/2} = ik'De^{ik'a/2} - ik'Fe^{-ik'a/2}. \tag{6.120}$$

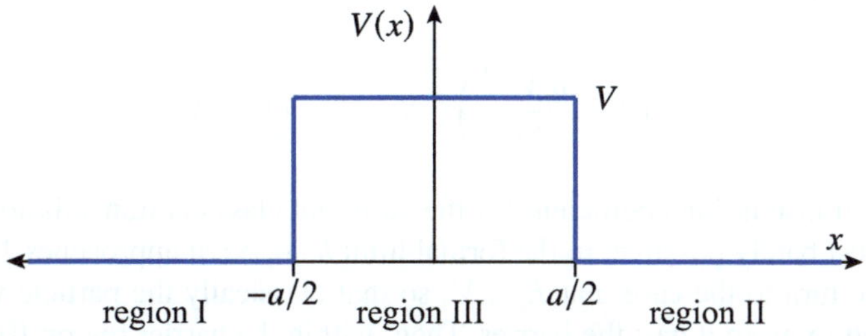

Fig. 6.3 A one-dimensional scattering potential of the type in (6.113). The potential can be either a barrier (if $V > 0$, as shown) or a well (if $V < 0$).

It is convenient to first solve (6.119) and (6.120) for D and F in terms of C, and plug the results into (6.117) and (6.118) which then involve only B and C as unknowns. The resulting (6.117) and (6.118) then combine to give

$$B = i\frac{k'^2 - k^2}{2kk'} \sin(k'a)\, C. \tag{6.121}$$

Now, since $k_T = k$, we also know $R + T = |B|^2 + |C|^2 = 1$, so we can solve to get

$$T = |C|^2 = \left[1 + \left(\frac{k'^2 - k^2}{2kk'}\right)^2 \sin^2(k'a)\right]^{-1} \tag{6.122}$$

for the transmission coefficient. Then, R is just $1 - T$.

To express the result directly in terms of the incident energy and the height of the potential barrier, we can now plug in $k' = k\sqrt{1 - V/E}$, to get

$$T = \left[1 + \frac{V^2}{4E(E - V)} \sin^2\left(\frac{a}{\hbar}\sqrt{2m(E - V)}\right)\right]^{-1}, \qquad (E > V). \tag{6.123}$$

As a check, if $V = 0$, then there is no barrier, and $T = 1$ and $R = 0$. In the high-energy limit, $E \gg V$ gives $T \approx 1$, and the barrier is almost transparent, as the potential is too weak to have much effect on the incident particles. It is perhaps more surprising that even if V and E are comparable, we also get $T = 1$ whenever $\sin(k'a) = 0$, which occurs if $k' = n\pi/a$ for any integer n. This shows that the barrier is transparent to the incident monoenergetic particle beam if its width a is an integer multiple of half the de Broglie wavelength $2\pi/k'$ inside the barrier. The discrete energies at which this occurs, called **scattering resonances**, are

$$E = E_n + V, \qquad E_n = \frac{\hbar^2\pi^2 n^2}{2ma^2}, \qquad (n = 1, 2, 3, \ldots). \tag{6.124}$$

These E_n coincide with the energy eigenvalues for a particle in a box of width a, but with walls of infinite potential height, as found in Sect. 6.4. Finally, note that if $E \approx V$, then we have $\sin^2\left(a\sqrt{2m(E - V)}/\hbar\right) \approx 2a^2 m(E - V)/\hbar^2$, so

$$T \approx \left(1 + \frac{mVa^2}{2\hbar^2}\right)^{-1}, \qquad (E \approx V). \tag{6.125}$$

This is the transmission coefficient for the case that classical transmission past the barrier is just barely possible; in the formal limit $\hbar \to \infty$, it approaches 1.

Now we turn to the case that $E < V$, so that classically the particle would not be expected to make it past the barrier. Then, within the barrier region III, we have instead of (6.115),

$$\psi_{\text{III}}(x) = De^{-\kappa x} + Fe^{\kappa x}, \tag{6.126}$$

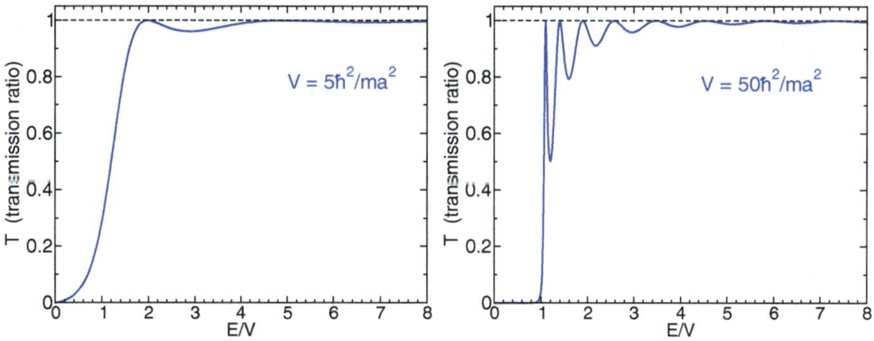

Fig. 6.4 Transmission ratio T for scattering from a one-dimensional rectangular barrier of width a and height $V = 5\hbar^2/ma^2$ (left) or $V = 50\hbar^2/ma^2$ (right), as a function of the energy E of the incident particles of mass m. For $E/V < 1$, the transmission is classically forbidden, and the nonzero T is an example of quantum tunneling. For $E/V > 1$, the barrier is perfectly transparent if its width is an integer multiple of half of the particle's de Broglie wavelength inside the barrier. Transparency also occurs in the limit $E \gg V$

where

$$\kappa = \sqrt{2m(V-E)}/\hbar. \tag{6.127}$$

All of the subsequent algebra is the same, but with $k' \to i\kappa$, so that the sine function is replaced by a hyperbolic sine, and

$$T = \left[1 + \frac{V^2}{4E(V-E)} \sinh^2\left(\frac{a}{\hbar}\sqrt{2m(V-E)}\right)\right]^{-1}, \qquad (E < V). \tag{6.128}$$

In this case, $T < 1$ always, but it is never 0. This is an example of **quantum tunneling**. Classically, there would be no transmission at all, but the Schrödinger equation gives a nonzero probability for the particle to enter the classically forbidden region and emerge on the right side of the barrier. For $E \ll V$, one finds that $T \ll 1$; the transmission due to tunneling becomes exponentially small in the limit of an incident energy much smaller than the barrier potential height. The results for the transmission coefficient T are shown as a function of the incident particle energy E in Fig. 6.4, for two different values of the barrier height V.

So far, we have assumed that the potential barrier height in Fig. 6.3 is positive, with $V > 0$. Now let us consider the case of scattering from a potential well, so that $V < 0$ in (6.113). In that case, all of the same algebra goes through as before, with $V = -|V|$. So, we have

$$T = \left[1 + \frac{V^2}{4E(E+|V|)} \sin^2\left(\frac{a}{\hbar}\sqrt{2m(E+|V|)}\right)\right]^{-1}, \qquad (E > 0 > V). \tag{6.129}$$

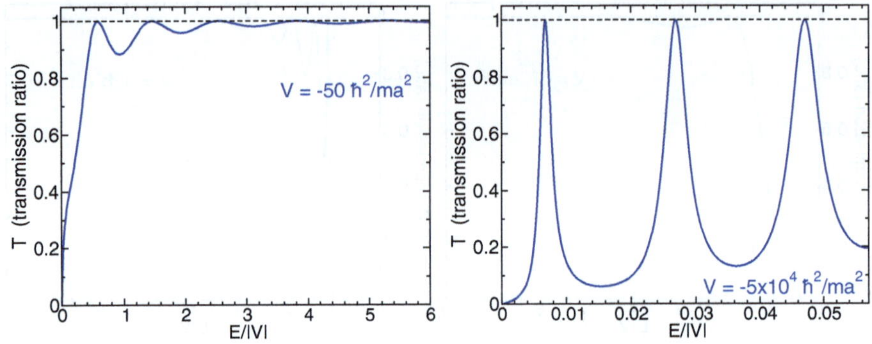

Fig. 6.5 Transmission ratio T for scattering from a one-dimensional well of width a and $V = -50\hbar^2/ma^2$ (left) or $V = -5 \times 10^4\,\hbar^2/ma^2$ (right). In the very low energy limit, the particles are entirely reflected. Transparency occurs in the limit of large E, and also when $E + |V|$ is equal to a resonance energy, given by the eigenvalues $E_n = \frac{\hbar^2\pi^2 n^2}{2ma^2}$ of a particle confined to a box of width a

Classically, there would never be reflection, as the particle incident from the left would have enough momentum and energy to follow a trajectory that takes it unavoidably to $x = +\infty$. The prediction of quantum mechanics in the low-energy limit is very different, as (6.129) gives $T \to 0$ for $E \to 0$. In the high-energy limit, one finds $T \to 1$ for $E \to \infty$, in agreement with the classical expectation. Just as we found for $V > 0$, there are scattering resonances that occur when $E = E_n - |V|$, where $E_n = \hbar^2\pi^2 n^2/2ma^2$ are the energies of bound states in a box (with walls at infinite potential) of width a. The results for the transmission coefficient T are shown as a function of the incident particle energy E in Fig. 6.5, for two different values of the well depth $|V|$. For larger negative V, the resonance energies are more sharply defined.

More generally, including in three-dimensional problems, an attractive potential can become almost transparent to scattering for certain resonance energies, which are determined by the geometry of the potential and the de Broglie wavelengths of the incident particles. This phenomenon is known as the **Ramsauer–Townsend effect**, as it was first observed independently by Carl Ramsauer and John S. Townsend in 1921, in the scattering of electrons from noble gas atoms Ar, Kr, and Xe, before its subsequent explanation by quantum mechanics. The qualitative explanation for this is that close to the positively charged nucleus, the scattering electrons see a spherically symmetric attractive potential well that is cut off at larger distances by the screening of the nuclear charge due to the atomic electrons. This geometry leads to a strong suppression of the scattering cross section, corresponding to near transparency for incident electron energies $E \approx 0.7\,\text{eV}$. The preceding results for the first peak in T in the one-dimensional well scattering problem with large negative V are a rough qualitative model for this phenomenon.

6.7 Particle Acted on by a Constant Force

Consider a particle acted on by a constant force. (For example, this could be a charged particle in a uniform electric field.) Let us take the force to have magnitude f, and to point in the negative x direction, so that the classical potential energy is $V = fx$ with positive f. We will treat this as a one-dimensional problem; restoring the effects of the y and z degrees of freedom just adds a separate additive contribution to the energy. Our Hamiltonian operator is therefore

$$H = \frac{P^2}{2m} + fX. \tag{6.130}$$

Our goal is to find the stationary states.

This problem is much easier to solve by starting in the momentum representation rather than the position representation. To do it, act with $\langle p|$ on the time-independent Schrödinger equation $H|\psi_E\rangle = E|\psi_E\rangle$. According to the momentum-representation rules $|\psi_E\rangle \to \widetilde{\psi}_E(p)$ and $P \to p$ and $X \to i\hbar d/dp$, as in Table 2.1, one obtains the first-order differential equation

$$\left(\frac{p^2}{2m} + i\hbar f\frac{d}{dp} - E\right)\widetilde{\psi}_E(p) = 0, \tag{6.131}$$

which can be rearranged into

$$\frac{d}{dp}\ln\left(\widetilde{\psi}_E(p)\right) = \frac{i}{\hbar f}\left(\frac{p^2}{2m} - E\right). \tag{6.132}$$

Integrating both sides and then exponentiating gives the solution, for any real energy E,

$$\widetilde{\psi}_E(p) = \widetilde{\psi}_E(0)\exp\left[\frac{i}{\hbar f}\left(\frac{p^3}{6m} - Ep\right)\right], \tag{6.133}$$

where the arbitrary normalization factor $\widetilde{\psi}_E(0)$ arises as the constant of integration.

Let us choose to fix the normalization factor so as to satisfy the Dirac orthonormality condition with respect to energy. To that end, consider

$$\langle\psi_{E'}|\psi_E\rangle = \int_{-\infty}^{\infty} dp\, \widetilde{\psi}^*_{E'}(p)\, \widetilde{\psi}_E(p). \tag{6.134}$$

Using (6.133), this can be rewritten as

$$\langle\psi_{E'}|\psi_E\rangle = \widetilde{\psi}^*_{E'}(0)\widetilde{\psi}_E(0)\int_{-\infty}^{\infty} dp\, \exp\left[\frac{ip}{\hbar f}(E' - E)\right], \tag{6.135}$$

where the p^3 terms in the exponential have conveniently canceled. The integral can be evaluated in terms of a Dirac delta function using (2.34), giving

$$\langle \psi_{E'} | \psi_E \rangle = \widetilde{\psi}^*_{E'}(0)\, \widetilde{\psi}_E(0)\, 2\pi\hbar f\, \delta(E' - E). \tag{6.136}$$

We now see that to achieve Dirac orthonormality of the energy eigenstates, $\langle \psi_{E'} | \psi_E \rangle = \delta(E' - E)$, we should choose the normalization $\widetilde{\psi}_E(0) = 1/\sqrt{2\pi\hbar f}$.

Having successfully found the momentum wavefunction, we can get the position wavefunction by taking the inverse Fourier transform. Applying (2.236),

$$\psi_E(x) = \frac{1}{\sqrt{2\pi\hbar}} \int_{-\infty}^{\infty} dp\, e^{ipx/\hbar}\, \widetilde{\psi}_E(p). \tag{6.137}$$

The integral over negative p is the complex conjugate of the integral over positive p, so

$$\psi_E(x) = \frac{1}{\pi\hbar\sqrt{f}} \int_0^{\infty} dp\, \cos\left(\frac{1}{\hbar f} \left[\frac{p^3}{6m} - Ep + fxp \right] \right). \tag{6.138}$$

This integral is not trivial, but can be put into a standard form by defining a dimensionless integration variable $s = p/(2\hbar m f)^{1/3}$ and a shifted and rescaled dimensionless position coordinate

$$y = \left(\frac{2mf}{\hbar^2} \right)^{1/3} (x - E/f). \tag{6.139}$$

The resulting integral over s is proportional to the **Airy function**, defined by

$$\mathrm{Ai}(y) = \frac{1}{\pi} \int_0^{\infty} ds\, \cos(sy + s^3/3). \tag{6.140}$$

This improper integral can be shown to converge, in the sense that it can be defined more formally either as $\mathrm{Ai}(y) = \lim_{N\to\infty} \frac{1}{\pi} \int_0^N ds\, \cos(sy + s^3/3)$, or by means of a convergence factor as $\mathrm{Ai}(y) = \lim_{\epsilon\to 0} \frac{1}{\pi} \int_0^{\infty} ds\, e^{-\epsilon s} \cos(sy + s^3/3)$.

The Airy function is the solution that is bounded as $y \to \infty$ of the Airy differential equation

$$\left(\frac{d^2}{dy^2} - y \right) \mathrm{Ai}(y) = 0. \tag{6.141}$$

It is named for George B. Airy, who first used it to describe optical caustics (like rainbows). The final result for the position wavefunction of the stationary state with energy E is

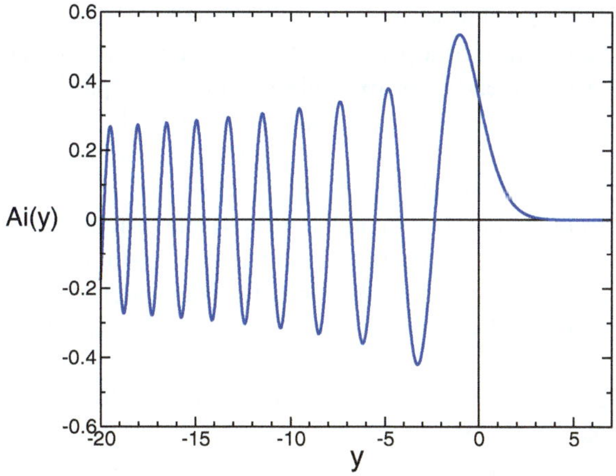

Fig. 6.6 The Airy function Ai(y), proportional to the position wavefunction for a particle in a linear potential with classically allowed region $y < 0$, classically forbidden region $y > 0$, and turning point $y = 0$

$$\psi_E(x) = \left(\frac{2m}{\hbar^2\sqrt{f}}\right)^{1/3} \mathrm{Ai}(y), \tag{6.142}$$

with y defined in (6.139). The point $y = 0$ (corresponding to $x = E/f$) is the classical turning point. This means that a classical particle of energy E coming from the left will follow a trajectory confined to $y < 0$, because it turns around at $y = 0$ where the momentum vanishes. The quantum wavefunction we have found can be thought of as the solution to a special kind of scattering problem in which particles incident from $x = -\infty$ are always reflected, effectively giving $T = 0$ and $R = 1$.

The Airy function is depicted in Fig. 6.6. It decreases rapidly in the classically forbidden region $y > 0$, and for large positive y approaches

$$\mathrm{Ai}(y) \approx \frac{1}{2\sqrt{\pi}\, y^{1/4}} \exp\left(-\frac{2}{3} y^{3/2}\right), \qquad (y \gg 0). \tag{6.143}$$

In the classically allowed region, it instead oscillates, but with a variable wavelength and amplitude that both decrease for larger $-y$. The asymptotic form for large negative y is

$$\mathrm{Ai}(y) \approx \frac{1}{\sqrt{\pi}(-y)^{1/4}} \sin\left(\frac{2}{3}(-y)^{3/2} + \frac{\pi}{4}\right), \qquad (y \ll 0). \tag{6.144}$$

The approximate values of the first four zeros of Ai(y) and its first derivative are

$$y \approx -2.33811,\ -4.08795,\ -5.52056,\ -6.78671, \ldots \text{[zeros of Ai}(y)], \tag{6.145}$$

$$y \approx -1.01879,\ -3.24820,\ -4.82010,\ -6.16331, \ldots \text{[zeros of Ai}'(y)]. \tag{6.146}$$

The solution that we have found is useful for understanding the tunneling penetration of a particle into a classically forbidden region in which the potential grows linearly, for example for a charged particle confined by a constant electric field. The exponentially falling form of the wavefunction is given by (6.143). Of course, this solution is only an idealization, because in reality the potential energy is always bounded from above. However, if the cutoff of the potential is far in the forbidden region, then this makes little difference.

A common practical application is to stitch the solutions of the Airy differential equation, including the linearly independent solution $\mathrm{Bi}(y)$, onto other solutions associated with other potentials in adjacent regions, by matching the wavefunction (and its first derivative, if the potential is finite). The function $\mathrm{Bi}(y)$ grows exponentially for large y, and so is excluded in the region that includes $y = \infty$, but is needed for the general solution in other regions. For example, this is used in one approach to the WKB approximation method, named for Gregor Wentzel, Hendrik Kramers, and Léon Brillouin, which is discussed in detail in other books.

Exercises

6.1 A particle of mass m moves in one dimension in the presence of an attractive delta-function potential $V(x) = -a\delta(x)$, where a is a positive constant with units of energy times length.

(a) Show that there is always exactly one bound-state solution, with energy $E = -ma^2/2\hbar^2$, and obtain its normalized position and momentum wavefunctions $\psi(x)$ and $\widetilde{\psi}(p)$.
(b) For the bound state, sketch $\psi(x)$, and compute $\langle X \rangle$, $\langle X^2 \rangle$, and ΔX.
(c) For the bound state, sketch $\widetilde{\psi}(p)$, and compute $\langle P \rangle$, $\langle P^2 \rangle$, and ΔP.
(d) What is $(\Delta X)(\Delta P)$? How does it compare to the result for a Gaussian wavefunction?

6.2 For the particle in a one-dimensional box of length L discussed in Sect. 6.4, consider the wavefunction $\psi_n(x)$ for each energy level n.

(a) Find the probability that the particle will be found within a distance $L/4$ of the center of the box. (Your answer should contain no trigonometric functions, and may have a different form for even and odd n.) For which n is this probability largest, and for which is it smallest?
(b) Find the uncertainties ΔX and ΔP for each energy level n. Show that their product is always consistent with the uncertainty principle.

6.3 Consider a particle of mass m moving in a potential that has a power-law behavior $V(x) = \lambda x^n$ at large positive x, where λ and n are positive constants. Show that the wavefunction for a stationary state falls off asymptotically at large x as $\psi \sim x^q \exp(-cx^\nu)$, where ν and c are constants that you will find in terms of the quantities n, λ, m, and $\hbar$, independent of the constant q, which depends on the energy eigenvalue. (The special case $n = 1$ was treated in Sect. 6.7, and the case $n = 2$ will be treated in Chap. 7. In the case of a three-dimensional spherically symmetric potential $V(r) = \lambda r^n$, the exponential scaling at large r has the same form with x replaced by r.)

6.4 Consider scattering of particles with mass m and energy E from a potential $V(x) = V_0 \delta(x/a)$, where a is a length scale and V_0 is a constant with units of energy.

(a) Find the transmission and reflection coefficients T and R by matching stationary-state wavefunctions at $x = 0$.
(b) Check your result for T by comparing to the behavior of (6.128) for small a.

6.5 Consider a particle of mass m moving in one dimension in a potential

$$V(x) = -aV_0\big[\delta(x-a) + \delta(x+a)\big], \tag{6.147}$$

where $a > 0$ and $V_0 > 0$. (Note that V_0 has units of energy.)

(a) What is the form of the wavefunction for a bound stationary state with even parity? Normalize your answer so that $\psi(x) = e^{-\kappa x}$ for large positive x.
(b) Find an equation that determines the energies for even-parity bound states, and determine graphically how many such states there are. [Hint: the equation can be written in the form $\kappa a =$ (polynomial in $e^{-\kappa a}$). Sketch the shape of the right-hand side as a function of κa.]
(c) Repeat parts (a) and (b) for odd-parity bound states. For what values of V_0 are there no such states? [Hint: again your equation should have the form $\kappa a =$ (polynomial in $e^{-\kappa a}$). Sketch the right-hand side, and consider its slope at $\kappa a = 0$.]
(d) Solve for the even-parity bound state energy analytically in the limit of $V_0 \ll \hbar^2/ma^2$.
(e) Find the even- and odd-parity bound-state energies in the limit $V_0 \gg \hbar^2/ma^2$. (They are equal to each other in that limit.)

6.6 For the potential in Exercise 6.5, let us look for stationary scattering states with $E = \hbar^2 k^2/2m > 0$ and wavefunctions of the form

$$\psi(x) = \begin{cases} e^{ikx} + Be^{-ikx} & (x \le -a) \\ De^{ikx} + Fe^{-ikx} & (-a \le x \le a) \\ Ce^{ikx} & (x \ge a). \end{cases} \tag{6.148}$$

(a) Derive four equations that relate the coefficients B, C, D, and F. Simplify the notation for your work below by writing them in terms of the dimensionless quantity $n = maV_0/\hbar^2 k$.

(b) Solve for C, and use it to find the transmission coefficient T. Some partial results: $C = [1 - iN_1 n + n^2(e^{iN_2 ka} - N_3)]^{-1}$, where N_1, N_2, and N_3 are certain positive integers that you will discover, and $T = [P_1 + P_2 \cos(N_2 ka) + P_3 \sin(N_2 ka)]^{-1}$ where P_1, P_2, and P_3 are certain polynomials in n with integer coefficients.

(c) Check that your transmission coefficient has the expected behavior when the energy E becomes very large. How does it behave when E is very small?

6.7 Consider a quantum mechanical "bouncing ball", consisting of a particle of mass m moving only vertically in the potential

$$V(x) = \begin{cases} mgx & \text{for } x > 0, \\ \infty & \text{for } x \leq 0, \end{cases} \tag{6.149}$$

where x is the height above an impenetrable surface at $x = 0$, and g is the acceleration due to a uniform gravitational field.

(a) Find the stationary-state wavefunctions $\psi_n(x)$ and energies E_n, in terms of the Airy function $\mathrm{Ai}(y)$ and its zeros y_n for $n = 1, 2, 3, \ldots$.

(b) For the ground state, estimate the numerical energy and characteristic height, defined as the point where $\psi_1(x)$ is maximum, when $m = 1$ gram and when $m = 1.7 \times 10^{-24}$ grams (roughly the mass of a neutron, proton, or hydrogen atom). Use (6.145) and (6.146) and $g = 9.8\ \mathrm{m/s^2}$.

The Harmonic Oscillator

7

7.1 The Unreasonable Effectiveness of the Harmonic Oscillator

A harmonic oscillator is any system in which the restoring force is proportional to the displacement from equilibrium. Consider the case of a particle of mass m moving in one dimension. In terms of energy, the classical harmonic oscillator is defined by

$$\text{kinetic energy} = \frac{p^2}{2m}, \qquad \text{potential energy} = V(x) = \frac{1}{2}m\omega^2 x^2. \tag{7.1}$$

Here the angular frequency ω has units of 1/time and parameterizes the strength of the restoring force, according to

$$\text{force} = -\frac{\partial V}{\partial x} = -m\omega^2 x. \tag{7.2}$$

The classical Hamiltonian is the sum of the kinetic and potential energies,

$$H = \frac{p^2}{2m} + \frac{1}{2}m\omega^2 x^2, \tag{7.3}$$

which leads to the equations of motion

$$\dot{x} = \frac{\partial H}{\partial p} = p/m, \qquad \dot{p} = -\frac{\partial H}{\partial x} = -m\omega^2 x, \tag{7.4}$$

with the general solution

$$x(t) = x_0 \cos(\omega t + \phi_0), \qquad p(t) = -m\omega x_0 \sin(\omega t + \phi_0), \tag{7.5}$$

S. P. Martin, *Quantum Mechanics*, Graduate Texts in Physics,
https://doi.org/10.1007/978-3-032-21119-4_7

where x_0 and ϕ_0 are constants determined by the initial conditions.

A wise theorist, Sidney Coleman, was fond of remarking that the career of a young physicist "consists of treating the harmonic oscillator in ever-increasing levels of abstraction". This may be a slight exaggeration, but there are at least three reasons why the quantum harmonic oscillator is particularly worthy of study.

First, the harmonic oscillator is a problem that can actually be solved exactly in closed form in quantum mechanics. Most problems in quantum mechanics do not have this property.

Second, it often arises as a good approximation to more complicated problems. To understand this, consider a more general potential $V(x)$, which we assume to have a local minimum at some point $x = x_0$, and to be smooth in a neighborhood of that point. Then we can expand it in a Taylor series for small $x - x_0$,

$$\begin{aligned} V(x) &= V(x_0) + (x - x_0)\frac{dV}{dx}\bigg|_{x=x_0} + \frac{1}{2}(x - x_0)^2\frac{d^2V}{dx^2}\bigg|_{x=x_0} \\ &\quad + \frac{1}{6}(x - x_0)^3\frac{d^3V}{dx^3}\bigg|_{x=x_0} + \cdots . \end{aligned} \tag{7.6}$$

The first term is a constant, and so can be absorbed into a redefinition of the zero point of energy; as a constant part of the Hamiltonian, it gives the same overall phase to all states, and therefore can be ignored. The second term is zero, since by definition dV/dx vanishes at a minimum of the potential. Thus, if the terms in the expansion are getting smaller, the leading behavior is captured by the quadratic term. After redefining coordinates by shifting x by the distance x_0, we can hope to approximate

$$V(x) \approx \frac{x^2}{2}\left(\frac{d^2V}{dx^2}\bigg|_{x=0}\right) \tag{7.7}$$

where the constant quantity in parentheses is defined to be $m\omega^2$.

The same argument generalizes to a particle moving in a three-dimensional potential. Let us suppose that the x, y, z coordinates have already been redefined so that the minimum of the potential is at the origin $(x, y, z) = (0, 0, 0)$. Then we have

$$V(x, y, z) = V(0, 0, 0) + \frac{1}{2}x_a x_b \mathcal{V}_{ab} + \cdots , \tag{7.8}$$

where by convention repeated indices $a, b, \ldots$ are implicitly summed over $1, 2, 3$, with $(x_1, x_2, x_3) = (x, y, z)$, and

$$\mathcal{V}_{ab} = \frac{\partial^2 V}{\partial x_a \partial x_b}\bigg|_{x_c=0} \tag{7.9}$$

is a real symmetric matrix. A linear term involving $\frac{\partial V}{\partial x_a}\big|_{x_b=0}$ vanishes because we are at the minimum of the potential. A theorem in linear algebra says that a real symmetric matrix can always be diagonalized by some orthogonal matrix O, according to

$$\mathcal{V} = O^T \tilde{\mathcal{V}} O, \tag{7.10}$$

where $\tilde{\mathcal{V}} = \text{diag}(\tilde{\mathcal{V}}_1, \tilde{\mathcal{V}}_2, \tilde{\mathcal{V}}_3)$, so that

$$\mathcal{V}_{ab} = O_{ca} \tilde{\mathcal{V}}_c O_{cb}, \tag{7.11}$$

and the orthogonality condition on O is written as

$$O_{ac} O_{bc} = O_{ca} O_{cb} = \delta_{ab}. \tag{7.12}$$

Dropping the constant $V(0, 0, 0)$, and assuming the higher-order terms in the expansion can be neglected, the quantum mechanical Hamiltonian can therefore be approximated as

$$H = \frac{1}{2m} P_a P_a + \frac{1}{2} O_{ca} \tilde{\mathcal{V}}_c O_{cb} X_a X_b. \tag{7.13}$$

Now we can do a change of variables, by defining

$$\tilde{X}_c = O_{cb} X_b, \qquad \tilde{P}_c = O_{cb} P_b. \tag{7.14}$$

A key feature of this change of variables is that $\tilde{X}_a$ and $\tilde{P}_a$ have canonical commutators, since

$$[\tilde{X}_a, \tilde{P}_b] = O_{ac} O_{bd} [X_c, P_d] = i\hbar O_{ac} O_{bc} = i\hbar \delta_{ab}, \tag{7.15}$$

where the second equality used the canonical commutation relation for the original variables, $[X_c, P_d] = i\hbar\delta_{cd}$, and (7.12) was used to get the last equality. Also,

$$\tilde{P}_a \tilde{P}_a = O_{ab} P_b O_{ac} P_c = \delta_{bc} P_b P_c = P_b P_b. \tag{7.16}$$

Therefore, the Hamiltonian in terms of the new canonical variables is

$$H = \sum_{a=1}^{3} \left(\frac{\tilde{P}_a^2}{2m} + \frac{1}{2} m \omega_a^2 \tilde{X}_a^2 \right), \tag{7.17}$$

where $m\omega_a^2 = \tilde{\mathcal{V}}_a$. This shows that H is the sum of three independent harmonic oscillator Hamiltonians, with possibly different natural frequencies. The general case with no special relationship between ω_1, ω_2, and ω_3 is called the anisotropic three-dimensional harmonic oscillator. The special case $\omega_1 = \omega_2 = \omega_3$ is the isotropic three-dimensional harmonic oscillator; we will study it in more detail later in Sect. 10.5, using its spherical coordinate wavefunctions.

The preceding illustrates why many systems can be treated as if they were, effectively, systems of harmonic oscillators, with various higher-order effects that one may hope to either neglect or treat as perturbations. Of course, there are some important problems that are not approximated well by the harmonic oscillator. The free particle has no restoring force at all. The potential of a particle in a box or a square well is not close to its Taylor series expansion. The hydrogen atom has a potential with minimum $-\infty$ at $r = 0$, and again the Taylor series expansion fails. Fortunately, those three cases are also exactly solvable, and one can say that, together with the harmonic oscillator, they are the four important examples of potential problems in nonrelativistic quantum mechanics that should be familiar to a well-educated physicist.

A third reason for the importance of the harmonic oscillator is that, as we will see, it has the unique property of equally spaced energy levels. This means that its operator algebra is suited for describing multiple energy excitations that can be added independently of each other. This turns out to have special relevance in quantum field theories, where free-particle modes with different momenta can be described using an infinite collection of harmonic oscillators, with couplings between them that describe particle interactions.

The energy eigenvalue problem of a harmonic oscillator in three dimensions can be solved as the tensor product of three one-dimensional harmonic oscillators. In the following two sections, we will use two very different methods to derive the solutions of the one-dimensional harmonic oscillator with Hamiltonian

$$H = \frac{P^2}{2m} + \frac{1}{2}m\omega^2 X^2. \tag{7.18}$$

But first, we make a general observation: the eigenstates of this H must all have positive energy. This should be intuitively plausible, since both contributions to the classical energy are manifestly non-negative. To prove it formally in the quantum theory, note that in a stationary state $|\psi\rangle$ with energy E,

$$E = \langle\psi|H|\psi\rangle = \frac{1}{2m}\langle\psi|P^2|\psi\rangle + \frac{1}{2}m\omega^2\langle\psi|X^2|\psi\rangle \tag{7.19}$$

$$= \frac{1}{2m}\langle P\psi|P\psi\rangle + \frac{1}{2}m\omega^2\langle X\psi|X\psi\rangle > 0. \tag{7.20}$$

Here we have applied the fact that P and X are Hermitian, so that for example $\langle\psi|P = \langle P^\dagger\psi| = \langle P\psi|$, and then used the positivity of the inner product for non-null kets. Strict positivity holds because there can be no state satisfying both $X|\psi\rangle = 0$ and $P|\psi\rangle = 0$, since $[X, P] = i\hbar$.

7.2 Position and Momentum Representations: The Differential Equations Approach

In this section, we will solve for the energy eigenstates of the harmonic oscillator in the position representation. This involves solving a differential equation for the wavefunction. It must be admitted that this procedure is less elegant than the algebraic (energy representation) approach given in the following section. So why do we bother with it? Besides the goal of building character in the student, the differential equation method is important to learn because of its greater applicability; it can be used to solve the eigenvalue problem for many other Hamiltonians for which algebraic approaches are not available.

The position-representation version of the time-independent Schrödinger equation for the harmonic oscillator $H|\psi\rangle = E|\psi\rangle$ is

$$\langle x| \left(\frac{P^2}{2m} + \frac{1}{2} m\omega^2 X^2 \right) |\psi\rangle = E\langle x|\psi\rangle, \tag{7.21}$$

or, in terms of the wavefunction $\psi(x) = \langle x|\psi\rangle$, after using $X \to x$ and $P \to -i\hbar d/dx$,

$$\frac{d^2\psi}{dx^2} + \frac{2m}{\hbar^2} \left(E - \frac{1}{2} m\omega^2 x^2 \right) \psi = 0. \tag{7.22}$$

We will now solve this as an eigenvalue problem for E and $\psi(x)$.

It is convenient to introduce dimensionless variables corresponding to the position and energy,

$$y = x/b, \qquad \mathcal{E} = E/\hbar\omega, \tag{7.23}$$

where we have defined a constant length scale

$$b = \sqrt{\hbar/m\omega}. \tag{7.24}$$

In terms of y and $\mathcal{E}$, (7.22) becomes

$$\frac{d^2\psi}{dy^2} + (2\mathcal{E} - y^2)\psi = 0. \tag{7.25}$$

Since H commutes with the parity operator Π, and bound energy eigenstates in one dimension never have degeneracies (as we discovered at the end of Sect. 6.3), our solutions will be simultaneous eigenstates of energy and parity.

To help us further understand the character of the solutions, first consider the behavior at large $|y|$, where we can neglect the constant $\mathcal{E}$ compared to y^2, so that $d^2\psi/dy^2 \approx y^2\psi$. Let us try an approximate solution of the form

$$\psi = Ay^n e^{-\alpha y^2}, \tag{7.26}$$

where n and α are constants. Then

$$\frac{d\psi}{dy} = A\left(ny^{n-1} - 2\alpha y^{n+1}\right)e^{-\alpha y^2} \approx -2\alpha y\psi, \tag{7.27}$$

since at large $|y|$ the first term in parentheses is negligible compared to the second. Repeating this yields

$$\frac{d^2\psi}{dy^2} \approx (2\alpha y)^2\psi. \tag{7.28}$$

Therefore the guess (7.26) indeed works for large y, with $\alpha = \pm 1/2$. We can reject $\alpha = -1/2$ on physical grounds, since the wavefunction would blow up at large $|y|$ and would not be normalizable. Such a purported wavefunction must be unphysical because it would imply that no matter how far from the origin you look, the probability that the particle would be found farther away grows exponentially.

This motivates trying a solution of the form

$$\psi = u(y)\,e^{-y^2/2}, \quad \text{with} \quad u = \sum_{j=0}^{\infty} c_j y^j, \tag{7.29}$$

where the c_j are constants to be determined. No negative powers of y are included in this guess, because we expect that ψ will be well-behaved as $y \to 0$, where the potential vanishes.

Plugging (7.29) into (7.25) gives

$$\sum_{j=0}^{\infty} c_j\left[j(j-1)y^{j-2} + (2\mathcal{E} - 1 - 2j)y^j\right] = 0. \tag{7.30}$$

Now we use a trick on the first term. Since it vanishes for $j = 0$ and $j = 1$, we rewrite it as $\sum_{j=2}^{\infty} c_j j(j-1)y^{j-2}$, and then let $j = k+2$ so that it becomes $\sum_{k=0}^{\infty} c_{k+2}(k+2)(k+1)y^k$, and finally rename $k \to j$, so that it becomes $\sum_{j=0}^{\infty} c_{j+2}(j+2)(j+1)y^j$. Now it can be nicely reunited with the second term, combining like powers of y to give

$$\sum_{j=0}^{\infty}\left[c_{j+2}(j+1)(j+2) + c_j(2\mathcal{E} - 1 - 2j)\right]y^j = 0. \tag{7.31}$$

Because each power y^j in the sum is independent, their coefficients must vanish separately for each j, so we have a recurrence relation,

$$c_{j+2} = \frac{2j+1-2\mathcal{E}}{(j+1)(j+2)} c_j. \tag{7.32}$$

This shows that we only need to know two constants c_0 and c_1, because if c_0 is known, then so are $c_2, c_4, c_6, \ldots$, and if c_1 is known, then so are c_3, c_5, c_7, etc. So, the solution is

$$\begin{aligned} u(y) &= c_0 \left[1 + \frac{1}{2}(1-2\mathcal{E})y^2 + \frac{1}{24}(1-2\mathcal{E})(5-2\mathcal{E})y^4 + \cdots \right] \\ &+ c_1 \left[y + \frac{1}{6}(3-2\mathcal{E})y^3 + \frac{1}{120}(3-2\mathcal{E})(7-2\mathcal{E})y^5 + \cdots \right], \end{aligned} \tag{7.33}$$

which, naively, appears to be an infinite series.

However, there is something horribly wrong with the solution if it is really an infinite series. If arbitrarily large powers j are present, the function $u(y)$ will necessarily grow too fast as $|y| \to \infty$. To see this, note that for large j, (7.32) implies

$$\frac{c_j}{c_{j+2}} = j/2 + 5/4 + \mathcal{E}/2 + O(1/j). \tag{7.34}$$

Consider for comparison the function $f(y) = y^p e^{y^2}$, with series expansion

$$f(y) = \sum_{k=0}^{\infty} \frac{y^{2k+p}}{k!}. \tag{7.35}$$

Now, writing $j = 2k + p$, this series has terms $C_j y^j$ where $C_j/C_{j+2} = k + 1 = j/2 + 1 - p/2$. Comparing to (7.34), it would appear that $u(y) \sim y^p e^{y^2}$ for $p = -\mathcal{E} - 1/2$, giving a large $|y|$ behavior

$$\psi(y) \sim (y^p e^{y^2}) e^{-y^2/2} = y^p e^{y^2/2}. \tag{7.36}$$

These are recognized as the unphysical solutions with $\alpha = -1/2$ that we already rejected.

For a sensible physical solution, there is only one way out: the series for $u(y)$ must terminate. To find a basis of such solutions, first consider $c_0 \neq 0$ with $c_1 = 0$, so that only even-index coefficients are present. If one of them vanishes, $c_{2k} = 0$ for some k, then all higher coefficients c_j with $j > 2k$ will also vanish, according to the recurrence relation. Likewise, we can take $c_1 \neq 0$ with $c_0 = 0$, so that only odd-index coefficients are present. If one of them vanishes, $c_{2k+1} = 0$ for some k, then all c_j with $j > 2k + 1$ will also vanish.

In either case, the condition for the series in $u(y)$ to terminate, yielding a physical solution, is that the numerator in the recurrence relation (7.32) must vanish for some non-negative integer $j = n$. Therefore, the allowed energy eigenvalues are $\mathcal{E} = n + 1/2$, or

$$E_n = (n + 1/2)\hbar\omega, \qquad (n = 0, 1, 2, \ldots). \tag{7.37}$$

It follows that $u(y)$ is a polynomial of degree n in y, and contains only even (odd) powers of y if n is even (odd). The wavefunctions are related to the energy eigenstate kets $|n\rangle$ of the harmonic oscillator by

$$\langle x|n\rangle = \psi_n(x), \tag{7.38}$$

with

$$H|n\rangle = \hbar\omega(n + 1/2)|n\rangle. \tag{7.39}$$

This shows that the energy levels of the one-dimensional harmonic oscillator are equally spaced and non-degenerate, with a quantum of energy given by $\Delta E = \hbar\omega$. There is also a **zero-point energy** of the ground state, $E_0 = \hbar\omega/2$, in agreement with our earlier proof that the energies had to be positive. The states with even n have even parity, and those with odd n have odd parity.

For any given n, the polynomials $u(y)$ can be constructed from the recurrence relation, (7.32), up to an overall multiplicative constant given by either c_0 or c_1. The resulting $u(y) = H_n(y)$ for a given n are called **Hermite polynomials**. From (7.25) and (7.29) with $\mathcal{E} = n + 1/2$, they satisfy the differential equation

$$\left(\frac{d^2}{dy^2} - 2y\frac{d}{dy} + 2n\right) H_n = 0. \tag{7.40}$$

A general expression for the Hermite polynomials is

$$H_n(y) = e^{y^2/2}\left(y - \frac{d}{dy}\right)^n e^{-y^2/2}. \tag{7.41}$$

This can be verified by plugging it into the differential equation (7.40). Here we are relying on the fact that stationary bound states in one-dimensional problems are non-degenerate, as we found at the end of Sect. 6.3, and therefore unique (up to a normalization constant) for a given n. An even nicer derivation of (7.41) will be found near the end of the Sect. 7.3.

The first few Hermite polynomials are

$$H_0(y) = 1, \qquad H_1(y) = 2y, \tag{7.42}$$

$$H_2(y) = -2 + 4y^2, \qquad H_3(y) = -12y + 8y^3, \tag{7.43}$$

$$H_4(y) = 12 - 48y^2 + 16y^4, \qquad H_5(y) = 120y - 160y^3 + 32y^5. \tag{7.44}$$

Using (7.41), one can establish the identities

$$H_n(-y) = (-1)^n H_n(y), \tag{7.45}$$

$$\frac{d}{dy} H_n(y) = 2n H_{n-1}(y), \tag{7.46}$$

$$H_{n+1}(y) = 2y H_n(y) - 2n H_{n-1}(y). \tag{7.47}$$

It is useful to define the function $F(y,t) = \sum_{n=0}^{\infty} \frac{t^n}{n!} H_n(y)$, which can be evaluated as follows. First, using (7.46), one finds the differential equation $\partial F/\partial y = 2tF$, which has the solution $F(y,t) = F(0,t)e^{2ty}$. Then using (7.47), one finds $\partial F/\partial t = 2(y-t)F$, which yields $\partial F(0,t)/\partial t = -2tF(0,t)$, which has the solution $F(0,t) = F(0,0)e^{-t^2}$. Since $F(0,0) = H_0(0) = 1$, we arrive at the **generating function for Hermite polynomials**,

$$F(y,t) = \sum_{n=0}^{\infty} \frac{t^n}{n!} H_n(y) = \exp(2yt - t^2). \tag{7.48}$$

One use of the generating function is to find the orthonormality properties of the Hermite polynomials. Consider the product of e^{-y^2} and two generating functions,

$$e^{-y^2} \sum_{n=0}^{\infty} \frac{t^n}{n!} H_n(y) \sum_{m=0}^{\infty} \frac{s^m}{m!} H_m(y) = e^{-y^2+2y(t+s)-s^2-t^2}. \tag{7.49}$$

Now integrating both sides with respect to y, we get

$$\sum_{n=0}^{\infty}\sum_{m=0}^{\infty} \frac{t^n}{n!}\frac{s^m}{m!} \left(\int_{-\infty}^{\infty} dy\, e^{-y^2} H_n(y) H_m(y) \right) = \sqrt{\pi} e^{2ts}. \tag{7.50}$$

The right side can be expanded as

$$\sqrt{\pi} e^{2ts} = \sqrt{\pi} \sum_{n=0}^{\infty} \frac{2^n t^n s^n}{n!} = \sqrt{\pi} \sum_{n=0}^{\infty}\sum_{m=0}^{\infty} \delta_{nm} \frac{2^n t^n s^m}{n!}. \tag{7.51}$$

Comparing the coefficients of $t^n s^m$, we arrive at the orthonormality condition

$$\int_{-\infty}^{\infty} dy\, e^{-y^2} H_n(y) H_m(y) = \delta_{nm} \sqrt{\pi}\, 2^n\, n!. \tag{7.52}$$

The orthonormal energy basis wavefunctions for the harmonic oscillator can now be written

$$\psi_n(x) = \left(\frac{m\omega}{\pi\hbar} \right)^{1/4} \sqrt{\frac{1}{2^n n!}} H_n(y) e^{-y^2/2}, \tag{7.53}$$

where $y = x/b = x\sqrt{m\omega/\hbar}$, and the multiplicative constant has been chosen so that

$$\int_{-\infty}^{\infty} dx\, \left(\psi_n(x)\right)^* \psi_m(x) = \delta_{nm}. \tag{7.54}$$

(Actually, the wavefunctions are all real, so the complex conjugation does nothing in this case.) In particular, the normalized ground state ($n = 0$) wavefunction is a pure Gaussian,

$$\psi_0(x) = \left(\frac{m\omega}{\pi\hbar}\right)^{1/4} \exp(-m\omega x^2/2\hbar). \tag{7.55}$$

The wavefunctions ψ_n and the corresponding probability densities $|\psi_n|^2$ for $n = 0, 1, 2, 3, 4$, and 16 are depicted in Fig. 7.1.

Another interesting relation satisfied by the harmonic oscillator wavefunctions, and thus the Hermite polynomials, is

$$\sum_{n=0}^{\infty} \psi_n(x')^* \psi_n(x) = \delta(x - x'). \tag{7.56}$$

This can be derived immediately from $\langle x|x'\rangle = \delta(x - x')$ by applying the completeness relation $\sum_{n=0}^{\infty} |n\rangle \langle n| = 1$.

Having found the position wavefunctions for the harmonic oscillator, we now turn to the momentum representation. One way to evaluate the momentum wavefunctions of the energy eigenstates $|n\rangle$ is to use the completeness relation:

$$\begin{aligned} \widetilde{\psi}_n(p) = \langle p|n\rangle &= \int_{-\infty}^{\infty} dx\, \langle p|x\rangle\langle x|n\rangle \\ &= \int_{-\infty}^{\infty} dx\, \frac{1}{\sqrt{2\pi\hbar}} e^{-ipx/\hbar}\, \psi_n(x), \end{aligned} \tag{7.57}$$

but it may not be immediately obvious how to evaluate the integral for general n. Fortunately, we can gain some insight by noting that the Hamiltonian in (7.18) has the very special property of being invariant under the simultaneous substitutions

$$X \leftrightarrow P, \qquad m\omega \leftrightarrow \frac{1}{m\omega}. \tag{7.58}$$

Since the energy eigenstates are non-degenerate, their momentum wavefunctions must be given, up to a phase $e^{i\varphi_n}$ to be determined, by making these same substitutions in the $\psi_n(x)$ given in (7.53). Therefore, defining a dimensionless variable proportional to momentum,

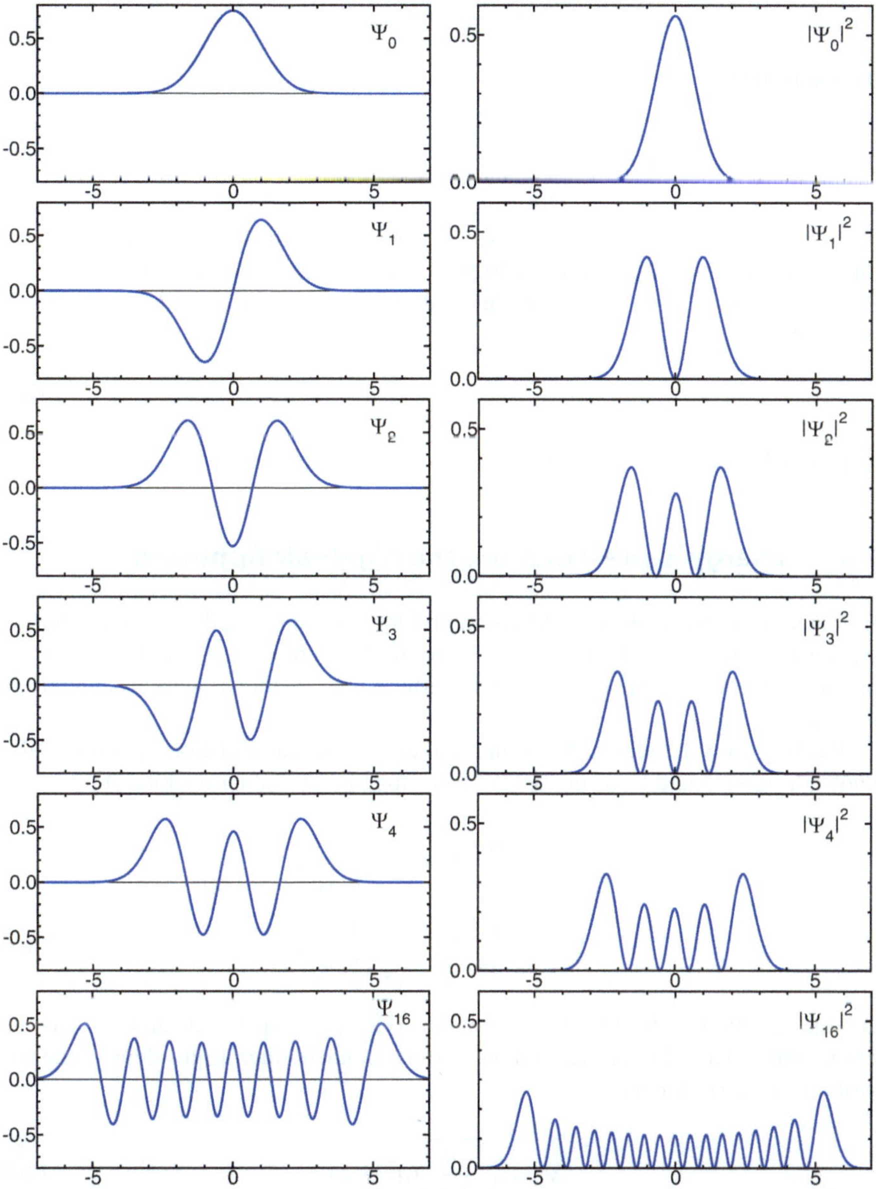

Fig. 7.1 Harmonic oscillator stationary-state wavefunctions ψ_n in units of $(m\omega/\hbar)^{1/4}$ (left column) and probability densities $|\psi_n|^2$ in units of $\sqrt{m\omega/\hbar}$ (right column) for $n = 0, 1, 2, 3, 4$, and 16, as functions of $y = x\sqrt{m\omega/\hbar}$

$$v = p/\sqrt{\hbar\omega m}, \tag{7.59}$$

it must be that

$$\widetilde{\psi}_n(p) = e^{i\varphi_n}\frac{1}{(\pi\hbar\omega m)^{1/4}}\sqrt{\frac{1}{2^n n!}}H_n(v)e^{-v^2/2}. \tag{7.60}$$

The phase factor $e^{i\varphi_n}$ must be chosen consistently with the phase convention of the ket $|n\rangle$ that has already been fixed by (7.53) and (7.38). For the lowest few $n = 0, 1, 2, 3, \ldots$, you can do the integral in (7.57) to check that (7.60) is indeed true with

$$e^{i\varphi_n} = (-i)^n, \tag{7.61}$$

and we will prove it for general n at the end of the next section.

7.3 Energy Representation: The Algebraic Approach

In this section, we will use a different, and more elegant, method to solve for the stationary states of the harmonic oscillator. In this approach, due to Dirac, we use operator algebraic methods rather than differential equations, working directly in the energy basis.

We begin by defining dimensionless creation and annihilation operators (also known as raising and lowering operators, or together as ladder operators) by

$$a = \sqrt{\frac{m\omega}{2\hbar}}X + i\frac{1}{\sqrt{2\hbar\omega m}}P, \tag{7.62}$$

$$a^\dagger = \sqrt{\frac{m\omega}{2\hbar}}X - i\frac{1}{\sqrt{2\hbar\omega m}}P. \tag{7.63}$$

As the notation indicates, these are not Hermitian operators and so are not observables, but rather are Hermitian adjoints of each other. Equivalently, one can write the relationship as

$$X = \sqrt{\frac{\hbar}{2m\omega}}(a^\dagger + a), \tag{7.64}$$

$$P = i\sqrt{\frac{\hbar\omega m}{2}}(a^\dagger - a). \tag{7.65}$$

Using the canonical commutation relations $[X, P] = i\hbar$, we find that

$$[a, a^\dagger] = 1. \tag{7.66}$$

We can also compute

$$a^\dagger a = \frac{m\omega}{2\hbar}X^2 + \frac{1}{2\hbar\omega m}P^2 + \frac{i}{2\hbar}[X, P] = H/\hbar\omega - 1/2,$$

so that the Hamiltonian is simply[1]

$$H = \hbar\omega(a^\dagger a + 1/2). \tag{7.67}$$

Using (7.66), one then obtains

$$[H, a] = -\hbar\omega\, a, \qquad [H, a^\dagger] = \hbar\omega\, a^\dagger. \tag{7.68}$$

Now suppose that we have an energy eigenstate $|E\rangle$, with energy eigenvalue E. Consider the Hamiltonian acting on the state $a^\dagger|E\rangle$:

$$H\left(a^\dagger|E\rangle\right) = a^\dagger H|E\rangle + [H, a^\dagger]|E\rangle = (E + \hbar\omega)a^\dagger|E\rangle. \tag{7.69}$$

This shows that $a^\dagger|E\rangle$ is an eigenstate of H with energy $E + \hbar\omega$. Repeating this n times, the state $(a^\dagger)^n|E\rangle$ must be an energy eigenstate with energy $E + n\hbar\omega$. Similarly,

$$H\,(a|E\rangle) = (E - \hbar\omega)\, a|E\rangle, \tag{7.70}$$

so the state $a^n|E\rangle$ is apparently an energy eigenstate with energy $E - n\hbar\omega$.

Thus, naively it might appear that, given $|E\rangle$, we can construct an infinite chain of energy eigenstates with both arbitrarily low and arbitrarily high energies,

$$\cdots, \quad |E-2\hbar\omega\rangle, \quad |E-\hbar\omega\rangle, \quad |E\rangle, \quad |E+\hbar\omega\rangle, \quad |E+2\hbar\omega\rangle, \quad \cdots. \tag{7.71}$$

But this cannot be true, since we proved at the end of Sect. 7.1 that all of the energy eigenvalues of the harmonic oscillator are positive. The only way out is that all of the kets in the chain with negative energy are actually the null ket. Let us rename the state with lowest non-negative energy as $|0\rangle$, where the label 0 is not the energy, but instead signifies that this is the ground state. It must satisfy

$$a|0\rangle = 0, \tag{7.72}$$

[1] One could also write the equivalent form $H = \hbar\omega(a^\dagger a + aa^\dagger)/2$. In choosing to write (7.67), we have followed the systematic protocol known as **normal ordering**. To "normal order" an operator means to rewrite it by moving all a operators to the right and all $a^\dagger$ operators to the left, using $aa^\dagger = a^\dagger a + 1$ (which is equivalent to the commutation relation) as many times as necessary.

so that all of the kets $a^n|0\rangle = 0$ for $n \geq 1$ are null and do not actually exist as physical states. Since $a^\dagger a|0\rangle = 0$, we have

$$(H/\hbar\omega - 1/2)\,|0\rangle = 0, \tag{7.73}$$

so the ground state must have $E_0 = \hbar\omega/2$, a result that we had also found in the previous section using the differential equation approach.

In Sect. 6.3, we proved that bound states in one-dimensional quantum mechanics never have degenerate energies. This implies that there is a unique state $|0\rangle$ with $E_0 = \hbar\omega/2$, and unique states with $E_n = \hbar\omega(n + 1/2)$ for $n = 0, 1, 2, \ldots$, just as we had found by the differential equations method in (7.37). Up to normalization, the energy eigenstates $|n\rangle$ are proportional to $(a^\dagger)^n|0\rangle$. Since they are non-degenerate eigenkets of a Hermitian operator (H), they can be normalized to form an orthobasis,

$$\langle k|n\rangle = \delta_{nk}. \tag{7.74}$$

Let $|0\rangle$ have norm 1. Then we can define the other orthobasis kets by recurrence, using

$$|n\rangle = c_n a^\dagger |n-1\rangle, \tag{7.75}$$

where c_n are normalization constants to be determined. We have

$$\langle n|n\rangle = |c_n|^2 \langle n-1|aa^\dagger|n-1\rangle = n|c_n|^2 \langle n-1|n-1\rangle, \tag{7.76}$$

where the second equality makes use of

$$aa^\dagger = a^\dagger a + 1 = H/\hbar\omega + 1/2 \tag{7.77}$$

and then $H|n-1\rangle = \hbar\omega(n - 1/2)|n-1\rangle$. Equation (7.76) shows that $\langle n|n\rangle = \langle n-1|n-1\rangle = 1$ requires $c_n = 1/\sqrt{n}$, where we have made an arbitrary choice of phase. From this, we use (7.75) and (7.77) to deduce that

$$a^\dagger|n\rangle = \sqrt{n+1}\,|n+1\rangle, \tag{7.78}$$

$$a|n\rangle = \sqrt{n}\,|n-1\rangle, \tag{7.79}$$

so that $a^\dagger$ raises the energy of the state (or creates an energy quantum), and a lowers the energy (or destroys an energy quantum). Taking the Hermitian conjugate gives

$$\langle n|\,a = \sqrt{n+1}\;\langle n+1|\,, \tag{7.80}$$

$$\langle n|\,a^\dagger = \sqrt{n}\;\langle n-1|\,. \tag{7.81}$$

It follows that the matrix elements of $a^\dagger$ and a in the energy eigenstate orthobasis are

$$\langle k|a^\dagger|n\rangle = \sqrt{n+1}\,\delta_{k,n+1}, \qquad \langle k|a|n\rangle = \sqrt{n}\,\delta_{k,n-1}. \tag{7.82}$$

Another consequence is that the nth excited state can be written in terms of n creation operators acting on the ground state,

$$|n\rangle = \frac{(a^\dagger)^n}{\sqrt{n!}}|0\rangle. \tag{7.83}$$

The Hamiltonian is sometimes written as $H = \hbar\omega(N + 1/2)$, where $N = a^\dagger a$ is called the **number operator**. It is clearly Hermitian, and satisfies

$$N|n\rangle = n|n\rangle, \tag{7.84}$$

so it is the observable that measures the number of energy quanta in the state.

The matrix elements of the position and momentum operators in the energy eigenstate basis follow immediately from (7.64) and (7.65) using (7.82), with the results

$$\langle k|X|n\rangle = \sqrt{\frac{\hbar}{2m\omega}}\left(\sqrt{n+1}\,\delta_{k,n+1} + \sqrt{n}\,\delta_{k,n-1}\right), \tag{7.85}$$

$$\langle k|P|n\rangle = i\sqrt{\frac{\hbar\omega m}{2}}\left(\sqrt{n+1}\,\delta_{k,n+1} - \sqrt{n}\,\delta_{k,n-1}\right). \tag{7.86}$$

In particular, their expectation values in any energy eigenstate $|n\rangle$ vanish,

$$\langle n|X|n\rangle = 0, \qquad \langle n|P|n\rangle = 0. \tag{7.87}$$

This is an example of the parity selection rule (Theorem 5.2), since X and P each have odd parity. One can also compute, using (7.78) and (7.79),

$$\begin{aligned} X^2|n\rangle &= \frac{\hbar}{2m\omega}(a^\dagger + a)(a^\dagger + a)|n\rangle \qquad (7.88)\\ &= \frac{\hbar}{2m\omega}\left(\sqrt{(n+1)(n+2)}|n+2\rangle + (2n+1)|n\rangle + \sqrt{n(n-1)}|n-2\rangle\right), \end{aligned}$$

with the consequence that

$$\langle n|X^2|n\rangle = \frac{\hbar}{m\omega}(n + 1/2). \tag{7.89}$$

In a very similar way,

$$\langle n|P^2|n\rangle = \hbar\omega m(n+1/2). \tag{7.90}$$

As a check,

$$\langle n|H|n\rangle = \frac{1}{2m}\langle n|P^2|n\rangle + \frac{1}{2}m\omega^2\langle n|X^2|n\rangle = \hbar\omega(n+1/2), \tag{7.91}$$

in agreement with the result for E_n that we found in (7.37). Furthermore, the uncertainties in position and momentum each grow with n, proportionally to $\sqrt{E_n}$:

$$\Delta X = \sqrt{\langle n|X^2|n\rangle - \langle n|X|n\rangle^2} = \left(\frac{\hbar}{m\omega}(n+1/2)\right)^{1/2}, \tag{7.92}$$

$$\Delta P = \sqrt{\langle n|P^2|n\rangle - \langle n|P|n\rangle^2} = \left(\hbar\omega m(n+1/2)\right)^{1/2}. \tag{7.93}$$

To further illustrate the power and convenience of the algebraic method, suppose that for some reason we needed to evaluate the matrix element $\langle 3|X^3|2\rangle$. We have, using (7.64),

$$\begin{aligned}\langle 3|X^3|2\rangle &= \left(\frac{\hbar}{2m\omega}\right)^{3/2}\langle 3|(a^\dagger + a)^3|2\rangle \\ &= \left(\frac{\hbar}{2m\omega}\right)^{3/2}\langle 3|\left(\cancel{a^{\dagger 3}} + a^{\dagger 2}a + a^\dagger a a^\dagger + a a^{\dagger 2} + \cancel{a^\dagger a^2} + \cancel{a a^\dagger a} + \cancel{a^2 a^\dagger} + \cancel{a^3}\right)|2\rangle.\end{aligned} \tag{7.94}$$

Here, we have crossed out terms that can be immediately seen to give no contribution by simply counting quanta created and destroyed. For the first term, we start in the ket on the far right with 2 quanta, and create 3 more, so it is proportional to $|5\rangle$, which by orthonormality has vanishing inner product with the bra $\langle 3|$ on the left. Likewise, the fifth, sixth, and seventh terms cannot contribute because we start with 2 quanta and then (in various orders) create 1 more but destroy 2, resulting in a state with 1, which again has vanishing inner product with the 3-quanta bra. And, the last term immediately vanishes because $a^3|2\rangle = 0$. The remaining three terms do not vanish, but can be easily evaluated with nothing but simple arithmetic, by applying (7.78) and (7.79) repeatedly. In contrast, evaluation of the same matrix element in the position representation approach of Sect. 7.2 would look like

$$\begin{aligned}\langle 3|X^3|2\rangle = \int_{-\infty}^{\infty} dx &\left(\frac{m\omega}{\pi\hbar 2^6(3!)^2}\right)^{1/4} H_3\left(x\sqrt{m\omega/\hbar}\right)e^{-m\omega x^2/2\hbar}\,x^3 \\ &\left(\frac{m\omega}{\pi\hbar 2^4(2!)^2}\right)^{1/4} H_2\left(x\sqrt{m\omega/\hbar}\right)e^{-m\omega x^2/2\hbar}.\end{aligned} \tag{7.95}$$

This is certainly doable, but less pleasant.

Let us now see how to connect the energy and position representations, by writing a and $a^\dagger$ as differential operators in the latter. In the position representation, $X \to x$ and $P \to -i\hbar d/dx$, so from (7.62) and (7.63) we get

$$a = \frac{1}{\sqrt{2}}\left(y + \frac{d}{dy}\right), \qquad a^\dagger = \frac{1}{\sqrt{2}}\left(y - \frac{d}{dy}\right), \tag{7.96}$$

where $y = x/b = x\sqrt{m\omega/\hbar}$ as before. Therefore, the condition $a|0\rangle = 0$ gives

$$\left(y + \frac{d}{dy}\right)\psi_0(y) = 0. \tag{7.97}$$

This implies $d\psi_0/\psi_0 = -y\,dy$, or $d(\ln\psi_0) = -d(y^2/2)$, so that upon integration, $\ln\psi_0 = -y^2/2 + \ln A_0$, where $\ln A_0$ is a constant of integration. Thus,

$$\psi_0(y) = A_0 e^{-y^2/2} = A_0 e^{-m\omega x^2/2\hbar}, \tag{7.98}$$

in agreement with (7.55) after fixing the normalization constant $A_0 = (m\omega/\pi\hbar)^{1/4}$. Now, combining (7.83) and (7.96), we have

$$\psi_n(x) = \langle x|n\rangle = \frac{1}{\sqrt{n!}}\left[\frac{1}{\sqrt{2}}\left(y - \frac{d}{dy}\right)\right]^n \left(\frac{m\omega}{\pi\hbar}\right)^{1/4} e^{-y^2/2}. \tag{7.99}$$

Comparing this with (7.53) yields the general form for the Hermite polynomials given in (7.41), as promised.

The energy and momentum representations can be connected in a similar way. In the momentum representation, $P \to p$ and $X \to i\hbar d/dp$, from which one finds

$$a = \frac{i}{\sqrt{2}}\left(v + \frac{d}{dv}\right), \qquad a^\dagger = -\frac{i}{\sqrt{2}}\left(v - \frac{d}{dv}\right), \tag{7.100}$$

where $v = p/\sqrt{\hbar\omega m}$ is the dimensionless rescaled momentum, as in (7.59) and (7.60). Evaluating (7.57) for the special case $n = 0$, we then find

$$\widetilde{\psi}_0(p) = \frac{1}{(\pi\hbar\omega m)^{1/4}} e^{-v^2/2} \tag{7.101}$$

for the ground state. Using (7.100) in (7.83), we obtain

$$\widetilde{\psi}_n(p) = \langle p|n\rangle = \frac{1}{\sqrt{n!}}\left[-\frac{i}{\sqrt{2}}\left(v - \frac{d}{dv}\right)\right]^n \frac{1}{(\pi\hbar\omega m)^{1/4}} e^{-v^2/2}. \tag{7.102}$$

Now, comparing to (7.41), we obtain (7.60), with $e^{i\varphi_n} = (-i)^n$, as promised.

7.4 Coherent States of the Harmonic Oscillator

Consider a macroscopic harmonic oscillator of the type one might encounter in a lab experiment in an introductory physics course, with mass $m = 0.2\,\text{kg}$, $\omega = 10\,\text{rad/s}$, and amplitude $x_0 = 0.1\,\text{m}$. Classically, the energy can be obtained as the potential energy at the extremum of the displacement, when $p = 0$ and $x = x_0$, so that $E = m\omega^2 x_0^2/2 = 0.1\,\text{J}$. The energy quantum associated with the oscillator is $\hbar\omega = 1.055 \times 10^{-33}\,\text{J}$. Therefore, we expect macroscopic oscillator states to have enormous numbers of energy quanta, roughly $n = E/\hbar\omega \approx 10^{32}$ in this example.

Is the energy eigenstate $|n = 10^{32}\rangle$ classical-like? The answer is clearly no, since in any energy eigenstate, $\langle X\rangle(t) = 0$ and $\langle P\rangle(t) = 0$ for all times t by the parity selection rule, while the classical trajectory, (7.5), involves macroscopic oscillations. We would like to find quantum states that are approximately classical, by which we mean that $\langle X\rangle(t) \approx x_{\text{cl}}(t)$ and $\langle P\rangle(t) \approx p_{\text{cl}}(t)$ should be as close as possible to the solutions in (7.5). The states that accomplish this are called **coherent** (or **quasi-classical**) states, and the argument just given ensures that they cannot be eigenstates of energy or parity.

Instead, with the wisdom of foresight, let us try kets $|\alpha\rangle$ that are eigenstates of the lowering operator a with eigenvalue α,

$$a|\alpha\rangle = \alpha|\alpha\rangle. \tag{7.103}$$

Since a is not Hermitian, we have no reason to expect that the allowed eigenvalues α will be real, and indeed it will turn out to be very important that they are complex in general. For the same reason, we also cannot expect that the set of all coherent states $|\alpha\rangle$ for different α will form an orthobasis or even be orthogonal, and again they are not.

To construct the coherent states that satisfy (7.103), let us try an arbitrary linear combination of energy eigenstates,

$$|\alpha\rangle = \sum_{n=0}^{\infty} c_n |n\rangle, \tag{7.104}$$

with coefficients c_n to be determined. Then,

$$a|\alpha\rangle = \sum_{n=0}^{\infty} c_n a|n\rangle = \sum_{n=1}^{\infty} c_n \sqrt{n}|n-1\rangle, \tag{7.105}$$

where in the second equality we used (7.79), and started the sum from $n = 1$ by exploiting the fact that the $n = 0$ term vanishes. We now relabel $n \to n + 1$, and require that the result

$$a|\alpha\rangle = \sum_{n=0}^{\infty} c_{n+1} \sqrt{n+1}|n\rangle \tag{7.106}$$

is equal to $\alpha \sum_{n=0}^{\infty} c_n |n\rangle$. This gives the recurrence relation

$$c_{n+1} = \frac{\alpha}{\sqrt{n+1}} c_n. \tag{7.107}$$

Starting with c_0, we get $c_1 = \alpha c_0$, $c_2 = \alpha^2 c_0/\sqrt{2}$, etc., or in general $c_n = \alpha^n c_0/\sqrt{n!}$. Therefore,

$$|\alpha\rangle = c_0 \sum_{n=0}^{\infty} \frac{\alpha^n}{\sqrt{n!}} |n\rangle. \tag{7.108}$$

To fix the normalization c_0, we require $\langle\alpha|\alpha\rangle = 1$, or

$$1 = |c_0|^2 \sum_{m=0}^{\infty} \sum_{n=0}^{\infty} \frac{\alpha^{*m}}{\sqrt{m!}} \frac{\alpha^n}{\sqrt{n!}} \langle m|n\rangle = |c_0|^2 \sum_{n=0}^{\infty} \frac{|\alpha|^{2n}}{n!} = |c_0|^2 e^{|\alpha|^2}, \tag{7.109}$$

where the second equality used the orthonormality $\langle m|n\rangle = \delta_{nm}$ to collapse the sum over m. Therefore, $c_0 = e^{-|\alpha|^2/2}$ (up to the usual arbitrary global phase), and the normalized coherent state with complex eigenvalue α is

$$|\alpha\rangle = e^{-|\alpha|^2/2} \sum_{n=0}^{\infty} \frac{\alpha^n}{\sqrt{n!}} |n\rangle. \tag{7.110}$$

Using the expression for $|n\rangle$ in terms of $|0\rangle$ in (7.83), this can be rewritten as

$$|\alpha\rangle = e^{-|\alpha|^2/2} \sum_{n=0}^{\infty} \frac{(\alpha a^\dagger)^n}{n!} |0\rangle = e^{-|\alpha|^2/2} e^{\alpha a^\dagger} |0\rangle. \tag{7.111}$$

If a harmonic oscillator is in a coherent state $|\alpha\rangle$, and the energy is measured, all results $E_n = \hbar\omega(n + 1/2)$ can occur, with probabilities

$$\mathcal{P}_n = |\langle n|\alpha\rangle|^2 = \frac{|\alpha|^{2n}}{n!} e^{-|\alpha|^2}. \tag{7.112}$$

These do not depend on the phase of α, and form a Poisson distribution with mean $|\alpha|^2$. Since

$$\mathcal{P}_n = \frac{|\alpha|^2}{n} \mathcal{P}_{n-1}, \tag{7.113}$$

we see that for increasing n, the probability increases as long as $n < |\alpha|^2$, but decreases thereafter. Therefore, the most probable result of an energy measurement is of order $\hbar\omega|\alpha|^2$. This shows that for a macroscopic oscillator like the one posed at the beginning of this section, $|\alpha| \sim \sqrt{n} \sim 10^{16}$ in a coherent state. The expectation value of the energy is

$$\begin{aligned} \langle\alpha|H|\alpha\rangle &= \hbar\omega\Big[(\langle\alpha|\,a^\dagger)(a|\alpha\rangle) + 1/2\Big] = \hbar\omega\Big[(\langle\alpha|\,\alpha^*)(\alpha|\alpha\rangle) + 1/2\Big] \\ &= \Big(|\alpha|^2 + \frac{1}{2}\Big)\hbar\omega, \end{aligned} \tag{7.114}$$

giving a similar measure of the average energy.

Next, let us compute the uncertainty in the energy. First, we use $H^2 = \hbar^2\omega^2(a^\dagger a+1/2)^2 = \hbar^2\omega^2\left(a^\dagger a a^\dagger a + a^\dagger a + 1/4\right) = \hbar^2\omega^2\left(a^\dagger a^\dagger a a + 2a^\dagger a + 1/4\right)$ to get

$$\langle\alpha|H^2|\alpha\rangle = (\hbar\omega)^2\left(|\alpha|^4 + 2|\alpha|^2 + 1/4\right). \tag{7.115}$$

This leads to

$$\Delta H = \sqrt{\langle H^2\rangle - \langle H\rangle^2} = \hbar\omega|\alpha|. \tag{7.116}$$

This is a very small energy uncertainty compared to the energy expectation value, with

$$\Delta H/\langle H\rangle \approx 1/|\alpha| \ll 1, \tag{7.117}$$

since we found $1/|\alpha|$ of order 10^{-16} in our numerical example. Thus, the energy distribution is peaked extremely sharply about the classical value.

We can likewise compute the expectation value and uncertainty of the position. First,

$$\langle\alpha|X|\alpha\rangle = \sqrt{\frac{\hbar}{2m\omega}}\langle\alpha|(a^\dagger + a)|\alpha\rangle = \sqrt{\frac{\hbar}{2m\omega}}(\alpha^* + \alpha) = \sqrt{\frac{2\hbar}{m\omega}}\mathrm{Re}[\alpha]. \tag{7.118}$$

Also, one has

$$\begin{aligned} \langle\alpha|X^2|\alpha\rangle &= \frac{\hbar}{2m\omega}\langle\alpha|(a^\dagger + a)^2|\alpha\rangle = \frac{\hbar}{2m\omega}\langle\alpha|(a^{\dagger 2} + 2a^\dagger a + a^2 + 1)|\alpha\rangle \\ &= \frac{\hbar}{2m\omega}\left[(\alpha^* + \alpha)^2 + 1\right]. \end{aligned} \tag{7.119}$$

It follows that

$$\Delta X = \sqrt{\langle\alpha|X^2|\alpha\rangle - \langle\alpha|X|\alpha\rangle^2} = \sqrt{\frac{\hbar}{2m\omega}}. \tag{7.120}$$

Remarkably, this does not depend on α at all, and in fact is exactly the same as for the ground state, making it ridiculously tiny by macroscopic standards. The interpretation of $\Delta X/\langle X\rangle \ll 1$ is that the position wavefunction is extremely sharply peaked about its expectation value. In a similar way, one can compute the expectation value and the uncertainty of the momentum:

$$\langle\alpha|P|\alpha\rangle = \sqrt{2m\hbar\omega}\,\mathrm{Im}[\alpha], \tag{7.121}$$

$$\Delta P = \sqrt{\frac{m\hbar\omega}{2}}. \tag{7.122}$$

The latter is again independent of α and is the same as in the ground state. Putting these results together, the product of uncertainties in position and momentum is

$$(\Delta X)(\Delta P) = \hbar/2, \tag{7.123}$$

which is exactly as small as it could possibly be, consistent with the uncertainty relation.

The preceding results suggest that there is some close relation between the ground state and the coherent states, since they have exactly the same uncertainties in position and momentum. To see this, let us work out the wavefunction of the coherent state $|\alpha\rangle$. We start with

$$\psi_\alpha(x) = \langle x|\alpha\rangle = e^{-|\alpha|^2/2}\langle x|e^{\alpha a^\dagger}|0\rangle \tag{7.124}$$

from (7.111). Next, we write

$$\alpha a^\dagger = A + B \tag{7.125}$$

where, from the definition of $a^\dagger$ in (7.63),

$$A = \alpha\sqrt{\frac{m\omega}{2\hbar}}X, \qquad B = -i\frac{\alpha}{\sqrt{2\hbar\omega m}}P. \tag{7.126}$$

Since $[A, B] = \alpha^2/2$ is a constant, we can use the Baker–Campbell–Hausdorff formula (2.86), repeated here for convenience as $e^{A+B} = e^A e^B e^{-[A,B]/2}$, to obtain

$$\psi_\alpha(x) = e^{-|\alpha|^2/2}e^{-\alpha^2/4}\exp\left(\sqrt{\frac{m\omega}{2\hbar}}\alpha x\right)\langle x|\exp\left(-i\frac{\alpha}{\sqrt{2\hbar\omega m}}P\right)|0\rangle. \tag{7.127}$$

The next step is to notice that the exponential operator inside the matrix element has the same form as the translation operator $T(a)$ defined in (5.16), with the real distance a replaced by the (complex, in general) number $\alpha\sqrt{\hbar/2\omega m}$. Therefore, we can invoke (5.23) to find

$$\langle x|\exp\left(-i\frac{\alpha}{\sqrt{2\hbar\omega m}}P\right)|0\rangle = \psi_0\left(x-\alpha\sqrt{\frac{\hbar}{2\omega m}}\right), \tag{7.128}$$

where $\psi_0(x)$ is the ground-state wavefunction, found in (7.55). Using this in (7.127), and rearranging the exponentials, gives

$$\psi_\alpha(x) = e^{(\alpha^2-|\alpha|^2)/2}\left(\frac{m\omega}{\pi\hbar}\right)^{1/4}\exp\left[-\frac{m\omega}{2\hbar}\left(x-\alpha\sqrt{\frac{2\hbar}{\omega m}}\right)^2\right]. \tag{7.129}$$

Now, writing α in terms of its real and imaginary parts, and then expressing them in terms of the position and momentum expectation values using (7.118) and (7.121), the wavefunction takes the simple form

$$\psi_\alpha(x) = e^{i\theta}e^{i\langle P\rangle x/\hbar}\psi_0(x-\langle X\rangle), \tag{7.130}$$

where $\theta = -\text{Re}[\alpha]\text{Im}[\alpha]$. Up to the physically irrelevant global phase provided by θ, the coherent state wavefunction is the same as the ground state, but displaced by the position expectation value and multiplied by the position-dependent phase factor that encodes the momentum expectation value.

So far, we have considered a coherent state $|\alpha\rangle$ at a fixed time t. Now let us investigate the time evolution of the state and its properties. Since the Hamiltonian is independent of time, the solution of the Schrödinger equation tells us that at time t the state is

$$\begin{aligned}
e^{-iHt/\hbar}|\alpha\rangle &= e^{-iHt/\hbar}e^{-|\alpha|^2/2}\sum_{n=0}^{\infty}\frac{\alpha^n}{\sqrt{n!}}|n\rangle = e^{-|\alpha|^2/2}\sum_{n=0}^{\infty}\frac{\alpha^n}{\sqrt{n!}}e^{-i\omega(n+1/2)t}|n\rangle \\
&= e^{-i\omega t/2}e^{-|\alpha|^2/2}\sum_{n=0}^{\infty}\frac{(\alpha e^{-i\omega t})^n}{\sqrt{n!}}|n\rangle \\
&= e^{-i\omega t/2}|\alpha e^{-i\omega t}\rangle.
\end{aligned} \tag{7.131}$$

This shows that the coherent state evolves to another coherent state in which α has changed[2] to $\alpha e^{-i\omega t}$. So, in the time-dependent coherent state, the magnitude of α does not change, and the phase of α rotates linearly with time, at the natural angular frequency of the harmonic oscillator.

Suppose that at time $t = 0$, we start with $\alpha = |\alpha|e^{-i\phi_0}$, so that at a general time t,

$$\alpha = |\alpha|e^{-i(\omega t+\phi_0)}. \tag{7.132}$$

[2] The state ket has also acquired an irrelevant global phase factor $e^{-i\omega t/2}$. In contrast, the complex phase $e^{-i\omega t}$ multiplying α (inside the ket symbol) is certainly physically relevant, as is clear from (7.118) and (7.121).

Then, using the results from (7.118) and (7.121), we get

$$\langle X\rangle(t) = \sqrt{\frac{2\hbar}{m\omega}}\,\mathrm{Re}\left[|\alpha|e^{-i(\omega t+\phi_0)}\right] = x_0\cos(\omega t+\phi_0), \tag{7.133}$$

$$\langle P\rangle(t) = \sqrt{2m\hbar\omega}\,\mathrm{Im}\left[|\alpha|e^{-i(\omega t+\phi_0)}\right] = -m\omega x_0\sin(\omega t+\phi_0), \tag{7.134}$$

where we have defined

$$x_0 = \sqrt{\frac{2\hbar}{m\omega}}|\alpha|. \tag{7.135}$$

This establishes the connection between the classical trajectory of the harmonic oscillator and the parameter α of the coherent state that most closely resembles it. Not only is the coherent state for a harmonic oscillator as close as possible to the classical limit, but we see, by comparing (7.133) and (7.134) to (7.5), that the expectation values of the position and momentum evolve in exactly the same way as their classical counterparts. The uncertainties do not change with time. To summarize the properties of a coherent state,

$$\langle H\rangle = \hbar\omega(|\alpha|^2+1/2), \qquad \Delta H = \hbar\omega|\alpha|, \tag{7.136}$$

$$\langle X\rangle(t) = x_{\mathrm{cl}}(t), \qquad \Delta X = \sqrt{\hbar/2m\omega}, \tag{7.137}$$

$$\langle P\rangle(t) = p_{\mathrm{cl}}(t), \qquad \Delta P = \sqrt{m\hbar\omega/2}. \tag{7.138}$$

The harmonic oscillator potential evidently has a remarkable "focusing" property, such that the coherent state wavefunctions do not spread out at all as they evolve in time, unlike the case for the free particle Gaussian states studied in Sect. 6.2.

There is a simple (and even practical) way to prepare a coherent state for the harmonic oscillator. Suppose we temporarily apply a constant force f, acting in the positive x direction, so that the new Hamiltonian is

$$H_f = \frac{P^2}{2m} + \frac{1}{2}m\omega^2X^2 - fX. \tag{7.139}$$

By defining a shifted position operator

$$X' = X - \frac{f}{m\omega^2}, \tag{7.140}$$

the Hamiltonian can be rewritten as

$$H_f = \frac{P^2}{2m} + \frac{1}{2}m\omega^2X'^2 - \frac{f^2}{2m\omega^2}. \tag{7.141}$$

The constant shift in the position operator does not affect the commutation relation,

$$[X', P] = i\hbar, \tag{7.142}$$

so the solution of the eigenvalue problem for H_f proceeds exactly as for the Hamiltonian with $f = 0$. The energy eigenvalues will be the same, but lowered by the constant amount $-f^2/2m\omega^2$. More importantly for our present purposes, the ground state $|0\rangle_f$ of H_f is defined by the property that it is annihilated by the operator

$$a' = \sqrt{\frac{m\omega}{2\hbar}} X' + i \frac{1}{\sqrt{2\hbar\omega m}} P = a - \frac{f}{\sqrt{2\hbar\omega^3 m}}, \tag{7.143}$$

where a is the annihilation operator for H with $f = 0$. This implies that the state $|0\rangle_f$ is exactly a coherent state of the original Hamiltonian H:

$$a|0\rangle_f = \alpha|0\rangle_f, \tag{7.144}$$

where the eigenvalue

$$\alpha = \frac{f}{\sqrt{2\hbar\omega^3 m}} \tag{7.145}$$

is a real number.

Thus, a coherent state of the harmonic oscillator can be prepared as follows. First, apply a constant force f, shifting the equilibrium position of the mass to the point $x_0 = f/m\omega^2$. Next, allow the oscillator to settle into the ground state of the new Hamiltonian H_f, with the same uncertainties in position and momentum as the ground state of H. Finally, we release the mass by removing the force f. At the instant that the mass is released, it will find itself in a coherent state of H, with α initially real and given by (7.145). It will therefore evolve in time as we have already seen, remaining in a coherent state as α acquires a nontrivial phase $e^{-i\omega t}$. The subsequent behavior is as close as possible to what would happen in classical mechanics if we displaced the oscillator and then released it from rest.

7.5 Three-Dimensional Harmonic Oscillator

We now return to the case of a three-dimensional harmonic oscillator, with Hamiltonian given by (7.17). Rewriting it without the tildes (which just denoted a particular choice of rectangular coordinates), it is

$$H = H_x + H_y + H_z, \tag{7.146}$$

where

$$H_x = \frac{P_x^2}{2m} + \frac{1}{2} m\omega_x^2 X^2 = \hbar\omega_x(a_x^\dagger a_x + 1/2), \tag{7.147}$$

with exactly analogous expressions for H_y and H_z. Here, we have made use of the experience of Sect. 7.3 to write

$$a_x = \sqrt{\frac{m\omega_x}{2\hbar}} X + i \frac{1}{\sqrt{2\hbar\omega_x m}} P_x, \tag{7.148}$$

$$a_y = \sqrt{\frac{m\omega_y}{2\hbar}} Y + i \frac{1}{\sqrt{2\hbar\omega_y m}} P_y, \tag{7.149}$$

etc. Now H_x and H_y and H_z all commute with each other, and are Hermitian. One can therefore find an orthobasis consisting of their common eigenstates, denoted $|n_x, n_y, n_z\rangle$, where operators with the x subscript just ignore the n_y and n_z labels, and similarly for operators with the y and z subscripts. Thus,

$$\begin{aligned} H_x|n_x, n_y, n_z\rangle &= E_x|n_x, n_y, n_z\rangle, \qquad H_y|n_x, n_y, n_z\rangle = E_y|n_x, n_y, n_z\rangle, \\ H_z|n_x, n_y, n_z\rangle &= E_z|n_x, n_y, n_z\rangle. \end{aligned} \tag{7.150}$$

(This can also be viewed as the tensor product of three one-dimensional harmonic oscillators.) From Sect. 7.3, we have already solved the eigenvalue problem for each H_j, and so we know that the allowed energies for each of $j = x, y, z$ are

$$E_j = \hbar\omega_j(n_j + 1/2), \qquad n_j = 0, 1, 2, \ldots, \tag{7.151}$$

and that the corresponding eigenkets satisfy, for example:

$$a_x|n_x, n_y, n_z\rangle = \sqrt{n_x}\,|n_x - 1, n_y, n_z\rangle, \tag{7.152}$$

$$a_x^\dagger|n_x, n_y, n_z\rangle = \sqrt{n_x + 1}\,|n_x + 1, n_y, n_z\rangle, \tag{7.153}$$

$$a_y|n_x, n_y, n_z\rangle = \sqrt{n_y}\,|n_x, n_y - 1, n_z\rangle, \tag{7.154}$$

etc. Since the labels n_x, n_y, and n_z uniquely specify the states of an orthobasis, the set $\{H_x, H_y, H_z\}$ forms a CSCO for this problem.

The state $|n_x, n_y, n_z\rangle$ has total energy eigenvalue

$$E = E_x + E_y + E_z = \hbar\omega_x(n_x + 1/2) + \hbar\omega_y(n_y + 1/2) + \hbar\omega_z(n_z + 1/2), \tag{7.155}$$

and its wavefunction is just a product of the wavefunctions for the one-dimensional harmonic oscillator,

$$\psi_{n_x,n_y,n_z}(\boldsymbol{r}) = \langle \boldsymbol{r}|n_x, n_y, n_z\rangle = \psi_{n_x,\omega_x}(x)\,\psi_{n_y,\omega_y}(y)\,\psi_{n_z,\omega_z}(z), \tag{7.156}$$

where each of the ψ_{n_j,ω_j} functions can be obtained from (7.53) in the obvious way.

In the special case of the isotropic three-dimensional harmonic oscillator, $\omega_x = \omega_y = \omega_z = \omega$, the system is invariant under arbitrary rotations about the origin. This symmetry is reflected in a degeneracy of the energy spectrum, which is now

$$E_n = \hbar\omega(n + 3/2), \tag{7.157}$$

for states $|n, u\rangle$, where $n = n_x + n_y + n_z = 0, 1, 2, \ldots$ and $u = 1, \ldots, g_n$ is a degeneracy label for the energy eigenvalue E_n. To find the degeneracy g_n of each energy level, we need to know how many ways there are to choose three non-negative integers that add up to n. The ground state with $n = 0$ is unique, as it can only be achieved with the lone combination $n_x = n_y = n_z = 0$, so $g_0 = 1$. The first excited energy level can be achieved in three different ways, by taking one of n_x, n_y, n_z to be 1, and the others to be 0, so $g_1 = 3$. Similar straightforward counting reveals that $g_2 = 6$ and $g_3 = 10$. More generally, this combinatorics problem can be solved using the dots and lines trick: one arranges n dots in a row, and splits them into three groups by placing two vertical lines as shown below.

$$\underbrace{\bullet\bullet\cdots\bullet}_{n_x}|\underbrace{\bullet\bullet\cdots\bullet}_{n_y}|\underbrace{\bullet\bullet\cdots\bullet}_{n_z} \tag{7.158}$$

The degeneracy of the energy eigenvalue E_n is therefore the number of ways of arranging a row of $n + 2$ objects, of which n are identical and 2 are identical,

$$g_n = \frac{(n+2)!}{n!\,2!} = \frac{1}{2}(n+1)(n+2). \tag{7.159}$$

There is another natural choice of CSCO that one can use for the isotropic three-dimensional harmonic oscillator, consisting of the total Hamiltonian H and two observables associated with the angular momentum. We will discuss this approach in Sect. 10.5.

Exercises

7.1 For a particle of mass m in a one-dimensional harmonic oscillator potential with angular frequency ω and minimum at the origin, use ladder operators to calculate the following matrix elements for energy eigenstates $|n\rangle$ and $|k\rangle$ where n and k are non-negative integers. (Your answers should make use of the Kronecker delta symbol.)

$$\langle k|a|n\rangle, \quad \langle k|a^\dagger|n\rangle, \quad \langle k|X|n\rangle, \quad \langle k|P|n\rangle, \quad \langle k|X^2|n\rangle, \quad \langle k|P^2|n\rangle, \quad \langle k|H|n\rangle.$$

Make note of how the parity selection rules apply to each of your results.

7.2 Consider a particle of mass m in a one-dimensional harmonic oscillator potential with angular frequency ω and minimum at the origin.

(a) Use ladder operators to calculate $\langle n|X^k|n\rangle$ and $\langle n|P^k|n\rangle$, for $k = 0, 1, 2, 3, 4, 5$, with n arbitrary. Make note of how the parity selection rule applies to your results.
(b) Find the expectation values of the kinetic and potential energies for the state $|n\rangle$, and show that they are equal. [This illustrates the virial theorem; see (3.44).]
(c) What is $(\Delta X)(\Delta P)$ for the state $|n\rangle$?

7.3 Use the raising and lowering identities (7.78) and (7.79) and the representation of a and $a^\dagger$ in (7.96) to derive the Hermite polynomial identities (7.46) and (7.47).

7.4 Consider a particle of mass m in a one-dimensional harmonic oscillator potential with angular frequency ω and minimum at the origin. For the observable $A = PX + XP$, find:

(a) The expression in simplest form for A in terms of the usual ladder operators.
(b) The expectation value for A for the nth excited energy eigenstate $|n\rangle$.
(c) The uncertainty for A in the state $|n\rangle$.

7.5 At time $t = 0$, a harmonic oscillator is in the state $|\psi(0)\rangle = \frac{1}{\sqrt{2}}|0\rangle + \frac{1}{\sqrt{2}}|1\rangle$ in the usual energy basis.

(a) Find the state at later times, $|\psi(t)\rangle$, in the energy basis.
(b) By direct calculation using the results of part (a) (without appealing to Ehrenfest's Theorem), find the expectation values $\langle X\rangle$ and $\langle P\rangle$ as functions of time t.
(c) Check that your results from part (b) obey Ehrenfest's Theorem.
(d) At time t, compute the probability that a measurement of the position yields $x > 0$. Evaluate the minimum and the maximum values of this probability numerically.

7.6 At time $t = 0$, a harmonic oscillator has wavefunction $\psi(x, 0) = cx^2 \exp(-m\omega x^2/2\hbar)$, where c is a positive real constant.

(a) By requiring the wavefunction to be normalized, find the constant c.
(b) If the energy is measured, what are the possible outcomes and their probabilities?
(c) Find the state ket $|\psi(t)\rangle$ in the energy basis, as a function of time.
(d) Compute the expectation values $\langle X\rangle$, $\langle P\rangle$, $\langle X^2\rangle$, and $\langle P^2\rangle$, and the uncertainties ΔX and ΔP, each as a function of time.
(e) At time t, what is the probability that a measurement of the position yields $x > 0$?

7.7 Consider a particle of mass m moving in one dimension in the potential

$$V(x) = \begin{cases} \frac{1}{2}m\omega^2x^2 & \text{for } x > 0, \\ \infty & \text{for } x \leq 0. \end{cases} \tag{7.160}$$

Find all of the energy eigenvalues and the corresponding unit-normalized wavefunctions. [Hint: try to make use of calculations already done, rather than new ones.]

7.8 Consider the following Hamiltonian for a particle of mass m moving in one dimension:

$$H = \frac{1}{2m}P^2 + \frac{1}{2}m\omega_0^2X^2 + \frac{1}{2}\omega_1(XP + PX). \tag{7.161}$$

(a) Treating the system classically, find Hamilton's equations of motion and use them to find the general solution for the position as a function of time. What does this tell you about the nature of the system?
(b) Write H in terms of the usual creation and annihilation operators $a^\dagger$ and a.
(c) Let us define new creation and annihilation operators $b^\dagger$ and b by

$$a^\dagger = ub^\dagger + ivb, \tag{7.162}$$

$$a = ub - ivb^\dagger, \tag{7.163}$$

where u and v are real numbers. (In general, this type of linear mixing of creation and annihilation operators is called a **Bogoliubov transformation**.) From this definition, compute $[a, a^\dagger]$ assuming $[b, b^\dagger] = 1$, in order to find a condition that must be satisfied by u and v.
(d) Now rewrite H in terms of $b^\dagger$ and b. Assuming that $\omega_1 < \omega_0$, solve for u and v such that $H = \hbar\Omega(b^\dagger b + 1/2)$, and find Ω in simplest form. Compare your result to that of part (a). What happens, in the classical version, and in the quantum version, if $\omega_1 > \omega_0$?

7.9 Consider the coherent states of the harmonic oscillator in Sect. 7.4.

(a) For two different coherent states $|\alpha\rangle$ and $|\beta\rangle$, compute $\langle\beta|\alpha\rangle$ and $|\langle\beta|\alpha\rangle|^2$ in simplest form. (Your answers should be exponentials. Therefore, they cannot vanish, which shows that the coherent states do *not* satisfy orthogonality.)
(b) Suppose that the oscillator is in the state $|\alpha\rangle$ at time $t = 0$. Find, in simplest form, the probability $P(t)$ to find it again in the state $|\alpha\rangle$ at a later time t. At what times is $P(t) = 1$?
(c) Suppose that $|\alpha| \gg 1$, as for a macroscopic oscillator. For what small length of time after $t = 0$ does the probability $P(t)$ remain greater than 0.5? [Hint: expand $\cos(\omega t)$ to order t^2.]
(d) What is the minimum value of $P(t)$, and at what time or times is it achieved?

(e) Estimate your answers to parts (c) and (d) numerically for a macroscopic oscillator with $m = 0.2\,\text{kg}$, $\omega = 10\ \text{s}^{-1}$, and $x_0 = 0.1\,\text{m}$.

7.10 We found in Sect. 7.4 that the eigenstates of the lowering operator a are of great interest, so it is natural to inquire about the possible eigenstates of the raising operator. Show that $a^\dagger$ has no normalizable eigenstates.

7.11 Consider a spinless particle of mass m, moving in three dimensions, with

$$H = \frac{1}{2m}(P_x^2 + P_y^2 + P_z^2) + \frac{1}{2}m\omega^2\left(\frac{5}{2}X^2 + \frac{5}{2}Y^2 + 3XY + Z^2\right). \tag{7.164}$$

(a) Consider the change of coordinates $x = cx' + sy'$, $y = -sx' + cy'$, $z = z'$, and similarly for the corresponding operators, where c and s are the cosine and sine of an arbitrary rotation angle. Rewrite the Hamiltonian in terms of the operators X', Y', Z', $P_{x'}$, $P_{y'}$, and $P_{z'}$. (You may use the fact that this is an orthogonal rotation on the coordinate system, so $P_x^2 + P_y^2 + P_z^2 = P_{x'}^2 + P_{y'}^2 + P_{z'}^2$.)
(b) Choose c and s (remembering that $c^2 + s^2 = 1$), so that H will not contain a term proportional to $X'Y'$. Rewrite the Hamiltonian with this choice. (There is more than one valid choice here.)
(c) What are the three smallest energy eigenvalues of H, and what are their degeneracies?

7.12 Two particles labeled 1 and 2 (they are distinguishable, but happen to have the same mass) are governed by a coupled harmonic oscillator Hamiltonian in one dimension,

$$H = \frac{1}{2m}(P_1^2 + P_2^2) + \frac{1}{2}m\omega^2 X_1^2 + \frac{1}{2}m\omega^2 X_2^2 + \frac{1}{2}m\Omega^2(X_1 - X_2)^2, \tag{7.165}$$

where the constant ω parametrizes the restoring force for the particles to the origin, and Ω describes the attractive force between the particles. The operators X_1, P_1 and X_2, P_2 satisfy the canonical commutation relations $[X_1, P_1] = i\hbar$ and $[X_2, P_2] = i\hbar$ and $[X_1, P_2] = [X_2, P_1] = 0$ and $[X_j, X_k] = 0$ and $[P_j, P_k] = 0$ for all $j, k = 1, 2$. Consider the new operators

$$U = \frac{1}{\sqrt{2}}(X_1 + X_2), \qquad V = \frac{1}{\sqrt{2}}(X_1 - X_2), \tag{7.166}$$

$$P_u = \frac{1}{\sqrt{2}}(P_1 + P_2), \qquad P_v = \frac{1}{\sqrt{2}}(P_1 - P_2). \tag{7.167}$$

(a) Derive all of the commutation relations of all pairs of operators from the set U, V, P_u, P_v.
(b) Write the Hamiltonian in terms of the operators U, V, P_u, P_v.

(c) Define appropriate raising and lowering operators for the U, V, P_u, P_v system, so that the Hamiltonian has a simple form in terms of them. From this, infer the eigenvalues of the Hamiltonian, and write down a suitable notation for the energy eigenbasis kets.

(d) Obtain the wavefunction of the ground state in the u, v representation $\psi(u, v)$, and use it to obtain the normalized ground-state wavefunction $\psi(x_1, x_2) = \langle x_1, x_2|\psi\rangle$.

7.13 Consider the isotropic harmonic oscillator for a single particle in two dimensions,

$$H = \frac{1}{2m}(P_x^2 + P_y^2) + \frac{1}{2}m\omega^2(X^2 + Y^2). \tag{7.168}$$

This can be viewed as the sum of independent oscillators for the x and y directions.

(a) Express H and the angular momentum operator $L_z = XP_y - YP_x$ in simplest form in terms of the raising and lowering operators $a_x^\dagger$, a_x, $a_y^\dagger$, a_y for the independent x and y oscillations. Compute the commutator of L_z and H. Are they compatible?

(b) Consider the basis kets

$$|n_x, n_y\rangle = \frac{(a_x^\dagger)^{n_x}}{\sqrt{n_x!}}\frac{(a_y^\dagger)^{n_y}}{\sqrt{n_y!}}|0, 0\rangle, \tag{7.169}$$

where $|0, 0\rangle$ is the ground state. Show that the $|n_x, n_y\rangle$ are eigenstates of energy and parity, and give the corresponding eigenvalues. What is the degeneracy of the nth excited energy level?

(c) Show that the basis kets $|n_x, n_y\rangle$ are not eigenstates of L_z.

(d) Define new raising and lowering operators

$$a_+^\dagger = (a_x^\dagger + ia_y^\dagger)/\sqrt{2}, \qquad a_+ = (a_x - ia_y)/\sqrt{2}, \tag{7.170}$$

$$a_-^\dagger = (a_x^\dagger - ia_y^\dagger)/\sqrt{2}, \qquad a_- = (a_x + ia_y)/\sqrt{2}. \tag{7.171}$$

Compute all of the nonzero commutators involving these four operators. Express H and L_z in terms of them, in normal-ordered form.

(e) Consider the basis kets

$$|n_+, n_-\rangle' = \frac{(a_+^\dagger)^{n_+}}{\sqrt{n_+!}}\frac{(a_-^\dagger)^{n_-}}{\sqrt{n_-!}}|0, 0\rangle. \tag{7.172}$$

Show that these are eigenstates of H and L_z, and give the corresponding eigenvalues. For the lowest three energy levels, write the eigenvalues of H and L_z, and express the corresponding eigenstates as linear combinations of the kets $|n_x, n_y\rangle$. Do H and L_z form a CSCO?

Angular Momentum and Its Representations 8

8.1 The Eigenvalue Problem for Angular Momentum

At the end of Sect. 5.3, we introduced the commutation relation algebra for the components of a generic angular momentum operator $\boldsymbol{J} = \hat{x} J_x + \hat{y} J_y + \hat{z} J_z$,

$$[J_a,\ J_b] = i\hbar \epsilon_{abc} J_c, \qquad (a, b, c = x, y, z), \tag{8.1}$$

with c implicitly summed over. Let us now find the simultaneous eigenvectors and the corresponding eigenvalues of the compatible observables $J^2 = J_x^2 + J_y^2 + J_z^2$ and J_z.

We start by observing that the eigenvalues of J^2 must be non-negative. To prove it, note that for any non-null ket $|\psi\rangle$,

$$\langle\psi|J^2|\psi\rangle = \|J_x|\psi\rangle\|^2 + \|J_y|\psi\rangle\|^2 + \|J_z|\psi\rangle\|^2 \geq 0, \tag{8.2}$$

where $\||v\rangle\|$ denotes the norm of $|v\rangle$, and equality occurs only if all three of the kets $J_x|\psi\rangle$, $J_y|\psi\rangle$, and $J_z|\psi\rangle$ are null. Now, if

$$J^2|\psi\rangle = \lambda|\psi\rangle, \tag{8.3}$$

then it follows that $\langle\psi|J^2|\psi\rangle = \lambda\langle\psi|\psi\rangle \geq 0$, so $\lambda \geq 0$. For reasons to become clear shortly, it is convenient to give λ the name $\hbar^2 j(j+1)$, by defining $j = \sqrt{\lambda/\hbar^2 + 1/4} - 1/2$. Since $\lambda \geq 0$, it follows that $j \geq 0$ also.

Because J^2 and J_z are compatible observables, Theorem 2.19 says that there must be an orthobasis consisting of common eigenstates, $|j, m, u\rangle$, which have J^2 and J_z eigenvalues $\hbar^2 j(j+1)$ and $\hbar m$ respectively, and satisfy orthonormality and completeness relations

S. P. Martin, *Quantum Mechanics*, Graduate Texts in Physics,
https://doi.org/10.1007/978-3-032-21119-4_8

$$\langle j', m', u' | j, m, u \rangle = \delta_{jj'} \delta_{mm'} \delta_{uu'}, \tag{8.4}$$

$$\sum_j \sum_m \sum_u |j, m, u\rangle\langle j, m, u| = I. \tag{8.5}$$

Here u is a possible degeneracy label that will be important in particular cases, but plays no role in the following discussion and so will be suppressed for simplicity. We therefore seek to solve the eigenvalue problem

$$J^2 |j, m\rangle = \hbar^2 j(j+1) |j, m\rangle, \tag{8.6}$$

$$J_z |j, m\rangle = \hbar m |j, m\rangle \tag{8.7}$$

for the allowed eigenvalues j and m, recognizing that there may be more than one sector of such states, distinguished by the suppressed degeneracy label. In the special case of orbital angular momentum, we already found in Sect. 5.3, by requiring the position wavefunction to be single-valued, that the allowed eigenvalues of L_z are $\hbar m$, where m is an integer. However, spin (intrinsic) angular momentum is not associated with a position wavefunction, so that argument does not apply, and m need not be an integer, as we will see.

It is very useful (and not just for the present purpose of solving the eigenvalue problem) to define the **angular momentum raising and lowering operators**, also sometimes known as **ladder operators** for angular momentum,

$$J_+ = J_x + i J_y, \qquad J_- = J_x - i J_y, \tag{8.8}$$

which will play a role similar to $a^\dagger$ and a for the harmonic oscillator. They are not Hermitian, but instead are Hermitian conjugates of each other,

$$(J_+)^\dagger = J_-. \tag{8.9}$$

The inverse relations of (8.8) are

$$J_x = \frac{1}{2}(J_- + J_+), \qquad J_y = \frac{i}{2}(J_- - J_+). \tag{8.10}$$

From these definitions, one finds

$$[J_z, J_\pm] = \pm\hbar J_\pm, \tag{8.11}$$

$$[J_+, J_-] = 2\hbar J_z, \tag{8.12}$$

$$J_+ J_- = J^2 - J_z^2 + \hbar J_z, \tag{8.13}$$

$$J_- J_+ = J^2 - J_z^2 - \hbar J_z. \tag{8.14}$$

From the last two equations, we get

$$J^2 = \frac{1}{2}J_+J_- + \frac{1}{2}J_-J_+ + J_z^2. \tag{8.15}$$

Each of J_+, J_-, and J_z commutes with the total angular momentum squared,

$$[J^2, J_+] = [J^2, J_-] = [J^2, J_z] = 0. \tag{8.16}$$

We now derive some useful facts by studying the ket $J_+|j, m\rangle$. Since J^2 commutes with J_+,

$$J^2 (J_+|j, m\rangle) = \hbar^2 j(j+1) (J_+|j, m\rangle). \tag{8.17}$$

Also,

$$J_z (J_+|j, m\rangle) = [J_z, J_+]|j, m\rangle + J_+J_z|j, m\rangle = \hbar(m+1) (J_+|j, m\rangle). \tag{8.18}$$

Thus, $J_+|j, m\rangle$ is an eigenstate of both J^2 and J_z, with eigenvalues $\hbar^2 j(j+1)$ and $\hbar(m+1)$, respectively. Due to the fact that the kets $|j, m, u\rangle$ were chosen as an orthobasis, we can conclude that, within each sector labeled by u, either $J_+|j, m\rangle$ is proportional to $|j, m+1\rangle$ or else it is the null ket. This is why J_+ is called a raising operator; it raises the J_z eigenvalue. To find out whether $J_+|j, m\rangle$ is the null ket, we compute its squared norm,

$$\begin{aligned} \|J_+|j, m\rangle\|^2 &= \langle j, m|J_-J_+|j, m\rangle = \langle j, m|\big(J^2 - J_z^2 - \hbar J_z\big)|j, m\rangle \\ &= \langle j, m|\left[\hbar^2 j(j+1) - (\hbar m)^2 - \hbar(\hbar m)\right]|j, m\rangle \\ &= \hbar^2[j(j+1) - m(m+1)]. \end{aligned} \tag{8.19}$$

From this we learn two useful things. First, the general properties of the inner product require that the squared norm is non-negative, so allowed j, m must satisfy $j(j+1) - m(m+1) = (j-m)(j+m+1) \geq 0$, and since we learned at the beginning of this section that $j \geq 0$, we can conclude that for every non-null ket $|j, m\rangle$,

$$m \leq j. \tag{8.20}$$

The second useful result from (8.19) is the constant of proportionality between $J_+|j, m\rangle$ and the unit-normalized ket $|j, m+1\rangle$,

$$J_+|j, m\rangle = \hbar\sqrt{j(j+1) - m(m+1)}\,|j, m+1\rangle. \tag{8.21}$$

Here we had to make an arbitrary and unavoidable choice of phase; (8.21) can be taken as the definition of the relative phase between $|j, m+1\rangle$ and $|j, m\rangle$. It follows from the $m = j$ special case of (8.21) that $J_+|j, j\rangle$ is actually the null ket.

Everything in the previous paragraph can be repeated for $J_-|j,m\rangle$. We learn that, for every non-null ket $|j,m\rangle$,

$$m \geq -j \tag{8.22}$$

due to the requirement of positive squared norm, and that $J_-|j,-j\rangle$ is the null ket, and that

$$J_-|j,m\rangle = \hbar\sqrt{j(j+1)-m(m-1)}\,|j,m-1\rangle. \tag{8.23}$$

This justifies calling J_- the lowering operator for (the z component of) angular momentum.

In (8.23), we have made another choice of phase, and we should be careful to make sure that it is consistent with the phase choice made in (8.21). This can be checked by using (8.21) and (8.23) to compute $J_+J_-|j,m\rangle$ and $J_-J_+|j,m\rangle$, and noting that they are equivalent to the results obtained using (8.13) and (8.14), respectively. A nice feature of (8.21) and (8.23) is that the coefficients on the right-hand sides are real and non-negative.

Combining (8.20) and (8.22) tells us that, for any given j, the only possible values of m for physical states $|j,m\rangle$ are in the range

$$-j \leq m \leq j. \tag{8.24}$$

Intuitively, this is the quantum version of the classical statement that the magnitude of the z-component is less than or equal to the magnitude of the angular momentum vector.

We are now ready to prove that all physical states must have $j = n/2$ for some integer n. If we compute $(J_+)^p|j,m\rangle$, it must be proportional to $|j,m+p\rangle$, according to (8.21). For some finite p, this must be the ket $|j,j\rangle$, because otherwise we could keep acting with J_+ to find non-null kets with arbitrarily large J_z eigenvalue, which would contradict (8.24). (Here, we are relying on the fact that (8.21) tells us that $|j,j\rangle$ is the unique non-null ket that is annihilated by J_+.) Therefore, $j = m+p$ for some non-negative integer p. Similarly, if we compute $(J_-)^q|j,m\rangle$, it must be proportional to $|j,m-q\rangle$, according to (8.23). For some finite q, this must be the ket $|j,-j\rangle$; otherwise, we would again contradict (8.24). Therefore, $-j = m-q$ for some non-negative integer q. Putting the results together gives $2j = p+q$, a non-negative integer. The allowed values are

$$j = 0,\ 1/2,\ 1,\ 3/2,\ 2,\ \ldots. \tag{8.25}$$

The fact that $j = m+p$ for some integer p, together with (8.24), also tells us which values of m can give non-null kets $|j,m\rangle$. For each j, they are the $2j+1$ values

$$m = -j,\ -j+1,\ \ldots,\ j-1,\ j. \tag{8.26}$$

For the special case that $\boldsymbol{J} = \boldsymbol{L}$, we already found in Sect. 5.3 that m must be an integer, so in the case of orbital angular momentum the allowed values of $j = l$ are further restricted to the non-negative integers $0, 1, 2, \ldots$. The cases of half-integer[1] j must correspond to something other than orbital angular momenta.

For a given state $|j, m, u\rangle$, the operations of J^2, J_z, J_+, and J_-, given by (8.6), (8.7), (8.21), and (8.23), are independent of u, and can change m, but not j. This has two important implications.

First, if we have a single representative state with a certain j and u, then all other states with the same j and u but other values of m are obtained by acting repeatedly with J_+ or J_-. Typically, u represents the eigenvalue(s) of some observable(s) that, together with J^2 and J_z, form a CSCO for the quantum system. If the Hamiltonian is invariant under rotations, then H can be taken to be one of those observables. The set of $2j + 1$ orthobasis states with common j and u and varying $m = -j, \ldots, j$ is called an **angular momentum multiplet**.

The second observation is that, because the operations of the angular momentum operators J^2, J_z, J_+, and J_- (and J_x, J_y) in the orthobasis $|j, m, u\rangle$ do not depend on u at all, we can work out their matrix representations and operations on kets for each relevant value of j just once, and the results will be applicable to any quantum system with that j.

For a $j = 0$ subspace, we have $J^2 = 0$ and $J_z = J_+ = J_- = J_x = J_y = 0$, so the state space has only one non-null ket $|j = 0, m = 0\rangle$. All angular momentum operators acting on $|j = 0, m = 0\rangle$ give the null ket, so that they are all represented by the 1×1 matrix 0.

8.2 The $j = 1/2$ Representation: Pauli Matrices and Spin

For a $j = 1/2$ subspace, the orthobasis consists of two states with $m = \pm 1/2$, with corresponding two-component vector representations

$$|j=1/2,\ m=1/2\rangle \leftrightarrow \begin{pmatrix}1\\0\end{pmatrix}, \qquad |j=1/2,\ m=-1/2\rangle \leftrightarrow \begin{pmatrix}0\\1\end{pmatrix}. \tag{8.27}$$

The matrix representations of the angular momentum operators can now be constructed from (8.6), (8.7), (8.10), (8.21), and (8.23), with the results

$$J^2 \leftrightarrow \frac{3\hbar^2}{4}\begin{pmatrix}1 & 0\\0 & 1\end{pmatrix}, \qquad J_z \leftrightarrow \frac{\hbar}{2}\begin{pmatrix}1 & 0\\0 & -1\end{pmatrix}, \tag{8.28}$$

[1] The standard term "half-integer" means an odd integer divided by 2. It might be more logical to call this "half-odd-integer" or "integer-plus-half", but it is difficult to fight tradition.

$$J_x \leftrightarrow \frac{\hbar}{2}\begin{pmatrix} 0 & 1 \\ 1 & 0 \end{pmatrix}, \qquad J_y \leftrightarrow \frac{\hbar}{2}\begin{pmatrix} 0 & -i \\ i & 0 \end{pmatrix}. \tag{8.29}$$

$$J_+ \leftrightarrow \hbar\begin{pmatrix} 0 & 1 \\ 0 & 0 \end{pmatrix}, \qquad J_- \leftrightarrow \hbar\begin{pmatrix} 0 & 0 \\ 1 & 0 \end{pmatrix}. \tag{8.30}$$

Some special operator identities that hold only in the $j = 1/2$ case are

$$J_x^2 = J_y^2 = J_z^2 = \frac{\hbar^2}{4}I, \qquad J_+^2 = J_-^2 = 0, \qquad (j = 1/2). \tag{8.31}$$

As noted after (8.26), the fact that $m = \pm 1/2$ is not an integer implies that $\boldsymbol{J}$ cannot be a pure orbital angular momentum.

The case of purely intrinsic angular momentum for a single particle is called **spin**, and we write $\boldsymbol{J} = \boldsymbol{S}$ and $j = s$ in that case. For $s = 1/2$, we say that the particle has spin-1/2, and we often rewrite (8.27) using a simplified notation for the S_z eigenstates,

$$|\uparrow\rangle \leftrightarrow \begin{pmatrix} 1 \\ 0 \end{pmatrix}, \qquad |\downarrow\rangle \leftrightarrow \begin{pmatrix} 0 \\ 1 \end{pmatrix}, \tag{8.32}$$

referred to as spin-up and spin-down, with S_z eigenvalues $\hbar m_s$, where $m_s = 1/2$ and $-1/2$, respectively.

From (8.28) and (8.29), one can deduce that the components of the spin operator written in ket-bra form are

$$S_x = \frac{\hbar}{2}\Big(|\downarrow\rangle\langle\uparrow| + |\uparrow\rangle\langle\downarrow|\Big), \tag{8.33}$$

$$S_y = \frac{i\hbar}{2}\Big(|\downarrow\rangle\langle\uparrow| - |\uparrow\rangle\langle\downarrow|\Big), \tag{8.34}$$

$$S_z = \frac{\hbar}{2}\Big(|\uparrow\rangle\langle\uparrow| - |\downarrow\rangle\langle\downarrow|\Big). \tag{8.35}$$

The standard notation for their matrix representations is

$$\boldsymbol{S} \leftrightarrow \frac{\hbar}{2}\boldsymbol{\sigma}, \tag{8.36}$$

where the components of the vector $\boldsymbol{\sigma}$ are known as the **Pauli matrices**, defined by

$$\sigma_x = \begin{pmatrix} 0 & 1 \\ 1 & 0 \end{pmatrix}, \qquad \sigma_y = \begin{pmatrix} 0 & -i \\ i & 0 \end{pmatrix}, \qquad \sigma_z = \begin{pmatrix} 1 & 0 \\ 0 & -1 \end{pmatrix}. \tag{8.37}$$

The Pauli matrices obey

$$\mathrm{Tr}[\sigma_a] = 0, \tag{8.38}$$

$$\det[\sigma_a] = -1, \tag{8.39}$$

and

$$\sigma_a \sigma_b = \delta_{ab} I + i \epsilon_{abc} \sigma_c, \tag{8.40}$$

for $a, b, c = x, y, z$, where I is the 2×2 identity matrix. This implies the commutation and anticommutation relations

$$[\sigma_a, \sigma_b] = 2i \epsilon_{abc} \sigma_c, \tag{8.41}$$

$$\{\sigma_a, \sigma_b\} = 2\delta_{ab} I. \tag{8.42}$$

For any spatial vector $\boldsymbol{v} = \hat{x} v_x + \hat{y} v_y + \hat{z} v_z$, we have

$$\boldsymbol{v} \cdot \boldsymbol{\sigma} = \begin{pmatrix} v_z & v_x - i v_y \\ v_x + i v_y & -v_z \end{pmatrix}, \tag{8.43}$$

and a formula useful for simplifications,

$$(\boldsymbol{v} \cdot \boldsymbol{\sigma})(\boldsymbol{w} \cdot \boldsymbol{\sigma}) = \boldsymbol{v} \cdot \boldsymbol{w} I + i(\boldsymbol{v} \times \boldsymbol{w}) \cdot \boldsymbol{\sigma}. \tag{8.44}$$

In particular,

$$(\boldsymbol{v} \cdot \boldsymbol{\sigma})(\boldsymbol{v} \cdot \boldsymbol{\sigma}) = v^2 I. \tag{8.45}$$

For a spin-1/2 particle like an electron, the complete state can be written as a linear combination of kets with S_z eigenvalues $\hbar m_s = \hbar/2$ and $-\hbar/2$, as

$$|\psi\rangle = |\psi_\uparrow\rangle \otimes |\uparrow\rangle + |\psi_\downarrow\rangle \otimes |\downarrow\rangle, \tag{8.46}$$

where $|\psi_\uparrow\rangle$ and $|\psi_\downarrow\rangle$ represent the non-spin degrees of freedom of the particle. This state can also be represented as a two-component column vector of kets, called a **spinor**,

$$|\psi\rangle = \begin{pmatrix} |\psi_\uparrow\rangle \\ |\psi_\downarrow\rangle \end{pmatrix}. \tag{8.47}$$

Acting on these two-component spinor kets, the spin operator is again represented by $\boldsymbol{S} \leftrightarrow \frac{\hbar}{2}\boldsymbol{\sigma}$. The bra corresponding to (8.47) is a two-component row spinor,

$$\langle\psi| = \left(\langle\psi_\uparrow| \; \langle\psi_\downarrow|\right). \tag{8.48}$$

The inner product of two states $|\psi\rangle$ and $|\chi\rangle$ is

$$\langle\chi|\psi\rangle = \left(\langle\chi_\uparrow|\ \langle\chi_\downarrow|\right)\begin{pmatrix}|\psi_\uparrow\rangle \\ |\psi_\downarrow\rangle\end{pmatrix} = \langle\chi_\uparrow|\psi_\uparrow\rangle + \langle\chi_\downarrow|\psi_\downarrow\rangle, \tag{8.49}$$

and unit normalization means

$$1 = \langle\psi|\psi\rangle = \langle\psi_\uparrow|\psi_\uparrow\rangle + \langle\psi_\downarrow|\psi_\downarrow\rangle. \tag{8.50}$$

For example, the two-component spinor position wavefunction for an electron is

$$\langle\boldsymbol{r}|\psi\rangle = \begin{pmatrix}\langle\boldsymbol{r}|\psi_\uparrow\rangle \\ \langle\boldsymbol{r}|\psi_\downarrow\rangle\end{pmatrix} = \begin{pmatrix}\psi_\uparrow(\boldsymbol{r}) \\ \psi_\downarrow(\boldsymbol{r})\end{pmatrix}. \tag{8.51}$$

These correspond to orthobasis eigenstates of the CSCO consisting of the observables $(\boldsymbol{R},\ S_z)$. If we impose the usual unit normalization condition for the state, the two spin-component wavefunctions are required to satisfy

$$1 = \int d^3r\ \left(|\psi_\uparrow(\boldsymbol{r})|^2 + |\psi_\downarrow(\boldsymbol{r})|^2\right), \tag{8.52}$$

but they are otherwise independent, in general.

8.3 Matrix Representation for $j = 1$

For a $j = 1$ subspace, the orthobasis of J_z eigenstates consists of three states $|j, m\rangle$ with $m = 1, 0, -1$, with corresponding vector representations

$$|1,1\rangle \leftrightarrow \begin{pmatrix}1\\0\\0\end{pmatrix}, \qquad |1,0\rangle \leftrightarrow \begin{pmatrix}0\\1\\0\end{pmatrix}, \qquad |1,-1\rangle \leftrightarrow \begin{pmatrix}0\\0\\1\end{pmatrix}. \tag{8.53}$$

The matrix representations of the angular momentum operators can now be constructed from (8.6), (8.7), (8.10), (8.21), and (8.23), with the results

$$J^2 \leftrightarrow 2\hbar^2\begin{pmatrix}1&0&0\\0&1&0\\0&0&1\end{pmatrix}, \qquad J_z \leftrightarrow \hbar\begin{pmatrix}1&0&0\\0&0&0\\0&0&-1\end{pmatrix}, \tag{8.54}$$

$$J_+ \leftrightarrow \sqrt{2}\hbar\begin{pmatrix}0&1&0\\0&0&1\\0&0&0\end{pmatrix}, \qquad J_- \leftrightarrow \sqrt{2}\hbar\begin{pmatrix}0&0&0\\1&0&0\\0&1&0\end{pmatrix}, \tag{8.55}$$

$$J_x \leftrightarrow \frac{\hbar}{\sqrt{2}} \begin{pmatrix} 0 & 1 & 0 \\ 1 & 0 & 1 \\ 0 & 1 & 0 \end{pmatrix}, \qquad J_y \leftrightarrow \frac{\hbar}{\sqrt{2}} \begin{pmatrix} 0 & -i & 0 \\ i & 0 & -i \\ 0 & i & 0 \end{pmatrix}. \tag{8.56}$$

As an example of the use of these matrices, suppose that a system is in an eigenstate of J_z, and we want to know the possible results and their probabilities if J_x is measured. To answer this question, we first find the eigenvalues and eigenstates of J_x, expressed in the orthobasis of J_z eigenvectors of (8.53). Since there is nothing special about the x direction as opposed to the z direction, we know even without computing the characteristic equation for J_x that its eigenvalues must be the same as those of J_z, namely $\hbar$, 0, and $-\hbar$. Then, solving for the eigenvectors of the matrix representation for J_x, one finds the normalized kets

$$|J_x = \hbar\rangle = \tfrac{1}{2}|1, 1\rangle + \tfrac{1}{\sqrt{2}}|1, 0\rangle + \tfrac{1}{2}|1, -1\rangle, \tag{8.57}$$

$$|J_x = 0\rangle = \tfrac{1}{\sqrt{2}}|1, 1\rangle - \tfrac{1}{\sqrt{2}}|1, -1\rangle, \tag{8.58}$$

$$|J_x = -\hbar\rangle = \tfrac{1}{2}|1, 1\rangle - \tfrac{1}{\sqrt{2}}|1, 0\rangle + \tfrac{1}{2}|1, -1\rangle. \tag{8.59}$$

As a check, these are mutually orthogonal, as required (Theorem 2.14) by the fact that they are eigenkets of a Hermitian operator with different eigenvalues. Then, for example, the probabilities that a measurement in the state $|1, 1\rangle$ will yield the results $J_x = \hbar$, $J_x = 0$, and $J_x = -\hbar$ are, by applying Postulate 4,

$$\mathcal{P}_{J_x=\hbar} = |\langle J_x = \hbar|1, 1\rangle|^2 = 1/4, \tag{8.60}$$

$$\mathcal{P}_{J_x=0} = |\langle J_x = 0|1, 1\rangle|^2 = 1/2, \tag{8.61}$$

$$\mathcal{P}_{J_x=-\hbar} = |\langle J_x = -\hbar|1, 1\rangle|^2 = 1/4. \tag{8.62}$$

8.4 Matrix Representation for Arbitrary j

For a subspace with arbitrary j, the orthobasis of J_z eigenstates consists of $2j + 1$ states $|j, m\rangle$ with $m = j,\ j{-}1, \ldots, -j{+}1,\ -j$. The corresponding column vector representations of these states, in that order, are

$$|j,\ j\rangle \leftrightarrow \begin{pmatrix} 1 \\ 0 \\ \vdots \\ 0 \end{pmatrix}, \quad |j,\ j{-}1\rangle \leftrightarrow \begin{pmatrix} 0 \\ 1 \\ \vdots \\ 0 \end{pmatrix}, \quad \ldots, \quad |j,\ -j\rangle \leftrightarrow \begin{pmatrix} 0 \\ 0 \\ \vdots \\ 1 \end{pmatrix}. \tag{8.63}$$

Again, (8.6), (8.7), (8.21), and (8.23) provide the matrix elements of the angular momentum operators. The $(2j+1)\times(2j+1)$ matrix representations have the forms

$$J^2 \leftrightarrow \hbar^2 j(j+1)\begin{pmatrix} 1 & 0 & \dots & 0 \\ 0 & 1 & \dots & 0 \\ \vdots & \vdots & \ddots & \vdots \\ 0 & 0 & \dots & 1 \end{pmatrix}, \qquad J_z \leftrightarrow \hbar \begin{pmatrix} j & 0 & \dots & 0 \\ 0 & j-1 & \dots & 0 \\ \vdots & \vdots & \ddots & \vdots \\ 0 & 0 & \dots & -j \end{pmatrix}, \tag{8.64}$$

$$J_+ \leftrightarrow \hbar \begin{pmatrix} 0 & \bullet & 0 & \dots & 0 & 0 \\ 0 & 0 & \bullet & \dots & 0 & 0 \\ 0 & 0 & 0 & \dots & 0 & 0 \\ \vdots & \vdots & \vdots & \ddots & \vdots & \vdots \\ 0 & 0 & 0 & \dots & 0 & \bullet \\ 0 & 0 & 0 & \dots & 0 & 0 \end{pmatrix}, \qquad J_- \leftrightarrow \hbar \begin{pmatrix} 0 & 0 & 0 & \dots & 0 & 0 \\ \bullet & 0 & 0 & \dots & 0 & 0 \\ 0 & \bullet & 0 & \dots & 0 & 0 \\ \vdots & \vdots & \vdots & \ddots & \vdots & \vdots \\ 0 & 0 & 0 & \dots & 0 & 0 \\ 0 & 0 & 0 & \dots & \bullet & 0 \end{pmatrix}, \tag{8.65}$$

where the $\bullet$'s are the only nonzero elements, which lie just above and below the main diagonal for J_+ and J_-, respectively. Using (8.10), J_x and J_y have matrix elements

$$(J_x)_{m',m} = \frac{\hbar}{2}\Big[\sqrt{j(j+1)-m(m-1)}\,\delta_{m',m-1} + \sqrt{j(j+1)-m(m+1)}\,\delta_{m',m+1}\Big], \tag{8.66}$$

$$(J_y)_{m',m} = i\frac{\hbar}{2}\Big[\sqrt{j(j+1)-m(m-1)}\,\delta_{m',m-1} - \sqrt{j(j+1)-m(m+1)}\,\delta_{m',m+1}\Big], \tag{8.67}$$

and so have the forms

$$J_x \leftrightarrow \hbar \begin{pmatrix} 0 & \bullet & 0 & \dots & 0 & 0 \\ \bullet & 0 & \bullet & \dots & 0 & 0 \\ 0 & \bullet & 0 & \dots & 0 & 0 \\ \vdots & \vdots & \vdots & \ddots & \vdots & \vdots \\ 0 & 0 & 0 & \dots & 0 & \bullet \\ 0 & 0 & 0 & \dots & \bullet & 0 \end{pmatrix}, \qquad J_y \leftrightarrow i\hbar \begin{pmatrix} 0 & -\bullet & 0 & \dots & 0 & 0 \\ \bullet & 0 & -\bullet & \dots & 0 & 0 \\ 0 & \bullet & 0 & \dots & 0 & 0 \\ \vdots & \vdots & \vdots & \ddots & \vdots & \vdots \\ 0 & 0 & 0 & \dots & 0 & -\bullet \\ 0 & 0 & 0 & \dots & \bullet & 0 \end{pmatrix}, \tag{8.68}$$

where each pair of $\bullet$'s on opposite sides of the main diagonal is composed of equal real positive numbers, consistent with the Hermiticity of these operators.

8.5 Matrix Representations for Unitary Rotation Operators

In Sect. 5.3, we discussed the unitary operators that act on the Hilbert space to generate rotations parameterized by $\boldsymbol{\alpha} = \hat{n}\alpha$, where the unit vector $\hat{n}$ is the axis of rotation, and α is the magnitude of the rotation angle:

$$U(\boldsymbol{\alpha}) = \exp\left(-i\boldsymbol{\alpha} \cdot \boldsymbol{J}/\hbar\right). \tag{8.69}$$

The matrix representations for these unitary operators are called **rotation matrices**, and are traditionally denoted as $D^{(j)}(\boldsymbol{\alpha})$. They are defined by

$$\langle j', m'|U(\boldsymbol{\alpha})|j, m\rangle = \delta_{jj'} D^{(j)}_{m'm}(\boldsymbol{\alpha}) \qquad (m, m' = -j, \ldots, j). \tag{8.70}$$

By completeness of the angular momentum eigenstates,

$$U(\boldsymbol{\alpha})|j, m\rangle = \sum_{m'=-j}^{j} |j, m'\rangle D^{(j)}_{m'm}(\boldsymbol{\alpha}). \tag{8.71}$$

Note that each $D^{(j)}(\boldsymbol{\alpha})$ is a $(2j+1) \times (2j+1)$ matrix, which can mix different m values but keeps j fixed.

The set of matrices $D^{(j)}(\boldsymbol{\alpha})$ for fixed j forms an **irreducible representation** of the rotation group. If you do one rotation $\boldsymbol{\alpha}$ followed by another $\boldsymbol{\beta}$, the result is always some other rotation $\boldsymbol{\gamma}$,

$$U(\boldsymbol{\gamma}) = U(\boldsymbol{\beta})U(\boldsymbol{\alpha}). \tag{8.72}$$

Then, applying completeness gives

$$D^{(j)}_{m'm}(\boldsymbol{\gamma}) = \sum_{m''=-j}^{j} D^{(j)}_{m'm''}(\boldsymbol{\beta}) D^{(j)}_{m''m}(\boldsymbol{\alpha}). \tag{8.73}$$

The unitarity of the operator $U(\boldsymbol{\alpha})$ and the fact that $U(\boldsymbol{\alpha})^{-1} = U(-\boldsymbol{\alpha})$ imply

$$D^{(j)}_{m'm}(-\boldsymbol{\alpha}) = \left[D^{(j)}_{mm'}(\boldsymbol{\alpha})\right]^*. \tag{8.74}$$

For small j, the exponentiation in (8.69) can be done explicitly in the matrix representation. For the trivial $j = 0$ case, we have $D^{(0)}(\boldsymbol{\alpha}) = e^0 = 1$, the unit 1×1 matrix, independent of α. This corresponds to the fact that states with zero angular momentum are invariant under rotations.

For $j = 1/2$, in terms of the vector of Pauli matrices $\boldsymbol{\sigma}$,

$$D^{(1/2)}(\boldsymbol{\alpha}) = \exp(-i\boldsymbol{\alpha}\cdot\boldsymbol{\sigma}/2) = \sum_{k=0}^{\infty}\frac{1}{k!}(-i\alpha\hat{n}\cdot\boldsymbol{\sigma}/2)^k. \tag{8.75}$$

Using (8.45), we have $(\hat{n}\cdot\boldsymbol{\sigma})^2 = I$ (the 2×2 unit matrix), so that the terms in (8.75) have a recurring structure. This allows us to resum the even and odd terms of the infinite sum separately, giving

$$D^{(1/2)}(\boldsymbol{\alpha}) = \cos(\alpha/2)I - i\hat{n}\cdot\boldsymbol{\sigma}\sin(\alpha/2). \tag{8.76}$$

As an example, suppose we have an electron spin that has been measured to be along the $\hat{z}$ direction, so that the state ket is $|\psi\rangle = |\uparrow\rangle$, or in the matrix representation,

$$|\psi\rangle \leftrightarrow \begin{pmatrix}1\\0\end{pmatrix}. \tag{8.77}$$

Now let us rotate this state by an angle θ about the $\hat{y}$ axis. The matrix representation of the rotation is

$$D^{(1/2)}(\hat{y}\theta) = \cos(\theta/2)I - i\sigma_y\sin(\theta/2) = \begin{pmatrix}\cos(\theta/2) & -\sin(\theta/2)\\ \sin(\theta/2) & \cos(\theta/2)\end{pmatrix}, \tag{8.78}$$

and so the rotated state has

$$|\psi'\rangle \leftrightarrow \begin{pmatrix}\cos(\theta/2)\\ \sin(\theta/2)\end{pmatrix}, \tag{8.79}$$

or $|\psi'\rangle = \cos(\theta/2)|\uparrow\rangle + \sin(\theta/2)|\downarrow\rangle$. You can check that this is indeed an eigenstate of the rotated operator

$$S'_z = \cos\theta\, S_z + \sin\theta\, S_x \leftrightarrow \frac{\hbar}{2}\begin{pmatrix}\cos\theta & \sin\theta\\ \sin\theta & -\cos\theta\end{pmatrix}, \tag{8.80}$$

with eigenvalue $+\hbar/2$. If we prepare the system in a spin eigenstate, then the probability to find the spin oriented in a direction at an angle θ with respect to the original direction is

$$\mathcal{P} = |\langle\psi'|\psi\rangle|^2 = \cos^2(\theta/2). \tag{8.81}$$

As a check, this is 1 for $\theta = 0$ or 2π, and 0 for $\theta = \pi$. The probability to find the spin oriented in any particular direction at a right angle to the original spin direction is $\cos^2(\pi/4) = 1/2$.

Abstracting this to a slightly more general case for a spin-1/2 system, let $|\hat{n}\rangle$ be the eigenstate of an arbitrary spin component $\hat{n} \cdot \boldsymbol{S}$ with eigenvalue $\hbar/2$. Then the probabilities for outcomes of the measurement of a different spin component $\hat{n}' \cdot \boldsymbol{S}$ are

$$\mathcal{P}(\hat{n}' \cdot \boldsymbol{S} = +\hbar/2) = |\langle \hat{n}' | \hat{n} \rangle|^2 = \cos^2(\theta/2), \tag{8.82}$$

$$\mathcal{P}(\hat{n}' \cdot \boldsymbol{S} = -\hbar/2) = |\langle -\hat{n}' | \hat{n} \rangle|^2 = \sin^2(\theta/2), \tag{8.83}$$

where θ is the angle between $\hat{n}$ and $\hat{n}'$, so $\cos\theta = \hat{n} \cdot \hat{n}'$.

A surprising feature of (8.76) is that a rotation about any axis by an angle $\alpha = 2\pi$ gives, instead of the identity matrix,

$$D^{(1/2)}(\hat{n} 2\pi) = -I \tag{8.84}$$

for a spin-1/2 system. Thus, a $j = 1/2$ state always acquires a minus sign when continuously rotated by an angle $\alpha = 2\pi$, even though such a 2π rotation corresponds to no change at all for rotations of classical objects. If the state starts in an eigenstate of a particular component $\hat{n}' \cdot \boldsymbol{J}$, then it will still be an eigenstate with the same eigenvalue after the 2π rotation, but with a minus-sign phase factor change on the ket.

For general j, the unitary matrix for a rotation about the z axis is easy to obtain, because J_z has a diagonal matrix representation, making the exponentiation easy,

$$D^{(j)}(\hat{z}\alpha) = \exp\left[-i\alpha \begin{pmatrix} j & 0 & \cdots & 0 \\ 0 & j-1 & \cdots & 0 \\ \vdots & \vdots & \ddots & \vdots \\ 0 & 0 & \cdots & -j \end{pmatrix}\right] = \begin{pmatrix} e^{-i\alpha j} & 0 & \cdots & 0 \\ 0 & e^{-i\alpha(j-1)} & \cdots & 0 \\ \vdots & \vdots & \ddots & \vdots \\ 0 & 0 & \cdots & e^{i\alpha j} \end{pmatrix}. \tag{8.85}$$

For example, in the $j = 1$ case,

$$D^{(1)}(\hat{z}\alpha) = \mathrm{diag}(e^{-i\alpha},\ 1,\ e^{i\alpha}). \tag{8.86}$$

The special case of a rotation through an angle 2π gives

$$D^{(j)}(\hat{z} 2\pi) = (-1)^{2j} I, \tag{8.87}$$

which is equal to the identity matrix for integer j, but is equal to $-I$ for half-integer j, generalizing what was found for $j = 1/2$. For this reason, for half-integer j it is impossible to define a single-valued continuous scalar position wavefunction (like the spherical harmonics introduced in the next section for integer j).

For rotations about other axes $\hat{n} \neq \hat{z}$, the algebraic form of matrices $D^{(j)}(\boldsymbol{\alpha})$ for general j can be considerably more complicated. The resummation of the infinite sum in the exponential may depend on matrix recurrence relations of higher order. For an example of a harder case that can still be done straightforwardly in closed form, one can evaluate $D^{(1)}(\hat{x}\alpha)$ by using

$$\begin{pmatrix} 0 & 1 & 0 \\ 1 & 0 & 1 \\ 0 & 1 & 0 \end{pmatrix}^3 = 2 \begin{pmatrix} 0 & 1 & 0 \\ 1 & 0 & 1 \\ 0 & 1 & 0 \end{pmatrix} \tag{8.88}$$

to resum the exponential series, with the result:

$$D^{(1)}(\hat{x}\alpha) = \begin{pmatrix} \cos^2(\alpha/2) & -\frac{i}{\sqrt{2}}\sin(\alpha) & -\sin^2(\alpha/2) \\ -\frac{i}{\sqrt{2}}\sin(\alpha) & \cos(\alpha) & -\frac{i}{\sqrt{2}}\sin(\alpha) \\ -\sin^2(\alpha/2) & -\frac{i}{\sqrt{2}}\sin(\alpha) & \cos^2(\alpha/2) \end{pmatrix}. \tag{8.89}$$

Finding a similar expression for a rotation about the $\hat{y}$ axis for $j = 1$ is left to Exercise 8.5.

For general j, the numerical value of any $D^{(j)}(\boldsymbol{\alpha})$ can always be obtained by exponentiation of the $(2j+1) \times (2j+1)$ matrix corresponding to (8.69). There is also a remarkable formula, due to Eugene P. Wigner, for rotations about the y axis,

$$D^{(j)}_{m'm}(\hat{y}\beta) = \sum_k (-1)^{k-m+m'} [\cos(\beta/2)]^{2j-2k+m-m'} [\sin(\beta/2)]^{2k-m+m'} \frac{[(j+m)!\,(j-m)!\,(j+m')!\,(j-m')!]^{1/2}}{k!\,(k-m+m')!\,(j+m-k)!\,(j-m'-k)!}, \tag{8.90}$$

where the sum is over integers k for which all of the factorials in the denominator have non-negative arguments; in other words, k is summed from $\max(0, m-m')$ to $\min(j+m, j-m')$. The proof is omitted here.

The Wigner formula for the matrix $D^{(j)}(\hat{y}\beta)$ is especially useful because it can be used to construct a general rotation matrix. The reason is that a general rotation can always be built up as a product of three sequential rotations about two fixed orthogonal axes. Suppose that we first rotate by an angle α about the $\hat{z}$ axis, then about the $\hat{y}$ axis by an angle β, and finally by an angle γ about the $\hat{z}$ axis again.[2] The change in coordinates induced by this rotation is

[2] There are different conventional ways of defining the three **Euler angles** needed for a general rotation. In classical mechanics, it is traditional to choose the middle rotation to be about the $\hat{x}$ axis. In quantum mechanics, it is preferable to use the $\hat{y}$ axis, because $D^{(j)}(\hat{y}\beta)$ has purely real entries, as exhibited in (8.90). There are also differing conventions for whether the axes of rotations are absolutely fixed, or whether the second and third rotation axes are the body-fixed axes, obtained by

$$\begin{pmatrix} x' \\ y' \\ z' \end{pmatrix} = \begin{pmatrix} c_\gamma & -s_\gamma & 0 \\ s_\gamma & c_\gamma & 0 \\ 0 & 0 & 1 \end{pmatrix} \begin{pmatrix} c_\beta & 0 & s_\beta \\ 0 & 1 & 0 \\ -s_\beta & 0 & c_\beta \end{pmatrix} \begin{pmatrix} c_\alpha & -s_\alpha & 0 \\ s_\alpha & c_\alpha & 0 \\ 0 & 0 & 1 \end{pmatrix} \begin{pmatrix} x \\ y \\ z \end{pmatrix}, \tag{8.91}$$

where $c_\alpha = \cos\alpha$, $s_\alpha = \sin\alpha$, etc. The corresponding unitary rotation operator is

$$U(\alpha, \beta, \gamma) = U(\hat{z}\gamma)\, U(\hat{y}\beta)\, U(\hat{z}\alpha), \tag{8.92}$$

so that the rotation matrix for a multiplet with angular momentum j is

$$D^{(j)}(\alpha, \beta, \gamma) = D^{(j)}(\hat{z}\gamma)\, D^{(j)}(\hat{y}\beta)\, D^{(j)}(\hat{z}\alpha). \tag{8.93}$$

The matrices for the first and last rotations about the fixed z axis are simple, being just given by (8.85).

8.6 Spherical Harmonic Representation of Orbital Angular Momentum

The orbital angular momentum operators L_x, L_y, and L_z were introduced in Sect. 5.3. We now consider the representation of these operators in terms of the position wavefunctions in spherical coordinates. Recall that in the spherical coordinate position orthobasis,

$$\psi(r, \theta, \phi) = \langle r, \theta, \phi | \psi \rangle. \tag{8.94}$$

From the definitions of L_x, L_y, and L_z in (5.50), we get their rectangular coordinate representations as differential operators:

$$L_x \leftrightarrow -i\hbar \left(y\frac{\partial}{\partial z} - z\frac{\partial}{\partial y} \right), \tag{8.95}$$

$$L_y \leftrightarrow -i\hbar \left(z\frac{\partial}{\partial x} - x\frac{\partial}{\partial z} \right), \tag{8.96}$$

$$L_z \leftrightarrow -i\hbar \left(x\frac{\partial}{\partial y} - y\frac{\partial}{\partial x} \right). \tag{8.97}$$

Converting these to spherical coordinate differential operators gives

$$L_x \leftrightarrow i\hbar \left(\sin\phi \frac{\partial}{\partial\theta} + \frac{\cos\phi\cos\theta}{\sin\theta}\frac{\partial}{\partial\phi} \right), \tag{8.98}$$

the previous rotations of the original (fixed) y and z axes. Here, our rotation axes are the absolutely fixed ones.

$$L_y \leftrightarrow i\hbar\left(-\cos\phi\,\frac{\partial}{\partial\theta}+\frac{\sin\phi\cos\theta}{\sin\theta}\frac{\partial}{\partial\phi}\right), \tag{8.99}$$

$$L_z \leftrightarrow -i\hbar\frac{\partial}{\partial\phi}, \tag{8.100}$$

and for the raising and lowering operators $L_\pm = L_x \pm iL_y$ as defined by (8.8),

$$L_+ \leftrightarrow \hbar e^{i\phi}\left(\frac{\partial}{\partial\theta}+i\,\frac{\cos\theta}{\sin\theta}\frac{\partial}{\partial\phi}\right), \tag{8.101}$$

$$L_- \leftrightarrow \hbar e^{-i\phi}\left(-\frac{\partial}{\partial\theta}+i\,\frac{\cos\theta}{\sin\theta}\frac{\partial}{\partial\phi}\right). \tag{8.102}$$

Then, using $L^2 = L_x^2 + L_y^2 + L_z^2 = (L_+L_- + L_-L_+)/2 + L_z^2$ from (8.15), we get

$$L^2 \leftrightarrow -\hbar^2\left(\frac{\partial^2}{\partial\theta^2}+\frac{\cos\theta}{\sin\theta}\frac{\partial}{\partial\theta}+\frac{1}{\sin^2\theta}\frac{\partial^2}{\partial\phi^2}\right). \tag{8.103}$$

For future reference, we note that L^2 is closely related to the part of the Laplacian involving angular derivatives; this is useful because the kinetic energy term in the Hamiltonian involves ∇^2. More precisely,

$$\nabla^2 = \frac{\partial^2}{\partial r^2}+\frac{2}{r}\frac{\partial}{\partial r}-\frac{1}{\hbar^2 r^2}L^2, \tag{8.104}$$

which becomes

$$\nabla^2 = \frac{\partial^2}{\partial r^2}+\frac{2}{r}\frac{\partial}{\partial r}-\frac{l(l+1)}{r^2} \tag{8.105}$$

when acting on a wavefunction corresponding to an eigenstate of L^2 with eigenvalue $\hbar^2 l(l+1)$.

We now solve the eigenvalue problem for orbital angular momentum. In doing so, we can make use of a notable feature of (8.98)–(8.103): there is no r dependence in any of the angular momentum operator representations as differential operators on the wavefunctions. Thus we can consider wavefunctions for eigenstates of L^2 and L_z in which the r dependence is factored out, with an orthobasis of position eigenkets that are a tensor product of radial and angular parts,

$$|r,\theta,\phi\rangle = |r\rangle \otimes |\theta,\phi\rangle, \tag{8.106}$$

with Dirac orthonormality conditions

$$\langle r'|r\rangle = \frac{1}{r^2}\delta(r-r'), \tag{8.107}$$

$$\langle\theta', \phi'|\theta, \phi\rangle = \delta(\phi - \phi')\,\delta(\cos\theta - \cos\theta'). \tag{8.108}$$

These kets obey completeness relations

$$\int_0^\infty dr\, r^2 |r\rangle\, \langle r| = I_r, \tag{8.109}$$

$$\int d\Omega\, |\theta, \phi\rangle\langle\theta, \phi| = I_{\theta,\phi}, \tag{8.110}$$

where I_r and $I_{\theta,\phi}$ are the identity operators on the respective Hilbert spaces, with $I = I_r \otimes I_{\theta,\phi}$. In (8.110), and from here on, we define

$$d\Omega = d\phi\, d(\cos\theta) \tag{8.111}$$

as the differential of solid angle in spherical coordinates, so that

$$\int d\Omega \cdots = \int_0^{2\pi} d\phi \int_{-1}^{1} d(\cos\theta) \cdots . \tag{8.112}$$

The Hilbert space spanned by the orthobasis of angular eigenkets $|\theta, \phi\rangle$ is also spanned by an orthobasis of (L^2, L_z) eigenkets $|l, m\rangle$ satisfying, from (8.6), (8.7), (8.21), and (8.23),

$$L_z|l, m\rangle = \hbar m|l, m\rangle, \tag{8.113}$$

$$L_+|l, m\rangle = \hbar\sqrt{l(l+1) - m(m+1)}\,|l, m+1\rangle, \tag{8.114}$$

$$L_-|l, m\rangle = \hbar\sqrt{l(l+1) - m(m-1)}\,|l, m-1\rangle, \tag{8.115}$$

$$L^2|l, m\rangle = \hbar^2 l(l+1)|l, m\rangle, \tag{8.116}$$

for $l = 0, 1, 2, \ldots$ and $m = -l, \ldots, l$. The $|l, m\rangle$ also satisfy orthonormality and completeness relations

$$\langle l', m'|l, m\rangle = \delta_{ll'}\,\delta_{mm'}, \tag{8.117}$$

$$\sum_{l=0}^{\infty} \sum_{m=-l}^{l} |l, m\rangle\langle l, m| = I_{\theta,\phi}. \tag{8.118}$$

We now define the angular wavefunctions

$$\langle\theta, \phi|l, m\rangle = Y_l^m(\theta, \phi), \tag{8.119}$$

called **spherical harmonics**.

The actions of the differential operators L_z, L_+, L_-, and L^2 defined in (8.100)–(8.103) on the spherical harmonic wavefunctions are expressed as

$$L_z Y_l^m(\theta,\phi) = \hbar m\, Y_l^m(\theta,\phi), \tag{8.120}$$

$$L_+ Y_l^m(\theta,\phi) = \hbar\sqrt{l(l+1)-m(m+1)}\, Y_l^{m+1}(\theta,\phi), \tag{8.121}$$

$$L_- Y_l^m(\theta,\phi) = \hbar\sqrt{l(l+1)-m(m-1)}\, Y_l^{m-1}(\theta,\phi), \tag{8.122}$$

$$L^2 Y_l^m(\theta,\phi) = \hbar^2 l(l+1)\, Y_l^m(\theta,\phi). \tag{8.123}$$

These are the position representation differential operator versions of the Hilbert space operator equations (8.113)–(8.116). For simplicity, we are using the same symbols for the differential operators (acting on wavefunctions) and the corresponding Hilbert space operators (acting on kets). From (8.110) and (8.117), the spherical harmonics satisfy the orthonormality rule

$$\int d\Omega\, Y_{l'}^{m'}(\theta,\phi)^*\, Y_l^m(\theta,\phi) = \delta_{ll'}\delta_{mm'}, \tag{8.124}$$

and from (8.108) and (8.118), the completeness relation

$$\sum_{l=0}^{\infty}\sum_{m=-l}^{l} Y_l^m(\theta',\phi')^*\, Y_l^m(\theta,\phi) = \delta(\phi-\phi')\,\delta(\cos\theta-\cos\theta'). \tag{8.125}$$

The full Hilbert space is spanned by the orthobasis of tensor product kets

$$|r\rangle \otimes |l,m\rangle, \tag{8.126}$$

which describe states in which the particle is known to be at a distance r from the origin, and in which L^2 and L_z are known to be $\hbar^2 l(l+1)$ and $\hbar m$ respectively. These states form an orthobasis alternative to (8.106).

Equation (8.120) together with (8.100) gives the simple differential equation

$$-i\hbar\frac{\partial}{\partial\phi}Y_l^m = \hbar m Y_l^m, \tag{8.127}$$

which has the general solution

$$Y_l^m(\theta,\phi) = e^{im\phi}\, f_l^m(\theta), \tag{8.128}$$

where the functions $f_l^m(\theta)$ are now to be determined. We know already that the largest possible value of m is l, so let us start with that case. Equation (8.121) tells us that $L_+ Y_l^l = 0$, or

$$\left(\frac{\partial}{\partial \theta} + i \, \frac{\cos\theta}{\sin\theta} \frac{\partial}{\partial \phi} \right) e^{il\phi} f_l^l(\theta) = 0 \tag{8.129}$$

after using (8.101). This reduces to

$$\frac{df_l^l}{d(\sin\theta)} = l \frac{f_l^l}{\sin\theta}, \tag{8.130}$$

which has the general solution

$$f_l^l(\theta) = e^C (\sin\theta)^l, \tag{8.131}$$

where C is a constant of integration. Therefore, we can write

$$Y_l^l(\theta, \phi) = \frac{(-1)^l}{2^l \, l!} \sqrt{\frac{(2l+1)!}{4\pi}} \, e^{il\phi} (\sin\theta)^l, \tag{8.132}$$

where a multiplicative normalization factor has been included, with magnitude chosen in such a way that the orthonormality condition (8.124) holds with $m' = m = l' = l$. The $(-1)^l$ factor is a choice of convention.

The spherical harmonics for the remaining values $m < l$ can now be obtained by acting repeatedly with L_-, using (8.122), and comparing to (8.102). First,

$$Y_l^{l-1}(\theta, \phi) = -\frac{e^{-i\phi}}{\sqrt{2l}} \left(\frac{\partial}{\partial \theta} + l \, \frac{\cos\theta}{\sin\theta} \right) Y_l^l(\theta, \phi). \tag{8.133}$$

Continuing in the same way, it can be shown by recursion that, for general $-l \leq m \leq l$,

$$Y_l^m(\theta, \phi) = \frac{(-1)^l}{2^l \, l!} \sqrt{\frac{(2l+1)\,(l+m)!}{4\pi\,(l-m)!}} \, e^{im\phi} \, (\sin\theta)^{-m} \frac{d^{l-m}}{d(\cos\theta)^{l-m}} (\sin^2\theta)^l. \tag{8.134}$$

The normalization factor in (8.134) ensures that (8.124) and (8.125) are satisfied. The phase convention here is determined by the choices that we have already made in (8.21), (8.23), and (8.132), and is called the **Condon–Shortley phase convention**. (Other phase and normalization conventions for the spherical harmonics exist, so one must be careful when comparing expressions from different sources.)

An equivalent way of writing the spherical harmonics is

$$Y_l^m(\theta, \phi) = \sqrt{\frac{(2l+1)\,(l-m)!}{4\pi\,(l+m)!}} \, e^{im\phi} \, P_l^m(\cos\theta), \tag{8.135}$$

where the $P_l^m(x)$ are called associated Legendre functions. They are solutions to the differential equations

$$\left[(1-x^2)\frac{d^2}{dx^2} - 2x\frac{d}{dx} + l(l+1) - \frac{m^2}{1-x^2}\right] P_l^m(x) = 0. \tag{8.136}$$

For the special case $m = 0$, the solutions are the ordinary Legendre polynomials,

$$P_l(x) = P_l^0(x) = \frac{1}{2^l\, l!}\frac{d^l}{dx^l}(x^2-1)^l, \tag{8.137}$$

which have a generating function

$$\sum_{l=0}^{\infty} t^l P_l(x) = (1 - 2tx + t^2)^{-1/2}, \tag{8.138}$$

and satisfy the orthonormality relations

$$\int_{-1}^{1} dx\, P_{l'}(x) P_l(x) = \frac{2}{2l+1}\delta_{ll'}. \tag{8.139}$$

The first few Legendre polynomials are

$$P_0(x) = 1, \qquad P_1(x) = x, \qquad P_2(x) = (3x^2 - 1)/2, \tag{8.140}$$

$$P_3(x) = (5x^3 - 3x)/2, \qquad P_4(x) = (35x^4 - 30x^2 + 3)/8. \tag{8.141}$$

For non-negative m, the associated Legendre functions are then given by

$$P_l^m(x) = (-1)^m (1-x^2)^{m/2}\frac{d^m}{dx^m} P_l(x), \qquad (m \geq 0), \tag{8.142}$$

and those with negative m are determined by the rule

$$P_l^{-m}(x) = (-1)^m \frac{(l-m)!}{(l+m)!} P_l^m(x), \qquad (m > 0). \tag{8.143}$$

With the Condon–Shortley phase convention,

$$Y_l^{-m}(\theta, \phi) = (-1)^m Y_l^m(\theta, \phi)^*. \tag{8.144}$$

For large l and maximal $|m|$, $Y_l^{\pm l}(\theta, \phi) \propto (\sin\theta)^l$, which has maximum magnitude at $\theta = \pi/2$. Thus, states with large angular momentum about the z axis have probability densities that are peaked near the xy plane and are suppressed near the z axis.

Since $Y_l^m(\theta, \phi)$ is proportional to $e^{im\phi}$, the only spherical harmonics that are independent of ϕ are those with $m = 0$,

$$Y_l^0(\theta, \phi) = \sqrt{\frac{2l+1}{4\pi}} P_l(\cos\theta). \tag{8.145}$$

In the special case $\theta = 0$, the coordinate ϕ is arbitrary. This implies that $Y_l^m(0, \phi)$ cannot possibly depend on ϕ, so it can only be nonzero if $m = 0$. From (8.145) and $P_l(1) = 1$, we therefore obtain the special value

$$Y_l^m(0, \phi) = \delta_{m0}\sqrt{\frac{2l+1}{4\pi}}. \tag{8.146}$$

The spherical harmonics that are most commonly encountered in practical problems are those with $l \leq 2$, which are

$$Y_0^0 = \frac{1}{\sqrt{4\pi}}, \tag{8.147}$$

$$Y_1^0 = \sqrt{\frac{3}{4\pi}} \cos\theta, \qquad Y_1^{\pm 1} = \mp\sqrt{\frac{3}{8\pi}}\, e^{\pm i\phi} \sin\theta, \tag{8.148}$$

$$Y_2^0 = \sqrt{\frac{5}{16\pi}}(3\cos^2\theta - 1), \qquad Y_2^{\pm 1} = \mp\sqrt{\frac{15}{8\pi}}\, e^{\pm i\phi} \sin\theta\, \cos\theta, \tag{8.149}$$

$$Y_2^{\pm 2} = \sqrt{\frac{15}{32\pi}}\, e^{\pm 2i\phi} \sin^2\theta. \tag{8.150}$$

Note that Y_l^m is always $e^{im\phi}$ times a polynomial in $\cos\theta$ and $\sin\theta$ of total degree l.

Using completeness of the orthobasis $|l, m\rangle$ over the Hilbert space component corresponding to the angular coordinates, as expressed in (8.118), any wavefunction can be expanded as

$$\psi(r, \theta, \phi) = \sum_{l=0}^{\infty} \sum_{m=-l}^{l} \big(\langle r| \otimes \langle \theta, \phi|\big)\big(I_r \otimes |l, m\rangle\, \langle l, m|\big)|\psi\rangle \tag{8.151}$$

$$= \sum_{l=0}^{\infty} \sum_{m=-l}^{l} Y_l^m(\theta, \phi)\, \big(\langle r| \otimes \langle l, m|\big)\, |\psi\rangle. \tag{8.152}$$

Defining **radial wavefunctions** $F_{l,m}(r) = \big(\langle r| \otimes \langle l, m|\big)\, |\psi\rangle$, this can be rewritten as

$$\psi(r, \theta, \phi) = \sum_{l=0}^{\infty} \sum_{m=-l}^{l} F_{l,m}(r)\, Y_l^m(\theta, \phi). \tag{8.153}$$

To find the $F_{l,m}(r)$ for a given $\psi(r, \theta, \phi)$, multiply both sides of (8.153) by $Y_{l'}^{m'}(\theta, \phi)^*$, then integrate $d\Omega$, and use the orthonormality condition (8.124) to reduce the double sum to a single term with $l' = l$ and $m' = m$, and finally rename $(l', m') \to (l, m)$. The result is

$$F_{l,m}(r) = \int d\Omega \; Y_l^m(\theta, \phi)^* \, \psi(r, \theta, \phi). \tag{8.154}$$

Of course, (8.153) and (8.154) also apply if ψ and F have no dependence on r.

If the wavefunction $\psi(r, \theta, \phi)$ is normalized, so that $\int_0^\infty dr\, r^2 \int d\Omega |\psi(r, \theta, \phi)|^2 = 1$, then one can use the orthonormality condition (8.124) again to obtain

$$\sum_{l=0}^{\infty} \sum_{m=-l}^{l} \int_0^\infty dr\, r^2 |F_{l,m}(r)|^2 = 1. \tag{8.155}$$

It follows that the probability of simultaneously measuring L^2 and L_z and getting the results $\hbar^2 l(l+1)$ and $\hbar m$ is, by using Postulate 4 with r playing the role of the degeneracy label,

$$\mathcal{P}_{l,m} = \int_0^\infty dr\, r^2 \big|(\langle r| \otimes \langle l, m|)|\psi\rangle\big|^2 = \int_0^\infty dr\, r^2 |F_{l,m}(r)|^2. \tag{8.156}$$

As a corollary, the probability of measuring just L^2 and getting $\hbar^2 l(l+1)$ is obtained by summing this over m, so $\mathcal{P}_l = \sum_{m=-l}^{l} \mathcal{P}_{l,m}$. And, the probability of measuring L_z and getting the result $\hbar m$ is $\mathcal{P}_m = \sum_{l=|m|}^{\infty} \mathcal{P}_{l,m}$. Here, the sum over l starts at $|m|$, because for smaller l there are no states that have L_z eigenvalue $\hbar m$.

In many cases, it is not necessary to actually do the radial integral in (8.156) in order to evaluate the probabilities; instead one can make use of ratios between the various possibilities. As a simple but essential example, any wavefunction that is a function of r only (with no θ or ϕ dependence) is proportional to Y_0^0, so one has $l = m = 0$, and the measurements of the compatible orbital angular momentum observables L^2 and L_z are both certain to give 0.

For a slightly less trivial example, consider a wavefunction

$$\psi(\boldsymbol{r}) = C \sin^2\theta \, \cos^2\phi \; f(r), \tag{8.157}$$

where C is a constant. (Note that we are not providing any context about whether this state is related to any particular Hamiltonian.) If L^2 and/or L_z is measured, what are the possible outcomes and their probabilities? To answer this, we seek to write the wavefunction as a linear combination of spherical harmonics multiplied by functions of r only. A useful clue is that the wavefunction is quadratic and even in sines and cosines of θ and ϕ, so one should expect that it will involve $l = 2$ and $l = 0$. Indeed, applying (8.154), one finds that

$$\sin^2\theta \, \cos^2\phi = \sqrt{\pi}\left(\frac{2}{3}Y_0^0 - \frac{2}{3\sqrt{5}}Y_2^0 + \sqrt{\frac{2}{15}}Y_2^2 + \sqrt{\frac{2}{15}}Y_2^{-2}\right). \tag{8.158}$$

This implies that the measurements of (L^2, L_z) can yield only the four possible pairs $(0, 0)$ and $(6\hbar^2, 0)$ and $(6\hbar^2, 2\hbar)$ and $(6\hbar^2, -2\hbar)$, with probability ratios, respectively,

$$\mathcal{P}_{0,0} : \mathcal{P}_{2,0} : \mathcal{P}_{2,2} : \mathcal{P}_{2,-2} = \left|\frac{2}{3}\right|^2 : \left|-\frac{2}{3\sqrt{5}}\right|^2 : \left|\sqrt{\frac{2}{15}}\right|^2 : \left|\sqrt{\frac{2}{15}}\right|^2. \quad (8.159)$$

By requiring that the sum of the probabilities is 1, it follows that

$$\mathcal{P}_{0,0} = 5/9, \qquad \mathcal{P}_{2,0} = 1/9, \qquad \mathcal{P}_{2,2} = \mathcal{P}_{2,-2} = 1/6. \quad (8.160)$$

Note that the radial wavefunction here was irrelevant for the angular momentum measurement probabilities, since it is a common factor.

One often needs to consider wavefunctions for states that are pure eigenstates of L^2 and L_z, with fixed eigenvalues $\hbar^2 l(l+1)$ and $\hbar m$, respectively. These will have the form

$$\psi(r, \theta, \phi) = F(r) Y_l^m(\theta, \phi). \quad (8.161)$$

According to (8.105), the Laplacian acting on such a wavefunction is

$$\nabla^2 \left[F(r) Y_l^m(\theta, \phi)\right] = \left[\frac{d^2 F}{dr^2} + \frac{2}{r}\frac{dF}{dr} - \frac{l(l+1)}{r^2} F\right] Y_l^m(\theta, \phi). \quad (8.162)$$

As a consequence of the rotational invariance of the Laplacian operator, the result maintains the form of an eigenfunction of L^2 and L_z, with the same eigenvalues.

We now state and prove an extraordinarily useful formula:

Theorem 8.1 (Spherical Harmonics Addition Formula) *Consider any two unit vectors $\hat{n}$ and $\hat{n}'$, characterized by their spherical coordinate angles (θ, ϕ) and (θ', ϕ'), respectively. Call the angle between these vectors γ, so that*

$$\hat{n} \cdot \hat{n}' = \cos\gamma = \cos\theta\cos\theta' + \sin\theta\sin\theta'\cos(\phi - \phi'). \quad (8.163)$$

Then, for each l,

$$P_l(\cos\gamma) = \frac{4\pi}{2l+1} \sum_{m=-l}^{l} Y_l^m(\theta', \phi')^* \, Y_l^m(\theta, \phi). \quad (8.164)$$

Proof Since $P_l(\cos\gamma)$ is a function of θ and ϕ, it can be written as an expansion in spherical harmonics with arguments θ and ϕ. Now, the spherical harmonics involved in this linear combination all must have the same l. To see this, note that we can rotate to a coordinate system in which $\hat{n}'$ is the new z direction, so that the polar angular coordinate of $\hat{n}$ is γ. In that coordinate system, $P_l(\cos\gamma)$ is proportional to a spherical harmonic with $m = 0$, an eigenfunction of the rotationally invariant differential operator L^2 with eigenvalue $\hbar^2 l(l+1)$. Likewise, $P_l(\cos\gamma)$ is also a function of θ' and ϕ', so it can also be written as a linear combination of spherical harmonics of those angles, with the same l. Therefore, it must be that $P_l(\cos\gamma)$ is a sum of terms of the form $Y_l^{m'}(\theta', \phi')^* Y_l^m(\theta, \phi)$. Each of these terms is proportional to $e^{i(m\phi - m'\phi')}$. If one chooses $\phi' = \phi$, this factor is $e^{i(m-m')\phi}$, but $\cos\gamma$ is independent of ϕ in that special case. Therefore, only terms with $m' = m$ can contribute, and we must have

$$P_l(\cos\gamma) = \sum_{m=-l}^{l} a_{lm} Y_l^m(\theta', \phi')^* Y_l^m(\theta, \phi). \tag{8.165}$$

The remaining task is to evaluate the coefficients a_{lm}. Because $P_l(\cos\gamma)$ is real, and is unchanged under $(\theta, \phi) \leftrightarrow (\theta', \phi')$, the coefficients must satisfy $a_{lm} = a_{lm}^*$, so they are real.

To learn more, consider the special case $(\theta, \phi) = (\theta', \phi')$, so that $\cos\gamma = 1$. Then, since the Legendre polynomials satisfy $P_l(1) = 1$, (8.165) gives

$$1 = \sum_{m=-l}^{l} a_{lm} \left|Y_l^m(\theta, \phi)\right|^2. \tag{8.166}$$

Integrating with respect to $d\Omega$, and using the orthonormality condition (8.124), we get

$$4\pi = \sum_{m=-l}^{l} a_{lm}. \tag{8.167}$$

Next, consider the square of (8.165), which can be written

$$\left(P_l(\cos\gamma)\right)^2 = \sum_{m=-l}^{l} a_{lm} Y_l^m(\theta', \phi')^* Y_l^m(\theta, \phi) \sum_{m'=-l}^{l} a_{lm'} Y_l^{m'}(\theta', \phi') Y_l^{m'}(\theta, \phi)^*. \tag{8.168}$$

Let us integrate over all angles (θ', ϕ'). To evaluate the integral of the left side, it is convenient to again use coordinates such that γ is the polar angle, and making use of (8.139), we have

$$\int d\Omega' \left(P_l(\cos\gamma)\right)^2 = 2\pi \int_{-1}^{1} d(\cos\gamma) \left(P_l(\cos\gamma)\right)^2 = \frac{4\pi}{2l+1}. \tag{8.169}$$

Meanwhile, the $d\Omega'$ integral of the right side of (8.168) is evaluated by using the orthonormality condition (8.124) again, after which only the terms with $m' = m$ contribute in the double sum. Comparing the two sides, we get

$$\frac{4\pi}{2l+1} = \sum_{m=-l}^{l} a_{lm}^2 Y_l^m(\theta,\phi) Y_l^m(\theta,\phi)^*. \tag{8.170}$$

Integrating this with respect to $d\Omega$, and using orthonormality once again, we find

$$\frac{(4\pi)^2}{2l+1} = \sum_{m=-l}^{l} a_{lm}^2. \tag{8.171}$$

We now have enough information to solve for the coefficients. From (8.167) and (8.171), we discover that

$$\sum_{m=-l}^{l} \left(a_{lm} - \frac{4\pi}{2l+1} \right)^2 = 0. \tag{8.172}$$

Since the left side is a sum of squares, the unique solution is obtained when each term vanishes, so that $a_{lm} = 4\pi/(2l+1)$ for all l, m. This concludes the proof of the spherical harmonics addition formula. □

As an application, consider the identity

$$\frac{1}{|\boldsymbol{r} - \boldsymbol{r}'|} = \sum_{l=0}^{\infty} \frac{r_{\min}^l}{r_{\max}^{l+1}} P_l(\cos\gamma), \tag{8.173}$$

with

$$r_{\max} = \max(r, r'), \qquad r_{\min} = \min(r, r'). \tag{8.174}$$

Equation (8.173) follows from the generating function (8.138), and may be familiar from the multipole expansion for problems with azimuthal symmetry in electrostatics. The spherical harmonics addition formula (8.164) allows us to rewrite it as

$$\frac{1}{|\boldsymbol{r} - \boldsymbol{r}'|} = \sum_{l=0}^{\infty} \frac{4\pi}{2l+1} \frac{r_{\min}^l}{r_{\max}^{l+1}} \sum_{m=-l}^{l} Y_l^m(\theta',\phi')^* \, Y_l^m(\theta,\phi), \tag{8.175}$$

a form that will be useful to us later.

8.7 Parity of Angular Momentum Eigenstates

As we noted in Sect. 5.3, angular momentum operators have even parity, which is another way of saying that they commute with the parity operator Π. According to Theorem 2.19, this means that the eigenstates of the angular momentum operators J^2 and J_z can also be chosen to be parity eigenstates.

First, let us work out the parity eigenvalues of the orbital angular momentum eigenstates $|l, m\rangle$. If $(x, y, z) \to (-x, -y, -z)$, the spherical coordinate transformations are

$$r \to r, \qquad \theta \to \pi - \theta, \qquad \phi \to \phi + \pi, \tag{8.176}$$

so that $\cos\theta \to -\cos\theta$, and $\sin\theta \to \sin\theta$, and $e^{im\phi} \to (-1)^m e^{im\phi}$. Using these in the definition of the spherical harmonics, (8.134), gives

$$Y_l^m(\pi - \theta, \phi + \pi) = (-1)^l Y_l^m(\theta, \phi). \tag{8.177}$$

Therefore,

$$\begin{aligned}\langle\theta, \phi|\Pi|l, m\rangle &= \langle\pi - \theta, \phi + \pi|l, m\rangle = (-1)^l Y_l^m(\theta, \phi)\\ &= (-1)^l \langle\theta, \phi|l, m\rangle.\end{aligned} \tag{8.178}$$

Since this is true for all $\langle\theta, \phi|$, we must have

$$\Pi|l, m\rangle = (-1)^l |l, m\rangle. \tag{8.179}$$

This shows that orbital angular momentum eigenstates are always even under parity if the quantum number l is even, and odd under parity if l is odd. This holds regardless of the quantum number m, which can be understood from the fact that the states $|l, m\rangle$ are all obtained from $|l, l\rangle$ by acting with L_-, which commutes with Π. [Therefore, the parity eigenvalue $(-1)^l$ could actually have been obtained by considering the spherical harmonics for only one sample value of m for each l, for example $m = l$ using (8.132), rather than the more general formula of (8.134).] Since space inversion in spherical coordinates does not change the radial coordinate, the parity eigenvalue of an angular momentum eigenstate also does not depend on the radial wavefunction or radial quantum numbers.

Let us next consider the parity of eigenkets $|s, m_s\rangle$ of intrinsic angular momentum operators S^2 and S_z. From the fact that the lowering operator S_- commutes with parity and relates kets with different m_s, we know that the parity eigenvalue of $|s, m_s\rangle$ cannot depend on m_s. We can therefore write

$$\Pi|s, m_s\rangle = \eta|s, m_s\rangle, \tag{8.180}$$

where η is known as the **intrinsic parity** of the particle in question. For any given particle type, the intrinsic parity can be chosen to be either $+1$ or -1, as an arbitrary convention. This conventional choice can have no observable consequences if the number of particles of each type does not change. This is because a change in the conventional choice of intrinsic parity just amounts to flipping the parity for every bra and every ket, therefore not affecting matrix elements. So, when the number of particles does not change, one might as well just always choose $\eta = +1$, and the parity of a single-particle L^2 eigenstate is $(-1)^l$, regardless of the spin. For a system of N particles with individual orbital angular momentum quantum numbers l_i, the parity eigenvalue is, from (8.179),

$$\pi_{\text{total}} = (-1)^{\sum_{i=1}^{N} l_i}. \tag{8.181}$$

This is the case for electrons in an atom, where the parity and the angular momenta are all defined with respect to the origin chosen to be the location of the fixed nucleus.

However, if the Hamiltonian is invariant under parity and can cause changes in the numbers of particles, it is natural and useful to adopt a convention in which the intrinsic parities of particles and antiparticles are chosen in a consistent way so that parity is conserved. Such Hamiltonians arise in quantum field theory, where the electromagnetic and strong nuclear interactions conserve parity, while the weak nuclear interactions do not but can often be treated as a perturbation. A generally accepted convention is to choose $\eta = +1$ for spin-1/2 particles (the electron, proton, neutron, muon, and quarks, for example). Then the structure of kinetic terms in relativistic quantum field theory can be used to show that their antiparticles (the positron, antiproton, antineutron, antimuon, and antiquarks) must be assigned $\eta = -1$ in the same convention.

For integer-spin particles, things are different. In processes governed by the electromagnetic and strong nuclear forces, parity conservation is experimentally verified, and the consistency of the intrinsic parity assignments requires that the photon and the gluon have intrinsic parity -1 (due to their vector nature), while the Higgs particle has intrinsic parity $+1$. The W and Z bosons are not assigned a definite intrinsic parity, because they are the mediators of the parity-violating weak interactions.

For a composite particle c, made out of two particles a and b with intrinsic parities η_a and η_b in a state with relative angular momentum quantum number L, the consistent intrinsic parity assignment is[3]

$$\eta_c = (-1)^L \eta_a \eta_b. \tag{8.182}$$

Thus, mesons composed of a quark and an antiquark in an $L = 0$ bound state (for example, pions and kaons) have intrinsic parity $\eta = -1$. For a bound state of three or

[3] Note that this differs from (8.181), since $(-1)^L$ is not always equal to $(-1)^{l_a+l_b}$. This is because the angular momentum L and the intrinsic parity η_c in (8.182) are defined with respect to the origin as the center-of-mass position of the two particles, rather than a fixed origin position as in (8.181).

more particles, the situation is more complicated, but the intrinsic parities of bound states can always be defined if the Hamiltonian is invariant under space inversion.

8.8 Time Reversal of Angular Momentum Eigenstates and Kramers Degeneracy

In Sect. 5.7, we found that the motion-reversal transformation (which coincides with time reversal when it is a symmetry) is implemented by an operator Θ, which keeps the position the same but flips the momentum. This is seen in (5.150) for kets, and in operator form in (5.153). We will now work out the motion-reversal properties of angular momentum eigenstates. The analysis relies on the peculiar antilinear and antiunitary nature of Θ. Section 5.7 is therefore an important prerequisite to this section, but antilinear operators do not appear in this book outside of that section and this one.

For orbital angular momentum eigenstates, we could, in principle, work out the action of Θ starting from the position-eigenstate basis definition in (5.150) and using antilinearity. However, it is better to take a more general approach based on the angular momentum algebra, so that our results will also apply to spins.

We therefore start with the fact that any angular momentum operator $\boldsymbol{J}$ is odd under time reversal, as expressed in (5.159), which can be written as

$$\Theta \boldsymbol{J} = -\boldsymbol{J}\Theta. \tag{8.183}$$

This implies, for any eigenstate $|j,m\rangle$ of J^2 and J_z, that

$$J^2\Theta|j,m\rangle = \Theta J^2|j,m\rangle = \hbar^2 j(j+1)\Theta|j,m\rangle, \tag{8.184}$$

$$J_z\Theta|j,m\rangle = -\Theta J_z|j,m\rangle = -\hbar m\Theta|j,m\rangle. \tag{8.185}$$

Therefore, $\Theta|j,m\rangle$ is also an eigenstate of J^2 and J_z, with eigenvalues $\hbar^2 j(j+1)$ and $-\hbar m$ respectively, so it must be proportional to $|j,-m\rangle$. The kets $|j,m\rangle$ and $\Theta|j,m\rangle$ have the same norm, due to the antiunitarity of Θ, so we must have

$$\Theta|j,m\rangle = \eta_{j,m}|j,-m\rangle, \tag{8.186}$$

where $\eta_{j,m}$ are phase factors (satisfying $|\eta_{j,m}| = 1$) that we need to determine.

Now consider the angular momentum ladder operators $J_\pm = J_x \pm iJ_y$. Taking into account the antilinearity of Θ in the form $\Theta i = -i\Theta$, we find using (5.159) that

$$J_-\Theta = -\Theta J_+. \tag{8.187}$$

We now compute each side of this equation acting on $|j,m\rangle$, for any m in the range $-j \le m \le j-1$. First,

$$\begin{aligned} J_-\Theta|j,m\rangle &= \eta_{j,m} J_-|j,-m\rangle \\ &= \eta_{j,m}\hbar\sqrt{j(j+1)-m(m+1)}\,|j,-m-1\rangle, \end{aligned} \tag{8.188}$$

where we used (8.23) to evaluate $J_-|j,-m\rangle$. Second,

$$\begin{aligned} -\Theta J_+|j,m\rangle &= -\hbar\sqrt{j(j+1)-m(m+1)}\,\Theta|j,m+1\rangle \\ &= -\hbar\sqrt{j(j+1)-m(m+1)}\,\eta_{j,m+1}|j,-m-1\rangle, \end{aligned} \tag{8.189}$$

where we used (8.21) to evaluate $J_+|j,m\rangle$. Comparing these results, we find

$$\eta_{j,m} = -\eta_{j,m+1}. \tag{8.190}$$

It follows that for fixed j, the signs alternate as m is decremented, so

$$\eta_{j,m} = e^{i\alpha_j}(-1)^{j-m}, \tag{8.191}$$

where $e^{i\alpha_j} = \eta_{j,j}$ is a phase factor that can be chosen as a convention. Putting this into (8.186), we arrive at the important rule

$$\Theta|j,m\rangle = e^{i\alpha_j}(-1)^{j-m}|j,-m\rangle \tag{8.192}$$

for the motion reversal of a general angular momentum eigenstate.

Consider the example of an orbital angular momentum, with L^2 and L_z eigenstates $|l,m\rangle$, with wavefunctions $Y_l^m(\theta,\phi) = \langle\theta,\phi|l,m\rangle$. As we found in a general way in (5.161), the wavefunction for $\Theta|l,m\rangle$ is obtained by taking the complex conjugate, so

$$\langle\theta,\phi|\Theta|l,m\rangle = Y_l^m(\theta,\phi)^* = (-1)^m Y_l^{-m}(\theta,\phi) = (-1)^m\langle\theta,\phi|l,-m\rangle, \tag{8.193}$$

where the second equality used our choice of the Condon–Shortley phase convention in (8.144). Comparing to (8.192), we see that $e^{i\alpha_l} = (-1)^l$ is the correct choice for the motion-reversal (time-reversal, when it is a symmetry) phase to be consistent with the Condon–Shortley convention. It also follows from (8.192) that $\Theta^2|l,m\rangle = |l,m\rangle$.

Another important case is that of a spin-1/2 system. Here, it is convenient to choose the convention $e^{i\alpha_{1/2}} = 1$, so that (8.192) becomes

$$\Theta|\uparrow\rangle = |\downarrow\rangle, \qquad \Theta|\downarrow\rangle = -|\uparrow\rangle, \tag{8.194}$$

or more generally for any spin state obtained as a linear combination,

$$\Theta\big(c_1|\uparrow\rangle + c_2|\downarrow\rangle\big) = c_1^*|\downarrow\rangle - c_2^*|\uparrow\rangle. \tag{8.195}$$

It follows that, acting on a general spin-1/2 state, $\Theta^2 = -I$, so that $\Theta^{-1} = -\Theta$.

More generally, we can use (8.192) twice to compute the action of Θ^2 on an arbitrary angular momentum eigenstate:

$$\begin{aligned}\Theta^2|j,m\rangle &= \Theta e^{i\alpha_j}(-1)^{j-m}|j,-m\rangle = e^{-i\alpha_j}(-1)^{j-m}\Theta|j,-m\rangle \\ &= e^{-i\alpha_j}(-1)^{j-m}e^{i\alpha_j}(-1)^{j+m}|j,m\rangle = (-1)^{2j}|j,m\rangle. \quad (8.196)\end{aligned}$$

So, acting on any space of states with fixed j,

$$\Theta^2 = (-1)^{2j}I. \quad (8.197)$$

This is always either I or $-I$, depending only on whether the angular momentum quantum number j is an integer or half-integer, which is a fixed property of the system.

The last result has an interesting practical consequence. Consider any system that has an odd number of spin-1/2 particles, so that the total angular momentum has half-integer j, and therefore $\Theta^2 = -I$. Take the general antiunitarity relation $\langle\psi|\chi\rangle = \langle\Theta\chi|\Theta\psi\rangle$, and plug in $|\chi\rangle = |\Theta\psi\rangle$. This gives

$$\langle\psi|\Theta\psi\rangle = \langle\Theta^2\psi|\Theta\psi\rangle = -\langle\psi|\Theta\psi\rangle. \quad (8.198)$$

Since $\langle\psi|\Theta\psi\rangle$ is equal to minus itself, it must be 0. Therefore, $|\psi\rangle$ and $\Theta|\psi\rangle$ are always orthogonal for a system with $\Theta^2 = -I$. Now suppose that the system has time-reversal invariance, so that H commutes with Θ. Then for any energy eigenstate $|\psi_E\rangle$ with energy eigenvalue E, we have $H\Theta|\psi_E\rangle = \Theta H|\psi_E\rangle = E\Theta|\psi_E\rangle$, so $\Theta|\psi_E\rangle$ is an eigenstate of H with the same energy eigenvalue E. We conclude that the orthogonal (and thus physically distinct) states $|\psi_E\rangle$ and $\Theta|\psi_E\rangle$ are degenerate. This result is called **Kramers degeneracy** after Hans Kramers: any system with time-reversal invariance and an odd number of spin-1/2 particles must have even degeneracy for each energy level. This is a powerful rule because many systems, even complicated ones, are time-reversal invariant or very nearly so. Kramers degeneracy is broken if we introduce a term in the Hamiltonian that violates time-reversal symmetry. A common way to do this is to turn on a static external magnetic field, so that H includes a term proportional to $\boldsymbol{S}\cdot\boldsymbol{B}$.

Exercises

8.1 Consider an eigenstate $|j,m\rangle$ of J^2 and J_z, as in (8.6)–(8.7). Find the expectation value and the uncertainties of J_x and J_y in this state. For which state or states $|j,m\rangle$ are these uncertainties maximized? For which are they minimized? Show that the uncertainty relation (3.19) is satisfied for $A = J_x$ and $B = J_y$.

8.2 Consider a spin-1/2 particle, and let $\hat{n} = \hat{x}\sin\beta + \hat{z}\cos\beta$ be a unit vector specified in terms of a fixed angle β.

(a) Construct the matrix representation of $\hat{n}\cdot\boldsymbol{S}$, working in the basis of S_z eigenstates denoted $|\uparrow\rangle$ and $|\downarrow\rangle$ for eigenvalues $+\hbar/2$ and $-\hbar/2$ respectively. Find its eigenvalues and eigenvectors.
(b) Suppose that the spin is in the state $|\uparrow\rangle$. What is the probability that the measurement of $\hat{n}\cdot\boldsymbol{S}$ yields the result $+\hbar/2$?
(c) Suppose that the measurement in part (b) has been carried out and the result for $\hat{n}\cdot\boldsymbol{S}$ was indeed $+\hbar/2$. Immediately afterwards, S_z is measured. What is the probability that the measurement yields $-\hbar/2$?
(d) Check that your results for parts (b) and (c) make sense for each of $\beta = 0, \pi$, and $\pi/2$.

8.3 Consider a spinless particle in a state with L^2 eigenvalue $2\hbar^2$ and L_z eigenvalue $\hbar$. As in the previous problem, let $\hat{n} = \hat{x}\sin\beta + \hat{z}\cos\beta$.

(a) Suppose that the angular momentum along the direction $\hat{n}$ is measured. What are the possible results, and their probabilities?
(b) For each of the possible results in part (a), suppose that L_z is then measured. What are the possible results, and their probabilities?
(c) Check that your results make sense for each of $\beta = 0, \pi$, and $\pi/2$.

8.4 Consider a particle in a state with spherical coordinate wavefunction of the form $\psi(r,\theta,\phi) = f(r)\cos^2\theta$.

(a) Write the wavefunction in terms of spherical harmonics $Y_l^m(\theta,\phi)$.
(b) If L^2 and L_z are measured simultaneously, find the possible results and their probabilities.

8.5 Find the unitary matrix for rotations about the $\hat{y}$ axis of $j = 1$ states, $D^{(1)}(\hat{y}\beta)$, without making use of (8.90). Then use (8.90) to check your answer.

8.6 Professor Bumble measured the spin component of a spin-1/2 system at time $t = 0$ and recorded the result as definitely $+\hbar/2$, but forgot whether it was S_x, S_y, or S_z. Assigning equal probabilities to these possibilities, what is the density operator ρ for the resulting mixed state at $t = 0$, in the basis of S_z eigenstates? What is the von Neumann entropy σ? If the Hamiltonian is $H = \omega S_z$, use the unitary time evolution operator to find the probability of getting the result $+\hbar/2$ for a measurement of S_x at time t.

Charged Particle in a Magnetic Field

9

9.1 Spin Precession in a Constant Magnetic Field

Consider a spin carrying a magnetic moment in a constant magnetic field. Let us assume that the particle carrying the spin has a fixed position, so that we do not need to worry about its kinetic energy or any position-dependent potentials. Specializing (4.48), the Hamiltonian is therefore simply

$$H = -\gamma \boldsymbol{B} \cdot \boldsymbol{S}, \tag{9.1}$$

where γ is the gyromagnetic ratio of the particle. Since we only care about the magnetic field at the point where the particle is located, we can orient our coordinate system so that $\boldsymbol{B} = B\hat{z}$, where B is a constant. In this section, we will study the time evolution of the spin state.

An orthobasis of states for this system consists of the eigenstates of S_z with eigenvalues $\hbar/2$ and $-\hbar/2$, denoted $|\uparrow\rangle$ for spin up and $|\downarrow\rangle$ for spin down. Then the Hamiltonian can be expressed as the spectral decomposition

$$H = \frac{1}{2}\hbar\omega_B\Big(|\uparrow\rangle\langle\uparrow| - |\downarrow\rangle\langle\downarrow|\Big), \tag{9.2}$$

where we have defined the **Larmor precession frequency**

$$\omega_B = -\gamma B. \tag{9.3}$$

(With this sign choice, ω_B is positive for the electron when B is positive, since the electron's gyromagnetic ratio $\gamma = -g_e e/2m_e c$ is a negative number. The same is true for a muon or a neutron. For a proton, ω_B is negative when B is positive.) Thus the matrix representation is

S. P. Martin, *Quantum Mechanics*, Graduate Texts in Physics,
https://doi.org/10.1007/978-3-032-21119-4_9

$$H = \frac{1}{2}\hbar\omega_B \begin{pmatrix} 1 & 0 \\ 0 & -1 \end{pmatrix}, \tag{9.4}$$

with energy eigenstates $|\uparrow\rangle$ with $E_\uparrow = \hbar\omega_B/2$, and $|\downarrow\rangle$ with $E_\downarrow = -\hbar\omega_B/2$, represented by $\begin{pmatrix} 1 \\ 0 \end{pmatrix}$ and $\begin{pmatrix} 0 \\ 1 \end{pmatrix}$, respectively. To decide which of these is the ground state requires knowing the sign of γB. In any case, a transition between the two stationary states typically involves emitting or absorbing a photon with energy $|E_\uparrow - E_\downarrow| = \hbar|\omega_B|$. The Planck–Einstein relation, discussed in Sect. 1.3, says that the wavelength of the photon will be $\lambda = 2\pi c/|\omega_B| = 2\pi c/|\gamma B|$, providing a way of determining the gyromagnetic ratio of the particle.

Another way of measuring the gyromagnetic ratio (or equivalently, the magnetic moment) is provided by the precession of the spin when the state is a linear combination of energy eigenstates. Suppose that at time $t = 0$ the normalized initial state is

$$|\psi(0)\rangle = a\,|\uparrow\rangle + b\,|\downarrow\rangle, \tag{9.5}$$

with complex coefficients satisfying $|a|^2 + |b|^2 = 1$. The subsequent time evolution can be obtained either by solving the time-dependent Schrödinger equation, or more directly by applying the unitary operator $U(t) = e^{-itH/\hbar}$, with the result

$$|\psi(t)\rangle = ae^{-i\omega_B t/2}\,|\uparrow\rangle + be^{i\omega_B t/2}\,|\downarrow\rangle. \tag{9.6}$$

If S_z is then measured, the possible results are the eigenvalues $\pm\hbar/2$, with probabilities

$$\mathcal{P}(S_z = \hbar/2) = |\langle\uparrow|\psi(t)\rangle|^2 = |a|^2, \tag{9.7}$$

$$\mathcal{P}(S_z = -\hbar/2) = |\langle\downarrow|\psi(t)\rangle|^2 = |b|^2, \tag{9.8}$$

which do not depend on time. (As a check, they sum to 1 because of the normalization condition.) It follows that the expectation value of S_z is

$$\langle S_z\rangle = \frac{\hbar}{2}(|a|^2 - |b|^2), \tag{9.9}$$

which is also independent of time.

If, instead, we measure S_x, then the possible results are again $\pm\hbar/2$, and the evaluation of the probabilities makes use of the corresponding eigenstates

$$|\uparrow_x\rangle = \frac{1}{\sqrt{2}}(|\uparrow\rangle + |\downarrow\rangle), \qquad (\text{for } S_x = \hbar/2), \tag{9.10}$$

$$|\downarrow_x\rangle = \frac{1}{\sqrt{2}}\big(|\uparrow\rangle - |\downarrow\rangle\big), \qquad \text{(for } S_x = -\hbar/2\text{)}. \tag{9.11}$$

We therefore find

$$\begin{aligned}\mathcal{P}(S_x = \hbar/2) &= |\langle\uparrow_x|\psi(t)\rangle|^2 = \left|\frac{1}{\sqrt{2}}\left(ae^{-i\omega_B t/2} + be^{i\omega_B t/2}\right)\right|^2 \\ &= \frac{1}{2}\left(|a|^2 + |b|^2 + a^*be^{i\omega_B t} + ab^*e^{-i\omega_B t}\right) \\ &= \frac{1}{2} + \mathrm{Re}\big(a^*be^{i\omega_B t}\big).\end{aligned} \tag{9.12}$$

Similarly (or, just by requiring that the probabilities add to 1), one finds

$$\mathcal{P}(S_x = -\hbar/2) = |\langle\downarrow_x|\psi(t)\rangle|^2 = \frac{1}{2} - \mathrm{Re}\big(a^*be^{i\omega_B t}\big). \tag{9.13}$$

The expectation value of S_x can be obtained either from the probability-weighted sum of eigenvalues as

$$\langle S_x\rangle = \frac{\hbar}{2}\mathcal{P}(S_x = \hbar/2) - \frac{\hbar}{2}\mathcal{P}(S_x = -\hbar/2), \tag{9.14}$$

or from matrix evaluation, using $S_x = (\hbar/2)\sigma_x$, as

$$\langle S_x\rangle = \begin{pmatrix} a^*e^{i\omega_B t/2} & b^*e^{-i\omega_B t/2}\end{pmatrix}\frac{\hbar}{2}\begin{pmatrix}0 & 1\\ 1 & 0\end{pmatrix}\begin{pmatrix}ae^{-i\omega_B t/2}\\ be^{i\omega_B t/2}\end{pmatrix} = \hbar\,\mathrm{Re}\big(a^*be^{i\omega_B t}\big). \tag{9.15}$$

For a measurement of S_y, the probabilities are evaluated using the corresponding eigenstates

$$|\uparrow_y\rangle = \frac{1}{\sqrt{2}}\big(|\uparrow\rangle + i\,|\downarrow\rangle\big), \qquad \text{(for } S_y = \hbar/2\text{)}, \tag{9.16}$$

$$|\downarrow_y\rangle = \frac{1}{\sqrt{2}}\big(|\uparrow\rangle - i\,|\downarrow\rangle\big), \qquad \text{(for } S_y = -\hbar/2\text{)}, \tag{9.17}$$

with the results

$$\mathcal{P}(S_y = \hbar/2) = |\langle\uparrow_y|\psi(t)\rangle|^2 = \frac{1}{2} + \mathrm{Im}\big(a^*be^{i\omega_B t}\big), \tag{9.18}$$

$$\mathcal{P}(S_y = -\hbar/2) = |\langle\downarrow_y|\psi(t)\rangle|^2 = \frac{1}{2} - \mathrm{Im}\big(a^*be^{i\omega_B t}\big), \tag{9.19}$$

and the resulting expectation value is

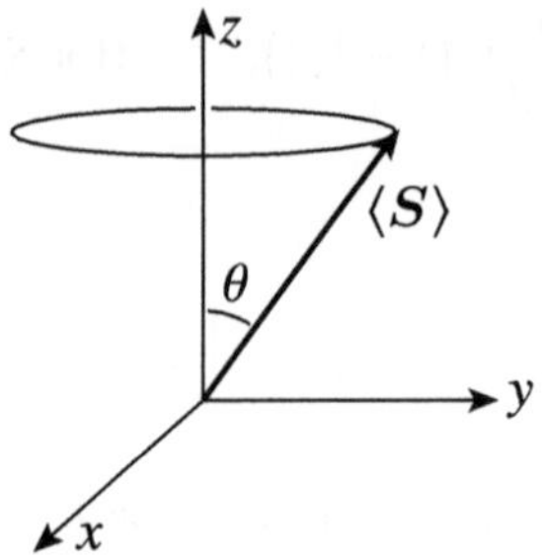

Fig. 9.1 The expectation value of a spin precesses about a magnetic field $\boldsymbol{B} = B\hat{z}$, with a period $2\pi/|\omega_B| = 2\pi/|\gamma B|$, where γ is the gyromagnetic ratio of the particle. The polar angle θ remains constant as the spin precesses, and is equal to $\arctan(2ab/(a^2 - b^2))$ for an initial state $|\psi\rangle = a\,|\uparrow\rangle + b\,|\downarrow\rangle$ with real a and b

$$\langle S_y \rangle = \hbar \, \mathrm{Im}\big(a^* b e^{i\omega_B t}\big). \tag{9.20}$$

The oscillation of these quantities describes precession of $\boldsymbol{S}$ about the $\boldsymbol{B}$ direction.

For example, take a and b to be real for simplicity. Then our results for the expectation values of the spin components are summarized as

$$\langle \boldsymbol{S} \rangle = \frac{\hbar}{2}\Big(2ab\left[\cos(\omega_B t)\hat{x} + \sin(\omega_B t)\hat{y}\right] + (a^2 - b^2)\hat{z}\Big). \tag{9.21}$$

At time $t = 0$, the spin expectation value lies in the xz plane. As time evolves, the angle $\theta = \arctan(2ab/(a^2 - b^2))$ between $\langle \boldsymbol{S} \rangle$ and $\hat{z}$ remains constant. (See Fig. 9.1.) The rotation of $\langle \boldsymbol{S} \rangle$ is counterclockwise (as seen looking down from positive z) if γB is negative, because then the angle $\phi = \omega_B t$ of the expectation value of the spin vector projected onto the xy plane increases at the constant rate $\omega_B = -\gamma B$. This precession frequency is independent of the initial orientation of the spin, provided only that $ab \neq 0$. Thus, the rate of precession of the spin in a known magnetic field provides another way of measuring the gyromagnetic ratio. This effect has many practical applications, including (in more sophisticated setups) the measurement with exquisite accuracy of the magnetic moments of the electron and the muon and various nuclei.

It is useful to describe the spin precession more directly in terms of the state ket, rather than the expectation value, which after all is merely a statistical quantity. Inspired by the result of (9.21), consider the time-dependent unit vector

$$\hat{n} = 2ab[\cos(\omega_B t)\hat{x} + \sin(\omega_B t)\hat{y}] + (a^2 - b^2)\hat{z}, \tag{9.22}$$

still taking a and b to be real. Note that $\hat{n} \cdot \hat{n} = 1$ follows from the normalization condition $a^2 + b^2 = 1$. Now define the spin component along $\hat{n}$, as the Hermitian operator

$$S_{\hat{n}} = \hat{n} \cdot \boldsymbol{S} = \frac{\hbar}{2} \begin{pmatrix} a^2 - b^2 & 2abe^{-i\omega_B t} \\ 2abe^{i\omega_B t} & b^2 - a^2 \end{pmatrix}. \tag{9.23}$$

The last expression is really the matrix representation in the S_z-eigenstate basis, obtained using the Pauli matrices according to $\boldsymbol{S} = (\hbar/2)\boldsymbol{\sigma}$. It is now straightforward to check that the state ket $|\psi(t)\rangle$ of (9.6), represented by the column vector

$$\begin{pmatrix} ae^{-i\omega_B t/2} \\ be^{i\omega_B t/2} \end{pmatrix}, \tag{9.24}$$

satisfies the eigenvalue equation

$$S_{\hat{n}} \, |\psi(t)\rangle = \frac{\hbar}{2} \, |\psi(t)\rangle \, . \tag{9.25}$$

In other words, the time-dependent state is simply the one in which measurement of $S_{\hat{n}}$ is always certain to give the result $\hbar/2$.

9.2 Magnetic Spin Resonance and Rabi Oscillations

As discussed in the previous section, the energy eigenvalues of a spin-1/2 particle with gyromagnetic ratio γ are split by a static magnetic field. The spin of a general state precesses in a way that does not change its expectation value along the magnetic field direction. One way to directly access the energy splitting is through emission or absorption of photons of the right energy. In this section, we will discuss another way. The idea is to supplement the main static component B of the magnetic field with a time-oscillating component $\widetilde{B}$ in a different direction. This oscillatory component drives transitions between the spin eigenstates defined with respect to the static field direction. As we will see, this effect is maximally enhanced (resonant) if the driving frequency for $\widetilde{B}$ is chosen equal to the Larmor precession frequency for B.

Specifically, let the magnetic field be

$$\boldsymbol{B} = B\hat{z} + \widetilde{B}\left[\cos(\omega t)\hat{x} + \sin(\omega t)\hat{y}\right]. \tag{9.26}$$

The oscillatory component rotates in the xy plane at the driving frequency ω, which can be adjusted independently by the experimentalist. Now define

$$\omega_B = -\gamma B, \qquad \Gamma = -\gamma \widetilde{B}, \tag{9.27}$$

both with units of angular frequency. Note that ω_B is the same as the Larmor precession frequency (see the previous section) for the constant component of the magnetic field. The Hamiltonian matrix in the S_z-eigenstate basis is

$$H = \frac{1}{2}\hbar\omega_B \begin{pmatrix} 1 & 0 \\ 0 & -1 \end{pmatrix} + \frac{1}{2}\hbar\Gamma \begin{pmatrix} 0 & e^{-i\omega t} \\ e^{i\omega t} & 0 \end{pmatrix}. \tag{9.28}$$

As an aside, this Hamiltonian can also serve as a model for a variety of other two-state physical systems, for which the static part of the Hamiltonian has been diagonalized by a choice of basis, and there is an off-diagonal contribution varying harmonically with time. Thus, the results below have more general interest and applicability.

Suppose that $\omega_B > 0$, so that the ground state (neglecting the oscillating part of H) is $|\downarrow\rangle$. We start in this ground state, and want to find the probability to find the system in the excited state $|\uparrow\rangle$, as a function of time. To accomplish this goal, we will find the time-dependent state

$$|\psi(t)\rangle = c_1(t)\,|\uparrow\rangle + c_2(t)\,|\downarrow\rangle \tag{9.29}$$

that solves the Schrödinger equation for the full H, subject to the initial condition

$$c_1(0) = 0, \qquad c_2(0) = 1. \tag{9.30}$$

In matrix form, the time-dependent Schrödinger equation is

$$i\hbar\frac{d}{dt}\begin{pmatrix} c_1 \\ c_2 \end{pmatrix} = \frac{\hbar}{2}\begin{pmatrix} \omega_B & \Gamma e^{-i\omega t} \\ \Gamma e^{i\omega t} & -\omega_B \end{pmatrix}\begin{pmatrix} c_1 \\ c_2 \end{pmatrix}. \tag{9.31}$$

It is helpful to define new coefficients $a_1(t)$ and $a_2(t)$, by

$$c_1 = e^{-i\omega t/2}a_1, \qquad c_2 = e^{i\omega t/2}a_2. \tag{9.32}$$

This is convenient because the resulting coupled first-order differential equations for a_1 and a_2 have coefficients with no explicit time dependence:

$$i\dot{a}_1 + (\omega - \omega_B)a_1/2 - \Gamma a_2/2 = 0, \tag{9.33}$$

$$i\dot{a}_2 - (\omega - \omega_B)a_2/2 - \Gamma a_1/2 = 0. \tag{9.34}$$

Mindful of the initial conditions in (9.30), we now try for a harmonic solution of the form

$$a_1 = b_1\sin(\Omega t), \tag{9.35}$$

$$a_2 = \cos(\Omega t) + b_2\sin(\Omega t), \tag{9.36}$$

where Ω, b_1, and b_2 are constants to be determined. Plugging this guess in, we can require the vanishing of the coefficients of $\cos(\Omega t)$ and $\sin(\Omega t)$ separately in each differential equation. Success is found provided that

$$\Omega^2 = \left[\Gamma^2 + (\omega - \omega_B)^2\right]/4 \tag{9.37}$$

and

$$b_1 = -i\Gamma/2\Omega, \qquad b_2 = i(\omega_B - \omega)/2\Omega. \tag{9.38}$$

Putting everything together, the coefficients in $|\psi(t)\rangle$ are

$$c_1 = -i\frac{\Gamma}{2\Omega}\sin(\Omega t)e^{-i\omega t/2}, \tag{9.39}$$

$$c_2 = \left[\cos(\Omega t) + i\frac{\omega_B - \omega}{2\Omega}\sin(\Omega t)\right]e^{i\omega t/2}. \tag{9.40}$$

The probability to find the spin in the excited state is therefore

$$\mathcal{P}(\uparrow) = |\langle\uparrow|\psi(t)\rangle|^2 = |c_1|^2 = \frac{\Gamma^2}{\Gamma^2 + (\omega - \omega_B)^2}\sin^2(\Omega t). \tag{9.41}$$

This shows resonance behavior, for if the driving frequency ω is close to the Larmor precession frequency ω_B, the system periodically oscillates into the higher state with probability close to 1, as depicted in Fig. 9.2. These **Rabi oscillations** (named after Isidor Isaac Rabi, who both measured and explained them theoretically) occur with angular frequency 2Ω, which near resonance is close in magnitude to $|\Gamma| = |\gamma\widetilde{B}|$, where $\widetilde{B}$ is the amplitude of the oscillating magnetic field. Away from the resonance, the probability for a spin flip is nonzero, but smaller, and the frequency of the observed Rabi oscillations is larger.

In real-world applications, the driving magnetic field is rarely of the form we assumed in (9.26), because it is not so easy to make a substantial magnetic field

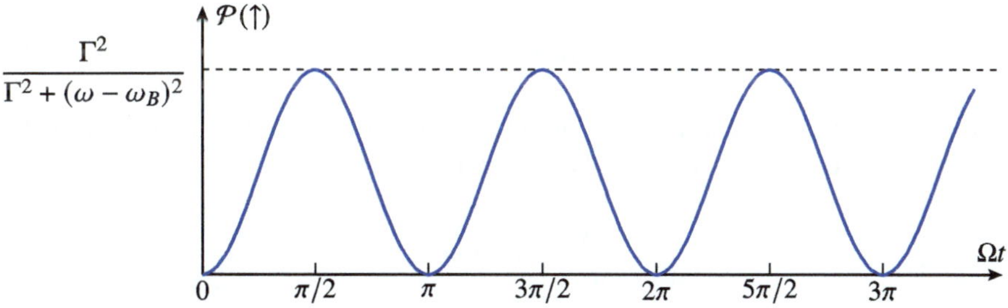

Fig. 9.2 The probability to find a spin-1/2 system in the higher-energy state due to magnetic spin resonance, as a function of time, according to (9.41). The peak probability for the spin to flip reaches unity for the resonant condition that the driving frequency ω is equal to the Larmor precession frequency $\omega_B = -\gamma B$, where γ is the gyromagnetic ratio of the particle and B is the constant part of the magnetic field

whose direction rotates at a high frequency. Instead, an oscillating magnetic field along a fixed direction, like $\widetilde{B}\cos(\omega t)\hat{x}$, is typically used for the driving field. This is more difficult to solve exactly, but has essentially the same behavior near resonance. The reason is that we can decompose it as

$$\widetilde{B}\cos(\omega t)\hat{x} = \frac{1}{2}\widetilde{B}[\cos(\omega t)\hat{x} + \sin(\omega t)\hat{y}] + \frac{1}{2}\widetilde{B}[\cos(\omega t)\hat{x} - \sin(\omega t)\hat{y}], \quad (9.42)$$

which is a superposition of two rotating fields with angular frequencies $+\omega$ and $-\omega$. When one of these is resonant (close to ω_B), the other is far from resonant and therefore has little effect.

This phenomenon has many applications, not the least of which is **magnetic resonance imaging**, commonly used in medical settings to safely produce maps of tissue using the magnetic moments of hydrogen nuclei (protons) affected in different signature ways by the fields of neighboring atoms in different molecules. In other applications, **nuclear magnetic resonance** is made more complicated and interesting by the fact that the nuclear spin s_N need not be $1/2$, in which case the number of states involved in the resonance is not 2, but $2s_N + 1$.

9.3 Landau Levels for a Charged Particle in a Constant Uniform Magnetic Field

Classically, a charged particle in a constant uniform magnetic field undergoes cyclotron motion, following a helical trajectory. This problem is exactly solvable in quantum mechanics, too, and we will now study it in detail.

First, let us gain some insight by reviewing the classical solution. To be concrete, let us take the particle to be an electron with mass m and charge $q = -e$. Recall that e is defined by (1.2) to be the proton's charge in the cgs unit system. In this section we orient the coordinate system so that the magnetic field points in the negative z direction,

$$\boldsymbol{B} = -B\hat{z}. \quad (9.43)$$

Then the Lorentz force law, (4.32), amounts to a first-order differential equation for the velocity vector $\boldsymbol{v} = d\boldsymbol{r}/dt$ of the electron,

$$\frac{d\boldsymbol{v}}{dt} = \omega_c\, \boldsymbol{v} \times \hat{z}, \quad (9.44)$$

or $\ddot{x} = \omega_c\dot{y}$ and $\ddot{y} = -\omega_c\dot{x}$ and $\ddot{z} = 0$ in rectangular coordinates, where

$$\omega_c = \frac{eB}{mc} \quad (9.45)$$

is called the **cyclotron angular frequency**.[1] Since the force has no component along $\hat{z}$, the velocity in that direction is constant. The general solution for the classical trajectory of the electron is a helix, which we take to have radius R and center (x_c, y_c) in the xy plane,

$$x(t) = x_c + R\sin(\omega_c t + \phi_0), \tag{9.46}$$

$$y(t) = y_c + R\cos(\omega_c t + \phi_0), \tag{9.47}$$

$$z(t) = z_0 + v_z t, \tag{9.48}$$

where ϕ_0, z_0, and v_z are the other constants needed to specify the initial conditions. It is important that the cyclotron frequency is independent of all 6 of these parameters. Since the classical motion is harmonic with angular frequency ω_c, it should not be too surprising that we will be able to solve the corresponding quantum mechanical problem using the algebraic technology that we developed for harmonic oscillators in Sect. 7.3, and that the energy quantum for orbital excitations of a free electron in a magnetic field will turn out to be

$$\hbar\omega_c = 1.15768 \times 10^{-4}\,\text{eV}\left(\frac{B}{\text{Tesla}}\right). \tag{9.49}$$

This is twice the product of the magnetic field and the Bohr magneton defined in (4.46).

The expression for the Hamiltonian depends on the choice of gauge for the vector potential. One convenient choice maintains manifest rotational symmetry about the $\hat{z}$ axis,

$$\boldsymbol{A} = \frac{1}{2}B(y\hat{x} - x\hat{y}) = -\frac{1}{2}Br\hat{\phi}, \tag{9.50}$$

where (r, ϕ, z) are the cylindrical coordinates. From the general form of (4.48), the total Hamiltonian operator is then

$$H_{\text{total}} = \frac{1}{2m}\left(\Pi_x^2 + \Pi_y^2 + P_z^2\right) - e\Phi(z) + \frac{g_e e}{2mc}\boldsymbol{B}\cdot\boldsymbol{S}, \tag{9.51}$$

where we have included spin and allowed for the possibility of a z-dependent electric potential, and the kinetic momentum operators

[1] For an isolated electron, the cyclotron frequency is nearly the same as the Larmor frequency $\omega_B = (g_e/2)\omega_c$ defined in the context of spin precession in Sect. 9.1. However, for electrons that are not isolated, m in (9.45) should be interpreted as the effective mass of the electron moving in its environment, which can differ greatly from the mass 0.511 MeV/c^2 of a free electron. This can give a very different numerical value for ω_c in practical applications involving electrons in materials, sometimes modifying (9.49) dramatically.

$$\Pi_x = P_x + \frac{eB}{2c}Y, \qquad \Pi_y = P_y - \frac{eB}{2c}X \tag{9.52}$$

obey the commutation relation

$$\left[\Pi_x, \Pi_y\right] = i\frac{e\hbar}{c}B. \tag{9.53}$$

Since $\boldsymbol{S}$ and P_z and $\Phi(z)$ commute with Π_x and Π_y, it makes sense to separate the spin and z-dependent parts of the Hamiltonian as $H_{\text{total}} = H + H_s + H_z$, where

$$H_s = -\frac{g_e eB}{2mc}S_z, \qquad H_z = \frac{P_z^2}{2m} - e\Phi(z), \tag{9.54}$$

but our primary interest is

$$H = \frac{1}{2m}\left(\Pi_x^2 + \Pi_y^2\right), \tag{9.55}$$

the Hamiltonian for the two-dimensional motion in the xy plane. The eigenvalues of H_s are just $E_s = \mp g_e\hbar eB/4mc$, for spin-up states with $S_z = \hbar/2$ and spin-down states with $S_z = -\hbar/2$. Thus, each spin-up state is lower than its counterpart spin-down state by an energy splitting $g_e\hbar eB/2mc \approx \hbar\omega_c$. We assume that the eigenvalues E_z of H_z have also been found. For example, if $\Phi = 0$, the H_z eigenstates are plane waves with momentum p_z along the z-direction and $E_z = p_z^2/2m$. In the opposite extreme, it might be that the potential $\Phi(z)$ confines the electron to the vicinity of a plane of constant z, by requiring a large energy gap for excitations along the z direction. In any case, we can concentrate on the remaining two-dimensional problem with Hamiltonian H, to which the constants E_s and E_z can be added to find the total energy.

Since H is quadratic in two Hermitian operators with a constant commutator, we are inspired to follow the logical path of the harmonic oscillator by defining raising and lowering operators

$$a^\dagger = \frac{-i}{\sqrt{2\hbar\omega_c m}}\left(\Pi_x - i\Pi_y\right), \qquad a = \frac{i}{\sqrt{2\hbar\omega_c m}}\left(\Pi_x + i\Pi_y\right). \tag{9.56}$$

You can check that they obey the usual ladder operator commutation relation

$$[a, a^\dagger] = 1, \tag{9.57}$$

and that the Hamiltonian H has the simple form

$$H = \hbar\omega_c(a^\dagger a + 1/2). \tag{9.58}$$

So, the energy levels are the same as for a one-dimensional harmonic oscillator with frequency ω_c,

$$E_n = \hbar\omega_c(n + 1/2), \tag{9.59}$$

where we will soon verify that $n = 0, 1, 2, \ldots$. These are called the **Landau levels** for a charged particle in a magnetic field, after Lev Landau. The energy spacing between them grows linearly with the applied magnetic field.

At this point, one might naively suppose that the Landau level states obey $a\,|n\rangle = \sqrt{n}\,|n-1\rangle$ and $a^\dagger\,|n\rangle = \sqrt{n+1}\,|n+1\rangle$. However, one must take into account the fact that in the present two-dimensional problem the states at each level n are not unique, unlike the one-dimensional harmonic oscillator. In fact, each Landau level is infinitely degenerate. This corresponds to the classical property that the energy is determined by the radius R of the helix, but one can always translate it so that it is centered about any desired point in the xy plane, without changing the energy.

To understand the quantum version of this degeneracy, let us first define two operators corresponding to the center of the classical trajectory. Solving (9.46) and (9.47) for (x_c, y_c) in terms of $x(t)$ and $y(t)$ and their time derivatives $\dot{x} = \pi_x/m$ and $\dot{y} = \pi_y/m$, and then promoting these objects to operators, we define

$$X_c = X + \frac{1}{m\omega_c}\Pi_y, \qquad Y_c = Y - \frac{1}{m\omega_c}\Pi_x, \tag{9.60}$$

which can also be written as

$$X_c = \frac{1}{2}X + \frac{1}{m\omega_c}P_y, \qquad Y_c = \frac{1}{2}Y - \frac{1}{m\omega_c}P_x. \tag{9.61}$$

You can check that X_c and Y_c are each constants of the motion, meaning that $[H, X_c] = [H, Y_c] = 0$. However, they do not commute with each other, as one finds

$$[X_c, Y_c] = -i\frac{\hbar c}{eB}. \tag{9.62}$$

Therefore, a CSCO can include either one of them (or any linear combination), but not both. For example, a valid CSCO choice would be H and X_c. In that case, the degeneracy of the orthobasis kets with fixed n is uncountably infinite, and corresponds to the eigenvalue of X_c.

However, let us make another choice, which we will see involves X_c and Y_c in a different way. Since the problem has symmetry under rotations about the z axis in this gauge, we know (and can explicitly check) that the angular momentum component

$$L_z = XP_y - YP_x \tag{9.63}$$

commutes with H. Therefore, we take the CSCO to consist of H and L_z, and denote the corresponding orthonormal energy eigenstate basis elements as $|n, \lambda\rangle$, where

$$H\,|n, \lambda\rangle = \hbar\omega_c(n + 1/2)\,|n, \lambda\rangle\,, \tag{9.64}$$

$$L_z\,|n, \lambda\rangle = \hbar\lambda\,|n, \lambda\rangle\,. \tag{9.65}$$

From the general theory of rotations, in Sect. 5.3, we know that λ must be an integer, but we will soon find that in the present problem it must satisfy the further restriction of being bounded from below (but not from above). This asymmetry is related to the sign of the magnetic field in (9.43). In this basis, the infinite degeneracy of the Landau levels is countable, since L_z has discrete eigenvalues, unlike X_c.

From the operator definitions, you can check that

$$[H, a] = -\hbar\omega_c a, \qquad [L_z, a] = \hbar a, \tag{9.66}$$

$$[H, a^\dagger] = \hbar\omega_c a^\dagger, \qquad [L_z, a^\dagger] = -\hbar a^\dagger. \tag{9.67}$$

Acting on $|n, \lambda\rangle$ with (9.66), we find that $a\,|n, \lambda\rangle$ must be an eigenstate of H and L_z with eigenvalues $\hbar\omega_c(n - 1/2)$ and $\hbar(\lambda + 1)$, respectively. It follows that a not only lowers n by 1, but must simultaneously raise λ by 1, so $a\,|n, \lambda\rangle = c\,|n-1, \lambda+1\rangle$ for some normalization constant c. Using orthonormality of the basis kets, the norm of the right side is $|c|^2$, while the norm of the left side is $\langle n, \lambda|a^\dagger a|n, \lambda\rangle = n$, where we have made use of (9.58) and the definition of n in (9.64). Therefore, we must have $c = \sqrt{n}$, where an arbitrary phase convention has been chosen. Doing a similar computation for $a^\dagger\,|n, \lambda\rangle$, we arrive at

$$a^\dagger\,|n, \lambda\rangle = \sqrt{n+1}\,|n+1, \lambda-1\rangle\,, \tag{9.68}$$

$$a\,|n, \lambda\rangle = \sqrt{n}\,|n-1, \lambda+1\rangle\,. \tag{9.69}$$

Equation (9.69) shows that n must be a non-negative integer, just as in the case of the ordinary harmonic oscillator, because acting with a on any of the states $|0, \lambda\rangle$ results in the null ket. The infinitely degenerate set of states with $n = 0$ and varying λ is called the **lowest Landau level.**

Acting with a or $a^\dagger$ changes both the energy level and the angular momentum. We will now construct raising and lowering operators for the angular momentum that leave the energy fixed. One clue to this construction is the fact, already noted, that X_c and Y_c commute with H. A second clue is that X_c and Y_c have a constant commutator with each other. We are therefore inspired to define a second pair of raising and lowering operators,

$$b^\dagger = \frac{1}{\sqrt{2}\ell_B}(X_c + iY_c)\,, \qquad b = \frac{1}{\sqrt{2}\ell_B}(X_c - iY_c)\,, \tag{9.70}$$

where we have defined the **magnetic length,**

$$\ell_B = \sqrt{\frac{\hbar c}{eB}} = 2.56556 \times 10^{-8}\ \mathrm{m} \left(\frac{\mathrm{Tesla}}{B}\right)^{1/2}. \tag{9.71}$$

These operators obey the usual ladder commutation relation

$$[b,\, b^\dagger] = 1, \tag{9.72}$$

and are independent in the sense that they commute with the $a, a^\dagger$ operators,

$$[a,\, b] = [a,\, b^\dagger] = [a^\dagger,\, b] = [a^\dagger,\, b^\dagger] = 0. \tag{9.73}$$

Furthermore, you are invited to check from the operator definitions that

$$b^\dagger b = \frac{1}{\hbar\omega_c} H + \frac{1}{\hbar} L_z - 1/2, \tag{9.74}$$

and that the counterparts of (9.66) and (9.67) are

$$[H,\, b] = 0, \qquad [L_z,\, b] = -\hbar b, \tag{9.75}$$

$$[H,\, b^\dagger] = 0, \qquad [L_z,\, b^\dagger] = \hbar b^\dagger. \tag{9.76}$$

With this information, it is now a short exercise, very similar to the derivation of (9.68) and (9.69), to check that (up to the usual arbitrary phase choice),

$$b^\dagger\, |n, \lambda\rangle = \sqrt{\lambda + n + 1}\, |n, \lambda{+}1\rangle\,, \tag{9.77}$$

$$b\, |n, \lambda\rangle = \sqrt{\lambda + n}\, |n, \lambda{-}1\rangle\,. \tag{9.78}$$

This shows how the infinite degeneracy of each Landau level n arises, as the operators $b^\dagger$ and b change the angular momentum within each level. However, (9.78) also shows that λ is restricted to be not less than $-n$, as attempting to lower it further with b would result in the null ket.

The complete set of eigenstates of H and L_z can now be constructed, from the state with $n = \lambda = 0$, as

$$|n, \lambda\rangle = \frac{\left(b^\dagger\right)^{\lambda+n}}{\sqrt{(\lambda+n)!}} \frac{\left(a^\dagger\right)^{n}}{\sqrt{n!}}\, |0, 0\rangle\,, \tag{9.79}$$

where $n = 0, 1, 2, \ldots$ is a non-negative integer, and $\lambda = -n, -n{+}1, -n{+}2, \ldots$ so that $\lambda + n$ is also a non-negative integer. This is a complete orthobasis for the two-dimensional degrees of freedom of a charged particle in a magnetic field.

To construct the corresponding wavefunctions $\psi_{n,\lambda}(x, y) = \langle x, y|n, \lambda\rangle$, it is convenient to define dimensionless complex coordinates z, z^* by

$$z = \frac{x + iy}{\sqrt{2}\ell_B} = \frac{re^{i\phi}}{\sqrt{2}\ell_B}, \qquad z^* = \frac{x - iy}{\sqrt{2}\ell_B} = \frac{re^{-i\phi}}{\sqrt{2}\ell_B}, \tag{9.80}$$

with partial derivatives defined by

$$\frac{\partial}{\partial z} = \frac{\ell_B}{\sqrt{2}}\left(\frac{\partial}{\partial x} - i\frac{\partial}{\partial y}\right), \qquad \frac{\partial}{\partial z^*} = \frac{\ell_B}{\sqrt{2}}\left(\frac{\partial}{\partial x} + i\frac{\partial}{\partial y}\right), \tag{9.81}$$

so that

$$\frac{\partial z}{\partial z} = \frac{\partial z^*}{\partial z^*} = 1, \qquad \frac{\partial z^*}{\partial z} = \frac{\partial z}{\partial z^*} = 0. \tag{9.82}$$

Then the position representations of the ladder operators are simply

$$a = \frac{\partial}{\partial z^*} + z/2, \qquad a^\dagger = -\frac{\partial}{\partial z} + z^*/2, \tag{9.83}$$

$$b = \frac{\partial}{\partial z} + z^*/2, \qquad b^\dagger = -\frac{\partial}{\partial z^*} + z/2. \tag{9.84}$$

Now, an eigenstate of L_z with eigenvalue $\hbar\lambda$ must have wavefunction proportional to $e^{i\lambda\phi}$, as we found on general grounds in Sect. 5.3. Therefore, $\psi_{0,0}$ cannot depend on ϕ, and so must depend on position only through the combination zz^*. Writing $\psi_{0,0} = f(zz^*)$, and noting that it must be annihilated by a (and also by b), we find $f'(zz^*) + f(zz^*)/2 = 0$, which has the solution $f(zz^*) = ce^{-zz^*/2}$, with a normalization constant c. The other states in the lowest Landau level ($n = 0$) are found by acting with $(b^\dagger)^\lambda/\sqrt{\lambda!}$ using (9.84). This gives the elegant result

$$\psi_{0,\lambda} = \frac{c}{\sqrt{\lambda!}} z^\lambda e^{-zz^*/2}. \tag{9.85}$$

Rewriting this in cylindrical coordinates, and requiring $\int_0^{2\pi} d\phi \int_0^\infty dr\, r|\psi_{0,0}|^2 = 1$ to fix c, we obtain the normalized wavefunctions

$$\psi_{0,\lambda} = \frac{1}{\sqrt{2\pi\lambda!}\,\ell_B}\left(\frac{r\,e^{i\phi}}{\sqrt{2}\ell_B}\right)^\lambda \exp\left(-r^2/4\ell_B^2\right). \tag{9.86}$$

The wavefunctions for the higher Landau levels are more complicated, but can now be obtained by acting with $(a^\dagger)^n/\sqrt{n!}$.

The wavefunction $\psi_{0,0}$ is sharply peaked at the origin $r = 0$, which is arbitrarily chosen. For nonzero λ, the probability density $|\psi_{0,\lambda}|^2$ instead has a ring-like profile, vanishing at the origin, and maximum at $r = R_\lambda = \sqrt{2\lambda}\ell_B$. For very large angular momentum λ, the radius of the ring R_λ is much larger than its width (defined as the range of r near R_λ for which the wavefunction is appreciable), which is always comparable to ℓ_B, independent of λ.

If we now suppose that the electron is confined to a disk of radius R centered at the origin in the xy plane, then the infinite degeneracy of each Landau level is reduced to a finite degeneracy. Intuitively, valid states must have $R_\lambda \lesssim R$, implying $\lambda \lesssim R^2/2\ell_B^2$, in order for the wavefunction ring to "fit" inside the disk. This is not exact, because confining the electron to $r < R$ introduces a boundary condition that modifies the stationary state wavefunctions in a complicated way. However, in a macroscopic disk with $R \gg \ell_B$, only the tiny fraction of states with $R - R_\lambda$ comparable to ℓ_B will be distorted significantly. For the purposes of counting the number of states in a macroscopic area, this makes almost no difference. So, we estimate that the number N of states in the lowest Landau level should be (one more than) the maximum allowed λ,

$$N \approx \lambda_{\text{max}} \approx \frac{R^2}{2\ell_B^2} = \frac{e\Phi_m}{2\pi\hbar c} = \frac{\Phi_m}{\Phi_0}, \tag{9.87}$$

where $\Phi_m = \pi R^2 B$ is the magnetic flux through the disk, and

$$\Phi_0 \equiv \frac{2\pi\hbar c}{e} = 4.13567 \times 10^{-7}\ \text{gauss} \cdot \text{cm}^2 = 4.13567 \times 10^{-15}\ \text{Tesla} \cdot \text{m}^2 \tag{9.88}$$

is the natural quantum[2] of magnetic flux. Thus, by varying the applied magnetic field, one can dial the number of electrons that can fit in the lowest Landau level, and in each higher energy level. This affects the conductivity properties of materials, and in particular is an important ingredient in the quantum Hall effect.

It is instructive to treat the same problem in a different gauge. The magnetic field $\boldsymbol{B} = -B\hat{z}$ can also be obtained as $\boldsymbol{\nabla} \times \boldsymbol{A}$ where we now choose

$$\boldsymbol{A} = -Bx\hat{y}. \tag{9.89}$$

This vector potential is invariant under translations in the y direction, rather than maintaining manifest invariance under rotations about the z axis. The Hamiltonian in this gauge is

$$H = \frac{1}{2m}P_x^2 + \frac{1}{2m}\left(P_y - \frac{eB}{c}X\right)^2. \tag{9.90}$$

Since this version of H commutes with P_y due to our gauge choice, we use them as our CSCO, and look for an orthobasis of common eigenstates $|n, k\rangle$ that satisfy

$$H\,|n, k\rangle = \hbar\omega_c(n + 1/2)\,|n, k\rangle\,, \qquad (n = 0, 1, 2, \ldots), \tag{9.91}$$

[2] In the context of superconductivity, a conventional definition of Φ_0 is used that is half as big, corresponding to a Cooper pair of electrons with charge that is twice as large.

$$P_y\,|n,k\rangle = \hbar k\,|n,k\rangle\,, \qquad (k = \text{real}). \tag{9.92}$$

Here we have used the fact that we already know the energy eigenvalues, since they are measurable and so cannot depend on our choice of gauge. The infinite degeneracy in the Landau levels is now represented by the freedom to choose any P_y eigenvalue $\hbar k$, independently of n. Substituting $P_y = \hbar k$, and defining

$$x_0 = \hbar k/m\omega_c = k\ell_B^2, \tag{9.93}$$

the Hamiltonian becomes

$$H = \frac{1}{2m}P_x^2 + \frac{1}{2}m\omega_c^2(X - x_0)^2. \tag{9.94}$$

This is simply a one-dimensional harmonic oscillator with angular frequency ω_c describing motion in the x direction, and displaced by the constant distance x_0 depending on the y momentum eigenvalue. The orthobasis position representation wavefunctions are, up to normalization,

$$\psi_{n,k}(x,y) = \langle x,y|n,k\rangle = \psi_n(x - x_0)e^{iky}, \tag{9.95}$$

where $\psi_n(x)$ are the harmonic oscillator wavefunctions of Sect. 7.2, with $\omega = \omega_c$ and length scale $b = \ell_B$. Thus, our new choice of gauge achieves a clean separation of x and y wavefunctions. The more general lesson is that different gauge choices are useful for highlighting different aspects of a physical system.

Suppose now that our electron is confined to a rectangular area, $0 < x < d_x$ and $0 < y < d_y$, with macroscopic lengths d_x and d_y. We could try to impose a boundary condition that the wavefunction vanishes outside of this rectangle, but then the H, P_y eigenvalue equations would be difficult to solve exactly. Fortunately, for many purposes we can still use the solutions found above, to a good approximation. For the vast majority of states in the lowest Landau levels, the boundary conditions at $x = 0$ and $x = d_x$ have almost no effect because the wavefunctions are only significant within a distance comparable to ℓ_B. For example, the lowest Landau level ($n = 0$) states have wavefunctions (up to normalization)

$$\psi_{0,k}(x,y) = \exp[-(x - x_0)^2/2\ell_B^2]\,e^{iky}. \tag{9.96}$$

Because of the exponential suppression, the distortion of this wavefunction resulting from the existence of the $x = 0$ and $x = d_x$ boundaries is very small except for the tiny fraction of states for which x_0 or $d_x - x_0$ is comparable to ℓ_B. This argument does not apply for the $y = 0$ and $y = d_y$ boundaries, because the wavefunction factor e^{iky} has constant magnitude. However, we can use the trick of imposing periodic boundary conditions in the y direction, so that $y = 0$ is identified with $y = d_y$. Since d_y is a macroscopic length, the microscopic physics should not depend very much on whether we identify the two distant sides, or more generally on what boundary conditions we impose, as long as they are consistent. The point is that imposing periodic boundary conditions in y is particularly easy to do in a consistent way for the wavefunctions proportional to e^{iky} (and in particular is much easier than trying to impose $\psi = 0$ at $y = 0$ and $y = d_y$). It just restricts the allowed eigenvalues to

$$k = 2\pi n_y / d_y, \qquad (n_y = \text{ integer}). \tag{9.97}$$

We can now repeat the estimate of the number of states per unit area in the lowest Landau level. To a good approximation, the valid states will be the ones for which $x_0 = \ell_B^2 2\pi n_y / d_y$ fits between 0 and d_x. This means that the allowed n_y are the integers in the range $0 \le n_y \le d_x d_y / 2\pi \ell_B^2$, resulting in a degeneracy for the lowest Landau level of

$$N \approx \frac{eBd_x d_y}{2\pi \hbar c} = \frac{\Phi_m}{\Phi_0}. \tag{9.98}$$

This agrees with our previous result (9.87) obtained using the rotationally symmetric gauge choice corresponding to the H, L_z orthobasis.

Exercises

9.1 Show that the spin expectation value precession result of (9.21) obeys Ehrenfest's Theorem, (3.39), with $\boldsymbol{S}$ playing the role of A.

9.2 A spin-1/2 particle with gyromagnetic ratio γ is in a uniform but time-varying magnetic field $\boldsymbol{B} = B(t)\hat{z}$ so that the Hamiltonian is $H = -\gamma \boldsymbol{B} \cdot \boldsymbol{S}$. At time $t = 0$, the spin of the particle along the direction $\hat{n} = (\hat{x} + \hat{z})/\sqrt{2}$ is measured to be $+\hbar/2$. You should express your answers in terms of the definite integral $I(t) = \int_0^t dt' B(t')$. [Hint: in this problem the Hamiltonian is not constant in time, so you need to solve a differential equation for the components of the state vector. That's where the definite integral comes in.]

(a) If the $\hat{z}$ component of the spin is measured at time $t > 0$, what is the probability that the result will be $S_z = -\hbar/2$?
(b) If, instead, the $\hat{y}$ component of the spin is measured at time $t > 0$, what is the probability that the result will be $S_y = +\hbar/2$?
(c) Find the expectation value of S_x as a function of time $t > 0$.

9.3 For the Rabi oscillation solution of (9.39) and (9.40), find the expectation value of the spin operator, $\langle \boldsymbol{S} \rangle$, as a function of time. Simplify the result as much as possible, and show that it obeys Ehrenfest's Theorem, (3.39) with $\boldsymbol{S}$ playing the role of A.

9.4 Consider the magnetic spin resonance system treated in Sect. 9.2, but now assume that the system is initially in an eigenstate of S_x with eigenvalue $\hbar/2$, so that the boundary condition for the differential equations (9.33) and (9.34) is $a_1(0) = a_2(0) = 1/\sqrt{2}$. Find the Rabi oscillation solution for $a_1(t)$ and $a_2(t)$ with the same $\Omega^2 = [\Gamma^2 + (\omega - \omega_B)^2]/4$. Show that the probability to find the system in the state $|\uparrow\rangle$ is $\mathcal{P}(\uparrow) = 1/2 + \eta \sin^2(\Omega t)$, where η is a quantity that you will find. What happens to this probability at resonance ($\omega = \omega_B$)? Under what conditions can $|\eta| = 1/2$, so that the probability periodically reaches 0 and 1?

9.5 Prove each of (9.72)–(9.78) from the operator definitions given earlier.

9.6 Find the $n = 1$ and $n = 2$ Landau level state wavefunctions $\psi_{1,\lambda}$ and $\psi_{2,\lambda}$ in the H, L_z eigenstate basis. Write your answers in terms of z and z^* defined in (9.80).

9.7 Consider an electron moving in the xy plane in the presence of uniform constant fields $\boldsymbol{B} = -B\hat{z}$ and $\boldsymbol{E} = E\hat{x}$. Use the same gauge choice as in (9.89) and (9.90). The presence of the electric field just adds a term eEX to the Hamiltonian.

(a) Show that there are stationary states with energies $E_{n,k} = \hbar\omega_c(n + 1/2) + c_1 kE/B + c_2 E^2/B^2$, where c_1 and c_2 are constants that you will determine, with corresponding wavefunctions $\psi_{n,k}(x, y) = \psi_n(x - x_0')e^{iky}$ where ψ_n are the standard one-dimensional harmonic oscillator wavefunctions, and x_0' is another constant that you will find.
(b) A wavepacket superposition of states with nearly the same k will have a velocity in the y direction, with dispersion because the energy depends on the wavenumber. Find the group velocity $v_g = (1/\hbar)\partial E_{n,k}/\partial k$. (This is equal to the time-averaged velocity for the cycloid motion of the corresponding classical problem.)

Examples with Spherical Symmetry 10

10.1 Stationary States with Spherical Symmetry

Consider a Hamiltonian of the form

$$H = \frac{P^2}{2\mu} + V(R). \tag{10.1}$$

In this chapter, we will write μ for the particle's mass, to avoid confusion with the eigenvalue $\hbar m$ of L_z, and R is the operator corresponding to the classical spherical coordinate r. The potential $V(R)$ is spherically symmetric, with no dependence on θ or ϕ. Likewise, the momentum squared operator

$$P^2 = P_x^2 + P_y^2 + P_z^2 = -\hbar^2\nabla^2 \tag{10.2}$$

is rotationally invariant, so there is no preferred direction associated with the Hamiltonian. It follows that H, L^2, and L_z are compatible observables, and we can look for an orthobasis of common eigenstates

$$|E, l, m\rangle, \tag{10.3}$$

where E is the energy of the state, the eigenvalue of the Hamiltonian. Because of (8.179), these are also parity eigenstates, with eigenvalue $(-1)^l$.

In the following, we will work in the position wavefunction representation with

$$\psi_{E,l,m}(r, \theta, \phi) = \langle r, \theta, \phi | E, l, m\rangle = R_{E,l}(r) Y_l^m(\theta, \phi), \tag{10.4}$$

where $R_{E,l}(r)$ is a radial wavefunction. Note that $R_{E,l}(r)$ will depend on both E and l in general, but it will not depend on m, because the operators L_+ and L_- raise and lower m without changing l or the radial dependence of the wavefunction. The

S. P. Martin, *Quantum Mechanics*, Graduate Texts in Physics,
https://doi.org/10.1007/978-3-032-21119-4_10

time-independent Schrödinger equation in this basis is

$$\left[-\frac{\hbar^2\nabla^2}{2\mu}+V(r)\right]\psi_{E,l,m}(r,\theta,\phi) = E\psi_{E,l,m}(r,\theta,\phi). \tag{10.5}$$

Now, using (8.105), this becomes

$$-\frac{\hbar^2}{2\mu}\left[\frac{d^2}{dr^2}+\frac{2}{r}\frac{d}{dr}-\frac{l(l+1)}{r^2}\right]R_{E,l}+V(r)R_{E,l} = ER_{E,l}, \tag{10.6}$$

where the spherical harmonics have been factored out.

In many cases, we will find that the radial wavefunction $R_{E,l}(r)$ has a power-law behavior r^p near the origin, for an integer p. By requiring the total probability to be finite, it is clear that p cannot be less than $-3/2$; otherwise, the normalization integral $\int dr\, r^2|R_{E,l}(r)|^2$ would diverge even for a finite volume. But in almost all cases, one can make the even stronger statement that $R_{E,l}(r)$ must be finite as $r \to 0$. This is because $\nabla^2(1/r) = -4\pi\delta^{(3)}(\boldsymbol{r})$, so that $p = -1$ would necessarily imply the presence of a delta function contribution to the potential $V(r)$. In the examples to follow, we will often make use of the finiteness of the radial wavefunction.

It is often useful to define a function

$$U_{E,l}(r) = rR_{E,l}(r), \tag{10.7}$$

in which one power of the radial coordinate has been factored out. The eigenvalue differential equation then becomes

$$\left[-\frac{\hbar^2}{2\mu}\frac{d^2}{dr^2}+\frac{\hbar^2 l(l+1)}{2\mu r^2}+V(r)\right]U_{E,l} = EU_{E,l}, \tag{10.8}$$

with the nice feature that there is no term with a single r derivative; that is the reason for sometimes using $U_{E,l}$ rather than $R_{E,l}$.

Indeed, (10.8) for $U_{E,l}$ is very similar to the one-dimensional time-independent Schrödinger equation (6.43) with x replaced by r. However, there are two important differences. First, the domain of the independent variable is now limited to non-negative values,

$$0 \le r < \infty, \tag{10.9}$$

unlike the one-dimensional Schrödinger equation. Second, the potential has effectively been modified to

$$V_{\text{eff}}(r) = V(r)+\frac{\hbar^2 l(l+1)}{2\mu r^2}, \tag{10.10}$$

where the second term is a repulsive centrifugal contribution that blows up at $r = 0$, and therefore makes the wavefunction vanish there unless $l = 0$. In fact, if $V(r)$ is finite at $r = 0$, one finds from (10.8) that for $l \neq 0$, $R_{E,l}$ must scale like r^l at small r.

With these caveats, the problem of stationary states for a particle in a spherically symmetric potential has been reduced to a particular type of one-dimensional problem. The same results that we derived in Sect. 6.3 [see the discussion surrounding (6.46)–(6.48)] for matching wavefunctions at special points hold here as for the one-dimensional problems. In particular, the radial wavefunction is always continuous, and its first derivative with respect to r is continuous at a point if the potential is finite (not necessarily continuous) there.

10.2 Free Particle in Spherical Coordinates

We have already discussed the position and momentum eigenstates and wavefunctions for a free particle moving in three dimensions, in Sect. 2.8. For position eigenstates, the CSCO used was (X, Y, Z), while for momentum eigenstates, it was (P_x, P_y, P_z). Now we will discuss the same physical problem using the CSCO (H, L^2, L_z) with eigenvalues $(E, \hbar^2 l(l+1), \hbar m)$, starting from (10.8) with $V(r) = 0$. We will solve the time-independent Schrödinger equation for fixed (E, l, m) in the position wavefunction representation with coordinates (r, θ, ϕ), where the domain may consist of only a sub-volume of the whole space. The resulting solutions can then be applied, using matching of the wavefunction, to problems in which a particle moves freely only in that sub-volume. They can also be applied to scattering problems in three dimensions.

To simplify things, define

$$k^2 = 2\mu E/\hbar^2, \qquad \rho = kr. \tag{10.11}$$

From the discussion in Sect. 2.8, we already know that the allowed eigenvalues E are non-negative, so that k is a real number with units of 1/length. Thus, for E positive, ρ is a radial coordinate rescaled to make it dimensionless. In terms of it, (10.8) becomes

$$\left(\frac{d^2}{d\rho^2} + 1 - \frac{l(l+1)}{\rho^2}\right) U_l(\rho) = 0. \tag{10.12}$$

Note that this equation and the forms of its solutions $U_l(\rho)$ do not depend on E, because the dependence has been hidden in ρ.

Let us start with the case $l = 0$. Then (10.12) is a familiar equation, with the familiar solutions $U_0(\rho) = \sin(\rho)$ or $\cos(\rho)$. Of these, the first solution $U_0 = \sin(\rho)$ gives a constant for $R(r)$ as $r \to 0$. However, the second solution $U_0 = \cos(\rho)$ gives a divergence $R(r) \sim 1/r$ for small r. This renders it physically unacceptable if the origin is included in the spatial region under consideration, because it would require a delta function potential $V(r)$, since $\nabla^2(1/r) = -4\pi\delta^{(3)}(\boldsymbol{r})$. Nevertheless, it is acceptable if we are solving for the free-particle wavefunction only in a region that does not include the origin. We therefore have two solutions for $L^2 = 0$, labeled A and B,

$$U_0^A(kr) = \sin(kr), \qquad U_0^B(kr) = -\cos(kr). \tag{10.13}$$

The minus sign is a phase choice for later convenience, and the B solution is acceptable if, and only if, the origin is excluded. These solutions for $l = 0$ can be used as seeds to find solutions for $l > 0$, as we will now see.

Writing the solution for general l in the form

$$U_l(\rho) = \rho^{l+1} f_l(\rho), \tag{10.14}$$

the differential equation (10.12) becomes

$$\left(\frac{d^2}{d\rho^2} + \frac{2(l+1)}{\rho}\frac{d}{d\rho} + 1\right) f_l(\rho) = 0. \tag{10.15}$$

Now we note that if $f_l(\rho)$ satisfies the differential equation for l, then

$$f_{l+1}(\rho) \equiv -\frac{1}{\rho}\frac{df_l}{d\rho} \tag{10.16}$$

will satisfy it for $l+1$, as can be proved by computing $d^2 f_{l+1}/d\rho^2$ and $df_{l+1}/d\rho$ and plugging into (10.15). So, for each of our two $l = 0$ solutions from (10.13),

$$f_0^A(\rho) = \frac{\sin\rho}{\rho}, \qquad f_0^B(\rho) = -\frac{\cos\rho}{\rho}, \tag{10.17}$$

by recursion we will have a solution

$$f_l = \left(-\frac{1}{\rho}\frac{d}{d\rho}\right)^l f_0. \tag{10.18}$$

Therefore, for each value of l, we have two linearly independent solutions for $R(r)$,

$$R_l^A(\rho) = \rho^l f_l^A = \rho^l \left(-\frac{1}{\rho}\frac{d}{d\rho}\right)^l \left(\frac{\sin\rho}{\rho}\right) \equiv j_l(\rho), \tag{10.19}$$

$$R_l^B(\rho) = \rho^l f_l^B = \rho^l \left(-\frac{1}{\rho}\frac{d}{d\rho}\right)^l \left(-\frac{\cos\rho}{\rho}\right) \equiv n_l(\rho). \tag{10.20}$$

The functions $j_l(\rho)$ are called the **spherical Bessel functions**, and $n_l(\rho)$ are called the **spherical Neumann functions**. The ones with $l \leq 2$ are

$$j_0(\rho) = \frac{\sin\rho}{\rho}, \qquad n_0(\rho) = -\frac{\cos\rho}{\rho}, \tag{10.21}$$

$$j_1(\rho) = \frac{\sin\rho}{\rho^2} - \frac{\cos\rho}{\rho}, \qquad n_1(\rho) = -\frac{\cos\rho}{\rho^2} - \frac{\sin\rho}{\rho}, \tag{10.22}$$

$$j_2(\rho)=\left(\frac{3}{\rho^3}-\frac{1}{\rho}\right)\sin\rho-\frac{3}{\rho^2}\cos\rho, \quad n_2(\rho)=\left(\frac{1}{\rho}-\frac{3}{\rho^3}\right)\cos\rho-\frac{3}{\rho^2}\sin\rho. \tag{10.23}$$

For small ρ, they can be shown to behave like

$$j_l(\rho) \sim \frac{\rho^l}{(2l+1)!!}, \tag{10.24}$$

$$n_l(\rho) \sim -\frac{(2l-1)!!}{\rho^{l+1}}, \tag{10.25}$$

where the double factorial notation means

$$(2l+1)!! = (2l+1)(2l-1)\cdots(1) = \frac{(2l+1)!}{2^l\, l!}, \tag{10.26}$$

with the special value $(-1)!! = 1$. Thus the j_l solutions are well-behaved at the origin, but the n_l solutions are not. For large ρ, they both fall off like $1/\rho$ multiplied by oscillating functions,

$$j_l(\rho) \sim \frac{1}{\rho}\sin(\rho - l\pi/2), \tag{10.27}$$

$$n_l(\rho) \sim -\frac{1}{\rho}\cos(\rho - l\pi/2). \tag{10.28}$$

The functions $n_l(\rho)$ are not allowed for the case of a completely free particle (one that has $V = 0$ everywhere including $r = 0$). However, they are useful in situations where we match the free-particle solutions in regions not including the origin to other solutions that do include the origin, as we will do in Sect. 10.4.

For example, suppose we have a potential that is spherically symmetric and piecewise constant within radial intervals, so that

$$V(r) = \begin{cases} V_0 & (0 \le r < a_1), \\ V_1 & (a_1 < r < a_2), \\ \cdots & \cdots \\ V_N & (a_N < r < \infty). \end{cases} \tag{10.29}$$

In each of these regions, the constant potential V_n can be absorbed into the constant energy, so the possible stationary-state solution wavefunctions are

$$\left(A_l^{(n)}\, j_l(k_n r) + B_l^{(n)}\, n_l(k_n r)\right)\, Y_l^m(\theta,\phi), \qquad \text{for } a_n < r < a_{n+1}, \tag{10.30}$$

for some constants $A_l^{(n)}$ and $B_l^{(n)}$, but now with

$$k_n = \sqrt{2\mu(E - V_n)}/\hbar, \tag{10.31}$$

and we define $a_0 = 0$ and $a_{N+1} = \infty$. The coefficients $A_l^{(n)}$ and $B_l^{(n)}$ can be determined by matching the wavefunctions, and their first derivative with respect to r, at each of the points a_n. This procedure also simultaneously determines the energy eigenvalue E. In the region $0 \le r < a_1$, the Neumann function solutions are not allowed, so one must have $B_l^{(0)} = 0$, but in all other regions $B_l^{(n)}$ is allowed to be nonzero. Note that there is a degeneracy $2l + 1$ for each of these stationary states, due to the fact that the energy does not depend on $m = -l, \ldots, l$. One can also match such wavefunctions to regions in which the potential is something more complicated.

Specific linear combinations of j_l and n_l are useful in certain kinds of problems. The **spherical Hankel functions** defined by

$$h_l^{(1)}(\rho) = j_l(\rho) + i n_l(\rho) = -i\rho^l \left(-\frac{1}{\rho}\frac{d}{d\rho}\right)^l \left(\frac{e^{i\rho}}{\rho}\right), \tag{10.32}$$

$$h_l^{(2)}(\rho) = j_l(\rho) - i n_l(\rho) = i\rho^l \left(-\frac{1}{\rho}\frac{d}{d\rho}\right)^l \left(\frac{e^{-i\rho}}{\rho}\right), \tag{10.33}$$

correspond to outgoing and ingoing spherical waves, respectively. To understand this, note that their behavior for large $\rho = kr$ is

$$h_l^{(1)}(\rho) \approx (-i)^{l+1} e^{i\rho}/\rho, \qquad h_l^{(2)}(\rho) \approx i^{l+1} e^{-i\rho}/\rho, \tag{10.34}$$

so that the time dependence of the corresponding wavefunctions is

$$\langle \boldsymbol{r}\,|\psi^{(1)}(t)\rangle \propto e^{-itE/\hbar} e^{ikr}/r, \tag{10.35}$$

$$\langle \boldsymbol{r}\,|\psi^{(2)}(t)\rangle \propto e^{-itE/\hbar} e^{-ikr}/r. \tag{10.36}$$

These maintain constant phase at $kr = \omega t + \text{constant}$ and $kr = -\omega t + \text{constant}$, respectively, where $\omega = E/\hbar$. (The phase velocity is therefore $\omega/k = E/p = \hbar k/2\mu$, but recall from the discussion leading to (6.26) that superpositions of wave packets with a finite spread of ω will have a group velocity twice this, $\hbar k/\mu$, which is the classical value.)

For problems in which $E < V_n$ in at least one region, k will be imaginary in that region. Then the solutions will involve $j_l(kr)$ and $n_l(kr)$ or $h_l^{(1)}(kr)$ and $h_l^{(2)}(kr)$, where now $k = i\sqrt{2\mu(V_n - E)}/\hbar$ is a purely imaginary number. In many such cases, a more convenient basis of solutions is provided by the **modified spherical**

Bessel functions,

$$i_l(\rho) = j_l(i\rho)/i^l = \rho^l \left(\frac{1}{\rho}\frac{d}{d\rho}\right)^l \left(\frac{\sinh\rho}{\rho}\right), \tag{10.37}$$

$$k_l(\rho) = -i^l h_l^{(1)}(i\rho) = \rho^l \left(-\frac{1}{\rho}\frac{d}{d\rho}\right)^l \left(\frac{e^{-\rho}}{\rho}\right), \tag{10.38}$$

where the names i_l and k_l are unfortunate but traditional, and

$$\rho = \kappa r, \qquad \kappa = \sqrt{2\mu(V_n - E)}/\hbar. \tag{10.39}$$

The function $i_l(\rho)$ is well-defined and useful for regions that include $r = 0$, while the function $k_l(\rho)$ is useful for regions that extend to $r = \infty$.

For the rest of this section, we consider a completely free particle whose domain includes all space, including the origin. The (H, L^2, L_z) eigenstate orthobasis wavefunctions are

$$\langle \boldsymbol{r}|k, l, m\rangle = \psi_{k,l,m}(\boldsymbol{r}) = R_{k,l}(r) Y_l^m(\theta, \phi), \qquad (E = \hbar^2 k^2/2\mu), \tag{10.40}$$

where the radial wavefunctions $R_{k,l}(r) = A_{k,l}\, j_l(kr)$ contain a normalization constant $A_{k,l}$ to be fixed. To do so, we can use the orthonormality relations for the spherical Bessel functions,[1]

$$\int_0^\infty dr\, r^2 j_l(kr) j_l(k'r) = \frac{\pi}{2k^2}\delta(k - k'), \tag{10.41}$$

for each l. It follows that if we choose $A_{k,l} = \sqrt{2/\pi}$, so that the radial wavefunctions are

$$R_{k,l}(r) = \sqrt{\frac{2}{\pi}}\, j_l(kr), \tag{10.42}$$

then we have the orthonormality relations,

$$\int dr\, r^2 R_{k',l}(r) R_{k,l}(r) = \frac{1}{k^2}\delta(k - k'), \tag{10.43}$$

and for the full eigenstates and their wavefunctions,

[1] Equation (10.41) is technically ill-defined, because the integration does not converge, even for $k \neq k'$. It should therefore be viewed as a formal relation, to be used within expressions where k or k' is integrated over. This is similar to the technically ill-defined nature of the Fourier integral for the delta function in (2.34).

$$
\begin{aligned}
\langle k', l', m' | k, l, m \rangle &= \int d^3\boldsymbol{r}\ \psi_{k',l',m'}(\boldsymbol{r})^*\ \psi_{k,l,m}(\boldsymbol{r}) \\
&= \frac{1}{k^2}\delta(k - k')\,\delta_{ll'}\delta_{mm'} . \qquad (10.44)
\end{aligned}
$$

With our choice of normalization in (10.42)–(10.44), we also have the completeness of the orthobasis in terms of an integral over k,

$$
\int_0^\infty dk\, k^2 \sum_{l=0}^{\infty} \sum_{m=-l}^{l} |k, l, m\rangle\langle k, l, m| = I, \qquad (10.45)
$$

from which follows

$$
\int_0^\infty dk\, k^2 \sum_{l=0}^{\infty} \sum_{m=-l}^{l} \psi_{k,l,m}(\boldsymbol{r}')^* \psi_{k,l,m}(\boldsymbol{r}) = \delta^{(3)}(\boldsymbol{r} - \boldsymbol{r}'). \qquad (10.46)
$$

A perfectly good alternative normalization choice would be to express the same results in terms of "energy normalization" kets

$$
|E, l, m\rangle = \frac{\sqrt{\mu k}}{\hbar} |k, l, m\rangle , \qquad (10.47)
$$

and the corresponding radial wavefunctions

$$
R_{E,l}(r) = \sqrt{\frac{2\mu k}{\pi \hbar^2}}\, j_l(kr). \qquad (10.48)
$$

If one uses $|E, l, m\rangle$, $R_{E,l}(r)$, and $\psi_{E,l,m}(\boldsymbol{r})$, then the Dirac orthonormality and completeness relations will have the same form as (10.43)–(10.46), but with $\delta(E - E')$ replacing $\delta(k - k')/k^2$ and $\int_0^\infty dE$ replacing $\int_0^\infty dk\, k^2$. This follows from $\delta(k - k') = \delta(E - E')\hbar^2 k/\mu$.

Let us now find the relation between the plane wavefunctions associated with momentum eigenstates $|\boldsymbol{p}\rangle$ and the spherical waves associated with the (H, L^2, L_z) eigenstates $|k, l, m\rangle$. Using the completeness relation, we have

$$
\langle \boldsymbol{r} | \boldsymbol{p} \rangle = \int_0^\infty dk\, k^2 \sum_{l=0}^{\infty} \sum_{m=-l}^{l} \langle \boldsymbol{r} | k, l, m\rangle\langle k, l, m | \boldsymbol{p}\rangle . \qquad (10.49)
$$

Now, both $|\boldsymbol{p}\rangle$ and $|k, l, m\rangle$ are eigenstates of H with energy eigenvalues $E = |\boldsymbol{p}|^2/2\mu$ and $\hbar^2 k^2/2\mu$, respectively. Therefore, Theorem 2.14 says that the last inner product in (10.49) must vanish unless $|\boldsymbol{p}| = \hbar k$, so it must be of the form

$$\langle k, l, m | \boldsymbol{p} \rangle = \frac{1}{(2\pi\hbar)^{3/2}} \frac{1}{k^2} \delta(k - |\boldsymbol{p}|/\hbar) \sqrt{\frac{\pi}{2}}\, C_{l,m}(\theta_{\boldsymbol{p}}, \phi_{\boldsymbol{p}}), \tag{10.50}$$

for some functions $C_{l,m}$ that depend only on l, m, and the angular coordinates of the vector $\boldsymbol{p}$. (The $C_{l,m}$ cannot depend on the magnitude k, because they are dimensionless and there is no other dimensionful quantity on which they could depend.) Since $k = |\boldsymbol{p}|/\hbar$ is enforced by the delta function, we can define $\boldsymbol{k} = \boldsymbol{p}/\hbar$ and use $\langle \boldsymbol{r} | \boldsymbol{k} \rangle$ from (2.256) and $\langle \boldsymbol{r} | k, l, m \rangle$ from (10.40) to find that (10.49) reads

$$\frac{e^{i\boldsymbol{k}\cdot\boldsymbol{r}}}{(2\pi\hbar)^{3/2}} = \frac{1}{(2\pi\hbar)^{3/2}} \sum_{l=0}^{\infty} \sum_{m=-l}^{l} C_{l,m}(\theta_{\boldsymbol{k}}, \phi_{\boldsymbol{k}})\, j_l(kr) Y_l^m(\theta, \phi). \tag{10.51}$$

The claim is that the coefficient functions turn out to be

$$C_{l,m}(\theta_{\boldsymbol{k}}, \phi_{\boldsymbol{k}}) = 4\pi\, i^l\, Y_l^m(\theta_{\boldsymbol{k}}, \phi_{\boldsymbol{k}})^*, \tag{10.52}$$

so that we have the remarkable identity relating plane waves to spherical waves,

$$e^{i\boldsymbol{k}\cdot\boldsymbol{r}} = \sum_{l=0}^{\infty} \sum_{m=-l}^{l} 4\pi i^l\, Y_l^m(\theta_{\boldsymbol{k}}, \phi_{\boldsymbol{k}})^*\, j_l(kr) Y_l^m(\theta, \phi). \tag{10.53}$$

Note that $(\theta_{\boldsymbol{k}}, \phi_{\boldsymbol{k}})$ are the spherical coordinate angles for the vector $\boldsymbol{k}$, while (θ, ϕ) are the angles for the vector $\boldsymbol{r}$ in the same coordinate system. The interpretation of (10.53) is that a plane wave with momentum $\boldsymbol{p} = \hbar\boldsymbol{k}$ consists of a superposition of spherical waves with all allowed (quantized) values of orbital angular momentum.

To prove (10.53), we first consider the case that $\boldsymbol{k} = k\hat{z}$. Then $\theta_{\boldsymbol{k}} = 0$, and $\boldsymbol{k}\cdot\boldsymbol{r} = kr\cos\theta$, so $e^{i\boldsymbol{k}\cdot\boldsymbol{r}} = e^{ikr\cos\theta}$. Because this does not depend on ϕ at all, its expansion in terms of spherical harmonics will only include the $m = 0$ functions $Y_l^0(\theta, \phi)$, which are proportional to the ordinary Legendre polynomials $P_l(\cos\theta)$. So, in this case the expansion in (10.53) has the simpler form

$$e^{ikr\cos\theta} = \sum_{l=0}^{\infty} c_l\, j_l(kr) P_l(\cos\theta), \tag{10.54}$$

where c_l are some coefficients that we need to determine. Now we can multiply both sides of (10.54) by $P_{l'}(\cos\theta)$, and integrate with respect to $\cos\theta$, making use of the identity

$$\int_{-1}^{1} du\, P_l(u) P_{l'}(u) = \frac{2}{2l+1}\delta_{ll'}. \tag{10.55}$$

The result, after relabeling $l' \to l$, is

$$c_l\, j_l(kr) = \frac{2l+1}{2} \int_{-1}^{1} d(\cos\theta)\, P_l(\cos\theta) e^{ikr\cos\theta}. \tag{10.56}$$

This must hold for all r, so we can consider the leading behavior as $r \to 0$ in particular. Using $j_l(kr) \approx (kr)^l/(2l+1)!!$ from (10.24), this gives

$$c_l \frac{(kr)^l}{(2l+1)!!} + O(r^{l+1}) = \frac{2l+1}{2} \int_{-1}^{1} du\, P_l(u) e^{ikru}. \tag{10.57}$$

This shows that the right side evidently must be proportional to r^l in the limit of small r. Expanding e^{ikru} in a power series in r, using $e^{ikru} = \sum_{n=0}^{\infty}(ikru)^n/n!$, this implies the identities

$$\int_{-1}^{1} du\, u^n P_l(u) = 0 \qquad \text{(for integer } n < l\text{)}, \tag{10.58}$$

and the result we need (from the $n = l$ term),

$$c_l \frac{k^l}{(2l+1)!!} = \frac{2l+1}{2} \frac{(ik)^l}{l!} \int_{-1}^{1} du\, u^l P_l(u). \tag{10.59}$$

Using (8.137) for the Legendre polynomials, one can obtain the integral

$$\int_{-1}^{1} du\, u^l P_l(u) = \frac{2(l!)}{(2l+1)!!}, \tag{10.60}$$

by integrating by parts l times. Thus, (10.59) simplifies to

$$c_l = (2l+1) i^l, \tag{10.61}$$

so that

$$e^{ikr\cos\theta} = \sum_{l=0}^{\infty} (2l+1) i^l\, j_l(kr) P_l(\cos\theta). \tag{10.62}$$

Now for the case of $\boldsymbol{k}$ in an arbitrary direction, substitute $kr\cos\theta \to \boldsymbol{k}\cdot\boldsymbol{r}$ in the preceding, or

$$\cos\theta \to \hat{k}\cdot\hat{r}. \tag{10.63}$$

Using the spherical harmonic addition identity, (8.164), one finally arrives at (10.53).

10.3 Particle Confined Inside a Sphere

As an application of the results of the previous section, consider a particle of mass μ confined within a sphere of radius a, but otherwise free, so that the potential is

$$V(r) = \begin{cases} 0 & (r < a), \\ \infty & (r \geq a). \end{cases} \tag{10.64}$$

The stationary-state wavefunctions are

$$\psi_{E,l,m} = \begin{cases} C_{E,l,m}\, j_l(kr) Y_l^m(\theta, \phi) & (r \leq a), \\ 0 & (r \geq a), \end{cases} \tag{10.65}$$

where $C_{E,l,m}$ are normalization constants and $E = \hbar^2 k^2/2\mu$. Continuity of the wavefunction at $r = a$ requires that

$$j_l(ka) = 0, \tag{10.66}$$

and this boundary condition determines the allowed quantized energy levels $E_{n,l}$. Let us see how this works for $l = 0, 1, 2$.

For $l = 0$, (10.66) becomes simply $\sin(ka)/ka = 0$, so $ka = n\pi$, where n is a positive integer. The energies are therefore

$$E_{n,0} = \frac{\hbar^2 k^2}{2\mu} = \frac{\hbar^2 \pi^2 n^2}{2\mu a^2}, \tag{10.67}$$

with corresponding wavefunctions for $l = m = 0$,

$$\psi_{n,0,0} = C_{n,0,0} \frac{\sin(n\pi r/a)}{n\pi r/a}. \tag{10.68}$$

Note that these wavefunctions approach a nonzero constant at $r = 0$. The number of zeros of the radial wavefunction, including the one at $r = a$, is n. The constant can be fixed by requiring the unit normalization condition $1 = 4\pi \int_0^a dr\, r^2 |\psi_{n,0,0}|^2$, which yields $C_{n,0,0} = n\sqrt{\pi/2a^3}$.

For $l = 1$, the boundary condition (10.66) gives

$$\tan(ka) = ka, \tag{10.69}$$

which is a transcendental equation that can be understood graphically and then solved numerically for $X = ka$ as shown in Fig. 10.1. The lowest three energy solutions have $ka/\pi \approx (1.4303,\ 2.4590,\ 3.4709,\ \ldots)$, so

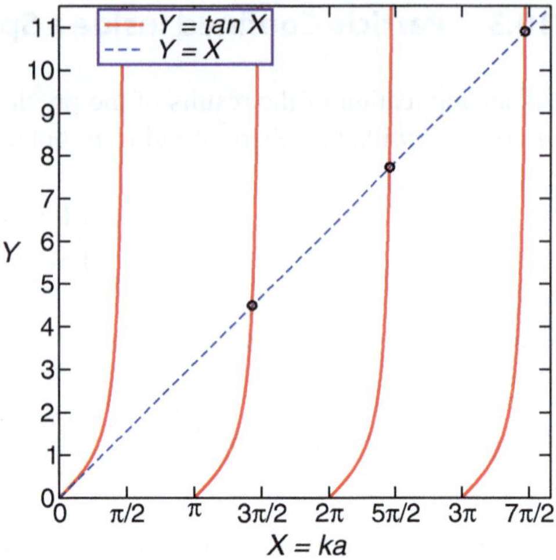

Fig. 10.1 The solutions of the transcendental equation $\tan X = X$ for positive $X = ka$ provide the eigenvalues for k for the $l = 1$ states of a particle confined to a ball of radius a. This graph shows the lowest three solutions $X \approx \pi(1.4303,\ 2.4590,\ 3.4709) = (4.4934,\ 7.7253,\ 10.9041)$, obtained as the intersections of $Y = \tan X$ with the line $Y = X$

$$E_{n,1} \approx \frac{\hbar^2\pi^2}{2\mu a^2}(2.0458,\ 6.0468,\ 12.0471,\ \ldots), \qquad \text{for } n = (2, 3, 4, \ldots), \tag{10.70}$$

where the label n is again the number of zeros of the radial wavefunction. In the $l = 1$ case, $r = 0$ and $r = a$ are always both zeros, so $n \geq 2$. Each of the $l = 1$ energy levels has degeneracy 3, because the quantum number $m = -1, 0, 1$ does not affect the energy.

For $l = 2$, the boundary condition (10.66) becomes

$$\tan(ka) = \frac{ka}{1 - (ka)^2/3}. \tag{10.71}$$

The solutions can again be found numerically, with the results $ka/\pi \approx (1.8346,\ 2.8950,\ 3.9225,\ \ldots)$, so

$$E_{n,2} \approx \frac{\hbar^2\pi^2}{2\mu a^2}(3.3656,\ 8.3812,\ 15.3861,\ \ldots), \qquad \text{for } n = (2, 3, 4, \ldots), \tag{10.72}$$

where again the label n is the number of zeros of the radial wavefunction $j_2(kr)$. Each of these energy levels has degeneracy 5, corresponding to $m = -2, -1, 0, 1, 2$.

The energy eigenvalues for higher angular momentum quantum number l can be numerically solved for in a similar way, and have increasingly higher energies. The degeneracy of each energy level $E_{n,l}$ is $2l + 1$, corresponding to the allowed L_z eigenvalues $\hbar m$. The lowest few energy levels, with $E_{n,l} < 10\hbar^2\pi^2/\mu a^2$ and $l \leq 7$, are depicted in Fig. 10.2.

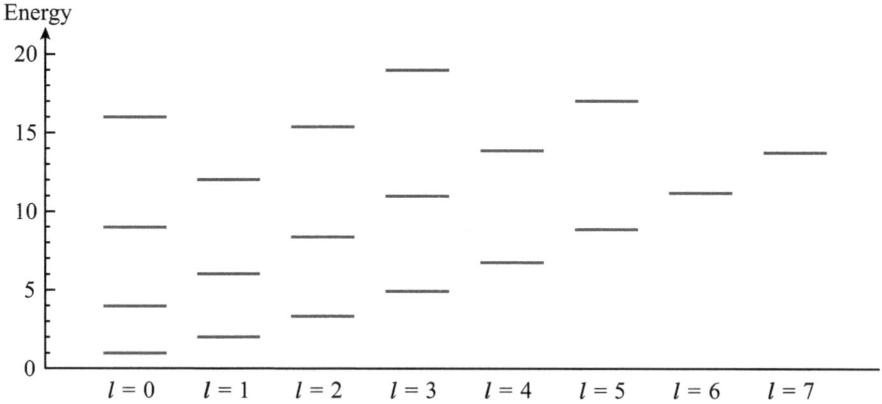

Fig. 10.2 The lowest energy levels, in units of the ground-state energy $E_{1,0} = \hbar^2\pi^2/2\mu a^2$, for a particle of mass μ confined inside a sphere of radius a, labeled by the angular momentum quantum number l. All energy eigenvalues with $E_{n,l} < 10\hbar^2\pi^2/\mu a^2$ and $l \leq 7$ are shown

10.4 Particle in a Spherical Potential Well

Consider a particle of mass μ in a spherical potential well of radius a and finite depth V_0,

$$V(r) = \begin{cases} -V_0 & (r < a), \\ 0 & (r \geq a). \end{cases} \tag{10.73}$$

We will consider bound-state solutions, which have $E < 0$. As discussed in Sect. 10.2, the stationary-state solution for the region $r < a$ involves the ordinary spherical Bessel function $j_l(kr)$, but now with

$$k = \sqrt{2\mu(E + V_0)}/\hbar. \tag{10.74}$$

The Neumann functions $n_l(\rho)$ are not finite at $\rho = 0$ and so do not appear for $r < a$. The solution for $r > a$ uses the modified spherical Bessel function $k_l(\kappa r)$ [see (10.38)] with

$$\kappa = \sqrt{-2\mu E}/\hbar, \tag{10.75}$$

because $k_l(\rho)$ is the linear combination that is well-behaved (vanishing exponentially, in fact) as $\rho \to \infty$. The stationary-state wavefunctions are therefore

$$\psi_{E,l,m}(\mathbf{r}) = \begin{cases} A\, j_l(kr)\, Y_l^m(\theta,\phi) & (r \leq a), \\ B\, k_l(\kappa r)\, Y_l^m(\theta,\phi) & (r \geq a), \end{cases} \tag{10.76}$$

where A and B are normalization constants, and

$$k^2 + \kappa^2 = 2\mu V_0/\hbar^2 \tag{10.77}$$

from (10.74) and (10.75).

At $r = a$, the wavefunction and its first derivative with respect to r are continuous, since the potential is finite there. This gives

$$A\, j_l(ka) = B\, k_l(\kappa a), \tag{10.78}$$

$$kA\, j_l'(ka) = \kappa B\, k_l'(\kappa a). \tag{10.79}$$

Taking the ratio of these to eliminate A and B yields

$$\frac{k\, j_l'(ka)}{j_l(ka)} = \frac{\kappa\, k_l'(\kappa a)}{k_l(\kappa a)}, \tag{10.80}$$

which is a transcendental equation that can be used together with (10.77) to solve for the allowed eigenvalues k, κ, and thus E. As always for a spherically symmetric potential, the allowed energies depend on l, but not m. Either equation (10.78) or (10.79) then also allows for the ratio A/B to be found for each E, l. The remaining unknown corresponds to the overall magnitude of the constants A and B, which can be fixed by requiring unit normalization of the wavefunction.

For example, if $l = 0$, (10.80) yields

$$-k \cot(ka) = \kappa. \tag{10.81}$$

Writing dimensionless positive variables $X = ka$ and $Y = \kappa a$, (10.77) and (10.81) give

$$X^2 + Y^2 = 2\mu V_0 a^2/\hbar^2, \tag{10.82}$$

$$-X \cot X = Y. \tag{10.83}$$

Fortunately, these are exactly the same equations we encountered for the odd-parity solutions of the one-dimensional square well problem, with $L \to 2a$. [Compare to (6.84) and (6.90).] The same graphical and numerical analysis therefore applies. In particular, if we label the $l = 0$ stationary states by $n = 1, 2, 3, \ldots$, then the condition for the bound state $|\psi_n\rangle$ to exist is

$$2\mu V_0 a^2/\hbar^2 > \pi^2(n - 1/2)^2. \tag{10.84}$$

Recall that in the case of a particle in a one-dimensional square well, there is always at least one bound state, but for a sufficiently shallow potential well only the even-parity ground state exists as a bound state. For the three-dimensional spherical well, since only the analogs of the odd-parity one-dimensional square-well states exist, (10.84) tells us that the existence of a bound state requires the potential to be sufficiently deep, $V_0 > \pi^2\hbar^2/8\mu a^2$. The more general lesson is that three-dimensional potentials may have no bound states if they are not sufficiently attractive.

10.5 Isotropic Three-Dimensional Harmonic Oscillator

In Sect. 7.5, we have already encountered the isotropic three-dimensional harmonic oscillator as a special case of the general anisotropic version, and obtained its stationary-state energy levels and degeneracies. We did this using a CSCO of (H_x, H_y, H_z), the individual Hamiltonians for excitations in the x, y, and z directions. In this section, we will solve the problem again, this time with a CSCO consisting of (H, L^2, L_z). This is possible because, in the isotropic case,

$$H = \frac{P^2}{2\mu} + \frac{1}{2}\mu\omega^2 R^2 \tag{10.85}$$

is invariant under all rotations and therefore commutes with L^2 and L_z. Note that we are now using μ as the symbol for the mass of the particle.

Before proceeding, let us anticipate what sort of answer we expect, given the results of Sect. 7.5. At an energy level $E = \hbar\omega(n + 3/2)$, the wavefunctions must be linear combinations of functions of the form

$$\langle x, y, z|n_x, n_y, n_z\rangle \propto H_{n_x}(x/b)H_{n_y}(y/b)H_{n_z}(z/b)\, e^{-(x^2+y^2+z^2)/2b^2}, \tag{10.86}$$

with $n_x + n_y + n_z = n$, where

$$b = \sqrt{\hbar/\mu\omega} \tag{10.87}$$

is the length scale in the problem, and the H_{n_a} are the Hermite polynomials. After translating to spherical coordinates, such a linear combination that is also an eigenstate of L^2 and L_z must be of the form

$$\langle r, \theta, \phi\,|n, l, m\rangle = R_{n,l}(r)\, Y_l^m(\theta, \phi), \tag{10.88}$$

where

$$R_{n,l}(r) = (\text{polynomial in } r \text{ of degree } n)\, e^{-r^2/2b^2}. \tag{10.89}$$

Our goal is to solve for the functions $R_{n,l}(r)$.

In Sect. 10.1, we found the differential equation for $U_{n,l} = r R_{n,l}$, which in the present case becomes

$$\left[-\frac{\hbar^2}{2\mu}\frac{d^2}{dr^2} + \frac{1}{2}\mu\omega^2 r^2 + \frac{\hbar^2 l(l+1)}{2\mu r^2} - E\right] U_{n,l} = 0. \tag{10.90}$$

Inspired by (10.89), we define dimensionless quantities x and $y(x)$ by

$$x = r/b, \qquad U_{n,l} = y(x)e^{-x^2/2}. \tag{10.91}$$

When plugged into (10.90), this gives

$$y'' - 2xy' + \left[\mathcal{E} - 1 - l(l+1)/x^2\right] y = 0, \tag{10.92}$$

where

$$\mathcal{E} = 2\mu b^2 E/\hbar^2 = 2E/\hbar\omega \tag{10.93}$$

is a dimensionless combination proportional to the energy eigenvalue. We already know from (7.157) that the allowed eigenvalues are $\mathcal{E} = 2n + 3$ for non-negative integers n, and from (10.89) that $y(x)$ is a polynomial, but in the following derivation we will proceed as if these facts were not known.

We now try a series solution for $y(x)$, of the form

$$y = x^q \sum_{p=0}^{\infty} c_p x^p. \tag{10.94}$$

Here q and the c_p are constants, with $c_0 \neq 0$ by definition. (Otherwise, we would adjust the value of q to make it so.) Now we prepare to plug into (10.92) by computing

$$y'' = \sum_{p=0}^{\infty} c_p(p+q)(p+q-1)x^{p+q-2}, \tag{10.95}$$

$$y/x^2 = \sum_{p=0}^{\infty} c_p x^{p+q-2}, \tag{10.96}$$

$$xy' = \sum_{p=0}^{\infty} c_p(p+q)x^{p+q} = \sum_{p=0}^{\infty} c_{p-2}(p+q-2)x^{p+q-2}, \tag{10.97}$$

$$y = \sum_{p=0}^{\infty} c_{p-2} x^{p+q-2}, \tag{10.98}$$

where in the last equality of (10.97) and in (10.98) we have used the trick of relabeling $p \to p-2$ and defining $c_{-2} = c_{-1} = 0$. The motivation behind this relabeling trick is that now all of the summands have the same powers of x, and so (10.92) becomes

$$\sum_{p=0}^{\infty} x^{p+q-2} \Big\{ c_p\big[(p+q)(p+q-1) - l(l+1)\big] \\ - c_{p-2}\big[2(p+q-2)+1-\mathcal{E}\big]\Big\} = 0. \tag{10.99}$$

For this equation to be satisfied for all x, each coefficient of a given power of x must vanish, so the quantity in braces must vanish for each p.

From the first term $p = 0$, using $c_{-2} = 0$ we find

$$c_0\big[q(q-1) - l(l+1)\big] = 0. \tag{10.100}$$

Since $c_0 \neq 0$, the possible solutions are $q = l+1$ and $q = -l$. However, the latter can be rejected on physical grounds, since it would imply that $y \sim x^{-l}$ for small x, which would mean $R_{n,l} \sim 1/r^{l+1}$ for small r, and the wavefunction would diverge at the point $r = 0$. Therefore,

$$q = l+1, \tag{10.101}$$

which implies that $R_{n,l} \sim r^l$ for small r. This suppression for small r is consistent with the existence of the repulsive $\hbar^2 l(l+1)/2\mu r^2$ centrifugal barrier contribution to the effective potential for nonzero angular momentum.

The second term $p = 1$ gives us

$$c_1\big[q(q+1) - l(l+1)\big] = 0, \tag{10.102}$$

because $c_{-1} = 0$. Now, plugging in $q = l+1$, this becomes $2(l+1)c_1 = 0$. Our knowledge of angular momentum eigenvalues tells us that $l+1$ cannot vanish, so we must have $c_1 = 0$.

For all larger p, the vanishing of (10.99) implies (after using $q = l+1$) that

$$c_p\, p(p+2l+1) = c_{p-2}\,(2p+2l-1-\mathcal{E}). \tag{10.103}$$

Since $c_1 = 0$, it follows that $c_p = 0$ for all odd p. Defining $C_j = c_{2j}$, we have

$$y = x^{l+1} \sum_{j=0}^{\infty} C_j x^{2j}, \tag{10.104}$$

with, from (10.103) by taking $p = 2(j+1)$, the recurrence relation

$$C_{j+1} = \frac{4j + 2l + 3 - \mathcal{E}}{2(j+1)(2j + 2l + 3)} C_j. \tag{10.105}$$

For large j, this gives $C_j \approx C_{j-1}/j$. So, if the series does not terminate, it would behave asymptotically like $y \sim x^{l+1} \sum_j (x^2)^j/j! \sim x^{l+1} e^{x^2}$, which would imply that $R_{n,l} \sim r^l e^{-r^2/2b^2} e^{r^2/b^2} \sim r^l e^{r^2/2b^2}$. As in the case of the one-dimensional harmonic oscillator, such a solution that blows up exponentially as $r \to \infty$ is unphysical, since it cannot be normalized.

We can therefore conclude that the series must terminate, and $y(x)$ is actually a polynomial, in agreement with (10.89). Then the recurrence relation (10.105) implies that physically valid solutions for $y(x)$ must have $C_{k+1} = 0$ for some integer $k \geq 0$, and so satisfy

$$\mathcal{E} = 4k + 2l + 3, \tag{10.106}$$

or, using (10.93),

$$E = \hbar\omega(2k + l + 3/2). \tag{10.107}$$

This shows that the energy eigenkets $|E, l, m\rangle$ are linear combinations of the kets $|n_x, n_y, n_z\rangle$, where

$$n_x + n_y + n_z = n = 2k + l. \tag{10.108}$$

An interesting feature of these results is that, unlike the particle-in-a-sphere and particle-in-a-spherical-well examples of Sects. 10.3 and 10.4, here the degeneracies in the energy levels are not entircly due to the rotational invariance of the problem. For a given l, there are $2l + 1$ degenerate states with $m = -l, \ldots, l$ that have the same energy. But there are also "accidental" energy degeneracies between states with different angular momentum magnitudes, for example $k = 0, l = 2$ and $k = 1, l = 0$.

Summarizing what we have learned, the three-dimensional isotropic harmonic oscillator stationary-state wavefunctions with definite (L^2, L_z) are

$$\psi_{k,l,m}(r, \theta, \phi) = A\, R_{k,l}(r)\, Y_l^m(\theta, \phi), \tag{10.109}$$

where A is a normalization constant, and the radial wavefunctions are of the form

$$R_{k,l}(r) = (r/b)^l P_{k,l}(r^2/b^2) e^{-r^2/2b^2}, \tag{10.110}$$

where, with $z = x^2 = r^2/b^2$,

$$P_{k,l}(z) = \sum_{j=0}^{k} C_j z^j. \tag{10.111}$$

Using (10.106), the recurrence relation (10.105) becomes

$$C_{j+1} = \frac{2(j-k)}{(j+1)(2j+2l+3)} C_j. \tag{10.112}$$

For the lowest few values $k = 0, 1, 2, 3$, we have (leaving the coefficient C_0 as an arbitrary normalization for the moment):

$$P_{0,l}(z) = C_0, \tag{10.113}$$

$$P_{1,l}(z) = C_0\left(1 - \frac{2}{2l+3}z\right), \tag{10.114}$$

$$P_{2,l}(z) = C_0\left(1 - \frac{4}{2l+3}z + \frac{4}{(2l+3)(2l+5)}z^2\right), \tag{10.115}$$

$$P_{3,l}(z) = C_0\left(1 - \frac{6}{2l+3}z + \frac{12}{(2l+3)(2l+5)}z^2 - \frac{8}{(2l+3)(2l+5)(2l+7)}z^3\right). \tag{10.116}$$

These polynomials are proportional to **associated Laguerre polynomials**, for which there are unfortunately several different notational conventions, so one must be careful in comparing difference sources. Here, we adopt the definition of the associated Laguerre polynomial $L_k^\alpha(z)$ used by the popular computer program Mathematica,

$$P_{k,l}(z) = L_k^{l+1/2}(z), \tag{10.117}$$

corresponding to the choice

$$C_0 = \frac{(2l+2k+1)!!}{2^k\, k!\, (2l+1)!!} = \frac{(2l+2k+1)!\, l!}{2^{2k}\, (2l+1)!\, (l+k)!\, k!}. \tag{10.118}$$

More generally, our conventional normalization is such that, at $z = 0$ and for any α,

$$L_k^\alpha(0) = \frac{\Gamma(k+\alpha+1)}{k!\, \Gamma(\alpha+1)}. \tag{10.119}$$

Here, the **Gamma function** is defined by

$$\Gamma(z) = \int_0^\infty dt\, t^{z-1} e^{-t}, \tag{10.120}$$

and satisfies $\Gamma(z) = (z-1)\Gamma(z-1)$. If n is a non-negative integer, then $n! = \Gamma(n+1)$. Another useful value is $\Gamma(1/2) = \sqrt{\pi}$.

An explicit, general form for the associated Laguerre polynomials is

$$L_N^\alpha(z) = \frac{1}{N!} z^{-\alpha} e^z \frac{d^N}{dz^N}\left(e^{-z} z^{N+\alpha}\right). \tag{10.121}$$

Note that the lower index N is always an integer equal to the degree of the polynomial, but the raised index α need not be, as in the present application where $\alpha = l + 1/2$. They satisfy the differential equation

$$\left[z\frac{d^2}{dz^2} + (1 - z + \alpha)\frac{d}{dz} + N\right] L_N^\alpha(z) = 0, \tag{10.122}$$

and the orthogonality relation

$$\int_0^\infty dz\, e^{-z} z^\alpha L_N^\alpha(z) L_{N'}^\alpha(z) = \frac{\Gamma(N+\alpha+1)}{N!}\delta_{NN'}. \tag{10.123}$$

The degree N is also equal to the number of zeros (in the present case, $N = k$), which all occur for positive real z.

Putting everything together, the wavefunctions for the stationary states of the three-dimensional isotropic harmonic oscillator with energies $E = \hbar\omega(2k+l+3/2)$ are

$$\psi_{k,l,m}(r,\theta,\phi) = \sqrt{\frac{2(k!)}{b^3\,\Gamma(k+l+3/2)}} \left(\frac{r}{b}\right)^l L_k^{l+1/2}(r^2/b^2)\, e^{-r^2/2b^2}\, Y_l^m(\theta,\phi). \tag{10.124}$$

The normalization factor A in (10.109) has been chosen, with the aid of (10.123), so that orthonormality holds,

$$\begin{aligned} \langle k', l', m' | k, l, m\rangle &= \int_0^\infty dr\, r^2 \int d\Omega\, \left[\psi_{k',l',m'}(r,\theta,\phi)\right]^* \psi_{k,l,m}(r,\theta,\phi) \\ &= \delta_{kk'}\delta_{ll'}\delta_{mm'}. \end{aligned} \tag{10.125}$$

Exercises

10.1 Consider a particle of mass μ in a spherically symmetric potential $V(R)$ in a stationary bound state with energy E and zero angular momentum, so that the wavefunction $\psi(r)$ depends only on r. The quantity $|\psi(0)|^2$ is useful for evaluating "contact" quantities that depend on the probability for the particle to be found at the potential source, the origin. (An example is the Darwin term of the hydrogen atom fine structure.) Show that in general

$$|\psi(0)|^2 = \frac{\mu}{2\pi\hbar^2}\langle V'(R)\rangle, \tag{10.126}$$

with the standard wavefunction normalization such that $4\pi\int_0^\infty dr\, r^2\, |\psi(r)|^2 = 1$. [Hint: multiply the Schrödinger equation $-\frac{\hbar^2}{2\mu}\left(\psi'' + \frac{2}{r}\psi'\right) + [V(r) - E]\psi = 0$ by $\psi'(r)$, and integrate by parts.] Check this result in the special cases of the $l = 0$ states of a particle confined to the interior of a sphere, and the ground state of the isotropic harmonic oscillator.

10.2 Use the spherical Bessel and Neumann function forms in (10.21)–(10.23) to verify the small-distance forms of (10.24) and (10.25) for $l = 0, 1, 2$.

10.3 Consider a particle of mass μ trapped inside a ball of radius b that has a hard core of radius a, so that the potential in spherical coordinates is

$$V(r) = \begin{cases} \infty & (\text{for } r < a \text{ and for } r > b), \\ 0 & (\text{for } a < r < b). \end{cases} \tag{10.127}$$

This means that the eigenstates of H, L^2, and L_z have wavefunctions of the form $\Psi_{E,l,m}(r, \theta, \phi) = [Aj_l(kr) + Bn_l(kr)]Y_l^m(\theta, \phi)$ for the region $a < r < b$.

(a) Find all of the allowed energy eigenstates and eigenvalues for $l = 0$. [Hint: use boundary conditions to solve for the ratio B/A twice, and require the two expressions to be equal.]
(b) For the case $l = 1$, find a transcendental equation whose solutions will yield the energy eigenvalues. Put your equation into the form $\tan[k(b - a)] = \{$an expression not involving sines or cosines$\}$. [Hint: first put the equation into a form that is polynomial in ka, kb, and their sines and cosines; then use trigonometric identities for $\sin(kb - ka)$ and $\cos(kb - ka)$.]
(c) For the special case $l = 1$ and $b = 2a$, write your transcendental equation in the form $\tan X = X/(1 + NX^2)$, where $X = ka$ and N is a certain integer that you will discover. Solve for X numerically to at least 3 digits of accuracy, and obtain the lowest energy for $l = 1$. How does it compare to the lowest energy for $l = 0$ that you found in part (a)?

10.4 Consider the isotropic three-dimensional harmonic oscillator problem, with potential $V(x, y, z) = \frac{1}{2}\mu\omega^2(x^2 + y^2 + z^2)$. The Hamiltonian H can be written as the sum of $H_x = \hbar\omega(a_x^\dagger a_x + 1/2)$, $H_y = \hbar\omega(a_y^\dagger a_y + 1/2)$, and $H_z = \hbar\omega(a_z^\dagger a_z + 1/2)$, which form a CSCO with corresponding orthonormal eigenbasis $|n_x, n_y, n_z\rangle$. Another choice of CSCO is H, L^2, and L_z, with corresponding eigenbasis $|n, l, m\rangle'$, where $n = n_x + n_y + n_z$. (The prime distinguishes the two types of orthobasis elements, since they both have three integer labels.)

(a) Show that the angular momentum component operators L_x, L_y, and L_z can be written as

$$L_b = -i\hbar\epsilon_{bcd}a_c^\dagger a_d. \tag{10.128}$$

(b) Construct the operator L^2 in terms of the ladder operators. Write the answer in normal-ordered form, which means that the commutation relations have been used to ensure that no creation operator appears to the right of an annihilation operator, as

$$\begin{aligned} L^2 = \hbar^2\big[&N_1(a_x^{\dagger 2}a_y^2 + a_x^{\dagger 2}a_z^2 + a_y^{\dagger 2}a_x^2 + a_y^{\dagger 2}a_z^2 + a_z^{\dagger 2}a_x^2 + a_z^{\dagger 2}a_y^2) \\ &+N_2(a_x^\dagger a_y^\dagger a_x a_y + a_x^\dagger a_z^\dagger a_x a_z + a_y^\dagger a_z^\dagger a_y a_z) \\ &+N_3(a_x^\dagger a_x + a_y^\dagger a_y + a_z^\dagger a_z)\big], \end{aligned} \tag{10.129}$$

where N_1, N_2, and N_3 are certain integers that you will discover. Note that L_x, L_y, L_z, and L^2 all give 0 acting on the ground state with $n = 0$.

(c) For the subspace of states $|1, 0, 0\rangle$, $|0, 1, 0\rangle$, and $|0, 0, 1\rangle$ with $n = 1$, find the action of L^2 and L_z. What are their matrix representations in that basis? Find the eigenvalues and eigenvectors of L_z within this subspace.

(d) For the subspace of states with $n = 2$, find the action of L^2 on each of the $|n_x, n_y, n_z\rangle$ basis states. Using these results, and using the ordering $|2, 0, 0\rangle$, $|0, 2, 0\rangle$, $|0, 0, 2\rangle$, $|1, 1, 0\rangle$, $|1, 0, 1\rangle$, $|0, 1, 1\rangle$, find the 6×6 matrix representation for L^2. Find the eigenvalues and normalized eigenvectors of L^2 for the $n = 2$ subspace in that basis.

(e) Compute the action of L_z on each of the simultaneous eigenvectors of H, L^2 found in the previous part. Within each sub-subspace of fixed $n = 2$ and fixed l, find the eigenvalues and eigenvectors of L_z, and so conclude by writing the six $|2, l, m\rangle'$ orthobasis states as linear combinations of the six $|n_x, n_y, n_z\rangle$ eigenstates.

Coulomb Potential and Hydrogen-Like Atoms

11

11.1 Bound States for Hydrogen-Like Atoms

One of the most important exactly solvable problems in quantum mechanics is that of a particle moving in a Coulomb potential. This includes the hydrogen atom, and more generally hydrogen-like ions, which consist of a single electron with mass m_e and charge $-e$ [with the normalization as defined in Gaussian (cgs) system units, see (1.1) and (1.2)], and a heavy nucleus with mass m_N and charge Ze, where Z is an integer. As discussed in a general context in Sect. 4.2, the problem can be separated into center-of-mass and relative degrees of freedom, where the latter uses a reduced mass $\mu = m_e m_N/(m_e + m_N)$, which is very close to the electron mass. In this chapter, we will simply write the mass as m_e, with the understanding that in a more precise numerical treatment one should replace $m_e \to \mu$ everywhere in the obvious way. For the ordinary (lightest isotope) hydrogen atom, m_N is the proton mass and the numerical relationship is $\mu = 0.999456 m_e$. We also neglect the small effects of special relativity and electron and nuclear spins, which will be treated in Chap. 17.

So, our Hamiltonian is $H = P^2/2m_e + V(R)$, where the potential energy in the position representation is

$$V(r) = -Ze^2/r. \tag{11.1}$$

The special case $Z = 1$ is the hydrogen atom. Because the potential is spherically symmetric, we choose the CSCO to be (H, L^2, L_z), and look for an orthobasis of eigenstates $|E, l, m\rangle$ with wavefunctions

$$\psi_{E,l,m}(r, \theta, \phi) = \langle r, \theta, \phi\,|E, l, m\rangle = R_{E,l}(r)\, Y_l^m(\theta, \phi). \tag{11.2}$$

The goal is to find the energy eigenvalues and the radial wavefunctions $R_{E,l}(r)$.

In this section, we will work out the properties of the bound-state solutions, for which the energy eigenvalues are discrete and satisfy $E < V(\infty) = 0$, and

S. P. Martin, *Quantum Mechanics*, Graduate Texts in Physics,
https://doi.org/10.1007/978-3-032-21119-4_11

$|R_{E,l}|$ decreases exponentially as $r \to \infty$. Unlike the case of the three-dimensional harmonic oscillator, there are also unbound energy eigenstates, for which E is continuous and non-negative, and $rR_{E,l}$ oscillates with an amplitude approaching a constant for large r. The unbound-state solutions will be found in Sect. 11.2.

The time-independent Schrödinger equation for the radial wavefunction is

$$\left[-\frac{\hbar^2}{2m_e}\left(\frac{d^2}{dr^2}+\frac{2}{r}\frac{d}{dr}-\frac{l(l+1)}{r^2}\right)-\frac{Ze^2}{r}-E\right]R_{E,l}(r) = 0. \quad (11.3)$$

A good first step is to replace r by a dimensionless variable. Since $-\hbar^2/2m_eE$ has units of [length]^2 and is a positive number for the bound states, we define a rescaled radial coordinate

$$s = 2r/b, \qquad b = \hbar/\sqrt{-2m_eE}. \quad (11.4)$$

With this change of variables, (11.3) becomes

$$\left(\frac{d^2}{ds^2}+\frac{2}{s}\frac{d}{ds}-\frac{l(l+1)}{s^2}+\frac{n}{s}-\frac{1}{4}\right)R_{E,l} = 0, \quad (11.5)$$

where we have introduced

$$n = \frac{Ze^2}{\hbar}\sqrt{\frac{m_e}{-2E}}, \quad (11.6)$$

a dimensionless quantity that parameterizes the energy eigenvalue. Note that n is a real number for $E < 0$, and is purely imaginary for $E > 0$. We will soon find out [see (11.15)] that n must be a positive integer for a bound-state solution, but we do not know that yet.

The next part of our strategy is to identify, and factor out, the large-distance and short-distance behaviors of $R_{E,l}$. For $s \to \infty$, (11.5) becomes $d^2R/ds^2 \approx R/4$, which has two linearly independent solutions, $e^{s/2}$ and $e^{-s/2}$. The first of these is unphysical, as it blows up for $s \to \infty$. In the opposite limit $s \to 0$, the last two terms of (11.5) can be neglected compared to the others, and there is a power-law solution for R proportional to s^l. We therefore write

$$R_{E,l} = s^l e^{-s/2} f(s), \quad (11.7)$$

which factors out the leading behavior in the two limits. Plugging this into (11.5) gives

$$s\frac{d^2f}{ds^2}+(2l+2-s)\frac{df}{ds}+(n-l-1)f = 0, \quad (11.8)$$

which we must now solve simultaneously for the eigenfunctions $f(s)$ and the corresponding eigenvalues n.

Equation (11.8) is a special case of a famous differential equation, called the **confluent hypergeometric equation**, which in its traditional general form is

$$x\frac{d^2F}{dx^2} + (c - x)\frac{dF}{dx} - aF = 0, \tag{11.9}$$

where a and c are constants. This has a unique (up to a multiplicative constant) solution that is finite as $x \to 0$, the confluent hypergeometric function, which has a series expansion

$$F(a, c, x) = 1 + \frac{a}{c}x + \frac{a(a+1)}{c(c+1)}\frac{x^2}{2!} + \frac{a(a+1)(a+2)}{c(c+1)(c+2)}\frac{x^3}{3!} + \cdots, \tag{11.10}$$

as can be verified by direct substitution into the differential equation. This series converges for all finite $|x|$, even if a, c, and x are complex, provided that $c \neq 0, -1, -2, \ldots$. One can check (either from the series solution, or by direct substitution into the differential equation) that it has an integral representation

$$F(a, c, x) = \frac{\Gamma(c)}{\Gamma(a)\Gamma(c-a)}\int_0^1 dt\, e^{tx}\, t^{a-1}(1-t)^{c-a-1}, \tag{11.11}$$

provided that $\text{Re}[c] > \text{Re}[a] > 0$. For real x large and positive, it has the asymptotic form

$$F(a, c, x) \approx \frac{\Gamma(c)}{\Gamma(a)}\, x^{a-c}\, e^x \qquad (\text{large real } x, \text{ and } a \neq 0, -1, -2, \ldots). \tag{11.12}$$

However, as we are about to see, the special case relevant for our present purposes will arise when a is a non-positive integer; in that case, $F(a, c, x)$ is instead a polynomial in x of degree $-a$, because the series in (11.10) terminates.

For the present application to the bound states of the hydrogen-like atom, we have $x = s$ and $a = l + 1 - n$ and $c = 2l + 2$, by comparing (11.8) to (11.9). Therefore, $f(s)$ in (11.7) is equal to $F(l{+}1{-}n,\ 2l{+}2,\ s)$, up to normalization, so

$$R_{n,l} = C\, s^l e^{-s/2} F(l{+}1{-}n,\ 2l{+}2,\ s). \tag{11.13}$$

Here we have replaced the subscript label E by the label n, which contains the same information due to (11.6), and C is a normalization constant to be chosen later. In the large-distance limit $s \to \infty$, the asymptotic form of (11.12) would seem to tell us

$$R_{n,l} \propto s^{-n-1} e^{s/2} \qquad (n{-}l{-}1 \neq 0, 1, 2, \ldots). \tag{11.14}$$

However, the exponential factor $e^{s/2}$ is the form that we already rejected as non-normalizable and therefore physically unacceptable; it would imply that no matter how far from the nucleus you look, the electron must have infinitely larger probability to be farther away. The only way to get a physically sensible bound state is to arrange for (11.14) not to apply. In other words, it is necessary that the series solution (11.10) for $F(l{+}1{-}n,\ 2l{+}2,\ s)$ terminates, so that instead of being proportional to e^s, it is actually a polynomial in s with degree that we will call k. This implies that

$$n = k + l + 1. \tag{11.15}$$

Since k, being the degree of the polynomial, is a non-negative integer, and $l + 1$ is always a positive integer, n must be a positive integer, called the **principal quantum number** of the hydrogen atom bound state.

Inverting (11.6), the allowed energy eigenvalues are

$$E_n = -\left(\frac{e^4 m_e}{2\hbar^2}\right)\frac{Z^2}{n^2} = -\left(\frac{e^2}{2a_0}\right)\frac{Z^2}{n^2}, \tag{11.16}$$

which depend only on n, not on k and l individually. To write the last expression we have defined[1] the **Bohr radius** (named after Niels Bohr),

$$a_0 = \frac{\hbar^2}{e^2 m_e} = 5.292 \times 10^{-11} \text{ meters} = \text{Bohr radius}. \tag{11.17}$$

The scale of energy is therefore

$$\frac{e^4 m_e}{2\hbar^2} = \frac{e^2}{2a_0} = \frac{\hbar^2}{2m_e a_0^2} = 2.180 \times 10^{-18} \text{ Joules} = 13.606 \text{ eV} = 1 \text{ Rydberg}, \tag{11.18}$$

named after Johannes Rydberg. The ground state has $n = 1$ and $k = l = 0$. More generally, the lowest possible energy for each l is obtained for $k = 0$, so that $n = l + 1$.

The energy levels of the hydrogen atom are depicted in Fig. 11.1. At each bound-state energy level n, the values of l that can occur are the integers from 0 to

[1] The precise numerical definitions of the Bohr radius and Rydberg energy units as listed here use the infinite nuclear mass limit, with $\mu = m_e$. An alternative definition uses, instead of the electron mass m_e, the reduced mass μ for the lightest isotope of hydrogen, which is smaller by a factor $m_p/(m_e + m_p) \approx 0.999456$.

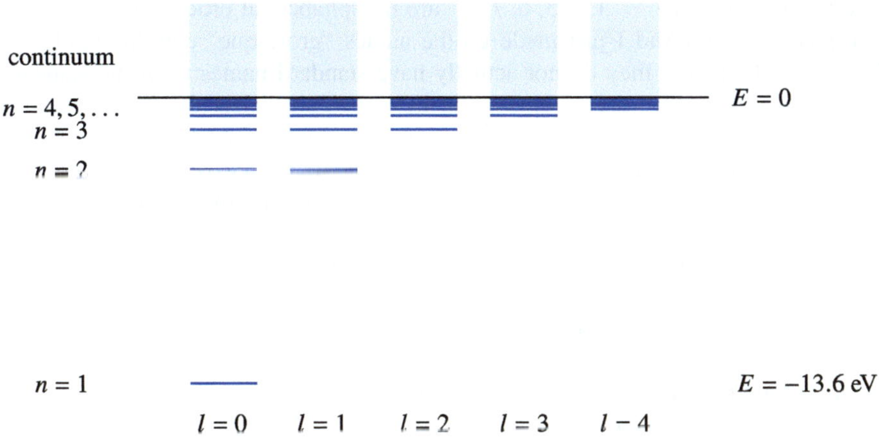

Fig. 11.1 Stationary-state energy levels of the hydrogen atom for $l \leq 4$. The bound state levels with $E_n = -13.6\,\text{eV}/n^2$ for $n = 1, 2, 3, \ldots$ are discrete and have degeneracy n^2, with $0 \leq l \leq n-1$. For each non-negative integer l, there are also unbound continuum energy eigenstates with $E \geq 0$

$n - 1$, each with degeneracy $2l + 1$, which comes from $m = -l, \ldots, l$. The total degeneracy for each energy level E_n is therefore

$$g_n = \sum_{l=0}^{n-1}(2l+1) = n^2. \tag{11.19}$$

Because this includes states with different angular momenta, the energy degeneracy is accidental, meaning that it cannot be explained by rotational invariance alone. This was also the case for the three-dimensional isotropic harmonic oscillator. For hydrogen-like atoms, this degeneracy is slightly broken by spin and relativistic effects, as we will discuss in detail in Chap. 17.

The distinct values of l are given letter codes that go back to the early history of spectroscopic analysis, specifically the visual appearance of photon emission lines from alkali metals. This **spectroscopic notation** is as follows:

$$\begin{aligned}
l = 0 &\leftrightarrow s \text{ "sharp"},\\
l = 1 &\leftrightarrow p \text{ "principal"},\\
l = 2 &\leftrightarrow d \text{ "diffuse"},\\
l = 3 &\leftrightarrow f \text{ "fundamental"},\\
l = 4 &\leftrightarrow g \text{ "grotesque"},\\
l = 5 &\leftrightarrow h \text{ "horrendous"},\\
\cdots &\quad \cdots
\end{aligned} \tag{11.20}$$

The letter codes for $l = 3, 4, 5, 6, 7, \ldots$ are in alphabetical order $f, g, h, i, k, \ldots$, but j is omitted. (And I just made up the names "grotesque" and "horrendous". Unlike the first four, they do not actually have standard names to fit the standard letters.) A supposedly useful mnemonic is "sober **p**hysicists **d**on't **f**ind **g**iraffes **h**iding **i**n **k**itchens". The hydrogen atom states are often referred to in the notation nl, but with $l = 0, 1, 2, 3, 4, 5, \ldots$ replaced by the spectroscopic code[2] letter $s, p, d, f, g, h \ldots$, so that the ground state is called $1s$ and the first excited states are $2s$ and $2p$, and the second excited states are $3s$, $3p$, and $3d$.

Returning to the radial wavefunction, (11.13) tells us that

$$R_{n,l} = C\, s^l e^{-s/2} F_{n,l}(s), \tag{11.21}$$

where we are now adopting the shorter notation

$$F_{n,l}(s) = F(l{+}1{-}n,\ 2l{+}2, s), \tag{11.22}$$

which is a polynomial of degree $k = n - l - 1$. Recalling that $s = 2r/b$ from (11.4), the exponential factor tells us that the wavefunctions are only significant within a distance set, in terms of the Bohr radius, by the length scale

$$b = na_0/Z. \tag{11.23}$$

Equations (11.16) and (11.23) show that the magnitude of the binding energy scales like Z^2, while the characteristic length scales like $1/Z$. In the classical limit $\hbar \to 0$, the Bohr radius goes to 0 and the binding energy of the ground state goes to $-\infty$, in accord with the discussion of the classical instability of atoms in Sect. 1.1.

For the lowest few energy states, the polynomials are, from (11.10),

$$F_{1,0} = 1, \tag{11.24}$$

$$F_{2,1} = 1, \quad F_{2,0} = 1 - \frac{s}{2}, \tag{11.25}$$

$$F_{3,2} = 1, \quad F_{3,1} = 1 - \frac{s}{4}, \quad F_{3,0} = 1 - s + \frac{s^2}{6}, \tag{11.26}$$

$$F_{4,3} = 1, \quad F_{4,2} = 1 - \frac{s}{6}, \quad F_{4,1} = 1 - \frac{s}{2} + \frac{s^2}{20}, \quad F_{4,0} = 1 - \frac{3s}{2} + \frac{s^2}{2} - \frac{s^3}{24}. \tag{11.27}$$

More generally, in terms of the associated Laguerre polynomials $L_N^{\alpha}(x)$ already defined explicitly by (10.121), it can be shown that

[2] In some other contexts, the letter codes for orbital angular momentum in (11.20) are capitalized.

$$F_{n,l} = \frac{(n-l-1)!\,(2l+1)!}{(n+l)!}\,L_{n-l-1}^{2l+1}(s). \tag{11.28}$$

The lower index on L_N^α is always equal to the degree of the associated Laguerre polynomial, which is also the number of its zeros, in this case $N = k$. Note, however, that in the case of the isotropic three-dimensional harmonic oscillator the upper index α was always half-integer [see (10.117)], while here $\alpha = 2l + 1$ is a positive odd integer.

From Theorem 2.14, we know that the kets $|n, l, m\rangle$ and $|n', l', m'\rangle$ are orthogonal whenever any of n, l, or m differ from n', l', or m', respectively. The orthogonality conditions $\delta_{ll'}$ and $\delta_{mm'}$ are already enforced by the proportionality of the wavefunctions to spherical harmonics. It follows that the associated Laguerre polynomials must also satisfy an orthogonality relation of the form (taking $x = ns = 2Zr/a_0$)

$$\int_0^\infty dx\, x^{2l+2}\, L_{n-l-1}^{2l+1}(x/n)\, L_{n'-l-1}^{2l+1}(x/n')\, e^{-x(1/n+1/n')/2} = \delta_{nn'}\, B_{n,l}, \tag{11.29}$$

for $n, n' = 1, 2, 3, \ldots$ and $l = 0, 1, \ldots, \min(n, n') - 1$. Although not obvious, it can be checked that this is true, with

$$B_{n,l} = \frac{2\,(n+l)!\,n^{2l+4}}{(n-l-1)!}. \tag{11.30}$$

Note that (11.29) is a quite different orthogonality relation for associated Laguerre polynomials than the one that was useful for the three-dimensional harmonic oscillator, (10.123).

Putting together (11.4), (11.21), (11.23), and (11.28), we obtain the radial wavefunctions of the hydrogen-like atom bound states,

$$R_{n,l}(r) = A_{n,l}\, e^{-Zr/na_0} \left(\frac{2Zr}{na_0}\right)^l L_{n-l-1}^{2l+1}(2Zr/na_0), \tag{11.31}$$

where we have introduced a new normalization constant factor

$$A_{n,l} = \left(\frac{2Z}{a_0}\right)^{3/2} \frac{n^l}{\sqrt{B_{n,l}}} = \left(\frac{Z}{a_0}\right)^{3/2} \frac{2}{n^2}\sqrt{\frac{(n-l-1)!}{(n+l)!}}, \tag{11.32}$$

chosen so that

$$\int_0^\infty dr\, r^2\, R_{n,l}(r) R_{n',l}(r) = \delta_{nn'}. \tag{11.33}$$

The full wavefunction [obtained by including the $Y_l^m(\theta, \phi)$ factor according to (11.2)] is

$$\psi_{n,l,m}(r,\theta,\phi) = R_{n,l}(r)Y_l^m(\theta,\phi). \tag{11.34}$$

An arbitrary choice of phase has been made in $A_{n,l}$, as usual. With this sensible choice, the radial wavefunctions $R_{n,l}$ are all real, and are positive for small r.

We now have everything necessary to evaluate the wavefunctions for the lowest few energy levels. For the ground state,

$$R_{1,0}(r) = \left(\frac{Z}{a_0}\right)^{3/2} 2e^{-Zr/a_0}, \tag{11.35}$$

and for the first excited states,

$$R_{2,0}(r) = \left(\frac{Z}{a_0}\right)^{3/2} \frac{1}{2\sqrt{2}}\left(2 - \frac{Zr}{a_0}\right) e^{-Zr/2a_0}, \tag{11.36}$$

$$R_{2,1}(r) = \left(\frac{Z}{a_0}\right)^{3/2} \frac{1}{2\sqrt{6}}\frac{Zr}{a_0} e^{-Zr/2a_0}, \tag{11.37}$$

and for the second excited states,

$$R_{3,0}(r) = \left(\frac{Z}{a_0}\right)^{3/2} \frac{2}{81\sqrt{3}}\left[27 - 18\frac{Zr}{a_0} + 2\left(\frac{Zr}{a_0}\right)^2\right] e^{-Zr/3a_0}, \tag{11.38}$$

$$R_{3,1}(r) = \left(\frac{Z}{a_0}\right)^{3/2} \frac{2\sqrt{2}}{81\sqrt{3}}\left[6\frac{Zr}{a_0} - \left(\frac{Zr}{a_0}\right)^2\right] e^{-Zr/3a_0}, \tag{11.39}$$

$$R_{3,2}(r) = \left(\frac{Z}{a_0}\right)^{3/2} \frac{2\sqrt{2}}{81\sqrt{15}}\left(\frac{Zr}{a_0}\right)^2 e^{-Zr/3a_0}. \tag{11.40}$$

These radial wavefunctions, and the corresponding radial probability densities $r^2|R_{n,l}|^2$, are graphed in Fig. 11.2 for $Z = 1$. For future reference, we also note that the radial wavefunctions at $r = 0$ can be evaluated, using (11.28), (11.31), and (11.32), as

$$R_{n,l}(0) = 2\left(\frac{Z}{na_0}\right)^{3/2} \delta_{l0}. \tag{11.41}$$

In particular, it is only nonzero for states with orbital angular momentum $l = 0$.

Putting in the spherical harmonics with the Condon–Shortley phase convention, the wavefunctions for the lowest few energy levels are

$$\psi_{1,0,0} = \left(\frac{Z}{a_0}\right)^{3/2} \frac{1}{\sqrt{\pi}} e^{-Zr/a_0}, \tag{11.42}$$

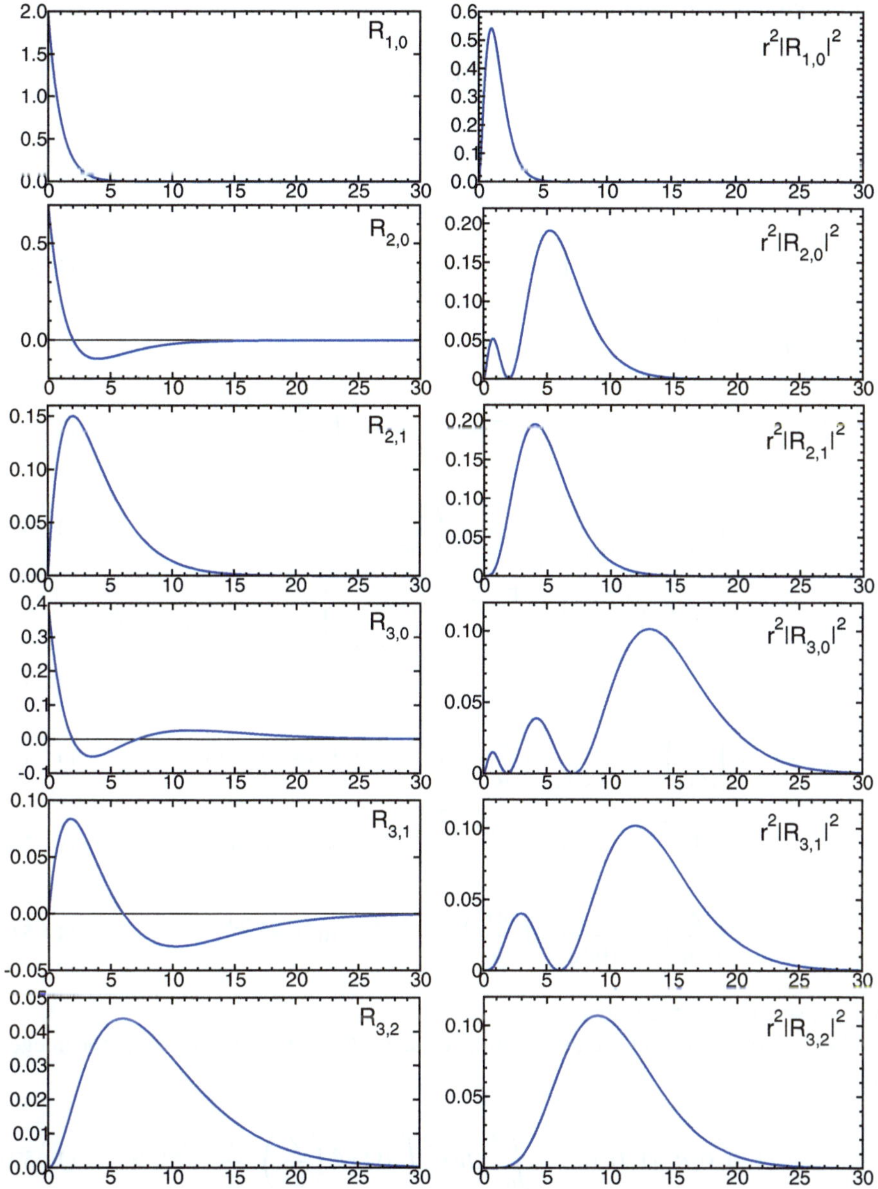

Fig. 11.2 Radial wavefunctions $R_{n,l}$ in units of $1/a_0^{3/2}$ (left column) and radial probability densities $r^2|R_{n,l}|^2$ in units of $1/a_0$ (right column) for the $Z = 1$ hydrogen atom stationary states with $n = 1, 2, 3$, as functions of r/a_0. Note the differing vertical scales

for the ground state, and

$$\psi_{2,0,0} = \left(\frac{Z}{a_0}\right)^{3/2} \frac{1}{4\sqrt{2\pi}} \left(2 - \frac{Zr}{a_0}\right) e^{-Zr/2a_0}, \tag{11.43}$$

$$\psi_{2,1,0} = \left(\frac{Z}{a_0}\right)^{3/2} \frac{1}{4\sqrt{2\pi}} \frac{Zr}{a_0} e^{-Zr/2a_0} \cos\theta, \tag{11.44}$$

$$\psi_{2,1,\pm1} = \mp\left(\frac{Z}{a_0}\right)^{3/2} \frac{1}{8\sqrt{\pi}} \frac{Zr}{a_0} e^{-Zr/2a_0} \sin\theta\, e^{\pm i\phi}, \tag{11.45}$$

for the first excited states, and

$$\psi_{3,0,0} = \left(\frac{Z}{a_0}\right)^{3/2} \frac{1}{81\sqrt{3\pi}} \left[27 - 18\frac{Zr}{a_0} + 2\left(\frac{Zr}{a_0}\right)^2\right] e^{-Zr/3a_0}, \tag{11.46}$$

$$\psi_{3,1,0} = \left(\frac{Z}{a_0}\right)^{3/2} \frac{\sqrt{2}}{81\sqrt{\pi}} \left[6\frac{Zr}{a_0} - \left(\frac{Zr}{a_0}\right)^2\right] e^{-Zr/3a_0} \cos\theta, \tag{11.47}$$

$$\psi_{3,1,\pm1} = \mp\left(\frac{Z}{a_0}\right)^{3/2} \frac{1}{81\sqrt{\pi}} \left[6\frac{Zr}{a_0} - \left(\frac{Zr}{a_0}\right)^2\right] e^{-Zr/3a_0} \sin\theta\, e^{\pm i\phi}, \tag{11.48}$$

$$\psi_{3,2,0} = \left(\frac{Z}{a_0}\right)^{3/2} \frac{1}{81\sqrt{6\pi}} \left(\frac{Zr}{a_0}\right)^2 e^{-Zr/3a_0} (3\cos^2\theta - 1), \tag{11.49}$$

$$\psi_{3,2,\pm1} = \mp\left(\frac{Z}{a_0}\right)^{3/2} \frac{1}{81\sqrt{\pi}} \left(\frac{Zr}{a_0}\right)^2 e^{-Zr/3a_0} \sin\theta \cos\theta\, e^{\pm i\phi}, \tag{11.50}$$

$$\psi_{3,2,\pm2} = \left(\frac{Z}{a_0}\right)^{3/2} \frac{1}{162\sqrt{\pi}} \left(\frac{Zr}{a_0}\right)^2 e^{-Zr/3a_0} \sin^2\theta\, e^{\pm 2i\phi}, \tag{11.51}$$

for the second excited states.

Let us now work out the expectation value of R^p in the ground state, for integer p,

$$\langle 1, 0, 0|R^p|1, 0, 0\rangle = \int_0^\infty dr\, r^2 \int d\Omega\, r^p \left(\frac{Z}{a_0}\right)^3 \frac{1}{\pi} e^{-2Zr/a_0}. \tag{11.52}$$

Using $\int d\Omega = 4\pi$, this evaluates to

$$\langle 1, 0, 0|R^p|1, 0, 0\rangle = \begin{cases} \frac{1}{2}(p+2)! \left(\frac{a_0}{2Z}\right)^p & (p \geq -2), \\ \infty & (p \leq -3). \end{cases} \tag{11.53}$$

In particular, plugging in $p = 0$ gives $\langle 1,0,0|1|1,0,0\rangle = 1$ (confirming the correct normalization of the wavefunction), while $p = 1$ gives

$$\langle 1,0,0|R|1,0,0\rangle = \langle R\rangle = \frac{3a_0}{2Z}, \tag{11.54}$$

and $p = -1$ gives

$$\langle 1,0,0|\, 1/R\, |1,0,0\rangle = Z/a_0. \tag{11.55}$$

(Note that this is not equal to $1/\langle R\rangle$.) We can also directly compute, for the ground state,

$$\langle 1,0,0|P^2|1,0,0\rangle = \int_0^\infty dr\, r^2 \int d\Omega\, \psi^*_{1,0,0}(-\hbar^2\nabla^2)\psi_{1,0,0} = \hbar^2 Z^2/a_0^2. \tag{11.56}$$

As a check, the expectation value of the Hamiltonian in the ground state is

$$\langle H\rangle = \langle P^2\rangle/2m_e - Ze^2\langle 1/R\rangle = -Z^2e^2/2a_0 = -Z^2\,\text{Rydberg}, \tag{11.57}$$

in agreement with the energy eigenvalue.

Later, we will want more general expressions for the expectation values $\langle R^p\rangle$ in arbitrary hydrogen-like atom stationary bound states $|n,l,m\rangle$, for various integer powers p. A helpful tool to find these for all integer $p \geq -1$, and for all integer $p \leq -3$ in terms of the single seed case $p = -2$, is the **Kramers–Pasternack recurrence formula**, which relates the expectation values for any three consecutive integer powers of the radial coordinate R. It is

$$\frac{q+1}{n^2}\langle R^q\rangle - (2q+1)\frac{a_0}{Z}\langle R^{q-1}\rangle + \frac{q}{4}\left[(2l+1)^2 - q^2\right]\frac{a_0^2}{Z^2}\langle R^{q-2}\rangle = 0, \tag{11.58}$$

valid for all $q > -2l-1$.

The proof of (11.58) is far from obvious, but goes as follows. Start from the differential equation for the radial wavefunction $R_{n,l}(s)$ with $s = 2Zr/na_0$, as given in (11.5). Multiply by $4s^{q+3}R'_{n,l} + 2(1-q)s^{q+2}R_{n,l}$ (this is perhaps the most non-obvious part), and then integrate with respect to s. Then, expand the integrand and eliminate all derivatives of $R_{n,l}$ using the integration-by-parts identities

$$\int_0^\infty ds\,\frac{d}{ds}\left(s^{q+3}R'^2_{n,l}\right) = 0, \tag{11.59}$$

$$\int_0^\infty ds\,\frac{d}{ds}\left(s^{q+2}R_{n,l}R'_{n,l}\right) = 0, \tag{11.60}$$

$$\int_0^\infty ds\,\frac{d}{ds}\left(s^{p}R^2_{n,l}\right) = 0, \tag{11.61}$$

for each of $p = q+1, q+2$, and $q+3$ in the last equation. The boundary terms in (11.59)–(11.61) vanish for $q > -2l - 1$, because $R_{n,l}$ scales like s^l for $s \to 0$, and like $e^{-s/2}$ for $s \to \infty$. The result is

$$\int_0^\infty ds \left\{(q+1)s^{q+2} - 2(2q+1)ns^{q+1} + q\left[(2l+1)^2 - q^2\right]s^q\right\} R_{n,l}^2 = 0. \tag{11.62}$$

Finally, using the change-of-integration-variable relation,

$$\langle R^p \rangle = \int_0^\infty dr\, r^{p+2}\left[R_{n,l}(r)\right]^2 = (na_0/2Z)^{p+3}\int_0^\infty ds\, s^{p+2}\left[R_{n,l}(s)\right]^2, \tag{11.63}$$

from (11.4), and (11.23), we arrive at (11.58).

Using $q = 0$ in the Kramers–Pasternack formula of (11.58), and $\langle R^0 \rangle = \langle 1 \rangle = 1$, we immediately find

$$\langle 1/R \rangle = \frac{Z}{n^2 a_0}, \tag{11.64}$$

in agreement with the $n = 1$ special case in (11.55). Then, using $q = 1$, one obtains

$$\langle R \rangle = \frac{a_0}{2Z}[3n^2 - l(l+1)]. \tag{11.65}$$

Using this as a measure of the "size" of the state $|n, l, m\rangle$, we note that it does not depend on the magnetic quantum number m, and

- shrinks with larger nuclear charge Z,
- grows with larger n, for fixed l,
- shrinks with larger l, for fixed n.

The last two of these features can be compared visually to the probability density distributions in the right column of Fig. 11.2.

Applying (11.58) with $q = 2$, we get

$$\langle R^2 \rangle = \frac{a_0^2}{Z^2}\frac{n^2}{2}\left[5n^2 - 3l(l+1) + 1\right]. \tag{11.66}$$

One can apply (11.58) recursively to find $\langle R^p \rangle$ for any desired positive integer p. However, for negative p, one finds an obstacle, that $\langle 1/R^2 \rangle$ cannot be determined by the recurrence relation alone. In Sect. 15.6, we will use another method to find

$$\langle 1/R^2 \rangle = \frac{Z^2}{a_0^2 n^3 (l + 1/2)}. \tag{11.67}$$

With this as a seed, all results for $\langle R^p \rangle$ with $p \le -3$ can then be determined by the recurrence relation. In particular, using $q = -1$ in (11.58) yields[3]

$$\langle 1/R^3 \rangle = \frac{Z^3}{a_0^3 n^3 l(l+1)(l+1/2)}. \tag{11.68}$$

The expectation value of P^2 can also be found by relating it to the Hamiltonian,

$$\langle P^2 \rangle = 2m_e\big(\langle H \rangle + Ze^2\langle 1/R \rangle\big) = m_e\left(\frac{e^2}{2a_0}\right)\frac{2Z^2}{n^2} = \frac{\hbar^2 Z^2}{a_0^2 n^2}, \tag{11.69}$$

where we have used the known energy eigenvalues and (11.64). This result can also be obtained from the virial theorem, (3.110).

The characteristic size of the hydrogen atom with $Z = 1$ is set by the Bohr radius $a_0 = \hbar^2/m_e e^2$. This could have been estimated by a dimensional analysis construction from the available quantities $\hbar, e, m_e$ in the problem, by requiring it to have units of length. Note that it vanishes in each of the following three extreme limits: (1) the classical limit $\hbar \to 0$, and (2) the limit that the electron mass m_e is taken very large, and (3) the limit of large e^2 so that the attractive electrical force between the nucleus and the electron is large.

However, if we add the speed of light c to our toolbox of quantities, then we can form a dimensionless quantity out of the available parameters in the problem, namely the **fine-structure constant**,

$$\alpha = \frac{e^2}{\hbar c} \approx 1/137.036. \tag{11.70}$$

So, we can construct another significant length scale, the **reduced**[4] **Compton wavelength** of the electron,

$$\lambda\!\!\!^{-}_e = \alpha a_0 = \frac{\hbar}{m_e c} \approx 3.862 \times 10^{-13}\,\text{m}. \tag{11.71}$$

Named for Arthur Compton's studies of the scattering of high-energy photons by electrons, this is the (very small) length scale at which virtual electron-positron pair production becomes important for understanding the energy levels of electron bound states. To see this, note that according to Einstein's famous formula relating energy and mass, the amount of energy needed to make such a pair is $\Delta E = 2m_e c^2$. For ultra-relativistic particles, energy is related to momentum by $\Delta E \sim c\Delta p$, so $e^- e^+$ pair production becomes important for $\Delta p > 2m_e c$. From the uncertainty principle

[3] Note that $\langle 1/R^3 \rangle$ diverges for $l = 0$. More generally, $\langle R^p \rangle = \int_0^\infty dr\, r^{2+p}|R_{n,l}(r)|^2$ diverges from the lower limit of integration if $p \le -2l - 3$, because $|R_{nl}| \propto r^l$ for small r.

[4] The ordinary Compton wavelength of the electron is defined as $\lambda_e = h/m_e c = 2\pi \lambda\!\!\!^{-}_e$.

$(\Delta x)(\Delta p) \geq \hbar/2$, confinement of the electron to a size $\Delta x \sim \hbar/4m_e c \sim \lambda\!\!\!^{-}_e/4$ will result in large enough fluctuations Δp to cause virtual e^-e^+ pair production to have a substantial impact. Furthermore, if we call $v = \sqrt{\langle P^2\rangle}/m_e$ the order of magnitude of the velocity of electrons in a hydrogen-like atom state with principal quantum number n, then

$$v \sim \frac{Z\hbar}{na_0 m_e} = Z\alpha c/n. \tag{11.72}$$

These considerations show that the nonrelativistic approximation is a good one only because the fine-structure constant α is small. That is what ensures $\lambda\!\!\!^{-}_e \ll a_0$ and $v \ll c$, so that electron-positron pair production and other relativistic corrections to the Hamiltonian can be neglected, to first approximation. However, for atoms with very large Z, the speeds of atomic electrons increase, and relativistic effects become important.

A related point is that the binding energy of the hydrogen atom is very small compared to $m_e c^2 = 0.511$ MeV, the rest energy of the electron. In terms of the fine-structure constant,

$$1 \text{ Rydberg} = \frac{e^2}{2a_0} = \frac{e^4 m_e}{2\hbar^2} = \frac{1}{2}\alpha^2 m_e c^2. \tag{11.73}$$

To estimate (very roughly) the size of the relativistic corrections to the energies, we can take the total classical relativistic energy of an electron with momentum p, and subtract off the rest energy, then expand in small p,

$$E_{\text{relativistic}} = \sqrt{m_e^2 c^4 + p^2 c^2} - m_e c^2 = \frac{p^2}{2m_e} - \frac{p^4}{8m_e^3 c^2} + \cdots, \tag{11.74}$$

where we have used the binomial expansion $\sqrt{1+x} = 1 + x/2 - x^2/8 + \cdots$ for small $x = p^2/m_e^2 c^2$. The first term is just the nonrelativistic kinetic energy that we have been using in the quantum Hamiltonian. So, we can estimate that relativistic effects should make a difference of order

$$\Delta E_{\text{relativistic}} \sim -\frac{\langle P^2\rangle^2}{8m_e^3 c^2} = -\frac{\alpha^2 Z^4}{4n^4} \text{ Rydberg}. \tag{11.75}$$

However, it turns out that a correct relativistic analysis must take into account spin. We will carry this out in Sect. 17.1, arriving at a complete result in (17.26). We will do it again in Sect. 28.5, using a manifestly relativistic approach. Our crude estimate in (11.75) is parametrically correct in the sense that the fine-structure effects indeed modify the binding energies by amounts that are suppressed by α^2 and by powers of n, but we will see that the numerical details are quite different than the naive estimate of (11.75).

When an electron transitions between states of the hydrogen atom, it releases or absorbs a photon. The photon energy is equal to the difference in energy levels, $\Delta E = E_n - E_{n'}$, with $n' < n$. This is equal to $\hbar\omega$ where ω is the angular frequency of the photon. Therefore, taking $Z = 1$ for the remainder of this section,

$$\omega_{n,n'} = \frac{13.6\,\text{eV}}{\hbar}\left(\frac{1}{n'^2} - \frac{1}{n^2}\right), \tag{11.76}$$

or equivalently in terms of wavelength,

$$\lambda_{n,n'} = (9.11 \times 10^{-8}\,\text{m})\frac{n'^2}{1 - n'^2/n^2}. \tag{11.77}$$

This formula was found empirically by Rydberg in 1888, generalizing work by Johann Balmer. For transitions between the ground state $n' = 1$ and the states with $n \geq 2$, these spectral lines are called the Lyman series, after Theodore Lyman. They are all in the ultraviolet range,

$$\text{Lyman series:} \quad \omega_{n,1} = (2.067 \times 10^{16}\,\text{s}^{-1})\left(1 - \frac{1}{n^2}\right), \quad (n = 2, 3, 4, \ldots), \tag{11.78}$$

with wavelengths from $\lambda_{2,1} = 1.216 \times 10^{-7}$ meters to $\lambda_{\infty,1} = 9.11 \times 10^{-8}$ meters. The latter wavelength, corresponding to a photon emitted when a free electron is captured to the ground state of a hydrogen atom, or absorbed in the process of ionizing a hydrogen atom that started in its ground state, is called the Lyman limit. The spectral line with $n' = 1$ and $n = 2$ is called the Lyman-alpha line, and is important in astronomy as an absorption feature when observing distant sources, for example as a tool to learn about intervening gas.

The Balmer series comes from transitions between $n' = 2$ and all higher-energy states $n \geq 3$,

$$\text{Balmer series:} \quad \omega_{n,2} = (5.168 \times 10^{15}\,\text{s}^{-1})\left(1 - \frac{4}{n^2}\right), \quad (n = 3, 4, 5, \ldots), \tag{11.79}$$

with wavelengths ranging from $\lambda_{3,2} = 6.56 \times 10^{-7}$ meters to $\lambda_{\infty,2} = 3.64 \times 10^{-7}$ meters. The Balmer lines are particularly interesting because four of them can be seen by the human eye. For $n = 3$ the Balmer line is red, which is responsible for the reddish color of supernova remnants and star-forming regions, including the Orion nebula. The $n = 4$, 5, and 6 Balmer lines are greenish-blue, blue, and violet, respectively, while $n = 7, \ldots, \infty$ are in the near ultraviolet. The remaining named series are all entirely in the infrared, and are called Paschen (transitions between $n' = 3$ and $n \geq 4$), Brackett (transitions between $n' = 4$ and $n \geq 5$), Pfund (transitions between $n' = 5$ and $n \geq 6$), and Humphreys (transitions between $n' = 6$ and $n \geq 7$), with higher n' series not named. There is significant overlap between the infrared series of spectral lines. Within each series, the spectral lines are called α, β, etc. in order of decreasing wavelength.

11.2 Unbound States of Coulomb Potentials

Now consider the unbound stationary states of the Coulomb problem, with $E \geq 0$. These include the ionized states of the hydrogen-like atom, describing an electron that is influenced by the nucleus but not localized near it. We again factor the electron's energy-eigenstate wavefunctions into the product of a radial wavefunction and a spherical harmonic,

$$\langle \boldsymbol{r}|\Phi_{E,l,m}\rangle = \Phi_{E,l,m}(r,\theta,\phi) = \widetilde{R}_{E,l}(r)Y_l^m(\theta,\phi) \qquad (E \geq 0). \quad (11.80)$$

We write $\Phi_{E,l,m}$ and $\widetilde{R}_{E,l}$ here to distinguish them from the bound-state wavefunctions $\psi_{n,l,m}$ and $R_{n,l}$ of the preceding section. For the unbound states, each $E \geq 0$ can have any non-negative integer value of l, as already depicted in Fig. 11.1.

The differential equation satisfied by the radial wavefunction for unbound states is the same as (11.5), with the important differences that the rescaled radial coordinate s and the quantity n, defined in (11.4) and (11.6) respectively, are imaginary for $E > 0$. Accordingly, we write

$$n = -i\frac{Z}{ka_0}, \quad (11.81)$$

which defines a continuous real wavenumber parameter k with units of inverse length. Comparing to the definition of n in (11.6) gives

$$ka_0 = \frac{\hbar}{e^2}\sqrt{\frac{2E}{m_e}}, \quad (11.82)$$

or, equivalently,

$$E = \frac{\hbar^2 k^2}{2m_e} = (ka_0)^2 \text{ Rydberg}. \quad (11.83)$$

Then, from (11.4), the rescaled radial coordinate used in the previous section becomes

$$s = 2ikr. \quad (11.84)$$

It follows that, up to normalization, we can obtain the solution in terms of the confluent hypergeometric function defined in (11.10), by simply re-using (11.13) with the preceding substitutions for n and s. Since s is now imaginary, there is no problem with having an asymptotic behavior like $e^{s/2}$ for large r, and therefore no restriction that the confluent hypergeometric function must be a polynomial; that is why there is no discrete quantization of k. The result is

$$\widetilde{R}_{E,l}(r) = C_{k,l}\,(2kr)^l\,e^{-ikr}\,F(l+1+i\frac{Z}{ka_0},\ 2l+2,\ 2ikr), \tag{11.85}$$

where $C_{k,l}$ is a normalization constant to be determined, which we will choose to be real.

Despite the appearance of the phase factor e^{-ikr} and the complex arguments of the confluent hypergeometric function, the radial wavefunction $\widetilde{R}_{E,l}(r)$ in (11.85) is real. This follows immediately from a property of the confluent hypergeometric function,

$$F(a,c,z) = e^z F(c-a,c,-z), \tag{11.86}$$

which can in turn be proved quickly from the integral representation of (11.11) by using the change of integration variable $t \to 1-t$.

Let us now consider the behavior of the radial wavefunction $\widetilde{R}_{E,l}$ for large r. This can be done by using the asymptotic form for the confluent hypergeometric function for complex arguments,

$$F(a,c,z) \approx \frac{\Gamma(c)}{\Gamma(c-a)}(-z)^{-a} + \frac{\Gamma(c)}{\Gamma(a)}z^{a-c}e^z \qquad (\text{large } |z|), \tag{11.87}$$

which generalizes (11.12) for real arguments. The Γ function for complex arguments was defined in (10.120). Applying this to (11.85), the two terms turn out to give contributions that are complex conjugates of each other, as required by the previous paragraph. After some simplification, one finds the asymptotic form for large kr,

$$\widetilde{R}_{E,l}(r) \approx C_{k,l}\,\frac{(2l+1)!\,e^{-\pi Z/2ka_0}}{|\Gamma(l+1+iZ/ka_0)|}\,\frac{1}{kr}\,\sin\Big(kr + \frac{Z}{ka_0}\ln(2kr) - \frac{\pi}{2}l - \delta\Big), \tag{11.88}$$

where

$$\delta = \mathrm{Arg}\big[\Gamma\big(l+1+iZ/ka_0\big)\big]. \tag{11.89}$$

The radial probability amplitude $r\widetilde{R}_{E,l}(r)$ approaches a sinusoidal oscillation with a constant amplitude at large distances, but with a phase shift that depends logarithmically on kr.

In order to determine the normalization constant, we first define the orthonormality properties of the states. Since the energy $E \geq 0$ is continuous, we will use a Dirac orthonormality condition for unbound states $|E,l,m\rangle$, along with the ordinary discrete orthonormality of the bound states $|n, l, m\rangle$ of the previous section. So,

$$\langle E', l', m'|E, l, m\rangle = \delta(E - E')\,\delta_{ll'}\,\delta_{mm'}, \tag{11.90}$$

$$\langle n', l', m'|E, l, m\rangle = 0, \tag{11.91}$$

$$\langle n', l', m'|n, l, m\rangle = \delta_{nn'}\,\delta_{ll'}\,\delta_{mm'}. \tag{11.92}$$

In terms of the radial wavefunctions for each l, these become

$$\int_0^\infty dr\, r^2\, \widetilde{R}_{E',l}(r)\widetilde{R}_{E,l}(r) = \delta(E - E'), \tag{11.93}$$

$$\int_0^\infty dr\, r^2\, \widetilde{R}_{E,l}(r) R_{n,l}(r) = 0, \tag{11.94}$$

$$\int_0^\infty dr\, r^2\, R_{n',l}(r) R_{n,l}(r) = \delta_{nn'}. \tag{11.95}$$

Now, for the purposes of normalizing the radial wavefunction in (11.88), note that when $E = E'$, one need only consider the asymptotic form at very large r, because contributions to $\int_0^\infty dr\, r^2\, |\widetilde{R}_{E,l}(r)|^2$ from any finite range in r contribute only an infinitesimal fraction of the total, which will have a Dirac delta function interpretation. In the very large r limit, the logarithmic variation in the phase shift can be neglected, and the normalization problem is the same as for the simpler radial wavefunctions

$$f_k(r) = A_k \frac{\sin(kr + \beta)}{kr}, \tag{11.96}$$

for some phase shift β, where comparison with (11.88) gives

$$C_{k,l} = \frac{|\Gamma(l + 1 + iZ/ka_0)|\, e^{\pi Z/2ka_0}}{(2l + 1)!} A_k. \tag{11.97}$$

The wavefunction norm is infinite, but integrating over a finite range $0 < r < D$ gives

$$\int_0^D dr\, r^2\, f_{k'}(r) f_k(r) = \frac{A_k A_{k'}}{2kk'} \left\{ \frac{\sin[(k - k')D]}{k - k'} + \frac{\sin[2\beta] - \sin[2\beta + (k + k')D]}{k + k'} \right\}. \tag{11.98}$$

In the formal limit $D \to \infty$, one can interpret the first term in the braces as $\pi\delta(k - k')$ [see (2.37)]. The remaining part remains bounded for all k and k', and oscillates rapidly as D increases. So, when inserted in any expression in which k or k' is integrated over,

$$\int_0^\infty dr\, r^2\, f_{k'}(r) f_k(r) = A_k A_{k'} \frac{\pi}{2k^2} \delta(k-k') = A_k^2 \frac{\pi \hbar^2}{2m_e k} \delta(E-E'), \tag{11.99}$$

where $dE/dk = \hbar^2 k/m_e$ from (11.83) was used in the last equality. Thus we can adopt the energy normalization for Dirac orthonormality, by taking $A_k^2 = 2m_e k/\pi\hbar^2$, allowing us to find $C_{k,l}$ from (11.97). Using this in (11.85), the final result for the unbound energy-eigenstate radial wavefunction is

$$\widetilde{R}_{E,l}(r) = \sqrt{\frac{2m_e k}{\pi \hbar^2}} \frac{|\Gamma(l+1+iZ/ka_0)| e^{\pi Z/2ka_0}}{(2l+1)!} (2kr)^l\, e^{-ikr}$$
$$\times F\left(l+1+\frac{iZ}{ka_0}, 2l+2, 2ikr\right), \tag{11.100}$$

where k is related to E by (11.83). Although this is not the simplest result one might have hoped for, we again remark that at least it is real, despite naive appearances.

For positive Z, the completeness relation for the energy eigenstates of (11.90)–(11.92) contains both a sum over bound states and an integral over unbound states:

$$\sum_{n=1}^{\infty} \sum_{l=0}^{n-1} \sum_{m=-l}^{l} |n,l,m\rangle\langle n,l,m| + \int_0^\infty dE \sum_{l=0}^{\infty} \sum_{m=-l}^{l} |E,l,m\rangle\langle E,l,m| = I. \tag{11.101}$$

This combination of ordinary and Dirac orthonormality and completeness will be crucial in understanding the evaluation of the ground-state energy of the hydrogen atom in an electric field (the quadratic Stark effect), in Sect. 15.7.

For the case of a repulsive Coulomb potential with $Z < 0$, there is just one slight difference: the bound states $|n,l,m\rangle$ do not exist, and the orthonormality and completeness relations therefore include only the unbound states with continuous positive E.

Exercises

11.1 Classically, for a hydrogen atom with $Z = 1$, what would be the maximum allowed distance of the electron from the proton, if the energy is -1 Rydberg? In the quantum ground state, compute the probability for the electron to be found farther away from the proton than this distance.

11.2 In this problem, we will reconstruct the Coulomb potential from the functional form of its ground-state wavefunction. Suppose that a particle in a spherically symmetric potential $V(r)$ has a stationary state wavefunction $\psi_E(r,\theta,\phi) =$

$Ne^{-r/b}$, where N and b are constants. Assume that $V(r)$ vanishes as $r \to \infty$. Use this to find the energy eigenvalue E, by matching leading terms in Schrödinger's equation as $r \to \infty$. Now that you have found E, consider finite r and find the potential $V(r)$.

11.3 For the ground state of the hydrogen atom, find the uncertainties of the rectangular coordinate Z and its momentum component P_z. Check that the uncertainty principle is satisfied by your answers.

11.4 For the ground state and the first excited states of the hydrogen atom, find all of the matrix elements of the rectangular coordinate operators for every pair of such states, $\langle n', l', m'|X|n, l, m\rangle$ and $\langle n', l', m'|Y|n, l, m\rangle$ and $\langle n', l', m'|Z|n, l, m\rangle$, with $n = 1, 2$ and $n' = 1, 2$. (Hint: many of them are zero.)

11.5 Use the Kramers–Pasternack formula (11.58) to derive the expectation values $\langle n, l, m|R^p|n, l, m\rangle$ for all integers $-3 \le p \le 3$. As seeds, you may use the obvious fact $\langle R^0\rangle = 1$, and also $\langle 1/R^2\rangle = 2Z^2/(a_0^2 n^3(2l + 1))$, which will be obtained in Sect. 15.6 using another trick.

11.6

(a) For the ground state of the hydrogen atom, by taking the Fourier transform as in (2.258), show that the momentum-representation wavefunction is

$$\widetilde{\psi}(\boldsymbol{p}) = \frac{C}{(p^2 + \hbar^2/a_0^2)^N}, \tag{11.102}$$

where N is a certain integer and C is a normalization factor, both of which you will discover. [Hint: since the ground state is spherically symmetric, the momentum wavefunction cannot depend on the direction of $\boldsymbol{p}$. So, for the purposes of computing the integral, you can take $\boldsymbol{p} = p\hat{z}$.]

(b) Use your result from the previous part to compute $\langle P^2\rangle$ and $\langle (P^2)^2\rangle$. The first of these should agree with the $n = 1$, $Z = 1$ special case of (11.69).

(c) Taking the result of part (a) at face value (ignoring the reality of relativity), compute numerically the probability for the magnitude of the electron's momentum to exceed $m_e c = \hbar/\alpha a_0$.

11.7 A particle of mass μ moves in the potential $V(R) = -a/R + \hbar^2 b/2\mu R^2$, where a and b are positive constants. Consider stationary-state wavefunctions of the form $\psi(\boldsymbol{r}) = f(r)Y_l^m(\theta, \phi)$. Find the bound-state energies and wavefunctions for orbital angular momentum l. (Hint: the radial effective potential has the same form as for the hydrogen atom, but with a modification of the angular momentum contribution.)

Addition of Angular Momenta

12

12.1 Statement of the Problem

In quantum mechanical problems, we often deal with two or more distinct angular momenta. The sum of two independent angular momentum operators,

$$\boldsymbol{J} = \boldsymbol{J}_1 + \boldsymbol{J}_2, \tag{12.1}$$

is also an angular momentum operator. This means that if all components of $\boldsymbol{J}_1$ commute with all components of $\boldsymbol{J}_2$, and if the components of $\boldsymbol{J}_1$ and $\boldsymbol{J}_2$ each satisfy the defining commutator algebra (5.81), then so will the components of $\boldsymbol{J}$.

Since J_1^2, J_2^2, J_{1z}, and J_{2z} are compatible operators, we can find an orthobasis of common eigenkets for them, with eigenvalues $\hbar^2 j_1(j_1+1)$, $\hbar^2 j_2(j_2+1)$, $\hbar m_1$, and $\hbar m_2$, respectively labeled

$$|j_1\, j_2\, m_1\, m_2\rangle \equiv |j_1\, m_1\rangle \otimes |j_2\, m_2\rangle. \tag{12.2}$$

We call this the **product orthobasis** for two angular momenta, because its elements consist of the tensor products of elements of the orthobases for the individual angular momenta. Here, we have suppressed any degeneracy labels, which might correspond to different radial wavefunctions, for example. For fixed j_1 and j_2, there are $2j_1 + 1$ allowed values of m_1, and $2j_2 + 1$ allowed values of m_2,

$$m_1 = j_1,\ j_1-1,\ \ldots,\ -j_1, \tag{12.3}$$

$$m_2 = j_2,\ j_2-1,\ \ldots,\ -j_2. \tag{12.4}$$

So, there are $(2j_1 + 1)(2j_2 + 1)$ orthobasis kets of the form (12.2) for fixed j_1 and j_2.

S. P. Martin, *Quantum Mechanics*, Graduate Texts in Physics,
https://doi.org/10.1007/978-3-032-21119-4_12

Another set of compatible operators is J_1^2, J_2^2, J^2, and J_z. They also have a set of common eigenkets that form a different orthobasis, with eigenvalues $\hbar^2 j_1(j_1{+}1)$, $\hbar^2 j_2(j_2{+}1)$, $\hbar^2 j(j{+}1)$, and $\hbar m$, and corresponding eigenstates that we can write as

$$|j_1\, j_2\, j\, m\rangle. \tag{12.5}$$

This is called the **total angular momentum orthobasis**. As we will see in the following sections, for fixed j_1 and j_2, the possible values of j range from a maximum of j_1+j_2 to a minimum of $|j_1-j_2|$, with integer increments,

$$j = j_1{+}j_2,\ j_1{+}j_2{-}1,\ \ldots,\ |j_1{-}j_2|. \tag{12.6}$$

Intuitively, the extremes for j correspond to the two angular momenta aligned in the same and in opposite directions, respectively. Then, for each j, there are $2j+1$ allowed values

$$m = j,\ j{-}1,\ \ldots,\ -j. \tag{12.7}$$

As a check, the number of total angular momentum basis states for fixed j_1 and j_2 is

$$\sum_{j=|j_1-j_2|}^{j_1+j_2} (2j+1) = (2j_1+1)(2j_2+1), \tag{12.8}$$

matching the result for the number of product basis kets.

Here, we face a potentially annoying problem of notation, because both orthobases are labeled by four numbers. In many cases, this will not cause confusion, but what if it does? In the common case that one of j_1 or j_2 is equal to $1/2$ and describes a spin, then in the product basis we will replace the corresponding magnetic quantum number $1/2$ or $-1/2$ by $\uparrow$ or $\downarrow$, respectively, signifying that the product basis is being used. In other cases when there is a chance of confusion, we will use a colon to separate the last two (magnetic quantum number) labels for the product angular momentum orthobasis. All other labels in both the product and total angular momentum bases are separated by no punctuation, or by a comma when it is typographically convenient. Also, there will be many occasions in which the labels j_1 and j_2 that are common to both orthobases are fixed, and understood from context. In that case, we can suppress those labels and call the total angular momentum orthobasis kets $|jm\rangle$ or $|j, m\rangle$, and the product orthobasis kets $|m_1 m_2\rangle$ or $|m_1, m_2\rangle$, but use the notation $|m_1 : m_2\rangle$ if there is a significant chance of confusion with the total angular momentum orthobasis.

A frequently encountered situation is that we might have to evaluate matrix elements involving the dot product of two angular momentum operators. If we are working in the product basis, the most straightforward way to do this is to write the dot product in terms of raising and lowering operators,

$$\boldsymbol{J}_1 \cdot \boldsymbol{J}_2 = \frac{1}{2}\left(J_{1+}J_{2-} + J_{1-}J_{2+}\right) + J_{1z}J_{2z}, \tag{12.9}$$

which follows from (8.10). Then each of J_{1+}, J_{1-}, and J_{1z} can be evaluated by their actions on the $|j_1 m_1\rangle$ component using the rules of (8.7), (8.21), (8.23), and similarly for the actions of J_{2+}, J_{2-}, and J_{2z} on the $|j_2 m_2\rangle$ part. However, it is often more convenient to use instead the total angular momentum basis. This is because one can use $J^2 = (\boldsymbol{J}_1 + \boldsymbol{J}_2)^2 = J_1^2 + J_2^2 + 2\boldsymbol{J}_1 \cdot \boldsymbol{J}_2$ to write

$$\boldsymbol{J}_1 \cdot \boldsymbol{J}_2 = \frac{1}{2}\left(J^2 - J_1^2 - J_2^2\right). \tag{12.10}$$

Now, when acting on the total angular momentum basis kets $|j_1\, j_2\, j\, m\rangle$, this operator just evaluates to a number,

$$\boldsymbol{J}_1 \cdot \boldsymbol{J}_2 = \frac{\hbar^2}{2}\left[j(j+1) - j_1(j_1+1) - j_2(j_2+1)\right], \tag{12.11}$$

independent of m. Solving problems in quantum mechanics is often the art of turning operators into numbers, and using the correct choice of basis certainly makes things easier. The specific lesson here is that the appearance of a dot product of angular momenta should immediately suggest using the total angular momentum basis.

Because they are both bases, each element of the total angular momentum orthobasis must be a linear combination of the product orthobasis kets, and vice versa. In general, the problem of **addition of angular momenta** is to evaluate the coefficients appearing in these linear combinations.

We will start with two useful and common special cases before taking on the most general case. First, in Sect. 12.2, we consider the case of two spins $s_1 = s_2 = 1/2$, for two particles with orbital angular momenta absent or disregarded. Next, in Sect. 12.3, we will take up the example of a single particle with arbitrary orbital angular momentum l and spin $s = 1/2$, which can be combined to form the total angular momentum of the particle. In Sects. 12.4 and 12.5 we will discuss the general case of addition of two arbitrary angular momenta. These results can be applied recursively to combine any number of angular momenta.

12.2 Addition of Two Spins

Consider two spins with $s_1 = s_2 = 1/2$. Spin magnitudes are always fixed, so we suppress those labels, and write the four product orthobasis kets as

$$|\uparrow\uparrow\rangle, \quad |\uparrow\downarrow\rangle, \quad |\downarrow\uparrow\rangle, \quad |\downarrow\downarrow\rangle. \tag{12.12}$$

Here the first $\uparrow$ or $\downarrow$ label on each ket stands for the eigenvalue $\hbar/2$ or $-\hbar/2$ of S_{1z}, and the second label similarly stands for the eigenvalue of S_{2z}. We say that the individual spins in this basis are either "up" or "down" relative to our choice of the $\hat{z}$ direction. Now, define the total spin operator by

$$\boldsymbol{S} = \boldsymbol{S}_1 + \boldsymbol{S}_2. \tag{12.13}$$

The product orthobasis kets are also eigenkets of S_z, since

$$S_z|m_1m_2\rangle = S_{1z}|m_1m_2\rangle + S_{2z}|m_1m_2\rangle = \hbar(m_1+m_2)|m_1m_2\rangle, \tag{12.14}$$

so that $S_z|\uparrow\uparrow\rangle = \hbar|\uparrow\uparrow\rangle$, and $S_z|\uparrow\downarrow\rangle = S_z|\downarrow\uparrow\rangle = 0$, and $S_z|\downarrow\downarrow\rangle = -\hbar|\downarrow\downarrow\rangle$. Choosing a representation in which

$$|\uparrow\uparrow\rangle \leftrightarrow \begin{pmatrix}1\\0\\0\\0\end{pmatrix}, \quad |\uparrow\downarrow\rangle \leftrightarrow \begin{pmatrix}0\\1\\0\\0\end{pmatrix}, \quad |\downarrow\uparrow\rangle \leftrightarrow \begin{pmatrix}0\\0\\1\\0\end{pmatrix}, \quad |\downarrow\downarrow\rangle \leftrightarrow \begin{pmatrix}0\\0\\0\\1\end{pmatrix}, \tag{12.15}$$

we have the matrix representation

$$S_z \leftrightarrow \hbar \begin{pmatrix}1&0&0&0\\0&0&0&0\\0&0&0&0\\0&0&0&-1\end{pmatrix}. \tag{12.16}$$

However, the product orthobasis kets are not all eigenstates of S^2. The operation of S^2 on them can be obtained from

$$\begin{aligned} S^2 &= (\boldsymbol{S}_1+\boldsymbol{S}_2)\cdot(\boldsymbol{S}_1+\boldsymbol{S}_2) = S_1^2 + S_2^2 + 2\boldsymbol{S}_1\cdot\boldsymbol{S}_2 \\ &= \frac{3}{2}\hbar^2 + S_{1+}S_{2-} + S_{1-}S_{2+} + 2S_{1z}S_{2z}, \end{aligned} \tag{12.17}$$

where in the last equality we have taken advantage of the fact that all of the kets in the state space under consideration are eigenkets of both S_1^2 and S_2^2 with eigenvalues $3\hbar^2/4$, and then applied (12.9) to write the result in terms of the angular momentum raising and lowering operators. Now, recall from (8.21) and (8.23) that, for any $j = 1/2$ system,

$$J_+|\uparrow\rangle = 0, \qquad J_+|\downarrow\rangle = \hbar|\uparrow\rangle, \tag{12.18}$$

$$J_-|\downarrow\rangle = 0, \qquad J_-|\uparrow\rangle = \hbar|\downarrow\rangle. \tag{12.19}$$

Applying these for each of $J = S_1$ and S_2, we obtain from (12.17),

$$S^2|\uparrow\uparrow\rangle = 2\hbar^2|\uparrow\uparrow\rangle, \tag{12.20}$$

$$S^2|\uparrow\downarrow\rangle = \hbar^2|\uparrow\downarrow\rangle + \hbar^2|\downarrow\uparrow\rangle, \tag{12.21}$$

$$S^2|\downarrow\uparrow\rangle = \hbar^2|\uparrow\downarrow\rangle + \hbar^2|\downarrow\uparrow\rangle, \tag{12.22}$$

$$S^2|\downarrow\downarrow\rangle = 2\hbar^2|\downarrow\downarrow\rangle. \tag{12.23}$$

In matrix representation form, this reads

$$S^2 \leftrightarrow \hbar^2 \begin{pmatrix} 2 & 0 & 0 & 0 \\ 0 & 1 & 1 & 0 \\ 0 & 1 & 1 & 0 \\ 0 & 0 & 0 & 2 \end{pmatrix}. \tag{12.24}$$

This matrix has eigenvalues $2\hbar^2, 2\hbar^2, 2\hbar^2$, and 0, which correspond to $S^2 = \hbar^2 s(s+1)$ with $s = 1$ and $s = 0$. The addition of two spins with $s_1 = s_2 = 1/2$ thus gives a triplet of states $|1, m\rangle$ with total spin $s = 1$ and $m = -1, 0, 1$, plus a singlet state with total spin $s = 0$, denoted $|0, 0\rangle$. By finding the eigenkets corresponding to the eigenvalue pairs (s, m), we can summarize the information about the addition of two spin-1/2 systems as

$$|1, 1\rangle = |\uparrow\uparrow\rangle, \tag{12.25}$$

$$|1, 0\rangle = \frac{1}{\sqrt{2}}\left(|\uparrow\downarrow\rangle + |\downarrow\uparrow\rangle\right), \tag{12.26}$$

$$|1, -1\rangle = |\downarrow\downarrow\rangle, \tag{12.27}$$

$$|0, 0\rangle = \frac{1}{\sqrt{2}}\left(|\uparrow\downarrow\rangle - |\downarrow\uparrow\rangle\right), \tag{12.28}$$

with total spin basis kets $|s, m\rangle$ on the left and product basis kets on the right. Both orthobases have 4 members, and we can of course invert the relationship for the $m = 0$ states, to find

$$|\uparrow\downarrow\rangle = \frac{1}{\sqrt{2}}\left(|1, 0\rangle + |0, 0\rangle\right), \tag{12.29}$$

$$|\downarrow\uparrow\rangle = \frac{1}{\sqrt{2}}\left(|1, 0\rangle - |0, 0\rangle\right). \tag{12.30}$$

As a way of expressing the fact that the tensor product of two spin-1/2 systems gives a sum of angular momenta 0 and 1, we write

$$\tfrac{1}{2} \otimes \tfrac{1}{2} = 0_A \oplus 1_S. \tag{12.31}$$

The subscripts A and S in this notation are a reminder that the $s = 0$ singlet state is antisymmetric under exchange of the two spins, while each of the $s = 1$, $m = 1, 0, -1$ triplet states is symmetric, as can be seen in (12.25)–(12.28).

An important practical question is: which orthobasis should we use? The answer depends on the problem under consideration. As a general rule, it is convenient to use a basis in which the operators most important to us are diagonal. For example, suppose the Hamiltonian is

$$H = b_1 S_{1z} + b_2 S_{2z}, \tag{12.32}$$

which could occur if both spins are interacting with a magnetic field but not with each other. This Hamiltonian is diagonal in the product basis, where the kets are already eigenstates of S_{1z} and S_{2z}, but it is diagonal in the total angular momentum basis only if $b_1 = b_2$. On the other hand, suppose that the Hamiltonian is of the form

$$H = a\mathbf{S}_1 \cdot \mathbf{S}_2, \tag{12.33}$$

which comes from the magnetic moments of the particles interacting with each other. In that case, we can use the standard trick of writing

$$\mathbf{S}_1 \cdot \mathbf{S}_2 = \frac{1}{2}\left(S^2 - S_1^2 - S_2^2\right). \tag{12.34}$$

The right side shows that this operator is diagonal in the total spin basis, where it evaluates to

$$\mathbf{S}_1 \cdot \mathbf{S}_2 = \frac{\hbar^2}{2}\left[s(s+1) - 3/2\right], \tag{12.35}$$

with $s = 1$ when acting on the triplet states and $s = 0$ for the singlet state. In other cases, a judgment can be made about which is the most convenient basis, often based on which part of the Hamiltonian, or some other observable of interest, is the most important.

As a classic example, the **hyperfine** splitting of the hydrogen atom ground state arises from the interaction of the spins of the electron and the proton. The hyperfine Hamiltonian H_{hf} has exactly the form of (12.33), with a constant a that is positive, and very small compared to the Rydberg energy scale divided by $\hbar^2$. Before taking into account fine or hyperfine effects, the hydrogen atom energy eigenstates can be given in the product spin basis labeled as

$$|\underbrace{n, l, m_l}_{\text{orbital}}, \overbrace{s_e, m_{s_e}}^{\text{electron spin}}, \underbrace{s_p, m_{s_p}}_{\text{proton spin}}\rangle = |n, l, m_l, m_{s_e}, m_{s_p}\rangle, \tag{12.36}$$

or we can use the total spin basis,

$$|n, l, m_l, s_e, s_p, s, m_s\rangle = |n, l, m_l, s, m_s\rangle. \tag{12.37}$$

In both cases, the individual spin labels $s_e, s_p = 1/2$ are completely fixed and therefore can be suppressed, as indicated. Our present interest is the effect on the ground state with $n = 1$ and $l = 0$. Since $H_{\text{orbital}} = P^2/2m_e - e^2/R$ trivially commutes with

$$H_{\text{hf}} = a\boldsymbol{S}_e \cdot \boldsymbol{S}_p, \tag{12.38}$$

they have an orthobasis of common eigenstates. The eigenvalues of H_{hf} are found immediately in the total spin basis, using (12.35), as

$$E_{\text{hf}} = \begin{cases} a\hbar^2/4 & (s = 1), \\ -3a\hbar^2/4 & (s = 0), \end{cases} \tag{12.39}$$

for the triplet and singlet total spin states, respectively. The energy splitting between these states is therefore $a\hbar^2$, which for the ground state of the hydrogen atom is, numerically,

$$\Delta E_{\text{hf}} = a\hbar^2 = 5.87 \times 10^{-6}\,\text{eV}, \tag{12.40}$$

corresponding to a wavelength $\lambda = 2\pi c/a\hbar = 0.211$ meters for photons emitted or absorbed in transitions between the $s = 1$ and $s = 0$ states. This is the famous 21 centimeter line of radio astronomy. The rate for such transitions turns out to be highly suppressed (for reasons to be discussed in Sect. 23.5), but space is big and mostly cold, so it is a very useful observational tool in astrophysics and experimental cosmology.

There is a special consideration when the two spin-1/2 particles are identical. The **Pauli exclusion principle** says that two identical particles with half-integer spin cannot be in the same quantum state, and more generally that the quantum state describing them must be antisymmetric under their exchange. Such particles are called **fermions**, after Enrico Fermi, while particles with integer spin are called **bosons**, after Satyendra Nath Bose. For example, suppose that the orbital wavefunctions of two spin-1/2 fermions are $\psi_a(\boldsymbol{r})$ and $\psi_b(\boldsymbol{r})$, corresponding to single-particle kets (neglecting spins for the moment) $|\psi_a\rangle$ and $|\psi_b\rangle$. From these, one can form symmetric and antisymmetric combinations

$$|\psi_a, \psi_b\rangle_S = \frac{1}{\sqrt{2}}\Big(|\psi_a, \psi_b\rangle + |\psi_b, \psi_a\rangle\Big), \tag{12.41}$$

$$|\psi_a, \psi_b\rangle_A = \frac{1}{\sqrt{2}}\Big(|\psi_a, \psi_b\rangle - |\psi_b, \psi_a\rangle\Big), \tag{12.42}$$

where, on the right-hand sides, the first and second entries in each ket correspond to particle labels 1 and 2, respectively. Then the allowed states constructed from the tensor product of the orbital and total spin states must be either symmetric in orbital kets and antisymmetric in spin kets, or vice versa,

$$|\psi_a, \psi_b\rangle_S \otimes \frac{1}{\sqrt{2}}\Big(|\uparrow\downarrow\rangle - |\downarrow\uparrow\rangle\Big) \qquad (s = 0), \tag{12.43}$$

$$|\psi_a, \psi_b\rangle_A \otimes |\uparrow\uparrow\rangle \qquad (s = 1,\ m_s = 1), \tag{12.44}$$

$$|\psi_a, \psi_b\rangle_A \otimes \frac{1}{\sqrt{2}}\Big(|\uparrow\downarrow\rangle + |\downarrow\uparrow\rangle\Big) \qquad (s = 1,\ m_s = 0), \tag{12.45}$$

$$|\psi_a, \psi_b\rangle_A \otimes |\downarrow\downarrow\rangle \qquad (s = 1,\ m_s = -1). \tag{12.46}$$

Other kets, such as $|\psi_a, \psi_b\rangle_S \otimes |\uparrow\uparrow\rangle$, do not exist as physical states. For example, the ground state of helium, with two electrons, has an orbital part that is symmetric under interchange of the positions of the two electrons. The spin part of the state is therefore the antisymmetric $s = 0$ combination.

12.3 Addition of Orbital Angular Momentum and Spin

In this section, we consider the combination of two angular momenta j_1 and j_2, with j_1 arbitrary and $j_2 = 1/2$. This applies, for example, to the case where we are combining the orbital $\boldsymbol{L}$ and intrinsic $\boldsymbol{S}$ angular momenta for a single spin-1/2 particle such as the electron. We will frame our discussion in that context.

As we saw in Sect. 8.2, the spin states of an electron can be associated with a two-component spinor, with the spin operator $\boldsymbol{S}$ represented by 2×2 matrices $\hbar\boldsymbol{\sigma}/2$. In the position-wavefunction spinor representation corresponding to the CSCO $(\boldsymbol{R}, S^2, S_z)$, the orbital angular momentum operators $\boldsymbol{L}$ and L^2 are represented by differential operators proportional to the unit matrix in the spin sector, for example [compare (8.100) and (8.101)]

$$L_z \leftrightarrow -i\hbar \begin{pmatrix} 1 & 0 \\ 0 & 1 \end{pmatrix} \frac{\partial}{\partial\phi}, \qquad L_+ \leftrightarrow \hbar e^{i\phi} \begin{pmatrix} 1 & 0 \\ 0 & 1 \end{pmatrix} \left(\frac{\partial}{\partial\theta} + i\frac{\cos\theta}{\sin\theta}\frac{\partial}{\partial\phi}\right), \tag{12.47}$$

etc. However, it is often more useful in problems with spherical symmetry to replace $\boldsymbol{R}$ in the CSCO with the radial coordinate operator R and the operators L^2 and L_z, so that all of the angular momentum operators are represented by matrices rather than differential operators on wavefunctions.

The angular momentum product orthobasis consists of common eigenkets of the compatible operators (L^2, S^2, L_z, S_z). The corresponding basis kets can be labeled $|l\, s\, m_l\, m_s\rangle$, where we have suppressed degeneracy labels corresponding to the radial wavefunction. Then, also suppressing the fixed labels l, s, the product basis states are

$$|m_l,\ \uparrow\rangle \qquad \text{and} \qquad |m_l,\ \downarrow\rangle, \tag{12.48}$$

where $m_s = 1/2$ and $-1/2$ are represented by $\uparrow$ and $\downarrow$, respectively. Using $\boldsymbol{J} = \boldsymbol{L} + \boldsymbol{S}$, we have another choice of compatible operators, $(L^2,\ S^2,\ J^2,\ J_z)$, associated with the total angular momentum orthobasis eigenkets

$$|l\,s\,j\,m\rangle \;=\; |j,\ m\rangle. \tag{12.49}$$

Our goal is to express the kets of (12.49) as linear combinations of the product basis kets (12.48), for each l.

If $l = 0$, then things are very easy; the product basis kets $|0,\ \uparrow\rangle$ and $|0,\ \downarrow\rangle$ are already eigenstates of J^2 with eigenvalue $3\hbar^2/4$, and of J_z with eigenvalues $\pm\hbar/2$, respectively. Therefore, the total angular momentum basis kets have $j = 1/2$ and $m = \pm 1/2$, and are $|\frac{1}{2},\ \frac{1}{2}\rangle = |0,\ \uparrow\rangle$ and $|\frac{1}{2},\ -\frac{1}{2}\rangle = |0,\ \downarrow\rangle$.

For $l \neq 0$, we begin with some preliminary counting, in order to know what to expect. There are $(2l + 1)2$ product basis kets $|m_l,\ \uparrow\rangle$ and $|m_l,\ \downarrow\rangle$. They are all eigenkets of $J_z = L_z + S_z$, and the largest eigenvalue of J_z is $\hbar m = \hbar(l+1/2)$. This implies that there must be, in the total angular momentum basis, a multiplet with $j = l+1/2$, which will have $2(l+1/2)+1 = 2l+2$ basis elements. Exactly one of those has $m = l-1/2$, but we also know that there are two linearly independent states with that eigenvalue in the product basis, namely $|l,\ \downarrow\rangle$ and $|l-1,\ \uparrow\rangle$. Therefore, there must also be a multiplet with $j = l - 1/2$, which has $2(l - 1/2) + 1 = 2l$ orthobasis members. Since we have accounted for all $4l + 2$ linearly independent kets, we infer that the tensor product of angular momentum l with angular momentum $1/2$ must consist of states with total angular momenta $j = l + 1/2$ and $l - 1/2$, and no others. In notation similar to (12.31), this is expressed as

$$l \otimes \tfrac{1}{2} \;=\; (l - \tfrac{1}{2}) \oplus (l + \tfrac{1}{2}). \tag{12.50}$$

Unlike (12.31), there are no A or S subscripts here, because there is no way to symmetrize or antisymmetrize with respect to angular momenta that are not the same.

To construct a complete map between the total and product orthobases, let us start with the state $|m_l, m_s\rangle = |l, \uparrow\rangle$. Acting on this with J_z gives

$$J_z|l, \uparrow\rangle \;=\; (L_z + S_z)|l, \uparrow\rangle \;=\; \hbar(l + 1/2)|l, \uparrow\rangle. \tag{12.51}$$

This is the unique eigenstate of J^2 and J_z with quantum numbers $j = l + 1/2$ and $m = l + 1/2$. So, up to a phase that we can set arbitrarily,

$$|l+\tfrac{1}{2},\ l+\tfrac{1}{2}\rangle \;=\; |l, \uparrow\rangle. \tag{12.52}$$

Now we use this to construct all of the other states with $j = l + 1/2$, by acting repeatedly with the lowering operator $J_- = L_- + S_-$. From (8.23),

$$(S_- + L_-)|l, \uparrow\rangle = J_-|l+\tfrac{1}{2}, l+\tfrac{1}{2}\rangle = \hbar\sqrt{2l+1}|l+\tfrac{1}{2}, l-\tfrac{1}{2}\rangle. \tag{12.53}$$

Using (8.23) again to evaluate the action of each of S_- and L_-, we get

$$|l+\tfrac{1}{2}, l-\tfrac{1}{2}\rangle = \frac{1}{\sqrt{2l+1}}\Big(|l, \downarrow\rangle + \sqrt{2l}|l-1, \uparrow\rangle\Big). \tag{12.54}$$

This is our second total angular momentum orthobasis eigenket. Acting with $J_- = L_- + S_-$ on it in a similar way gives

$$|l+\tfrac{1}{2}, l-\tfrac{3}{2}\rangle = \frac{1}{\sqrt{2l+1}}\Big(\sqrt{2}|l-1, \downarrow\rangle + \sqrt{2l-1}|l-2, \uparrow\rangle\Big). \tag{12.55}$$

Repeating this strategy of applying J_-, we obtain all of the $j = l+1/2$ kets by induction,

$$|l+\tfrac{1}{2}, m\rangle = \frac{1}{\sqrt{2l+1}}\Big(\sqrt{l-m+1/2}|m+\tfrac{1}{2}, \downarrow\rangle + \sqrt{l+m+1/2}|m-\tfrac{1}{2}, \uparrow\rangle\Big) \tag{12.56}$$

for all $m = -l-1/2, \ldots, l+1/2$.

Having found all the states with total angular momentum $j = l+1/2$, now we find the states with $j = l-1/2$. Each state with $j = l-1/2$ must be a linear combination of the form

$$|l-\tfrac{1}{2}, m\rangle = a|m+\tfrac{1}{2}, \downarrow\rangle + b|m-\tfrac{1}{2}, \uparrow\rangle. \tag{12.57}$$

We know this because the product orthobasis kets on the right side are the only two that have the correct eigenvalue m of $J_z/\hbar$. Now, we can appeal to Theorem 2.14, which assures us that the state $|l-\tfrac{1}{2}, m\rangle$ must be orthogonal to the state $|l+\tfrac{1}{2}, m\rangle$, since they have different eigenvalues of the Hermitian operator J^2. Carrying out the inner product of (12.56) with (12.57), we learn that

$$a\sqrt{l-m+1/2} = -b\sqrt{l+m+1/2}. \tag{12.58}$$

As an arbitrary phase convention, we choose a real and positive. Requiring unit normalization of the ket then uniquely determines that the states $|j, m\rangle$ with $j = l-1/2$ are

$$|l-\tfrac{1}{2}, m\rangle = \frac{1}{\sqrt{2l+1}}\Big(\sqrt{l+m+1/2}|m+\tfrac{1}{2}, \downarrow\rangle - \sqrt{l-m+1/2}|m-\tfrac{1}{2}, \uparrow\rangle\Big), \tag{12.59}$$

for $m = -l+1/2, \ldots, l-1/2$.

That concludes our task. To recapitulate, in (12.56) and (12.59), we have obtained the J^2, J_z orthobasis kets as linear combinations of the product orthobasis kets.

As a concrete example, which we will have occasion to use later (at the end of Sect. 17.3), consider the addition of an angular momentum $l = 1$ to a spin-1/2,

$$1 \otimes \tfrac{1}{2} = \tfrac{3}{2} \oplus \tfrac{1}{2}. \tag{12.60}$$

Applying (12.56), we have the $j = 3/2$ total angular momentum basis states

$$\left|\tfrac{3}{2}, \tfrac{3}{2}\right\rangle = |1, \uparrow\rangle, \tag{12.61}$$

$$\left|\tfrac{3}{2}, \tfrac{1}{2}\right\rangle = \sqrt{\tfrac{1}{3}}\,|1, \downarrow\rangle + \sqrt{\tfrac{2}{3}}\,|0, \uparrow\rangle, \tag{12.62}$$

$$\left|\tfrac{3}{2}, -\tfrac{1}{2}\right\rangle = \sqrt{\tfrac{2}{3}}\,|0, \downarrow\rangle + \sqrt{\tfrac{1}{3}}\,|-1, \uparrow\rangle, \tag{12.63}$$

$$\left|\tfrac{3}{2}, -\tfrac{3}{2}\right\rangle = |-1, \downarrow\rangle, \tag{12.64}$$

and applying (12.59) we get the $j = 1/2$ total angular momentum basis states

$$\left|\tfrac{1}{2}, \tfrac{1}{2}\right\rangle = \sqrt{\tfrac{2}{3}}\,|1, \downarrow\rangle - \sqrt{\tfrac{1}{3}}\,|0, \uparrow\rangle, \tag{12.65}$$

$$\left|\tfrac{1}{2}, -\tfrac{1}{2}\right\rangle = \sqrt{\tfrac{1}{3}}\,|0, \downarrow\rangle - \sqrt{\tfrac{2}{3}}\,|-1, \uparrow\rangle. \tag{12.66}$$

12.4 The General Case and Clebsch–Gordan Coefficients

Now consider the general case of addition of angular momenta $\boldsymbol{J}_1$ and $\boldsymbol{J}_2$. In this section, the labels j_1, j_2 on kets $|j_1 j_2 m_1 {:} m_2\rangle$ and $|j_1 j_2\, j m\rangle$ and the corresponding bras are always the same, and so are routinely (but not always) omitted. Product angular momentum basis kets will always be distinguished by use of a colon between the magnetic quantum numbers.

Let us apply reasoning similar to the counting that led to (12.50). The largest eigenvalue of $J_z = J_{1z} + J_{2z}$ is $\hbar(j_1 + j_2)$, and there is only one such state, $|j_1 : j_2\rangle$, so there must be exactly one total angular momentum multiplet with $j = j_1 + j_2$. The subspace with J_z eigenvalue $\hbar(j_1 + j_2 - 1)$ has dimension 2, spanned by the two states in the product basis $|j_1 - 1 : j_2\rangle$ and $|j_1 : j_2 - 1\rangle$. One linear combination of these is found in the $j = j_1 + j_2$ multiplet, so to accommodate the other there must also be exactly one total angular momentum multiplet with $j = j_1 + j_2 - 1$. Similarly, the subspace with J_z eigenvalue $\hbar(j_1 + j_2 - 2)$ has dimension 3, spanned by the product basis kets $|j_1 - 2 : j_2\rangle$ and $|j_1 - 1 : j_2 - 1\rangle$ and $|j_1 : j_2 - 2\rangle$. Two linear combinations of these will occur in the $j = j_1 + j_2$ and $j = j_1 + j_2 - 1$ multiplets that we already know about, so there must also be exactly one total angular momentum multiplet with $j = j_1 + j_2 - 2$.

Continuing in this way, one finds that the tensor product of a multiplet with J_1^2 eigenvalue $\hbar^2 j_1(j_1+1)$ and a multiplet with J_2^2 eigenvalue $\hbar^2 j_2(j_2+1)$ must consist of a sum of multiplets with $J^2 = \hbar^2 j(j+1)$, with j taking on the values from $|j_1 - j_2|$ to $j_1 + j_2$, with integer increments,

$$j_1 \otimes j_2 = |j_1 - j_2| \oplus \cdots \oplus (j_1 + j_2). \tag{12.67}$$

Each of the total angular momentum basis kets $|j_1\, j_2\, j, m\rangle$ is a linear combination of the product basis kets $|j_1\, j_2\, m_1\!:\!m_2\rangle$. By the completeness of the latter, we can write

$$|j_1\, j_2\, j, m\rangle = \sum_{m_1=-j_1}^{j_1} \sum_{m_2=-j_2}^{j_2} |j_1\, j_2\, m_1\!:\!m_2\rangle\, \langle j_1\, j_2\, m_1\!:\!m_2|j_1\, j_2\, j, m\rangle \tag{12.68}$$

$$= \sum_{m_1=-j_1}^{j_1} \sum_{m_2=-j_2}^{j_2} |j_1\, j_2\, m_1\!:\!m_2\rangle\, C^{j_1\, j_2\, j}_{m_1 m_2 m} \tag{12.69}$$

where the inner products

$$C^{j_1\, j_2\, j}_{m_1 m_2 m} \equiv \langle j_1\, j_2\, m_1\!:\!m_2|j_1\, j_2\, j, m\rangle \tag{12.70}$$

are known as **Clebsch–Gordan coefficients**, after mathematicians Alfred Clebsch and Paul Gordan. Various notations and conventions for them appear in the literature. We will usually use the C notation for these inner products, as a way of saving space, with commas inserted between the superscripts or subscripts when it helps to make the meaning clear.

The Clebsch–Gordan coefficients obey selection rules; they can be nonzero only if

$$|j_1 - j_2| \le j \le j_1 + j_2, \quad \text{(triangle condition)}, \tag{12.71}$$

$$j_1 + j_2 - j \text{ is an integer}, \tag{12.72}$$

$$m = m_1 + m_2. \tag{12.73}$$

The first two of these simply restate (12.67), and the last follows immediately from evaluating the matrix element of $J_z = J_{1z} + J_{2z}$ between $\langle j_1\, j_2\, m_1\!:\!m_2|$ and $|j_1\, j_2\, j, m\rangle$.

In general, the Clebsch–Gordan coefficients are subject to ambiguities, because the normalized states $|j_1\, j_2\, j, m\rangle$ are determined only up to a phase. These phase ambiguities are resolved here by adopting a convention that $C^{j_1\, j_2\, j}_{m_1 m_2 m}$ is real and positive when $m_1 = j_1$ and $m = j$. (Note that this convention gives a special role to the first of the angular momenta being combined.) The coefficient is nonzero only when $m_2 = j - j_1$, so we can specify the convention as

$$\text{phase convention:} \quad C^{j_1\, j_2\, j}_{j_1,\, j-j_1,\, j} \text{ is real and positive.} \tag{12.74}$$

Then, because the actions of J_- and J_{1-} and J_{2-} each involve only real coefficients, as given by (8.23), we find, by acting on (12.69) repeatedly with J_- on the left side and its equivalent $J_{1-} + J_{2-}$ on the right side, the convenient and useful result that all of the Clebsch–Gordan coefficients will be real.

One can also use completeness with respect to the orthobasis $|j_1\, j_2\, j, m\rangle$ to obtain the inverse relation to (12.69), which gives each product orthobasis ket as a linear combination of the total angular momentum orthobasis kets,

$$|j_1\, j_2\, m_1 : m_2\rangle = \sum_{j=|j_1-j_2|}^{j_1+j_2} \sum_{m=-j}^{j} |j_1\, j_2\, j, m\rangle\, \langle j_1\, j_2\, j, m | j_1\, j_2\, m_1 : m_2\rangle \tag{12.75}$$

$$= \sum_{j=|j_1-j_2|}^{j_1+j_2} \sum_{m=-j}^{j} |j_1\, j_2\, j, m\rangle\, C^{j_1\, j_2\, j}_{m_1\, m_2\, m}. \tag{12.76}$$

The matrix elements on the right are actually the complex conjugates of the Clebsch–Gordan coefficients, but in our phase convention, all of them are real anyway.

As (12.69) and (12.76) demonstrate, results for total angular momentum orthobasis states in terms of product states, or vice versa, can always be expressed in terms of Clebsch–Gordan coefficients. For example, the content of the results for $l \otimes \frac{1}{2} = (l - \frac{1}{2}) \oplus (l + \frac{1}{2})$ that we found in (12.56) and (12.59) can be expressed, writing $m = m_l \pm 1/2$ as appropriate, as

$$C^{l,\, \frac{1}{2},\, l+\frac{1}{2}}_{m_l,\, -\frac{1}{2},\, m_l-\frac{1}{2}} = \sqrt{\frac{l+1-m_l}{2l+1}}, \qquad C^{l,\, \frac{1}{2},\, l+\frac{1}{2}}_{m_l,\, \frac{1}{2},\, m_l+\frac{1}{2}} = \sqrt{\frac{l+1+m_l}{2l+1}}, \tag{12.77}$$

$$C^{l,\, \frac{1}{2},\, l-\frac{1}{2}}_{m_l,\, -\frac{1}{2},\, m_l-\frac{1}{2}} = \sqrt{\frac{l+m_l}{2l+1}}, \qquad C^{l,\, \frac{1}{2},\, l-\frac{1}{2}}_{m_l,\, \frac{1}{2},\, m_l+\frac{1}{2}} = -\sqrt{\frac{l-m_l}{2l+1}}, \tag{12.78}$$

from which the results in (12.61)–(12.66) follow as the special case with $l = 1$. Although we claimed to be interested in the case that l was an orbital angular quantum number, there is no difference in the arithmetic for any angular momentum, and l can even be taken to be a half integer in these formulas.

As another useful example, consider the angular momentum addition problem $j \otimes 1 = (j-1) \oplus j \oplus (j+1)$. By a similar strategy, to be formalized below in the general case, the relevant nonzero Clebsch–Gordan coefficients are found to be

$$C^{j,\,1,\,j+1}_{m,\,\pm 1,\,m\pm 1} = \sqrt{\frac{(j+1\pm m)(j+2\pm m)}{2(j+1)(2j+1)}}, \quad C^{j,\,1,\,j+1}_{m,\,0,\,m} = \sqrt{\frac{(j+1-m)(j+1+m)}{(j+1)(2j+1)}}, \tag{12.79}$$

$$C^{j,\,1,\,j}_{m,\,\pm 1,\,m\pm 1} = \mp\sqrt{\frac{(j+1\pm m)(j\mp m)}{2j(j+1)}}, \qquad C^{j,\,1,\,j}_{m,\,0,\,m} = \sqrt{\frac{1}{j(j+1)}}\,m, \tag{12.80}$$

$$C^{j,\,1,\,j-1}_{m,\,\pm 1,\,m\pm 1} = \sqrt{\frac{(j-1\mp m)(j\mp m)}{2j(2j+1)}}, \qquad C^{j,\,1,\,j-1}_{m,\,0,\,m} = -\sqrt{\frac{(j+m)(j-m)}{j(2j+1)}}. \tag{12.81}$$

One can easily find published or online tables of Clebsch–Gordan coefficients, and software implementations of them, but it is a good idea to read the fine print to make sure of the phase convention being used.

We now point out some important identities satisfied by Clebsch–Gordan coefficients in general. First, by using (8.21) and (8.23) to evaluate

$$\langle m_1\!:\!m_2|J_-|j,m\rangle = \big(\langle j,m|(J_{1+}+J_{2+})|m_1\!:\!m_2\rangle\big)^*, \tag{12.82}$$

one finds

$$\begin{aligned}\sqrt{j(j+1)-m(m-1)}\,C^{j_1\,j_2\,j}_{m_1,m_2,m-1} &= \sqrt{j_1(j_1+1)-m_1(m_1+1)}\,C^{j_1\,j_2\,j}_{m_1+1,m_2,m}\\ &\quad+\sqrt{j_2(j_2+1)-m_2(m_2+1)}\,C^{j_1\,j_2\,j}_{m_1,m_2+1,m}.\end{aligned} \tag{12.83}$$

Similarly, from evaluating matrix elements of J_+,

$$\begin{aligned}\sqrt{j(j+1)-m(m+1)}\,C^{j_1\,j_2\,j}_{m_1,m_2,m+1} &= \sqrt{j_1(j_1+1)-m_1(m_1-1)}\,C^{j_1\,j_2\,j}_{m_1-1,m_2,m}\\ &\quad+\sqrt{j_2(j_2+1)-m_2(m_2-1)}\,C^{j_1\,j_2\,j}_{m_1,m_2-1,m}.\end{aligned} \tag{12.84}$$

From the total angular momentum orthonormality relations, using completeness of the product basis and the reality of the Clebsch–Gordan coefficients in our chosen convention, we also obtain

$$\sum_{m_1=-j_1}^{j_1}\sum_{m_2=-j_2}^{j_2} C^{j_1\,j_2\,j}_{m_1\,m_2\,m}C^{j_1\,j_2\,j'}_{m_1\,m_2\,m'} = \delta_{jj'}\,\delta_{mm'}, \tag{12.85}$$

for all $j, j' = |j_1-j_2|,\ldots,j_1+j_2$ with $m=-j,\ldots,j$ and $m'=-j',\ldots,j'$. Similarly, from the product basis orthonormality relations and completeness of the total angular momentum orthobasis,

$$\sum_{j=|j_1-j_2|}^{j_1+j_2} \sum_{m=-j}^{j} C^{j_1\,j_2\,j}_{m_1\,m_2\,m} C^{j_1\,j_2\,j}_{m_1'\,m_2'\,m} = \delta_{m_1 m_1'}\,\delta_{m_2 m_2'}, \tag{12.86}$$

for all $m_1, m_1' = -j_1, \dots j_1$ and $m_2, m_2' = -j_2, \dots, j_2$.

Consider (12.84) in the special case $m = j$, so that the left side vanishes. This gives

$$C^{j_1\,j_2\,j}_{m_1-1,\,m_2,\,j} = -\left(\frac{j_2(j_2+1) - m_2(m_2-1)}{j_1(j_1+1) - m_1(m_1-1)}\right)^{1/2} C^{j_1\,j_2\,j}_{m_1,\,m_2-1,\,j}. \tag{12.87}$$

This relation can be used to lower m_1 and raise m_2 in unit steps, starting from $m_1 = j_1$, giving a sign flip each time. Therefore, in our convention specified in (12.74),

$$\operatorname{sign}\left(C^{j_1\,j_2\,j}_{m_1,\,j-m_1,\,j}\right) = (-1)^{j_1-m_1}. \tag{12.88}$$

In particular, taking $m_1 = j - j_2$, we must have

$$\operatorname{sign}\left(C^{j_1\,j_2\,j}_{j-j_2,\,j_2,\,j}\right) = (-1)^{j_1+j_2-j}. \tag{12.89}$$

Meanwhile, directly from our convention in (12.74), we also have

$$\operatorname{sign}\left(C^{j_2\,j_1\,j}_{j_2,\,j-j_2,\,j}\right) = 1. \tag{12.90}$$

The states $|j_1\,j_2\,j, j\rangle$ and $|j_2\,j_1\,j, j\rangle$ obtained by interchanging the roles of $\boldsymbol{J}_1$ and $\boldsymbol{J}_2$ are really the same state physically, so they can only differ by a phase, which in our convention must be simply a sign. Let us define the product orthobasis kets to be invariant under interchange of the roles of the two angular momenta,

$$|j_1\,j_2\,m_1\!:\!m_2\rangle = |j_2\,j_1\,m_2\!:\!m_1\rangle = |j_1, m_1\rangle \otimes |j_2, m_2\rangle. \tag{12.91}$$

Then, comparing (12.89) and (12.90), we see that the total angular momentum basis kets must obey $|j_1\,j_2\,j, j\rangle = (-1)^{j_1+j_2-j}|j_2\,j_1\,j, j\rangle$. Now we act on this equation repeatedly with J_- to get the relation between the more general states $|j_1\,j_2\,j, m\rangle$ and $|j_2\,j_1\,j, m\rangle$. Doing so cannot change the relative sign, because the operator $J_- = J_{1-} + J_{2-}$ does not know or care which of the two angular momenta we chose to be the first one, so

$$|j_1\,j_2\,j, m\rangle = (-1)^{j_1+j_2-j}|j_2\,j_1\,j, m\rangle. \tag{12.92}$$

We therefore arrive at the rule for Clebsch–Gordan coefficients when we interchange the two angular momenta being added,

$$C^{j_1\,j_2\,j}_{m_1 m_2 m} = (-1)^{j_1+j_2-j}\, C^{j_2\,j_1\,j}_{m_2 m_1 m}. \tag{12.93}$$

Although we used a convention choice in the intermediate steps of deriving it, (12.93) is actually independent of the convention choice. A similar sort of strategy (with details omitted here) can be used to show that if one simultaneously flips the signs of m_1, m_2, and m, then the Clebsch–Gordan coefficients must also satisfy

$$C^{j_1\,j_2\,j}_{m_1 m_2 m} = (-1)^{j_1+j_2-j}\, C^{j_1\,j_2\,j}_{-m_1,-m_2,-m} \tag{12.94}$$

in our phase convention.

The following recipe can be used to construct the total angular momentum orthobasis states, and thus the Clebsch–Gordan coefficients, in a general case. Assume that $j_1 \geq j_2$. (Otherwise, one simply reverses the roles of j_1 and j_2.) We start with the state with the highest possible m. Since there is only one product orthobasis state with $J_{1z} + J_{2z}$ eigenvalue equal to $\hbar(j_1 + j_2)$, we have, using our phase convention choice,

$$|j_1{+}j_2,\ j_1{+}j_2\rangle = |j_1 : j_2\rangle. \tag{12.95}$$

Now we follow the strategy of repeatedly acting with the lowering operator $J_- = J_{1-} + J_{2-}$ to find new states $|j_1 + j_2, m\rangle$. Using (8.23) gives

$$J_-|j_1{+}j_2,\ j_1{+}j_2\rangle = \hbar\sqrt{2(j_1 + j_2)}\,|j_1{+}j_2,\ j_1{+}j_2{-}1\rangle, \tag{12.96}$$

which also can be evaluated as

$$J_{1-}|j_1 : j_2\rangle + J_{2-}|j_1 : j_2\rangle = \hbar\sqrt{2j_1}\,|j_1{-}1 : j_2\rangle + \hbar\sqrt{2j_2}\,|j_1 : j_2{-}1\rangle. \tag{12.97}$$

Therefore,

$$|j_1{+}j_2,\ j_1{+}j_2{-}1\rangle = \frac{1}{\sqrt{j_1 + j_2}}\Big(\sqrt{j_2}\,|j_1 : j_2{-}1\rangle + \sqrt{j_1}\,|j_1{-}1 : j_2\rangle\Big). \tag{12.98}$$

[The special cases $j_1 = j_2 = 1/2$ and $j_1 = l$, $j_2 = 1/2$ were previously found in (12.26) and (12.54), respectively.] Continuing to act with J_-, we can similarly find all of the orthobasis kets $|j_1{+}j_2, m\rangle$. Eventually we will obtain $m = -j_1 - j_2$, which is then annihilated by J_-.

Next, we proceed to obtain the states that have total $j = j_1 + j_2 - 1$. We start by writing the most general possible expression for the state with the largest possible m,

$$|j_1{+}j_2{-}1,\ j_1{+}j_2{-}1\rangle = a\,|j_1 : j_2{-}1\rangle + b\,|j_1{-}1 : j_2\rangle. \tag{12.99}$$

The coefficients a and b can be uniquely identified by requiring that this ket is orthogonal to $|j_1+j_2,\ j_1+j_2-1\rangle$, that the ket is normalized so that $|a|^2+|b|^2=1$, and that a is real and positive according to our phase convention. This gives

$$|j_1+j_2-1,\ j_1+j_2-1\rangle \;=\; \frac{1}{\sqrt{j_1+j_2}}\Big(\sqrt{j_1}|j_1:j_2-1\rangle - \sqrt{j_2}|j_1-1:j_2\rangle\Big). \quad (12.100)$$

[Again, we had previously derived this in the special cases $j_1 = j_2 = 1/2$ and $j_1 = l$, $j_2 = 1/2$, in (12.28) and (12.59), respectively.] Now, we again apply the operator J_- repeatedly, to obtain all of the other states $|j_1+j_2-1,\ m\rangle$.

Continuing on our vaunted quest, we proceed to the states with $j = j_1 + j_2 - 2$, starting again with the maximum value of m. This must be some linear combination of the three product orthobasis kets with $m = m_1 + m_2 = j_1 + j_2 - 2$, which we write as

$$|j_1+j_2-2,\ j_1+j_2-2\rangle \;=\; a|j_1:j_2-2\rangle + b|j_1-1:j_2-1\rangle + c|j_1-2:j_2\rangle. \quad (12.101)$$

The three unknown coefficients a, b, c can be obtained by demanding that this is orthogonal to both of the kets $|j_1+j_2,\ j_1+j_2-2\rangle$ and $|j_1+j_2-1,\ j_1+j_2-2\rangle$, as required by Theorem 2.14 (since all three are eigenvectors of J^2 with different eigenvalues), and that the ket is normalized, so $|a|^2 + |b|^2 + |c|^2 = 1$, and that a is real and positive according to our phase convention. The remaining states $|j_1+j_2-2,\ m\rangle$ are then obtained by acting repeatedly with J_-.

The preceding illustrates the general procedure, which is shown schematically in Fig. 12.1. Each column represents the repeated action of J_- on the states with fixed j, starting at the top with $m = j$ and ending at the bottom with $m = -j$. After completing each column, we move to the next column by first writing a candidate ket $|j,\ j\rangle$ as a linear combination of the product orthobasis kets that have $m_1 + m_2 = m = j$. The coefficients are determined by requiring orthogonality to all of the previously obtained total angular momentum orthobasis kets with that value of m and total angular momentum $j + 1$ or larger. Demanding that the ket be normalized and obey our phase convention, the state is then uniquely determined, and the rest of the states $|j,\ m\rangle$ follow by applying J_- repeatedly. The process iterates by moving to the next column with j lower by 1. Eventually, we will finish the last column of orthobasis kets, which will have $j = |j_1 - j_2|$, and there will be none more.

Consider the special case of adding two angular momenta $j_1 = 1$ and $j_2 = 1$. The resulting allowed values of j are 0, 1, and 2. By following the procedure summarized in Fig. 12.1, one finds the $j = 2$ states

$$|2, 2\rangle \;=\; |1:1\rangle, \quad (12.102)$$

$$|2, 1\rangle \;=\; \big(|1:0\rangle + |0:1\rangle\big)/\sqrt{2}, \quad (12.103)$$

$$|2, 0\rangle \;=\; \big(|1:-1\rangle + 2|0:0\rangle + |-1:1\rangle\big)/\sqrt{6}, \quad (12.104)$$

$$
\begin{array}{lllll}
\text{Start: } |\mathcal{J},\mathcal{J}\rangle & & & & \\
\quad\downarrow J_- & & & & \\
|\mathcal{J},\mathcal{J}-1\rangle & \overset{\text{orthogonal}}{\longrightarrow} & |\mathcal{J}-1,\mathcal{J}-1\rangle & & \\
\quad\downarrow J_- & & \quad\downarrow J_- & & \\
|\mathcal{J},\mathcal{J}-2\rangle & & |\mathcal{J}-1,\mathcal{J}-2\rangle & \overset{\text{orthogonal}}{\longrightarrow} & |\mathcal{J}-2,\mathcal{J}-2\rangle \\
\quad\downarrow J_- & & \quad\downarrow J_- & & \quad\downarrow J_- \quad \cdots \\
\quad\vdots & & \quad\vdots & & \quad\vdots \quad\quad \cdots \\
\quad\downarrow J_- & & \quad\downarrow J_- & & \quad\downarrow J_- \\
|\mathcal{J},-\mathcal{J}+2\rangle & & |\mathcal{J}-1,-\mathcal{J}+2\rangle & & |\mathcal{J}-2,-\mathcal{J}+2\rangle \\
\quad\downarrow J_- & & \quad\downarrow J_- & & \\
|\mathcal{J},-\mathcal{J}+1\rangle & & |\mathcal{J}-1,-\mathcal{J}+1\rangle & & \\
\quad\downarrow J_- & & & & \\
|\mathcal{J},-\mathcal{J}\rangle & & & &
\end{array}
$$

Fig. 12.1 A plan to compute the total angular momentum orthobasis kets $|j_1\, j_2\, j, m\rangle$, abbreviated here as $|j, m\rangle$, in terms of product orthobasis kets, when two angular momenta j_1 and j_2 are combined. In this figure, $\mathcal{J} \equiv j_1 + j_2$ is the maximum value of j. The process starts at the upper left with $j = m = \mathcal{J}$. Each column has fixed j, and is constructed from top to bottom using J_-. Then the next column is started by constructing the state with maximum m, by requiring it to be orthogonal to all of the previously found kets with that same m. The process ends after the column with $j = |j_1 - j_2|$ is finished, with $m = -|j_1 - j_2|$. This process provides the Clebsch–Gordan coefficients $\langle m_1\!:\!m_2|j, m\rangle = \langle j_1\, j_2\, m_1\!:\!m_2|j_1\, j_2\, j, m\rangle = C^{j_1\, j_2\, j}_{m_1\, m_2\, m}$

$$|2,-1\rangle = \big(|0\!:\!-1\rangle + |-1\!:\!0\rangle\big)/\sqrt{2}, \tag{12.105}$$

$$|2,-2\rangle = |-1\!:\!-1\rangle, \tag{12.106}$$

which are all symmetric under $m_1 \leftrightarrow m_2$, followed by the $j = 1$ states

$$|1,1\rangle = \big(|1\!:\!0\rangle - |0\!:\!1\rangle\big)/\sqrt{2}, \tag{12.107}$$

$$|1,0\rangle = \big(|1\!:\!-1\rangle - |-1\!:\!1\rangle\big)/\sqrt{2}, \tag{12.108}$$

$$|1,-1\rangle = \big(|0\!:\!-1\rangle - |-1\!:\!0\rangle\big)/\sqrt{2}, \tag{12.109}$$

which are each antisymmetric under the same exchange, and finally the $j = 0$ state

$$|0,0\rangle = \big(|1\!:\!-1\rangle - |0\!:\!0\rangle + |-1\!:\!1\rangle\big)/\sqrt{3}, \tag{12.110}$$

which is symmetric. [These results can be checked using the general Clebsch–Gordan coefficients for $j \otimes 1$ in (12.79)–(12.81).] To summarize what has been done, we write

$$1 \otimes 1 = 2_S \oplus 1_A \oplus 0_S, \tag{12.111}$$

where the S and A subscripts indicate the symmetry or antisymmetry of the total angular momentum basis kets under exchange of the two product basis eigenvalues m_1 and m_2.

More generally, for the combination of two equal angular momenta $j_1 = j_2 = j$, the symmetry and antisymmetry properties for exchange of m_1 and m_2 are summarized by

$$j \otimes j = (2j)_S \oplus (2j-1)_A \oplus (2j-2)_S \oplus \cdots \oplus 0, \tag{12.112}$$

with alternating S and A, so that the singlet on the far right is symmetric if j is an integer and antisymmetric if j is a half-integer. These symmetry and antisymmetry properties follow immediately from (12.93).

The simplest nontrivial example that cannot be obtained from (12.77)–(12.78) or from (12.79)–(12.81) is $j_1 = j_2 = 3/2$. Call the resulting total angular momentum quantum number J. Then you can show that the $J = 3, m_J = 3$ and $J = 3, m_J = 0$ and $J = 0, m_J = 0$ states are, in terms of product basis states $|m_1 : m_2\rangle$,

$$|3, 3\rangle = \left|\tfrac{3}{2} : \tfrac{3}{2}\right\rangle, \tag{12.113}$$

$$|3, 0\rangle = \frac{1}{\sqrt{20}}\left(\left|\tfrac{3}{2} : -\tfrac{3}{2}\right\rangle + \left|-\tfrac{3}{2} : \tfrac{3}{2}\right\rangle + 3\left|\tfrac{1}{2} : -\tfrac{1}{2}\right\rangle + 3\left|-\tfrac{1}{2} : \tfrac{1}{2}\right\rangle\right), \tag{12.114}$$

$$|0, 0\rangle = \frac{1}{2}\left(\left|\tfrac{3}{2} : -\tfrac{3}{2}\right\rangle - \left|-\tfrac{3}{2} : \tfrac{3}{2}\right\rangle - \left|\tfrac{1}{2} : -\tfrac{1}{2}\right\rangle + \left|-\tfrac{1}{2} : \tfrac{1}{2}\right\rangle\right), \tag{12.115}$$

respectively. It is left as Exercise 12.6 to work out the other total angular momentum orthobasis states $|J, m_J\rangle$ in this example.

One can also combine three (or more) angular momenta, to obtain an orthobasis of kets that are eigenstates of the observables J^2 and J_z for the total angular momentum $\boldsymbol{J} = \boldsymbol{J}_1 + \boldsymbol{J}_2 + \boldsymbol{J}_3$. This can be done recursively, by first combining $\boldsymbol{J}_1$ and $\boldsymbol{J}_2$ into an angular momentum operator $\boldsymbol{J}_{12}$, and then combining the results with $\boldsymbol{J}_3$. For example, combining three spin-1/2 angular momenta results in

$$\tfrac{1}{2} \otimes \tfrac{1}{2} \otimes \tfrac{1}{2} = (1 \oplus 0) \otimes \tfrac{1}{2} = (1 \otimes \tfrac{1}{2}) \oplus (0 \otimes \tfrac{1}{2}) \tag{12.116}$$

$$= (3/2)_S \oplus (1/2)_a \oplus (1/2)_b. \tag{12.117}$$

The subscript S on the $j = 3/2$ multiplet indicates that the states in it are symmetric under exchange of any two of the three individual spins, as we will soon check. The result also has two $j = 1/2$ multiplets, distinguished by degeneracy labels a and b. To check the multiplicities of states, note that on the left side of (12.117) there are $2^3 = 8$ product orthobasis states, since each individual spin has 2 values of $m_s = \pm 1/2$. On the right, the multiplicities of the total angular momentum states are $4 + 2 + 2 = 8$. To explicitly construct the total angular momentum orthobasis kets, one can first use (12.25)–(12.28) to construct the J_{12}^2, J_{12z} eigenstate kets. Combining the resulting $j_{12} = 1$ states with the third spin, one finds, by using (12.56), that the $j = 3/2$ states are

$$|\tfrac{3}{2}, \tfrac{3}{2}\rangle = |\uparrow\uparrow\uparrow\rangle, \tag{12.118}$$

$$|\tfrac{3}{2}, \tfrac{1}{2}\rangle = \frac{1}{\sqrt{3}}\big(|\uparrow\uparrow\downarrow\rangle + |\uparrow\downarrow\uparrow\rangle + |\downarrow\uparrow\uparrow\rangle\big), \tag{12.119}$$

$$|\tfrac{3}{2}, -\tfrac{1}{2}\rangle = \frac{1}{\sqrt{3}}\big(|\uparrow\downarrow\downarrow\rangle + |\downarrow\uparrow\downarrow\rangle + |\downarrow\downarrow\uparrow\rangle\big), \tag{12.120}$$

$$|\tfrac{3}{2}, -\tfrac{3}{2}\rangle = |\downarrow\downarrow\downarrow\rangle, \tag{12.121}$$

and, by using (12.59), the $j = 1/2$ states

$$|\tfrac{1}{2}, \tfrac{1}{2}, a\rangle = \frac{1}{\sqrt{6}}\big(2|\uparrow\uparrow\downarrow\rangle - |\uparrow\downarrow\uparrow\rangle - |\downarrow\uparrow\uparrow\rangle\big), \tag{12.122}$$

$$|\tfrac{1}{2}, -\tfrac{1}{2}, a\rangle = \frac{1}{\sqrt{6}}\big(|\uparrow\downarrow\downarrow\rangle + |\downarrow\uparrow\downarrow\rangle - 2|\downarrow\downarrow\uparrow\rangle\big). \tag{12.123}$$

From combining the $j_{12} = 0$ states from (12.28) with the third spin, we have the additional $j = 1/2$ total angular momentum basis states

$$|\tfrac{1}{2}, \tfrac{1}{2}, b\rangle = \frac{1}{\sqrt{2}}\big(|\uparrow\downarrow\uparrow\rangle - |\downarrow\uparrow\uparrow\rangle\big), \tag{12.124}$$

$$|\tfrac{1}{2}, -\tfrac{1}{2}, b\rangle = \frac{1}{\sqrt{2}}\big(|\uparrow\downarrow\downarrow\rangle - |\downarrow\uparrow\downarrow\rangle\big). \tag{12.125}$$

The explicit forms for the $j = 3/2$ states show that they are indeed each symmetric under exchange of any two of the three individual spins. For the $j = 1/2$ states, the ones labeled a are symmetric, and the ones labeled b are antisymmetric, under exchange of the first two spins. However, these $j = 1/2$ states do not have a complete symmetry or antisymmetry under exchange of every pair of spins. Also note that there can be no totally antisymmetric combination of three or more spin 1/2 states, simply because in the product basis there are only two available values $\uparrow$ and $\downarrow$.

Similarly, combining four spin-1/2 angular momenta gives

$$\tfrac{1}{2} \otimes \tfrac{1}{2} \otimes \tfrac{1}{2} \otimes \tfrac{1}{2} = (1 \oplus 0) \otimes (1 \oplus 0) = 2_S \oplus 1 \oplus 1 \oplus 1 \oplus 0 \oplus 0. \tag{12.126}$$

Here the counting is that the number of product orthobasis states on the left is $2^4 = 16$, while the counting of multiplicities of total angular momentum orthobasis states on the right is $5 + 3 + 3 + 3 + 1 + 1 = 16$. Only the total spin 2 multiplet is totally symmetric. The other multiplets on the right side of (12.126) have mixed symmetry properties under exchange of the spins, and there is no totally antisymmetric multiplet.

In the case of N electrons in an atom, the angular momentum eigenstates are often given in spectroscopic notation, defined as follows. First, combine all of the individual spin operators $\boldsymbol{S}_i$ to obtain the total spin angular momentum operator $\boldsymbol{S} = \sum_i \boldsymbol{S}_i$. The eigenvalues of the operator $S^2 = \boldsymbol{S} \cdot \boldsymbol{S}$ are then denoted by $\hbar^2 S(S+1)$ where S is a number.[1] Clearly, if N is even, then the allowed numbers S must be integers with $0 \leq S \leq N/2$. If N is odd, then the number S must be half-integer, with $1/2 \leq S \leq N/2$. Next, combine all of the individual orbital angular momenta operators $\boldsymbol{L}_i$ to obtain the total orbital angular momentum operator $\boldsymbol{L} = \sum_i \boldsymbol{L}_i$. The eigenvalues of the operator $L^2 = \boldsymbol{L} \cdot \boldsymbol{L}$ are likewise denoted $\hbar^2 L(L+1)$, where the quantum number L is always an integer. Finally, the operators $\boldsymbol{S}$ and $\boldsymbol{L}$ are combined to form the total angular momentum operator $\boldsymbol{J}$, and the operators J^2 and J_z have eigenvalues denoted $\hbar^2 J(J+1)$ and $\hbar m_J$, respectively. The traditional notation for a multiplet of common eigenstates of the observables S^2, L^2, and J^2 is then

$$ {}^{2S+1}L_J, \tag{12.127} $$

where S, L, and J are the quantum numbers, but with L replaced by the capital letter code $S, P, D, F, G, \ldots$, according to whether the number L is $0, 1, 2, 3, 4, \ldots$, as indicated in (11.20). The degeneracy, or multiplicity, of each group of states denoted by ${}^{2S+1}L_J$ is $2J+1$, since the eigenvalue m_J can take on the values $-J, -J+1, \ldots, J-1, J$.

For example, a single electron always has $S = 1/2$. For $L = 0$, it has $J = 1/2$, and for $L \geq 1$ it can have $J = L \pm 1/2$, as we saw in Sect. 12.3. So, the list of total angular momentum multiplets for a single electron in an atom is

$$ {}^2S_{1/2},\ {}^2P_{1/2},\ {}^2P_{3/2},\ {}^2D_{3/2},\ {}^2D_{5/2},\ {}^2F_{5/2},\ {}^2F_{7/2},\ {}^2G_{7/2},\ {}^2G_{9/2} \cdots \tag{12.128} $$

For two electrons, the possible total spin quantum numbers are $S = 0$ and 1, as we saw in Sect. 12.2. The orbital angular momenta $\boldsymbol{L}_1$ and $\boldsymbol{L}_2$ with quantum numbers l_1 and l_2 can be combined into $L = |l_1 - l_2|, \ldots, l_1 + l_2$. Therefore, the possible total angular momentum eigenstates for two electrons include

$$ {}^1S_0,\ {}^1P_1,\ {}^1D_2,\ {}^1F_3,\ {}^1G_4 \cdots \tag{12.129} $$

for $S = 0$, and

$$ {}^3S_1,\ {}^3P_0,\ {}^3P_1,\ {}^3P_2,\ {}^3D_1,\ {}^3D_2,\ {}^3D_3,\ {}^3F_2,\ {}^3F_3,\ {}^3F_4 \cdots \tag{12.130} $$

[1] Following a common but potentially confusing notation for combined angular momenta, capital letters are often used both for the names of the operators and the corresponding quantum numbers.

for $S = 1$. However, as we will discuss in more detail in Sect. 18.1, Fermi–Dirac statistics requires that the total state must be antisymmetric under exchange of the two electrons. In the special case that the electrons have the same radial wavefunction and the same orbital angular momentum quantum number $l_1 = l_2$, then (12.112) implies that if L is odd (an antisymmetric position wavefunction) then only the symmetric spin combination $S = 1$ is allowed, and if L is even (a symmetric position wavefunction), then only the antisymmetric spin combination $S = 0$ is allowed.

12.5 How Spherical Harmonics Combine

The spherical harmonics are the position representations of the orbital angular momentum eigenstates of L^2 and L_z. In this section, we will learn how the addition of angular momentum technology is realized in terms of the spherical harmonics.

We start by considering fixed l_1, l_2, l, and m. Let us define the following function of θ and ϕ, as a linear combination of products of spherical harmonics weighted by the Clebsch–Gordan coefficients defined by (12.70),

$$\Psi_l^m(\theta,\phi) \equiv \sum_{m_1,m_2} Y_{l_1}^{m_1}(\theta,\phi)\, Y_{l_2}^{m_2}(\theta,\phi)\, C_{m_1 m_2 m}^{l_1 l_2 l}, \tag{12.131}$$

where the sums are over all m_1 and m_2 such that the Clebsch–Gordan coefficient does not vanish (equivalently, over m_1, m_2 with $m_1 + m_2 = m$). Our goal is to find another simple expression for $\Psi_l^m(\theta,\phi)$. We claim that, as its labeling suggests, this wavefunction is an eigenstate of the differential operators L^2 and L_z, with eigenvalues $\hbar^2 l(l+1)$ and $\hbar m$, respectively. This implies that it must be proportional to $Y_l^m(\theta,\phi)$.

To prove this claim, we will check the action of the differential operators L_z, L_-, and L_+ on it. First, by applying (8.120), we have

$$L_z \Psi_l^m = \sum_{m_1,m_2} (\hbar m_1 + \hbar m_2)\, Y_{l_1}^{m_1}\, Y_{l_2}^{m_2}\, C_{m_1 m_2 m}^{l_1 l_2 l}. \tag{12.132}$$

The Clebsch–Gordan coefficient enforces that all nonzero contributions have $m_1 + m_2 = m$, so

$$L_z \Psi_l^m = \hbar m \Psi_l^m. \tag{12.133}$$

Next, applying (8.122) we find

$$\begin{aligned} L_- \Psi_l^m = \hbar \sum_{m_1,m_2} \Big[&\sqrt{l_1(l_1+1) - m_1(m_1-1)}\, Y_{l_1}^{m_1-1}\, Y_{l_2}^{m_2} \\ &+ \sqrt{l_2(l_2+1) - m_2(m_2-1)}\, Y_{l_1}^{m_1}\, Y_{l_2}^{m_2-1} \Big] C_{m_1 m_2 m}^{l_1 l_2 l}. \end{aligned} \tag{12.134}$$

Since we are summing over all m_1 and m_2, we can use the trick of relabeling $m_1 \to m_1 + 1$ in the first term and $m_2 \to m_2 + 1$ in the second term, to obtain

$$L_- \Psi_l^m = \hbar \sum_{m_1, m_2} \left[\sqrt{l_1(l_1+1) - m_1(m_1+1)}\, C^{l_1 l_2 l}_{m_1+1, m_2, m} + \sqrt{l_2(l_2+1) - m_2(m_2+1)}\, C^{l_1 l_2 l}_{m_1, m_2+1, m} \right] Y_{l_1}^{m_1} Y_{l_2}^{m_2}. \qquad (12.135)$$

Now the Clebsch–Gordan recurrence relation (12.83) neatly turns this into

$$L_- \Psi_l^m = \hbar \sum_{m_1, m_2} \sqrt{l(l+1) - m(m-1)}\, C^{l_1 l_2 l}_{m_1, m_2, m-1} Y_{l_1}^{m_1} Y_{l_2}^{m_2}, \qquad (12.136)$$

or, using the definition (12.131), simply,

$$L_- \Psi_l^m = \hbar \sqrt{l(l+1) - m(m-1)}\, \Psi_l^{m-1}. \qquad (12.137)$$

In an exactly analogous way, one finds

$$L_+ \Psi_l^m = \hbar \sqrt{l(l+1) - m(m+1)}\, \Psi_l^{m+1}. \qquad (12.138)$$

Taken together, (12.133), (12.137) and (12.138), combined with $L^2 = (L_+ L_- + L_- L_+)/2 + L_z^2$ from (8.15), show that this wavefunction also obeys

$$L^2 \Psi_l^m = \hbar^2 l(l+1) \Psi_l^m. \qquad (12.139)$$

Since the spherical harmonic Y_l^m is (up to an overall multiplicative constant) the unique function of (θ, ϕ) with L^2 eigenvalue $\hbar^2 l(l+1)$ and L_z eigenvalue $\hbar m$, (12.133) and (12.139) show that Ψ_l^m must be proportional to it. Therefore,

$$\Psi_l^m(\theta, \phi) = c_{l_1 l_2 l m}\, Y_l^m(\theta, \phi) \qquad (12.140)$$

as claimed, for some constant $c_{l_1 l_2 l m}$ (independent of θ, ϕ).

Furthermore, by acting with L_- on both sides of the previous equation, we obtain

$$L_- \Psi_l^m = c_{l_1 l_2 l m}\, L_- Y_l^m = c_{l_1 l_2 l m} \hbar \sqrt{l(l+1) - m(m-1)}\, Y_l^{m-1}, \qquad (12.141)$$

but also from (12.137) this is equal to

$$L_- \Psi_l^m = \hbar \sqrt{l(l+1) - m(m-1)}\, c_{l_1 l_2 l, m-1}\, Y_l^{m-1}. \qquad (12.142)$$

Comparing these tells us that $c_{l_1 l_2 l m} = c_{l_1 l_2 l, m-1}$ for all $-l + 1 \le m \le l$, so the constant of proportionality does not actually depend on m, and we therefore drop that label.

All of the preceding just served to prove that we can write

$$\sum_{m_1,m_2} Y_{l_1}^{m_1}(\theta,\phi)\,Y_{l_2}^{m_2}(\theta,\phi)\,C_{m_1m_2m}^{l_1l_2l} = c_{l_1l_2l}\,Y_l^m(\theta,\phi) \tag{12.143}$$

for some proportionality constant $c_{l_1l_2l}$. Our remaining task is to identify this constant. Fortunately, there is a trick to do this very easily; just consider the special case $\theta = 0$. Using the fact found in (8.146),

$$Y_l^m(0,\phi) = \delta_{m0}\sqrt{\frac{2l+1}{4\pi}}, \tag{12.144}$$

the double sum in (12.143) collapses to a single term with $m_1 = m_2 = 0$, and it reads

$$\sqrt{\frac{2l_1+1}{4\pi}}\sqrt{\frac{2l_2+1}{4\pi}}\,C_{000}^{l_1l_2l} = c_{l_1l_2l}\sqrt{\frac{2l+1}{4\pi}}. \tag{12.145}$$

Solving this for $c_{l_1l_2l}$, (12.143) becomes

$$\sum_{m_1,m_2} Y_{l_1}^{m_1}(\theta,\phi)\,Y_{l_2}^{m_2}(\theta,\phi)\,C_{m_1m_2m}^{l_1l_2l} = \sqrt{\frac{(2l_1+1)(2l_2+1)}{4\pi(2l+1)}}\,C_{000}^{l_1l_2l}\,Y_l^m(\theta,\phi). \tag{12.146}$$

This accomplishes our goal.

To derive an interesting and useful identity, start with (12.146), multiply by $C_{m_1'm_2'm}^{l_1l_2l}$, then sum over all l, m, then apply the orthogonality of the Clebsch–Gordan coefficients of (12.86) on the left side, and finally relabel $m_1' \to m_1$ and $m_2' \to m_2$. The result is

$$Y_{l_1}^{m_1}(\theta,\phi)Y_{l_2}^{m_2}(\theta,\phi) = \sum_{l=|l_1-l_2|}^{l_1+l_2}\sqrt{\frac{(2l_1+1)(2l_2+1)}{4\pi(2l+1)}}\,C_{000}^{l_1l_2l}\,C_{m_1m_2m}^{l_1l_2l}\,Y_l^m(\theta,\phi), \tag{12.147}$$

where $m = m_1 + m_2$ on the right side. The product of any two spherical harmonics is thus an appropriately weighted sum over the spherical harmonics selected by the corresponding addition of angular momenta.

Another closely related identity follows. Start from (12.147), multiply by $[Y_{l'}^{m'}(\theta,\phi)]^*$, integrate over the angular coordinates, then apply the orthonormality of the spherical harmonics using (8.124) on the right side, and finally relabel $l' \to l$ and $m' \to m$. The result is

$$\int d\Omega \left[Y_l^m(\theta,\phi)\right]^* Y_{l_1}^{m_1}(\theta,\phi) Y_{l_2}^{m_2}(\theta,\phi) = \sqrt{\frac{(2l_1+1)(2l_2+1)}{4\pi(2l+1)}}\, C_{000}^{l_1 l_2 l} C_{m_1 m_2 m}^{l_1 l_2 l}. \tag{12.148}$$

This is the **Wigner–Eckart formula for spherical harmonics**. It is a special case of the general Wigner–Eckart theorem (a statement about matrix elements of operators, and applicable not just to orbital angular momentum) to be derived in Sect. 13.3. Remarkably, the integral depends on the magnetic quantum numbers m_1, m_2, and m only through the coefficient $C_{m_1 m_2 m}^{l_1 l_2 l}$, so the Clebsch–Gordan selection rules (12.71)–(12.73) also govern which integrals of this type can be nonzero. This formula can be used to evaluate the integral of any product of three spherical harmonics (with or without complex conjugates), simply by making use of (8.144).

Exercises

12.1 An atomic electron is in a state with orbital angular momentum $l = 2$, with L_z and S_z eigenvalues 0 and $+\hbar/2$, respectively. If its total angular momentum squared is measured to be $J^2 = \hbar^2 j(j+1)$, what are the possible results for j, and their probabilities?

12.2 Two particles both have spin 1/2, and one of them is in a p-wave ($l = 1$) orbital angular momentum state while the other is in a d-wave ($l = 2$) orbital angular momentum state. If $\boldsymbol{J}$ is the total angular momentum operator, what are the allowed eigenvalues of J^2, and their degeneracies? [Hint: start by figuring out the total number of basis states by multiplying together the individual angular momentum multiplicities. You should find 60, so that should be the sum of the degeneracies.]

12.3 Consider a quantum system with two independent spin-1/2 operators, $\boldsymbol{S}_1$ and $\boldsymbol{S}_2$, so that the state space is spanned by an orthobasis of S_{1z} and S_{2z} eigenstates $|\uparrow\uparrow\rangle$, $|\uparrow\downarrow\rangle$, $|\downarrow\uparrow\rangle$, and $|\downarrow\downarrow\rangle$. In each ket, the first entry labels states with S_{1z} eigenvalue $\pm\hbar/2$, and the second entry labels states with S_{2z} eigenvalue $\pm\hbar/2$. At time $t = 0$, the state is

$$|\psi(0)\rangle = \frac{1}{2}|\uparrow\uparrow\rangle + \frac{1}{2}|\uparrow\downarrow\rangle + \frac{1}{\sqrt{2}}|\downarrow\downarrow\rangle. \tag{12.149}$$

(a) At time $t = 0$, we simultaneously measure S^2 and S_z, where $\boldsymbol{S} = \boldsymbol{S}_1 + \boldsymbol{S}_2$ is the total spin operator. What are the possible outcomes, and their probabilities?
(b) Suppose that instead of the above measurements, we let the system evolve until time t, with the Hamiltonian $H = \omega_1 S_{1z} + \omega_2 S_{2z}$, where ω_1 and ω_2 are constants. What is the state at time t? If we measure S_{1z} at time t, what

are the possible outcomes, and their probabilities? Use these results to find the expectation value $\langle S_{1z} \rangle$ as a function of time.

(c) Now suppose instead that the Hamiltonian of the system is $H = a\boldsymbol{S}_1 \cdot \boldsymbol{S}_2$, where a is a constant. What is the state at time t, and what are the possible outcomes and probabilities for a measurement of S_{1z}? Use these results to find $\langle S_{1z} \rangle$ as a function of time. [Hint: to find the state at time t, it is best to use the total angular momentum basis, in which H is diagonal.]

12.4 Consider the hyperfine splitting of the $n = 1$ level of the hydrogen atom in the presence of a constant external magnetic field $\boldsymbol{B} = B\hat{z}$. Since we are only treating the ground state, there is no orbital angular momentum $\boldsymbol{L}$. Let the electron and proton spin operators be $\boldsymbol{S}$ and $\boldsymbol{I}$, respectively, and let their total be $\boldsymbol{J} = \boldsymbol{S} + \boldsymbol{I}$. The Hamiltonian is

$$H = \frac{E_\gamma}{\hbar^2}\boldsymbol{S} \cdot \boldsymbol{I} + \frac{2\mu_B}{\hbar}\boldsymbol{B} \cdot \boldsymbol{S}, \tag{12.150}$$

where μ_B is the Bohr magneton and E_γ is the energy of the 21.1 centimeter line. (There is also an interaction term $-(g_p\mu_N/\hbar)\boldsymbol{B} \cdot \boldsymbol{I}$, but it is neglected because $\mu_N \ll \mu_B$.)

(a) Evaluate each of the two operators $\boldsymbol{S} \cdot \boldsymbol{I}$ and $\boldsymbol{B} \cdot \boldsymbol{S}$ acting on each of the product basis states labeled by the eigenvalues of S_z and I_z, and find the matrix representation of H in that basis.

(b) Repeat part (a), but this time use the total angular momentum basis labeled by the eigenvalues of J^2 and J_z.

(c) In the limit that B is so large that E_γ can be neglected, find the energy eigenvalues, and the corresponding energy eigenstates in the product orthobasis.

(d) In the limit that B is so small that it can be neglected, find the energy eigenvalues and the corresponding energy eigenstates in the total angular momentum basis.

(e) Find the energy eigenvalues for general B. Expand them for small B and for large B to show agreement with the special limits obtained in parts (c) and (d). (You do not need to find the energy eigenstates.)

(f) Graph the energy eigenvalues in units of μeV as a function of the magnetic field strength B for $0 < B < 1200$ gauss. Label the energies near $B = 0$ with $j = 0$ and $j = 1$, and the energies for large B with m_S and m_I.

12.5 The electron and its antiparticle, the positron, can form bound states before they annihilate, called positronium. The positronium states with lowest energy and zero orbital angular momentum have a Hamiltonian

$$H = \frac{a}{\hbar^2}\boldsymbol{S}_e \cdot \boldsymbol{S}_p + \frac{b}{\hbar}(S_{ez} - S_{pz}), \tag{12.151}$$

where $\boldsymbol{S}_e$ and $\boldsymbol{S}_p$ are the spin operators for the electron and positron, respectively, and $a = 8.41 \times 10^{-4}$ eV, and $b = \hbar e B / m_e c$ where B is an external magnetic field in the $\hat{z}$ direction.

(a) Working in the total spin basis of eigenstates of S^2 and S_z where $\boldsymbol{S} = \boldsymbol{S}_e + \boldsymbol{S}_p$, find the Hamiltonian as a 4×4 matrix in terms of a and b, and find its eigenvalues and eigenvectors.

(b) The ${}^1S_0$ singlet (parapositronium) state decays rather quickly by annihilation to two photons, with a lifetime of about 1.25×10^{-10} seconds, while the ${}^3S_1$ triplet (orthopositronium) states are relatively stable, with a lifetime of about 1.42×10^{-7} seconds. (Assume that these statements do not depend on the applied magnetic field.) Estimate the magnetic field B required to make the ground state lifetime longer by 10%.

12.6 For the addition of two angular momenta $j_1 = j_2 = 3/2$, find all of the total angular momentum states in terms of the product basis states, completing (12.113)–(12.115). Summarize the results as a list of the nonzero Clebsch–Gordan coefficients.

12.7 General expressions for the Clebsch–Gordan coefficients for the addition of angular momenta l and 1 were given in (12.79)–(12.81). Use the results to write each of $Y_l^m(\theta, \phi) \cos\theta$ and $Y_l^m(\theta, \phi) \sin\theta \cos\phi$ and $Y_l^m(\theta, \phi) \sin\theta \sin\phi$ as linear combinations of other spherical harmonics, with constant coefficients.

Tensor Operators and Useful Rules They Obey

13

13.1 Irreducible Tensor Operators

Consider a finite spatial rotation generated by a unitary operator, so that a state ket $|\psi\rangle$ is transformed to $|\psi'\rangle = U(\boldsymbol{\alpha})\,|\psi\rangle$, as discussed in Sect. 5.3. Recall, from the discussion surrounding (2.112)–(2.117), that for any operator A we can consider the corresponding rotated version defined by $A' = U(\boldsymbol{\alpha})AU(\boldsymbol{\alpha})^\dagger$, so that any matrix element $\langle\chi|A|\psi\rangle$ is equal to the matrix element of the rotated operator for the rotated states, $\langle\chi'|A'|\psi'\rangle$. We will now study operators that transform under rotations in a simple and useful way, specifically, as irreducible representations of the rotation group.

An **irreducible tensor operator** (also known as a **spherical tensor operator**) of order j is a collection of $2j+1$ component operators labeled by an index m,

$$T_m^{(j)}, \qquad (m = -j, \ldots, j), \tag{13.1}$$

which, by definition, are required to obey the following transformation rule for every unitary rotation operator $U(\boldsymbol{\alpha})$,

$$U(\boldsymbol{\alpha})\, T_m^{(j)}\, U(\boldsymbol{\alpha})^\dagger = \sum_{m'=-j}^{j} T_{m'}^{(j)}\, D_{m'm}^{(j)}(\boldsymbol{\alpha}), \tag{13.2}$$

with the rotation matrices $D^{(j)}(\boldsymbol{\alpha})$ defined in (8.70). Using the form of $U(\boldsymbol{\alpha})$ for an infinitesimal rotation in terms of the total angular momentum operator $\boldsymbol{J}$ in (8.69), and the actions of J_z, J_+, and J_- on angular momentum eigenstates as given in (8.7), (8.21), and (8.23), which are reproduced here for convenience:

$$J_z\,|jm\rangle = \hbar m\,|jm\rangle\,, \tag{13.3}$$

S. P. Martin, *Quantum Mechanics*, Graduate Texts in Physics,
https://doi.org/10.1007/978-3-032-21119-4_13

$$J_+ \, |jm\rangle = \hbar\sqrt{j(j+1) - m(m+1)} \; |j, m{+}1\rangle , \tag{13.4}$$

$$J_- \, |jm\rangle = \hbar\sqrt{j(j+1) - m(m-1)} \; |j, m{-}1\rangle , \tag{13.5}$$

one can show (see Exercise 13.1) that the components of an irreducible tensor operator must obey the commutation relations

$$\left[J_z, \, T_m^{(j)}\right] = \hbar m \, T_m^{(j)}, \tag{13.6}$$

$$\left[J_+, \, T_m^{(j)}\right] = \hbar\sqrt{j(j+1) - m(m+1)} \, T_{m+1}^{(j)}, \tag{13.7}$$

$$\left[J_-, \, T_m^{(j)}\right] = \hbar\sqrt{j(j+1) - m(m-1)} \, T_{m-1}^{(j)}. \tag{13.8}$$

Equations (13.6)–(13.8) serve as an equivalent necessary and sufficient requirement for $T_m^{(j)}$ to form an irreducible tensor operator. Their similarity to (13.3)–(13.5) suggests that the operators $T_m^{(j)}$ "carry" angular momentum as indicated by the quantum number labels j and m. This intuition is valid, as we will explain in the following, with the most general and precise statement to be given in (13.71).

The simplest type of irreducible tensor operator has $j = 0$ with a single component, and is called a **scalar operator**. A scalar operator S obeys $U(\boldsymbol{\alpha}) \, S \, U(\boldsymbol{\alpha})^\dagger = S$, or equivalently it commutes with the total angular momentum operators J_z, J_+, J_-, and therefore also with J_x, J_y, and J^2. Consider a state $|\alpha, j, m\rangle$ that is an eigenstate of J^2 and J_z with eigenvalues $\hbar^2 j(j+1)$ and $\hbar m$, with α a degeneracy label. It follows that the state $S \, |\alpha, j, m\rangle$ is also an eigenstate of J^2 and J_z with the same eigenvalues. In this sense, a scalar operator carries no total angular momentum.

Irreducible tensor operators with $j = 1/2$, or more generally with half-integer j, are called **spinor operators**. However, they cannot be observables, because they are necessarily double-valued; under a rotation by 2π they acquire a minus sign, just as we saw for matrix representations of states with half-integer angular momentum in (8.84) and (8.87). Observables can be constructed by taking the product of an even number of spinor operators, using Theorem 13.1 below. We will not discuss them further, although they do play an important role in quantum field theories with fermions.

An irreducible tensor operator $V_m^{(1)}$ with order $j = 1$ and three components $m = -1, 0, 1$ is called a **vector operator**. The familiar rectangular x, y, z components of $\boldsymbol{V} = \hat{x} V_x + \hat{y} V_y + \hat{z} V_z$ can then be defined as follows:

$$V_x = \left(V_{-1}^{(1)} - V_1^{(1)}\right)/\sqrt{2}, \tag{13.9}$$

$$V_y = i\left(V_{-1}^{(1)} + V_1^{(1)}\right)/\sqrt{2}, \tag{13.10}$$

$$V_z = V_0^{(1)}. \tag{13.11}$$

These can be shown, using (13.6)–(13.8) with $j = 1$, to obey the commutator rule

$$[J_a, V_b] = i\hbar\epsilon_{abc}V_c, \qquad (a, b, c = x, y, z), \tag{13.12}$$

which is often taken as the defining property of a vector operator in rectangular coordinates. For example, the rectangular coordinate position operators found in $\boldsymbol{R} = \hat{x}X + \hat{y}Y + \hat{z}Z$ can be expressed in terms of components of an irreducible tensor operator $R^{(1)}$, as

$$R_1^{(1)} = -(X + iY)/\sqrt{2}, \qquad R_{-1}^{(1)} = (X - iY)/\sqrt{2}, \tag{13.13}$$

$$R_0^{(1)} = Z, \tag{13.14}$$

and you can check that the requisite commutation relations (13.6)–(13.8) are indeed satisfied. (See also Exercise 5.1.) Besides the position operator, other examples of vector operators are the momentum operator $\boldsymbol{P}$, the orbital angular momentum $\boldsymbol{L}$, the spin $\boldsymbol{S}$, and the total angular momentum operator itself, $\boldsymbol{J} = \boldsymbol{L} + \boldsymbol{S}$. The angular momentum raising and lowering operators defined by (8.8) are given in the irreducible tensor component formalism by $J_+ = -\sqrt{2}J_1^{(1)}$ and $J_- = \sqrt{2}J_{-1}^{(1)}$, respectively, while $J_z = J_0^{(1)}$.

Given any two irreducible tensor operators, one can multiply them to construct others. The way to do this is specified by the following theorem.

Theorem 13.1 (Product Rules for Tensor Operators) *If $T^{(j_1)}$ and $S^{(j_2)}$ are irreducible tensor operators of orders j_1 and j_2 respectively, and $C_{m_1\,m_2\,m}^{j_1\,j_2\,j}$ are the Clebsch–Gordan coefficients defined in (12.70), then for $j = |j_1 - j_2|, \ldots, j_1 + j_2$,*

$$W_m^{(j)} = \sum_{m_1=-j_1}^{j_1} \sum_{m_2=-j_2}^{j_2} T_{m_1}^{(j_1)} S_{m_2}^{(j_2)} C_{m_1\,m_2\,m}^{j_1\,j_2\,j} \qquad (\textit{for } m = -j, \ldots, j) \tag{13.15}$$

is an irreducible tensor operator of order j. Furthermore,

$$T_{m_1}^{(j_1)} S_{m_2}^{(j_2)} = \sum_{j=|j_1-j_2|}^{j_1+j_2} W_m^{(j)} C_{m_1\,m_2\,m}^{j_1\,j_2\,j} \qquad (\textit{with } m = m_1 + m_2), \tag{13.16}$$

in a convention in which the Clebsch–Gordan coefficients are real.

The proof is left as Exercise 13.3.

As an example, consider two vector operators $V^{(1)}$ and $U^{(1)}$. The results of (12.102)–(12.110) can be used to extract the necessary Clebsch–Gordan coefficients for $1 \otimes 1 = 0 \oplus 1 \oplus 2$. Then (13.15) allows us to construct the following three tensor operators, of orders 0, 1, and 2:

$$W_0^{(0)} = \left(V_1^{(1)}U_{-1}^{(1)} + V_{-1}^{(1)}U_1^{(1)} - V_0^{(1)}U_0^{(1)}\right)/\sqrt{3}, \tag{13.17}$$

and

$$W_{\pm 1}^{(1)} = \pm\left(V_{\pm 1}^{(1)}U_0^{(1)} - V_0^{(1)}U_{\pm 1}^{(1)}\right)/\sqrt{2}, \tag{13.18}$$

$$W_0^{(1)} = \left(V_1^{(1)}U_{-1}^{(1)} - V_{-1}^{(1)}U_1^{(1)}\right)/\sqrt{2}, \tag{13.19}$$

and

$$W_{\pm 2}^{(2)} = V_{\pm 1}^{(1)}U_{\pm 1}^{(1)}, \tag{13.20}$$

$$W_{\pm 1}^{(2)} = \left(V_{\pm 1}^{(1)}U_0^{(1)} + V_0^{(1)}U_{\pm 1}^{(1)}\right)/\sqrt{2}, \tag{13.21}$$

$$W_0^{(2)} = \left(V_1^{(1)}U_{-1}^{(1)} + V_{-1}^{(1)}U_1^{(1)} + 2V_0^{(1)}U_0^{(1)}\right)/\sqrt{6}. \tag{13.22}$$

Now, $W_0^{(0)}$ is a scalar operator, equal to $-\boldsymbol{V}\cdot\boldsymbol{U}/\sqrt{3}$. It is also not hard to check that the components $W_m^{(1)}$ form a vector operator, and using (13.9)–(13.11), the corresponding rectangular components are those of $i(\boldsymbol{V}\times\boldsymbol{U})/\sqrt{2}$. The $j=2$ tensor operator can be rewritten, in terms of the rectangular components of $\boldsymbol{V}$ and $\boldsymbol{U}$, as

$$W_{\pm 2}^{(2)} = (V_x \pm iV_y)(U_x \pm iU_y)/2, \tag{13.23}$$

$$W_{\pm 1}^{(2)} = \left[(\mp V_x - iV_y)U_z + V_z(\mp U_x - iU_y)\right]/2, \tag{13.24}$$

$$W_0^{(2)} = \left(2V_zU_z - V_xU_x - V_yU_y\right)/\sqrt{6}. \tag{13.25}$$

A different basis for these $j=2$ operators consists of the five operators

$$\begin{array}{lll} V_xU_y + V_yU_x, & V_xU_z + V_zU_x, & V_yU_z + V_zU_y, \\ V_xU_x - V_yU_y, & 2V_zU_z - V_xU_x - V_yU_y, & \end{array} \tag{13.26}$$

which are also a basis for the entries of the traceless symmetric 3×3 matrix

$$\mathcal{W}_{ab} = (V_aU_b + V_bU_a)/2 - \delta_{ab}\boldsymbol{V}\cdot\boldsymbol{U}/3, \qquad (a,b = x,y,z). \tag{13.27}$$

Therefore, $W_m^{(2)}$ and $\mathcal{W}_{ab}$ are the same operators in different guises. In terms of rectangular components, the product of V_a and U_b decomposes into a linear combination of the irreducible tensor operators with $j=0,1,2$, as

$$V_aU_b = \delta_{ab}\boldsymbol{V}\cdot\boldsymbol{U}/3 + \epsilon_{abc}(\boldsymbol{V}\times\boldsymbol{U})_c/2 + \mathcal{W}_{ab}. \tag{13.28}$$

This can be checked using $\epsilon_{abc}(\boldsymbol{V} \times \boldsymbol{U})_c = V_a U_b - V_b U_a$. Thus $V_a U_b$ is an example of a **reducible tensor operator**.[1]

As a further specialization, we can take $\boldsymbol{V} = \boldsymbol{U} = \boldsymbol{R}$, the position operator. The resulting $j = 2$ spherical tensor is called the **quadrupole moment operator**,

$$Q^{(2)}_{\pm 2} = (X \pm iY)^2/2, \tag{13.29}$$

$$Q^{(2)}_{\pm 1} = (\mp X - iY)Z, \tag{13.30}$$

$$Q^{(2)}_{0} = (2Z^2 - X^2 - Y^2)/\sqrt{6}. \tag{13.31}$$

This five-component operator has the same content as the five rectangular components of the traceless symmetric tensor

$$Q_{ab} = R_a R_b - \delta_{ab} R^2/3, \tag{13.32}$$

in the sense that each consists of (complex) linear combinations of the other. This operator will appear in our study of absorption and emission of electromagnetic radiation, in Sect. 23.5.

It is straightforward to check that the position representation of the quadrupole moment operator can be expressed in terms of spherical harmonics as

$$Q^{(2)}_{m} = \sqrt{\frac{8\pi}{15}}\, r^2 Y_2^m(\theta, \phi). \tag{13.33}$$

This is not an accident, as any $(2l + 1)$-component operator with position representation of the form $f(r) Y_l^m(\theta, \phi)$ can be shown to satisfy the requirements (13.6)–(13.8) for a tensor operator of order l. In particular, consider the **multipole moment operator** with position representation

$$T^{(l)}_{m} = r^l Y_l^m(\theta, \phi). \tag{13.34}$$

As we have just seen, the quadrupole moment operator is the special case $l = 2$ (up to a normalization). The dipole moment operator with $l = 1$ is equivalent to the position vector operator, as in (13.13) and (13.14). The octupole moment operator has $l = 3$, etc. To justify the terminology, recall that for a classical charge distribution $\rho(\boldsymbol{r})$, the electrostatic potential is

$$\Phi(\boldsymbol{r}) = \int d^3 r' \, \frac{\rho(\boldsymbol{r}')}{|\boldsymbol{r} - \boldsymbol{r}'|}. \tag{13.35}$$

[1] A **Cartesian tensor operator** $T_{a_1,\ldots,a_n}$ with $n \geq 2$ indices $a_i = x, y, z$ is generically (if no special symmetry, antisymmetry, or trace conditions are imposed) reducible, in the sense that its entries can be written as linear combinations of irreducible tensor operators with more than one j.

Applying the expansion formula (8.175) for large r, this can be rewritten as

$$\Phi(\boldsymbol{r}) = \sum_{l=0}^{\infty} \frac{4\pi}{2l+1} \sum_{m=-l}^{l} q_{lm}^{*} \frac{Y_l^m(\theta,\phi)}{r^{l+1}}, \tag{13.36}$$

where

$$q_{lm} = \int d^3\boldsymbol{r}'\, \rho(\boldsymbol{r}')\, T_m^{(l)}(r',\theta',\phi'). \tag{13.37}$$

Thus, for a particle of charge q and wavefunction $\psi(\boldsymbol{r})$, the q_{lm} can be obtained by integrating $T_m^{(l)}$ weighted by the charge probability distribution $\rho(\boldsymbol{r}) = q|\psi(\boldsymbol{r})|^2$. In bra-ket form,

$$q_{lm} = q\langle\psi|T_m^{(l)}|\psi\rangle, \tag{13.38}$$

the expectation value of the multipole moment operator.

13.2 Selection Rules for Scalar and Vector Operators and the Landé Projection Formula

As remarked in the previous section, scalar operators do not change the total angular momentum quantum numbers of the states they act on. Furthermore, their matrix elements between total angular momentum (J^2, J_z) eigenstates do not depend on the magnetic quantum number m. Both properties are incorporated in the following:

Theorem 13.2 (Scalar Operator Selection Rule) *If S is a scalar operator, and $|\alpha jm\rangle$ are eigenstates of the total angular momentum operators J^2 and J_z, with additional labels $\alpha, \beta, \ldots$, then*

$$\langle\beta j'm'|\, S\,|\alpha jm\rangle = \delta_{j'j}\delta_{m'm}\,\langle\beta j\|\, S\,\|\alpha j\rangle, \tag{13.39}$$

where $\langle\beta j\|\, S\,\|\alpha j\rangle$ does not depend on m.

As terminology, the object $\langle\beta j\|\, S\,\|\alpha j\rangle$ is called a **reduced matrix element**. This is not a matrix element in the usual sense, but rather just a symbol denoting a number that depends on α, β, and j and on the choice of the operator S, but not on m or m'.

To prove Theorem 13.2, consider $\langle\beta j'm'|[J_z, S]|\alpha jm\rangle$, which is 0 from the definition of a scalar operator. Evaluating each J_z acting on the adjacent bra or ket, this immediately tells us that $(\hbar m' - \hbar m)\langle\beta j'm'|S|\alpha jm\rangle = 0$, so the matrix element in question can only be nonzero if $m' = m$. Similarly, the matrix element of $[J^2, S] = 0$ tells us that $[\hbar^2 j'(j'+1) - \hbar^2 j(j+1)]\langle\beta j'm'|S|\alpha jm\rangle = 0$, so the matrix element can only be nonzero if $j' = j$, since the other root $j' = -j-1$

is impossible because j and j' are both non-negative. It remains to show that the matrix element is independent of m, which follows from

$$\langle \beta j, m+1 | S | \alpha j, m+1 \rangle = \langle \beta jm | S | \alpha jm \rangle \tag{13.40}$$

for $-j \leq m \leq j-1$. To establish this, start with $[S, J_+] = 0$ from the definition of a scalar operator, which immediately implies $\langle \beta j, m+1 | [S, J_+] | \alpha jm \rangle = 0$. Then (13.40) follows from the facts that $J_+ |\alpha jm\rangle = c_{jm} |\alpha j, m+1\rangle$ and $\langle \beta j, m+1 | J_+ = c_{jm} \langle \beta jm|$, with the *same* nonzero proportionality factor $c_{jm} = \hbar\sqrt{j(j+1) - m(m+1)}$.

Theorem 13.2 is a powerful selection rule that can be applied, for example, if S is a term in the Hamiltonian, which is often invariant under rotations and therefore a scalar operator. (Exceptions include a charged particle in the presence of an external electric or magnetic field, which will always pick out a special direction that destroys the rotational symmetry. For example, if there is a constant external electric field $\boldsymbol{E}$, the Hamiltonian will contain a term proportional to $\boldsymbol{E} \cdot \boldsymbol{R}$, which is not a scalar operator. This is because $\boldsymbol{E}$, being constant, does *not* transform as a vector operator in the sense we have defined.) This theorem will even be useful to us in the seemingly trivial case that S is the identity operator, where it reads

$$\langle \beta j' m' | \alpha jm \rangle = \delta_{jj'} \delta_{mm'} \langle \beta jm | \alpha jm \rangle. \tag{13.41}$$

We now turn our attention to vector operators, which in this section we will treat in terms of their rectangular coordinate components V_a for $a = x, y, z$ with $\boldsymbol{V} = \hat{x} V_x + \hat{y} V_y + \hat{z} V_z$. By definition, these satisfy

$$[J_a, V_b] = i\hbar \epsilon_{abc} V_c. \tag{13.42}$$

One useful observation is that the operator $\boldsymbol{J} \cdot \boldsymbol{V} = \boldsymbol{V} \cdot \boldsymbol{J}$ does not have an operator-ordering problem, since each rectangular component J_a commutes with the corresponding component V_a. Furthermore, it is a scalar operator:

$$\left[J_a, \boldsymbol{J} \cdot \boldsymbol{V}\right] = 0. \tag{13.43}$$

One can also check straightforwardly that

$$\left[J^2, V_a\right] = i\hbar \epsilon_{abc} (V_b J_c + J_c V_b). \tag{13.44}$$

From this, and the identity $\epsilon_{abc}\epsilon_{dec} = \delta_{ad}\delta_{be} - \delta_{ae}\delta_{bd}$, one can establish the double-commutator relation

$$[J^2, [J^2, V_a]] = 2\hbar^2 \left(J^2 V_a + V_a J^2 - 2J_a (\boldsymbol{J} \cdot \boldsymbol{V}) \right). \tag{13.45}$$

We now exploit (13.45) by evaluating both sides as a matrix element between orthobasis states $\langle\beta, j', m'|$ and $|\alpha, j, m\rangle$, where α and β are degeneracy labels. Each instance of J^2 can be immediately evaluated as either $\hbar^2 j(j+1)$ or $\hbar^2 j'(j'+1)$, leading to

$$\hbar^4\Big([j(j+1) - j'(j'+1)]^2 - 2[j(j+1) + j'(j'+1)]\Big)\langle\beta, j', m'|V_a|\alpha, j, m\rangle = \\ -4\hbar^2\langle\beta, j', m'|J_a(\boldsymbol{J}\cdot\boldsymbol{V})|\alpha, j, m\rangle. \tag{13.46}$$

The matrix element on the right can be separated, by using completeness, to get

$$\langle\beta, j', m'|J_a(\boldsymbol{J}\cdot\boldsymbol{V})|\alpha, j, m\rangle = \\ \sum_{\gamma, j'', m''}\langle\beta, j', m'|J_a|\gamma j'' m''\rangle\langle\gamma j'' m''|\,\boldsymbol{J}\cdot\boldsymbol{V}\,|\alpha, j, m\rangle. \tag{13.47}$$

Here, the first matrix element vanishes unless $\gamma = \beta$ and $j'' = j'$. According to Theorem 13.2 the second matrix element vanishes unless $j'' = j$ and $m'' = m$, since $\boldsymbol{J}\cdot\boldsymbol{V}$ is a scalar operator. Therefore, only one term in the completeness sum contributes, and (13.46) simplifies to a remarkably useful formula,

$$\hbar^2(j + j' + 2)(j + j')[(j - j')^2 - 1]\langle\beta, j', m'|V_a|\alpha, j, m\rangle = \\ -4\delta_{jj'}\langle j, m'|J_a|j, m\rangle\langle\beta, j, m|\,\boldsymbol{J}\cdot\boldsymbol{V}\,|\alpha, j, m\rangle, \tag{13.48}$$

where the irrelevant label β has been removed from the first matrix element on the right.

One important application of (13.48) is the special case $j' = j$. Then it immediately reduces to the following result due to Alfred Landé:

Theorem 13.3 (Vector Operator Projection Rule) *If an operator* $\boldsymbol{V}$ *transforms as a vector with respect to rotations generated by the angular momentum operator* $\boldsymbol{J}$*, then the* ***Landé projection formula*** *holds for* $j \neq 0$*:*

$$\langle\beta, j, m'|\,V_a\,|\alpha, j, m\rangle = \langle j, m'|\,J_a\,|j, m\rangle\frac{\langle\beta, j\|\,\boldsymbol{J}\cdot\boldsymbol{V}\,\|\alpha, j\rangle}{\hbar^2 j(j+1)}, \tag{13.49}$$

where the reduced matrix element

$$\langle\beta, j\|\,\boldsymbol{J}\cdot\boldsymbol{V}\,\|\alpha, j\rangle \equiv \langle\beta, j, m|\,\boldsymbol{J}\cdot\boldsymbol{V}\,|\alpha, j, m\rangle \tag{13.50}$$

does not depend on m *or* m'*. For the special case* $j = 0$*,*

$$\langle\beta, 0, 0|V_a|\alpha, 0, 0\rangle = 0. \tag{13.51}$$

Because $\boldsymbol{J} \cdot \boldsymbol{V}$ is a scalar operator, Theorem 13.2 reminds us that, as claimed, the right side of (13.50) does not actually depend on m (or m'). So, in applications, we can pick any convenient value for m to do the evaluation of $\langle \beta, j \| \boldsymbol{J} \cdot \boldsymbol{V} \| \alpha, j \rangle$.

Of course, if $j = 0$ then (13.49) fails, but in that case $j = m = m' = 0$, and the vanishing of $\langle \beta, 0, 0 | \boldsymbol{V} | \alpha, 0, 0 \rangle$ follows from spherical symmetry. The $j = 0$ states have no special direction along which the matrix element of a vector operator could point.

Intuitively, the Landé projection formula says that, within each subspace of fixed j, all vector operators are proportional to each other, and in particular to the angular momentum operator. As a check, if we take $\boldsymbol{V} = \boldsymbol{J}$, then $\boldsymbol{J} \cdot \boldsymbol{V} = J^2$, and (13.49) is satisfied. The reason for the word "projection" in the name is that it implies that fixed-j matrix elements of $\boldsymbol{V}$ are equal to those of $\boldsymbol{J}(\boldsymbol{J}\cdot\boldsymbol{V})/J^2$, the formal geometric projection of $\boldsymbol{V}$ onto the $\boldsymbol{J}$ direction. The projection formula is a particularly useful special case of the Wigner–Eckart theorem discussed in the next section.

Another important application of (13.48) occurs if we take $\boldsymbol{V} = \boldsymbol{R}$, the position operator. As a notable example, this will be useful for evaluating absorption and emission of electromagnetic radiation in the electric dipole approximation (in Sect. 23.4). For this reason, the conditions on the matrix elements of the components of $\boldsymbol{R}$ that we are about to derive are often called the **dipole selection rules**.

When considering matrix elements of $\boldsymbol{R}$, we can take $\boldsymbol{J}$ to be just the orbital angular momentum operator $\boldsymbol{L}$, since that is what generates rotations for $\boldsymbol{R}$; the spin is not relevant. Further, $\boldsymbol{L} \cdot \boldsymbol{R} = (\boldsymbol{R} \times \boldsymbol{P}) \cdot \boldsymbol{R} = 0$, so the right side of (13.48) vanishes, giving

$$(l + l' + 2)(l + l')[(l - l')^2 - 1] \left\langle \beta, l', m' \right| \boldsymbol{R} \left| \alpha, l, m \right\rangle = 0. \tag{13.52}$$

Therefore, for the matrix element to be nonzero, we need

$$(l + l' + 2)(l + l')[(l - l')^2 - 1] = 0. \tag{13.53}$$

Now, $l + l' + 2$ cannot vanish, because l and l' are always non-negative. Also, the condition $l + l' = 0$ can only be satisfied if $l = l' = 0$, but then $\langle \beta, l', m' | \boldsymbol{R} | \alpha, l, m \rangle$ can only be $\langle \beta, 0, 0 | \boldsymbol{R} | \alpha, 0, 0 \rangle$, which must vanish anyway due to the parity selection rule, since $l = 0$ states have the same (even) parity and $\boldsymbol{R}$ has odd parity. From (13.53), the only other possibility allowing a non-vanishing matrix element is $(l - l')^2 = 1$. Thus, we have the first dipole selection rule,

$$\langle \beta, l', m' | \boldsymbol{R} | \alpha, l, m \rangle \neq 0 \quad \text{requires } l' - l = \pm 1. \tag{13.54}$$

Since this holds for the whole vector $\boldsymbol{R}$, it is also true for the matrix elements of each of its components, X, Y, and Z. The remaining dipole selection rules are

$$\langle \beta, l', m' | Z | \alpha, l, m \rangle \neq 0 \quad \text{requires } m' = m, \tag{13.55}$$

$$\langle \beta, l', m' | (X + iY) | \alpha, l, m \rangle \neq 0 \quad \text{requires } m' = m + 1, \tag{13.56}$$

$$\langle \beta, l', m' | (X - iY) | \alpha, l, m \rangle \neq 0 \quad \text{requires } m' = m - 1. \tag{13.57}$$

A simple proof uses the ϕ-dependence of the position representations of the spherical harmonics and X, Y, and Z. Suppressing the labels α, β, which play no role,

$$\langle l', m' | Z | l, m \rangle \propto \int_0^{2\pi} d\phi \left(e^{-im'\phi}\right)(1)\left(e^{im\phi}\right) \propto \delta_{mm'}, \tag{13.58}$$

$$\langle l', m' | (X \pm iY) | l, m \rangle \propto \int_0^{2\pi} d\phi \left(e^{-im'\phi}\right)\left(e^{\pm i\phi}\right)\left(e^{im\phi}\right) \propto \delta_{m',m\pm 1}. \tag{13.59}$$

Equations (13.56) and (13.57) also imply that the corresponding matrix elements of X and Y vanish unless $m' = m \pm 1$.

As yet another application of (13.48), consider matrix elements $\langle \beta, j, m' | \boldsymbol{V} | \alpha, j, m \rangle$ of a general vector operator $\boldsymbol{V}$. Compared to the special case $\boldsymbol{V} = \boldsymbol{R}$, the only difference in the arguments that led to (13.54) is that now we must admit the possibility that the right side of (13.48) does not vanish for $j = j'$. Therefore,

$$\langle \beta, j', m' | \boldsymbol{V} | \alpha, j, m \rangle \neq 0 \quad \text{requires } j' - j = 0, \pm 1. \tag{13.60}$$

Also, by computing the matrix elements of $[J_z, V_0^{(1)}] = 0$ and $[J_z, V_{\pm 1}^{(1)}] = \pm\hbar V_{\pm 1}^{(1)}$, evaluating J_z on the adjacent bra or ket, one finds

$$\langle \beta, j', m' | V_z | \alpha, j, m \rangle \neq 0 \quad \text{requires } m' = m, \tag{13.61}$$

$$\langle \beta, j', m' | (V_x + iV_y) | \alpha, j, m \rangle \neq 0 \quad \text{requires } m' = m + 1, \tag{13.62}$$

$$\langle \beta, j', m' | (V_x - iV_y) | \alpha, j, m \rangle \neq 0 \quad \text{requires } m' = m - 1. \tag{13.63}$$

These are the general vector operator selection rules.

13.3 The Wigner–Eckart Theorem and Selection Rules for Tensor Operators

In this section, we will state and prove the **Wigner–Eckart theorem** for matrix elements of irreducible tensor operators. Developed independently by Eugene Wigner and Carl Eckart, this is a generalization of three simpler results that we have already given: (12.148) for spherical harmonics, Theorem 13.2 for scalar operators, and Theorem 13.3 for vector operators. The reader may wish to contemplate the parallels between the following derivations and those used in Sect. 12.5 for spherical harmonics.

Consider an irreducible tensor operator $T^{(j_1)}_{m_1}$ with $m_1 = -j_1, \ldots, j_1$, and a multiplet of angular momentum eigenstates $|\alpha j_2 m_2\rangle$ with $m_2 = -j_2, \ldots, j_2$, where α is a degeneracy label. Now we construct the states

$$|\gamma jm\rangle = \sum_{m_1,m_2} T^{(j_1)}_{m_1} |\alpha j_2 m_2\rangle \, C^{j_1\, j_2\, j}_{m_1 m_2 m}, \tag{13.64}$$

for $j = |j_1 - j_2|, \ldots, j_1 + j_2$ and $m = -j, \ldots, j$, where $C^{j_1\, j_2\, j}_{m_1 m_2 m}$ are Clebsch–Gordan coefficients as defined by (12.70). We claim that, as the labeling suggests, (13.64) is an eigenstate of J^2 and J_z with eigenvalues $\hbar^2 j(j+1)$ and $\hbar m$, respectively. (It is not necessarily normalized to unity.) To prove this, it suffices to show that

$$J_z\,|\gamma jm\rangle = \hbar m\,|\gamma jm\rangle\,, \tag{13.65}$$

$$J_-\,|\gamma jm\rangle = \hbar\sqrt{j(j+1) - m(m-1)}\,|\gamma j, m-1\rangle\,, \tag{13.66}$$

$$J_+\,|\gamma jm\rangle = \hbar\sqrt{j(j+1) - m(m+1)}\,|\gamma j, m+1\rangle\,, \tag{13.67}$$

as then the result for $J^2 = (J_+J_- + J_-J_+)/2 + J_z^2$ follows.

To prove (13.65), we use $[J_z, T^{(j_1)}_{m_1}] = \hbar m_1 T^{(j_1)}_{m_1}$ from (13.6) to obtain

$$J_z\,|\gamma jm\rangle = \sum_{m_1,m_2} T^{(j_1)}_{m_1}(\hbar m_1 + J_z)\,|\alpha j_2 m_2\rangle \, C^{j_1\, j_2\, j}_{m_1 m_2 m}. \tag{13.68}$$

Then, use $J_z\,|\alpha j_2 m_2\rangle = \hbar m_2\,|\alpha j_2 m_2\rangle$, and (13.65) follows by noting that the Clebsch–Gordan coefficient enforces that only terms with $m_1 + m_2 = m$ contribute in the sum.

To prove (13.66), we similarly use the commutator $[J_-, T^{(j_1)}_{m_1}]$ in (13.8) and the J_- ladder operator rule in (13.5) to find

$$\begin{aligned} J_-\,|\gamma jm\rangle = \hbar \sum_{m_1,m_2} C^{j_1\, j_2\, j}_{m_1 m_2 m}\Big(&\sqrt{j_1(j_1+1) - m_1(m_1-1)}\, T^{(j_1)}_{m_1-1}\,|\alpha j_2 m_2\rangle \\ &+ \sqrt{j_2(j_2+1) - m_2(m_2-1)}\, T^{(j_1)}_{m_1}\,|\alpha j_2, m_2-1\rangle\Big). \end{aligned} \tag{13.69}$$

Now, since we are summing over all m_1 and m_2, we can use the trick of relabeling indices according to $m_1 \to m_1 + 1$ in the first term and $m_2 \to m_2 + 1$ in the second term. The result is

$$\begin{aligned} J_-\,|\gamma jm\rangle = \hbar \sum_{m_1,m_2} T^{(j_1)}_{m_1}\,|\alpha j_2 m_2\rangle \Big(&\sqrt{j_1(j_1+1) - m_1(m_1+1)}\, C^{j_1\, j_2\, j}_{m_1+1, m_2, m} \\ &+ \sqrt{j_2(j_2+1) - m_2(m_2+1)}\, C^{j_1\, j_2\, j}_{m_1, m_2+1, m}\Big). \end{aligned} \tag{13.70}$$

Equation (13.66) then follows immediately from using the Clebsch–Gordan recurrence relation (12.83). The proof of (13.67) is completely analogous.

Having established that the states $|\gamma jm\rangle$ defined by (13.64) are really eigenstates of J^2 and J_z, we can now invert the relationship by solving for each $T^{(j_1)}_{m_1}\,|\alpha j_2 m_2\rangle$ in terms of $|\gamma jm\rangle$. Starting from (13.64), multiply by $C^{j_1\,j_2\,j}_{m'_1\,m'_2\,m}$, sum over all j and m, and then use the Clebsch–Gordan orthonormality relation (12.86) to collapse the m_1, m_2 sums to a single term, and finally relabel $m'_1 \to m_1$ and $m'_2 \to m_2$. The result is

$$T^{(j_1)}_{m_1}\,|\alpha j_2 m_2\rangle = \sum_{j,m} |\gamma jm\rangle\; C^{j_1\,j_2\,j}_{m_1\,m_2\,m}. \tag{13.71}$$

This is the precise version of the intuitive statement that $T^{(j_1)}_{m_1}$ carries angular momentum quantum numbers j_1 and m_1, and that it imparts them to the states it acts on, yielding a sum of states that occur in the addition of angular momenta j_1 and j_2.

Now acting on (13.71) with $\langle\beta j'm'|$, then using the orthogonality condition (13.41) to collapse the sum on the right, and finally relabeling $j' \to j$ and $m' \to m$, we obtain

$$\langle\beta jm|\,T^{(j_1)}_{m_1}\,|\alpha j_2 m_2\rangle = \langle\beta jm|\gamma jm\rangle C^{j_1\,j_2\,j}_{m_1\,m_2\,m}. \tag{13.72}$$

By its definition in (13.64), the state $|\gamma jm\rangle$ depends on α, j, j_1, j_2, and the choice of tensor operator $T^{(j_1)}$. However, the matrix element $\langle\beta jm|\gamma jm\rangle$ clearly does not depend on m_1 or m_2, and from Theorem 13.2 with S taken to be the identity operator, we see that it cannot actually depend on m either. Therefore, we have proved a powerfully general result:

Theorem 13.4 (Wigner–Eckart) *If $T^{(j_1)}_{m_1}$ are the components of an irreducible tensor operator, and $|\alpha j_2 m_2\rangle$ and $|\beta jm\rangle$ are eigenstates of J^2 and J_z with eigenvalues as labeled, then*[2]

$$\langle\beta jm|\,T^{(j_1)}_{m_1}\,|\alpha j_2 m_2\rangle = \langle\beta j\|\,T^{(j_1)}\,\|\alpha j_2\rangle\; C^{j_1\,j_2\,j}_{m_1\,m_2\,m}, \tag{13.73}$$

where the reduced matrix element $\langle\beta j\|\,T^{(j_1)}\,\|\alpha j_2\rangle$ does not depend on m, m_1, or m_2.

For $j_1 = m_1 = 0$, so that $T^{(0)}_0 = S$ is a scalar operator, we can use the fact that $C^{0\,j\,j'}_{0\,m\,m'} = \delta_{jj'}\delta_{mm'}$ to recover Theorem 13.2 as a special case of the Wigner–

[2] Some sources define the reduced matrix element with other normalization conventions, most commonly by including a factor of $1/\sqrt{2j+1}$ (and sometimes a different sign) on the right side of (13.73).

Eckart theorem. This is hardly surprising, since we just used the former in the proof of the latter. The Landé projection formula, (13.49), can be shown to be equivalent to the special case obtained for vector operators when $j_1 = 1$. The triple spherical harmonic integral relation (12.148) is equivalent to the special case in which the generator of rotations $\boldsymbol{J}$ is taken to be the orbital angular momentum $\boldsymbol{L}$, and the tensor operator is taken to be defined by the position representation $T_{m_1}^{(l_1)} = Y_{l_1}^{m_1}(\theta, \phi)$.

One way to use the Wigner–Eckart theorem is as a labor-saving device. If we know the matrix element on the left side of (13.73) for just one instance (m_1, m_2, m) for which it is nonzero, then we can infer the reduced matrix element. Once that is done, all other cases of (m_1, m_2, m) with the same α, β, j_1, j_2, and j will also be known.

The Wigner–Eckart theorem also provides selection rules, since it implies that the matrix element can only be nonzero if the Clebsch–Gordan coefficient $C_{m_1\,m_2\,m}^{j_1\,j_2\,j}$ is nonzero. This means that the same selection rules apply, so that $\left\langle \beta j m | T_{m_1}^{(j_1)} | \alpha j_2 m_2 \right\rangle$ can be nonzero only if

$$j = |j_1 - j_2|, \;\ldots,\; j_1 + j_2, \tag{13.74}$$

$$m = m_1 + m_2. \tag{13.75}$$

These **tensor operator selection rules** generalize the results in (13.60)–(13.63) that we obtained for the special case of vector operators.

Exercises

13.1 Prove the angular momentum commutation relations for tensor operators, (13.6)–(13.8), from earlier results.

13.2 Suppose that $\boldsymbol{V}$ and $\boldsymbol{W}$ are rectangular-coordinate vector operators, obeying (13.12). Show that their cross product $\boldsymbol{V} \times \boldsymbol{W}$ is also a vector operator by the same definition, and that their dot product $\boldsymbol{V} \cdot \boldsymbol{W}$ is a scalar operator (commutes with $\boldsymbol{J}$).

13.3 Prove Theorem 13.1, which allows tensor operators to be constructed from products of other tensor operators.

13.4 Consider angular momentum eigenstates $|\alpha j m_j\rangle$ with J^2 and J_z eigenvalues $\hbar^2 j(j+1)$ and $\hbar m_j$, respectively. For what values of j, m_j is the matrix element $\langle \alpha j m_j | (X^2 - Y^2) | \beta\, 0\, 0 \rangle$ possibly nonzero? What about $\langle \alpha j m_j | XY | \beta\, 0\, 0 \rangle$ and $\langle \alpha j m_j | XZ | \beta\, 0\, 0 \rangle$?

13.5 Show that for any angular momentum eigenstate $|\alpha j m_j\rangle$, the expectation values of the components of an irreducible tensor operator $T_m^{(l)}$ vanish unless $l \leq 2j$

and $m = 0$. This applies in particular to the multipole moment operator defined in (13.34). Thus, the only nonzero quadrupole moment component [as defined in (13.38)] for an angular momentum eigenstate is q_{20}, and only for states with $j \geq 1$. A spin-0 or spin-1/2 particle cannot have a quadrupole moment, even if it is composite.

13.6 The deuteron is a composite particle with electric charge $+e$ (i.e. $Q = 1$ in units of e) and spin one, so that $j = 1$ in its rest frame. Apply the Wigner–Eckart theorem to find the ratios of the electric quadrupole moment operator expectation values q_{20}, defined in (13.38), for each of the three states with $m_j = 1, 0, -1$.

13.7 Consider a basis of angular momentum eigenstates $|\alpha jm\rangle$ for a particle of unit charge, for fixed α and j, and $m = -j, \ldots, j$. Suppose that the electric dipole and quadrupole moments, defined by the expectation values in the highest-m state,

$$p = \langle \alpha jj|Z|\alpha jj\rangle, \tag{13.76}$$

$$q = \langle \alpha jj|(2Z^2 - X^2 - Y^2)|\alpha jj\rangle/\sqrt{6}, \tag{13.77}$$

are known. Use the Wigner–Eckart theorem to find, in terms of p and q and appropriate Clebsch–Gordan coefficients,

(a) all of the matrix elements $\langle \alpha jm'|A|\alpha jm\rangle$ for each of $A = X$ and Y and Z.
(b) all matrix elements $\langle \alpha jm'|A|\alpha jm\rangle$ for each of the operators $A = XY$ and XZ and YZ and $X^2 - Y^2$ and $(2Z^2 - X^2 - Y^2)/\sqrt{6}$. For which m, m' can the results be nonzero?

Entanglement, the EPR Problem, Hidden Variables, and Bell Inequalities

14

14.1 The Einstein–Podolsky–Rosen Problem

Although Einstein played an important role in the early development of quantum mechanics, like many others he was troubled by some of its implications. This was in part due to his discomfort with the reliance of Postulates 4 and 5 on probabilities for the result of a measurement and the subsequent state, rather than completely deterministic time evolution. Even after he grudgingly accepted the possibility of non-determinism, Einstein found the implications of entanglement of quantum states difficult to accept, and doubted whether the standard formulation of quantum mechanics should be regarded as complete.

In 1935, Einstein, Boris Podolsky, and Nathan Rosen described a thought experiment that highlighted the perceived weirdness associated with measurements involving entangled states. This problem is often called the EPR paradox, but it should be emphasized at the outset that although it may be troubling to humans accustomed to classical ways of thinking, in the end we will find that there is really nothing paradoxical about it. (Indeed, EPR themselves did not call it a "paradox".) EPR described the problem in terms of measurements of positions and momenta, but David Bohm later distilled the basic ideas into a simpler form in terms of measurements of components of two entangled spins, and that is the version we will now describe.

Consider a system of two spin-1/2 particles, labeled 1 and 2, in a state with the total spin $\boldsymbol{S} = \boldsymbol{S}_1 + \boldsymbol{S}_2$ equal to 0,

$$|S = 0\rangle = \frac{1}{\sqrt{2}}\Big(|\uparrow\downarrow\rangle - |\downarrow\uparrow\rangle\Big) = \frac{1}{\sqrt{2}}\Big(\big|+\hat{z}, -\hat{z}\big\rangle - \big|-\hat{z}, +\hat{z}\big\rangle\Big). \tag{14.1}$$

The notation is such that, for example, $|\uparrow\downarrow\rangle = \big|+\hat{z}, -\hat{z}\big\rangle \equiv \big|+\hat{z}\big\rangle_1 \otimes \big|-\hat{z}\big\rangle_2$ is an eigenstate of both S_{1z} and S_{2z}, with eigenvalues $+\hbar/2$ and $-\hbar/2$, respectively. The

S. P. Martin, *Quantum Mechanics*, Graduate Texts in Physics,
https://doi.org/10.1007/978-3-032-21119-4_14

state of the system is thus represented by a ket in the tensor product of the Hilbert spaces for the two individual particles. In the following, we will purposely avoid the $\uparrow$ and $\downarrow$ notation, because we will want to consider components of the spins other than the z component. In fact, since the $S = 0$ state has no preferred direction, there is nothing special about $\hat{z}$, and up to an irrelevant phase we could just as easily write the state in (14.1) as

$$|S=0\rangle = \frac{1}{\sqrt{2}}\Big(|+\hat{n}, -\hat{n}\rangle - |-\hat{n}, +\hat{n}\rangle\Big) \tag{14.2}$$

for any convenient unit vector direction $\hat{n}$. Importantly, there is no way to write this total spin-0 ket as a single product of kets in the individual spin-1/2 particle Hilbert spaces; no matter the choice of bases for the individual spins, it is always a nontrivial linear combination of such products. Thus, it is an example of an entangled state as defined in Sect. 2.9.

The particles also have spatial wavefunctions that we need not write explicitly. The only thing important for us is that they describe localized wave packets moving away from each other. The two spins, although entangled, are therefore separated by a large distance at late times. This is sometimes called **nonlocal entanglement**.

One way to realize such a state would be an electron-positron pair from the decay of a neutral pion at rest,[1]

$$\pi^0 \to e^- e^+. \tag{14.3}$$

The pion has spin 0, and since we are in its rest frame, the total angular momentum is also 0. We imagine that this experiment will be conducted many times, with two ideal observers Alice and Bob located on opposite sides of the point where the pions decay, and far away, as shown in Fig. 14.1. They each have a small detector capable of measuring any desired component of the spin of a particle moving through it. Since the detectors are far apart, so that the spatial wavefunctions for the two particles have no overlap, Alice can only measure one of the spins, and Bob can only measure the other. Let us choose the labeling of the particles so that Alice always makes measurements on particle 1 (which on an event-by-event basis might be either the electron or the positron), and Bob always on particle 2. Each can measure any component of that spin, or choose to make no measurement. For simplicity, we assume that the Hamiltonian is just that of free particles, so that the spin states have no Hamiltonian time evolution. Throughout this chapter, we will use spin operators with a factor of $\hbar/2$ extracted,

$$\boldsymbol{\sigma}_k = \frac{2}{\hbar}\boldsymbol{S}_k \tag{14.4}$$

[1] This is a very rare decay, occurring for less than 10^{-7} of all neutral pions. But this is a thought experiment; we only care that it is possible in principle, not about such practical difficulties.

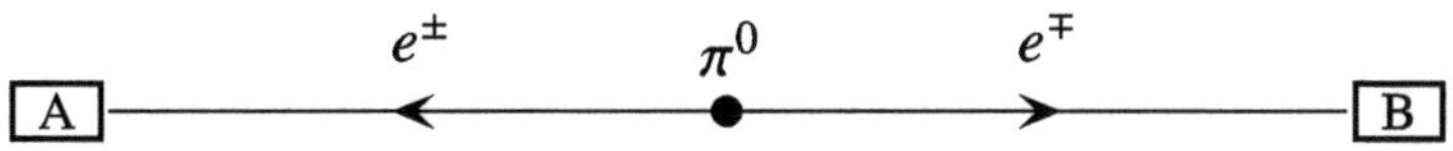

Fig. 14.1 The setup for Bohm's version of the EPR problem. Decays of neutral pions at rest produce two spin-1/2 particles moving in opposite directions, in an entangled spin state with total spin 0, and with spatial wavefunctions that have no overlap at late times. Two distant observers Alice and Bob can each measure any component they choose of the spin of the particle that arrives at their respective detector A or B

for particles $k = 1, 2$, represented by the Pauli matrices. All three components of $\boldsymbol{\sigma}_1$ commute with all three components of $\boldsymbol{\sigma}_2$, because they operate on distinct one-particle Hilbert spaces.

Now suppose that Alice measures σ_{1z}. The result will be either $+1$ or -1, with equal probability. If the result was $+1$, then if Bob measured σ_{2z} the result will necessarily be -1. This follows from the form of the state (14.1). We can say that Alice has collapsed the state to $|+\hat{z}, -\hat{z}\rangle$ by her measurement of σ_{1z}, so that the only possible outcome for Bob is $\sigma_{2z} = -1$. Conversely, if Alice instead measured σ_{1z} to be -1, then if Bob measures σ_{2z} the result must be $+1$. There is a perfect anticorrelation between their measurements.

This perfect anticorrelation in Alice's and Bob's measured results for σ_{1z} and σ_{2z} should not be at all surprising. In fact, it is actually much more general, in the sense that it does not even rely on quantum mechanics being correct. It would inevitably be true in any alternative theory, just as long as angular momentum is conserved. There is a similar perfect anticorrelation between the measured electric charges; if Alice sees that particle 1 is an electron, then she can immediately be sure that Bob's particle 2 will be a positron, due to conservation of charge. As an analogy, suppose that Charlie at the central point had taken a pair of shoes and shipped one to Alice and one to Bob, concealed in boxes. Before opening her box, Alice would know nothing about the shoe Bob received, but if she opens her box to find a left shoe, she would immediately gain the information that Bob's box has a right shoe. This has nothing to do with quantum mechanics, of course.

However, spins are much more interesting than electric charges or shoes, and can reveal nontrivial aspects of measurement correlations in quantum mechanics. This is because spins can point in arbitrary directions, and Alice and Bob could choose to make measurements of different spin components. For example, suppose Alice measures σ_{1x} but Bob still measures σ_{2z}. If Alice obtains the result $\sigma_{1x} = +1$, then from (14.2) with $\hat{n} = \hat{x}$, we can say that the total state collapses to $|+\hat{x}, -\hat{x}\rangle$. Then, since $|-\hat{x}\rangle_2 = (|+\hat{z}\rangle_2 - |-\hat{z}\rangle_2)/\sqrt{2}$, Bob will obtain the results $\sigma_{2z} = \pm 1$ randomly, and with equal probability. The same random results will occur for Bob's measurement of σ_{2z} if the result of Alice's measurement was instead $\sigma_{1x} = -1$, or if she decides to make no measurement at all.

To be more general, suppose that Alice and Bob measure arbitrary spin components $\hat{a} \cdot \boldsymbol{\sigma}_1$ and $\hat{b} \cdot \boldsymbol{\sigma}_2$, respectively, for unit vectors $\hat{a}$ and $\hat{b}$. Neither can predict with certainty what the other will measure on an event-by-event basis (unless $\hat{a}$ and $\hat{b}$ are parallel), but there is a statistical correlation between their measurements, which can be expressed as the expectation value

$$C(\hat{a}, \hat{b}) = \langle \hat{a} \cdot \boldsymbol{\sigma}_1 \, \hat{b} \cdot \boldsymbol{\sigma}_2 \rangle \tag{14.5}$$

in the state $|S = 0\rangle$. A labor-saving trick for evaluating this is to note that $(\boldsymbol{\sigma}_1 + \boldsymbol{\sigma}_2)\,|S = 0\rangle = 0$, so one can replace $\hat{b} \cdot \boldsymbol{\sigma}_2$ with $-\hat{b} \cdot \boldsymbol{\sigma}_1$ when acting on the state $|S = 0\rangle$, followed by use of the Pauli matrix identity $(\hat{a} \cdot \boldsymbol{\sigma}_1)(\hat{b} \cdot \boldsymbol{\sigma}_1) = \hat{a} \cdot \hat{b} + i(\hat{a} \times \hat{b}) \cdot \boldsymbol{\sigma}_1$ from (8.44). Then, since the expectation value of $\boldsymbol{\sigma}_1$ in the $S = 0$ state vanishes, the prediction of quantum mechanics is

$$C(\hat{a}, \hat{b}) = -\hat{a} \cdot \hat{b}. \tag{14.6}$$

This agrees with the special cases already discussed; for $\hat{a} = \hat{b}$, the measurements are perfectly anticorrelated, and when $\hat{a}$ and $\hat{b}$ are orthogonal there is no correlation. More generally, $C(\hat{a}, \hat{b})$ is always in the range from -1 (perfect anticorrelation) to $+1$ (perfect correlation).

The key issue that bothered EPR is that the results obtained by Bob would seem to depend on the choices made by Alice, and vice versa, even if they are so far apart that there is no possibility of communication between them when they each choose which spin measurement to make. They could even make their individual decisions about which spin components to measure randomly for each individual pion decay after it has occurred and while the electron and positron are already in flight. At the last moment before particle 1 arrives, Alice could decide to measure σ_{1z}. Then, from the results of Alice's measurement alone, she can be immediately certain of the result if Bob measures σ_{2z}, a measurement occurring far away and quite out of her control. Or, she could decide instead to measure σ_{1x}, in which case (regardless of the outcome of her own measurement) she would correctly assign a probability of 50% to the outcome of a measurement by Bob of σ_{2z}. The same follows if she decides to make no measurement at all. This is all despite the fact that, because of the finite speed of light, no signal carrying the news of her last-moment decision could reach Bob before his measurement occurs. Einstein referred to this sort of thing as 'spooky action at a distance', and argued that it pointed to an incompleteness of the standard quantum theory as a description of reality. However, Nature does not care whether we, or even Einstein, find a phenomenon peculiar or spooky. There are two things to check to make sure that there is not a problem with the quantum theory, let alone a true paradox.

The first thing to check is that the predictions of quantum mechanics are really consistent and unambiguous. In the preceding, we have phrased things in terms of Alice's measurement collapsing the state, with implications for what Bob will see. But the situation is symmetrical; we could just as easily describe any pair of measurements in terms of Bob collapsing the state, with implications for what Alice will see. For the case in which Alice measured $\sigma_{1z} = +1$ and Bob measured

$\sigma_{2z} = -1$, we could choose to say that Bob has collapsed the total state to $|+\hat{z}, -\hat{z}\rangle$, which means that the only possible result for Alice's measurement was, indeed, $\sigma_{1z} = +1$. This outcome is fully consistent with the interpretation in which Alice's measurement collapses the state first. A third way is to view the collapse symmetrically, by taking Alice's and Bob's measurements to be a single joint measurement. (This is possible because all components of $\boldsymbol{\sigma}_1$ commute with all components of $\boldsymbol{\sigma}_2$.) Again, the state collapses to $|+\hat{z}, -\hat{z}\rangle$. For any given outcome of both measurements, quantum mechanics indeed always gives the same result for the collapse of the state regardless of what order we assign to the measurements.[2] To see this more generally, suppose that Alice measures $\hat{a} \cdot \boldsymbol{\sigma}_1$ to be $+1$ and Bob measures $\hat{b} \cdot \boldsymbol{\sigma}_2$ to be $+1$. As you will verify when you do Exercise 14.1, the three ways of describing the state collapse give the same probabilities for this outcome, and the same final state.

The second thing to check is that quantum mechanics does not provide any instantaneous communication between the distant Alice and Bob, whose measurements are events occurring at a spacelike separation in the language of special relativity. In general, an instantaneous (or even faster-than-light) communication between spacelike separated points in one inertial reference frame would appear to travel backwards in time in some other inertial frame that is moving with respect to the first. This would imply causality paradoxes, because signals by combinations of actors in different inertial frames can be constructed in such a way that a message could be received on a time-like path before it is sent from a previous point on the same path. For example, you could use instantaneous communications to convince your parents not to meet, thus paradoxically preventing your own birth, as illustrated in Fig. 14.2.

To see that Alice and Bob's experiments do not actually allow instantaneous communication, we need only note that no matter what each of them chooses to measure, the results of their individual measurements will always appear to be completely random, until they meet to compare them, or at least send a conventional signal at a speed not exceeding that of light. Even if Alice and Bob agree to seek maximum anticorrelation by measuring σ_{1z} and σ_{2z} respectively (and even assuming that they do not break their promises!), they will each record their data locally as a random sequence of $+1$ and -1, containing no information. *After* communicating by conventional means, they can check that their results do have correlations in agreement with quantum mechanics, but the key point is that no information is transmitted in either direction until that happens.

Why, then, did EPR view the situation as disturbing? The essential issue is that one must not fall into the trap of thinking of the state of a system as always defined only by local information, a viewpoint sometimes called **local realism**. If you try to

[2] This is necessary, because if the measurements are spacelike separated events, special relativity says that there is no invariant sense of which occurred first. There is an inertial frame in which they are simultaneous, but in some other frames, Alice's measurement was first, and in others Bob's was first.

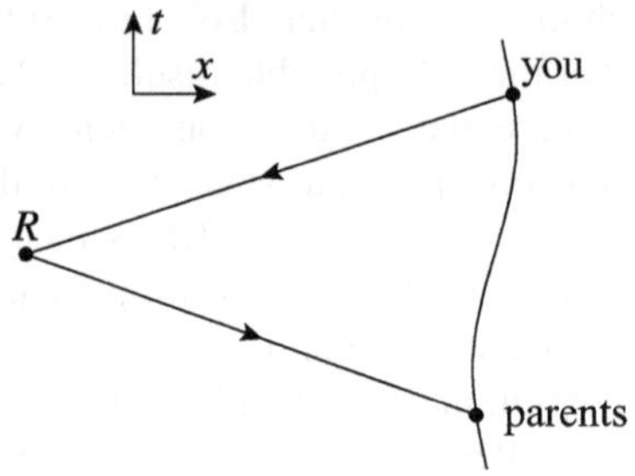

Fig. 14.2 If instantaneous communication were possible, then you could send a signal to a relay R which would be instantaneous (at constant time) in some other reference frame, but would arrive at an earlier time in your own reference frame. The relay could then send a signal, instantaneous in a different reference frame, which might prevent your parents from meeting, preventing your own birth. Fortunately, instantaneous communication is not a consequence of quantum mechanics

think about particle 2 only in terms of non-entangled kets such as $|+\hat{z}\rangle_2$ or $|-\hat{x}\rangle_2$, then it might indeed appear paradoxical that Alice's distant measurement of σ_{1z} can affect what Bob measures. However, that way of thinking about the state of particle 2 is explicitly rejected by quantum mechanics, which insists that we take into account the full nonlocal entanglement in order to get correct and consistent predictions. The quantum state describes both particles at the same time, not one particle or the other. The real value of the EPR problem is that it forces us to recognize this fundamental truth about how quantum mechanics works.

One sometimes sees grand pronouncements, inspired by the EPR problem, along the lines of "quantum mechanics is nonlocal". However, one must be careful about what this implies, because the word "nonlocal" has several completely different meanings in physics. For example, it is possible to define a Hamiltonian that has nonlocal dynamics; that is what quantum field theorists usually mean when they talk about the possibility of nonlocality. That is certainly not the case in the preceding discussion, as we took $H = 0$ for the spin degrees of freedom, and even the H describing the time evolution of the spatial wavefunctions was (implicitly) just that of free particle propagation. The entanglement of the state occurred due to perfectly local dynamics. The rejection of local realism in favor of nonlocal entanglement of states does not imply that we need to accept a nonlocal Hamiltonian, for which there is absolutely no experimental evidence.

14.2 Hidden Variables and Bell's Inequality

Having learned in the previous section that standard quantum mechanics rejects local realism in favor of allowing nonlocal entanglement, it must be recognized that this is a falsifiable hypothesis to be experimentally tested against alternatives. There is a general class of alternatives based on **hidden variables**, which attempt to incorporate local realism in a way consistent with experiment. In the remaining sections of this chapter, we will consider local hidden-variables theories, see how they make predictions that are incompatible with quantum mechanics, and understand why experiments show that they cannot be correct.

Hidden variables are supposed to be quantities that we cannot directly measure or control, but are necessary to completely characterize the state of a system. They could be part of some completely deterministic alternative to quantum mechanics, or could be an extra part of a framework that otherwise looks like conventional quantum mechanics. In the latter version, it is not just a ket $|\psi\rangle$ that describes the state of the system, but rather the state ket together with the hidden variable(s), which we will generically call λ. They are supposed to be distributed in a way that is either truly random, or just seems to be random because of our lack of knowledge of them. For any given system, the distribution of the hidden variables can be described by a probability density function $\mu(\lambda)$ with the properties

$$\int d\lambda\,\mu(\lambda) = 1, \tag{14.7}$$

$$\mu(\lambda) \geq 0, \tag{14.8}$$

so that $\mu(\lambda_0)\,d\lambda$ is the probability that λ will be found between λ_0 and $\lambda_0 + d\lambda$. [The following arguments do not require us to specify the allowed domain for λ, or the corresponding limits of the definite integral in (14.7). Also, the arguments are unaffected if λ takes on discrete values, with $\int d\lambda$ replaced by $\sum_\lambda$.] The function $\mu(\lambda)$ might depend on the particulars of the system and its history.

The presence of λ is supposed to be responsible for the illusion of random outcomes of measurements. For some observable A, and a state fully characterized by non-hidden properties ψ and hidden variables λ, the hypothesis is that the outcome of a measurement is not random, but rather is determined as some function

$$\text{measured value of } A = f_{A,\psi}(\lambda). \tag{14.9}$$

The function $f_{A,\psi}(\lambda)$ can be chosen so that as λ runs over all possible values, it returns all allowed values predicted by quantum mechanics for the measurement of A in the state ψ. Because we have no way of knowing what λ is, the results of individual measurements appear random. So, the mean value of many measurements of A for apparently identical states ψ will be

$$\overline{A} = \int d\lambda\,\mu(\lambda)\,f_{A,\psi}(\lambda). \tag{14.10}$$

This is the counterpart of the expectation value $\langle A\rangle = \langle\psi|A|\psi\rangle$ in standard quantum theory.

For example, let us consider how this would work for a system with two spin-1/2's combined to have total spin $S = 0$ in an entangled state coming from π^0 decay, as considered in Sect. 14.1. To be general, consider the measurements by Alice and Bob of arbitrary components of the spins of particles 1 and 2, say $\hat{a}\cdot\boldsymbol{\sigma}_1$

and $\hat{b} \cdot \boldsymbol{\sigma}_2$, where $\hat{a}$ and $\hat{b}$ are unit vectors. Then the outcomes for Alice's and Bob's measurements are respectively determined by some functions (suppressing the ψ label from now on):

$$\hat{a} \cdot \boldsymbol{\sigma}_1 = f_A(\hat{a}, \lambda) = +1 \text{ or } -1, \tag{14.11}$$

$$\hat{b} \cdot \boldsymbol{\sigma}_2 = f_B(\hat{b}, \lambda) = +1 \text{ or } -1. \tag{14.12}$$

To incorporate local realism, the value of λ is assumed to be the same for both particles 1 and 2, and is fixed at the moment that the parent pion decays. Thus, in each π^0 decay event, λ has an unknown, but specific, value, which already determines what the ensuing measurement of any component of the e^+ and e^- spins will be, through the functions f_A and f_B.

In the case of spins, the functions determining the measurement outcomes must satisfy some constraints. First, a measured value for $\hat{n} \cdot \boldsymbol{\sigma}_1$ is the same as the negative of a measured value for $-\hat{n} \cdot \boldsymbol{\sigma}_1$, so $f_A(\hat{n}, \lambda) = -f_A(-\hat{n}, \lambda)$, and similarly $f_B(\hat{n}, \lambda) = -f_B(-\hat{n}, \lambda)$, for any $\hat{n}$. Also, to be consistent with angular momentum conservation, it must be true that whenever $\hat{a} = \hat{b} = \hat{n}$, the measurement results are perfectly anticorrelated in the case of the EPR setup. Thus,

$$f_B(\hat{n}, \lambda) = -f_A(\hat{n}, \lambda) \tag{14.13}$$

so that the measurements are consistent with total spin $S = 0$.

We can now write down the hidden-variables prediction for the statistical correlation of Alice's and Bob's measurements. It is the product of the possible measurement results, integrated over the hidden variables weighted by their probability density,

$$C(\hat{a}, \hat{b}) = \int d\lambda\, \mu(\lambda)\, f_A(\hat{a}, \lambda) f_B(\hat{b}, \lambda). \tag{14.14}$$

This hidden-variables result should be contrasted with the prediction of quantum mechanics that we found in (14.6), which was $C(\hat{a}, \hat{b})_{\text{QM}} = -\hat{a} \cdot \hat{b}$.

The hidden-variables result for $C(\hat{a}, \hat{b})$ is clearly less specific than the quantum mechanics prediction, because we have not committed to a particular form for the functions μ, f_A, and f_B. Nevertheless, it is possible to draw some specific conclusions. In the case that $\hat{b} = \hat{a}$, the hidden-variables prediction is

$$\begin{aligned} C(\hat{a}, \hat{a}) &= \int d\lambda\, \mu(\lambda)\, f_A(\hat{a}, \lambda) f_B(\hat{a}, \lambda) = -\int d\lambda\, \mu(\lambda)\, f_A(\hat{a}, \lambda)^2 \\ &= -\int d\lambda\, \mu(\lambda) \\ &= -1, \end{aligned} \tag{14.15}$$

where we used (14.13) to get the second equality, then the fact that $f_A(\hat{a}, \lambda) = \pm 1$, and finally (14.7). In this special case, the hidden-variables prediction does agree with that of quantum mechanics. This had to be true, simply because in our hidden-variables theory we built in the perfect anticorrelation required by angular momentum conservation, through (14.13).

However, for more general $\hat{a}$ and $\hat{b}$, it is not immediately clear whether the hidden-variables predictions for $C(\hat{a}, \hat{b})$ can be made to always agree with the predictions of quantum mechanics. In 1964, John S. Bell answered the question by discovering that the correlations predicted by hidden-variables theories must satisfy an inequality that is clearly violated by the correlations predicted by quantum mechanics. Surprisingly, this is true for *any* choices of the functions $\mu(\lambda)$, $f_A(\hat{a}, \lambda)$, and $f_B(\hat{b}, \lambda)$ in the hidden-variables theory. This provides a way for experiment to decisively settle the question of whether the general hidden-variables idea or standard quantum mechanics is true.

Bell's inequality is remarkably simple to derive. We start with

$$C(\hat{a}, \hat{b}) - C(\hat{a}, \hat{c}) = \int d\lambda\, \mu(\lambda) \left[f_A(\hat{a}, \lambda) f_B(\hat{b}, \lambda) - f_A(\hat{a}, \lambda) f_B(\hat{c}, \lambda) \right] \tag{14.16}$$

$$= -\int d\lambda\, \mu(\lambda) \left[f_A(\hat{a}, \lambda) f_A(\hat{b}, \lambda) - f_A(\hat{a}, \lambda) f_A(\hat{c}, \lambda) \right], \tag{14.17}$$

where we have used (14.13) to replace f_B in favor of f_A. Multiply the second term on the right side by $1 = [f_A(\hat{b}, \lambda)]^2$, and rearrange to get

$$C(\hat{a}, \hat{b}) - C(\hat{a}, \hat{c}) = -\int d\lambda\, \mu(\lambda) \left[1 - f_A(\hat{b}, \lambda) f_A(\hat{c}, \lambda) \right] f_A(\hat{a}, \lambda) f_A(\hat{b}, \lambda). \tag{14.18}$$

Now, take the absolute value of both sides, and use the general fact that the absolute value of any integral is always less than or equal to the integral of the magnitude of the integrand:

$$\left| \int d\lambda\, F(\lambda) \right| \leq \int d\lambda\, |F(\lambda)|. \tag{14.19}$$

Let $F(\lambda)$ be the integrand on the right side of (14.18). Then since $f_A(\hat{a}, \lambda) f_A(\hat{b}, \lambda)$ is always equal to ± 1, and both $\mu(\lambda)$ and $1 - f_A(\hat{b}, \lambda) f_A(\hat{c}, \lambda)$ are always non-negative, we have $|F(\lambda)| = \mu(\lambda) \left[1 - f_A(\hat{b}, \lambda) f_A(\hat{c}, \lambda) \right]$. Therefore we obtain

$$\left| C(\hat{a}, \hat{b}) - C(\hat{a}, \hat{c}) \right| \leq \int d\lambda\, \mu(\lambda) \left[1 - f_A(\hat{b}, \lambda) f_A(\hat{c}, \lambda) \right], \tag{14.20}$$

or, using (14.7) and (14.13) and (14.14) again,

$$\left|C(\hat{a},\hat{b}) - C(\hat{a},\hat{c})\right| \leq 1 + C(\hat{b},\hat{c}). \tag{14.21}$$

This is Bell's original inequality governing correlations of spin-1/2 measurements in a total spin $S = 0$ state in hidden-variables theories.

To see that Bell's inequality is predicted to be violated by quantum mechanics, consider what happens if we substitute the quantum prediction of the correlation from (14.6) into (14.21),

$$\left|\hat{a}\cdot\hat{b} - \hat{a}\cdot\hat{c}\right| \leq 1 - \hat{b}\cdot\hat{c}. \tag{14.22}$$

This reduces the question to a purely geometric claim about three arbitrary unit vectors. A single example will be sufficient. Take $\hat{a} = \hat{x}$, $\hat{b} = \hat{y}$, and $\hat{c} = (\hat{x} + \hat{y})/\sqrt{2}$. The left side of (14.22) is $1/\sqrt{2}$, while the right side is $1 - 1/\sqrt{2}$, which is less, so the inequality is violated. Thus, quantum mechanics makes a specific prediction that cannot be satisfied in any local hidden-variables theory.

14.3 Quantum Mechanics vs. Hidden Variables, Without Inequalities

There are various other Bell-type inequalities, which show that quantum mechanics and hidden-variables theories can be distinguished by measuring statistical correlations. In this section we will discuss another way, which is theoretically even sharper because it does not rely on inequalities or statistics, although in practical terms it is somewhat harder to realize experimentally. The result in this example is that quantum mechanics and hidden-variables theories make predictions that disagree, not just for correlations on a statistical basis, but for certain measurements on an event-by-event basis with 100% probability.

Consider the following entangled state involving three spin-1/2 particles,[3]

$$|\psi\rangle = \frac{1}{\sqrt{2}}\Big(\left|+\hat{z},+\hat{z},+\hat{z}\right\rangle - \left|-\hat{z},-\hat{z},-\hat{z}\right\rangle\Big). \tag{14.23}$$

We assume that the three particles, labeled 1, 2, and 3, start from a central point and are spatially separated without disturbing the spins. The particles are observed by Alice, Bob, and Charlie at three remote sites. It is agreed that each of them will independently measure a random choice of either the x or y component of the spin, so that Alice always measures either σ_{1x} or σ_{1y}, while Bob always measures

[3] This three-spin state was proposed in 1990 by N. David Mermin, closely following a similar four-spin state given by Daniel Greenberger, Michael Horne, and Anton Zeilinger. Examples of this type are commonly called **GHZ states**.

σ_{2x} or σ_{2y}, and Charlie always measures σ_{3x} or σ_{3y}. The experiment is repeated many times, always starting with the same state $|\psi\rangle$, and then the observers meet to compare their results.

Let us first analyze the situation according to the standard rules of quantum mechanics. The state $|\psi\rangle$ is not an eigenstate of any of the individual spin operators σ_{1x}, σ_{1y}, σ_{2x}, σ_{2y}, σ_{3x}, or σ_{3y}, and each of the individual measurements of Alice, Bob, and Charlie will have a random result ± 1 with equal probability. However, you can check that $|\psi\rangle$ has the remarkable property that it is an eigenstate of all four of the operators of the form

$$\Omega_{abc} = \sigma_{1a}\sigma_{2b}\sigma_{3c}, \tag{14.24}$$

where $(a, b, c) = (x, x, x)$, (x, y, y), (y, x, y), and (y, y, x). For the operator Ω_{xxx} the eigenvalue is -1, and for each of the other three operators Ω_{xyy}, Ω_{yxy}, and Ω_{yyx}, the eigenvalue is $+1$.

Before meeting, Alice, Bob, and Charlie have not learned anything except that their own individual measurement results appear completely random. However, consider what they find when they finally get together to compare their results. They decide to first look only at the results for cases in which two of them had chosen to measure the y component of the spin, and the other had measured the x component of the spin. In those cases, we can equivalently say that collectively they had measured one of the observables Ω_{xyy} or Ω_{yxy} or Ω_{yyx}, and because

$$\Omega_{xyy}\,|\psi\rangle = \Omega_{yxy}\,|\psi\rangle = \Omega_{yyx}\,|\psi\rangle = |\psi\rangle\,, \tag{14.25}$$

they must find that the product of their measured results was definitely $+1$, every time. (Sometimes all three of them will have measured the result to be $+1$, and sometimes two of them will have obtained -1 and the other will have obtained $+1$.) On the other hand, in the cases where all three had chosen to measure the x component of the spin, we can equivalently say that their choices amounted to measuring Ω_{xxx}. Now, since

$$\Omega_{xxx}\,|\psi\rangle = -\,|\psi\rangle\,, \tag{14.26}$$

it must be that the product of their three measurements is -1. (Sometimes they will all get the result -1, and sometimes two will get $+1$ and the other will get -1.)

So far, we have analyzed the experiment assuming the predictions of quantum mechanics. Now let us try to explain these results in terms of a general local hidden-variables theory, in which the measurement of σ_{1x} is supposed to be some $f_A(\hat{x}, \lambda) = \pm 1$ and the measurement of σ_{2y} is $f_B(\hat{y}, \lambda) = \pm 1$, etc. The hidden variable(s) λ are supposed to be different for each repetition of the experiment, giving the illusion of randomness. Then, the observed results for Ω_{xyy}, Ω_{yxy}, Ω_{yyx} would imply that, respectively,

$$f_A(\hat{x}, \lambda)\, f_B(\hat{y}, \lambda)\, f_C(\hat{y}, \lambda) = 1, \tag{14.27}$$

$$f_A(\hat{y},\lambda)\, f_B(\hat{x},\lambda)\, f_C(\hat{y},\lambda) = 1, \tag{14.28}$$

$$f_A(\hat{y},\lambda)\, f_B(\hat{y},\lambda)\, f_C(\hat{x},\lambda) = 1, \tag{14.29}$$

for every λ. Now, multiplying these three equations together, and using $f_A(\hat{y},\lambda)^2 = f_B(\hat{y},\lambda)^2 = f_C(\hat{y},\lambda)^2 = 1$, we discover that

$$f_A(\hat{x},\lambda)\, f_B(\hat{x},\lambda)\, f_C(\hat{x},\lambda) = 1. \tag{14.30}$$

This tells us that in those cases where they had all measured the x component of spin, the product of their measurements should have been $+1$, with 100% certainty, according to the hidden-variables theory. This is in direct contradiction to the quantum mechanics prediction.

To recapitulate: if we stipulate that the measurement of Ω_{xyy}, Ω_{yxy}, or Ω_{yyx} always gives $+1$, then, according to any local hidden-variables theory, measurement of Ω_{xxx} must always give $+1$ as well. However, the prediction of quantum mechanics, for the given state $|\psi\rangle$, is that Ω_{xxx} always gives -1. The hidden-variables and quantum mechanics hypotheses make contradictory predictions. One of them must be wrong. The experimental verdict will be revealed at the end of the next section.

14.4 Aspect's Experiments and the Demise of Local Hidden Variables

In practice, the most decisive real-world experimental tests, many first conducted by Alain Aspect and collaborators, involve polarizations of photons in entangled states, rather than spin-1/2 systems. In a series of increasingly sensitive experiments, the Bell-type inequalities of local hidden-variables theories have been put to the test and found to be clearly violated, and in just the way predicted by standard quantum mechanics. In this section, we will describe a Bell-type inequality for hidden variables, and the corresponding quantum prediction, as tested in a 1982 experiment by A. Aspect, J. Dalibard, and G. Roger, referred to in the following as the ADR experiment.

Consider observables $A(\hat{a})$ and $B(\hat{b})$ that can take only the values ± 1, and depend on unit vectors $\hat{a}$ and $\hat{b}$, and are associated with spatially separated detectors. These generalize the spin-1/2 observables $\hat{a}\cdot\boldsymbol{\sigma}_1$ and $\hat{b}\cdot\boldsymbol{\sigma}_2$ of Alice and Bob in the previous two sections. The experiment can change $\hat{a}$ and $\hat{b}$ independently at any time. According to the hidden-variables proposal, the measured values are supposed to be determined by some functions

$$A(\hat{a}) = f_A(\hat{a},\lambda) = \pm 1, \tag{14.31}$$

$$B(\hat{b}) = f_B(\hat{b},\lambda) = \pm 1. \tag{14.32}$$

Table 14.1 Possible measurement results pertaining to the ADR experiment, as dictated by (14.31)–(14.33), in a theory of local hidden variables. In addition to the eight possibilities shown, there are eight more in which all measurements have the opposite signs. The [result] in the last column is the quantity inside the square brackets on the right side of (14.33). It is 2 or −2 in all cases

$f_A(\hat{a}, \lambda)$	$f_A(\hat{a}', \lambda)$	$f_B(\hat{b}, \lambda)$	$f_B(\hat{b}', \lambda)$	[result]
1	1	1	1	2
1	1	1	−1	2
1	1	−1	1	−2
1	−1	1	1	−2
1	1	−1	−1	−2
1	−1	1	−1	2
1	−1	−1	1	−2
1	−1	−1	−1	2

The hidden variables λ are assumed to be local, implying that they are shared by A and B, since they will have originated from a single location and event. Their probability density distribution is some non-negative function $\mu(\lambda)$, just as in (14.7). The correlation between measurement results $C(\hat{a}, \hat{b})$ is again given by the formula (14.14). However, since the observables are not necessarily components of angular momenta (and will not be in the ADR experiment), we will not assume or use (14.13). This means that the Bell inequality we are about to derive applies more generally than the original one.

Consider the combination of correlations

$$
\begin{aligned}
C(\hat{a}, \hat{b}) - C(\hat{a}, \hat{b}') + C(\hat{a}', \hat{b}) + C(\hat{a}', \hat{b}') &= \\
\int d\lambda\, \mu(\lambda) \Big[f_A(\hat{a}, \lambda) f_B(\hat{b}, \lambda) - f_A(\hat{a}, \lambda) f_B(\hat{b}', \lambda) \\
+ f_A(\hat{a}', \lambda) f_B(\hat{b}, \lambda) + f_A(\hat{a}', \lambda) f_B(\hat{b}', \lambda) \Big], \qquad (14.33)
\end{aligned}
$$

for arbitrary unit vectors $\hat{a}$ and $\hat{a}'$ associated with observable A, and $\hat{b}$ and $\hat{b}'$ associated with observable B. Since each of the four terms is bounded by -1 and $+1$, in any conceivable theory the left side must be in the range from -4 to 4, inclusive. However, in hidden-variables theories, a stronger statement holds. This is because a simple brute-force enumeration of all possible results for the expression inside the square brackets on the right side shows that it can only be 2 or -2. (See Table 14.1.) Using (14.7), this implies that the magnitude of the integral on the right side of (14.33) is bounded by 2. So, we have another Bell-type result, the CHSH inequality, named for John Clauser, Michael Horne, Abner Shimony, and Richard Holt:

$$\left|C(\hat{a},\hat{b}) - C(\hat{a},\hat{b}') + C(\hat{a}',\hat{b}) + C(\hat{a}',\hat{b}')\right| \le 2, \tag{14.34}$$

which must be obeyed in any local hidden-variables theory.

In the ADR experiment, an excited state of a calcium atom with total angular momentum $J = 0$ decays to an intermediate state with $J = 1$ by emitting one photon in an electric dipole transition, and then to another $J = 0$ state by emitting a second photon in another electric dipole transition. The photons travel in opposite directions to distant detectors A and B, which, by a choice of coordinates, are located on the positive and negative z-axis, respectively. The joint state of the two photons assigns them both circular polarizations, with $S_{1z} = \pm\hbar$ and $S_{2z} = \mp\hbar$, in an entangled state $|\psi\rangle$ with total $S_z = 0$ by angular momentum conservation. If one photon is in an L-circular polarization state, then the other also has L-circular polarization, because both its momentum direction and its spin direction are opposite to those of the first. Similarly, the two photons could both have R-circular polarization.

However, the detectors A and B measure the linear polarizations of the photons, not circular polarizations, as follows. A linear polarizer sends each photon into one photomultiplier if the linear polarization is along a selected direction, and into a different one if the linear polarization is in the orthogonal direction. (The allowed polarizations are always orthogonal to the photon momentum direction.) If detector A measures the linear polarization along the chosen direction $\hat{a}$, then it records $A(\hat{a}) = +1$, and if it instead detects the polarization perpendicular to $\hat{a}$, then it records $A(\hat{a}) = -1$. Similarly, if B sees the linear polarization along the chosen direction $\hat{b}$, then it records $B(\hat{b}) = +1$, and if it detects the polarization perpendicular to $\hat{b}$, then it records $B(\hat{b}) = -1$. The experiment is designed so that detector A can switch between two choices $\hat{a}$ and $\hat{a}'$, and detector B can switch between $\hat{b}$ and $\hat{b}'$. The switching is done randomly and quickly, with times shorter than the light travel time between the two detectors. This avoids possible biasing of the results by some hypothetical communication mechanism that might relate the choices made by the detectors.

Let us now work out the quantum-mechanical prediction for the correlations of measurements of $A(\hat{a})$ and $B(\hat{b})$. As already noted, the photons are produced in a state where the circular polarizations are the same, but could be either both L or both R. Since the atomic transitions have equal amplitudes for these two possibilities due to spherical symmetry, we can write

$$|\psi\rangle = \frac{1}{\sqrt{2}}\big(|L, L\rangle + |R, R\rangle\big), \tag{14.35}$$

where the first and second ket labels refer to the photons that are seen by detectors A and B, respectively. The single-photon orthobasis states obey $\langle R|R\rangle = \langle L|L\rangle = 1$ and $\langle L|R\rangle = \langle R|L\rangle = 0$. To make contact with the measurements made by the experiment, it is convenient to rewrite the state using a linear polarization orthobasis. Our chosen coordinates imply that the momenta of the photons that reach detectors

A and B point in the directions $\hat{k}_A = \hat{z}$ and $\hat{k}_B = -\hat{z}$. Then the transformations from the circular polarization orthobases to the linear polarization orthobases are

$$|L\rangle_A = \frac{1}{\sqrt{2}}\big(|\hat{x}\rangle_A - i\,|\hat{y}\rangle_A\big), \qquad |R\rangle_A = \frac{1}{\sqrt{2}}\big(|\hat{x}\rangle_A + i\,|\hat{y}\rangle_A\big), \tag{14.36}$$

$$|L\rangle_B = \frac{1}{\sqrt{2}}\big(|\hat{x}\rangle_B + i\,|\hat{y}\rangle_B\big), \qquad |R\rangle_B = \frac{1}{\sqrt{2}}\big(|\hat{x}\rangle_B - i\,|\hat{y}\rangle_B\big), \tag{14.37}$$

where $|\hat{x}\rangle$ and $|\hat{y}\rangle$ are states with linear polarizations along the x and y axes respectively, and $\langle\hat{x}|\hat{x}\rangle = \langle\hat{y}|\hat{y}\rangle = 1$ and $\langle\hat{x}|\hat{y}\rangle = \langle\hat{y}|\hat{x}\rangle = 0$. Using (14.36)–(14.37) in (14.35),

$$|\psi\rangle = \frac{1}{\sqrt{2}}\big(|\hat{x},\hat{x}\rangle + |\hat{y},\hat{y}\rangle\big). \tag{14.38}$$

This state has even parity, in agreement with the fact that the transition involves two even-parity atomic states.

A state $|\hat{n}\rangle$ describing a photon with linear polarization $\hat{n}$ has, up to an arbitrary phase,

$$\langle\hat{n}|\hat{x}\rangle = \hat{n}_x, \qquad \langle\hat{n}|\hat{y}\rangle = \hat{n}_y. \tag{14.39}$$

So, if we use $|\hat{a},\hat{b}\rangle$ to denote the state in which the linear polarizations are $\hat{a}$ and $\hat{b}$, then

$$\langle\hat{a},\hat{b}|\psi\rangle = \frac{1}{\sqrt{2}}\left(\hat{a}_x\hat{b}_x + \hat{a}_y\hat{b}_y\right), \tag{14.40}$$

again up to an irrelevant phase. So, the probability for the linear polarizations to be observed as $A(\hat{a}) = 1$, $B(\hat{b}) = 1$ is

$$\frac{1}{2}(\hat{a}\cdot\hat{b})^2 = \frac{1}{2}\cos^2\theta_{ab}, \tag{14.41}$$

where θ_{ab} is the angle between $\hat{a}$ and $\hat{b}$. The same result is obtained for the probability to observe the polarizations both orthogonal to $\hat{a}$ and $\hat{b}$, so that $A(\hat{a}) = -1$, $B(\hat{b}) = -1$. The probabilities to observe $A(\hat{a}) = 1$, $B(\hat{b}) = -1$ and $A(\hat{a}) = -1$, $B(\hat{b}) = 1$ are both

$$\frac{1}{2}|\hat{a}\times\hat{b}|^2 = \frac{1}{2}\sin^2\theta_{ab}. \tag{14.42}$$

Weighting each of the outcomes for $A(\hat{a})B(\hat{b})$ by these respective probabilities, we arrive at

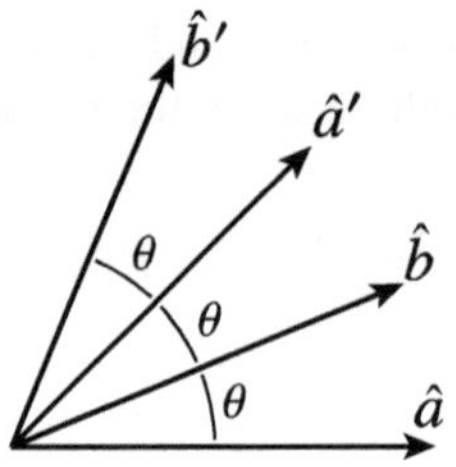

Fig. 14.3 The configuration for the unit vectors $\hat{a}$, $\hat{a}'$, $\hat{b}$, and $\hat{b}'$ that provides the maximum violation of the Bell-type inequality (14.34) in the ADR photon polarization experiment, using the quantum prediction of (14.43). The three angles labeled θ are all equal to $\pi/8$, so that $\theta_{ab} = \theta_{a'b} = \theta_{a'b'} = \pi/8$, and $\theta_{ab'} = 3\pi/8$

$$\begin{aligned} C(\hat{a}, \hat{b}) &= \langle A(\hat{a}) B(\hat{b}) \rangle = \frac{1}{2}\cos^2\theta_{ab} + \frac{1}{2}\cos^2\theta_{ab} - \frac{1}{2}\sin^2\theta_{ab} - \frac{1}{2}\sin^2\theta_{ab} \\ &= \cos(2\theta_{ab}). \end{aligned} \tag{14.43}$$

This is the quantum-mechanical prediction for the correlation.

The local hidden-variables Bell inequality, (14.34), is incompatible with the quantum-mechanical result of (14.43) for many choices of $\hat{a}$, $\hat{a}'$, $\hat{b}$, and $\hat{b}'$. (Due to the geometry of the experiment and the transverse polarization of photons, these four unit vectors must all lie in the xy plane.) The most extreme violation of the inequality, used in the actual ADR experiment, occurs if one chooses the unit vectors as depicted in Fig. 14.3, with

$$C(\hat{a}, \hat{b}) = C(\hat{a}', \hat{b}) = C(\hat{a}', \hat{b}') = \cos(\pi/4) = 1/\sqrt{2}, \tag{14.44}$$

$$C(\hat{a}, \hat{b}') = \cos(3\pi/4) = -1/\sqrt{2}. \tag{14.45}$$

The (idealized) quantum-mechanical prediction for that particular geometry is therefore

$$\left| C(\hat{a}, \hat{b}) - C(\hat{a}, \hat{b}') + C(\hat{a}', \hat{b}) + C(\hat{a}', \hat{b}') \right| = 2\sqrt{2} \approx 2.828. \tag{14.46}$$

The expected quantum-mechanical prediction for the ADR experiment as performed was slightly lower at 2.70 ± 0.05, due to experimental non-ideal realities. This was in excellent agreement with the actual experimental result 2.697 ± 0.015 reported by ADR. This result is also completely incompatible with the upper bound of 2 from (14.34) in all theories in which the probabilities are determined by local hidden variables.

Further experiments have sharpened the result that local hidden variables cannot explain observed phenomena, and that nonlocal entanglement of states as predicted by quantum mechanics is both consistent and supported by the experimental evidence. In particular, these experiments address a perverse loophole in Bell's original argument that we have not mentioned yet. It is natural to expect that the detector settings made in each trial (in the Aspect experiment case, the choice of $\hat{a}$ or $\hat{a}'$, and the choice of $\hat{b}$ or $\hat{b}'$) should not be correlated in any special way with the hidden variables λ of the decaying particle. But what if they are, because of some influence in their shared past? This could mean that event statistics are somehow biased by the detector setting choices in just such a very special way as to allow a hidden-variables theory to not only violate Bell-type inequalities but also to closely mimic the standard quantum mechanics predictions. Experiments of various types have taken great pains to choose the detector settings randomly in very different (and often quite convoluted) ways to discount this already implausible conspiracy theory scenario. Furthermore, the striking perfect-correlation quantum mechanics prediction for a state of the GHZ type discussed in Sect. 14.3 (but with photon polarizations rather than spins) has been confirmed within experimental uncertainties. The experimental verdict is that "local realism" is not real.

Since local hidden-variables theories do not agree with experiment, one might entertain the possibility of "nonlocal hidden variables", which could allow interactions between spacelike separated points. This is hard to rule out in general because it could predict anything, including causality violation; one might as well explain experiments by invoking sorcery. Furthermore, nonlocal hidden variables abandon the original motivation of local realism, and in any case are a cure for a disease that does not exist.

Exercises

14.1 Consider the nonlocally entangled EPR state $|S = 0\rangle$ of Sect. 14.1. Suppose that Alice and Bob measure $\hat{a} \cdot \boldsymbol{\sigma}_1$ and $\hat{b} \cdot \boldsymbol{\sigma}_2$, respectively. Let θ be the angle between $\hat{a}$ and $\hat{b}$. The four possible outcomes of their measurements are denoted $(+1, +1)$, $(+1, -1)$, $(-1, +1)$, and $(-1, -1)$.

(a) Suppose Alice's measurement was first. For the outcome $\hat{a} \cdot \boldsymbol{\sigma}_1 = +1$, the associated projection operator can be written as

$$P_{\hat{a}\cdot\boldsymbol{\sigma}_1=+1} = |\hat{a}, \hat{n}\rangle\langle\hat{a}, \hat{n}| + |\hat{a}, -\hat{n}\rangle\langle\hat{a}, -\hat{n}| \tag{14.47}$$

for any choice of unit vector $\hat{n}$. It is convenient to choose $\hat{n} = \hat{a}$, both in (14.47) and in the expression for the starting state $|S = 0\rangle$ in (14.2). Apply Postulates 4 and 5 to show that the probability of this outcome is $1/2$, and that the normalized state after the measurement is $|\hat{a}, -\hat{a}\rangle$. Now consider Bob's subsequent measurement. The projection operator associated with Bob's outcome $\hat{b} \cdot \boldsymbol{\sigma}_2 = +1$ is

$$P_{\hat{b}\cdot\boldsymbol{\sigma}_2=+1} = |\hat{n}, \hat{b}\rangle\langle\hat{n}, \hat{b}| + |-\hat{n}, \hat{b}\rangle\langle-\hat{n}, \hat{b}|, \tag{14.48}$$

where again $\hat{n}$ is any unit vector. It is again convenient to choose $\hat{n} = \hat{a}$. Apply Postulates 4 and 5 again to show that the probability for the outcome $(+1, +1)$ is $\mathcal{P}(+1, +1) = \frac{1}{2} \sin^2(\theta/2)$, and that the state after the measurements is $|\hat{a}, \hat{b}\rangle$.

(b) Repeat part (a), but this time take Bob's measurement first. Show that the final state and the probability for the outcome $(+1, +1)$ are the same as found in part (a).

(c) Repeat parts (a) and (b), but this time assuming the measurements are simultaneous, using a projection operator

$$P_{\hat{a}\cdot\boldsymbol{\sigma}_1=+1,\,\hat{b}\cdot\boldsymbol{\sigma}_2=+1} = |\hat{a}, \hat{b}\rangle\langle\hat{a}, \hat{b}|. \tag{14.49}$$

Check that the results agree with those found in parts (a) and (b).

(d) Repeat the analysis of parts (a), (b), and (c) for each of the other three outcomes $(+1, -1)$, $(-1, +1)$, and $(-1, -1)$. As checks, your four probabilities should add to 1, and can be used to obtain the expectation value $\langle \hat{a} \cdot \boldsymbol{\sigma}_1 \, \hat{b} \cdot \boldsymbol{\sigma}_2 \rangle$ given in (14.6).

14.2 Show that Mermin's 3-spin GHZ-type state in (14.23) satisfies the eigenvalue equations (14.25) and (14.26) for the operators Ω_{xyy}, Ω_{yxy}, Ω_{yyx} and Ω_{xxx} defined in terms of the individual spin components in (14.24).

Stationary-State Perturbation Theory 15

15.1 Perturbative Expansion for Energy Eigenstates

Given a Hamiltonian H that does not depend on time, we would like to solve the time-independent Schrödinger equation,

$$H\,|\psi_n\rangle = E_n\,|\psi_n\rangle\,, \tag{15.1}$$

as an eigenvalue problem for the stationary states $|\psi_n\rangle$ and their energies E_n. Unfortunately, in many cases this is too hard to do exactly. However, suppose that it is possible to split the Hamiltonian into two parts,

$$H = H_0 + \lambda W, \tag{15.2}$$

where H_0 is a simpler Hamiltonian for which we already know how to solve the eigenvalue problem, and λ is a small parameter, so that W can be treated as a perturbation. Then we can hope to find an approximate solution for the full Hamiltonian H, by expanding in λ. We will now work out how to do this. This is called **time-independent perturbation theory** or **stationary-state perturbation theory**.

To begin, assume that we have found the complete set of eigenstates $|n\rangle$ and eigenvalues $\mathcal{E}_n$ for H_0, so that

$$H_0\,|n\rangle = \mathcal{E}_n\,|n\rangle\,. \tag{15.3}$$

In general, the states $|n\rangle$ for a given $\mathcal{E}_n$ might be degenerate, but for now we assume they are not. For each unperturbed stationary state $|n\rangle$ and energy $\mathcal{E}_n$, we want to solve for the corresponding $|\psi_n\rangle$ and energy E_n. The connection between the unperturbed and perturbed stationary-state solutions is assumed to be smoothly

S. P. Martin, *Quantum Mechanics*, Graduate Texts in Physics,
https://doi.org/10.1007/978-3-032-21119-4_15

continuous in λ in order for the following strategy to work. In particular, there should be a one-to-one correspondence between the known $|n\rangle$ and the desired $|\psi_n\rangle$ for each λ, as long as λ is not too large.

The perturbation theory strategy is to write the unknown full eigenstates and energy eigenvalues each as a power series in λ,

$$|\psi_n\rangle = |\psi_n^{(0)}\rangle + \lambda|\psi_n^{(1)}\rangle + \lambda^2|\psi_n^{(2)}\rangle + \lambda^3|\psi_n^{(3)}\rangle + \cdots, \tag{15.4}$$

$$E_n = E_n^{(0)} + \lambda E_n^{(1)} + \lambda^2 E_n^{(2)} + \lambda^3 E_n^{(3)} + \cdots. \tag{15.5}$$

Now, $\lambda = 0$ must recover the unperturbed solutions, so

$$|\psi_n^{(0)}\rangle = |n\rangle, \qquad E_n^{(0)} = \mathcal{E}_n. \tag{15.6}$$

The unperturbed stationary states are a complete orthobasis, with

$$\sum_n |n\rangle\langle n| = I, \tag{15.7}$$

$$\langle n'|n\rangle = \delta_{nn'}. \tag{15.8}$$

This follows because H_0 is assumed to be an observable.

We could also choose unit normalization for the eigenkets $|\psi_n\rangle$, but we won't. It is much more convenient to choose the following normalization condition for them instead:

$$\langle n|\psi_n\rangle = 1 \qquad \text{(for each } n\text{)}. \tag{15.9}$$

This will result in $\langle\psi_n|\psi_n\rangle \neq 1$, but after obtaining the solutions for $|\psi_n\rangle$, we can renormalize them later, as a last step. Plugging (15.4) into (15.9), we have

$$\langle n|\left(|n\rangle + \lambda|\psi_n^{(1)}\rangle + \lambda^2|\psi_n^{(2)}\rangle + \cdots\right) = 1. \tag{15.10}$$

Since $\langle n|n\rangle = 1$, in order to have a solution for general λ, it is necessary and sufficient that the coefficient of λ^k in (15.10) vanishes for each $k \geq 1$, which implies

$$\langle n|\psi_n^{(k)}\rangle = 0 \qquad \text{(for } k = 1, 2, 3, \ldots\text{)}. \tag{15.11}$$

Thus, the normalization choice (15.9) ensures that all of the corrections to each state are orthogonal to the corresponding unperturbed state. This is extremely important, and will be used repeatedly in the following.

The time-independent Schrödinger equation (15.1) now reads

$$
\begin{aligned}
&(H_0 + \lambda W)\left(|n\rangle + \lambda|\psi_n^{(1)}\rangle + \lambda^2|\psi_n^{(2)}\rangle + \cdots\right) \\
&\quad = \left(\mathcal{E}_n + \lambda E_n^{(1)} + \lambda^2 E_n^{(2)} + \cdots\right)\left(|n\rangle + \lambda|\psi_n^{(1)}\rangle + \lambda^2|\psi_n^{(2)}\rangle + \cdots\right).
\end{aligned}
\tag{15.12}
$$

Expanding both sides, and matching the coefficient of each power λ^k, we get

$$\lambda^0: \quad H_0\,|n\rangle = \mathcal{E}_n\,|n\rangle\,, \tag{15.13}$$

$$\lambda^1: \quad H_0|\psi_n^{(1)}\rangle + W\,|n\rangle = \mathcal{E}_n|\psi_n^{(1)}\rangle + E_n^{(1)}\,|n\rangle\,, \tag{15.14}$$

$$\lambda^2: \quad H_0|\psi_n^{(2)}\rangle + W|\psi_n^{(1)}\rangle = \mathcal{E}_n|\psi_n^{(2)}\rangle + E_n^{(1)}|\psi_n^{(1)}\rangle + E_n^{(2)}\,|n\rangle\,, \tag{15.15}$$

and, in general,

$$\lambda^k: \quad H_0|\psi_n^{(k)}\rangle + W|\psi_n^{(k-1)}\rangle = \mathcal{E}_n|\psi_n^{(k)}\rangle + \sum_{j=1}^{k-1} E_n^{(j)}|\psi_n^{(k-j)}\rangle + E_n^{(k)}\,|n\rangle \tag{15.16}$$

Equation (15.13) is, of course, just a repetition of (15.3).

Consider first the effect of the perturbation at order λ^1. Taking the inner product of (15.14) with $\langle n|$ gives

$$\langle n|H_0|\psi_n^{(1)}\rangle + \langle n|W|n\rangle = \mathcal{E}_n\,\langle n|\psi_n^{(1)}\rangle + E_n^{(1)}\,\langle n|n\rangle\,. \tag{15.17}$$

Now using $\langle n|\,H_0 = \mathcal{E}_n\,\langle n|$ on the first term, we see that it and the first term on the right side are equal, and both vanish due to (15.11). Therefore, using the orthonormality condition (15.8) on the last term, (15.17) simplifies to

$$E_n^{(1)} = \langle n|W|n\rangle\,, \tag{15.18}$$

so that

$$E_n = \mathcal{E}_n + \lambda\,\langle n|W|n\rangle + \cdots\,. \tag{15.19}$$

Thus, to first order in the expansion in λ, the energy shift in the state $|n\rangle$ is simply the expectation value of the Hamiltonian perturbation. This is the most important and commonly used result from stationary-state perturbation theory.

To find the first-order correction to the energy eigenkets, we note that completeness of the H_0 eigenstates allows us to write

$$|\psi_n^{(1)}\rangle = \sum_{m\neq n} |m\rangle\langle m|\psi_n^{(1)}\rangle, \tag{15.20}$$

where we have made good use of (15.11) to dispense with the $m = n$ term. We now need to find the coefficients $\langle m|\psi_n^{(1)}\rangle$ for $m \neq n$. To do so, take $\langle m|$ acting on (15.14),

$$\langle m|H_0|\psi_n^{(1)}\rangle + \langle m|W|n\rangle = \mathcal{E}_n \langle m|\psi_n^{(1)}\rangle + E_n^{(1)} \langle m|n\rangle . \tag{15.21}$$

The last term vanishes due to the orthonormality of the unperturbed stationary states (15.8), and the first term on the left can be simplified using $\langle m| H_0 = \mathcal{E}_m \langle m|$. Therefore,

$$(\mathcal{E}_n - \mathcal{E}_m) \langle m|\psi_n^{(1)}\rangle = \langle m|W|n\rangle \qquad (m \neq n). \tag{15.22}$$

Now, assuming that there are no degeneracies, so that $\mathcal{E}_m \neq \mathcal{E}_n$ for $m \neq n$, we can solve for $\langle m|\psi_n^{(1)}\rangle$, and (15.20) becomes

$$|\psi_n^{(1)}\rangle = \sum_{m\neq n} |m\rangle \frac{\langle m|W|n\rangle}{\mathcal{E}_n - \mathcal{E}_m}. \tag{15.23}$$

We have thus obtained the perturbed energy eigenstates at first order in the expansion in λ,

$$|\psi_n\rangle = |n\rangle + \lambda \sum_{m\neq n} |m\rangle \frac{\langle m|W|n\rangle}{\mathcal{E}_n - \mathcal{E}_m} + \cdots . \tag{15.24}$$

We can now see why the normalization condition (15.9) was so useful and important; it gave us (15.11), which in turn allowed us to exclude $m = n$ from the sum, which avoided a term with a disastrous vanishing denominator[1] in $|\psi_n^{(1)}\rangle$, and in similar expressions to follow.

Having solved the perturbation theory problem at order λ^1, we can continue iteratively to obtain the perturbed energies and their eigenstates at any desired order in λ. We will now find the necessary recurrence relations in general. At order λ^k, the inner product of $\langle n|$ with (15.16) is

$$\langle n|H_0|\psi_n^{(k)}\rangle + \langle n|W|\psi_n^{(k-1)}\rangle = \mathcal{E}_n \langle n|\psi_n^{(k)}\rangle + \sum_{j=1}^{k-1} E_n^{(j)} \langle n|\psi_n^{(k-j)}\rangle + E_n^{(k)}. \tag{15.25}$$

The first term on the left vanishes because of (15.11), after using $\langle n| H_0 = \langle n| \mathcal{E}_n$. On the right side of the equality, using (15.11) again ensures that all of the terms

[1] Of course, the denominator is still a disaster if the unperturbed states have energy degeneracies, so that $\mathcal{E}_m = \mathcal{E}_n$ for some $m \neq n$. This issue is addressed by "degenerate perturbation theory", in Sect. 15.6.

vanish except the last one, where we have used $\langle n|n\rangle = 1$. So, we find the remarkable result

$$E_n^{(k)} = \langle n|W|\psi_n^{(k-1)}\rangle\,, \tag{15.26}$$

that the λ^k correction to each energy eigenvalue is always obtained in terms of the λ^{k-1} correction to the corresponding ket.

Also, using completeness, we can write

$$|\psi_n^{(k)}\rangle = \sum_{m\neq n} |m\rangle\langle m|\psi_n^{(k)}\rangle, \tag{15.27}$$

where once again (15.11) has been used to eliminate the $m = n$ term. To find the coefficients $\langle m|\psi_n^{(k)}\rangle$ for $m \neq n$, we take $\langle m|$ acting on (15.16), which gives

$$(\mathcal{E}_n - \mathcal{E}_m)\,\langle m|\psi_n^{(k)}\rangle = \langle m|W|\psi_n^{(k-1)}\rangle - \sum_{j=1}^{k-1} E_n^{(j)}\,\langle m|\psi_n^{(k-j)}\rangle\,, \tag{15.28}$$

where $\langle m|n\rangle = 0$ has been used to eliminate the term proportional to $E_n^{(k)}$. The right-hand side only involves expressions from orders less than k. Therefore, (15.26), (15.27), and (15.28) can be used to solve iteratively for $E_n^{(k)}$ and then $|\psi_n^{(k)}\rangle$ at each successive level k.

Let us apply the preceding for $k = 2$. Evaluating (15.26) using (15.23) gives

$$E_n^{(2)} = \sum_{m\neq n} \frac{|\langle m|W|n\rangle|^2}{\mathcal{E}_n - \mathcal{E}_m}, \tag{15.29}$$

the second-order corrections to the energies. Also, (15.28) gives

$$\langle m|\psi_n^{(2)}\rangle = \frac{1}{\mathcal{E}_n - \mathcal{E}_m}\Big(\langle m|W|\psi_n^{(1)}\rangle - E_n^{(1)}\,\langle m|\psi_n^{(1)}\rangle\Big), \tag{15.30}$$

or, using (15.18) and (15.23),

$$\langle m|\psi_n^{(2)}\rangle = \sum_{p\neq n} \frac{\langle m|W|p\rangle\,\langle p|W|n\rangle}{(\mathcal{E}_n - \mathcal{E}_p)(\mathcal{E}_n - \mathcal{E}_m)} - \frac{\langle m|W|n\rangle\,\langle n|W|n\rangle}{(\mathcal{E}_n - \mathcal{E}_m)^2} \tag{15.31}$$

for $n \neq m$. Summarizing the results to order λ^2:

$$E_n = \mathcal{E}_n + \lambda\,\langle n|W|n\rangle + \lambda^2 \sum_{m\neq n} \frac{|\langle m|W|n\rangle|^2}{\mathcal{E}_n - \mathcal{E}_m} + \cdots\,, \tag{15.32}$$

and

$$|\psi_n\rangle = |n\rangle + \lambda \sum_{m\neq n} |m\rangle \frac{\langle m|W|n\rangle}{\mathcal{E}_n - \mathcal{E}_m}$$
$$+ \lambda^2 \sum_{m\neq n} |m\rangle \left(\sum_{p\neq n} \frac{\langle m|W|p\rangle \langle p|W|n\rangle}{(\mathcal{E}_n - \mathcal{E}_p)(\mathcal{E}_n - \mathcal{E}_m)} - \frac{\langle m|W|n\rangle \langle n|W|n\rangle}{(\mathcal{E}_n - \mathcal{E}_m)^2} \right) + \cdots . \tag{15.33}$$

Don't forget that, by design, this ket $|\psi_n\rangle$ does not have unit norm, but now it can be renormalized as $|\psi_n\rangle/\sqrt{\langle\psi_n|\psi_n\rangle}$, if desired. In doing so, it often makes sense to expand in λ and drop the terms of order λ^3 or higher. From (15.33), we find using the orthonormality of the unperturbed kets that

$$\langle\psi_n|\psi_n\rangle = 1 + \lambda^2 \sum_{m\neq n} \frac{|\langle m|W|n\rangle|^2}{(\mathcal{E}_n - \mathcal{E}_m)^2} + \cdots , \tag{15.34}$$

so that, working consistently to second order in λ, the normalized eigenstate can be obtained by just multiplying the first term $|n\rangle$ in (15.33) by

$$1/\sqrt{\langle\psi_n|\psi_n\rangle} = 1 - \frac{\lambda^2}{2} \sum_{m\neq n} \frac{|\langle m|W|n\rangle|^2}{(\mathcal{E}_n - \mathcal{E}_m)^2} + \cdots . \tag{15.35}$$

It is also worth noting that, using (15.26), the energy eigenvalue to all orders in λ is

$$E_n = \mathcal{E}_n + \sum_{k=1}^{\infty} \lambda^k \langle n|W|\psi_n^{(k-1)}\rangle = \mathcal{E}_n + \lambda \langle n|W \sum_{k=1}^{\infty} \lambda^{k-1} |\psi_n^{(k-1)}\rangle , \tag{15.36}$$

which can be rewritten

$$E_n = \mathcal{E}_n + \lambda \langle n|W|\psi_n\rangle . \tag{15.37}$$

The all-orders energy E_n is therefore known exactly in terms of the all-orders state ket $|\psi_n\rangle$. Of course, the latter may well be only known partially as a perturbative expansion.

So far, we have used the parameter λ as a way of keeping track of the order of perturbation theory. In any given term, the power of λ is the same as the number of matrix elements of W, so we can simply absorb λ into the definition of W without losing any information. (The convergence of perturbation theory still relies on the assumption that the properties of W allow it to be treated as a small correction. For example, W may contain one or more small dimensionless parameters that effectively play the role of λ.) Equivalently, taking $\lambda = 1$, we have

$$E_n = \mathcal{E}_n + E_n^{(1)} + E_n^{(2)} + E_n^{(3)} + \cdots, \quad (15.38)$$

$$|\psi_n\rangle = |n\rangle + |\psi_n^{(1)}\rangle + |\psi_n^{(2)}\rangle + |\psi_n^{(3)}\rangle + \cdots. \quad (15.39)$$

The preceding results can be expressed more compactly with the following simplifying notation,

$$W_{mn} = \langle m|W|n\rangle, \quad (15.40)$$

$$\mathcal{E}_{nm} = \mathcal{E}_n - \mathcal{E}_m. \quad (15.41)$$

Then, the energy eigenvalue corrections are summarized as

$$E_n^{(1)} = W_{nn}, \quad (15.42)$$

$$E_n^{(2)} = \sum_{m\neq n} \frac{|W_{mn}|^2}{\mathcal{E}_{nm}}, \quad (15.43)$$

$$E_n^{(3)} = \sum_{m\neq n}\sum_{p\neq n} \frac{W_{nm}W_{mp}W_{pn}}{\mathcal{E}_{np}\mathcal{E}_{nm}} - W_{nn}\sum_{m\neq n}\frac{|W_{mn}|^2}{\mathcal{E}_{nm}^2}, \quad (15.44)$$

etc., and for the (unnormalized) energy eigenstates,

$$|\psi_n^{(1)}\rangle = \sum_{m\neq n} |m\rangle \frac{W_{mn}}{\mathcal{E}_{nm}}, \quad (15.45)$$

$$|\psi_n^{(2)}\rangle = \sum_{m\neq n} |m\rangle \left(\sum_{p\neq n} \frac{W_{mp}W_{pn}}{\mathcal{E}_{np}\mathcal{E}_{nm}} - \frac{W_{nn}W_{mn}}{\mathcal{E}_{nm}^2} \right), \quad (15.46)$$

etc. All higher-order corrections likewise involve powers of matrix elements of W in the numerators, and powers of unperturbed energy differences $\mathcal{E}_{nm}$'s in the denominators. This means that the perturbative expansion, as we have constructed it here, may fail to converge for a state $|n\rangle$ if there is some other state $|m\rangle$ such that $\mathcal{E}_{nm}$ is zero, or small compared to the matrix elements of W. To treat such cases, we will need to use degenerate perturbation theory (Sect. 15.6), or almost-degenerate perturbation theory (Sect. 15.8), respectively.

15.2 Simple Examples of Perturbation Theory

Consider the one-dimensional harmonic oscillator with a linear perturbation:

$$H_0 = \frac{P^2}{2m} + \frac{1}{2}m\omega^2 X^2, \qquad W = -fX. \quad (15.47)$$

Here f is a constant force, which plays the role of the expansion parameter λ in the discussion of Sect. 15.1. Before solving this as a problem in perturbation theory for small f, we recall that it is special in that the stationary states can actually be found exactly. As already noted at the end of Sect. 7.4, this is because the total Hamiltonian can be rewritten, by completing the square, as

$$H = \frac{P^2}{2m} + \frac{1}{2}m\omega^2\Big(X - \frac{f}{m\omega^2}\Big)^2 - \frac{f^2}{2m\omega^2}, \tag{15.48}$$

so that after a shift $X \to X + f/m\omega^2$, the new H will have the same m, ω, and energy eigenvalues, except that the latter will be offset by a constant $-f^2/2m\omega^2$. Thus we have

$$E_n = \hbar\omega(n + 1/2) - f^2/2m\omega^2, \tag{15.49}$$

exactly. Furthermore, the corresponding energy eigenstates must be

$$|\psi_n\rangle = T(f/m\omega^2)\,|n\rangle\,, \tag{15.50}$$

where $T(a)$ is the translation operator of (5.16), and $|n\rangle$ are the stationary states of H_0. We can expand this exact result to linear order in f, using (7.65) for the momentum operator in terms of harmonic oscillator creation and annihilation operators, to get

$$T(f/m\omega^2) = I + \frac{f}{\sqrt{2m\hbar\omega^3}}(a^\dagger - a) + O(f^2), \tag{15.51}$$

so that the eigenstates of H are

$$|\psi_n\rangle = |n\rangle + \frac{f}{\sqrt{2m\hbar\omega^3}}\Big(\sqrt{n+1}\,|n{+}1\rangle - \sqrt{n}\,|n{-}1\rangle\Big) + O(f^2), \tag{15.52}$$

in terms of the eigenstates of H_0.

Now let us solve the same problem using the technology of perturbation theory that we have developed. The unperturbed energy eigenvalues and eigenstates are $\mathcal{E}_n = \hbar\omega(n{+}1/2)$ and $|n\rangle$. The general matrix elements needed for the perturbative expansion are

$$\begin{aligned} W_{n'n} &= \langle n'|W|n\rangle = -f\,\langle n'|X|n\rangle = -f\sqrt{\frac{\hbar}{2m\omega}}\,\langle n'|(a^\dagger + a)|n\rangle \\ &= -f\sqrt{\frac{\hbar}{2m\omega}}\Big(\sqrt{n+1}\,\delta_{n',n+1} + \sqrt{n}\,\delta_{n',n-1}\Big). \end{aligned} \tag{15.53}$$

Now, since this vanishes for $n' = n$, the first-order corrections to the energies are all 0. From (15.42) and (15.43), we get, through second order,

$$E_n^{(1)} = 0, \tag{15.54}$$

$$E_n^{(2)} = \frac{\left|-f\sqrt{\hbar(n+1)/2m\omega}\right|^2}{\mathcal{E}_n - \mathcal{E}_{n+1}} + \frac{\left|-f\sqrt{\hbar n/2m\omega}\right|^2}{\mathcal{E}_n - \mathcal{E}_{n-1}}, \tag{15.55}$$

where only two terms in the sum contributed. Now $\mathcal{E}_n - \mathcal{E}_{n+1} = -\hbar\omega$ and $\mathcal{E}_n - \mathcal{E}_{n-1} = \hbar\omega$, so (15.55) simplifies to

$$E_n^{(2)} = -f^2/2m\omega^2, \tag{15.56}$$

and we have

$$E_n = \hbar\omega(n + 1/2) - f^2/2m\omega^2 + \cdots . \tag{15.57}$$

This agrees with the exact result of (15.49). In fact, this comparison shows that the possible higher-order terms (indicated by the ellipsis here) must actually conspire to vanish. The first-order perturbative correction to the stationary state $|n\rangle$, applying (15.45), also only has two terms in the sum,

$$|\psi_n^{(1)}\rangle = -f\sqrt{\frac{\hbar}{2m\omega}}\left(\frac{\sqrt{n+1}\,|n{+}1\rangle}{-\hbar\omega} + \frac{\sqrt{n}\,|n{-}1\rangle}{\hbar\omega}\right) \tag{15.58}$$

$$= \frac{f}{\sqrt{2m\hbar\omega^3}}\left(\sqrt{n+1}\,|n{+}1\rangle - \sqrt{n}\,|n{-}1\rangle\right). \tag{15.59}$$

Again this agrees with the exact result in (15.50), after the expansion for linear order in f as in (15.52).

As a second example, less amenable to an easy exact calculation, consider the potential

$$V(x) = \begin{cases} V_0 \ (|x| < a/2), \\ 0 \ \ (a/2 < |x| < L/2), \\ \infty \ (|x| > L/2), \end{cases} \tag{15.60}$$

as in Fig. 15.1. This is an infinite square well with a perturbation bump in the middle, so we define $H = H_0 + W$ where, using a hybrid operator/position-representation notation,

$$H_0 = \frac{P^2}{2m} + \begin{cases} 0 & (|x| < L/2), \\ \infty & (|x| > L/2), \end{cases} \tag{15.61}$$

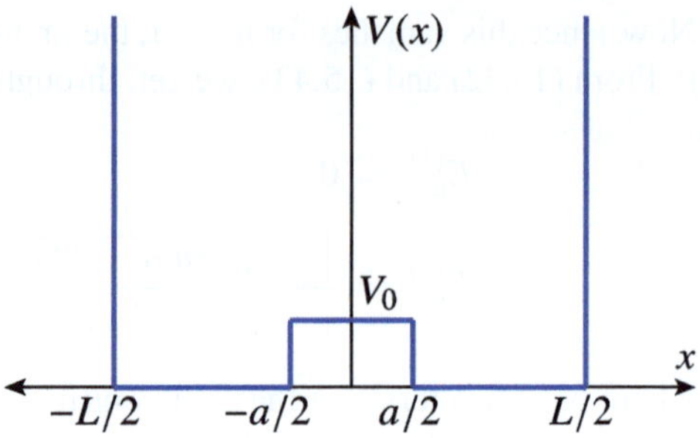

Fig. 15.1 An infinite one-dimensional square well potential with width L, with a symmetric bump of height V_0 and width a to be treated as a perturbation

and the perturbation Hamiltonian is the bump potential

$$W = \begin{cases} V_0 \ (|x| < a/2), \\ 0 \quad (|x| > a/2). \end{cases} \tag{15.62}$$

The infinite square well problem defined by H_0 was solved in Sect. 6.4. The wavefunctions $\langle x|n\rangle = \phi_n(x)$ for the unperturbed H_0 eigenstates $|n\rangle$ were found to be

$$\phi_n(x) = \sqrt{\frac{2}{L}} \cos(n\pi x/L) \qquad (n = 1, 3, 5, \ldots), \tag{15.63}$$

$$\phi_n(x) = \sqrt{\frac{2}{L}} \sin(n\pi x/L) \qquad (n = 2, 4, 6, \ldots), \tag{15.64}$$

with, in both cases, energies

$$\mathcal{E}_n = \frac{\hbar^2\pi^2 n^2}{2mL^2}. \tag{15.65}$$

Applying (15.18) or its equivalent (15.42), the leading correction to the energy eigenvalue for the state n is

$$E_n^{(1)} = V_0 \int_{-a/2}^{a/2} dx\, |\phi_n(x)|^2. \tag{15.66}$$

For the ground state with $n = 1$, the energy eigenvalue to first order in V_0 evaluates to

$$E_1 = \frac{\hbar^2\pi^2}{2mL^2} + V_0 \left(\frac{a}{L} + \frac{1}{\pi} \sin(\pi a/L) \right). \tag{15.67}$$

As a check, note that if $a = L$, then the perturbation simply adds V_0 to the energy eigenvalue, independent of L. As expected, this is just the constant shift in the Hamiltonian.

For the change in the ground-state wavefunction, we apply (15.45). This requires us to find $W_{k1}/\mathcal{E}_{1k}$, which for odd $k \neq 1$ is

$$\frac{\langle k|W|1\rangle}{\mathcal{E}_1 - \mathcal{E}_k} = \frac{2mL^2}{\hbar^2\pi^2(1-k^2)}\frac{2V_0}{L}\int_{-a/2}^{a/2} dx\, \cos\left(\frac{\pi k x}{L}\right)\cos\left(\frac{\pi x}{L}\right) \tag{15.68}$$

$$= \frac{4V_0 mL^2}{\hbar^2\pi^3(1-k^2)}\left[\frac{\sin(\pi a(k-1)/2L)}{k-1} + \frac{\sin(\pi a(k+1)/2L)}{k+1}\right]. \tag{15.69}$$

For even k we have instead

$$\frac{\langle k|W|1\rangle}{\mathcal{E}_1 - \mathcal{E}_k} = 0, \tag{15.70}$$

because of

$$\int_{-a/2}^{a/2} dx\, \sin\left(\frac{\pi k x}{L}\right)\cos\left(\frac{\pi x}{L}\right) = 0. \tag{15.71}$$

This reflects (not sorry about the pun) the parity selection rule of (5.105). Since the perturbation Hamiltonian has even parity $\pi_W = 1$, and $\pi_k = (-1)^{k-1}$ for a state $|k\rangle$, the parity selection rule tells us that $\langle k|W|n\rangle$ vanishes if $k + n$ is odd, and in particular that $\langle k|W|1\rangle = 0$ for even k. It follows that the corrected ground-state wavefunction is

$$\psi_1(x) = \sqrt{\frac{2}{L}}\cos\left(\frac{\pi x}{L}\right) + \psi_1^{(1)}(x), \tag{15.72}$$

with

$$\psi_1^{(1)}(x) = \sum_{k=3,5,7,\ldots} \frac{\langle k|W|1\rangle}{\mathcal{E}_1 - \mathcal{E}_k}\phi_k(x). \tag{15.73}$$

The corrections to the energies and wavefunctions of the other stationary states are left as an exercise.

For a third example, let us estimate the effect of the proton size on the energy levels of the hydrogen atom. In doing so, we will treat the proton (not completely realistically) as a ball of uniform charge density with radius a_p. For all $r \geq a_p$, the electric field is the same as if the proton were a point charge. For $r \leq a_p$, an elementary application of Gauss' law in classical electrostatics shows that the electric field rises linearly with r, so that the potential energy of the electron is

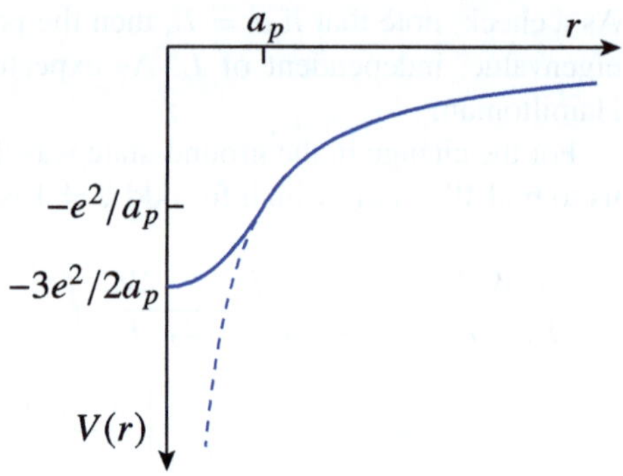

Fig. 15.2 The solid line shows the shape of the potential energy for the electron in the hydrogen atom, with the proton modeled as a sphere of radius a_p with uniform charge density. The dashed line shows the potential with the proton modeled as a point, as in Sect. 11.1

$$V(r) = \begin{cases} -e^2\left(\dfrac{3}{2a_p} - \dfrac{r^2}{2a_p^3}\right), & (r \leq a_p), \\ -\dfrac{e^2}{r}, & (r \geq a_p), \end{cases} \tag{15.74}$$

as shown in Fig. 15.2.

Therefore, compared to our treatment in Sect. 11.1, the perturbation to the Hamiltonian is

$$W = \begin{cases} e^2\left(\dfrac{1}{r} - \dfrac{3}{2a_p} + \dfrac{r^2}{2a_p^3}\right), & (r \leq a_p), \\ 0, & (r \geq a_p), \end{cases} \tag{15.75}$$

in the position representation. So, the first-order correction to the ground-state energy is

$$E^{(1)}_{1,0,0} = \langle 1,0,0|W|1,0,0\rangle \tag{15.76}$$

$$= 4\pi \int_0^{a_p} dr\, r^2 \left(\frac{1}{\sqrt{\pi a_0^3}} e^{-r/a_0}\right)^2 e^2 \left(\frac{1}{r} - \frac{3}{2a_p} + \frac{r^2}{2a_p^3}\right), \tag{15.77}$$

where the 4π factor comes from the angular integration. Within the radial integral, $e^{-2r/a_0} \approx 1$ is an excellent approximation, since $a_p \ll a_0$. The integration then yields

$$E^{(1)}_{1,0,0} = \frac{2e^2 a_p^2}{5a_0^3} = \frac{4a_p^2}{5a_0^2}\,\text{Rydberg} \approx 2.1 \times 10^{-10}\ \text{Rydberg}, \tag{15.78}$$

where we have used $a_p \approx 8.5 \times 10^{-16}$ meters in the last approximation.

In the preceding, we used the small size of the proton compared to the Bohr radius to effectively replace the radial wavefunction by its (very nearly constant) value at $r = 0$, with $|R_{1,0}(r)|^2 \to |R_{1,0}(0)|^2$. For a general n, l, m state, the same procedure will give

$$E^{(1)}_{n,l,m} = |R_{n,l}(0)|^2 \int d\Omega \, |Y_l^m(\theta,\phi)|^2 \int_0^{a_p} dr \, r^2 \, e^2 \left(\frac{1}{r} - \frac{3}{2a_p} + \frac{r^2}{2a_p^3}\right). \quad (15.79)$$

Now, $\int d\Omega |Y_l^m|^2 = 1$, and $R_{n,l}(0) = 0$ unless $l = 0$, which also implies $m = 0$. From (11.41), we know that the hydrogen atom radial wavefunction evaluated at $r = 0$ is $R_{n,l}(0) = \delta_{l0} 2/(na_0)^{3/2}$. So, we find

$$E^{(1)}_{n,l,m} = \delta_{l0} \, \delta_{m0} \, \frac{4a_p^2}{5a_0^2 n^3} \text{ Rydberg}. \quad (15.80)$$

Thus the finite proton radius only affects the energies of the states with no angular momentum, and with decreasing importance for larger principal quantum number n.

The preceding calculation is just an estimate, as the proton is certainly not a ball of uniform charge density. A more sophisticated calculation can be used to define the effective charge radius of the proton, which can then be compared to various experimental determinations of the same quantity. This has recently been the subject of some interest and controversy, because experimental determinations of the proton charge radius inferred from electron scattering seem to be slightly larger than those obtained from measuring the energy levels of muonic hydrogen, in which a muon replaces the electron. The muon is used because its much larger mass gives it a smaller Bohr radius and larger binding energy than the electron. This leads to a larger effect in $E^{(1)}_{n,l,m}$ by a factor $m_\mu^3/m_e^3 \approx 8.84 \times 10^6$; the heavier muon is much more likely to be found inside the proton than an electron, and therefore provides a better probe.

15.3 Helium Atom Ground State from First-Order Perturbation Theory

In this section, we consider the helium atom consisting of two electrons, each with mass m_e and charge $-e$, and a heavy nucleus with charge $Ze = +2e$ at the origin, as pictured in Fig. 15.3. The Hamiltonian is[2]

---

[2] For simplicity, we ignore small effects due to spin interactions and the finite mass of the nucleus. Also, we ignore the fact that the electrons are identical; this issue turns out not to affect the following treatment of the ground state because the electron spins will be oriented oppositely. It does affect the excited states of helium, as we will discuss later, in Sect. 18.3.

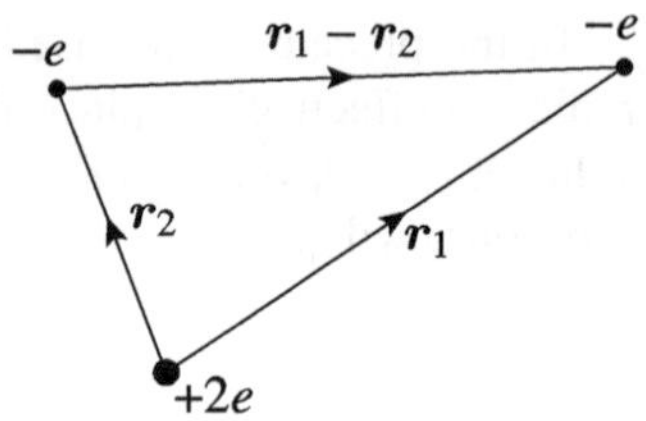

Fig. 15.3 The helium atom consists of a very heavy nucleus of charge $+2e$ at the origin, and two electrons with mass m_e and charge $-e$ at variable positions $\boldsymbol{r}_1$ and $\boldsymbol{r}_2$

$$H = H_1 + H_2 + H_{12}, \tag{15.81}$$

in which, in the position representation,

$$H_1 = -\frac{\hbar^2}{2m_e}\nabla_1^2 - \frac{Ze^2}{r_1}, \qquad H_2 = -\frac{\hbar^2}{2m_e}\nabla_2^2 - \frac{Ze^2}{r_2}, \tag{15.82}$$

$$H_{12} = \frac{e^2}{|\boldsymbol{r}_1 - \boldsymbol{r}_2|}, \tag{15.83}$$

where $Z = 2$, and $\boldsymbol{r}_1$ and $\boldsymbol{r}_2$ describe the position coordinates of the two electrons, and ∇_1^2 and ∇_2^2 are the corresponding Laplacians. If the electrons did not interact with each other, then H_1 and H_2 would just be the Hamiltonians for hydrogen-like atoms with $Z = 2$. The electron-electron interaction term H_{12} is the source of complications, and in this section we treat it as the perturbation W, although it is not particularly small.

The lowest energy eigenstate of the unperturbed Hamiltonian $H_0 = H_1 + H_2$ is simply the tensor product of two hydrogen-like atomic ground states with $(n, l, m) = (1, 0, 0)$,

$$|(1,0,0),(1,0,0)\rangle = |1,0,0\rangle_1 \otimes |1,0,0\rangle_2 . \tag{15.84}$$

The position wavefunction for the unperturbed state is

$$\psi(\boldsymbol{r}_1, \boldsymbol{r}_2) = \langle \boldsymbol{r}_1|1,0,0\rangle_1 \langle \boldsymbol{r}_2|1,0,0\rangle_2 = \psi_{1,0,0}(\boldsymbol{r}_1)\,\psi_{1,0,0}(\boldsymbol{r}_2), \tag{15.85}$$

where, as we found in (11.42),

$$\psi_{1,0,0}(\boldsymbol{r}) = \sqrt{\frac{Z^3}{\pi a_0^3}}\, e^{-Zr/a_0}. \tag{15.86}$$

The corresponding unperturbed total energy eigenvalue for $H_1 + H_2$ is, including the contributions of both electrons,

$$\mathcal{E}_0 = 2\left(-\frac{Z^2 e^2}{2a_0}\right) = -8\,\text{Rydberg} = -108.85\ \text{eV}. \tag{15.87}$$

This is almost 30 eV lower than the experimental value −79.01 eV (the minimum energy needed to completely ionize the helium atom). The reason is that the electron-electron Coulomb repulsive potential H_{12} raises the energy, by an amount that we now estimate.

At first order in perturbation theory, the ground-state energy correction is

$$E^{(1)} = \langle (1,0,0),(1,0,0)|H_{12}|(1,0,0),(1,0,0)\rangle \tag{15.88}$$

$$= \int d^3\boldsymbol{r}_1 \int d^3\boldsymbol{r}_2\, |\psi_{1,0,0}(\boldsymbol{r}_1)|^2 |\psi_{1,0,0}(\boldsymbol{r}_2)|^2 \frac{e^2}{|\boldsymbol{r}_1 - \boldsymbol{r}_2|}. \tag{15.89}$$

This can be written as

$$E^{(1)} - e^2 \left(\frac{Z^3}{\pi a_0^3}\right)^2 I, \tag{15.90}$$

with

$$I = \int_0^\infty dr_1\, r_1^2 \int d\Omega_1 \int_0^\infty dr_2\, r_2^2 \int d\Omega_2\, e^{-2Zr_1/a_0} e^{-2Zr_2/a_0} \frac{1}{|\boldsymbol{r}_1 - \boldsymbol{r}_2|}. \tag{15.91}$$

Our challenge now is to evaluate the integral I. To do so, recall the identity [rewritten from (8.175)]

$$\frac{1}{|\boldsymbol{r}_1 - \boldsymbol{r}_2|} = \sum_{l=0}^{\infty} \frac{4\pi}{2l+1} \frac{r_{\min}^l}{r_{\max}^{l+1}} \sum_{m=-l}^{l} Y_l^m(\theta_2,\phi_2)^* \, Y_l^m(\theta_1,\phi_1), \tag{15.92}$$

in which

$$r_{\max} = \max(r_1, r_2), \qquad r_{\min} = \min(r_1, r_2). \tag{15.93}$$

Using (15.92) in the integral I, we have $\int d\Omega_1\, Y_l^m(\theta_1,\phi_1) = 0$ unless $m = l = 0$, and similarly for the $d\Omega_2$ integral. Since only that one term survives, we can replace

$$\frac{1}{|\boldsymbol{r}_1 - \boldsymbol{r}_2|} \to \frac{4\pi}{r_{\max}} Y_0^0(\theta_2,\phi_2)^* \, Y_0^0(\theta_1,\phi_1) = 1/r_{\max}, \tag{15.94}$$

eliminating all of the angular dependences of the integrand. So, using $\int d\Omega_1 = \int d\Omega_2 = 4\pi$,

$$I = (4\pi)^2 \int_0^\infty dr_1\, r_1^2 \int_0^\infty dr_2\, r_2^2\, e^{-2Zr_1/a_0} e^{-2Zr_2/a_0} \frac{1}{r_{\max}}. \tag{15.95}$$

Separating the dr_2 integral into two parts, according to whether $r_2 < r_1$ or $r_2 > r_1$, now gives

$$I = (4\pi)^2 \int_0^\infty dr_1\, e^{-2Zr_1/a_0} \left(\int_0^{r_1} dr_2\, r_1 r_2^2 e^{-2Zr_2/a_0} + \int_{r_1}^\infty dr_2\, r_1^2 r_2 e^{-2Zr_2/a_0} \right), \tag{15.96}$$

which can be done straightforwardly. The two parts turn out to contribute equally to the total, which is

$$I = \frac{5\pi^2 a_0^5}{8Z^5}. \tag{15.97}$$

Putting this into (15.90), the first-order correction to the ground-state energy is

$$E^{(1)} = \frac{5Ze^2}{8a_0}. \tag{15.98}$$

Combining this with (15.87), and plugging in $Z = 2$, the total ground-state energy for helium estimated from first-order perturbation theory is

$$E = \mathcal{E}_0 + E^{(1)} = (-8 + 5/2) \text{ Rydberg} = -74.83 \text{ eV}. \tag{15.99}$$

Note that the first-order correction is not suppressed by an adjustable small parameter compared to the unperturbed energy; it is $5/16$ as large in magnitude. This might cause worry that perhaps the perturbative expansion is not converging fast enough to be trusted. Nevertheless, our final result differs by only about 5.3% from the experimental result of −79.01 eV. In Sect. 16.4, we will use a non-perturbative technique, the variational method, to do even better.

15.4 Brillouin–Wigner Perturbation Theory

The standard organization of stationary-state perturbation theory summarized in (15.38)–(15.46) has an alternative, called **Brillouin–Wigner perturbation theory** after Léon Brillouin and Wigner, which sometimes has better numerical convergence behavior. To derive it, we start with the eigenvalue problem in the form

$$(H_0 + W - E_n)|\psi_n\rangle = 0, \tag{15.100}$$

using the same notations as in Sect. 15.1, but with $\lambda = 1$ from the start. Now, act with the unperturbed bra $\langle m|$, then use $\langle m|H_0 = \langle m|\mathcal{E}_m$, and rearrange the result to get

$$\langle m|\psi_n\rangle = \frac{\langle m|W|\psi_n\rangle}{E_n - \mathcal{E}_m}. \tag{15.101}$$

Note the simultaneous presence in the denominator of the full energy eigenvalue E_n and the unperturbed energy $\mathcal{E}_m$. From completeness of the unperturbed states $|m\rangle$, we also have

$$|\psi_n\rangle = \sum_m |m\rangle \langle m|\psi_n\rangle = |n\rangle \langle n|\psi_n\rangle + \sum_{m\neq n} |m\rangle \langle m|\psi_n\rangle . \qquad (15.102)$$

As in Sect. 15.1, we choose the normalization condition $\langle n|\psi_n\rangle = 1$, and use (15.101) in (15.102) to obtain

$$|\psi_n\rangle = |n\rangle + \sum_{m\neq n} |m\rangle \frac{\langle m|W|\psi_n\rangle}{E_n - \mathcal{E}_m} . \qquad (15.103)$$

This has a straightforward iterative solution,

$$\begin{aligned} |\psi_n\rangle &= |n\rangle + \sum_{m\neq n} |m\rangle \frac{1}{E_n - \mathcal{E}_m} \langle m|W|n\rangle \\ &+ \sum_{m\neq n}\sum_{p\neq n} |m\rangle \frac{1}{E_n - \mathcal{E}_m} \langle m|W|p\rangle \frac{1}{E_n - \mathcal{E}_p} \langle p|W|n\rangle + \cdots , \end{aligned} \qquad (15.104)$$

as can be checked by plugging it in. To find an equation for the corresponding energy eigenvalues, we now use $E_n = \mathcal{E}_n + \langle n|W|\psi_n\rangle$ from (15.37) to obtain

$$\begin{aligned} E_n &= \mathcal{E}_n + \langle n|W|n\rangle + \sum_{m\neq n} \langle n|W|m\rangle \frac{1}{E_n - \mathcal{E}_m} \langle m|W|n\rangle \\ &+ \sum_{m\neq n}\sum_{p\neq n} \langle n|W|m\rangle \frac{1}{E_n - \mathcal{E}_m} \langle m|W|p\rangle \frac{1}{E_n - \mathcal{E}_p} \langle p|W|n\rangle + \cdots . \end{aligned} \qquad (15.105)$$

Equations (15.104) and (15.105) summarize Brillouin–Wigner perturbation theory. They have a simpler structure than the standard perturbation theory of Sect. 15.1, but have the disadvantage that they are not yet a complete solution, since the full energy E_n appears on both sides of (15.105).

At first order in Brillouin–Wigner perturbation theory, only the first two terms on the right-hand side of (15.105) are included, and the solution for E_n is the same as in standard perturbation theory. At second order or beyond, each E_n is obtained by solving a nonlinear algebraic equation. The results are often more accurate than those found from the standard perturbation theory of Sect. 15.1 at the same order. One way of thinking about this is that, at a given order in perturbation theory, the E_n appearing in the denominators of Brillouin–Wigner perturbation theory contain more complete information than the $\mathcal{E}_n$ appearing in the denominators of standard perturbation theory. Once E_n has been obtained by solving (15.105) truncated at a specific order in the expansion, it can be plugged into (15.104) to obtain the corresponding energy eigenstate.

15.5 Dalgarno–Lewis Method for Simplifying Perturbation Theory

The formulas of stationary-state perturbation theory found in Sect. 15.1 for the state, and for the energy beyond first order, contain sums over all unperturbed states. For example, the second-order correction to the energy contains the sum

$$E_n^{(2)} = \sum_{m \neq n} \frac{|\langle m|W|n\rangle|^2}{\mathcal{E}_n - \mathcal{E}_m}. \tag{15.106}$$

While this is a well-defined formal solution, in practice it is sometimes hard to compute such sums directly. It is especially difficult if some of the unperturbed states form a continuum, in which case the corresponding part of the summation will actually be an integration. Fortunately, there is an elegant method, due to Alexander Dalgarno and John T. Lewis, that allows these sums to be recast into a form that can often be evaluated exactly. Even when an exact result for the summation cannot be obtained, the **Dalgarno–Lewis method** still allows for a systematic approximation, often by numerically solving a differential equation.

Suppose that, given $|n\rangle$, one can find an operator A that satisfies

$$[A, H_0]\,|n\rangle = (W - c)\,|n\rangle\,, \tag{15.107}$$

where H_0 and W are the unperturbed and perturbation parts of the Hamiltonian, respectively, and c is a constant. Actually, c is not arbitrary, because after acting on both sides of (15.107) with $\langle n|$, the left side vanishes, which implies $c = \langle n|W|n\rangle = E_n^{(1)}$. Also, since only the commutator of A appears in (15.107), one can always add a constant multiple of the identity operator to it, to ensure that

$$\langle n|A|n\rangle = 0. \tag{15.108}$$

Note that the required operator A is different for each $|n\rangle$. Now, for every other unperturbed orthobasis state $|m\rangle$, (15.107) implies $\langle m|W|n\rangle = \langle m|[A, H_0]|n\rangle + c\,\langle m|n\rangle$, or, since orthonormality of the H_0 eigenstates eliminates the c term,

$$\langle m|W|n\rangle = (\mathcal{E}_n - \mathcal{E}_m)\,\langle m|A|n\rangle\,, \qquad (m \neq n). \tag{15.109}$$

The Dalgarno–Lewis idea is to use (15.109) to cancel the energy denominators in the results of perturbation theory, by expressing matrix elements of W in terms of matrix elements of A. So, if we can find A, the expressions in perturbation theory can then be simplified using the completeness relation for the unperturbed states.

For example, from (15.23), the first-order correction to the state is

$$|\psi_n^{(1)}\rangle = \sum_{m\neq n} |m\rangle \frac{\langle m|W|n\rangle}{\mathcal{E}_n - \mathcal{E}_m} = \sum_{m\neq n} |m\rangle\langle m|A|n\rangle = \sum_m |m\rangle\,\langle m|A|n\rangle. \quad (15.110)$$

The last equality relies on the fact that the additional term with $m = n$ vanishes, because of (15.108). Now, from the completeness relation $\sum_m |m\rangle\langle m| = I$, we conclude that the first-order correction to the state $|n\rangle$ is simply

$$|\psi_n^{(1)}\rangle = A\,|n\rangle\,. \quad (15.111)$$

In words, the desired operator A maps the unperturbed state to its first-order correction. Similarly, the second-order correction to the energy is, from (15.106),

$$E_n^{(2)} = \sum_{m\neq n} \langle n|W|m\rangle\,\langle m|A|n\rangle = \sum_m \langle n|W|m\rangle\,\langle m|A|n\rangle = \langle n|WA|n\rangle\,, \quad (15.112)$$

where we have again used (15.108) to account for the vanishing of the additional term with $m = n$, and then used completeness. As a check, this can be rewritten as

$$E_n^{(2)} = \langle n|W|\psi_n^{(1)}\rangle\,, \quad (15.113)$$

in accord with (15.26) with $k = 2$. In a similar way, you can show that even the third-order correction to the energy can be written as

$$E_n^{(3)} = \langle\psi_n^{(1)}|W|\psi_n^{(1)}\rangle - \langle\psi_n^{(1)}|\psi_n^{(1)}\rangle\langle n|W|n\rangle\,, \quad (15.114)$$

without energy-difference denominators.

Of course, this wonderful simplification does not come for free. To make use of the Dalgarno–Lewis method, one must either find the operator A, or equivalently (and more directly) solve for the ket $A|n\rangle = |\psi_n^{(1)}\rangle$. Rewriting (15.107), the key equation to be solved is

$$(\mathcal{E}_n - H_0)\,|\psi_n^{(1)}\rangle = W\,|n\rangle - |n\rangle\,\langle n|W|n\rangle\,. \quad (15.115)$$

In favorable cases, this can be solved exactly for $|\psi_n^{(1)}\rangle$ or the corresponding wavefunction. We will see an example of this in Sect. 15.7 when we treat the ground state of the hydrogen atom in an electric field (the quadratic Stark effect). Even when this is not feasible, it is often possible to cast (15.115) in the position representation as a differential equation that can be solved approximately, perhaps numerically. This is often faster and simpler than trying to do the sums over all other states $|m\rangle$ with energy denominators $\mathcal{E}_n - \mathcal{E}_m$, especially if there is an infinite number of such states. The great advantage of the Dalgarno–Lewis method is that (15.115) only involves the unperturbed state $|n\rangle$ that one is considering, rather than infinite sums over other states. Once (15.115) has been solved to find $|\psi_n^{(1)}\rangle$, the second and third-order corrections to the energy follow immediately from (15.113) and (15.114), respectively.

15.6 Degenerate Perturbation Theory

As noted in Sect. 15.1, a special problem arises when unperturbed states $|n\rangle$ have degeneracies, because then the energy differences $\mathcal{E}_{nm} = \mathcal{E}_n - \mathcal{E}_m$ appearing in the denominators of (15.43)–(15.46) will vanish. In this section, we show how to deal with this difficulty.

The bad-denominator problem is related to the fact that perturbation theory requires a smooth one-to-one correspondence between the unperturbed states and the perturbed states as the expansion parameter λ varies. Specifically, this is what allowed us to demand that the unperturbed state $|n\rangle$ and the corrections to it $|\psi_n^{(k)}\rangle$ are orthogonal, as we did in (15.11). This in turn was used to exclude the terms in the sums that would otherwise have had vanishing energy-difference denominators. In the degenerate case, the essential problem is that for a given perturbed state $|\psi_n\rangle$ there is an ambiguity in deciding which of the degenerate unperturbed states $|n\rangle$ corresponds to it and therefore can be excluded from the sums, since arbitrary linear combinations of them have the same H_0 eigenvalue $\mathcal{E}_n$. This is illustrated in Fig. 15.4, which compares the situations for non-degenerate and degenerate unperturbed states.

To resolve the problem, we need to answer the following question: if a group of unperturbed states $\{|n\rangle\}$ has the same unperturbed energy $\mathcal{E}_n$, then which linear combinations of them are the $|\psi_n^{(0)}\rangle$? The problem of avoiding bad (vanishing)

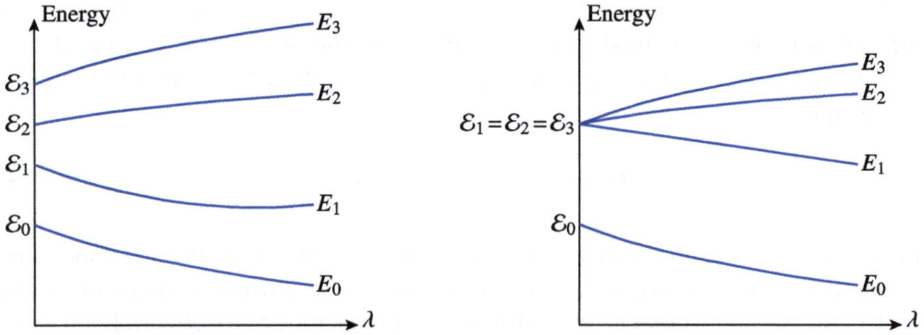

Fig. 15.4 Perturbed energies as a function of the expansion parameter λ, for non-degenerate (left) and degenerate (right) perturbation theory. In degenerate perturbation theory, for each of the perturbed states $|\psi_1\rangle$, $|\psi_2\rangle$, $|\psi_3\rangle$ there is an ambiguity in choosing the corresponding unperturbed states $|1\rangle$, $|2\rangle$, $|3\rangle$, since arbitrary linear combinations of them have the same H_0 eigenvalue $\mathcal{E}_1 = \mathcal{E}_2 = \mathcal{E}_3$. This ambiguity is resolved (at least in part) by choosing the unperturbed states to diagonalize the perturbation Hamiltonian W within each subspace of degenerate $\mathcal{E}_n$

denominators in perturbation theory is the same as the problem of choosing the correct linear combinations of unperturbed energy states.

Consider what happens when we return to (15.22), rewritten here for convenience,

$$(\mathcal{E}_n - \mathcal{E}_m)\,\langle m|\psi_n^{(1)}\rangle = \langle m|W|n\rangle \qquad (m \neq n), \tag{15.116}$$

but now allowing for the possibility that some unperturbed states have degenerate unperturbed energies. Whenever $\mathcal{E}_m \neq \mathcal{E}_n$, we can still solve

$$\langle m|\psi_n^{(1)}\rangle = \frac{W_{mn}}{\mathcal{E}_{nm}}, \qquad (\mathcal{E}_m \neq \mathcal{E}_n). \tag{15.117}$$

However, for $\mathcal{E}_m = \mathcal{E}_n$, we instead learn only that $\langle m|W|n\rangle$ must vanish for $m \neq n$, in order to have a consistent perturbative solution. It is therefore required that the matrix elements of W form a diagonal matrix when restricted to each subspace with degenerate unperturbed energies,

$$\langle m|W|n\rangle = \delta_{nm} W_{nn} \qquad (\text{for } \mathcal{E}_m = \mathcal{E}_n). \tag{15.118}$$

Fortunately, this is always possible to arrange by a suitable choice of the orthobasis of stationary states $|n\rangle$. This is assured by Theorem 2.15, since W is Hermitian.

Equation (15.118) is a necessary (but not always sufficient, as we will see) requirement on the choice of unperturbed states. The point is that by choosing a "good" basis of unperturbed states in this way, the perturbation Hamiltonian matrix W_{mn} does not connect pairs of distinct unperturbed states for which the energy-difference denominators would vanish.

In practice, one must solve the eigenvalue problem for $g_{\mathcal{E}_n} \times g_{\mathcal{E}_n}$ matrices, formed by the matrix elements of the operator W restricted to each $g_{\mathcal{E}_n}$-dimensional degenerate subspace. Once we choose the states $|n\rangle$ as the resulting orthonormal eigenvectors, then it follows from (15.18) that the eigenvalues are the corresponding first-order corrections to the energies,

$$E_n^{(1)} = W_{nn}, \tag{15.119}$$

the same result as for the non-degenerate case. Note that it is only necessary to diagonalize W within the subspaces for degenerate $\mathcal{E}_n$. In particular, for each unperturbed state $|n\rangle$ that is non-degenerate, there is nothing to do; W restricted to that subspace is simply a 1×1 matrix, so $|n\rangle$ is automatically an eigenvector, with eigenvalue $W_{nn} = \langle n|W|n\rangle$.

Often, only the first-order energy corrections are required; then (15.119) is sufficient provided that the unperturbed energy basis has been chosen to satisfy (15.118), and in that case you can skip the complications of the next few paragraphs, from here to (15.130).

If one needs the first-order corrections to the state ket $|\psi_n^{(1)}\rangle$ and the second-order corrections to the energies $E_n^{(2)}$, then one must confront the fact that (15.116) provides no information about $\langle m|\psi_n^{(1)}\rangle$ for $\mathcal{E}_m = \mathcal{E}_n$. By completeness,

$$|\psi_n^{(1)}\rangle = \sum_m{}' |m\rangle \frac{W_{mn}}{\mathcal{E}_{nm}} + \sum_m{}'' |m\rangle\langle m|\psi_n^{(1)}\rangle, \tag{15.120}$$

where we have introduced the notations, always for fixed n = the state being perturbed,

$$\sum_m{}' = \text{sum over all } m \text{ such that } \mathcal{E}_m \neq \mathcal{E}_n, \tag{15.121}$$

$$\sum_m{}'' = \text{sum over all } m \text{ such that } m \neq n \text{ and } \mathcal{E}_m = \mathcal{E}_n, \tag{15.122}$$

and used (15.117) in the first summation. To solve for the remaining matrix elements $\langle m|\psi_n^{(1)}\rangle$ appearing in the last summation, one must use the equations governing the perturbative expansion at order λ^2, and possibly beyond.

To see how this goes, consider (15.28) with $k = 2$ and $\mathcal{E}_m = \mathcal{E}_n$, which yields

$$E_n^{(1)} \langle m|\psi_n^{(1)}\rangle = \langle m|W|\psi_n^{(1)}\rangle, \qquad (\text{for } m \neq n \text{ and } \mathcal{E}_m = \mathcal{E}_n). \tag{15.123}$$

We now define an operator $W^{(2)}$ by specifying its matrix elements

$$W_{mn}^{(2)} \equiv \sum_p{}' \frac{W_{mp}W_{pn}}{\mathcal{E}_{np}}. \tag{15.124}$$

Then, using $E_n^{(1)} = W_{nn}$, and (15.120) with m replaced by p, (15.123) becomes

$$W_{nn} \langle m|\psi_n^{(1)}\rangle = W_{mn}^{(2)} + \sum_p{}'' W_{mp} \langle p|\psi_n^{(1)}\rangle, \quad (m \neq n \text{ and } \mathcal{E}_m = \mathcal{E}_n). \tag{15.125}$$

In the last summation, we know that $W_{mp} = \delta_{mp}W_{mm}$, because of (15.118). Therefore, only the single term with $p = m$ in that sum survives, and (15.125) reduces to

$$(W_{nn} - W_{mm}) \langle m|\psi_n^{(1)}\rangle = W_{mn}^{(2)}, \quad (m \neq n \text{ and } \mathcal{E}_m = \mathcal{E}_n). \tag{15.126}$$

Now, for states that are non-degenerate at first order, so that $W_{mm} \neq W_{nn}$, we obtain

$$\langle m|\psi_n^{(1)}\rangle = \frac{W_{mn}^{(2)}}{W_{nn} - W_{mm}} \qquad (\mathcal{E}_m = \mathcal{E}_n \text{ and } W_{mm} \neq W_{nn}), \tag{15.127}$$

but in the opposite case we learn only that

$$W_{mn}^{(2)} = 0, \qquad (m \neq n \text{ and } \mathcal{E}_m = \mathcal{E}_n \text{ and } W_{mm} = W_{nn}). \tag{15.128}$$

This is a second condition that must be imposed on the choice of the unperturbed states, in addition to (15.118).

To summarize the preceding paragraph, we have found that we must choose the unperturbed kets $|n\rangle$ so that on subspaces where both H_0 and W are proportional to the identity, the matrix $W^{(2)}$ is also diagonal. Fortunately, this is always possible, because $W^{(2)}$ is a Hermitian matrix when restricted[3] to this subspace. Now, the second-order corrections to the energy eigenvalues are obtained from (15.26), and are simply

$$E_n^{(2)} = W_{nn}^{(2)} = \sum_m{}' \frac{|W_{mn}|^2}{\mathcal{E}_{nm}}, \tag{15.129}$$

where it is crucial that the unperturbed states satisfy the conditions given in both (15.118) and (15.128), and there is no vanishing denominator problem because of the fact that the sum excludes all terms with $\mathcal{E}_m = \mathcal{E}_n$.

There may still be some ambiguity in the matrix elements $\langle m|\psi_n^{(1)}\rangle$, if for some $m \neq n$ one has $\mathcal{E}_m = \mathcal{E}_n$ and $W_{mm} = W_{nn}$ and $W_{mm}^{(2)} = W_{nn}^{(2)}$, so that the degeneracy is not completely lifted even at second order. In that case, one must use the information from third order, that is, $k = 3$ in (15.16). One then finds a third condition on the choice of unperturbed states $|n\rangle$, that the operator $W^{(3)}$ defined by its matrix elements

$$W_{mn}^{(3)} \equiv \sum_q{}' \sum_p{}' \frac{W_{mq} W_{qp} W_{pn}}{\mathcal{E}_{nq}\mathcal{E}_{np}} - W_{nn} \sum_p{}' \frac{W_{mp} W_{pn}}{\mathcal{E}_{np}^2} \tag{15.130}$$

is diagonal on each subspace of states where the operators H_0, W, and $W^{(2)}$ are proportional to the identity. If the degeneracy of unperturbed states is still not removed at third order (that is, $W^{(3)}$ has degenerate eigenvalues when restricted to that subspace of states), one must continue to $k = 4$ in (15.16), etc. These considerations from $k \geq 3$ affect the determination of $|\psi_n^{(1)}\rangle$ (through $\langle m|\psi_n^{(1)}\rangle$) even though they seemingly involve higher orders in perturbation theory, but they do not affect the determination of $E_n^{(2)}$ given by (15.129).

Practical applications of degenerate perturbation theory often only need $E_n^{(1)}$ as given in (15.119), and therefore one only needs to worry about choosing the

[3] The operator $W^{(2)}$ is not Hermitian on the whole state space, because n and m are not treated symmetrically in the definition (15.124). However, it is Hermitian within the subspace of matrix elements with $\mathcal{E}_m = \mathcal{E}_n$, for which $W_{mn}^{(2)*} = W_{nm}^{(2)}$.

unperturbed states to satisfy (15.118). Examples will appear in Sects. 15.7 and 15.8 and in Chapter 17.

To conclude this section, we exploit first-order perturbation theory to infer the exact expectation value $\langle 1/R^2 \rangle$ for the states $|n, l, m\rangle$ of the hydrogen atom, fulfilling a promise made in Sect. 11.1. Consider the Hamiltonian $H_\lambda = H_0 + \lambda/R^2$, where $H_0 = P^2/2m_e - e^2/R$ is the unperturbed hydrogen atom Hamiltonian, and λ is a small parameter. In this case, the operator $1/R^2$ commutes with the angular momentum operators L^2 and L_z, so it is diagonal in the orthobasis of states $|n, l, m\rangle$ for fixed n and different l, m. Therefore $|n, l, m\rangle$ is a good basis for degenerate first-order perturbation theory in λ, and the first-order correction to the energy eigenvalue for each state is simply proportional to the expectation value that we seek,

$$E^{(1)}_{n,l,m} = \lambda \langle 1/R^2 \rangle_{n,l,m}. \tag{15.131}$$

Meanwhile, the exact energy eigenvalues $E_{\lambda,k,l,m}$ of H_λ are the solutions to the radial wavefunction eigenvalue problem

$$\left[-\frac{\hbar^2}{2m_e} \left(\frac{d^2}{dr^2} + \frac{2}{r}\frac{d}{dr} - \frac{l(l+1)}{r^2} \right) - \frac{e^2}{r} + \frac{\lambda}{r^2} \right] R_{k,l} = E_{\lambda,k,l,m} R_{k,l}, \tag{15.132}$$

where the integer k is the radial quantum number (the number of zeros of $R_{k,l}(r)$ for $0 < r < \infty$). Now we can define a quantity l' by

$$\frac{\hbar^2}{2m_e} l'(l'+1) = \frac{\hbar^2}{2m_e} l(l+1) + \lambda, \tag{15.133}$$

or, equivalently,

$$l' = \sqrt{(l+1/2)^2 + 2m_e\lambda/\hbar^2} - 1/2. \tag{15.134}$$

Note that l' need not be an integer, unlike l. Nevertheless, the analysis of the radial wavefunction eigenvalue problem that led to (11.16) goes through, but with l replaced everywhere by l', and n replaced by $k + l' + 1$. In particular, it is crucial that k is a non-negative integer, in order that the wavefunction is normalizable. The result is

$$E_{\lambda,k,l,m} = -\frac{e^2}{2a_0(k + l' + 1)^2}. \tag{15.135}$$

Expanding in small λ, for fixed integers k and l,

$$E_\lambda = E_{\lambda=0} + \lambda \frac{dE_\lambda}{d\lambda}\bigg|_{\lambda=0} + \cdots, \tag{15.136}$$

and by comparing to (15.131), we find

$$\langle 1/R^2\rangle_{n,l,m} = \frac{d}{d\lambda}\left(\frac{-e^2}{2a_0(k+l'+1)^2}\right)\bigg|_{\lambda=0} = \frac{e^2}{a_0(k+l+1)^3}\frac{dl'}{d\lambda}\bigg|_{\lambda=0} \tag{15.137}$$

From (15.134), we have

$$\frac{dl'}{d\lambda}\bigg|_{\lambda=0} = \frac{m_e}{\hbar^2(l+1/2)}. \tag{15.138}$$

Plugging (15.138) into (15.137), and using $n = k + l + 1$, we obtain

$$\langle 1/R^2\rangle_{n,l,m} = \frac{m_e e^2}{\hbar^2 a_0 n^3(l+1/2)} = \frac{1}{a_0^2 n^3(l+1/2)}. \tag{15.139}$$

Equation (11.67) follows by restoring the general nuclear charge Z, using the rule $a_0 \to a_0/Z$.

The same sort of perturbation theory trick can be used to check the result we found for $\langle 1/R\rangle_{n,l,m}$ in (11.64). Consider $H_Z = P^2/2m_e - Ze^2/R$, the Hamiltonian for general Z. Rewrite this as $H_Z = H_0 + W$, where $H_0 = P^2/2m_e - e^2/R$ is the $Z = 1$ Hamiltonian, and $W = (1-Z)e^2/R$ is treated as a perturbation by taking Z to be close to 1. Then, for an eigenstate of H_0 with principal quantum number n, the first-order energy correction is

$$E_n^{(1)} = (1-Z)e^2\langle 1/R\rangle_{n,l,m}, \tag{15.140}$$

which is proportional to the expectation value that we seek. (Again we are making use of the fact that the basis $|n,l,m\rangle$ is a good one for first-order degenerate perturbation theory, in the sense that the perturbation $1/R$ is diagonal for fixed n and different l, m.) On the other hand, the exact energies were found in (11.16) to be

$$E_n = -\frac{Z^2e^2}{2a_0n^2} = -\frac{e^2}{2a_0n^2}\left[1 + 2(Z-1) + (Z-1)^2\right]. \tag{15.141}$$

Comparing the terms linear in small $Z - 1$ in (15.140) and (15.141), we must have

$$\langle 1/R\rangle_{n,l,m} = \frac{1}{a_0n^2}. \tag{15.142}$$

This confirms (11.64) with $Z = 1$.

15.7 Hydrogen Atom in Electric Field (Stark Effect)

Consider a hydrogen atom in a static external electric field $\boldsymbol{E} = E\hat{z}$. The corresponding electrostatic potential seen by the electron is $\Phi(\boldsymbol{R}) = -EZ$, giving a potential energy contribution to the Hamiltonian,

$$W = -e\Phi = eEZ, \tag{15.143}$$

which we will treat as a perturbation. The calculation of shifts in the atomic energy levels, known as the Stark effect, provides a nice example of both non-degenerate second-order and degenerate first-order perturbation theory.

For the ground state $|n, l, m\rangle = |1, 0, 0\rangle$, the first-order correction to the energy vanishes,

$$E^{(1)}_{1,0,0} = \langle 1, 0, 0|W|1, 0, 0\rangle = eE\langle 1, 0, 0|Z|1, 0, 0\rangle = 0. \tag{15.144}$$

This can be understood as a consequence of the parity selection rule (5.105). More generally, the expectation value of Z in any parity eigenstate is always 0, because $\Pi Z \Pi = -Z$, so that $\pi_Z = -1$, giving $\pi_Z \pi^2_{n,l,m} = -1$. Another way to see that the matrix element vanishes is through the dipole selection rule (13.54).

Since the first-order correction to the ground-state energy is 0, we turn to the second-order correction. Applying the general formula of (15.43), and remembering that the unperturbed energies do not depend on l and m, we have

$$E^{(2)}_{1,0,0} = e^2E^2\left(\sum_{n=2}^{\infty}\sum_{l=0}^{n-1}\sum_{m=-l}^{l} \frac{|\langle n, l, m|Z|1, 0, 0\rangle|^2}{\mathcal{E}_1 - \mathcal{E}_n} + \int_0^{\infty} d\mathcal{E} \sum_{l=0}^{\infty}\sum_{m=-l}^{l} \frac{|\langle \mathcal{E}, l, m|Z|1, 0, 0\rangle|^2}{\mathcal{E}_1 - \mathcal{E}}\right). \tag{15.145}$$

This includes integration over the unbound states $|\mathcal{E}, l, m\rangle$ with continuous unperturbed energies $\mathcal{E} > 0$, as well as summation over the bound states $|n, l, m\rangle$ with discrete unperturbed energies $\mathcal{E}_n = -e^2/2a_0n^2$, excluding $n = 1$. Equation (15.145) simplifies considerably if we use the dipole selection rules of (13.54) and (13.55), which inform us that only the $l = 1$ and $m = 0$ matrix elements are nonzero. Therefore, it can be rewritten as

$$E^{(2)}_{1,0,0} = -e^2E^2\left(\sum_{n=2}^{\infty} \frac{|\langle n, 1, 0|Z|1, 0, 0\rangle|^2}{\mathcal{E}_n - \mathcal{E}_1} + \int_0^{\infty} d\mathcal{E}\, \frac{|\langle \mathcal{E}, 1, 0|Z|1, 0, 0\rangle|^2}{\mathcal{E} - \mathcal{E}_1}\right), \tag{15.146}$$

where the overall minus sign appears because the denominators have flipped sign to make them positive. Since both the sum and the integral are clearly positive,

(15.146) is negative; the presence of the electric field lowers the hydrogen atom ground-state energy.

The evaluation of (15.146) is not trivial, so we will study it using a series of tricks with instructive value. First, we will find a simple lower bound on its magnitude, then a simple upper bound, and then attempt a numerical approximation by including only the bound-state sum contribution. Finally, we will obtain the exact value of $E^{(2)}_{1,0,0}$ by exploiting the power of the Dalgarno–Lewis method of Sect. 15.5.

To obtain the lower bound, note that the magnitude of (15.146) is certainly greater than that of the $n = 2$ term alone, since all of the contributions have the same sign. So, using $\mathcal{E}_2 - \mathcal{E}_1 = 3e^2/8a_0$,

$$|E^{(2)}_{1,0,0}| > (n = 2 \text{ term only}) = \frac{8}{3}E^2 a_0 |\langle 2, 1, 0|Z|1, 0, 0\rangle|^2. \quad (15.147)$$

The relevant matrix element is

$$\begin{aligned}\langle 2, 1, 0|Z|1, 0, 0\rangle &= \int_0^{2\pi} d\phi \int_{-1}^{1} d(\cos\theta) \int_0^\infty dr\, r^2\, (R_{21}Y_1^0)^*\, r\cos\theta\, (R_{10}Y_0^0) \\ &= \frac{128\sqrt{2}}{243}a_0. \end{aligned} \quad (15.148)$$

Therefore, we have

$$|E^{(2)}_{1,0,0}| > \frac{2^{18}}{3^{11}}E^2 a_0^3 \approx 1.4798 E^2 a_0^3 \quad (15.149)$$

as a lower bound on the energy shift of the ground state due to the external electric field.

We also obtain a bound from the other direction, by noting that the denominators in (15.146) with $n \neq 2$ are all larger than they would be if each instance of $\mathcal{E}_n$ or $\mathcal{E}$ were replaced by $\mathcal{E}_2$. So,

$$|E^{(2)}_{1,0,0}| < \frac{8}{3}E^2 a_0 \left(\sum_{n=2}^{\infty} |\langle n, 1, 0|Z|1, 0, 0\rangle|^2 + \int_0^\infty d\mathcal{E} |\langle \mathcal{E}, 1, 0|Z|1, 0, 0\rangle|^2 \right). \quad (15.150)$$

With the annoying energy-difference denominators gone, we can evaluate the whole quantity in parentheses using the completeness identity of (11.101), remembering that the missing $l \neq 1$ and $m \neq 0$ terms vanish by the dipole selection rules. The result is

$$|E^{(2)}_{1,0,0}| < \frac{8}{3}E^2 a_0 \langle 1, 0, 0|Z^2|1, 0, 0\rangle. \quad (15.151)$$

Another trick: by rotational symmetry of the ground state, we can replace Z^2 in the expectation value by the average of X^2, Y^2, and Z^2, which is easier to compute. So

$$\begin{aligned}\langle 1,0,0|Z^2|1,0,0\rangle &= \frac{1}{3}\langle 1,0,0|(X^2+Y^2+Z^2)|1,0,0\rangle = \frac{1}{3}\langle 1,0,0|R^2|1,0,0\rangle \\ &= a_0^2,\end{aligned} \tag{15.152}$$

where, at the end, we have used (11.53) with $p = 2$. Thus we have

$$|E^{(2)}_{1,0,0}| < \frac{8}{3}E^2 a_0^3 \approx 2.6667 E^2 a_0^3 \tag{15.153}$$

as an upper bound on the energy shift to complement the lower bound in (15.149).

As an attempt to do better than the bounds in (15.149) and (15.153), one can add up the exact contributions from all bound states $n = 2, 3, 4, \ldots$ with $\mathcal{E}_n = -e^2/2a_0 n^2$ in (15.146). It is possible, but quite nontrivial, to show that

$$e^2 \frac{|\langle n,1,0|Z|1,0,0\rangle|^2}{\mathcal{E}_n - \mathcal{E}_1} = \frac{(2n)^9 (n-1)^{2n-6}}{3(n+1)^{2n+6}} a_0^3. \tag{15.154}$$

The sum over all integers $n \geq 2$ in (15.146) is then found to converge to

$$E^{(2),\,\text{partial}}_{1,0,0} \approx -1.8316 E^2 a_0^3, \tag{15.155}$$

which is indeed between the two bounds that we have already obtained. However, this turns out to be numerically far from the true answer; as we are about to discover, the integral over unbound states contributes significantly.

Fortunately, one can obtain the correct result while avoiding having to perform directly the difficult sum plus integration, by instead using the Dalgarno–Lewis method described in Sect. 15.5. To do this, we must first use (15.115) to solve for the first-order correction to the ground-state wavefunction, $\psi^{(1)}_{1,0,0}(\boldsymbol{r})$. In the position representation, this equation is

$$(\mathcal{E}_1 - H_0)\psi^{(1)}_{1,0,0} = eE\, r\cos\theta\, \psi^{(0)}_{1,0,0}, \tag{15.156}$$

where $\mathcal{E}_1 = -e^2/2a_0$ and $\psi^{(0)}_{1,0,0} = e^{-r/a_0}/\sqrt{\pi a_0^3}$ from the results of Sect. 11.1, and we have used the fact that $W = eEZ$ has vanishing expectation value in the unperturbed ground state, which simplifies the right side. The unperturbed Hamiltonian operator is

$$
\begin{aligned}
H_0 &= -\frac{\hbar^2\nabla^2}{2m_e} - \frac{e^2}{r} \\
&= -\frac{a_0 e^2}{2}\left[\frac{\partial^2}{\partial r^2} + \frac{2}{r}\frac{\partial}{\partial r} + \frac{1}{r^2 \sin\theta}\frac{\partial}{\partial\theta}\left(\sin\theta\frac{\partial}{\partial\theta}\right)\right] - \frac{e^2}{r},
\end{aligned}
\tag{15.157}
$$

where, in the second equality, we have used the fact that there is no ϕ dependence because of the symmetry of the problem about the z-axis, and $\hbar^2/2m_e = a_0 e^2/2$. Given the form of (15.156), it is reasonable to try for a solution of the form

$$
\psi^{(1)}_{1,0,0} = f(r)\cos\theta\, \psi^{(0)}_{1,0,0}, \tag{15.158}
$$

where $f(r)$ is to be determined. With this guess, the differential equation (15.156) becomes

$$
\frac{1}{2}f'' + \left(\frac{1}{r} - \frac{1}{a_0}\right)f' - \frac{1}{r^2}f = \frac{E}{e a_0}r, \tag{15.159}
$$

in which we are delighted to see that the θ dependence has disappeared. Since the inhomogeneous part is linear in r, we are inspired to make the further guess that f is a quadratic polynomial in r, and are rewarded with the simple solution

$$
f = -\frac{E}{e}(a_0 + r/2)r. \tag{15.160}
$$

[This corresponds to the Dalgarno–Lewis operator $A = -\frac{E}{e}(a_0 + R/2)Z$ as the solution to (15.107) and (15.108), although we do not directly need it.]

Having successfully found $\psi^{(1)}_{1,0,0}$, the rest is relatively easy. Using (15.113) in the position representation, we get

$$
E^{(2)}_{1,0,0} = 2\pi \int_{-1}^{1} d(\cos\theta) \int_0^\infty dr\, r^2 \left(eEr\cos\theta\right)\left(-\frac{E}{e}(a_0 + r/2)r\cos\theta\right)|\psi^{(0)}_{1,0,0}|^2, \tag{15.161}
$$

which evaluates finally to the correct, and surprisingly simple, result

$$
E^{(2)}_{1,0,0} = -\frac{9}{4}a_0^3 E^2 = -2.25 a_0^3 E^2. \tag{15.162}
$$

Because the leading correction to the ground-state energy comes from second order in perturbation theory, and is therefore quadratic in the applied electric field, this is called the **quadratic Stark effect**.

Let us now see how to connect (15.162) to an experimental measurement. In general, for a dipole moment $\boldsymbol{p}$ induced by an external electric field $\boldsymbol{E}$, the polarizability α is defined by

$$\boldsymbol{p} = \alpha \boldsymbol{E}. \tag{15.163}$$

As the electric field is increased, the change in the energy of the dipole is

$$dU = -\boldsymbol{p} \cdot d\boldsymbol{E} = -\alpha\, E\, dE, \tag{15.164}$$

so that the total energy of the induced dipole is

$$U = -\frac{1}{2}\alpha E^2. \tag{15.165}$$

Comparing to (15.162), we have $-\frac{9}{4}a_0^3 E^2 = -\frac{1}{2}\alpha E^2$, or

$$\alpha = \frac{9}{2}a_0^3 = 6.67 \times 10^{-31}\, m^3, \tag{15.166}$$

which agrees well with the experimental value for the polarizability of atomic hydrogen in the ground state.

Let us now turn to the Stark effect for the first excited ($n = 2$) stationary states of the hydrogen atom. The unperturbed states are

$$|n, l, m\rangle = |2, 0, 0\rangle\,,\ |2, 1, 0\rangle\,,\ |2, 1, 1\rangle\,,\ \text{and}\ |2, 1, -1\rangle\,, \tag{15.167}$$

which all have the same unperturbed energy $\mathcal{E}_2 = -e^2/8a_0$. We therefore need to apply the method of degenerate perturbation theory as outlined in Sect. 15.6. To start, we must choose the unperturbed orthobasis of $n = 2$ states as the eigenkets of the 4×4 matrix

$$W_{(l',m'),\,(l,m)} = \langle 2, l', m'|eEZ|2, l, m\rangle. \tag{15.168}$$

Using the dipole selection rule of (13.54) and (13.55), the only nonzero entries of this matrix come from

$$\langle 2, 1, 0|Z|2, 0, 0\rangle = \langle 2, 0, 0|Z|2, 1, 0\rangle, \tag{15.169}$$

which can be evaluated as

$$\begin{aligned}\langle 2, 1, 0|Z|2, 0, 0\rangle &= 2\pi \int_{-1}^{1} d(\cos\theta) \int_0^\infty dr\, r^2 \left(R_{21}Y_1^0\right)^* \left(r\cos\theta\right) \left(R_{20}Y_0^0\right) \\ &= \frac{1}{16}\int_{-1}^{1} d(\cos\theta)\, \cos^2\theta \int_0^\infty dr \left(\frac{r}{a_0}\right)^4 (2 - r/a_0)\, e^{-r/a_0} \\ &= -3a_0. \end{aligned} \tag{15.170}$$

So, in the orthobasis of (15.167), the perturbation Hamiltonian has the matrix representation

$$W \leftrightarrow -3a_0eE\begin{pmatrix}0&1&0&0\\1&0&0&0\\0&0&0&0\\0&0&0&0\end{pmatrix}, \tag{15.171}$$

which has eigenvalues $-3a_0eE$, $+3a_0eE$, 0, and 0. Thus, the energy eigenvalues to first order in the perturbing electric field and the corresponding unperturbed energy eigenstates are

$$\underline{\text{Energies}} \qquad\qquad \underline{\text{States}} \tag{15.172}$$

$$-\frac{e^2}{8a_0} - 3a_0eE \qquad \frac{1}{\sqrt{2}}\left(|2,0,0\rangle + |2,1,0\rangle\right), \tag{15.173}$$

$$-\frac{e^2}{8a_0} + 3a_0eE \qquad \frac{1}{\sqrt{2}}\left(|2,0,0\rangle - |2,1,0\rangle\right), \tag{15.174}$$

$$-\frac{e^2}{8a_0} \qquad |2,1,1\rangle, \qquad |2,1,-1\rangle. \tag{15.175}$$

Because the perturbed $n = 2$ energy levels vary linearly with the applied electric field, this is called the **linear Stark effect**.

Now that we have found the leading results in perturbation theory for the $n = 1$ and $n = 2$ energy levels, it is time to confess the awful truth that in this example, the perturbative expansion does not converge! The perturbation Hamiltonian $W = eEZ$ is negative with an arbitrarily large magnitude; it approaches $-\infty$ for $z \to -\infty$. This implies that in the idealized problem of a hydrogen atom in a uniform electric field filling infinite space, there are no bound states at all. Instead, the exact stationary-state solutions will have only continuous energies, with resonances near the bound state energies that we have obtained. For this reason, the perturbative expansion in powers of the electric field E actually cannot converge to the supposed atomic bound states.

Fortunately, however, in any real-world version of the Stark effect, the electric field is cut off at some large distance, and the perturbation Hamiltonian remains finite. So, our effort was not at all wasted. In fact, the practically relevant results really are the perturbative ones we found, not the "exact" calculation for unbound states based on an unattainable ideal.

15.8 Almost-Degenerate Perturbation Theory

The perturbative expansion can also face problems if the perturbed state is almost, but not quite, degenerate with some other states. This means that for some subset of stationary states $|n_1\rangle, |n_2\rangle, \ldots, |n_N\rangle$, the differences in unperturbed energies $\mathcal{E}_{n_j} - \mathcal{E}_{n_k}$ are all small compared to the magnitudes of the matrix elements of W. In this case, the terms in the perturbation expansions for E_{n_j} and $|\psi_{n_j}\rangle$ will be finite, but will not become smaller at higher orders, and convergence will fail. To get sensible results, we must formulate **almost-degenerate perturbation theory** to evaluate them, as follows.

For inspiration, we write the perturbation Hamiltonian as

$$W = \sum_n \sum_m |n\rangle \langle n|W|m\rangle \langle m| , \tag{15.176}$$

by using completeness twice. Consider a projection operator onto the subspace of almost-degenerate states that we wish to study,

$$P_{\text{deg}} = \sum_k |n_k\rangle\langle n_k| . \tag{15.177}$$

Now we define the perturbation Hamiltonian projected onto the almost-degenerate subspace,

$$W_{\text{deg}} = P_{\text{deg}} W P_{\text{deg}}, \tag{15.178}$$

which is equivalent to only including the orthobasis elements for the degenerate subspace in the summations in (15.176). Note that $P_{\text{deg}} W_{\text{deg}} P_{\text{deg}} = W_{\text{deg}}$, since $P_{\text{deg}}^2 = P_{\text{deg}}$, as always for projection operators.

The strategy is now to re-divide the full Hamiltonian as

$$H = H_0' + W', \tag{15.179}$$

where

$$H_0' = H_0 + W_{\text{deg}}, \tag{15.180}$$

$$W' = W - W_{\text{deg}}. \tag{15.181}$$

One now endeavors to solve the eigenvalue problem for H_0' exactly, and then treat W' as the perturbation. In favorable circumstances, for example if the dimension of the almost-degenerate subspace is small, this can be done exactly. In less favorable cases, one might have to solve the eigenvalue problem for H_0' in an approximation, before applying W' as the perturbation. In any case, let us refer to the unperturbed eigenstates of H_0' within the almost-degenerate subspace as $|n_j'\rangle$, with unperturbed energies $\mathcal{E}_{n'_j}$. The point is that all of the W' matrix elements within the almost-degenerate subspace vanish,

$$\langle n'_j | W' | n'_k \rangle = 0. \tag{15.182}$$

This follows because the left side can be rewritten, using $|n'_k\rangle = P_{\text{deg}}|n'_k\rangle$, as

$$\langle n'_j | P_{\text{deg}}(W - W_{\text{deg}}) P_{\text{deg}} | n'_k \rangle = \langle n'_j | (W_{\text{deg}} - W_{\text{deg}}) | n'_k \rangle = 0. \tag{15.183}$$

One can now fearlessly apply the usual non-degenerate perturbation theory results of (15.42)–(15.46) with H'_0 as the unperturbed Hamiltonian with eigenstates $|n'_j\rangle$, and W' as the perturbation, because small denominators $\mathcal{E}_{n'_j} - \mathcal{E}_{n'_k}$ can never appear in the sums, due to the selection rule (15.182).

As a minimal example of all three types of stationary-state perturbation theory (non-degenerate, degenerate, almost-degenerate), consider a system with only 2 states, with

$$H_0 = \begin{pmatrix} a & 0 \\ 0 & b \end{pmatrix}, \qquad W = \begin{pmatrix} 0 & c \\ c^* & 0 \end{pmatrix}, \tag{15.184}$$

where a, b are real numbers and c is complex. Of course, 2×2 matrices are easy, so the eigenvalues for $H = H_0 + W$ can be found exactly in this example, with the results

$$E_{1,2} = \frac{1}{2}\left(a + b \mp \sqrt{(b-a)^2 + 4|c|^2}\right). \tag{15.185}$$

If one applies non-degenerate perturbation theory, one readily obtains (assuming that $b > a$)

$$E_1 = a - |c|^2/(b-a) + \cdots, \qquad E_2 = b + |c|^2/(b-a) + \cdots, \tag{15.186}$$

with the perturbation c contributing only at quadratic order.

On the other hand, if $b = a$, then the previous expansion fails, but one can apply degenerate perturbation theory, with the result

$$E_1 = a - |c|, \qquad E_2 = a + |c| \tag{15.187}$$

at first order. In fact, this coincides with the exact result.

Finally, the application of almost-degenerate perturbation theory taking $a \approx b$ is just the exact result, since the 2×2 almost-degenerate subspace consists of the whole state space in this simple example. The exact eigenvalues behave quadratically with $|c|$ for small $|c| \ll |b-a|$, and transition to behaving linearly with $|c|$ for large $|c| \gg$

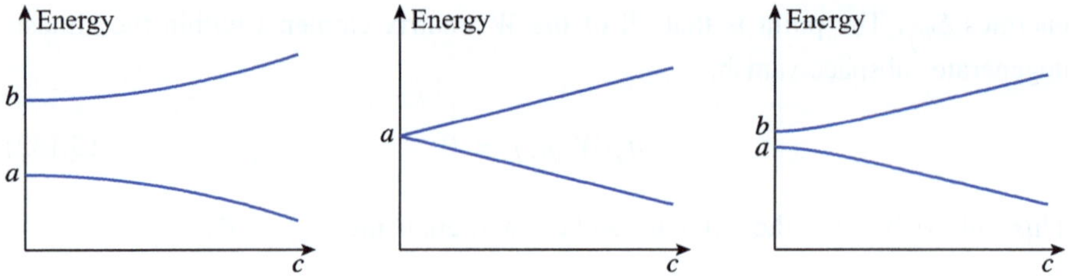

Fig. 15.5 Perturbation theory results at leading order for energy eigenvalues of a two-state system with Hamiltonian given by (15.184), for non-degenerate perturbation theory with $b-a \gg |c|$ (left), degenerate perturbation theory with $a = b$ (center), and almost-degenerate perturbation theory with $a \approx b$ (right). For a two-state system, almost-degenerate perturbation theory is the exact result, by definition. In all three cases, the energy eigenvalues are moved apart by the perturbation, illustrating level repulsion

$|b-a|$. These statements are illustrated in Fig. 15.5. The fact that increasing the off-diagonal perturbation moves the energy eigenvalues apart is called **level repulsion.**

As a less minimal example of almost-degenerate perturbation theory, consider a state space of dimension 3, with matrix representations for the unperturbed and perturbation Hamiltonians

$$H_0 = \begin{pmatrix} a-\Delta & 0 & 0 \\ 0 & a+\Delta & 0 \\ 0 & 0 & b \end{pmatrix}, \qquad W = \begin{pmatrix} 0 & \delta & \epsilon \\ \delta & 0 & 0 \\ \epsilon & 0 & 0 \end{pmatrix}, \tag{15.188}$$

where a, b, Δ, δ, and ϵ are all real constants. The exact eigenvalues of $H = H_0 + W$ can be obtained by solving a cubic equation, but the results are complicated. However, it is worth noting that, because of the theorem in linear algebra that the sum of the eigenvalues of a matrix is equal to the trace, the exact energies must satisfy a simple sum rule,

$$E_1 + E_2 + E_3 = 2a + b, \tag{15.189}$$

regardless of the values of the other parameters. Now let us see what perturbation theory says.

First, let us find the energy eigenvalues by applying non-degenerate perturbation theory. Because H_0 is diagonal, the unperturbed stationary states are the kets represented by the unit column vectors

$$|1\rangle = \begin{pmatrix} 1 \\ 0 \\ 0 \end{pmatrix}, \qquad |2\rangle = \begin{pmatrix} 0 \\ 1 \\ 0 \end{pmatrix}, \qquad |3\rangle = \begin{pmatrix} 0 \\ 0 \\ 1 \end{pmatrix}, \tag{15.190}$$

with

$$\mathcal{E}_1 = a - \Delta, \qquad \mathcal{E}_2 = a + \Delta, \qquad \mathcal{E}_3 = b. \tag{15.191}$$

From (15.42), the first-order corrections to the energies vanish, because W has no nonzero diagonal entries. Applying (15.43) gives the energy eigenvalues at second order in non-degenerate perturbation theory in δ, ϵ,

$$E_1 = a - \Delta + \epsilon^2/(a - b - \Delta) - \delta^2/2\Delta, \tag{15.192}$$

$$E_2 = a + \Delta + \delta^2/2\Delta, \tag{15.193}$$

$$E_3 = b + \epsilon^2/(b - a + \Delta). \tag{15.194}$$

As a check, these do satisfy the sum rule (15.189).

However, if Δ is small compared to δ, then the last term in each of (15.192) and (15.193) will be large, signaling a breakdown in the perturbative expansion because the unperturbed states $|1\rangle$ and $|2\rangle$ are too close in energy. In that case, one must use almost-degenerate perturbation theory. In this example,

$$P_{\text{deg}} = \begin{pmatrix} 1 & 0 & 0 \\ 0 & 1 & 0 \\ 0 & 0 & 0 \end{pmatrix}, \qquad W_{\text{deg}} = \begin{pmatrix} 0 & \delta & 0 \\ \delta & 0 & 0 \\ 0 & 0 & 0 \end{pmatrix}, \tag{15.195}$$

and so

$$H_0' = \begin{pmatrix} a - \Delta & \delta & 0 \\ \delta & a + \Delta & 0 \\ 0 & 0 & b \end{pmatrix}, \qquad W' = \begin{pmatrix} 0 & 0 & \epsilon \\ 0 & 0 & 0 \\ \epsilon & 0 & 0 \end{pmatrix}. \tag{15.196}$$

Note that W' is indeed 0 within the 2×2 subspace of almost-degenerate states. It is not too hard to find the unperturbed energies and eigenstates of H_0',

$$\mathcal{E}_{1'} = a - \sqrt{\delta^2 + \Delta^2}, \qquad |1'\rangle = \sqrt{\frac{r_+}{2}}\,|1\rangle - \sqrt{\frac{r_-}{2}}\,|2\rangle, \tag{15.197}$$

$$\mathcal{E}_{2'} = a + \sqrt{\delta^2 + \Delta^2}, \qquad |2'\rangle = \sqrt{\frac{r_-}{2}}\,|1\rangle + \sqrt{\frac{r_+}{2}}\,|2\rangle, \tag{15.198}$$

$$\mathcal{E}_{3'} = b, \qquad |3'\rangle = |3\rangle, \tag{15.199}$$

where

$$r_\pm = 1 \pm \frac{\Delta}{\sqrt{\delta^2 + \Delta^2}}. \tag{15.200}$$

We then apply (15.42) and (15.43) with H_0' and W', while being careful to note that the matrix representation of W' in (15.196) is in the original basis $|1\rangle$, $|2\rangle$, $|3\rangle$, so that

$$W' = \epsilon\,|1\rangle\langle 3| + \epsilon\,|3\rangle\langle 1|\,. \tag{15.201}$$

The first-order contributions to the energies vanish because W' is also off-diagonal in the basis $|1'\rangle$, $|2'\rangle$, $|3'\rangle$. The second-order energy corrections are, from (15.43),

$$E_{1'}^{(2)} = |\langle 3'|W'|1'\rangle|^2/(\mathcal{E}_{1'} - \mathcal{E}_{3'}), \tag{15.202}$$

$$E_{2'}^{(2)} = |\langle 3'|W'|2'\rangle|^2/(\mathcal{E}_{2'} - \mathcal{E}_{3'}), \tag{15.203}$$

$$E_{3'}^{(2)} = |\langle 1'|W'|3'\rangle|^2/(\mathcal{E}_{3'} - \mathcal{E}_{1'}) + |\langle 2'|W'|3'\rangle|^2/(\mathcal{E}_{3'} - \mathcal{E}_{2'}), \tag{15.204}$$

resulting in the energy eigenvalues at second-order in almost-degenerate perturbation theory,

$$E_{1'} = a - \sqrt{\delta^2 + \Delta^2} + \frac{\epsilon^2 r_+}{2(a - b - \sqrt{\delta^2 + \Delta^2})}, \tag{15.205}$$

$$E_{2'} = a + \sqrt{\delta^2 + \Delta^2} + \frac{\epsilon^2 r_-}{2(a - b + \sqrt{\delta^2 + \Delta^2})}, \tag{15.206}$$

$$E_{3'} = b + \frac{\epsilon^2 (b - a - \Delta)}{(b - a)^2 - (\delta^2 + \Delta^2)}, \tag{15.207}$$

now with ϵ as the sole expansion parameter. Again one can check that the sum rule (15.189) is indeed satisfied.

The almost-degenerate perturbation theory expansion in ϵ will be a good one as long as $|\epsilon/(\mathcal{E}_{1'} - \mathcal{E}_{3'})| = |\epsilon/(a - b - \sqrt{\delta^2 + \Delta^2})|$ and $|\epsilon/(\mathcal{E}_{2'} - \mathcal{E}_{3'})| = |\epsilon/(a - b + \sqrt{\delta^2 + \Delta^2})|$ are both small. Conversely, if they are not, then even this expansion will break down, and almost-degenerate perturbation theory would, unfortunately, consist of solving the original 3×3 problem exactly.

Exercises

15.1 Consider a harmonic oscillator with Hamiltonian $H = H_0 + W$, where $H_0 = P^2/2m + m\omega^2 X^2/2$ and $W = \lambda P^4$ is to be treated as a small perturbation.

(a) Compute the correction to the energy of the unperturbed state $|n\rangle$ at first order in λ. For fixed n, your answer is good for sufficiently small positive λ. However, you should find that it grows with n, so if $|\lambda|$ is fixed, no matter how small it is, the perturbation expansion will break down for sufficiently large n.

(b) The leading relativistic correction to the energy is realized by $\lambda = -1/8m^3c^2$, where c is the speed of light. The negative sign technically makes the energy unbounded from below, but disaster can be avoided by including even higher-order terms in H. Find the condition for first-order perturbation theory to give a small correction to the energy of the state $|n\rangle$, in the form $n \ll n_c$, where n_c is to be determined in terms of m and ω.

(c) Find the ground state in terms of the unperturbed states $|n\rangle$, at first order in λ.

15.2 The Hamiltonian for a rigid rotator in a magnetic field perpendicular to the y-axis is $H = aL^2 + bL_z + cL_x$, where a, b, and c are positive constants, and $\boldsymbol{L}$ is the angular momentum operator.

(a) Write down the exact energy eigenvalues. (This should be quick. Start by writing $bL_z + cL_x = b'\hat{n} \cdot \boldsymbol{L}$, where $\hat{n}$ is a unit vector; what is b'?)
(b) Now take $c \ll a, b$. Find the unperturbed energy eigenvalues with $c = 0$ for the eigenkets $|l, m\rangle$ of L^2 and L_z, and use them to obtain the energy eigenvalues at second order in the perturbation. Compare them to the exact answer from part (a). [Hint: the necessary matrix elements $\langle l', m'|L_x|l, m\rangle$ can be found in (8.66).]

15.3 A spinless particle of charge q and mass m is confined to a cubic box of side L with center at the origin. A weak uniform electric field E is applied, with direction parallel to one of the sides of the box, so that the perturbation Hamiltonian is $W = -qEZ$, where Z is the position coordinate.

(a) Write down the unperturbed energy eigenvalues and corresponding normalized wavefunctions, in terms of appropriate quantum numbers. [Hint: for a particle confined to a one-dimensional box, the wavefunctions were given in (6.63) and (6.64), for odd and even n, respectively. Here you may find it more convenient to use the unified expression $\varphi_n(x) = \sqrt{\frac{2}{L}} \sin\left(\frac{n\pi x}{L} + \frac{n\pi}{2}\right)$, which is equivalent up to irrelevant signs.]
(b) Find the change in the ground-state energy at second order in the electric field. Leave your answer in terms of an infinite sum of the form $\sum_{n=2,4,6,\ldots} \frac{n^p}{(n^2-1)^r}$, where p and r are certain integers that you will find. Evaluate the sum to three significant digits. How does the change in the ground-state energy scale with the size of the box?

15.4 A perturbed isotropic two-dimensional harmonic oscillator has Hamiltonian

$$H = \frac{1}{2m}(P_x^2 + P_y^2) + \frac{1}{2}m\omega^2(X^2 + Y^2) + \lambda X^2 Y^2. \tag{15.208}$$

(a) First treating $\lambda = 0$, what are the energies and the degeneracies of the three lowest energy levels?
(b) Now treating λ as a small perturbation, find the first-order corrections to the energies of the three lowest unperturbed energy levels.

15.5 Consider the Stark effect for the $n = 3$ states of the Hydrogen atom:

(a) For the unperturbed states $|n, l, m\rangle$ with $n = 3$, find all of the nonzero matrix elements $\langle 3, l', m'|Z|3, l, m\rangle$. (Use selection rules, and the fact that Z is Hermitian, to reduce the labor.)
(b) Include an electric field $\boldsymbol{E} = E\hat{z}$, treated as a perturbation. Obtain the energy eigenvalue corrections at first order in degenerate perturbation theory. Give the corresponding energy eigenstates to zeroth order in the perturbation, in terms of the unperturbed eigenstates $|3, l, m\rangle$.

15.6 Starting from Brillouin–Wigner perturbation theory at second order in the state (15.104) and third order in the energy (15.105), derive the corresponding results for standard perturbation theory, given at the end of Sect. 15.1.

15.7 Derive the formula for $E_n^{(3)}$ in terms of $|\psi_n^{(1)}\rangle$, as given in (15.114).

15.8 Consider a 3-state system with Hamiltonian $H = H_0 + W$, where in a certain basis

$$H_0 = \begin{pmatrix} \mathcal{E}_1 & 0 & 0 \\ 0 & \mathcal{E}_2 & 0 \\ 0 & 0 & \mathcal{E}_2 \end{pmatrix}, \qquad W = \begin{pmatrix} 0 & a & b \\ a & 0 & c \\ b & c & 0 \end{pmatrix}, \tag{15.209}$$

where $\mathcal{E}_1$ and $\mathcal{E}_2$ are distinct, and much larger than $|a|$, $|b|$, $|c|$.

(a) Assuming that $c \neq 0$, find the energy eigenvalues at second order in degenerate perturbation theory. Recall that this requires going to a new unperturbed basis, chosen so that W is diagonal within the 2×2 subspace corresponding to the degenerate H_0 eigenvalues.
(b) Now consider $c = 0$. Again compute the three energy eigenvalues at second order in degenerate perturbation theory. Note that, as explained in the discussion leading to (15.128), since the degeneracy is not lifted at first order, it is necessary to choose a different unperturbed basis, such that the matrix $W^{(2)}$ defined in (15.124) is diagonal within the 2×2 subspace corresponding to the degenerate H_0 eigenvalues. (It is not diagonal, or even Hermitian, on the whole 3×3 state space.) You should find that the two energy eigenvalues arising from $\mathcal{E}_2$ are *not* the same as the $c \to 0$ limit of part (a).
(c) Check that the sum of the three energy eigenvalues in both cases is equal to $\mathcal{E}_1 + 2\mathcal{E}_2$, which is required because that is the trace of H. For the case $c = 0$, compute the exact eigenvalues of H, and check that expanding them to second order in the perturbations reproduces the results of part (b).

The Variational Method

16

16.1 Estimate and Upper Bound on the Ground-State Energy

In Chap. 15, we studied how to use perturbation theory to find approximate solutions to the Hamiltonian eigenvalue problem. In this chapter, we will consider the **variational method** (also known as the **Rayleigh–Ritz method**), which applies even in situations for which it is not possible to make a sensible perturbation series expansion. Because the variational method is not based on a perturbative expansion, there is no need to know the exact solution to a simpler unperturbed problem.

The key idea comes from an inequality satisfied by the expectation value of the Hamiltonian in any state $|\psi\rangle$ of our choice. Although the eigenvalues E_n and eigenstates $|E_n\rangle$ of H are unknown, we can use completeness to write

$$\begin{aligned} \langle\psi|H|\psi\rangle &= \sum_n \langle\psi|H|E_n\rangle\langle E_n|\psi\rangle = \sum_n E_n \langle\psi|E_n\rangle\langle E_n|\psi\rangle \\ &\geq E_0 \sum_n \langle\psi|E_n\rangle\langle E_n|\psi\rangle\,, \end{aligned} \tag{16.1}$$

where we used $E_n \geq E_0$ to get the last line. (Possible degeneracies in the energies are handled by simply adding a degeneracy label and summing over them in the completeness relation, as usual.) Using completeness in reverse to rewrite $\sum_n \langle\psi|E_n\rangle\langle E_n|\psi\rangle = \langle\psi|\psi\rangle$, we obtain the bound

$$\frac{\langle\psi|H|\psi\rangle}{\langle\psi|\psi\rangle} \geq E_0. \tag{16.2}$$

(We have not assumed the normalization condition $\langle\psi|\psi\rangle = 1$, as this is sometimes not convenient.) Now we can make a guess, or a class of guesses, for $|\psi\rangle$, and then

S. P. Martin, *Quantum Mechanics*, Graduate Texts in Physics,
https://doi.org/10.1007/978-3-032-21119-4_16

evaluate the left side of (16.2) to obtain an estimate for E_0 that is guaranteed not to be lower than the true value. The guess is called a **trial state** or **trial wavefunction**.

In a typical application, we improve the quality of our guess by choosing a whole family of trial states $|\psi(a)\rangle$, where a denotes a continuously variable parameter (or, more generally, a set of several parameters). Now we compute

$$E(a) = \frac{\langle\psi(a)|H|\psi(a)\rangle}{\langle\psi(a)|\psi(a)\rangle}, \tag{16.3}$$

and minimize it with respect to a to obtain a_{min}, typically by solving

$$\frac{\partial E(a)}{\partial a} = 0. \tag{16.4}$$

Then $E(a_{\text{min}})$ is our corresponding best estimate and upper bound for E_0, and $|\psi(a_{\text{min}})\rangle$ is our best estimate for the ground state $|E_0\rangle$.

The variational method tends to forgive us if our guess is good, but not perfect. To see why, suppose the result of our best guess can be written as

$$|\psi(a_{\text{min}})\rangle = |E_0\rangle + \sum_{n\neq 0} \delta_n |E_n\rangle, \tag{16.5}$$

where the δ_n are some unknown, but hopefully small, numbers representing the fact that our guess was not perfect. Then we compute

$$\langle\psi|H|\psi\rangle = E_0 + \sum_{n\neq 0} E_n|\delta_n|^2, \tag{16.6}$$

$$\langle\psi|\psi\rangle = 1 + \sum_{n\neq 0} |\delta_n|^2, \tag{16.7}$$

which yields

$$\frac{\langle\psi|H|\psi\rangle}{\langle\psi|\psi\rangle} = E_0 + \frac{\sum_{n\neq 0}(E_n - E_0)|\delta_n|^2}{1 + \sum_{n\neq 0}|\delta_n|^2}. \tag{16.8}$$

The error in our estimate for E_0 is seen to be quadratic, involving $|\delta_n|^2$ rather than $|\delta_n|$. The lesson here is that the estimate for the energy is better than the estimate for the state, provided only that the best-estimate errors are not too horrible, obeying $|\delta_n| < 1$. A 10% error in the state ket typically leads to only a $\sim$ 1% error in the energy.

Still, we do want to make a good guess for our parameterized family of states $|\psi(a)\rangle$, so that the best of them has a chance to be as close as possible to the true ground state. We should choose trial wavefunctions $\psi(\boldsymbol{r})$ such that $|\psi(\boldsymbol{r})|$ is larger where the potential $V(\boldsymbol{r})$ is smaller. Furthermore, the wavefunction should have as few wiggles and nodes as possible, consistent with the given potential $V(\boldsymbol{r})$, in order to minimize the kinetic contribution to the Hamiltonian. In three-dimensional problems, if the Hamiltonian has rotational symmetry, minimizing the kinetic contribution likewise implies that we should make guesses that have L^2 eigenvalue 0.

16.2 Variational Method for Excited States

The variational method can also provide information about higher-energy states, by projecting out lower-energy states from the trial state guesses. For example, suppose that we want to obtain the energy eigenvalue for the first excited state(s) $|E_1\rangle$. If we somehow already knew the exact $|E_0\rangle$ (assumed to be nondegenerate), then we could define the projector

$$P = I - |E_0\rangle \langle E_0| , \tag{16.9}$$

which satisfies $P^2 = P$ and $P|E_0\rangle = 0$. Then, for all trial states $P|\chi\rangle$, one can prove the bound

$$\frac{\langle\chi|PHP|\chi\rangle}{\langle\chi|P|\chi\rangle} \geq E_1, \tag{16.10}$$

where E_1 is the first excited energy eigenvalue.

In practice, we may only have a variational estimate $|\psi_{\text{est}}\rangle$ for $|E_0\rangle$, for example $|\psi_{\text{est}}\rangle = |\psi(a_{\text{min}})\rangle$ from a previous variational estimate of the ground state. In that case, it might be that the best we can do is to instead define

$$\widetilde{P} = I - |\psi_{\text{est}}\rangle \langle\psi_{\text{est}}| , \tag{16.11}$$

(assuming that $|\psi_{\text{est}}\rangle$ has been normalized to unity) and then minimize the expectation value of the Hamiltonian in a parameterized family of states $\widetilde{P}|\chi(b)\rangle$, where $|\chi(b)\rangle$ is chosen based on a prior understanding of the properties of the first excited state(s), and depends on a variational parameter b. Unfortunately, the accuracy of the resulting estimate for E_1 will be limited by the goodness of the estimate $|\psi_{\text{est}}\rangle \approx |E_0\rangle$. To the extent that this is not a good approximation, there is some leakage from the ground state into the guess, and we do not have a strict bound on the result of the minimization of the Hamiltonian expectation value. For this reason, it is often computationally challenging to obtain good estimates for all but the lowest few energy eigenvalues.

Fortunately, strict bounds and variational estimates for some of the higher energy states can be obtained if the Hamiltonian is compatible with some other observable(s). Suppose that there is an observable Ω that satisfies

$$[H, \Omega] = 0. \tag{16.12}$$

This implies that one can choose an orthobasis of common eigenstates of H and Ω, so that the energy eigenstates can be sub-classified by their Ω eigenvalues, denoted ω. Common examples for Ω are the angular momentum operators J^2, J_z and/or the parity operator Π. In such cases, we can define a family of trial kets $|\psi_\omega(a)\rangle$ that are restricted to be eigenstates of Ω with a fixed eigenvalue ω, so

$$\Omega\,|\psi_\omega(a)\rangle = \omega\,|\psi_\omega(a)\rangle\,. \tag{16.13}$$

In most cases, it will be obvious how to do this, but if necessary we can define the projection operator for each ω,

$$P_\omega = \sum_u |\omega, u\rangle\,\langle\omega, u|\,, \tag{16.14}$$

where the u are degeneracy labels for ω, and then take $|\psi_\omega(a)\rangle = P_\omega\,|\psi(a)\rangle$, for some suitable $|\psi(a)\rangle$. In any case, given a family of trial states $|\psi_\omega(a)\rangle$, one finds

$$E(a, \omega) = \frac{\langle\psi_\omega(a)|H|\psi_\omega(a)\rangle}{\langle\psi_\omega(a)|\psi_\omega(a)\rangle} \geq E_{\text{min}}(\omega), \tag{16.15}$$

where $E_{\text{min}}(\omega)$ is the true minimum energy eigenvalue within the subspace of states that have $\Omega = \omega$. Thus, we obtain a separate variational estimate and bound on the lowest energy eigenstate within each ω subspace.

For example, if the Hamiltonian for a spinless particle has rotational symmetry, then we can take $\Omega = L^2$. Then, applying the variational principle in turn to the trial wavefunctions $\psi(r, \theta, \phi) = R_A(r)$ and $R_B(r)Y_1^0(\theta, \phi)$ and $R_C(r)Y_2^0(\theta, \phi)$, we will obtain estimates for the minimum energy eigenvalues and wavefunctions for the ground state with $l = 0$, and for the lowest energy state that has $l = 1$, and for the lowest energy state that has $l = 2$, respectively. (Recall that rotational invariance guarantees that the energy eigenvalues and the radial wavefunctions are actually independent of m, so we took $m = 0$ in each case just to be specific. Any other value of m would work just as well.) It is not necessary in this case to go to the trouble of defining the projection operator in (16.14), because the eigenstates of L^2 are known in terms of the spherical harmonics.

16.3 Examples of the Variational Method

As an easy example, consider a system for which we already know the exact answer: the harmonic oscillator in one dimension, with

$$H = \frac{P^2}{2m} + \frac{1}{2}m\omega^2 X^2. \tag{16.16}$$

We now use a family of trial wavefunctions

$$\psi(a) = e^{-x^2/a^2}. \tag{16.17}$$

This is a very fortunate guess, as it just so happens to include the correct ground-state wavefunction as a special case. Now, calculate

$$\langle\psi(a)|\psi(a)\rangle = \int_{-\infty}^{\infty} dx\, e^{-2x^2/a^2} = \sqrt{\frac{\pi}{2}}a, \tag{16.18}$$

and

$$\begin{aligned}\langle\psi(a)|H|\psi(a)\rangle &= \int_{-\infty}^{\infty} dx\, e^{-x^2/a^2}\left[-\frac{\hbar^2}{2m}\frac{d^2}{dx^2} + \frac{1}{2}m\omega^2x^2\right]e^{-x^2/a^2} \\ &= \int_{-\infty}^{\infty} dx \left[\frac{\hbar^2}{2m}\left(\frac{2}{a^2} - \frac{4x^2}{a^4}\right) + \frac{1}{2}m\omega^2x^2\right]e^{-2x^2/a^2} \\ &= \sqrt{\frac{\pi}{2}}\left(\frac{\hbar^2}{2ma} + \frac{1}{8}m\omega^2a^3\right).\end{aligned} \tag{16.19}$$

Using these in (16.3) gives

$$E(a) = \frac{\hbar^2}{2ma^2} + \frac{m\omega^2a^2}{8}. \tag{16.20}$$

The first term is the kinetic contribution, which wants a to be large, and the second term is the potential contribution, which prefers a to be small. Minimizing $E(a)$ with respect to a^2 gives

$$a^2_{\text{min}} = \frac{2\hbar}{m\omega}, \tag{16.21}$$

and so the result of the variational method is

$$E_0 \leq E(a_{\text{min}}) = \hbar\omega/2, \tag{16.22}$$

in agreement with the exact result. Of course, $\psi(a_{\text{min}})$ is also the exact ground-state wavefunction (up to the normalization factor), as a result of our fortunate guess.

For a less trivial example, consider the Hamiltonian

$$H = \frac{P^2}{2m} + \lambda X^4, \tag{16.23}$$

which is a cousin of the harmonic oscillator but with a quartic potential. Not feeling very imaginative at the moment, and remembering the forgiving nature of the variational principle, we again choose as the trial wavefunction

$$\psi(a) = e^{-x^2/a^2}, \tag{16.24}$$

which yields

$$E(a) = \frac{\hbar^2}{2ma^2} + \frac{3}{16}\lambda a^4. \tag{16.25}$$

(The computation of the first term is identical to that for the previous example, as the trial wavefunction and the $P^2/2m$ part of the Hamiltonian have not changed.) Now minimizing $E(a)$ with respect to a^2 gives

$$a^2_{\text{min}} = \left(\frac{4\hbar^2}{3\lambda m}\right)^{1/3} \approx 1.101\left(\frac{\hbar^2}{\lambda m}\right)^{1/3}, \tag{16.26}$$

and

$$E_{\text{min}} = \frac{3}{8}\left(\frac{6\hbar^4\lambda}{m^2}\right)^{1/3} \approx 0.681\left(\frac{\hbar^4\lambda}{m^2}\right)^{1/3} \tag{16.27}$$

as the estimate of the ground-state energy.

To do better, we invent a three-parameter trial wavefunction

$$\psi(a, b, c) = (1 + bx^2 + cx^4)e^{-x^2/a^2}. \tag{16.28}$$

We do not bother including odd powers of x, because we expect that the parity of the ground state is $+1$. Using a computer to minimize $E(a, b, c)$ with respect to all three of a, b, c gives

$$E(a_{\text{min}}, b_{\text{min}}, c_{\text{min}}) \approx 0.668\left(\frac{\hbar^4\lambda}{m^2}\right)^{1/3} \tag{16.29}$$

as a better estimate. As always, the more general trial wavefunction gives an estimate that is lower, and closer to the true ground-state energy eigenvalue.

The first excited state for H in (16.23) will have one node, at $x = 0$, and odd parity. To find it, we can therefore exploit (16.15), by choosing a simple trial wavefunction

$$\psi(a) = xe^{-x^2/a^2}. \tag{16.30}$$

A straightforward calculation of the Hamiltonian's expectation value and the norm gives

$$E(a) = \frac{3\hbar^2}{2ma^2} + \frac{15}{16}\lambda a^4, \tag{16.31}$$

with minimum at

$$a_{\min}^2 = \left(\frac{4\hbar^2}{5\lambda m}\right)^{1/3}, \tag{16.32}$$

leading to the estimate and upper bound for E_1 of

$$E(a_{\min}) = \frac{9}{4}\left(\frac{5\hbar^4\lambda}{4m^2}\right)^{1/3} \approx 2.424\left(\frac{\hbar^4\lambda}{m^2}\right)^{1/3}. \tag{16.33}$$

A somewhat better result could follow from using a multi-parameter trial wavefunction of the form $\psi(a, b, c, \ldots) = (x + bx^3 + cx^5 + \cdots)e^{-x^2/a^2}$. Note that it is important to be careful to only include odd-parity terms in this trial wavefunction, to avoid contamination from the even-parity ground state. The actual first excited state energy eigenvalue E_1 turns out to be

$$E_1 \approx 2.394\left(\frac{\hbar^4\lambda}{m^2}\right)^{1/3}, \tag{16.34}$$

as can be obtained from numerical minimization of, say, a 6-parameter trial wavefunction.

Next, consider as a three-dimensional example the hydrogen atom Hamiltonian, written in the spherical coordinate position representation as

$$H = T + V, \qquad T = -\frac{\hbar^2\nabla^2}{2m_e}, \qquad V = -\frac{e^2}{r}. \tag{16.35}$$

Although we know the exact wavefunction, let us ignore the temptation to use it, and instead try a Gaussian trial wavefunction with a variational parameter k,

$$\psi(k) = e^{-k^2r^2}/\sqrt{4\pi}. \tag{16.36}$$

To find the expectation value of the Hamiltonian, we first compute the kinetic contribution,

$$\langle\psi|T|\psi\rangle = -\frac{\hbar^2}{2m_e}\int_0^\infty dr\, r^2 e^{-k^2r^2}\left(\frac{d^2}{dr^2} + \frac{2}{r}\frac{d}{dr}\right)e^{-k^2r^2} = \frac{\hbar^2}{m_e k}\sqrt{\frac{\pi}{2}}\frac{3}{16}, \tag{16.37}$$

then the potential contribution,

$$\langle \psi | V | \psi \rangle = -\frac{e^2}{4k^2}, \tag{16.38}$$

and finally the normalization factor,

$$\langle \psi | \psi \rangle = \sqrt{\frac{\pi}{2}} \frac{1}{8k^3}. \tag{16.39}$$

Putting these together, we have

$$E(k) = \frac{3\hbar^2 k^2}{2m_e} - \sqrt{\frac{8}{\pi}} e^2 k. \tag{16.40}$$

The minimum is found from $\partial E(k)/\partial k = 0$ to be at

$$k_{\min} = \sqrt{\frac{8}{\pi}} \frac{e^2 m_e}{3\hbar^2} = \sqrt{\frac{8}{\pi}} \frac{1}{3a_0}. \tag{16.41}$$

Plugging this in gives the estimate of the ground-state energy

$$E(k_{\min}) = -\left(\frac{8}{3\pi}\right) \frac{e^2}{2a_0} \approx -0.849\,\text{Rydberg}. \tag{16.42}$$

Of course, the exact answer is -1 Rydberg. Even though our choice of one-parameter trial wavefunctions was not a very good one, because the harmonic oscillator is a poor approximation to the Coulomb potential, the result was within about 15% of the correct answer.

16.4 Helium Atom Ground State from the Variational Method

We conclude this chapter with a more practical example in which the variational method shines: the helium atom consisting of two electrons with charge $-e$ and a nucleus with charge $+2e$. The Hamiltonian for this system was already given in (15.81)–(15.83) of Sect. 15.3, where we treated it using first-order perturbation theory.

To use the variational method in the simplest way, let us choose a factorized trial wavefunction of the form

$$\Psi(\tilde{Z}, \boldsymbol{r}_1, \boldsymbol{r}_2) = \psi(\tilde{Z}, \boldsymbol{r}_1)\, \psi(\tilde{Z}, \boldsymbol{r}_2), \tag{16.43}$$

where, for each electron,

$$\psi(\tilde{Z}, \boldsymbol{r}) = \sqrt{\frac{\tilde{Z}^3}{\pi a_0^3}}\, e^{-\tilde{Z}r/a_0} \tag{16.44}$$

is the normalized ground-state wavefunction for a hydrogen-like atom with general atomic number $\tilde{Z}$ [see (11.42)]. However, here we do not set $\tilde{Z}$ equal to $Z = 2$ in the trial wavefunction, but instead treat it as a continuous variational parameter, adjusting it to find the minimum expectation value of the total H. In contrast, the Z that appears in the Hamiltonian is fixed to be equal to the actual number of protons in the nucleus. For helium, $Z = 2$, but we will leave it general for the following discussion, so that our results apply to any ion with 2 electrons. For the rest of this section, we leave the $\tilde{Z}$ argument on wavefunctions and kets implicit.

We need to compute, and then minimize

$$E(\tilde{Z}) = E_1(\tilde{Z}) + E_2(\tilde{Z}) + E_{12}(\tilde{Z}), \tag{16.45}$$

where $E_1(\tilde{Z}) = E_2(\tilde{Z}) = \langle\Psi|H_1|\Psi\rangle$, and $E_{12}(\tilde{Z}) = \langle\Psi|H_{12}|\Psi\rangle$, since we have arranged for $\langle\Psi|\Psi\rangle = 1$ in this example. Because the nuclear charge $+Ze$ is partly screened by the other electron, we anticipate that the value $\tilde{Z}_{\text{min}}$ that minimizes $E(\tilde{Z})$ will be in the range $Z - 1 < \tilde{Z}_{\text{min}} < Z$, so that for helium, $1 < \tilde{Z}_{\text{min}} < 2$.

To find the non-interaction contributions, we first compute

$$E_1(\tilde{Z}) = \int d^3\boldsymbol{r}_1\, \psi(\boldsymbol{r}_1)^* \left(-\frac{\hbar^2}{2m_e}\nabla_1^2 - \frac{Ze^2}{r_1}\right)\psi(\boldsymbol{r}_1) \int d^3\boldsymbol{r}_2\, |\psi(\boldsymbol{r}_2)|^2. \tag{16.46}$$

Note that the potential energy contribution is $-Ze^2/r_1$, not $-\tilde{Z}e^2/r_1$, because $\tilde{Z}$ is our variational parameter, not the actual nuclear charge. In this expression, $\int d^3\boldsymbol{r}_2\, |\psi(\boldsymbol{r}_2)|^2 = 1$, and the remaining integral factor evaluates simply, to give

$$E_1(\tilde{Z}) = \frac{e^2}{2a_0}\left(\tilde{Z}^2 - 2\tilde{Z}Z\right). \tag{16.47}$$

Of course, $E_2(\tilde{Z}) = E_1(\tilde{Z})$ follows from an identical calculation with the labels 1 and 2 interchanged. The more difficult part is the interaction contribution

$$E_{12}(\tilde{Z}) = e^2 \int d^3\boldsymbol{r}_1 \int d^3\boldsymbol{r}_2\, |\psi(\boldsymbol{r}_1)|^2\, |\psi(\boldsymbol{r}_2)|^2\, \frac{1}{|\boldsymbol{r}_1 - \boldsymbol{r}_2|}. \tag{16.48}$$

Fortunately, we have already evaluated this integral in (15.88)–(15.98), with the result

$$E_{12}(\tilde{Z}) = \frac{5\tilde{Z}e^2}{8a_0}. \tag{16.49}$$

Plugging this into (16.45), along with $E_1(\tilde{Z}) = E_2(\tilde{Z})$ from (16.47), gives

$$E(\tilde{Z}) = \tilde{Z}(\tilde{Z} - 2Z + 5/8)\frac{e^2}{a_0}. \tag{16.50}$$

Minimizing this with respect to our variational parameter $\tilde{Z}$ gives

$$\tilde{Z}_{\text{min}} = Z - 5/16 = 27/16, \tag{16.51}$$

where the last equality finally specializes to $Z = 2$ for helium. Note that indeed $Z - 1 < \tilde{Z}_{\text{min}} < Z$, as we had anticipated.

Our variational method estimate for the ground-state energy of the helium atom is therefore

$$E(\tilde{Z}_{\text{min}}) = -\tilde{Z}^2_{\text{min}}\frac{e^2}{a_0} = -\left(\frac{27}{16}\right)^2\frac{e^2}{a_0} = -5.695 \text{ Rydberg} = -77.49 \text{ eV}. \tag{16.52}$$

This can be compared to the experimental value (the negative of the minimum ionization energy needed to completely remove both electrons from the He atom in its ground state), which is

$$E_{\text{exp}} = -5.807 \text{ Rydberg} = -79.005 \text{ eV}. \tag{16.53}$$

As expected, our variational method estimate is higher than the true value, but by less than 2%. Recall that the first-order perturbation theory method in Sect. 15.3 gave instead

$$E_{\text{pert.}} = -5.5 \text{ Rydberg} = -74.83 \text{ eV}, \tag{16.54}$$

which is about 5.3% higher than the experimental value. The variational method, even with the simplest trial wavefunction, performs better than first-order perturbation theory. If one uses a more sophisticated non-factorized trial wavefunction with more parameters, and asks a computer to do the integrations and minimization numerically, then the variational method will win even more convincingly for a given amount of calculation effort expended. The variational method also scales well to more complex problems in atomic and molecular physics, where perturbation theory may have more difficulty.

Exercises

16.1 Consider a particle of mass m moving in one dimension in the confining potential $V(X) = \lambda X^6$. Using a Gaussian trial wavefunction, estimate the ground-state energy and wavefunction. What is the corresponding estimate for $\langle X^2 \rangle$? Do the same for the first excited state, using an appropriate trial wavefunction.

16.2 Consider the problem of a particle of mass m in one dimension, confined to a box $-L/2 < x < L/2$, as studied in Sect. 6.4.

(a) Your task is to find variational principle estimates for the first three energy levels E_1, E_2, and E_3. Start with the trial wavefunctions

$$\psi_1(x) = (x + L/2)(x - L/2), \tag{16.55}$$

$$\psi_2(x) = x(x + L/2)(x - L/2), \tag{16.56}$$

$$\psi_3(x) = (1 + ax + bx^2)(x + L/2)(x - L/2). \tag{16.57}$$

It would be a bad idea to treat a and b as variational parameters. Why? What is the correct way to choose a and b (pretending that you do not know the exact stationary states), and what are their resulting values? What are the resulting estimates for E_1, E_2, and E_3, and how do they compare to the exact values?

(b) Now try instead $\psi_1(x) = (1 + cx^2)(x + L/2)(x - L/2)$ for the ground state, where this time c really is a variational parameter. What is the resulting estimate for the ground-state energy, in simplest form? You should find that it involves $\sqrt{133}$, and that it comes extremely close to the exact value.

16.3 A particle of mass m moves in a spherically symmetric potential $V(R)$. Denote by E_l the lowest bound state energy with orbital angular momentum l [that is, with L^2 eigenvalue $\hbar^2 l(l+1)$]. Use the variational principle to prove that E_l increases monotonically with l. (It obviously follows that the ground state must have $l = 0$.) Compare to Figs. 10.2 and 11.1.

16.4 Consider a particle of mass m in the three-dimensional potential $V(R) = \lambda R$, corresponding to a constant restoring force. Use two different trial wavefunctions, $\psi = e^{-r/a}$ and $\psi = e^{-r^2/b^2}$, to estimate the ground-state energy. Which trial wavefunction does a better job?

16.5 Consider a particle of mass m moving in three dimensions in a Yukawa (also known as screened Coulomb) potential

$$V(R) = -\frac{g}{R}e^{-R/a}, \tag{16.58}$$

which behaves at short distances like a Coulomb potential and vanishes exponentially at large distances. Consider a trial wavefunction $\psi = e^{-r/\beta a}$, where β is a dimensionless variational parameter.

(a) Show that the energy expectation value for the trial state is

$$E(\beta) = \frac{\hbar^2}{ma^2}\left(\frac{1}{2\beta^2} - \frac{nG}{\beta(2+\beta)^2}\right), \tag{16.59}$$

where $G = gma/\hbar^2$, and n is a certain positive integer that you will discover.

(b) Find the condition (cubic in β, and linear in G) that gives the minimum of $E(\beta)$.
(c) Since the potential vanishes at large distances, the condition for a bound state is $E < 0$. This requires $G > G_c$, where G_c is a certain critical value for the strength of the potential. Use the condition you found in the previous part to eliminate G, together with $E(\beta) = 0$, to show that the estimate from this trial wavefunction for the critical value is $G_c = 1$. To illustrate this, make a graph of $E(\beta)$ for $0 < \beta < 3$, for the three cases $G = 1/2$, $G = 1$, and $G = 2$. What is β_{min} when $G = G_c = 1$? (By using a more carefully chosen trial wavefunction, it is possible to show that the condition for a bound state to exist is more accurately given by $gma/\hbar^2 > 0.8399$.)

16.6 Consider a particle of mass μ moving in the three-dimensional isotropic harmonic oscillator potential $V(R) = \frac{1}{2}\mu\omega^2 R^2$. Instead of solving exactly as we did in Sect. 10.5, let us see how well the variational principle works with a trial wavefunction $\psi_{l,m}(n,k) = r^n e^{-kr} Y_l^m(\theta,\phi)$ in spherical coordinates. The quantities n and k are the variational parameters. We could take n to vary continuously, but we will consider it to only take on integer values in this problem.

(a) Compute the energy function for the trial wavefunction. Write your answer in the form $E(n,k) = (\hbar^2 k^2/2\mu)A + (\mu\omega^2/k^2)B$, where you should find that A is a quadratic polynomial in l and n divided by a quadratic polynomial in n, and B is another quadratic polynomial in n.
(b) Minimize $E(n,k)$ with respect to k, to obtain $E(n,k_{\text{min}})$. Work in terms of the symbols A and B, and only plug in what they are in the next step.
(c) Estimate the ground-state energy for $l = 0$. Which integer n does the best job? (Plug in $n = 0, 1, 2, 3, 4, 5, 6$.) How does it compare to the exact answer?
(d) Repeat part (c) for the lowest state with $l = 1$, and for $l = 2$.

Fine, Hyperfine, and Magnetic Effects for the Hydrogen Atom

17

17.1 Relativistic Kinetic, Spin-Orbit, and Darwin Corrections

The hydrogen atom Hamiltonian was treated in Chap. 11 in a nonrelativistic approximation. There, it was argued that relativistic effects should alter the energy eigenvalues by an amount suppressed by the square of the fine-structure constant, $\alpha = 1/137.036\ldots$. Now it is time to find those corrections, using stationary-state perturbation theory.

An expansion of the Dirac equation, carried out in Sect. 28.4, shows that a more accurate approximation for the electron (with mass called m_e in this chapter) moving in a general spherically symmetric electrostatic potential[1] $\Phi(R)$ is given by

$$H = H_0 + H_{\text{rel}} + H_{\text{s.o.}} + H_{\text{Darwin}}, \tag{17.1}$$

where

$$H_0 = \frac{P^2}{2m_e} - e\Phi(R) \tag{17.2}$$

is the nonrelativistic approximation we have already used, with $P^2 = \boldsymbol{P} \cdot \boldsymbol{P}$, and

$$H_{\text{rel}} = -\frac{(P^2)^2}{8m_e^3c^2}, \tag{17.3}$$

$$H_{\text{s.o.}} = -\frac{e}{2m_e^2c^2}\frac{1}{R}\frac{d\Phi}{dR}\,\boldsymbol{S} \cdot \boldsymbol{L}, \tag{17.4}$$

$$H_{\text{Darwin}} = -\frac{\hbar^2 e}{8m_e^2c^2}\nabla^2\Phi \tag{17.5}$$

[1] For hydrogen-like atoms, $\Phi(R) = Ze/R$, but in (17.2)–(17.5) we choose to be more general.

S. P. Martin, *Quantum Mechanics*, Graduate Texts in Physics,
https://doi.org/10.1007/978-3-032-21119-4_17

are called the relativistic kinetic, spin-orbit, and Darwin terms, respectively. (The last is named for Charles G. Darwin, grandson of the Charles R. Darwin famous for evolutionary biology.) These three terms give numerical contributions that are of the same order, and they are collectively called the **fine-structure** corrections to the hydrogen atom energies. Each of them is a relativistic effect, as evidenced by the $1/c^2$ factors, which would make them formally vanish if the speed of light were infinite. In this section, we will calculate their effects using perturbation theory. Before doing so, we briefly comment on the physical reasons behind the fine-structure terms.

First, H_{rel} comes from the binomial expansion of the special relativistic expression for the energy of a massive particle with classical momentum p,

$$E = \sqrt{m_e^2 c^4 + p^2 c^2} = m_e c^2 + \frac{p^2}{2m_e} - \frac{(p^2)^2}{8m_e^3 c^2} + \cdots . \tag{17.6}$$

The first term is just a constant, Einstein's equivalence of rest mass and energy. We can ignore it, since it causes all states to be multiplied by the same time-dependent phase factor. The second term is the kinetic term in H_0, and the third gives H_{rel} when $\boldsymbol{p}$ is promoted from classical momentum to the quantum momentum operator $\boldsymbol{P}$.

The $H_{\text{s.o.}}$ term can be understood as the energy due to the interaction of the intrinsic magnetic moment of the electron with the magnetic field $\boldsymbol{B}$ that it experiences because it is moving. Naively, this gives a contribution to the Hamiltonian $-\boldsymbol{\mu}_e \cdot \boldsymbol{B}$, where the magnetic field as seen in the electron's rest frame moving with velocity $\boldsymbol{v}$ is, in Gaussian cgs units,

$$\boldsymbol{B} = -\frac{\boldsymbol{v}}{c} \times \boldsymbol{E}. \tag{17.7}$$

Here, we can use $\boldsymbol{v} = \boldsymbol{p}/m_e$ and, for a spherically symmetric potential,

$$\boldsymbol{E} = -\nabla\Phi = -\hat{r}\frac{d\Phi}{dr}, \tag{17.8}$$

so that, promoting $\boldsymbol{r}$ and $\boldsymbol{p}$ to quantum operators,

$$\boldsymbol{B} = \frac{1}{m_e c}\frac{1}{R}\frac{d\Phi(R)}{dR}(\boldsymbol{P} \times \boldsymbol{R}) = -\frac{1}{m_e c}\frac{1}{R}\frac{d\Phi(R)}{dR}\boldsymbol{L}. \tag{17.9}$$

There is no operator-ordering issue in writing the cross product $\boldsymbol{L} = \boldsymbol{R}\times\boldsymbol{P} = -\boldsymbol{P}\times\boldsymbol{R}$, because different rectangular coordinate components of $\boldsymbol{P}$ and $\boldsymbol{R}$ commute. The magnetic moment of the electron is opposite the direction of the spin, because the charge $-e$ is negative,

$$\boldsymbol{\mu}_e = -\frac{e}{m_e c}\boldsymbol{S}, \tag{17.10}$$

where we have used (4.38) and (4.39) with the (very good) approximation $g_e \approx 2$. So, naively we would have

$$H_{\text{s.o., naive}} = -\boldsymbol{\mu}_e \cdot \boldsymbol{B} = -\frac{e}{m_e^2 c^2} \frac{1}{R} \frac{d\Phi}{dR} \boldsymbol{S} \cdot \boldsymbol{L}. \tag{17.11}$$

The extra factor of 1/2 in the true $H_{\text{s.o.}}$ in (17.4) is known as the Thomas precession effect. Llewellyn Thomas explained it in 1925 as a classical relativistic effect of the electron being at rest in a rotating (not inertial) reference frame. Alternatively, the derivation from the Dirac equation in Sect. 28.4 gives the correct normalization for $H_{\text{s.o.}}$ more directly.

Finally, the physical origin of the Darwin term H_{Darwin} is related to the fact that, as we noted following (11.71), the behavior of the electron is modified by relativistic effects on a small distance scale comparable to its reduced Compton wavelength, $\lambda\!\!\!^{-}_e = \hbar/m_e c$. This effectively makes the replacement $\Phi(R) \to \Phi(R) + \frac{\lambda\!\!\!^{-2}_e}{8}\nabla^2\Phi$, resulting in the correction of (17.5). A proper derivation from the Dirac equation can again be found in Sect. 28.4.

Now we apply first-order perturbation theory to evaluate the H_{rel} contribution to the energies of the unperturbed states $|nlm\rangle$, specializing to the case of the hydrogen atom with $Z = 1$. These unperturbed states have degeneracies for $n > 1$, so we are really doing degenerate perturbation theory, but $W = H_{\text{rel}}$ is already diagonal in this basis, so we can directly apply (15.119). This gives

$$E_{nlm}^{(1),\,\text{rel}} = \langle H_{\text{rel}} \rangle_{nlm} = -\frac{1}{8 m_e^3 c^2} \langle (P^2)^2 \rangle_{nlm}, \tag{17.12}$$

where, here and in the following, expectation values $\langle nlm| \cdots |nlm\rangle$ are denoted by $\langle \cdots \rangle_{nlm}$. To evaluate this efficiently, we use the trick of rewriting it as $P^2 = 2m_e(H_0 + e^2/R)$,

$$E_{nlm}^{(1),\,\text{rel}} = -\frac{1}{2 m_e c^2} \langle (H_0 + e^2/R)^2 \rangle_{nlm} \tag{17.13}$$

$$= -\frac{1}{2 m_e c^2} \left(\mathcal{E}_n^2 + 2\mathcal{E}_n e^2 \langle 1/R \rangle_{nlm} + e^4 \langle 1/R^2 \rangle_{nlm} \right), \tag{17.14}$$

where $\mathcal{E}_n = -e^2/2a_0 n^2$ is the eigenvalue of H_0 for the state $|nlm\rangle$. The expectation values of $1/R$ and $1/R^2$ were given in (11.64) and (11.67), respectively. Plugging those results into (17.14), we find

$$E_{nlm}^{(1),\,\text{rel}} = \alpha^2 \frac{e^2}{2a_0} \frac{1}{n^3} \left(\frac{3}{4n} - \frac{1}{l + 1/2} \right), \tag{17.15}$$

where we have used $a_0 = \hbar^2/m_e e^2$ to write the result in terms of the fine-structure constant $\alpha = e^2/\hbar c \approx 1/137.036$, and $e^2/2a_0 = 1$ Rydberg ≈ 13.606 eV.

By itself, (17.15) is not very useful, because the spin-orbit and Darwin terms contribute at the same order. To evaluate the spin-orbit contribution, we note that for the special case of the hydrogen atom, the electrostatic potential is $\Phi = e/R$, so that (17.4) becomes

$$H_{\text{s.o.}} = \frac{e^2}{2m_e^2c^2}\frac{1}{R^3}\boldsymbol{S}\cdot\boldsymbol{L}. \tag{17.16}$$

This vanishes for all $l = 0$ states, including the ground state $n = 1$, since they have no orbital angular momentum. Because the $n > 1$ states have H_0 eigenvalue degeneracies, we are really doing degenerate perturbation theory, which means that we must use a basis in which $H_{\text{s.o.}}$ is diagonal. We use the standard trick for dot products of angular momentum operators,

$$\boldsymbol{S}\cdot\boldsymbol{L} = \frac{1}{2}(J^2 - L^2 - S^2) = \frac{\hbar^2}{2}\left[j(j+1) - l(l+1) - s(s+1)\right], \tag{17.17}$$

which is diagonal in the total angular momentum orthobasis $|nlsjm_j\rangle$. (In particular, we do *not* want to use the product orthobasis $|nlsm_lm_s\rangle$ here, because then $\boldsymbol{S}\cdot\boldsymbol{L}$ would not be diagonal, as required for degenerate perturbation theory.) Using $s = 1/2$, it follows that the first-order correction is

$$E_{nlj}^{(1),\,\text{s.o.}} = \frac{\hbar^2e^2}{4m_e^2c^2}\left[j(j+1) - l(l+1) - 3/4\right]\langle 1/R^3\rangle_{nl}, \qquad (l \neq 0). \tag{17.18}$$

The expectation value of $1/R^3$ was given in (11.68), reproduced here for convenience,

$$\langle 1/R^3\rangle_{nl} = \frac{1}{a_0^3n^3l(l+1)(l+1/2)}. \tag{17.19}$$

It follows that

$$E_{nlj}^{(1),\,\text{s.o.}} = \frac{\hbar^2e^2}{2m_e^2c^2a_0^3}\frac{j(j+1) - l(l+1) - 3/4}{n^3l(l+1)(2l+1)}, \qquad (l \neq 0). \tag{17.20}$$

Now, using $j = l \pm 1/2$, and writing the prefactor in terms of the fine-structure constant and the Rydberg energy,

$$E_{nlj}^{(1),\,\text{s.o.}} = \alpha^2\left(\frac{e^2}{2a_0}\right)\frac{-1 \pm (2l+1)}{2n^3l(l+1)(2l+1)}, \qquad (l \neq 0,\ j = l \pm 1/2), \tag{17.21}$$

to go along with $E_{n01/2}^{(1),\,\text{s.o.}} = 0$.

Now we turn to the Darwin term of (17.5), which for the hydrogen atom includes, in the position representation, a factor $\nabla^2\Phi = e\nabla^2(1/r) = -4\pi e\delta^{(3)}(\boldsymbol{r})$. It is left to Exercise 17.1 to compute the first-order energy correction

$$E^{(1),\text{Darwin}} = \delta_{l0}\,\alpha^2\left(\frac{e^2}{2a_0}\right)\frac{1}{n^3}, \tag{17.22}$$

which depends only on n, l and is nonzero only for $l = 0$.

We are finally ready to combine the three fine-structure contributions, $E^{\text{fine}} = E^{(1),\text{rel}} + E^{(1),\text{s.o.}} + E^{(1),\text{Darwin}}$, from (17.15), (17.21), and (17.22). This can be done in three separate cases,

$$(l=0,\ j=1/2):\quad E^{\text{fine}} = \frac{\alpha^2}{n^3}\left(\frac{e^2}{2a_0}\right)\left(\frac{3}{4n} - 2 + 0 + 1\right), \tag{17.23}$$

$$(l\neq 0,\ j=l+1/2):\quad E^{\text{fine}} = \frac{\alpha^2}{n^3}\left(\frac{e^2}{2a_0}\right)\left(\frac{3}{4n} - \frac{2}{2l+1} + \frac{1}{(l+1)(2l+1)} + 0\right), \tag{17.24}$$

$$(l\neq 0,\ j=l-1/2):\quad E^{\text{fine}} = \frac{\alpha^2}{n^3}\left(\frac{e^2}{2a_0}\right)\left(\frac{3}{4n} - \frac{2}{2l+1} - \frac{1}{l(2l+1)} + 0\right). \tag{17.25}$$

All three cases are seen to be encompassed by the simple formula

$$E^{\text{fine}}_{nj} = \frac{\alpha^2}{n^3}\left(\frac{e^2}{2a_0}\right)\left(\frac{3}{4n} - \frac{1}{j+1/2}\right), \tag{17.26}$$

which notably depends only on n and j, and not separately on l, as one might have guessed. This result can also be obtained by expanding the exact solution of the Dirac equation; this will be done in Sect. 28.5.

It is convenient to express the fine-structure energy shifts in terms of

$$E_f = \alpha^2 e^2/2a_0 = 7.245\times 10^{-4}\ \text{eV}. \tag{17.27}$$

The results for the lowest few energy levels are then as follows. For $n = 1$,

$$1S_{1/2}\ (n=1, l=0, j=1/2)\qquad E^{\text{fine}} = -E_f/4, \tag{17.28}$$

and for $n = 2$,

$$\left.\begin{array}{l} 2S_{1/2}\ (n=2, l=0, j=1/2) \\ 2P_{1/2}\ (n=2, l=1, j=1/2) \end{array}\right\}\quad E^{\text{fine}} = -5E_f/64, \tag{17.29}$$

$$2P_{3/2}\ (n=2, l=1, j=3/2)\qquad E^{\text{fine}} = -E_f/64, \tag{17.30}$$

where the $2S_{1/2}$ and $2P_{1/2}$ states stay degenerate, and for $n = 3$,

$$\left.\begin{array}{l} 3S_{1/2}\ \ (n=3, l=0, j=1/2) \\ 3P_{1/2}\ \ (n=3, l=1, j=1/2) \end{array}\right\} \ E^{\text{fine}} = -E_f/36, \tag{17.31}$$

$$\left.\begin{array}{l} 3P_{3/2}\ \ (n=3, l=1, j=3/2) \\ 3D_{3/2}\ \ (n=3, l=2, j=3/2) \end{array}\right\} \ E^{\text{fine}} = -E_f/108, \tag{17.32}$$

$$3D_{5/2}\ \ (n=3, l=2, j=5/2) \quad E^{\text{fine}} = -E_f/324. \tag{17.33}$$

Note that the $3S_{1/2}$ and $3P_{1/2}$ states stay degenerate, as do the $3P_{3/2}$ and $3D_{3/2}$ states. Including electron spin, each principal quantum number level n has $(2s + 1)\sum_{l=0}^{n-1}(2l+1) = 2n^2$ states. The fine structure splits each of those levels into n sub-levels, with $j = 1/2, 3/2, \ldots, n - 1/2$. The degeneracies are further enhanced by a factor of 2 for the proton spin, but then are further split by the hyperfine structure to be discussed next.

17.2 Hyperfine Structure of Hydrogen

The hyperfine structure takes into account that an atomic nucleus also has a magnetic dipole moment (along its spin direction) and an electric quadrupole moment (if the nuclear spin is at least 1; see Exercise 13.5). For the ^{1}H isotope of hydrogen with a nucleus consisting of one proton and no neutrons, the magnetic moment of the proton is related to its spin-1/2 operator, which we denote $\boldsymbol{S}_p$, by (see Sect. 4.3):

$$\boldsymbol{\mu}_p = \frac{g_p e}{2m_p c}\boldsymbol{S}_p, \tag{17.34}$$

where m_p is the proton's mass, and $g_p = 5.5856946893(16)$ is the proton's g-factor. The hyperfine contributions to the energy eigenvalues are numerically suppressed compared to the fine-structure contributions because $m_p \gg m_e$.

The interaction between the magnetic field created by the proton's magnetic moment and the moving electron introduces a proton spin-orbit coupling term in the Hamiltonian,

$$H_{\text{s.o.}}^{\text{proton}} = \frac{e}{m_e c R^3}\boldsymbol{\mu}_p \cdot \boldsymbol{L} = \frac{g_p e^2}{2m_e m_p c^2}\frac{1}{R^3}\boldsymbol{S}_p \cdot \boldsymbol{L}. \tag{17.35}$$

In addition, there is the energy of the magnetostatic interaction of the dipole moments of the proton and the electron,

$$H_{\text{dipole-dipole}} = \frac{1}{R^3}\left[\boldsymbol{\mu}_e \cdot \boldsymbol{\mu}_p - \frac{3}{R^2}(\boldsymbol{\mu}_e \cdot \boldsymbol{R})\,(\boldsymbol{\mu}_p \cdot \boldsymbol{R})\right] - \frac{8\pi}{3}\boldsymbol{\mu}_e \cdot \boldsymbol{\mu}_p\,\delta^{(3)}(\boldsymbol{R}), \tag{17.36}$$

in which the final term is called a contact term. These two contributions are of the same numerical order. So, using

$$\boldsymbol{\mu}_e = -\frac{g_e e}{2m_e c}\boldsymbol{S} \tag{17.37}$$

with $g_e \approx 2$ and $\boldsymbol{S}$ the electron's spin, the full hyperfine contribution to the Hamiltonian for hydrogen is

$$\begin{aligned} H_{\text{hf}} &= H_{\text{s.o.}}^{\text{proton}} + H_{\text{dipole-dipole}} \\ &= \frac{g_p e^2}{2m_e m_p c^2}\left[\frac{1}{R^3}(\boldsymbol{L}-\boldsymbol{S})\cdot\boldsymbol{S}_p + \frac{3}{R^5}(\boldsymbol{R}\cdot\boldsymbol{S})(\boldsymbol{R}\cdot\boldsymbol{S}_p) + \frac{8\pi}{3}\boldsymbol{S}\cdot\boldsymbol{S}_p\,\delta^{(3)}(\boldsymbol{R})\right]. \end{aligned} \tag{17.38}$$

We now treat this as a perturbation of the fine-structure results found in the previous section.

In the previous section, we found that even after including the fine structure, there are still energy degeneracies. Therefore, we again need to do degenerate perturbation theory by choosing a good set of unperturbed orthobasis states, so that H_{hf} is diagonal in that basis. The appropriate basis is again a total angular momentum basis, but now taking into account the proton's spin in addition to the electron's spin and orbital angular momentum. We define

$$\boldsymbol{J} = \boldsymbol{L} + \boldsymbol{S}, \tag{17.39}$$

$$\boldsymbol{F} = \boldsymbol{J} + \boldsymbol{S}_p = \boldsymbol{L} + \boldsymbol{S} + \boldsymbol{S}_p, \tag{17.40}$$

so that $\boldsymbol{F}$ is the total angular momentum including the proton's spin, and $\boldsymbol{J}$ is the total excluding the proton's spin. The eigenvalues for J^2 and J_z are denoted $\hbar^2 j(j+1)$ and $\hbar m_j$, where the allowed values for j are $l-1/2$ and $l+1/2$ when $l \neq 0$ and $j = 1/2$ when $l = 0$, and in each case m_j takes on the $2j+1$ values $-j, -j+1, \ldots, j-1, j$. Because $\boldsymbol{F}$ arises from adding the orbital angular momentum to two spin-1/2 angular momenta, the eigenvalues for F^2 and F_z are $\hbar^2 f(f+1)$ and $\hbar m_f$, with integer values $f = j - 1/2$ and $j + 1/2$, and in each case $m_f = -f, -f+1, \ldots, f-1, f$. The total angular momentum orthobasis states are constructed by first combining $\boldsymbol{L}$ and $\boldsymbol{S}$ eigenstates to obtain an orthobasis $|l\,s\,j\,m_j\rangle$, and then taking the tensor product with the proton spin eigenstates $|s_p m_{s_p}\rangle$ and going to the total angular momentum basis to obtain states labeled $|l\,s\,j\,s_p\,f\,m_f\rangle$. Here the s and s_p spin labels are trivial and can be omitted, as they are both fixed to be 1/2. So, including the principal quantum number n, the CSCO = $\{H_0, L^2, J^2, F^2, F_z\}$ eigenstate orthobasis kets can be labeled

$$|n\,l\,j\,f\,m_f\rangle. \tag{17.41}$$

We now want to show that H_{hf} is diagonal in this basis, and obtain its expectation values. It is convenient to treat separately the case $l = 0$ (including the ground state) and the case $l \neq 0$.

For $l = 0$, the first two terms in (17.38) do not contribute. To see this, note that the spherical symmetry of $l = 0$ states guarantees that the expectation value of these two terms is proportional to the angular integral of $3(\hat{r} \cdot \boldsymbol{S})(\hat{r} \cdot \boldsymbol{S}_p) - \boldsymbol{S} \cdot \boldsymbol{S}_p$, where we have used $\boldsymbol{L} = 0$ for $l = 0$. Such an integral is a scalar, is linear in each of the two spins, and depends on no other vectors, so it can only be proportional to their dot product. That is, we must have

$$\int d\Omega \left[3(\hat{r} \cdot \boldsymbol{S})(\hat{r} \cdot \boldsymbol{S}_p) - \boldsymbol{S} \cdot \boldsymbol{S}_p\right] = a\boldsymbol{S} \cdot \boldsymbol{S}_p \tag{17.42}$$

for some constant number a. Now $a = 0$ follows from evaluating the left side for $\boldsymbol{S} = \boldsymbol{S}_p = \hat{z}$, using $\hat{r} \cdot \hat{z} = \cos\theta$.

The last term in (17.38), proportional to $\delta^{(3)}(\boldsymbol{R})$, does contribute for $l = 0$, and only in that case. To see this, note that

$$\langle n l m_l | \delta^{(3)}(\boldsymbol{R}) | n l m_l \rangle = |\psi_{nlm_l}(0, 0, 0)|^2, \tag{17.43}$$

the square of the wavefunction at the origin. Using the radial wavefunction at $r = 0$ from (11.41), this is nonzero only if $l = 0$, which of course implies $m_l = 0$ also, and

$$\langle n l m_l | \delta^{(3)}(\boldsymbol{R}) | n l m_l \rangle = \delta_{l0}\, \delta_{m_l 0} \frac{1}{\pi a_0^3 n^3}. \tag{17.44}$$

Now, with $l = 0$, we have $\boldsymbol{F} = \boldsymbol{S} + \boldsymbol{S}_p$, so

$$\boldsymbol{S} \cdot \boldsymbol{S}_p = \frac{1}{2}(F^2 - S^2 - S_p^2). \tag{17.45}$$

This is indeed diagonal in the total angular momentum basis $|n l j f m_f\rangle$, with $F^2 - S^2 - S_p^2 = \hbar^2[f(f+1) - 3/2]$, where the possible values of f are 0 and 1. Putting together the factors,

$$\langle H_{\rm hf} \rangle = \frac{g_p e^2}{2m_e m_p c^2} \frac{8\pi}{3} \frac{\hbar^2}{2} [f(f+1) - 3/2] \frac{1}{\pi a_0^3 n^3} \qquad (l = 0), \tag{17.46}$$

so we arrive at the hyperfine energy correction for $l = 0$, $j = 1/2$ states,

$$E_{\rm hf}^{(l=0)} = \langle H_{\rm hf} \rangle = \frac{g_p m_e}{m_p} \frac{\alpha^2}{n^3} \left(\frac{e^2}{2a_0}\right) \times \begin{cases} 2/3 & \text{for } f = 1, \\ -2 & \text{for } f = 0. \end{cases} \tag{17.47}$$

In particular, the hyperfine interaction splits the $n = 1$ ground state by an amount

$$E_\gamma = E_{\rm hf}^{(f=1)} - E_{\rm hf}^{(f=0)} = \frac{8 g_p m_e}{3 m_p} \alpha^2 \left(\frac{e^2}{2a_0}\right) = 5.878 \times 10^{-6}\,\text{eV}. \tag{17.48}$$

We had already discussed this in (12.38)–(12.40); here we have shown how the numerical prefactor arises. The actual experimental value 5.87433×10^{-6} eV is very slightly less than (17.48), due to higher-order effects. The $f = 1$ state is higher in energy than the $f = 0$ state, and when it decays it emits a photon with this energy splitting, whose frequency is $\nu = \omega/2\pi = E_\gamma/2\pi\hbar = 1.42041 \times 10^9$ Hz, corresponding to a wavelength

$$\lambda = c/\nu = 21.106\,\text{cm}. \tag{17.49}$$

This is the 21 centimeter line of radio astronomy, which comes from a photon emission or absorption between states with different orientations of the electron and proton spins in the hydrogen atom ground state.

Now take the case $l \neq 0$. We will use the Landé projection formula (13.49) derived in Sect. 13.2 for a general vector operator, given in our present problem by the vector that appears dotted into $\boldsymbol{S}_p$ in (17.38),

$$\boldsymbol{V} = \frac{1}{R^3}(\boldsymbol{L} - \boldsymbol{S}) + \frac{3}{R^5}(\boldsymbol{R} \cdot \boldsymbol{S})\,\boldsymbol{R}, \tag{17.50}$$

so that

$$H_{\text{hf}}^{l\neq 0} = \frac{g_p e^2}{2m_e m_p c^2}\boldsymbol{V} \cdot \boldsymbol{S}_p. \tag{17.51}$$

Here we have used the fact that the $\delta^{(3)}(\boldsymbol{R})$ term does not contribute for $l \neq 0$, as noted in the calculation of (17.44). Because $\boldsymbol{V}$ itself does not involve $\boldsymbol{S}_p$ at all, for the purposes of evaluating its matrix elements we can temporarily ignore the f and m_f quantum numbers when using (13.49) and (13.50), and so take the total angular momentum operator to be just $\boldsymbol{J} = \boldsymbol{L} + \boldsymbol{S}$, the generator of rotations for the orbital and electron spin degrees of freedom.[2] Also, in our present problem, we are calculating an energy correction as an expectation value, so the degeneracy labels α and β appearing in (13.49) are the same, $\alpha = (n, l)$, and the Landé projection formula (13.49) gives

$$\langle \alpha\, j m_j' | \boldsymbol{V} | \alpha\, j m_j \rangle = \langle j m_j' | \boldsymbol{J} | j m_j \rangle \frac{\langle \alpha\, j \| \boldsymbol{J} \cdot \boldsymbol{V} \| \alpha\, j \rangle}{\hbar^2 j(j+1)}, \tag{17.52}$$

where the reduced matrix element is the scalar expectation value

$$\langle \alpha\, j \| \boldsymbol{J} \cdot \boldsymbol{V} \| \alpha\, j \rangle = \langle \alpha\, j m_j | \boldsymbol{J} \cdot \boldsymbol{V} | \alpha\, j m_j \rangle. \tag{17.53}$$

[2] If $\boldsymbol{V}$ had contained $\boldsymbol{S}_p$, then the Landé projection formula (13.49) would apply only if $\boldsymbol{J}$ were replaced by $\boldsymbol{F}$, the total angular momentum operator including the proton spin.

Recall that, because of Theorem 13.2, since $\boldsymbol{J}\cdot\boldsymbol{V}$ is a scalar operator, $\langle\alpha j\|\boldsymbol{J}\cdot\boldsymbol{V}\|\alpha j\rangle$ does not actually depend on m_j at all, despite its appearance on the right side of (17.53).

To evaluate $\langle\alpha j\|\boldsymbol{J}\cdot\boldsymbol{V}\|\alpha j\rangle$ we therefore need the expectation values for the operators $\boldsymbol{J}\cdot(\boldsymbol{L}-\boldsymbol{S})/R^3$ and $(\boldsymbol{R}\cdot\boldsymbol{J})(\boldsymbol{R}\cdot\boldsymbol{S})/R^5$ in a state $|\alpha j m_j\rangle$. For the first of these, we calculate:

$$\boldsymbol{J}\cdot(\boldsymbol{L}-\boldsymbol{S}) = (\boldsymbol{L}+\boldsymbol{S})\cdot(\boldsymbol{L}-\boldsymbol{S}) = L^2 - S^2 = \hbar^2[l(l+1)-3/4]. \tag{17.54}$$

For the second, we use $\boldsymbol{R}\cdot\boldsymbol{L} = \boldsymbol{R}\cdot(\boldsymbol{R}\times\boldsymbol{P}) = 0$, which implies $\boldsymbol{R}\cdot\boldsymbol{J} = \boldsymbol{R}\cdot\boldsymbol{S}$, and therefore

$$(\boldsymbol{R}\cdot\boldsymbol{J})(\boldsymbol{R}\cdot\boldsymbol{S}) = (\boldsymbol{R}\cdot\boldsymbol{S})(\boldsymbol{R}\cdot\boldsymbol{S}) = \hbar^2(\boldsymbol{R}\cdot\boldsymbol{\sigma})(\boldsymbol{R}\cdot\boldsymbol{\sigma})/4 = \hbar^2R^2/4, \tag{17.55}$$

where we have used the Pauli matrix representation of spin, and then the identity (8.45). Putting (17.54) and (17.55) into (17.53) gives

$$\langle\alpha j\|\boldsymbol{J}\cdot\boldsymbol{V}\|\alpha j\rangle = \hbar^2 l(l+1)\langle nl|\,1/R^3\,|nl\rangle \tag{17.56}$$

for the operator $\boldsymbol{V}$ defined in (17.50). So, we can write for (17.51), within matrix elements of states sharing the same n, l, j,

$$H_{\text{hf}}^{l\neq 0} = \frac{g_p e^2}{2m_e m_p c^2}\frac{l(l+1)}{j(j+1)}\,\boldsymbol{J}\cdot\boldsymbol{S}_p\,\frac{1}{R^3}. \tag{17.57}$$

In the orthobasis $|nljfm_f\rangle$, the angular momentum operator factor in (17.57) is diagonal, and evaluates to

$$\boldsymbol{J}\cdot\boldsymbol{S}_p = \frac{1}{2}(F^2 - J^2 - S_p^2) = \frac{\hbar^2}{2}(f(f+1) - j(j+1) - 3/4). \tag{17.58}$$

The radial part of the expectation value was found in (11.68),

$$\langle 1/R^3\rangle = \frac{1}{a_0^3 n^3 l(l+1)(l+1/2)}. \tag{17.59}$$

Using (17.58) and (17.59) to find the expectation value of (17.57), and eliminating j in favor of f, gives

$$E_{\text{hf}} = \langle H_{\text{hf}}\rangle = \frac{g_p m_e}{m_p}\frac{\alpha^2}{n^3}\left(\frac{e^2}{2a_0}\right)\frac{\pm 1}{(l+1/2)(2f+1)}, \tag{17.60}$$

where the + sign applies if $f = j+1/2$ and the − sign if $f = j-1/2$. Actually, this agrees with the $l = 0$, $j = 1/2$ formula in (17.47), so that (17.60) is an all-purpose result, correct for all states $|nljfm_f\rangle$ in the total angular momentum basis.

Note that the hyperfine energy contribution is independent of the quantum number m_f. This had to be true, because the energy eigenvalues for the hydrogen atom cannot possibly depend on the orientation of the total angular momentum with respect to an arbitrarily chosen $\hat{z}$ direction. Therefore, each state with total angular momentum quantum number f has a remaining exact degeneracy of $2f + 1$. The only way to break this degeneracy is to introduce some external effect that picks out a special direction in space, for example an external electric or magnetic field.

In terms of the total $n = 1$ hyperfine splitting E_γ defined in (17.48), the hyperfine splittings of the $n = 2$ states are

$$2S_{1/2}(l = 0, j = 1/2) : E_{\text{hf}} = \begin{cases} E_\gamma/32 & (f = 1) \\ -3E_\gamma/32 & (f = 0) \end{cases} \rightarrow \Delta E_{\text{hf}}^{2S_{1/2}} = E_\gamma/8, \tag{17.61}$$

$$2P_{1/2}(l = 1, j = 1/2) : E_{\text{hf}} = \begin{cases} E_\gamma/96 & (f = 1) \\ -E_\gamma/32 & (f = 0) \end{cases} \rightarrow \Delta E_{\text{hf}}^{2P_{1/2}} = E_\gamma/24, \tag{17.62}$$

$$2P_{3/2}(l = 1, j = 3/2) : E_{\text{hf}} = \begin{cases} E_\gamma/160 & (f = 2) \\ -E_\gamma/96 & (f = 1) \end{cases} \rightarrow \Delta E_{\text{hf}}^{2P_{3/2}} = E_\gamma/60. \tag{17.63}$$

Note that of the $2(2j + 1)$ states before the hyperfine splitting is taken into account, the $2j+2$ states with $f = j+1/2$ are raised by an amount multiplied by $1/(2j+2)$, while the $2j$ states with $f = j - 1/2$ are lowered by the same amount multiplied by $1/2j$. Therefore, the average hyperfine energy contribution to all of the states in any given n, l, j level is always 0.

So far, we have evaluated the fine and hyperfine contributions, by working at first order in degenerate perturbation theory. However, there is a second-order effect, called the **Lamb shift** after Willis E. Lamb Jr., which for $l = 0$ states is numerically larger than the hyperfine effect. The Lamb shift arises due to the electron emitting and reabsorbing virtual photons, quanta of the electromagnetic field. To calculate it in a systematic way requires the toolbox of quantum electrodynamics, including renormalization of ultraviolet divergent contributions and a careful treatment of infrared effects. Understanding these issues played a critical role in the development of relativistic quantum field theory, but here we can only summarize the results numerically. The Lamb shift for atomic hydrogen states $|n, l, j\rangle$ has been calculated to be approximately

$$\Delta E_{\text{Lamb}} = \frac{\alpha^3}{n^3}\left(\frac{e^2}{2a_0}\right) \times \begin{cases} 6.50 & (\text{for } l = 0,\ j = 1/2), \\ 0.026 \pm \dfrac{2}{\pi(2j+1)(2l+1)} & (\text{for } l \neq 0,\ j = l \pm 1/2), \end{cases} \tag{17.64}$$

where the decimal coefficients actually have a slight dependence on n. Note that the Lamb shift is much larger for s-wave ($l = 0$) states than for $l \neq 0$ states. The extra factor of α here (compared to the α^2 factor for fine and hyperfine energy corrections) is indicative of an effect of higher order in perturbation theory. Effects that are even higher order in $\alpha = 1/137.036$ slightly modify each of the fine-structure, hyperfine, and Lamb shift contributions, so each of the energy splittings that we have calculated here cannot be trusted beyond about the percent level. More heroic calculations, not reviewed here, have been done to improve beyond that accuracy.

The fine, hyperfine, and Lamb shift splitting effects on the $n = 2$ states of atomic hydrogen are summarized and illustrated in Fig. 17.1.

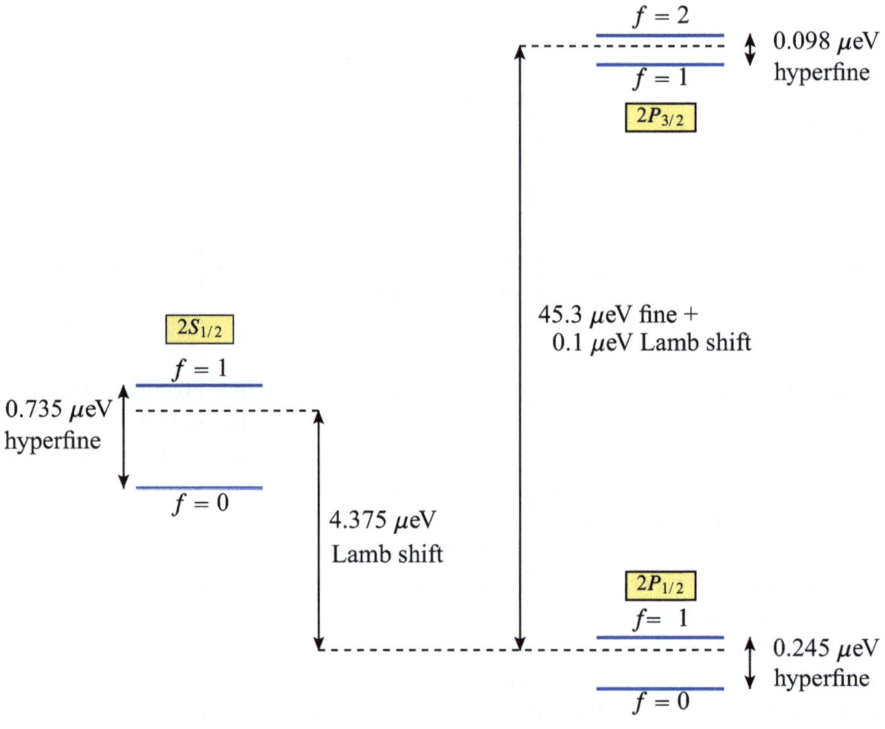

Fig. 17.1 The fine, hyperfine, and Lamb shift energy splittings in μeV for the $n = 2$ energy levels of the hydrogen atom, with nl_j spectroscopic notation, where l is the orbital angular momentum replaced by its (capitalized) letter code as in (11.20), j is the total angular momentum quantum number excluding the proton spin, and f is the total angular momentum quantum number including the proton spin. Relative energy spacings are indicated qualitatively, but not to scale. The Lamb-shift splitting between the $2S_{1/2}$ and $2P_{1/2}$ levels is the experimental one, which differs slightly from the calculation quoted in (17.64) due to higher-order effects. Each energy level shown has a degeneracy $2f + 1$

17.3 Hydrogen Atom in External Magnetic Field (Zeeman and Paschen–Back Effects)

Atomic and molecular state properties can be probed using an external magnetic field. In this section, we consider the energy levels of the hydrogen atom in the presence of a constant uniform magnetic field, which is taken, without loss of generality, to point in the $\hat{z}$ direction,

$$\boldsymbol{B} = B\hat{z}. \tag{17.65}$$

This is associated with the Coulomb-gauge vector potential in spherical coordinates,

$$\boldsymbol{A} = \frac{1}{2}Br\sin\theta\,\hat{\phi}. \tag{17.66}$$

We wish to apply this in the Coulomb-gauge Hamiltonian for a charged particle moving in an electromagnetic field, (4.55), which includes $\boldsymbol{A}\cdot\boldsymbol{P}$. Working in the position representation with $\boldsymbol{P} = -i\hbar\nabla$,

$$\boldsymbol{A}\cdot\boldsymbol{P} = \frac{1}{2}B\Big(-i\hbar\frac{\partial}{\partial\phi}\Big) = \frac{1}{2}BL_z, \tag{17.67}$$

where the last equality uses the representation of L_z of (8.100). So, from (4.55), the external magnetic field contribution to the electron's Hamiltonian is

$$\Delta H = \frac{eB}{2m_e c}\left(L_z + g_e S_z\right) + \frac{e^2B^2}{8m_e c^2}r^2\sin^2\theta. \tag{17.68}$$

Let us first get a rough idea of the order of magnitude of these contributions, for typical laboratory magnetic fields. Putting in the numerical values of the constants gives

$$\begin{aligned}\Delta H =\; &(5.8\times 10^{-5}\,\text{eV})\frac{1}{\hbar}\left(L_z + g_e S_z\right)\frac{B}{\text{Tesla}}\\ &+ (6.2\times 10^{-11}\,\text{eV})\frac{r^2}{a_0^2}\sin^2\theta\left(\frac{B}{\text{Tesla}}\right)^2.\end{aligned} \tag{17.69}$$

From this, we estimate that the term linear in B will contribute comparably to the fine-structure effects in atomic hydrogen if B is of order 1 Tesla. For fields typically accessible in laboratories (of order 10 Tesla or less), the term quadratic in B is much smaller, so we will neglect it. Taking $g_e = 2$, the new Hamiltonian contribution to be considered in this section is therefore

$$H_B = \mu_B B\left(L_z + 2S_z\right)/\hbar, \tag{17.70}$$

where the Bohr magneton μ_B was defined in (4.46). Our goal is to understand the energy splitting of hydrogen atomic states due to this Hamiltonian perturbation, as a function of the magnetic field strength.

Let us first consider the weak-field limit, in which $\mu_B B$ can be treated as a perturbation that is small compared to the fine-structure contributions, but still large compared to the hyperfine energy splittings, which we will neglect. This is called the **Zeeman effect**, after Pieter Zeeman. We apply degenerate first-order perturbation theory to the fine-structure eigenstates $|n\,l\,j\,m\rangle$, with the unperturbed energies now taken to be

$$E = \frac{e^2}{2a_0}\left[-\frac{1}{n^2} + \frac{\alpha^2}{n^3}\left(\frac{3}{4n} - \frac{1}{j+1/2}\right)\right]. \tag{17.71}$$

Following the general discussion of degenerate perturbation theory in Sect. 15.6, the energy shifts due to B will be the eigenvalues of the $(2j+1)\times(2j+1)$ perturbation matrix

$$(H_B)_{m',m} = \frac{\mu_B B}{\hbar}\langle n\,l\,j\,m'|(L_z + 2S_z)|n\,l\,j\,m\rangle, \tag{17.72}$$

for each energy level with fixed n, l, j. (There is no need to consider matrix elements between different values of l, because the perturbation $L_z + 2S_z$ cannot connect states with different l.) Now we write

$$L_z + 2S_z = J_z + S_z, \tag{17.73}$$

and replace J_z by its eigenvalue, using $\langle l\,j\,m'|J_z|l\,j\,m\rangle = \hbar m\delta_{mm'}$. To finish the calculation, we need the matrix elements $\langle l\,j\,m'|S_z|l\,j\,m\rangle$. Here, the Landé projcction formula of (13.49) and (13.50) comes to our aid, with $\boldsymbol{S}$ in the role of $\boldsymbol{V}$. It tells us that the matrix elements of S_z are proportional to those of J_z, with

$$\langle l\,j\,m'|S_z|l\,j\,m\rangle = \langle l\,j\,m'|J_z|l\,j\,m\rangle\frac{\langle l\,j\,m|\boldsymbol{J}\cdot\boldsymbol{S}|l\,j\,m\rangle}{\hbar^2 j(j+1)}, \tag{17.74}$$

where here, and from now on, the common label n is suppressed. Now, we again evaluate $\langle l\,j\,m'|J_z|l\,j\,m\rangle = \hbar m\delta_{mm'}$, and are delighted to see that the whole H_B is proportional to $\delta_{mm'}$, and so is actually a diagonal matrix on the subspace of degenerate total angular momentum basis states. It remains to calculate the matrix element $\langle l\,j\,m|\boldsymbol{J}\cdot\boldsymbol{S}|l\,j\,m\rangle$. This can be done using the standard trick for dot products of angular momenta,

$$\boldsymbol{J}\cdot\boldsymbol{S} = \frac{1}{2}\left[J^2 + S^2 - (\boldsymbol{J}-\boldsymbol{S})^2\right] = \frac{1}{2}\left[J^2 + S^2 - L^2\right], \tag{17.75}$$

which is replaced by its eigenvalue, $\hbar^2[j(j+1) + s(s+1) - l(l+1)]/2$, when acting on $|l\,j\,m\rangle$.

Putting the ingredients of (17.72) together, the energy shifts are

$$\Delta E_B = g\mu_B B m, \tag{17.76}$$

where m is the eigenvalue of $J_z/\hbar$, and

$$g = \frac{3}{2} + \frac{s(s+1) - l(l+1)}{2j(j+1)} \tag{17.77}$$

is called the **Landé g-factor**. These energy splittings are directly proportional to the magnetic quantum number m for the total angular momentum. Using the facts that $s = 1/2$ and $j = l \pm 1/2$, we can rewrite the Landé g-factor for the hydrogen atom as

$$g = \begin{cases} 1 + \frac{1}{2j} & \text{(for } j = l + 1/2\text{)}, \\ 1 - \frac{1}{2j+2} & \text{(for } j = l - 1/2\text{)}. \end{cases} \tag{17.78}$$

The weak-field Zeeman energy shifts for the lowest few values of j are therefore

$$\Delta E_{S_{1/2}} = 2\mu_B B m, \qquad (m = \pm 1/2), \tag{17.79}$$

$$\Delta E_{P_{1/2}} = \frac{2}{3}\mu_B B m, \qquad (m = \pm 1/2), \tag{17.80}$$

$$\Delta E_{P_{3/2}} = \frac{4}{3}\mu_B B m, \qquad (m = \pm 3/2,\ \pm 1/2), \tag{17.81}$$

$$\Delta E_{D_{3/2}} = \frac{4}{5}\mu_B B m, \qquad (m = \pm 3/2,\ \pm 1/2), \tag{17.82}$$

$$\Delta E_{D_{5/2}} = \frac{6}{5}\mu_B B m, \qquad (m = \pm 5/2,\ \pm 3/2,\ \pm 1/2), \tag{17.83}$$

etc. By applying a weak magnetic field, the energy degeneracies for each principal quantum number n are eliminated, and the observed splittings can be used to count the (previously degenerate) states and deduce their angular momentum quantum numbers. This type of phenomenon is the historical reason for calling m the magnetic quantum number.

Next, consider the opposite limit, called the **Paschen–Back** regime after Friedrich Paschen and Ernst Back, in which the effect of the external magnetic field is taken to be much stronger than the fine-structure effect, which will be treated as a perturbation. The Hamiltonian $H_0 = P^2/2m_e - e^2/R$ commutes with the magnetic field Hamiltonian H_B in (17.70), so the exact energy eigenstates of $H_0 + H_B$ are the product angular momentum states $|nlm_lm_s\rangle$, with energy eigenvalues $E = -e^2/2a_0n^2 + \Delta E_B$, where

$$\Delta E_B = \mu_B B(m_l + 2m_s). \tag{17.84}$$

Now we must apply the fine-structure Hamiltonian to these states as a perturbation. A key point is that when doing so, we cannot use the states $|nljm\rangle$ as the unperturbed states, as we did in the previous treatment of the fine structure with a vanishing or weak external magnetic field. This is because the strong magnetic field has eliminated the spherical symmetry and the degeneracy associated with it. Instead, degenerate perturbation theory tells us that the fine-structure energy shifts are obtained as the eigenvalues of the matrix of the fine-structure Hamiltonian H_{fine} on each degenerate subspace of product basis states $|nlm_lm_s\rangle$ with fixed n and fixed $m_l + 2m_s$. Fortunately, as we will soon see, when restricted to each of these degenerate subspaces, H_{fine} is diagonal in the product basis, so the problem amounts to finding its expectation values in the product basis states.

First, consider the easier case that $l = 0$. Then the spin-orbit Hamiltonian vanishes, and the fine-structure contributions to the energies are the same as found in (17.23) from the H_{rel} and H_{Darwin} terms. Since $m_l = 0$ in this case, the combined energy shift due to the strong external B field and the fine structure is

$$\Delta E^{l=0}_{B,\,\text{fine}} = 2\mu_B B m_s + \frac{\alpha^2}{n^3}\left(\frac{e^2}{2a_0}\right)\left(\frac{3}{4n} - 1\right). \tag{17.85}$$

Now consider the case $l \neq 0$. It is convenient to write the operator $\boldsymbol{S}\cdot\boldsymbol{L}$ in the spin-orbit Hamiltonian in (17.16) as

$$\boldsymbol{S}\cdot\boldsymbol{L} = S_z L_z + \frac{1}{2}(S_+L_- + S_-L_+). \tag{17.86}$$

The reason is that in the product basis states $|nlm_lm_s\rangle$, the S_zL_z part immediately evaluates to $\hbar^2 m_s m_l$, and the combination $S_+L_- + S_-L_+$ always changes $m_l + 2m_s$ by one unit, so that it has vanishing matrix elements within the degenerate subspace. Thus, $H_{\text{s.o.}}$ is diagonal on each degenerate subspace with fixed n and fixed $m_l + 2m_s$, and the energy corrections from first-order perturbation theory are just the expectation values

$$\langle nlm_lm_s|H_{\text{s.o.}}|nlm_lm_s\rangle = \frac{e^2}{2m_e^2c^2}\langle 1/R^3\rangle_{nl}\,\langle lm_lm_s|\,\boldsymbol{S}\cdot\boldsymbol{L}\,|lm_lm_s\rangle\,. \tag{17.87}$$

The expectation value $\langle 1/R^3\rangle_{nl}$ was already given in (17.19). Note the difference from the previously treated spin-orbit case with a weak magnetic field, where one must instead evaluate the expectation value of $\boldsymbol{L}\cdot\boldsymbol{S}$ in the total angular momentum basis states. Combining the spin-orbit result with the H_{rel} contribution from (17.15), and recalling from (17.22) that the Darwin contribution vanishes for $l \neq 0$, we obtain the total first-order perturbation contribution from the strong external B field and the fine-structure effects:

$$\begin{aligned}\Delta E^{l\neq 0}_{B,\text{fine}} &= \mu_B B(m_l + 2m_s)\\ &\quad+\alpha^2\frac{e^2}{2a_0}\frac{1}{n^3}\left(\frac{m_lm_s}{l(l+1)(l+1/2)} + \frac{3}{4n} - \frac{1}{l+1/2}\right).\end{aligned} \tag{17.88}$$

Finally, let us consider the intermediate case in which the effects of the fine structure and external-magnetic-field Hamiltonians are comparable to each other, so that $H_{\text{fine}} + H_B$ must be treated together as a single perturbation to H_0. The preceding results for weak B and for strong B can be obtained as special-case limits of the following analysis. As before, we will ignore the smaller hyperfine effects. For the unperturbed states, it is not a bad idea to choose[3] the total angular momentum basis elements $|nl\,jm\rangle$, so that the matrix for the fine-structure Hamiltonian H_{fine} is diagonal and we can make use of results already found for it in Sect. 17.1.

For $l = 0$ states, there is no distinction between S_z and J_z, so $H_B = 2\mu_B B m_s = 2\mu_B B m$ is also diagonal in the total angular momentum basis. This means that the energy shift for the $l = 0$ states is the same as we already found in (17.85).

For $l \neq 0$ states, things are more complicated, as H_B is not diagonal. Let us consider only the $n = 2$ level with $l = 1$, where the six degenerate unperturbed states are

$$|j, m\rangle = |\tfrac{3}{2}, \tfrac{3}{2}\rangle,\ |\tfrac{3}{2}, \tfrac{1}{2}\rangle,\ |\tfrac{3}{2}, -\tfrac{1}{2}\rangle,\ |\tfrac{3}{2}, -\tfrac{3}{2}\rangle,\ |\tfrac{1}{2}, \tfrac{1}{2}\rangle,\ |\tfrac{1}{2}, -\tfrac{1}{2}\rangle. \tag{17.89}$$

Following the rules of degenerate perturbation theory, the first-order energy corrections are the eigenvalues of the matrix for $W = H_B + H_{\text{fine}}$ on this subspace. Consider first the contribution from $H_B = \mu_B B(L_z + 2S_z)/\hbar$. Because it commutes with J_z, the only non-zero matrix elements of H_B in the $|j, m\rangle$ basis are between states with the same m. These matrix elements can be most efficiently evaluated by using (12.61)–(12.66) to express the $|j, m\rangle$ states in terms of the product orthobasis states $|m_l, m_s\rangle$, allowing the operator $(L_z + 2S_z)/\hbar$ to be replaced by the number $m_l + 2m_s$. The complete list of non-zero matrix elements in the $|j, m\rangle$ basis is:

$$\langle\tfrac{3}{2}, \tfrac{3}{2}|(L_z + 2S_z)/\hbar\,|\tfrac{3}{2}, \tfrac{3}{2}\rangle = 2, \qquad \langle\tfrac{3}{2}, -\tfrac{3}{2}|(L_z + 2S_z)/\hbar\,|\tfrac{3}{2}, -\tfrac{3}{2}\rangle = -2, \tag{17.90}$$

$$\langle\tfrac{3}{2}, \tfrac{1}{2}|(L_z + 2S_z)/\hbar\,|\tfrac{3}{2}, \tfrac{1}{2}\rangle = 2/3, \quad \langle\tfrac{3}{2}, -\tfrac{1}{2}|(L_z + 2S_z)/\hbar\,|\tfrac{3}{2}, -\tfrac{1}{2}\rangle = -2/3, \tag{17.91}$$

$$\langle\tfrac{1}{2}, \tfrac{1}{2}|(L_z + 2S_z)/\hbar\,|\tfrac{1}{2}, \tfrac{1}{2}\rangle = 1/3, \quad \langle\tfrac{1}{2}, -\tfrac{1}{2}|(L_z + 2S_z)/\hbar\,|\tfrac{1}{2}, -\tfrac{1}{2}\rangle = -1/3, \tag{17.92}$$

$$\langle\tfrac{3}{2}, \tfrac{1}{2}|(L_z + 2S_z)/\hbar\,|\tfrac{1}{2}, \tfrac{1}{2}\rangle = \langle\tfrac{1}{2}, \tfrac{1}{2}|(L_z + 2S_z)/\hbar\,|\tfrac{3}{2}, \tfrac{1}{2}\rangle = -\sqrt{2}/3, \tag{17.93}$$

$$\langle\tfrac{3}{2}, -\tfrac{1}{2}|(L_z + 2S_z)/\hbar\,|\tfrac{1}{2}, -\tfrac{1}{2}\rangle = \langle\tfrac{1}{2}, -\tfrac{1}{2}|(L_z + 2S_z)/\hbar\,|\tfrac{3}{2}, -\tfrac{1}{2}\rangle = -\sqrt{2}/3. \tag{17.94}$$

[3] One could instead choose the product basis states as the unperturbed eigenstates of H_0. Then H_B would be diagonal and H_{fine} would be off-diagonal, but the final results are of course the same.

For the H_{fine} contribution, we see from (17.26) that the nonzero matrix elements are

$$\langle \tfrac{3}{2}, m|H_{\text{fine}}|\tfrac{3}{2}, m\rangle = -\frac{1}{64}\alpha^2\frac{e^2}{2a_0}, \qquad (m = \pm 3/2, \pm 1/2), \tag{17.95}$$

$$\langle \tfrac{1}{2}, m|H_{\text{fine}}|\tfrac{1}{2}, m\rangle = -\frac{5}{64}\alpha^2\frac{e^2}{2a_0}, \qquad (m = \pm 1/2). \tag{17.96}$$

We can now put together these results to construct the matrix elements of $W = H_B + H_{\text{fine}}$. Using the notations

$$a = \frac{1}{64}\alpha^2\frac{e^2}{2a_0}, \qquad b = \mu_B B \tag{17.97}$$

for simplicity, and ordering the basis elements as $|\tfrac{3}{2}, \tfrac{3}{2}\rangle$, $|\tfrac{3}{2}, -\tfrac{3}{2}\rangle$, $|\tfrac{3}{2}, \tfrac{1}{2}\rangle$, $|\tfrac{1}{2}, \tfrac{1}{2}\rangle$, $|\tfrac{3}{2}, -\tfrac{1}{2}\rangle$, $|\tfrac{1}{2}, -\tfrac{1}{2}\rangle$, the matrix representation for the perturbation is

$$W = \begin{pmatrix} -a+2b & 0 & 0 & 0 & 0 & 0 \\ 0 & -a-2b & 0 & 0 & 0 & 0 \\ 0 & 0 & -a+2b/3 & -\sqrt{2}b/3 & 0 & 0 \\ 0 & 0 & -\sqrt{2}b/3 & -5a+b/3 & 0 & 0 \\ 0 & 0 & 0 & 0 & -a-2b/3 & -\sqrt{2}b/3 \\ 0 & 0 & 0 & 0 & -\sqrt{2}b/3 & -5a-b/3 \end{pmatrix}. \tag{17.98}$$

This contains only 1×1 and 2×2 nonzero blocks, with eigenvalues

$$\Delta E_{B,\,\text{fine}}^{n=2,\,l=1} = -a \pm 2b, \tag{17.99}$$

$$-3a + b/2 \pm \sqrt{4a^2 + 2ab/3 + b^2/4}, \tag{17.100}$$

$$-3a - b/2 \pm \sqrt{4a^2 - 2ab/3 + b^2/4}. \tag{17.101}$$

In the same notation, the $l = 0$ energy corrections from (17.85) are

$$\Delta E_{B,\,\text{fine}}^{n=2,\,l=0} = -5a \pm b. \tag{17.102}$$

These eight energy corrections to the $n = 2$ states are graphed in Fig. 17.2 as a function of the external magnetic field B, also including the small positive Lamb shift of 4.375 μeV for the $l = 0$ eigenstates. You are encouraged to confirm that expanding (17.99)–(17.102) in small b/a recovers the weak-field limit of (17.26) and (17.79)–(17.81), and expanding in small a/b recovers the strong-field limit found in (17.85) and (17.88).

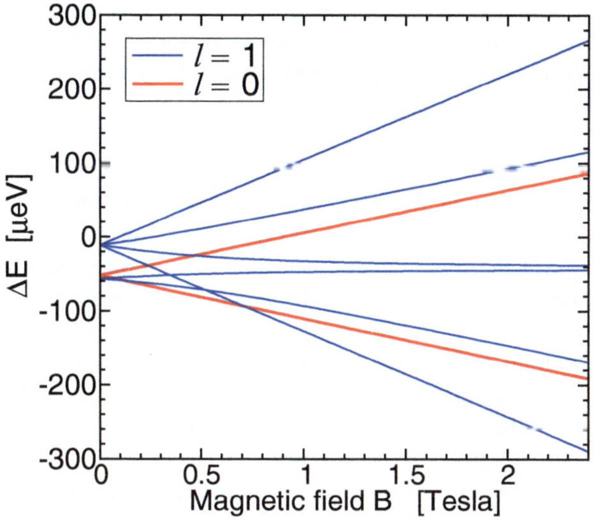

Fig. 17.2 Energy corrections (in μeV) to the $n = 2$ states of atomic hydrogen, due to the fine-structure Hamiltonian and an external magnetic field B (in Tesla), as found in (17.99)–(17.102). At small B, the weak-field Zeeman effect energy eigenstates approach total angular momentum eigenstates $|l, s, j, m\rangle$; the upper four states are $2P_{3/2}$, and the lower four states are $2P_{1/2}$ and $2S_{1/2}$. The small Lamb shift of 4.375 μeV has also been included for the $2S_{1/2}$ states, and is barely visible. At large B, the strong-field (Paschen–Back regime) energy eigenstates approach product angular momentum eigenstates $|l, s, m_l, m_s\rangle$, with $m_l + 2m_s = 2, 1, 0, -1$, and -2, from top to bottom

Exercises

17.1 Consider the normalized hydrogen atom stationary-state wavefunctions, which are given by $\psi_{n,l,m}(\boldsymbol{r}) = R_{n,l}(r)Y_l^m(\theta, \phi)$. For general n, l, m, compute the probability density for the electron to be at the origin, $|\psi_{n,l,m}(0)|^2$. [Hint: use (11.41).] Use the result to obtain the Darwin term energy correction in (17.22).

17.2 Make a diagram of the relative energies for the $n = 3$ level of atomic hydrogen, similar to Fig. 17.1 for the $n = 2$ level. Don't try to draw the energy splittings to scale, but include numerical values in μeV for the fine-structure, hyperfine, and Lamb shift splittings. Use spectroscopic notation to label the states before the hyperfine effect, and label the hyperfine-split states by their f (grand total angular momentum) quantum numbers.

17.3 Deuterium, also known as heavy hydrogen, is an atom with one electron and a nucleus consisting of a deuteron (a proton-neutron bound state) with spin 1. The magnetic moment operator of the deuteron is

$$\boldsymbol{\mu}_d = \frac{g_d e}{2m_d c}\boldsymbol{S}_d, \tag{17.103}$$

where $\boldsymbol{S}_d$ is the spin-1 operator, $m_d = 1875.6\ \text{MeV}/c^2$ is the deuteron mass, and $g_d = 1.713$. Work out the hyperfine structure of the $n = 1$ state of deuterium. What are the energy, frequency, and wavelength of the photon corresponding to the hyperfine transition of the ground state?

17.4 An electron is moving in a spherically symmetric but otherwise generic (no other special properties) potential $V(r)$, with a perturbation Hamiltonian of the spin-orbit type, $W = a\boldsymbol{L} \cdot \boldsymbol{S}$, where a is a positive constant, $\boldsymbol{L}$ is the orbital angular momentum, and $\boldsymbol{S}$ is the spin operator. Consider a set of unperturbed states that are degenerate in energy (before including W), and have L^2 eigenvalue $12\hbar^2$.

(a) Work in the total angular momentum basis $|j, m_j\rangle$ to answer the following questions. What are the energy corrections due to W? What are the remaining degeneracies of the corresponding energy eigenstates?
(b) Repeat part (a), but this time do your work in terms of the matrix elements of W in the product angular momentum basis $|m_l, m_s\rangle$. [Hint: you may want to use (17.86).]
(c) Repeat part (a), but this time pretend that you live in a universe where the electron has spin $3/2$ (instead of spin $1/2$, as in our universe).

17.5 Consider a hydrogen atom in a constant uniform magnetic field. Find the energy shifts for the $n = 3$ states [similar to (17.99)–(17.102) for the $n = 2$ states].

Identical Particles and Multi-electron Atoms 18

18.1 Intrinsic Indistinguishability of Identical Particles

Particles are said to be **identical** if there is no way of distinguishing them, even in principle. In particular, they must have the same mass, spin magnitude, charge, and any other intrinsic properties. Classically, one can keep track of individual particles, even if they are identical, by following their trajectories, which are determined by the equations of motion. However, in quantum mechanics, this is impossible.

For example, if we scatter two electrons off of each other, as illustrated in Fig. 18.1, we may choose to label the electrons in the initial state by $1, 2$, and the electrons in the final state by $3, 4$. However, it has no meaning to say that final-state electron 3 is uniquely the same as the initial-state electron 1. Both of the outgoing electrons 3 and 4 have an equal claim to being the same as the incoming electron 1 (or 2). Although we may choose a labeling scheme, and it is generally quite useful to do so, the physics results cannot depend at all on that choice.

One way to describe a state with N identical particles is to take a tensor product of N identical individual state spaces. Suppose that we have an orthobasis of single-particle kets $|\alpha_n\rangle$, where each α_n represents a whole set of CSCO labels for one particle. An orthobasis of kets for the N-particle Hilbert space is, mathematically,[1]

$$|\alpha_1\rangle \otimes |\alpha_2\rangle \otimes \cdots \otimes |\alpha_N\rangle. \tag{18.1}$$

For example, we could take $\alpha_n = (\boldsymbol{r}_n, m_n)$, and the orthobasis elements would be

$$|\boldsymbol{r}_1, m_1\rangle \otimes |\boldsymbol{r}_2, m_2\rangle \otimes \cdots \otimes |\boldsymbol{r}_N, m_N\rangle, \tag{18.2}$$

[1] We say "mathematically" here because the physical N-identical-particle Hilbert space is a proper subset of the mathematical N-identical-particle Hilbert space, as we will soon see.

S. P. Martin, *Quantum Mechanics*, Graduate Texts in Physics,
https://doi.org/10.1007/978-3-032-21119-4_18

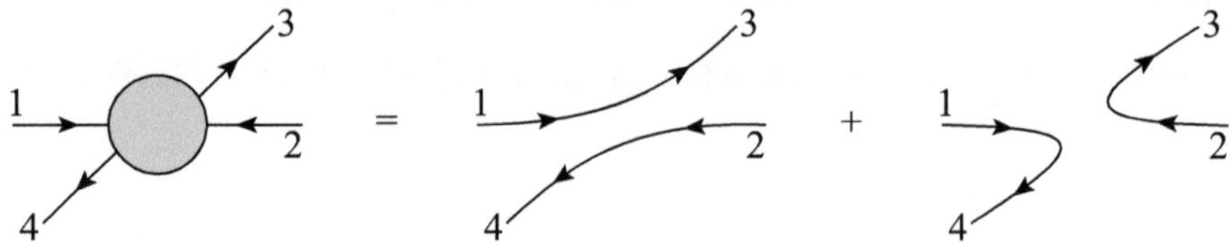

Fig. 18.1 In elastic scattering of indistinguishable particles (for example, electrons) $1, 2 \to 3, 4$, each of the final-state particles labeled 3 and 4 has an equal claim to be the same as each of the initial-state particles 1 and 2 of the same type. It is allowed and useful to label the identical particles, but physical results cannot depend on our labeling choice

where the interpretation of $|\boldsymbol{r}_n, m_n\rangle$ is that we are certain to find the particle at position $\boldsymbol{r}_n$ with spin component $S_z = \hbar m_n$. Alternatively, α_n could instead include the momentum $\boldsymbol{p}_n$ rather than the position, or it could include the energy, L^2, and L_z quantum numbers of the nth particle, or some other choice of CSCO eigenvalues. In any case, we can dispense with the tensor product notation and write (18.1) as a single ket for a given orthobasis element,

$$|\alpha_1, \alpha_2, \ldots, \alpha_N\rangle. \tag{18.3}$$

The fact that the particles are identical means that although we may choose to label them by integers $n = 1, 2, \ldots, N$, we cannot distinguish them, and so physical results cannot depend on our arbitrary choice of labels. In particular, this means that any observable operator must be unchanged if we exchange any pair of labels.

For example, the Hamiltonian for two identical particles of mass m might be

$$H = \frac{P_1^2}{2m} + \frac{P_2^2}{2m} + V(\boldsymbol{R}_1) + V(\boldsymbol{R}_2) + V_{\text{int}}(|\boldsymbol{R}_1 - \boldsymbol{R}_2|), \tag{18.4}$$

since this is invariant under the exchange of labels $1 \leftrightarrow 2$. It is important here that the potentials $V(\boldsymbol{r})$ for the individual identical particles 1 and 2 are the same function. The total momentum operator for N identical particles,

$$\boldsymbol{P} = \boldsymbol{P}_1 + \boldsymbol{P}_2 + \cdots + \boldsymbol{P}_N, \tag{18.5}$$

is likewise an observable, but the individual operators $\boldsymbol{P}_n$ appearing in it are not observables. Similarly, we can define the observable operator for the density at a fixed position $\boldsymbol{r}$ as

$$\rho(\boldsymbol{r}) = \sum_{n=1}^{N} \delta^{(3)}(\boldsymbol{r} - \boldsymbol{R}_n). \tag{18.6}$$

Here $\boldsymbol{r}$ is an ordinary 3-vector that labels which operator we are talking about, and $\boldsymbol{R}_n$ are the individual position operators for the identical particles.

Pair-exchange operators swap the labels of identical particles, and are defined by

$$P_{ij}|\ldots,\alpha_i,\ldots,\alpha_j,\ldots\rangle = |\ldots,\alpha_j,\ldots,\alpha_i,\ldots\rangle. \tag{18.7}$$

So, for example, $P_{13}|\alpha_1,\alpha_2,\alpha_3\rangle = |\alpha_3,\alpha_2,\alpha_1\rangle$ in the special case of a system of three identical particles. Note that P_{ij} is both unitary and Hermitian (like the parity operator), and $P_{ij} = P_{ji}$, and $(P_{ij})^2 = I$. It follows from the last property that if $P_{ij}|\psi\rangle = \lambda|\psi\rangle$, then $\lambda^2 = 1$ unless $|\psi\rangle$ is the null ket, so the only possible eigenvalues of P_{ij} are $\lambda = \pm 1$. Despite being Hermitian, P_{ij} is not itself an observable for $N > 2$, since, for example, P_{12} is not left unchanged if we exchange identical particles 1 and 3.

Now, since observables must be symmetric under the pair exchange of labels, we have

Theorem 18.1 *For any observable A, and any pair-exchange operator P_{ij} for identical particles,*

$$[P_{ij}, A] = 0. \tag{18.8}$$

Proof Write $A = A(a_1, a_2, \ldots, a_N)$, where the a_n are operators that act nontrivially only on orthobasis kets $|\alpha_n\rangle$ for the nth particle. Since A must be symmetric under exchange of labels,

$$A(\ldots,a_i,\ldots,a_j,\ldots) = A(\ldots,a_j,\ldots,a_i,\ldots). \tag{18.9}$$

It follows that

$$\begin{aligned} P_{ij}A|\ldots,\alpha_i,\ldots,\alpha_j,\ldots\rangle &= A|\ldots,\alpha_j,\ldots,\alpha_i,\ldots\rangle \\ &= AP_{ij}|\ldots,\alpha_i,\ldots,\alpha_j,\ldots\rangle, \end{aligned} \tag{18.10}$$

where the first equality made use of (18.9). Since the states $|\ldots,\alpha_i,\ldots,\alpha_j,\ldots\rangle$ are an orthobasis, the theorem follows by linearity since $P_{ij}A$ must equal AP_{ij} acting on every state. □

In particular, the Hamiltonian must commute with every P_{ij},

$$[H, P_{ij}] = 0. \tag{18.11}$$

Now suppose that $|E\rangle$ is an eigenket of H, with energy eigenvalue E. Then,

$$H\left(P_{ij}|E\rangle\right) = P_{ij}H|E\rangle = EP_{ij}|E\rangle, \tag{18.12}$$

so $P_{ij}|E\rangle$ is also an eigenket of H with the same energy eigenvalue E. This fact sometimes goes by the name of **exchange degeneracy**.

However, while mathematically useful, the exchange degeneracy is physically spurious, because it only applies to mathematical kets, and not to physical states. The reason is the **spin-statistics principle**: physical states are required to be eigenstates of all exchange operators P_{ij} for pairs of identical particles, with eigenvalue $+1$ if the identical particles being exchanged are bosons (have integer spin), and eigenvalue -1 if they are fermions (have spin $1/2$, $3/2$, ...). These are the only two possible eigenvalues, as noted after (18.7). Fermions are said to obey **Fermi–Dirac statistics**, and bosons obey **Bose–Einstein statistics**.

The spin-statistics principle is sometimes taken as a postulate, but it can be proved as a theorem in four-dimensional relativistic quantum field theory, the fundamental quantum mechanical framework that underlies our best understanding of the universe at small distance scales. Since it can be proved in that context, rather than assumed, it was not listed among the postulates of quantum mechanics in Sect. 3.1. On the other hand, if a quantum system is not assumed to obey special relativity, then the spin-statistics principle cannot be proved, but can be adopted (or not) as an assumption. Alternatively, one can view the spin-statistics principle simply as an experimentally observed fact, subject at any time to potential falsification.

The spin-statistics principle applies not just for the particles that we currently believe to be fundamental, including bosons (photons, gluons, W and Z bosons, and the Higgs particle) and fermions (electrons, muons, tau leptons, neutrinos, quarks), but even for composite bosons (for example, H atoms, deuterons, ^{4}He nuclei, pions) and composite fermions (protons, neutrons, ^{3}He nuclei). Whether a composite particle is a boson or a fermion is determined simply by whether its constituents include an even or odd number of fermions, respectively. For example, a ^{4}He nucleus consists of two protons and two neutrons. Doing a pair exchange of two ^{4}He nuclei therefore involves the simultaneous exchange of four fermions, resulting in a factor of $(-1)^4$. The proton can be viewed as consisting of $3+n$ quarks and n antiquarks, where naively $n=0$, but the presence of virtual particles inside the proton means that n is actually indeterminate. Nevertheless, n takes on integer values, so the pair exchange of two protons always results in a factor of $(-1)^{3+n}(-1)^n = -1$, making the proton a fermion.

The exchange degeneracy described by (18.12) is completely eliminated from physical states by the spin-statistics principle. For example, if there are two identical particles, then $|\alpha_1,\alpha_2\rangle$ and $|\alpha_2,\alpha_1\rangle$ are not eigenstates of P_{12} unless $\alpha_1=\alpha_2$. If $\alpha_1\neq\alpha_2$, then using

$$P_{12}|\alpha_1,\alpha_2\rangle = |\alpha_2,\alpha_1\rangle, \qquad P_{12}|\alpha_2,\alpha_1\rangle = |\alpha_1,\alpha_2\rangle, \tag{18.13}$$

one finds that the linear combinations

$$|\alpha_1, \alpha_2\rangle_S = \frac{1}{\sqrt{2}} \left(|\alpha_1, \alpha_2\rangle + |\alpha_2, \alpha_1\rangle \right), \tag{18.14}$$

$$|\alpha_1, \alpha_2\rangle_A = \frac{1}{\sqrt{2}} \left(|\alpha_1, \alpha_2\rangle - |\alpha_2, \alpha_1\rangle \right), \tag{18.15}$$

are normalized eigenkets of P_{12}, with eigenvalues $+1$ and -1, respectively. Therefore, according to the spin-statistics principle, only the linear combination $|\alpha_1, \alpha_2\rangle_S$ is a physical state if the two particles are identical bosons, and only $|\alpha_1, \alpha_2\rangle_A$ is a physical state if the two particles are identical fermions. In either case, if $|\alpha_1, \alpha_2\rangle$ is an energy eigenstate, then exactly one linear combination is physical, and there is no true energy degeneracy associated with the exchange of identical particles. (However, there may be energy degeneracies for other reasons.)

If instead $\alpha_1 = \alpha_2$, then $|\alpha_1, \alpha_1\rangle_S = |\alpha_1, \alpha_1\rangle$ is a physical state if the particles are bosons; note that only the normalization has changed compared to (18.14). But if the particles are fermions, then $|\alpha_1, \alpha_1\rangle_A$ is the null ket, so there is no such physical state. This is the **Pauli exclusion principle**: two identical fermions are not allowed to be in the same state. In particular, they are not allowed to have the same eigenvalues of a CSCO.

To construct physical states that satisfy the spin-statistics principle for $N > 2$ identical particles, we first construct general **permutation operators**, which are arbitrary products of pair-exchange operators, for example $P = P_{12}P_{13}P_{45}$. (Note that the order of the P_{ij} matters, because they do not always commute.) There are many different ways of writing a given permutation operator as a product in this way, because of identities like $P_{12}P_{23}P_{12} = P_{13}$, but the total number of distinct permutation operators is exactly $N!$. They can all be classified as odd or even, according to whether the number of pair-exchange operators in the product is even or odd. If P is an even permutation, we define $(-1)^P = +1$, and if P is an odd permutation, we write $(-1)^P = -1$.

We can then define the **symmetrization and antisymmetrization operators** as sums over all $N!$ distinct permutation operators for N identical particles,

$$\mathcal{S} = \frac{1}{N!} \sum_P P, \tag{18.16}$$

$$\mathcal{A} = \frac{1}{N!} \sum_P (-1)^P P. \tag{18.17}$$

These are projection operators onto the subspaces of completely symmetric and antisymmetric kets, respectively, and satisfy the properties $\mathcal{S}^2 = \mathcal{S}$ and $\mathcal{A}^2 = \mathcal{A}$ and $\mathcal{S}\mathcal{A} = \mathcal{A}\mathcal{S} = 0$. Furthermore, we have $P_{ij}\mathcal{S} = \mathcal{S}$ and $P_{ij}\mathcal{A} = -\mathcal{A}$ for all i, j, so that any ket acted on by $\mathcal{S}$ will be an eigenstate of every P_{ij} with eigenvalue 1, and any ket acted on by $\mathcal{A}$ will be an eigenstate of every P_{ij} with eigenvalue -1.

It follows that the physical orthobasis states for N identical fermions are proportional to $\mathcal{A}|\alpha_1, \alpha_2, \ldots, \alpha_N\rangle$,

$$|\alpha_1, \alpha_2, \ldots, \alpha_N\rangle_A = \frac{1}{\sqrt{N!}} \sum_P (-1)^P P |\alpha_1, \alpha_2, \ldots, \alpha_N\rangle. \quad (18.18)$$

These are the simultaneous eigenstates for every pair-exchange operator P_{ij} with eigenvalue -1, unique up to normalization, that one can build out of the single-particle orthobasis states $|\alpha_i\rangle$. The prefactor has been chosen to make the ket have unit norm if the $|\alpha_i\rangle$ are orthonormal and all distinct from each other. An equivalent way to write this is the **Slater determinant**, named after John C. Slater,

$$|\alpha_1, \alpha_2, \ldots, \alpha_N\rangle_A = \frac{1}{\sqrt{N!}} \begin{vmatrix} |\alpha_1\rangle_1 & |\alpha_1\rangle_2 & \ldots & |\alpha_1\rangle_N \\ |\alpha_2\rangle_1 & |\alpha_2\rangle_2 & \ldots & |\alpha_2\rangle_N \\ \vdots & \vdots & \ddots & \vdots \\ |\alpha_N\rangle_1 & |\alpha_N\rangle_2 & \ldots & |\alpha_N\rangle_N \end{vmatrix}, \quad (18.19)$$

where the products obtained from the determinant are to be understood as

$$|\alpha_i\rangle_1 \, |\alpha_j\rangle_2 \ldots |\alpha_k\rangle_N = |\alpha_i, \alpha_j, \ldots, \alpha_k\rangle. \quad (18.20)$$

If any two of the α_i are the same, then the determinant vanishes, and there is no corresponding physical state at all. This is the more general version of the Pauli exclusion principle.

Similarly, the physical orthobasis states for N identical bosons are proportional to completely symmetrized kets $\mathcal{S}|\alpha_1, \alpha_2, \ldots, \alpha_N\rangle$, because these are the simultaneous eigenstates for all of the pair-exchange operators P_{ij} with eigenvalue $+1$, unique up to normalization, that one can build out of the single-particle orthobasis states. Here, the normalization factor is complicated slightly by the fact that for bosons the α_i can be the same. If we write n_j for the number of times each distinct α_j is repeated in the list, and take the $|\alpha_j\rangle$ to be orthonormal, then the normalized physical orthobasis states are

$$|\alpha_1, \alpha_2, \ldots, \alpha_N\rangle_S = \sqrt{\frac{N!}{\prod_j (n_j!)}} \, \mathcal{S}|\alpha_1, \alpha_2, \ldots, \alpha_N\rangle. \quad (18.21)$$

This can also be written as

$$|\alpha_1, \alpha_2, \ldots, \alpha_N\rangle_S = \sqrt{\frac{\prod_j (n_j!)}{N!}} \sum_P{}' \, P|\alpha_1, \alpha_2, \ldots, \alpha_N\rangle, \quad (18.22)$$

where $\sum'_P$ means that the $\prod_j (n_j!)$ redundancies in the sum are eliminated, by including each distinct ket $P|\alpha_1, \alpha_2, \ldots, \alpha_N\rangle$ only once.

For example, consider $N = 3$ identical particles. Let us construct the physical states that can be made using a finite number n of orthobasis states for the individual particles, $|\alpha_i\rangle$ with $i = 1, 2, \ldots, n$. If the three identical particles are bosons, then we have $n(n+1)(n+2)/6$ physical orthobasis states, consisting of $n(n-1)(n-2)/6$ states with distinct i, j, k,

$$|\alpha_i, \alpha_j, \alpha_k\rangle_S = \frac{1}{\sqrt{6}}\Big(|\alpha_i, \alpha_j, \alpha_k\rangle + |\alpha_i, \alpha_k, \alpha_j\rangle + |\alpha_j, \alpha_k, \alpha_i\rangle + |\alpha_j, \alpha_i, \alpha_k\rangle + |\alpha_k, \alpha_i, \alpha_j\rangle + |\alpha_k, \alpha_j, \alpha_i\rangle\Big), \tag{18.23}$$

and $n(n-1)$ orthobasis states with distinct i and j,

$$|\alpha_i, \alpha_i, \alpha_j\rangle_S = \frac{1}{\sqrt{3}}\Big(|\alpha_i, \alpha_i, \alpha_j\rangle + |\alpha_i, \alpha_j, \alpha_i\rangle + |\alpha_j, \alpha_i, \alpha_i\rangle\Big), \tag{18.24}$$

and n orthobasis states,

$$|\alpha_i, \alpha_i, \alpha_i\rangle_S = |\alpha_i, \alpha_i, \alpha_i\rangle. \tag{18.25}$$

Each of these is an eigenstate of each of P_{12}, P_{13}, and P_{23}, with eigenvalue $+1$. For fermions, there are only the $n(n-1)(n-2)/6$ orthobasis states,

$$|\alpha_i, \alpha_j, \alpha_k\rangle_A = \frac{1}{\sqrt{6}}\Big(|\alpha_i, \alpha_j, \alpha_k\rangle - |\alpha_i, \alpha_k, \alpha_j\rangle + |\alpha_j, \alpha_k, \alpha_i\rangle - |\alpha_j, \alpha_i, \alpha_k\rangle + |\alpha_k, \alpha_i, \alpha_j\rangle - |\alpha_k, \alpha_j, \alpha_i\rangle\Big) \tag{18.26}$$

for distinct i, j, k.

An alternative way of specifying the orthobasis states for N identical particles is by giving the **occupation number** for each of the single-particle orthobasis states $|\alpha_i\rangle$. The occupation number n_i is the number of times that α_i appears in the symmetrized or antisymmetrized state ket. For bosons, the nonzero occupation numbers are the same as the numbers $n_1, n_2, \ldots$ appearing in (18.22). For fermions, each occupation number is always 0 or 1, due to the Pauli exclusion principle. Then, in both cases, the orthobasis states can be written using the occupation-number notation $|n_1, n_2, n_3, \ldots\rangle$, where some ordering has been chosen for the α_i. The individual occupation numbers can change with time, due to interactions. However, we will usually treat the total of the occupation numbers as fixed at $\sum_i n_i = N$. (Quantum field theory is what happens to you when you allow the number of particles to be variable.)

So far, we have considered orthobasis states for N identical particles, built out of the orthobasis states $|\alpha_i\rangle$ for the individual particles. It is crucial that the resulting states are indeed an orthobasis for the full Hilbert state space of N identical particles, whether they are bosons or fermions. This means that if we define any N linear combinations of single-particle states

$$|\phi_\mu\rangle = \sum_i c_{\mu i}|\alpha_i\rangle, \qquad (\mu = 1, \ldots, N), \tag{18.27}$$

which may be taken to have unit norm but are not necessarily elements of an orthobasis, then the allowed physical states for identical fermions can be written as

$$|\phi_1, \phi_2, \ldots, \phi_N\rangle_A = \mathcal{A}|\phi_1, \phi_2, \ldots, \phi_N\rangle, \tag{18.28}$$

and the allowed physical states for identical bosons are

$$|\phi_1, \phi_2, \ldots, \phi_N\rangle_S = \mathcal{S}|\phi_1, \phi_2, \ldots, \phi_N\rangle, \tag{18.29}$$

up to normalization. For fermions, a general (not necessarily normalized) Slater determinant ket

$$|\phi_1, \phi_2, \ldots, \phi_N\rangle_A = \begin{vmatrix} |\phi_1\rangle_1 & |\phi_1\rangle_2 & \cdots & |\phi_1\rangle_N \\ |\phi_2\rangle_1 & |\phi_2\rangle_2 & \cdots & |\phi_2\rangle_N \\ \vdots & \vdots & \ddots & \vdots \\ |\phi_N\rangle_1 & |\phi_N\rangle_2 & \cdots & |\phi_N\rangle_N \end{vmatrix} \tag{18.30}$$

is physical (provided that it is not null), and can be shown to be a unique linear combination of the orthobasis states $|\alpha_1, \alpha_2, \ldots, \alpha_N\rangle_A$ defined by (18.19). Here, the states $|\phi_\mu\rangle$ must be linearly independent, because otherwise the Slater determinant will vanish. Similarly, if the identical particles are bosons, then the (not necessarily normalized) state

$$|\phi_1, \phi_2, \ldots, \phi_N\rangle_S = \sum_P P|\phi_1, \phi_2, \ldots, \phi_N\rangle, \tag{18.31}$$

is an allowed physical state, even if the $|\phi_\mu\rangle$ are not orthobasis states or even linearly independent, and any state of this form can be written in a unique way as a linear combination of the orthobasis states $|\alpha_1, \alpha_2, \ldots, \alpha_N\rangle_S$ defined by (18.22).

Consider N identical particles, each of which has a single-particle Hamiltonian of the form

$$H_i = \frac{P_i^2}{2m} + V(\boldsymbol{R}_i). \tag{18.32}$$

For now, we are neglecting any kind of Hamiltonian interaction between the particles [like V_{int} in (18.4)]. Suppose that the single-particle energy-eigenstate position wavefunctions (neglecting spin) are

$$\phi_0(\boldsymbol{r}),\ \phi_1(\boldsymbol{r}),\ \phi_2(\boldsymbol{r}),\ \phi_3(\boldsymbol{r}),\ \ldots, \tag{18.33}$$

with individual H_i energy eigenvalues

$$\mathcal{E}_0 < \mathcal{E}_1 \leq \mathcal{E}_2 \leq \mathcal{E}_3 \leq \cdots . \tag{18.34}$$

We now consider the possible lowest energy states of the total unperturbed Hamiltonian $H_0 = \sum_{i=1}^{N} H_i$. Interactions between the particles can be treated later as a perturbation.

If $N = 2$, then the unperturbed ground state has $E = 2\mathcal{E}_0$, with position wavefunction

$$\psi(\boldsymbol{r}_1, \boldsymbol{r}_2) = \phi_0(\boldsymbol{r}_1)\phi_0(\boldsymbol{r}_2). \tag{18.35}$$

Since this is symmetric under exchange of the labels 1, 2, the spin part of the state must also be symmetric if the particles are bosons, or antisymmetric if the particles are fermions. Call the spin operators $\boldsymbol{S}_1$ and $\boldsymbol{S}_2$, so that the total spin is $\boldsymbol{S} = \boldsymbol{S}_1 + \boldsymbol{S}_2$, and denote the eigenvalue of the operator S^2 as $\hbar^2 s(s+1)$, where s is the total-spin quantum number. For spin-0 bosons, the total-spin quantum number is of course $s = 0$, and the (unique) spin state is trivially symmetric. For spin-1 bosons with position wavefunction (18.35), one must have either $s = 0$ or $s = 2$, because those are the symmetric combinations from the addition of angular momentum rule

$$1 \otimes 1 = 0_S \oplus 1_A \oplus 2_S. \tag{18.36}$$

For spin-1/2 fermions, the spin state accompanying (18.35) must be $s = 0$, the antisymmetric combination in

$$\tfrac{1}{2} \otimes \tfrac{1}{2} = 0_A \oplus 1_S. \tag{18.37}$$

We will make use of this in the next two sections.

For any number N of identical bosons, one can have a symmetric position wavefunction

$$\psi(\boldsymbol{r}_1, \boldsymbol{r}_2, \ldots, \boldsymbol{r}_N) = \phi_0(\boldsymbol{r}_1)\phi_0(\boldsymbol{r}_2)\phi_0(\boldsymbol{r}_3)\cdots\phi_0(\boldsymbol{r}_N), \tag{18.38}$$

together with a spin state that is symmetric under interchange of any two of the N spins. To minimize the energy, the occupation number of the single-particle ground state should be maximized, forming a **Bose–Einstein condensate** with ground-state energy

$$E = N\mathcal{E}_0. \tag{18.39}$$

In the occupation number notation, the ground state is $|N, 0, 0, \ldots\rangle$, where the ordering of single-particle state labels is the same as for the energy eigenvalues.

In contrast, for spin-1/2 fermions, at most two can be put into each spatial wavefunction, due to the Pauli exclusion principle. So, one can assign two fermions to ϕ_0 (with $s = 0$ for their antisymmetric combined spin state), two more to ϕ_1 (again with $s = 0$ for their spin state) and so on. If N is even, then the last two fermions in an $s = 0$ state will occupy the $\phi_{\frac{N}{2}-1}$ wavefunction, while if N is odd then the last fermion will have wavefunction $\phi_{\frac{N-1}{2}}$. The total unperturbed energy for N spin-1/2 fermions is therefore

$$E = \begin{cases} 2\mathcal{E}_0 + 2\mathcal{E}_1 + \cdots + 2\mathcal{E}_{\frac{N}{2}-1} & \text{(even } N\text{)}, \\ 2\mathcal{E}_0 + 2\mathcal{E}_1 + \cdots + 2\mathcal{E}_{\frac{N-3}{2}} + \mathcal{E}_{\frac{N-1}{2}} & \text{(odd } N\text{)}. \end{cases} \tag{18.40}$$

One application of this counting is to the shell model for electrons in atoms. If one ignores the electron-electron Coulomb repulsion, fine-structure, and hyperfine interactions, and just considers electrons as moving in a central potential, then each single-particle spatial state with (H, L^2, L_z) eigenvalues $(E_{n,l}, \hbar^2 l(l+1), \hbar m)$ for $m = -l, \ldots, l$ can be occupied by 2 electrons in an $s = 0$ spin state. So, in the ground state of the multi-electron atom, an unperturbed energy level $E_{n,l}$ can contain up to $2(2l+1)$ electrons. We will explore this in more detail in Sect. 18.4.

18.2 Wavefunctions and Spin for Two Identical Particles

In practice, the spin degrees of freedom often play a special role in specifying the orthobasis of states for identical particles. As an example, consider a system of two electrons. The total spin is $\boldsymbol{S} = \boldsymbol{S}_1 + \boldsymbol{S}_2$, and the operator S^2 has eigenvalues $\hbar^2 s(s+1)$ where $s = 0, 1$, while the S_z eigenvalue $\hbar m_S$ can have $m_S = 0$ (if $s = 0$) and $m_S = 1, 0, -1$ (if $s = 1$). The state can be specified in a hybrid form, with wavefunctions for the spatial degrees of freedom and kets for the spin degrees of freedom, as

$$\psi_{0,0}(\boldsymbol{r}_1, \boldsymbol{r}_2)\, \chi_{0,0} + \sum_{m_S=-1}^{1} \psi_{1,m_S}(\boldsymbol{r}_1, \boldsymbol{r}_2)\, \chi_{1,m_S}. \tag{18.41}$$

Here χ_{s,m_S} is the orthobasis ket for a state with total spin quantum numbers (s, m_S), so that in terms of the product orthobasis kets,

$$\chi_{0,0} = \frac{1}{\sqrt{2}} \left(|\uparrow\downarrow\rangle - |\downarrow\uparrow\rangle \right), \tag{18.42}$$

$$\chi_{1,1} = |\uparrow\uparrow\rangle, \tag{18.43}$$

$$\chi_{1,0} = \frac{1}{\sqrt{2}} \left(|\uparrow\downarrow\rangle + |\downarrow\uparrow\rangle \right), \tag{18.44}$$

$$\chi_{1,-1} = |\downarrow\downarrow\rangle. \tag{18.45}$$

The pair-exchange operator separates into spatial and spin parts, as

$$P_{12} = P_{12}^{\text{spatial}} P_{12}^{\text{spin}}. \tag{18.46}$$

The total spin singlet ($s = 0$) state is antisymmetric (eigenvalue -1 for P_{12}^{spin}), while the total spin triplet ($s = 1$) states are each symmetric (eigenvalue $+1$ for P_{12}^{spin}). Since the total state must be antisymmetric to describe identical fermions, we must have

$$\psi_{0,0}(\boldsymbol{r}_1, \boldsymbol{r}_2) = \psi_{0,0}(\boldsymbol{r}_2, \boldsymbol{r}_1), \tag{18.47}$$

$$\psi_{1,m_S}(\boldsymbol{r}_1, \boldsymbol{r}_2) = -\psi_{1,m_S}(\boldsymbol{r}_2, \boldsymbol{r}_1). \tag{18.48}$$

Thus $\psi_{0,0}$ is a symmetric spatial wavefunction for the total spin singlet, and ψ_{1,m_S} are three distinct antisymmetric spatial wavefunctions for the total spin triplets. In the context of the spatial wavefunctions, "symmetric" (or "antisymmetric") means eigenvalue $+1$ (or -1) for P_{12}^{spatial}. If the operator S^2 commutes with the Hamiltonian H, then they have common eigenstates, so one can label the energy levels as either $s = 1$ or $s = 0$.

Suppose the two electrons are in a potential $V(\boldsymbol{r})$ that is large in magnitude compared to their Coulomb interaction with each other and any spin interactions, which we therefore neglect. In that approximation,

$$H = \frac{1}{2m_e}(P_1^2 + P_2^2) + V(\boldsymbol{R}_1) + V(\boldsymbol{R}_2), \tag{18.49}$$

which could be the unperturbed Hamiltonian in a perturbative approach, and the Schrödinger equation for the wavefunction $\psi(\boldsymbol{r}_1, \boldsymbol{r}_2)$ is

$$\left[-\frac{\hbar^2}{2m_e}(\nabla_1^2 + \nabla_2^2) + V(\boldsymbol{r}_1) + V(\boldsymbol{r}_2) - E\right]\psi = 0. \tag{18.50}$$

Let $\phi_a(\boldsymbol{r})$ be the wavefunction solutions to the single-particle eigenvalue differential equation

$$\left[-\frac{\hbar^2}{2m_e}\nabla^2 + V(\boldsymbol{r}) - \mathcal{E}_a\right]\phi_a(\boldsymbol{r}) = 0. \tag{18.51}$$

Then, there are spin singlets (but not spin triplets) with energies $E = 2\mathcal{E}_a$, and total spatial wavefunctions

$$\psi_{0,0}(\boldsymbol{r}_1, \boldsymbol{r}_2) = \phi_a(\boldsymbol{r}_1)\phi_a(\boldsymbol{r}_2). \tag{18.52}$$

This will apply, for example, if $\mathcal{E}_a$ is the non-degenerate lowest eigenvalue for (18.51), so that, as already noted in the previous section, the ground state of the two-electron system is a spin singlet state. Also, for each pair of distinct eigenfunctions $\phi_a(\boldsymbol{r})$ and $\phi_b(\boldsymbol{r})$ with eigenvalues $\mathcal{E}_a$ and $\mathcal{E}_b$, we have four energy eigenstates with $E = \mathcal{E}_a + \mathcal{E}_b$, with spatial wavefunctions

$$\psi_{0,0}(\boldsymbol{r}_1, \boldsymbol{r}_2) = \frac{1}{\sqrt{2}} \left[\phi_a(\boldsymbol{r}_1)\phi_b(\boldsymbol{r}_2) + \phi_b(\boldsymbol{r}_1)\phi_a(\boldsymbol{r}_2)\right], \quad (s = 0), \tag{18.53}$$

$$\psi_{1,m_S}(\boldsymbol{r}_1, \boldsymbol{r}_2) = \frac{1}{\sqrt{2}} \left[\phi_a(\boldsymbol{r}_1)\phi_b(\boldsymbol{r}_2) - \phi_b(\boldsymbol{r}_1)\phi_a(\boldsymbol{r}_2)\right], \quad (s = 1;\ m_S = \pm 1,\ 0), \tag{18.54}$$

which are total spin singlets and triplets, respectively. The eigenvalues $\mathcal{E}_a$ may have degeneracies, so that in the notation we are using here, $\mathcal{E}_a$ and $\mathcal{E}_b$ could be equal even if $\phi_a(\boldsymbol{r})$ and $\phi_b(\boldsymbol{r})$ are distinct eigenfunctions.

For each of these possibilities, consider the probability to find one electron within a volume $d^3\boldsymbol{r}_1$ near $\boldsymbol{r}_1$ and the other within a volume $d^3\boldsymbol{r}_2$ near $\boldsymbol{r}_2$. From Postulate 4 in Sect. 3.1,

$$d\mathcal{P}(\boldsymbol{r}_1, \boldsymbol{r}_2) = d^3\boldsymbol{r}_1\, d^3\boldsymbol{r}_2\, |\psi(\boldsymbol{r}_1, \boldsymbol{r}_2)|^2, \tag{18.55}$$

where

$$|\psi(\boldsymbol{r}_1, \boldsymbol{r}_2)|^2 = \frac{1}{2}\Big(|\phi_a(\boldsymbol{r}_1)|^2|\phi_b(\boldsymbol{r}_2)|^2 + |\phi_b(\boldsymbol{r}_1)|^2|\phi_a(\boldsymbol{r}_2)|^2 \\ \pm 2\mathrm{Re}\left[\phi_a(\boldsymbol{r}_1)\phi_b^*(\boldsymbol{r}_1)\phi_b(\boldsymbol{r}_2)\phi_a^*(\boldsymbol{r}_2)\right]\Big), \tag{18.56}$$

with the $+$ sign for spin-singlet states and the $-$ sign for spin-triplet states. Note that for either sign, (18.56) is invariant under the exchange $\boldsymbol{r}_1 \leftrightarrow \boldsymbol{r}_2$.

The $\pm$ term in (18.56) is known as the **exchange density**. As a consequence of it, when the electrons are in the spin-triplet state, they avoid each other. To see this, note that the probability density vanishes when $\boldsymbol{r}_1 = \boldsymbol{r}_2$, and therefore (by continuity) is small when $\boldsymbol{r}_1 \approx \boldsymbol{r}_2$. Conversely, in the spin-singlet state, the two electrons "attract"; there is an enhanced probability density for $\boldsymbol{r}_1 \approx \boldsymbol{r}_2$ due to constructive interference. This attraction or repulsion is not due to any electromagnetic interaction or other Hamiltonian interaction between the identical fermions, but rather due solely to the Fermi–Dirac statistics.

Electrons in Argonne are identical to those in Batavia. So, it is natural to wonder why we do not need to worry about significant correlation effects from antisymmetrization of wavefunctions belonging to very distant electrons. Suppose that we define normalized single-particle stationary-state wavefunctions that are peaked in Argonne and Batavia, with very little overlap, so that $\phi_A(\boldsymbol{r})$ is insignificant outside of a small volume somewhere in Argonne, and $\phi_B(\boldsymbol{r})$ is insignificant outside of a small volume in Batavia, and therefore

$$\phi_A(\boldsymbol{r})\phi_B(\boldsymbol{r}) \approx 0 \tag{18.57}$$

everywhere. The total wavefunction is

$$\psi(\boldsymbol{r}_1, \boldsymbol{r}_2) = \frac{1}{\sqrt{2}}\big[\phi_A(\boldsymbol{r}_1)\phi_B(\boldsymbol{r}_2) \pm \phi_B(\boldsymbol{r}_1)\phi_A(\boldsymbol{r}_2)\big], \tag{18.58}$$

where the $+$ sign applies for spin singlets and the $-$ sign for spin triplet states. Note that there is no sense in which the electron labeled 1 is nearer to Argonne, or nearer to Batavia; it is always equally likely to be found in either place.

Now we can ask: what is the probability of seeing an electron within an infinitesimal volume $d^3\boldsymbol{r}$ near $\boldsymbol{r}$? The answer is

$$d\mathcal{P}(\boldsymbol{r}) = d^3\boldsymbol{r} \int d^3\boldsymbol{r}_2\, |\psi(\boldsymbol{r}, \boldsymbol{r}_2)|^2 + d^3\boldsymbol{r} \int d^3\boldsymbol{r}_1\, |\psi(\boldsymbol{r}_1, \boldsymbol{r})|^2. \tag{18.59}$$

The first term is the probability that the electron labeled 1 is near $\boldsymbol{r}$, summed over all possible electron 2 positions, and the second term is the same with $1 \leftrightarrow 2$. This can be rewritten as

$$\begin{aligned} d\mathcal{P}(\boldsymbol{r}) = d^3\boldsymbol{r} \bigg(&|\phi_A(\boldsymbol{r})|^2 \int d^3\boldsymbol{x}\, |\phi_B(\boldsymbol{x})|^2 + |\phi_B(\boldsymbol{r})|^2 \int d^3\boldsymbol{x}\, |\phi_A(\boldsymbol{x})|^2 \\ &\pm 2\text{Re}\left[\phi_A(\boldsymbol{r})\phi_B(\boldsymbol{r})^* \int d^3\boldsymbol{x}\, \phi_A(\boldsymbol{x})^*\phi_B(\boldsymbol{x})\right]\bigg). \end{aligned} \tag{18.60}$$

The important point is that the last term with $\pm$ is (doubly!) negligible, because of (18.57). In each of the first two terms, the integral is 1 by the assumed normalization, so

$$d\mathcal{P}(\boldsymbol{r}) = d^3\boldsymbol{r}\, \Big(|\phi_A(\boldsymbol{r})|^2 + |\phi_B(\boldsymbol{r})|^2\Big) \tag{18.61}$$

to a very good approximation. This shows that, despite the form of the wavefunction in (18.58), and despite the fact that our labeling 1, 2 of the electrons did not distinguish either one as being "the electron in Batavia", the probability density decouples into two separate additive contributions with negligible interference, for sufficiently separated identical particle wavefunctions. It is the lack of wavefunction overlap for single-particle (approximate) stationary states that enforces the decoupling, not the electron labels. The same type of argument applies to the decoupling of expectation values of observables, including the Hamiltonian. When in Batavia, we can usually ignore electrons in Argonne, and vice versa.

Now we shift our attention to the case of two identical bosons. Let us consider the cases of spin 0 and spin 1. For spin 0, the spin degrees of freedom are trivial, and in place of (18.41) we have simply a symmetric wavefunction

$$\psi_{0,0}(\boldsymbol{r}_1, \boldsymbol{r}_2) = \psi_{0,0}(\boldsymbol{r}_2, \boldsymbol{r}_1). \tag{18.62}$$

The case of spin 1 bosons is more interesting, because the two individual spins can combine to form nine total-spin orthobasis states χ_{s,m_S}, with $s = 0, m_S = 0$, and $s = 1, m_S = 1, 0, -1$, and $s = 2, m_S = 2, 1, 0, -1, -2$. Using the addition of angular momentum formalism discussed in Sect. 12.4, one can write these in terms of the product orthobasis kets $|m_{s1} : m_{s2}\rangle$, where the eigenvalues of the individual spin operators S_{1z} and S_{2z} are $\hbar m_{s1}$ and $\hbar m_{s2}$, with $m_{s1}, m_{s2} = -1, 0, 1$. The total-spin S^2, S_z eigenstates are, from (12.102)–(12.110),

$$\chi_{2,\pm 2} = |\pm 1 : \pm 1\rangle, \tag{18.63}$$

$$\chi_{2,\pm 1} = \big(|\pm 1 : 0\rangle + |0 : \pm 1\rangle\big)/\sqrt{2}, \tag{18.64}$$

$$\chi_{2,0} = \big(|1 : -1\rangle + 2|0 : 0\rangle + |-1 : 1\rangle\big)/\sqrt{6}, \tag{18.65}$$

$$\chi_{1,\pm 1} = \big(|\pm 1 : 0\rangle - |0 : \pm 1\rangle\big)/\sqrt{2}, \tag{18.66}$$

$$\chi_{1,0} = \big(|1 : -1\rangle - |-1 : 1\rangle\big)/\sqrt{2}, \tag{18.67}$$

$$\chi_{0,0} = \big(|1 : -1\rangle - |0 : 0\rangle + |-1 : 1\rangle\big)/\sqrt{3}. \tag{18.68}$$

These expressions show that χ_{s,m_S} has P_{12}^{spin} eigenvalue equal to $(-1)^s$. This same fact can be expressed in the shorthand form for addition of angular momentum

$$1 \otimes 1 = 2_S \oplus 1_A \oplus 0_S, \tag{18.69}$$

where the subscripts S and A indicate symmetry or antisymmetry under exchange of the two spins. Therefore, the total state for two identical spin-1 bosons is, again in a hybrid notation with wavefunctions for spatial degrees of freedom and kets for spin degrees of freedom,

$$\Psi = \psi_{0,0}(\boldsymbol{r}_1, \boldsymbol{r}_2)\chi_{0,0} + \sum_{m_S=-1}^{1} \psi_{1,m_S}(\boldsymbol{r}_1, \boldsymbol{r}_2)\chi_{1,m_S} + \sum_{m_S=-2}^{2} \psi_{2,m_S}(\boldsymbol{r}_1, \boldsymbol{r}_2)\chi_{2,m_S}. \tag{18.70}$$

This contains nine component wavefunctions satisfying

$$\psi_{s,m_S}(\boldsymbol{r}_1, \boldsymbol{r}_2) = (-1)^s \psi_{s,m_S}(\boldsymbol{r}_2, \boldsymbol{r}_1). \tag{18.71}$$

Under exchange of the boson labels, the position wavefunctions for the total spin 0 and 2 states are symmetric, and the position wavefunctions for total spin 1 are antisymmetric.

18.3 Excited States of the Helium Atom

The binding energy of the ground state of the two-electron helium atom was studied in Sects. 15.3 and 16.4. In doing so, we ignored the spin degrees of freedom, as well as the fact that the electrons are identical particles. This is now seen to be justified by the fact that the electrons can arrange themselves into a total spin $s = 0$ state. Then the total spatial wavefunction is the symmetric $\psi_{0,0}(\boldsymbol{r}_1, \boldsymbol{r}_2) = \psi_{0,0}(\boldsymbol{r}_2, \boldsymbol{r}_1)$, just as we had assumed (without bothering to justify it in detail) in Sects. 15.3 and 16.4. As we found in Sect. 16.4, the helium ground state spatial wavefunction is approximately just that of the product of two identical $1s$ hydrogen-like states with $Z_{\text{eff}} \approx 1.7$. The ground state of helium does not have a counterpart with total spin $s = 1$, because that would force the spatial wavefunction to be antisymmetric under exchange of the two electron labels, which is inconsistent with it being the product of two identical lowest-energy single-particle states.

However, for the excited states of helium, we must explicitly take into account both spin and the Fermi–Dirac statistics of the electrons. The Hamiltonian is once again given by (15.81)–(15.83). We start by neglecting the $H_{12} = e^2/|\boldsymbol{r}_1 - \boldsymbol{r}_2|$ interaction Hamiltonian, and later include it as a perturbation. Consider the spatial wavefunctions that can be formed from the individual $Z = 2$ hydrogen-like states, with one electron in the ground state ($1s$) and the other in a higher level, $2s$ or $2p$ or $3s$ or $3p$ or $3d$ or $4s$, etc.:

$$\psi(\boldsymbol{r}_1, \boldsymbol{r}_2) = \frac{1}{\sqrt{2}} \left[\phi_{1,0,0}(\boldsymbol{r}_1)\phi_{n,l,m}(\boldsymbol{r}_2) \pm \phi_{1,0,0}(\boldsymbol{r}_2)\phi_{n,l,m}(\boldsymbol{r}_1)\right]. \qquad (18.72)$$

Here, if the + sign applies, then the spatial wavefunction is symmetric, and so the total-spin state must have $s = 0$; these are called para-helium states. (The ground state is also a para-helium state.) Conversely, if the − sign applies, then the spatial wavefunction is antisymmetric, and so the total-spin state must be symmetric with $s = 1$; these are called ortho-helium states. For both of these, the unperturbed energy eigenvalues of $H_0 = H_1 + H_2$ are

$$\mathcal{E}_{n,l,m} = -\frac{Z^2 e^2}{2a_0}\left(1 + 1/n^2\right). \qquad (18.73)$$

Now, treating H_{12} as a perturbation, we obtain the first-order energy correction

$$E^{(1)}_{n,l,m} = \langle\psi|H_{12}|\psi\rangle = e^2\left(I_{nlm} \pm J_{nlm}\right), \qquad (18.74)$$

where the + sign applies to para-helium ($s = 0$) and the − sign to ortho-helium ($s = 1$), and

$$I_{nlm} = \int d^3\boldsymbol{r}_1 \int d^3\boldsymbol{r}_2\, |\phi_{1,0,0}(\boldsymbol{r}_1)|^2\, |\phi_{n,l,m}(\boldsymbol{r}_2)|^2\, \frac{1}{|\boldsymbol{r}_1 - \boldsymbol{r}_2|}, \qquad (18.75)$$

$$J_{nlm} = \int d^3\boldsymbol{r}_1 \int d^3\boldsymbol{r}_2\ \phi_{1,0,0}(\boldsymbol{r}_1)\phi_{n,l,m}(\boldsymbol{r}_1)^*\phi_{1,0,0}(\boldsymbol{r}_2)^*\phi_{n,l,m}(\boldsymbol{r}_2)\frac{1}{|\boldsymbol{r}_1 - \boldsymbol{r}_2|}. \tag{18.76}$$

Both integrals are manifestly real. Here I_{nlm}, which is manifestly positive, is called the "direct integral", while J_{nlm} is called the "exchange integral". We will not evaluate them explicitly, but it is worth knowing that J_{nlm} is also positive, but always smaller in magnitude than I_{nlm}. (The example of $n, l, m = 2, 0, 0$ is left to Exercise 18.4.) This shows that each ortho-helium ($s = 1$) state is always lower in energy than the corresponding para-helium ($s = 0$) state. A qualitative explanation for this is that in the $s = 0$ combination with an antisymmetric spin state, the symmetric spatial wavefunction results in a higher electrostatic repulsion between the electrons, due to their greater wavefunction overlap, than for the $s = 1$ combination with an antisymmetric spatial wavefunction.

The allowed set of total angular momentum quantum numbers j depends on s and l, the orbital angular momentum quantum number of the excited electron state. For $s = 0$ we must have $j = l$. For $s = 1$ with $l = 0$, the only allowed value is $j = 1$. For $s = 1$ and $l \geq 1$, one can instead have j equal to $l - 1$, l, or $l + 1$. A sketch of the observed energies for the lowest few states of the neutral helium atom is shown in Fig. 18.2, classified by their unperturbed quantum numbers on the left, and by their ${}^{2s+1}l_j$ angular momentum labels on the right. The excited states are each split into $s = 0$ para-helium (upper) and $s = 1$ ortho-helium (lower) levels. Transitions between para-helium states and ortho-helium states are highly suppressed, because they require spin flips, so to a good approximation they form two distinct sets of spectral lines. For this reason it was thought by early investigators that they might actually be two separate kinds of helium.

Each of the ortho-helium levels with $l \geq 1$ is split by small fine-structure effects into its separate $j = l - 1$, l and $l + 1$ states. These splittings are not shown in Fig. 18.2.

Regarding hyperfine effects, the helium nucleus exists in two stable isotopes. Almost all naturally occurring helium nuclei are ^{4}He, a boson with spin 0. A spin-0 particle has no special direction in its rest frame, and therefore cannot have a magnetic dipole moment or electric quadrupole moment, so ^{4}He atoms have no hyperfine splitting. A very tiny fraction of naturally occurring helium nuclei is ^{3}He, a fermion with spin 1/2. The ^{3}He atomic states with a given j are therefore further split by the hyperfine contribution.

18.4 Multi-electron Atoms

Consider the problem of finding the energy eigenstates for multi-electron atoms. To a good approximation, the Hamiltonian depends on the nonrelativistic kinetic energy of the electrons, their electrostatic attraction to the nucleus, their pairwise electrostatic repulsion, and the couplings between the electron spins and their orbital angular momenta. These effects alone are enough to provide a challenge

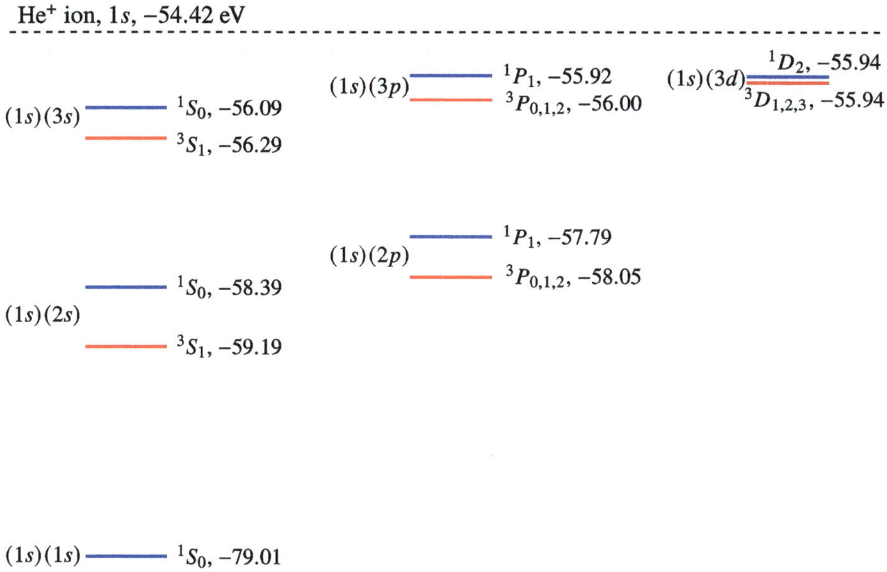

Fig. 18.2 The lowest few energy levels of the neutral helium atom, with electron configurations, $^{2s+1}l_j$ spectroscopic notations, and energies in eV. Relative energy spacings are indicated qualitatively, but not to scale. The ground state has both electrons in the lowest energy level $n = 1$ and opposite spins ($s = 0$, para-helium). The remaining states have one electron in the lowest level, $1s$, and one in an excited energy level $n > 1$, and are split into para-helium (upper, blue lines, $s = 0$) and ortho-helium (lower, red lines, $s = 1$) states. Would-be states with both electrons in an excited level are above the ground-state energy of the He$^+$ ion, shown as the dashed line, corresponding to one electron in the ground state and the other completely ionized away. Not shown are other states with $(1s)(4s)$ and $(1s)(4p)$ etc., which fall below the dashed line and asymptotically approach it as $n \to \infty$ for the excited electron. Small fine-structure effects, also not shown, split the ortho-helium levels $^3P_{0,1,2}$, $^3D_{1,2,3}$, etc., into their different j components, with larger j slightly lower in energy

that we cannot hope to solve exactly, and approximation methods must be used. In the following, we will not consider other, usually smaller, effects, including the contributions due to the relativistic correction to the electron kinetic energies, Darwin terms, and hyperfine contributions due to the magnetic dipole and electric quadrupole moments of the nucleus. These can be added as further perturbations.

One way to make progress is the **central-field method**. We invent a central potential $U(r)$, a rotationally invariant function (of the radial coordinate only) that is chosen to approximate the electrostatic potential seen by an electron at a given point due to the nucleus and the suitably averaged effects of the other electrons. Although it is not easy to give a rigorous definition of "suitably averaged", there are motivated techniques for estimating $U(r)$ by iteration (notably the Hartree and Hartree–Fock approximations, due to Douglas Hartree and improved by Vladimir Fock) that will not be discussed here. In any case, the choice of $U(r)$ is arbitrary, in principle, because we will add it to one part of the Hamiltonian (the unperturbed part, or the part used to define trial states for use with the variational principle) and subtract it from another (the perturbation part), so that the total Hamiltonian does not depend on it.

More specifically, we write the approximate Hamiltonian as

$$H = H_0 + W + H_{\text{s.o.}}, \tag{18.77}$$

where

$$H_0 = \sum_{i=1}^{N_e} \left(\frac{P_i^2}{2m_e} + U(R_i) \right), \tag{18.78}$$

$$W = \sum_{i=1}^{N_e} \left(-U(R_i) - \frac{Ze^2}{R_i} + \sum_{j=1}^{i-1} \frac{e^2}{|\boldsymbol{R}_i - \boldsymbol{R}_j|} \right), \tag{18.79}$$

$$H_{\text{s.o.}} = \sum_{i=1}^{N_e} \boldsymbol{S}_i \cdot \boldsymbol{\Lambda}_i. \tag{18.80}$$

This Hamiltonian is invariant under the exchange of any two electrons, as it must be. In the spin-orbit term (18.80),

$$\boldsymbol{\Lambda}_i = -\frac{e}{2m_e^2c^2} (\nabla \Phi_i) \times \boldsymbol{P}_i \tag{18.81}$$

where the electric potential seen by the ith electron is

$$\Phi_i(\boldsymbol{R}_i) = \frac{Ze}{R_i} - \sum_{j \neq i} \frac{e}{|\boldsymbol{R}_i - \boldsymbol{R}_j|}. \tag{18.82}$$

The contribution to the spin-orbit interaction $H_{\text{s.o.}}$ from each electron generalizes the case of a spherically symmetric potential in (17.4), and reduces, for hydrogen with $Z = 1$, to (17.16). Note that there is no operator-ordering problem associated with the placement of $\boldsymbol{P}_i$ in (18.81), because of the vector calculus identity $\nabla \times (\nabla \Phi) = 0$. We have distinguished the number of protons in the nucleus Z from the number of electrons N_e, so that the formulas apply to charged ions as well as neutral atoms.

One should aim to choose the function $U(r)$ in such a way that the eigenstates of H_0 are as close as possible to those of the full Hamiltonian. A reasonable choice might interpolate between

$$U(r) \approx \begin{cases} -Ze^2/r & \text{(small } r\text{)}, \\ -(Z - N_e + 1)e^2/r & \text{(large } r\text{)}. \end{cases} \tag{18.83}$$

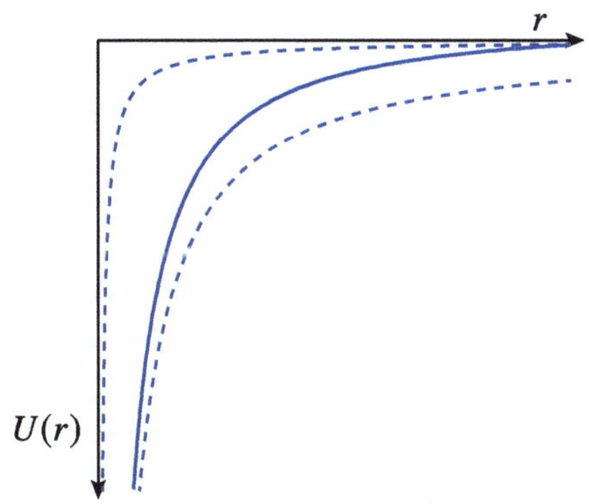

Fig. 18.3 The solid line shows the qualitative shape of a choice for the central-field potential energy function $U(r)$ for the electrons in a neutral atom with $N_e = Z = 10$. The upper dashed and lower dashed lines show the potentials $-e^2/r$ (maximally screened by the 9 other electrons closer to the nucleus) and $-10e^2/r$ (no screening, all 9 other electrons farther from the nucleus), respectively

The idea behind this choice is the principle from classical electrostatics, following from Gauss' law, that the electric field and electric potential at a point $\boldsymbol{r}$ in a spherically symmetric charge distribution are both determined only by the total charge contained within r. Thus, at small r there is no screening of the charge $+Ze$ nuclear potential due to other electrons, because they are likely to be found farther from the nucleus. Conversely, at large r the other $N_e - 1$ electrons are all likely to be closer to the nucleus and so the effective number of charges seen by an outermost very distant electron is only $Z - N_e + 1$. A sketch of a plausible choice of $U(r)$ is shown in Fig. 18.3, compared to the asymptotic limits of no screening at small r and maximum screening at large r. A choice of $U(r)$ should be justified *a posteriori* by the successful convergence of perturbation theory or the variational method.

The single-electron eigenstates of the central-field Hamiltonian H_0, called **orbital states** or **orbitals**, will have wavefunctions of the form

$$\phi_{n,l,m_l,m_s}(r,\theta,\phi) = R^U_{n,l}(r) Y^{m_l}_l(\theta,\phi)\,\chi_{m_s}, \tag{18.84}$$

where $\chi_{m_s} = |\uparrow\rangle$ or $|\downarrow\rangle$ for the spin states. The radial wavefunctions $R^U_{n,l}(r)$ will not be the same as for a hydrogen-like atom, because $U(r)$ is not proportional to $1/r$. However, they will be qualitatively similar. In particular, the label n is a positive integer, and then $0 \leq l \leq n-1$, with $n-l-1$ equal to the number of radial zeros for $r > 0$. In general, the orbital energy eigenvalues of H_0 will increase with n, but will also depend on l, unlike hydrogen-like atom energy eigenstates. The unperturbed eigenstates of H_0 will be Slater determinants formed out of N_e states of the form of (18.84). Then the effects of W and $H_{\text{s.o.}}$ can be added as perturbations.

A set of orbitals with the same values of n is called a **shell**, and those with the same values of both n and l form a **subshell**. Each subshell consists of

$$g_{n,l}^{\text{subshell}} = 2(2l+1) \tag{18.85}$$

states, due to the m_s and m_l quantum numbers. (This neglects any further degeneracy due to nuclear spins, which are different for distinct isotopes of the same element.) The orbital states and subshells are labeled nl, but with $l = 0, 1, 2, 3, \ldots$ replaced by the spectroscopic code letter $s, p, d, f, \ldots$, just as for the hydrogen atom states. The state of a given atom can be specified by giving the occupation numbers of each subshell, called the **electron configuration**, subject to the Pauli exclusion principle (enforced by the Slater determinant), which says that we cannot put two electrons in the same state due to their Fermi–Dirac statistics. Using (18.85), the maximum number of electrons in an s, p, d, f subshell is 2, 6, 10, 14, respectively. For a given l, smaller n tends to have lower energy, but this does not always mean that the subshells fill up from smaller to larger n, as we will soon see. For a given n, the states with smaller l tend to have lower energies. The reason is that wavefunctions with smaller l are enhanced closer to the nucleus, where the attractive force of the nucleus is less screened by the other electrons in the complete subshells.

The $1s$ shell can fit up to 2 electrons, and so is the only shell necessary for H and He, but is insufficient to accommodate larger Z atoms. The next to fill is the $2s$ subshell, which again fits up to 2 electrons, and so is the last subshell needed for Li and Be. The $2p$ subshell fills next, and can fit up to 6 electrons, so it is enough to accommodate the neutral atoms with Z up to 10, namely B, C, N, O, F, and Ne. Likewise, the $3s$ subshell accommodates the last electron for Na and Mg, while the $3p$ subshell is used for Al, Si, P, S, Cl, Ar. However, the $4s$ subshell tends to fill before the $3d$ subshell, in K, Ca, Sc, Ti, V, Mn, Fe, Co, Ni, and Zn. This tendency has two exceptions; within this sequence from $Z = 19$ to $Z = 30$, Cr and Cu break the trend, by having only one electron in the $4s$ subshell, with the $3d$ subshell containing five and ten electrons, respectively.

The subshell filling order for multi-electron atoms is, empirically,

$$1s,\ 2s,\ 2p,\ 3s,\ 3p,\ \begin{pmatrix}4s\\3d\end{pmatrix},\ 4p,\ \begin{pmatrix}5s\\4d\end{pmatrix},\ 5p,\ \begin{pmatrix}6s\\4f\\5d\end{pmatrix},\ 6p,\ \begin{pmatrix}7s\\5f\\6d\end{pmatrix}, \tag{18.86}$$

where the cases in parentheses correspond to ambiguities due to close energies, with the most common (but not universally followed) filling order being from top to bottom. The fact noted earlier that smaller l tends to give lower energy for a given n explains why $2s$ fills before $2p$, and why the $4s$ can compete with the $3d$, and the $5s$ can compete with the $4d$, etc.

Some example electron configurations are

$$\text{H: } 1s^1, \qquad \text{He: } 1s^2, \qquad \text{Li: } 1s^2\,2s^1, \qquad \text{C: } 1s^2\,2s^2\,2p^2,$$
$$\text{N: } 1s^2\,2s^2\,2p^3, \qquad \text{Na: } 1s^2\,2s^2\,2p^6\,3s^1,$$

$$\text{Hg:}\ 1s^2\,2s^2\,2p^6\,3s^2\,3p^6\,4s^2\,3d^{10}\,4p^6\,5s^2\,4d^{10}\,5p^6\,6s^2\,4f^{14}\,5d^{10}.$$

The chemical properties of a given element are mostly dependent on the electron configuration, and the similarities between elements in a given column of the periodic table are due to having similar configurations of electrons in the outermost (partially filled) subshells.

The inert (noble) gases are those whose highest energy electrons are in a full p subshell (in the cases of Ne, Ar, Kr, Xe, Rn) or just a full $1s$ shell (in the case of He), with no vacancies in previous subshells. This leads to a very low tendency to form chemical bonds. (The p subshells are special, because the next higher energy subshell is always s. This means a big energy gap to excite an electron out of a p subshell.) It is common to give the electron configuration for any element by just listing the filled orbitals in excess of that for the previous inert gas, which is indicated in square brackets. For example, for the inert gases themselves, one can write the electron configurations as

$$\text{He:}\ 1s^2,\quad \text{Ne:}\ [\text{He}]\,2s^2\,2p^6,\quad \text{Ar:}\ [\text{Ne}]\,3s^2\,3p^6,\quad \text{Kr:}\ [\text{Ar}]\,4s^2\,3d^{10}\,4p^6,$$
$$\text{Xe:}\ [\text{Kr}]\,5s^2\,4d^{10}\,5p^6,\qquad \text{Rn:}\ [\text{Xe}]\,6s^2\,4f^{14}\,5d^{10}\,6p^6.$$

For the alkali metals, there is an s subshell with only one electron,

$$\text{Li:}\ [\text{He}]\,2s^1,\quad \text{Na:}\ [\text{Ne}]\,3s^1,\quad \text{K:}\ [\text{Ar}]\,4s^1,$$
$$\text{Rb:}\ [\text{Kr}]\,5s^1,\quad \text{Cs:}\ [\text{Xe}]\,6s^1,\quad \text{Fr:}\ [\text{Rn}]\,7s^1,$$

while the halogens are all missing 1 electron in a p subshell,

$$\text{F:}\ [\text{He}]\,2s^2\,2p^5,\qquad \text{Cl:}\ [\text{Ne}]\,3s^2\,3p^5,\qquad \text{Br:}\ [\text{Ar}]\,3d^{10}\,4s^2\,4p^5,$$
$$\text{I:}\ [\text{Kr}]\,4d^{10}\,5s^2\,5p^5,\qquad \text{At:}\ [\text{Xe}]\,4f^{14}\,5d^{10}\,6s^2\,6p^5.$$

A good periodic table of the elements will give the electron configuration for the atomic ground state of each element.

The energy needed to remove a single electron from the ground state of a neutral atom is called the **first-ionization energy**. Experimental results for the first-ionization energy as a function of the atomic number Z are shown in Fig. 18.4, including as the first two data points 13.6 eV for H and 24.6 eV for He. (Note that the result for He can be obtained from numbers given in Fig. 18.2.) Because of the shielding of the nuclear charge by the innermost electrons, the results do not depend very strongly on Z; the outermost electrons are attracted to the nucleus as if they "see" a reduced net charge of order e rather than the full nuclear charge Ze. The first-ionization energies are largest for the inert gases, and are smallest for alkali metals, which each have a single, relatively weakly bound, s-subshell electron available for ionization.

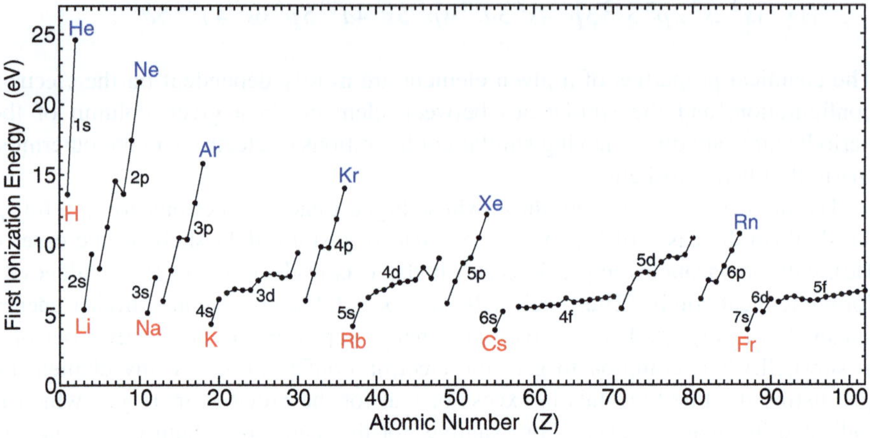

Fig. 18.4 The energy (in eV) needed to remove a single electron from the ground state of a neutral atom, as a function of the atomic number Z. Atoms with the same highest-energy subshells ($1s$, $2s$, $2p$, $3s$, ...) are connected by lines to guide the eye, as labeled. The inert gas elements (He, Ne, Ar, Kr, Xe, Rn) each have completely filled s- and p-subshells and thus a large first-ionization energy, while the alkali metal atoms (Li, Na, K, Rb, Cs, Fr) each have a lone s-subshell electron and a small first-ionization energy

A more complete specification of the atomic state will include information about the angular momentum quantum numbers. Suppose that we can treat the spin-orbit contribution $H_{\text{s.o.}}$ as small compared to H_0 and W in (18.77)–(18.80). Now, all three of the summed orbital, summed spin, and total angular momentum operators

$$\boldsymbol{L} = \sum_{i=1}^{N_e} \boldsymbol{L}_i, \qquad \boldsymbol{S} = \sum_{i=1}^{N_e} \boldsymbol{S}_i, \qquad \boldsymbol{J} = \boldsymbol{L} + \boldsymbol{S} \tag{18.87}$$

commute with H_0 (as well as W). One can therefore choose unperturbed orthobasis states that are not only eigenstates of H_0, but also eigenstates of the operators L^2, S^2, L_z and S_z with eigenvalues $\hbar^2 L(L+1)$, $\hbar^2 S(S+1)$, $\hbar m_L$, and $\hbar m_S$ respectively. These states also carry other quantum number labels (including the unperturbed H_0 energy eigenvalue), which we will symbolically denote as N, so that the states are labeled

$$|NLSm_L m_S\rangle. \tag{18.88}$$

For fixed N, L, S, the H_0 energy eigenvalues have degeneracy $g_{NLS} = (2S + 1)(2L + 1)$, due to the magnetic quantum numbers m_L, m_S. Alternatively, one can choose as an orthobasis the linear combinations of the same unperturbed states that are eigenstates of H_0, S^2, L^2, and J^2 [with eigenvalue $\hbar^2 J(J + 1)$], and J_z (with eigenvalue $\hbar m_J$),

$$|NLSJm_J\rangle = \sum_{m_L,m_S} C^{LSJ}_{m_Lm_Sm_J}|NLSm_Lm_S\rangle, \tag{18.89}$$

where $C^{LSJ}_{m_Lm_Sm_J}$ are the Clebsch–Gordan coefficients. The total number of such states for fixed N, L, S is again $(2S+1)(2L+1)$. The approximation in which the states constructed in this way from eigenstates of H_0 (or perhaps $H_0 + W$) are close to being eigenstates of the full Hamiltonian is called the **LS coupling** (or **Russell–Saunders coupling**, after Henry N. Russell and Frederick Saunders) approximation.[2]

We now investigate the spin-orbit fine-structure energy corrections, using degenerate perturbation theory. To do so, we must find the matrix elements of $H_{\text{s.o.}}$ on the degenerate subspaces of H_0, and diagonalize it. We start by working with the product orthobasis kets, which decompose into linear combinations of tensor products of orbital and spin angular momentum eigenkets,

$$|NLSm_Lm_S\rangle = \sum_{\alpha,\beta} c_{\alpha\beta}|\alpha NLm_L\rangle \otimes |\beta Sm_S\rangle. \tag{18.90}$$

Here α is a degeneracy label whose presence reflects the fact that there can be more than one state with eigenvalues L, m_L, and similarly β labels the different spin states with eigenvalues S, m_S. The whole state $|NLSm_Lm_S\rangle$ is a Slater determinant incorporating the antisymmetry under interchange of any two electrons. This puts constraints, which we will not explicitly identify or make use of, on the coefficients $c_{\alpha\beta}$.

In each subspace of degenerate H_0 eigenstates, the matrix elements of $H_{\text{s.o.}}$ from (18.80) are

$$\langle NLSm'_Lm'_S|H_{\text{s.o.}}|NLSm_Lm_S\rangle =$$
$$\sum_{\alpha,\beta}\sum_{\alpha',\beta'} c_{\alpha\beta}c^*_{\alpha'\beta'}\sum_{i=1}^{N_e}\langle\alpha' NLm'_L|\,\boldsymbol{\Lambda}_i\,|\alpha NLm_L\rangle\cdot\langle\beta' Sm'_S|\,\boldsymbol{S}_i\,|\beta Sm_S\rangle. \tag{18.91}$$

Electrons in complete subshells can be omitted, because pairs of opposite spin expectation values with the same orbital quantum numbers will always cancel in the sum over i. Now we will invoke the Landé projection formula, (13.49), twice. First, note that each $\boldsymbol{\Lambda}_i$ is a vector operator with respect to the total orbital angular momentum operator $\boldsymbol{L}$, without involving spin. Therefore, for $L \neq 0$, we have

[2] The LS coupling approximation is not so good if Z is very large, because then $H_{\text{s.o.}}$ cannot be treated as a small perturbation. This is because, unlike H_0 and W, the spin-orbit term $H_{\text{s.o.}}$ does not commute with $\boldsymbol{S}$ or $\boldsymbol{L}$ if there is more than one electron. We will return to this point near the end of this section.

$$\langle \alpha' N L m'_L | \mathbf{\Lambda}_i | \alpha N L m_L \rangle = \langle L m'_L | \mathbf{L} | L m_L \rangle \frac{\langle \alpha' N L \| \mathbf{L} \cdot \mathbf{\Lambda}_i \| \alpha N L \rangle}{\hbar^2 L(L+1)}. \tag{18.92}$$

As noted in Theorem 13.3, for the case $L = 0$, the matrix element simply vanishes, and there is no spin-orbit splitting. Similarly, each of the individual electron spin operators $\mathbf{S}_i$ is a vector with respect to rotations generated by the total spin angular momentum operator $\mathbf{S}$, without involving orbital angular momentum, so we can write, for $S \neq 0$,

$$\langle \beta' S m'_S | \mathbf{S}_i | \beta S m_S \rangle = \langle S m'_S | \mathbf{S} | S m_S \rangle \frac{\langle \beta' S \| \mathbf{S} \cdot \mathbf{S}_i \| \beta S \rangle}{\hbar^2 S(S+1)}. \tag{18.93}$$

while for $S = 0$ the matrix element again vanishes. Therefore, putting things together,

$$\langle N L S m'_L m'_S | H_{\text{s.o.}} | N L S m_L m_S \rangle = \zeta(N, L, S) \langle L S m'_L m'_S | \mathbf{S} \cdot \mathbf{L} | L S m_L m_S \rangle, \tag{18.94}$$

where the factor (for $L \neq 0$ and $S \neq 0$)

$$\zeta(N, L, S) = \sum_{\alpha,\beta} \sum_{\alpha',\beta'} c_{\alpha\beta} c^*_{\alpha'\beta'} \sum_{i=1}^{N_e} \frac{\langle \alpha' N L \| \mathbf{L} \cdot \mathbf{\Lambda}_i \| \alpha N L \rangle}{\hbar^2 L(L+1)} \frac{\langle \beta' S \| \mathbf{S} \cdot \mathbf{S}_i \| \beta S \rangle}{\hbar^2 S(S+1)} \tag{18.95}$$

may be quite difficult to evaluate exactly, but has the important property that it does not depend on the magnetic quantum numbers m_L, m_S, m'_L, m'_S at all, because the reduced matrix elements $\langle \alpha' N L \| \mathbf{L} \cdot \mathbf{\Lambda}_i \| \alpha N L \rangle$ and $\langle \beta' S \| \mathbf{S} \cdot \mathbf{S}_i \| \beta S \rangle$ do not. For a given atom, $\zeta(N, L, S)$ can, in principle, be evaluated, at least approximately. This is beyond our scope; we will merely consider it to be an empirical function of N, L, S.

Equation (18.94) shows that the matrix elements of $H_{\text{s.o.}}$ are proportional to those of $\mathbf{S} \cdot \mathbf{L} = (J^2 - L^2 - S^2)/2$, where J^2, L^2 and S^2 are understood to be operators. To properly conduct degenerate perturbation theory, we should switch to the orthobasis $|N L S J m_J\rangle$, in which this operator is diagonal. We therefore arrive at a formula for the spin-orbit correction to the energies, at first order in degenerate perturbation theory,

$$\Delta E(N, L, S, J) = \zeta(N, L, S) \hbar^2 \big[J(J+1) - L(L+1) - S(S+1) \big]/2, \tag{18.96}$$

where J, L, and S are now numbers. This spin-orbit splitting vanishes in the special cases $L = 0$ (which then implies $J = S$) and $S = 0$ (which then implies $J = L$), consistent with observations already made in the previous paragraph. Equation (18.96) shows that the degeneracy $(2S+1)(2L+1)$ is partially broken by $H_{\text{s.o.}}$, but states with a given J are not split, with remaining degeneracies $g_J = 2J + 1$, since $\Delta E(N, L, S, J)$ does not depend on m_J.

We can now use (18.96) to obtain a simple formula for the energy differences between successive eigenvalues $J-1$ and J, for fixed N, L, and S,

$$\Delta E(N, L, S, J) - \Delta E(N, L, S, J-1) = \hbar^2 \zeta(N, L, S) J. \qquad (18.97)$$

This result, that the spin-orbit fine-structure energy splitting between states J and $J-1$ with fixed N, L, S is proportional to J, is called the **Landé interval rule**, as observed in spectral lines from transitions between atomic states. It can be used as a check of the validity of the LS coupling scheme approximation.

The preceding discussion shows how, within the LS coupling approximation, energy eigenstates can be built by first adding all of the spins, then adding all of the orbital angular momenta, and then adding the two to get the total angular momentum. The angular momentum quantum numbers of these states are then traditionally specified with the **term symbol**

$$^{2S+1}L_J = \text{Russell–Saunders } (LS) \text{ spectroscopic notation}, \qquad (18.98)$$

where the eigenvalues of operators J^2, S^2, and L^2 are numbers $\hbar^2 J(J + 1)$, $\hbar^2 S(S + 1)$, and $\hbar^2 L(L + 1)$, respectively, and then the value of L is replaced by its spectroscopic code $S, P, D, F, G, H, \ldots$ for $L = 0, 1, 2, 3, 4, 5, \ldots$. Thus, for example, the hydrogen and helium atom ground states are given as $^2S_{1/2}$ and $^1S_0$, respectively. Many editions of the periodic table of elements list the Russell–Saunders notation for the atomic ground states along with the electron configuration.

The list of possible term symbols $^{2S+1}L_J$ of an atom can be obtained from looking only at the **valence electrons**, the ones in the incompletely filled subshells. This is because each electron in a filled subshell with a given m_s, m_l, and m_j always has a partner with the opposite values of those quantum numbers, giving no net contribution to S, L, and J. The parity of the state is useful for selection rules for transitions. From (8.181), it is $\pi = (-1)^{\sum_i l_i}$, where l_i are the individual orbital angular momenta, specified by the electronic configuration. [Note that for orbitals the parity is defined with respect to the fixed nucleus as the origin, not the center of mass of the valence electrons, so even for two electrons π is not necessarily equal to $(-1)^L$.]

For an atom with a partly filled subshell, there may be several candidates for the ground state term symbol. Three empirical rules, known as **Hund's Rules** after Friedrich Hund, can be used to predict which the ground state will be, given the electron configuration. They are:

Hund's Rule 1 The L, S, J multiplet with the lowest energy will have the largest S from among the candidates.

(Qualitative justification: for larger S, the spin wavefunction is more symmetric, which means the spatial wavefunction is more antisymmetric, which means the electrons are less often near each other, which gives less repulsive forces between them, and so lower energy. We have already seen this effect in action; it is why each ortho-helium $S = 1$ state is lower in energy than its para-helium $S = 0$ counterpart.)

Hund's Rule 2 If there are two or more candidates selected by the first rule that have the same largest S, then the lowest energy will have the largest L among them.

(Qualitative justification: larger L means that the orbital wavefunctions can be arranged to have less overlap, again leading to smaller repulsive forces and thus lower energy.)

Hund's Rule 3 If there are two or more candidates selected by the first two rules with the same S and L, then the spin-orbit coupling splits apart the different levels of $J = |L-S|, \ldots, L+S$. If an incomplete subshell is not more than half-filled, then the ground state has $J = |L-S|$, otherwise the ground state has $J = L+S$.

(Qualitative justification: the form of (18.96) immediately tells us that for fixed L, S, the minimum energy will occur for one of the two extremes for J, either $|L-S|$ or $L+S$. Hund's Rule 3 is equivalent to the empirical fact that $\zeta(N, L, S)$ in (18.96) turns out to be positive if the subshell is not more than half-filled, and negative otherwise.)

We now illustrate the prediction of term symbols and ground states, using Hund's rules where necessary, with examples from the beginning of the periodic table. For the first four elements, Hund's rules are not needed.

Hydrogen (H, $Z = 1$) has electron configuration $1s^1$. Because there is one electron with $S = 1/2$ and $L = 0$, the only possibility is $J = 1/2$, and the term symbol is ${}^2S_{1/2}$.

Helium (He, $Z = 2$) has electron configuration $1s^2$. As always when there are no partially filled subshells, the ground state is ${}^1S_0$. The same result is obtained for all inert (noble) gases.

Lithium (Li, $Z = 3$) has electron configuration $1s^2\,2s^1$. Just as for hydrogen, there is one electron that is not in a filled subshell, with $S = 1/2$ and $L = 0$, so the term symbol is ${}^2S_{1/2}$. The same applies for all alkali metals, the elements in the same periodic table column as Li, which have a single s subshell electron.

Beryllium (Be, $Z = 4$) has electron configuration $1s^2\,2s^2$, with only completely filled subshells, so the ground state is ${}^1S_0$. The same applies for all elements in the Be periodic table column.

Boron (B, $Z = 5$) has electron configuration $1s^2\,2s^2\,2p^1$. Here, the $1s$ and $2s$ subshells are irrelevant, since they are completely filled. In the incomplete $2p$ subshell, we have one electron with $L = 1$ and $S = 1/2$. Since this is the only choice, we do not need Hund's rules 1 or 2. The possible candidates have $J = 1/2$ and $J = 3/2$, with term symbols ${}^2P_{1/2}$ and ${}^2P_{3/2}$. Because the last incomplete subshell is less than half filled (1 out of 6), Hund's rule 3 tells us that the ground state has $J = |L - S| = 1/2$ and is ${}^2P_{1/2}$. The same result is obtained for all elements in the B column of the periodic table, which have a single p electron.

Carbon (C, $Z = 6$) has electron configuration $1s^2\,2s^2\,2p^2$. We start by finding the total number of allowed states associated with the incomplete $2p$ subshell. The

2 electrons have 6 states from which to choose. In general, the number of ways of choosing k things from n possibilities, without regard to order, is equal to the binomial coefficient

$$\binom{n}{k} = \frac{n!}{k!\,(n-k)!}. \tag{18.99}$$

So in our present case there are $(6 \cdot 5)/(2 \cdot 1) = 15$ completely antisymmetric states. Now, let us see how these 15 states are divided into ${}^{2S+1}L_J$ multiplets. Combining the two spin-1/2's gives $S = 0$ (the antisymmetric combination) and $S = 1$ (the symmetric combination). Combining the two orbital angular momenta, each with $l = 1$, gives $L = 0$ (symmetric), and $L = 1$ (antisymmetric), and $L = 2$ (symmetric). Therefore, requiring that the full state be antisymmetric under exchange of the two electrons, we have the following allowed combinations. First, there is $L = 2, S = 0$, which implies $J = 2$. This possibility ${}^1D_2$ consists of $2J + 1 = 5$ states. Second, there is $L = 1, S = 1$, which allows $J = 0, 1, 2$, so the terms are ${}^3P_0$, ${}^3P_1$, and ${}^3P_2$, with 1, 3, and 5 states respectively. Third, there is $L = 0, S = 0$, which of course allows only one state, ${}^1S_0$ with $J = 0$. As a check, the total of these possibilities is indeed 15. Now we are ready to use Hund's rules. Hund's rule 1 selects the largest possible value of S, which is $S = 1$. Hund's rule 2 tells us nothing, because the remaining competitors all have $L = 1$. Finally, for Hund's rule 3, we note that the $2p$ subshell is not more than half-filled (2 out of 6), so that $J = |L - S| = 0$ in the ground state. We therefore have $S = 1, L = 1$, $J = 0$, and the ground state of carbon is ${}^3P_0$. The same logic and result apply to all elements in the C column of the periodic table, which have two p electrons.

Nitrogen (N, $Z = 7$) has electron configuration $1s^2\,2s^2\,2p^3$. The total number of possible electron states associated with the three electrons in the $2p$ subshell is $(6{\cdot}5{\cdot}4)/(3{\cdot}2{\cdot}1) = 20$ antisymmetric states. To see which L, S, J combinations these correspond to, first consider the combination of the orbital angular momenta, which gives $1{\otimes}1{\otimes}1 = 0{\oplus}1{\oplus}1{\oplus}1{\oplus}2{\oplus}2{\oplus}3$ for the possible values of L. If $L = 3$, one can check that the spatial wavefunction constructed by addition of angular momentum would be totally symmetric under exchange of the electron labels, but there is no totally antisymmetric spin state for ≥ 3 electrons; we conclude from this that L cannot be 3. If $L = 0$, then the spatial wavefunction is instead totally antisymmetric, from which we conclude that the spin state must be totally symmetric, so $S = 3/2$. Conversely, if $S = 3/2$, then only the totally antisymmetric $L = 0$ combination is allowed. We conclude that the only (L, S) combinations that are allowed are $L = 0, S = 3/2$ with $J = 3/2$ (term ${}^4S_{3/2}$, 4 states), and $L = 1, S = 1/2$ with $J = 1/2$ and $3/2$ (terms ${}^2P_{1/2}$ and ${}^2P_{3/2}$, 6 states), and $L = 2, S = 1/2$ with $J = 3/2$ and $J = 5/2$ (terms ${}^2D_{3/2}$ and ${}^2D_{5/2}$, 10 states). As a check, these indeed add up to 20 states. Hund's rule 1 tells us that the ground state should have maximal S, which selects the $L = 0, S = 3/2, J = 3/2$ combination. This is the only one that realizes that maximal value of S, so the other two Hund's rules are not needed, and the ground state term is ${}^4S_{3/2}$. The same logic and result apply to all elements in the N column of the periodic table, which have three p electrons.

Oxygen (O, $Z = 8$) has electron configuration $1s^2\ 2s^2\ 2p^4$. As in the previous examples, we start by doing the counting of the total number of physical antisymmetric electron states for the incomplete subshell. Since there are 4 electrons and 6 available states, that number is $(6{\cdot}5{\cdot}4{\cdot}3)/(4{\cdot}3{\cdot}2{\cdot}1) = 15$. However, as a useful trick it is better to consider the 2 "holes", by which we mean the electrons omitted from the completely filled $2p$ subshell. Treating the holes as equivalent spin-1/2 particles, we again obtain $(6 \cdot 5)/(2 \cdot 1) = 15$ states. A complete $2p$ subshell would have $L = S = 0$, so we treat the 2 holes as carrying the same S and L as the 4 electrons. So, the allowed candidate multiplets are the same as we found for carbon with $2p^2$, namely $(L, S, J) = (2, 0, 2)$, $(1, 1, 0)$, $(1, 1, 1)$, $(1, 1, 2)$, and $(0, 0, 0)$, with term symbols ${}^1D_2$, ${}^3P_0$, ${}^3P_1$, ${}^3P_2$, and ${}^1S_0$. Hund's rule 1 says the ground state has $S = 1$, and Hund's rule 2 does not differentiate between the remaining candidates, which all have $L = 1$. To apply Hund's rule 3, we note that the $2p$ subshell is now more than half-filled with electrons (4 out of 6), so $J = L + S = 2$ for the ground state. (Note that this is the difference between carbon and oxygen; although the enumeration of the candidate terms is easier in terms of the holes, Hund's rule 3 for the selection of the ground state still applies to the actual electrons, not the holes.) So, the ground state term for oxygen is ${}^3P_2$. The same logic and result apply to all elements in the O column of the periodic table, with four p electrons.

Fluorine (F, $Z = 9$) has electron configuration $1s^2\ 2s^2\ 2p^5$. Using the trick introduced in the oxygen example, we get the allowed (L, S, J) multiplets by considering the $2p$ subshell to consist of just one hole, rather than 5 electrons. This means that, just as for boron, the only combinations are $L = 1$, $S = 1/2$ with $J = 1/2$ or $3/2$. Since these have the same L and S, Hund's rules 1 and 2 provide no information. To apply Hund's rule 3, we note that the $2p$ subshell is more than half full (5 out of 6), and so the ground state has $J = L + S = 3/2$, with term symbol ${}^2P_{3/2}$. The same applies to all halogens, the elements in the F column of the periodic table, with five p electrons.

Hund's rules correctly predict the ground state of all neutral atoms for which the ground state term symbol is unambiguously measured. They also often give the correct result for charged ions, and for neutral atoms with an electron configuration different from the ground state. For example, the singly ionized carbon ion C^+ has electron configuration $1s^2\ 2s^2\ 2p$, which is the same as boron, and has the same term symbol ${}^2P_{1/2}$, as correctly predicted by Hund's rules. Also, the neutral carbon atom has excited states with electron configuration $1s^2\,2s\,2p^3$. Among those excited states, Hund's rules correctly predict that the lowest energy term is ${}^5S_2$.

An important experimental tool for exposing the properties of atomic states is to observe the behavior of energy levels and transition spectra as a function of an applied magnetic field, called the Zeeman effect. In Sect. 17.3, we studied the Zeeman effect for atomic hydrogen, and much of that discussion can be adapted to the multi-electron case. Adding the contributions to the Hamiltonian linear in a constant, uniform magnetic field $\boldsymbol{B} = B\hat{z}$, one obtains a result of the same form as (17.70),

$$H_B = \mu_B B(L_z + 2S_z)/\hbar, \tag{18.100}$$

but now with L_z and S_z respectively equal to the total orbital and spin angular momentum operators, obtained as the sums of the operators for the individual electrons. Within the LS coupling scheme approximation, the analysis for the weak-field (small-B) Zeeman effect in multi-electron atoms just mimics what we already did for hydrogen from (17.72)–(17.77). So, atomic states characterized by N, L, S, J obtain energy splittings proportional to the total angular momentum quantum number m_J,

$$\Delta E_B = g\,\mu_B B\, m_J, \qquad (m_J = -J,\ -J+1, \ldots,\ J-1,\ J), \tag{18.101}$$

where, just as in (17.77), the Landé g-factor is

$$g = \frac{3}{2} + \frac{S(S+1) - L(L+1)}{2J(J+1)}. \tag{18.102}$$

(Of course, unlike the atomic hydrogen case, in general S need not be equal to 1/2, and J need not be equal to $L \pm 1/2$.) For weak external magnetic fields, the magnetic moment of the atom is therefore related to its total angular momentum by $-g\mu_B \boldsymbol{J}/\hbar$, and the splitting exposes the degeneracies and angular momentum quantum numbers for the multi-electron atomic states.

The strong magnetic field (Paschen–Back) limit for multi-electron atoms is also quite similar to the example of the hydrogen atom, as given in (17.84)–(17.88). This includes the use of $|NLSm_L m_S\rangle$, in which H_B is diagonal, as the degenerate unperturbed states to which $H_{\text{s.o.}}$ is applied as a perturbation. For multi-electron atoms, the counterpart of (17.87) is

$$\langle NLSm_L m_S|H_{\text{s.o.}}|NLSm_L m_S\rangle = \zeta(N, L, S)\langle NLSm_L m_S|\,\boldsymbol{S}\cdot\boldsymbol{L}\,|NLSm_L m_S\rangle \tag{18.103}$$

where $\zeta(N, L, S)$ is the quantity appearing in (18.95)–(18.97). Now $\boldsymbol{S}\cdot\boldsymbol{L} = \frac{1}{2}(S_+L_- + S_-L_+) + S_z L_z$ evaluates to $\hbar^2 m_S m_L$. Therefore, we have in the strong-field limit

$$\Delta E_{B,\,\text{fine}} = \mu_B B(m_L + 2m_S) + \hbar^2\zeta(N, L, S)m_S m_L, \tag{18.104}$$

where we have kept only the Zeeman term and the spin-orbit part of the fine-structure effect, since these are the contributions that contain the dependence on m_L and m_S. For the intermediate case that the fine-structure and external magnetic fields are comparable, things are more complicated and one should find the energy shift by treating $H_B + H_{\text{s.o.}}$ together as a single perturbation, to be diagonalized using either the basis of states $|NLSJm_J\rangle$ or the basis $|NLSm_L m_S\rangle$.

It is important to keep in mind that the LS coupling scheme approximation is based on the assumption that $H_{\text{s.o.}}$ can be treated as a perturbation. The point is that while the unperturbed Hamiltonian H_0 commutes with the total spin and total orbital angular momentum operators $\boldsymbol{S}$ and $\boldsymbol{L}$, the spin-orbit part $H_{\text{s.o.}}$ does not. As a consequence, the LS coupling scheme approximation is not as good for heavier atoms with large Z. To understand this, consider the limit in which the potential in the spin-orbit interaction is spherically symmetric, $\Phi(r) = Z_{\text{eff}}e/r$, where $Z_{\text{eff}} < Z$ roughly takes into account the effects of partial screening of the nuclear charge by the other electrons. Then we have a sum of terms like (17.4) for each electron,

$$H_{\text{s.o.}} = \sum_{i=1}^{N_e} \xi(R_i)\boldsymbol{S}_i \cdot \boldsymbol{L}_i, \qquad \xi(r) = -\frac{e}{2m_e^2c^2}\frac{1}{r}\frac{\partial \Phi}{\partial r}. \tag{18.105}$$

Now, as a qualitative approximation, suppose that the electrons are in hydrogen-like orbitals. Then $H_{\text{s.o.}}$ scales like Z_{eff}/R^3, and since the characteristic size of orbitals scales like $1/Z_{\text{eff}}$, as can be seen from (11.65), the matrix elements of $H_{\text{s.o.}}$ scale like Z_{eff}^4. This is in comparison to the Z_{eff}^2 scaling of the effects of H_0.

In the limit that the effects of $H_{\text{s.o.}}$ are too large to be treated as a perturbation to H_0, one can instead use the **jj coupling approximation**. The idea is that H_0 and $H_{\text{s.o.}}$ both commute with the total angular momentum of each electron, $\boldsymbol{J}_i = \boldsymbol{L}_i + \boldsymbol{S}_i$ and with their total $\boldsymbol{J} = \sum_{i=1}^{N_e} \boldsymbol{J}_i$. Thus, $H_0 + H_{\text{s.o.}}$ is diagonal in an orthobasis labeled by the quantum numbers l_i, s_i, and j_i associated with the operators L_i^2, S_i^2, and J_i^2. The W part of the Hamiltonian in (18.79) does not commute with the operators L_i^2 and J_i^2, so it is treated as a perturbation, which then splits the states into different values of the total angular momentum quantum number J, with a remaining degeneracy $2J + 1$ due to m_J (neglecting small hyperfine effects associated with electronic interactions with the nucleus). In the jj coupling scheme appropriate for large $H_{\text{s.o.}}$ and small W, the term symbols specifying the angular momenta are thus written as

$$(j_1, j_2, \ldots, j_n)_J, \tag{18.106}$$

where n is the number of electrons in the incomplete subshell.

The standard example of the jj coupling scheme is the lead atom (Pb, $Z = 82$). The electron configuration is [Xe]$4f^{14}\,5d^{10}\,6s^2\,6p^2$, so what matters is the two $l = 1$ electrons in the unfilled $6p$ subshell. Just as for carbon, the number of antisymmetric states that we can form from two p electrons in the same subshell is $6 \cdot 5/2 = 15$. Each electron can have total angular momentum $j = 1/2$ or $3/2$. Now we use

$$\tfrac{1}{2} \otimes \tfrac{1}{2} = 0_A \oplus 1_S \tag{18.107}$$

$$\tfrac{3}{2} \otimes \tfrac{3}{2} = 0_A \oplus 1_S \oplus 2_A \oplus 3_S, \tag{18.108}$$

$$\tfrac{1}{2} \otimes \tfrac{3}{2} = 1 \oplus 2 \tag{18.109}$$

to find that the possible jj coupling terms for two p electrons are

$$(j_1, j_2)_J = (\tfrac{1}{2}, \tfrac{1}{2})_0, \quad (\tfrac{1}{2}, \tfrac{3}{2})_1, \quad (\tfrac{1}{2}, \tfrac{3}{2})_2, \quad (\tfrac{3}{2}, \tfrac{3}{2})_2, \quad (\tfrac{3}{2}, \tfrac{3}{2})_0, \tag{18.110}$$

with degeneracies 1, 3, 5, 5, 1, respectively (which indeed add up to 15 as a check). Experimentally, it turns out that the order given in (18.110) is in increasing energy, so the ground state has $J = 0$. By way of comparison, the Hund's rules prediction for the LS-coupling term for Pb is $^3P_0$ (by an argument identical to the one for carbon), which also has $J = 0$, so the two schemes agree in that respect. However, the jj coupling scheme provides a somewhat better numerical approximation to the energy levels and other properties of the lowest few eigenstates of Pb than the LS coupling scheme does. For other atoms with high Z, neither the LS scheme nor the jj scheme is very accurate, and more complicated descriptions are appropriate.

Exercises

18.1 Two identical particles are moving in a three-dimensional isotropic harmonic oscillator potential with natural frequency ω. The only Hamiltonian interaction between the two particles is the spin-spin $W = a\boldsymbol{S}_1 \cdot \boldsymbol{S}_2$, where a is a small constant ($\hbar|a| \ll \omega$).

(a) If the particles are spin-1 bosons, find the energy, the spectroscopic term, and the degeneracy of the ground state. Give separate answers for positive and negative a.
(b) Repeat part (a) if the particles have spin 0.
(c) Repeat part (a) if the particles have spin 2.
(d) Repeat part (a) if the particles are fermions with spin 1/2.

18.2 Two identical spin-1/2 fermions are free to move inside a cubic box of side a with one corner at the origin. The potential is infinite outside of the box. The fermions do not have any Hamiltonian interaction with each other.

(a) The *one-particle* energy eigenstate wavefunctions are

$$\psi_{n_x,n_y,n_z}(x, y, z) = \left(\frac{2}{a}\right)^{3/2} \sin\left(\frac{n_x\pi x}{a}\right) \sin\left(\frac{n_y\pi y}{a}\right) \sin\left(\frac{n_z\pi z}{a}\right). \tag{18.111}$$

What are the corresponding one-particle energies?
(b) In the remainder of this problem, we will consider two-particle energy eigenstates, organized in terms of the total spin, $S = 0$ or $S = 1$. But notice that in this problem we do not have spherical symmetry, and the Hamiltonian does not

commute with L^2 or L_z. Therefore, we cannot organize energy eigenstates in terms of orbital angular momentum. Instead, consider states that are either of the form

$$\chi_{S,m_S}\,\psi_N(x_1, y_1, z_1)\psi_N(x_2, y_2, z_2), \tag{18.112}$$

when the two particles have the same spatial wavefunction ψ_N, or

$$\chi_{S,m_S}\,\frac{1}{\sqrt{2}}\left[\psi_N(x_1, y_1, z_1)\psi_M(x_2, y_2, z_2) \pm \psi_M(x_1, y_1, z_1)\psi_N(x_2, y_2, z_2)\right], \tag{18.113}$$

if they have different spatial wavefunctions, where χ_{S,m_S} is the spin ket for $S = 0$ or $S = 1$, and N or M represents (n_x, n_y, n_z) for a given single-particle wavefunction. Find the energy eigenvalues and the corresponding degeneracies for the lowest three energy levels of the two-fermion system.
[Hints: the one-particle energies simply add to give the two-particle energies, since we are not including any interactions between the particles. Recall that for $S = 0$, the spatial wavefunction is symmetric under interchange of the labels 1 and 2, while for $S = 1$, the spatial wavefunction must be antisymmetric. You should find that the total number of states in the lowest three energy levels is a number greater than 35.]

18.3 Consider three identical spin-1/2 fermions bound in a three-dimensional isotropic harmonic oscillator potential characterized by frequency ω. The fermions do not interact with each other.

(a) Use the Pauli exclusion principle to find the energies and the degeneracies of the ground-state energy level and the first excited energy level. Give the spectroscopic term notation ${}^{2S+1}L_J$ for the allowed states in each case. [Hint: combine two of the fermions, then add the third. For the ground-state energy level, you should find two distinct spectroscopic terms. For the first excited state energy level, there are 11 spectroscopic terms, some of which are the same.]
(b) A spin-orbit perturbation Hamiltonian is added, $W = a\boldsymbol{L} \cdot \boldsymbol{S}$, where $\boldsymbol{L}$ and $\boldsymbol{S}$ are the total orbital and spin angular momenta of the fermions, and $\hbar|a| \ll \omega$. What are the energies and spectroscopic term notations of the lowest three energy levels if a is positive? What if a is negative?
(c) Now suppose that in addition to the perturbation W, there is a much weaker perturbation $W' = bL_z$. Use degenerate perturbation theory to find the energy splitting of the lowest level you found in part (b), by finding the matrix representation for W' and finding its eigenvalues. The cases of positive a and negative a are different, so treat them separately. [Hint: you may find (12.61)–(12.66) useful.]

18.4 Compute the direct and exchange integrals I_{200} and J_{200}, as defined in (18.75) and (18.76), for the $(1s)(2s)$ excited states of the helium atom. Note that from (11.42) and (11.43), the one-electron wavefunctions with $Z = 2$ are $\phi_{1,0,0}(r) = e^{-2r/a_0}\sqrt{8/\pi a_0^3}$ and $\phi_{2,0,0}(r) = e^{-r/a_0}(1 - r/a_0)/\sqrt{\pi a_0^3}$. Check that $0 < J_{200} < I_{200}$. Use these results to estimate the energies of the ${}^1S_0$ and ${}^3S_1$ states, and their energy splitting $\Delta E = 2e^2 J_{200}$, at first order in perturbation theory. Compare to the results quoted in Fig. 18.2. [Hints: to set up the integrals, follow the example of the method used in (15.88)–(15.97). Before integrating, you are encouraged to change to dimensionless variables $u = r_1/a_0$ and $v = r_2/a_0$. The perturbation theory result for ΔE is far from a triumph, as you should find that it is numerically about a factor of 3 larger than the experimental value. However, it does have two important qualitative features: the correct sign, and a magnitude much smaller than the energy shift from the direct integral.]

18.5 For each of calcium ($Z = 20$), scandium ($Z = 21$), titanium ($Z = 22$), nickel ($Z = 28$), germanium ($Z = 32$) and selenium ($Z = 34$), use the shell model to find the electronic configuration. Assume that the $4s$ shell fills before the $3d$ shell; although this is not always the case, it is true for these examples. In the Russell–Saunders approximation, find the spectroscopic term notation for the electronic configurations you found, and apply Hund's rules to select the ground state in each case.

18.6 In the Bizarro Universe, far away and very long ago, everything was just like in our universe except that electrons were fermions with spin $3/2$ instead of spin $1/2$. Consider atomic states in the Bizarro Universe.

(a) How many Bizarro electrons fit into each s, p, and d subshell (that is, $l = 0, 1, 2$, respectively)?
(b) What are the atomic numbers of the lightest two Bizarro noble gases (the lightest has a filled $1s$ subshell, the next has a filled $2p$ subshell)? What about the lightest two Bizarro alkali metals (one s-subshell electron) and Bizarro halogens (one missing p-subshell electron)?
(c) Consider Bizarro neon, a neutral atom with $Z = 10$. What is the electron configuration, instead of $(1s)^2(2s)^2(2p)^6$ as in our universe? What are all of the ${}^{2S+1}L_J$ spectroscopic terms for the electron configuration you found, and which of them is the ground state?

19 Periodic Potentials

19.1 Bloch Waves

In this chapter we will discuss potentials that are periodic in space, which are useful for modeling the states of electrons in crystalline solids. Consider first the class of one-dimensional problems involving a Hamiltonian $H = P^2/2m + V(X)$, where the potential function obeys

$$V(x + a) = V(x) \tag{19.1}$$

for all x. The length a might be the distance between adjacent sites of a lattice, occupied by identical fixed atoms or molecules, which provide the potential seen by relatively light and mobile electrons whose stationary states and energies we wish to find. More generally, each range of length a is called a **primitive cell**. This potential has a symmetry that is a discrete subgroup of the translations, and is associated with the unitary operator

$$T(a) = e^{-iaP/\hbar}. \tag{19.2}$$

Recall from (5.30) and (5.31) that $T(a)XT(a)^{-1} = X - a$, and $T(a)PT(a)^{-1} = P$, from which it is easy to show that $T(a)HT(a)^{-1} = H$. This is equivalent to $[T(a), H] = 0$, showing that $T(a)$ is a symmetry of the system.

Like parity, the symmetry $T(a)$ is discrete, because $T(a)^n = T(na)$ is a symmetry of the potential only for integer n. Noether's principle applies only to continuous symmetries, so it does not give us a conserved quantity in this case. However, the symmetry still provides useful information about the stationary states and energies. Since H and $T(a)$ commute, they have an orthobasis of common eigenstates, with

$$H|\psi\rangle = E|\psi\rangle, \qquad T(a)|\psi\rangle = e^{-i\theta}|\psi\rangle. \tag{19.3}$$

S. P. Martin, *Quantum Mechanics*, Graduate Texts in Physics,
https://doi.org/10.1007/978-3-032-21119-4_19

The unitarity of $T(a)$ guarantees that its eigenvalues $e^{-i\theta}$ have modulus 1, so θ is always a real number. We now define a quantity

$$q = \theta/a, \tag{19.4}$$

where $\hbar q$ is sometimes called the **quasi-momentum** of the state. Note that we can restrict to the range $-\pi < \theta \leq \pi$, because values outside of this range are equivalent to values within the range, due to the invariance of $e^{-i\theta}$ under $\theta \to \theta + 2\pi$. Therefore, without loss of generality, we restrict to the **first Brillouin zone**

$$-\frac{\pi}{a} < q \leq \frac{\pi}{a}, \tag{19.5}$$

although the choice $0 \leq q < 2\pi/a$ would work just as well. In terms of the wavefunction $\psi(x) = \langle x|\psi\rangle$, we use $\langle x|T(a)|\psi\rangle = \psi(x-a)$ from (5.23) to get $\psi(x-a) = e^{-iqa}\psi(x)$, or equivalently

$$\psi(x+a) = e^{iqa}\psi(x). \tag{19.6}$$

Thus, the wavefunction itself is not periodic, but picks up a phase factor when we translate by an integer number of lattice spacings. The probability density $|\psi(x)|^2$ does satisfy periodicity, since the phase factor cancels.

It is useful to define a function $u(x)$ according to

$$\psi(x) = e^{iqx}u(x). \tag{19.7}$$

Substituting this into (19.6), we discover that $u(x)$ is periodic,

$$u(x+a) = u(x). \tag{19.8}$$

The form (19.7), which separates the lattice periodicity of $u(x)$ from the non-periodic modulation phase factor e^{iqx}, is called a **Bloch wave** after Felix Bloch. It is important that the solution for $u(x)$ will depend on the choice of q, and that there will be many such solutions with different energies, which we could label by an integer n. Therefore, it might be more appropriate to write $\psi_{n,q}(x)$ and $u_{n,q}(x)$, but we will not do so in order to avoid notational clutter.

The Bloch wave form is a useful construction because after selecting q, one can solve for $\psi(x)$ [or $u(x)$] in just the domain $0 \leq x \leq a$, and then extend it to the whole infinite (or very large) domain of x. We will study a simple example of this in the next section.

The Bloch wave idea generalizes to three dimensions in the following way. Suppose that we have a crystal lattice invariant under translations

$$\boldsymbol{r} \to \boldsymbol{r} + \boldsymbol{a}_j, \qquad (j = 1, 2, 3), \tag{19.9}$$

where $\boldsymbol{a}_1, \boldsymbol{a}_2$, and $\boldsymbol{a}_3$ are three linearly independent (but not necessarily orthogonal) vectors called the primitive lattice vectors. (In the special case of a simple cubic lattice, the vectors $\boldsymbol{a}_1$, $\boldsymbol{a}_2$ and $\boldsymbol{a}_3$ all have the same length and are mutually orthogonal, but in three dimensions there are many more interesting possibilities.) Assuming the potential $V(\boldsymbol{r})$ has the same periodicity as the lattice, the operators H, $T(\boldsymbol{a}_1)$, $T(\boldsymbol{a}_2)$, and $T(\boldsymbol{a}_3)$ are all compatible, guaranteeing the existence of an orthobasis of stationary states whose wavefunctions obey

$$\psi(\boldsymbol{r}+\boldsymbol{a}_j) = e^{i\theta_j}\psi(\boldsymbol{r}), \qquad (j = 1, 2, 3). \tag{19.10}$$

For each choice of $(\theta_1, \theta_2, \theta_3)$, we define a vector $\boldsymbol{q}$ according to

$$\boldsymbol{q}\cdot\boldsymbol{a}_j = \theta_j, \qquad (j = 1, 2, 3). \tag{19.11}$$

(In general, this requires solving three equations for the three components of $\boldsymbol{q}$, but we will provide the explicit solution very soon.) Then (19.10) implies that for every lattice vector of the form

$$\boldsymbol{a} \equiv n_1\boldsymbol{a}_1 + n_2\boldsymbol{a}_2 + n_3\boldsymbol{a}_3, \qquad (\text{integer } n_1, n_2, n_3), \tag{19.12}$$

we must have

$$\psi(\boldsymbol{r}+\boldsymbol{a}) = e^{i\boldsymbol{q}\cdot\boldsymbol{a}}\psi(\boldsymbol{r}). \tag{19.13}$$

Now define a function $u(\boldsymbol{r})$ by

$$\psi(\boldsymbol{r}) = e^{i\boldsymbol{q}\cdot\boldsymbol{r}}u(\boldsymbol{r}). \tag{19.14}$$

Plugging this into (19.13), we find that $u(\boldsymbol{r})$ satisfies periodicity under any lattice translation,

$$u(\boldsymbol{r}+\boldsymbol{a}) = u(\boldsymbol{r}). \tag{19.15}$$

Equation (19.14) is the Bloch wavefunction for the three-dimensional lattice.

To solve the three equations in (19.11) for $\boldsymbol{q}$, and for other purposes as well, it is useful to define the **reciprocal lattice**, as follows. First, we construct three primitive reciprocal lattice vectors $\boldsymbol{b}_j$ with $j = 1, 2, 3$, defined to satisfy

$$\boldsymbol{a}_j\cdot\boldsymbol{b}_{j'} = 2\pi\delta_{jj'}. \tag{19.16}$$

To construct them explicitly, consider first $\boldsymbol{b}_1$. It is orthogonal to $\boldsymbol{a}_2$ and $\boldsymbol{a}_3$, and therefore proportional to $\boldsymbol{a}_2\times\boldsymbol{a}_3$. The constant of proportionality is determined by requiring $\boldsymbol{a}_1\cdot\boldsymbol{b}_1 = 2\pi$. Doing the same for $\boldsymbol{b}_2$ and $\boldsymbol{b}_3$, we obtain

$$b_1 = 2\pi(a_2 \times a_3)/\mathcal{V}, \qquad b_2 = 2\pi(a_3 \times a_1)/\mathcal{V},$$
$$b_3 = 2\pi(a_1 \times a_2)/\mathcal{V}, \tag{19.17}$$

where the normalization factor

$$\mathcal{V} = a_1 \cdot (a_2 \times a_3) = a_2 \cdot (a_3 \times a_1) = a_3 \cdot (a_1 \times a_2) \tag{19.18}$$

is, up to a sign, the volume of the primitive cell parallelepiped with edges given by the vectors a_1, a_2, a_3.

We can now immediately write down the solution to (19.11):

$$q = \frac{\theta_1}{2\pi} b_1 + \frac{\theta_2}{2\pi} b_2 + \frac{\theta_3}{2\pi} b_3. \tag{19.19}$$

The factor $e^{iq\cdot a}$ in (19.13) is unchanged if $q \cdot a$ is shifted by an integer multiple of 2π. It follows that for a general reciprocal lattice vector

$$b = m_1 b_1 + m_2 b_2 + m_3 b_3, \qquad \text{(integer } m_1, m_2, m_3\text{)}, \tag{19.20}$$

there is no physical effect from the shift

$$q \to q + b. \tag{19.21}$$

Therefore, the vector q is defined modulo these shifts. So, as in the one-dimensional case, one can restrict q to avoid redundancy, and without loss of generality.

From (19.10), one obvious choice is to restrict to q such that $-\pi < \theta_j \leq \pi$, since we can always shift each θ_j by an integer multiple of 2π without changing anything. An equally good choice is $0 \leq \theta_j < 2\pi$. The second option is equivalent to

$$0 \leq q \cdot a_j < 2\pi, \qquad \text{(for } j = 1, 2, 3\text{)}, \tag{19.22}$$

and is called the **primitive reciprocal cell.**

An alternative way to restrict q to eliminate redundancy without losing generality is to instead require it to be within the first Brillouin zone, defined for three-dimensional lattices as the set of vectors q that satisfy the minimum length requirement

$$|q| \leq |q - b| \quad \text{for all non-zero } b \text{ in (19.20)}. \tag{19.23}$$

The first Brillouin zone is always a convex polyhedron in q-space, and it has the same volume as the primitive reciprocal cell (19.22), namely $\mathcal{V}_{\text{recip}} = |b_1 \cdot (b_2 \times b_3)| = (2\pi)^3/|\mathcal{V}|$. These two ways of restricting q are just different descriptions of the same physics.

19.2 Kronig–Penney Delta-Function Model

In this section, we will study a simple model for a periodic potential in one dimension, the **Kronig–Penney** model, named for Ralph Kronig and William Penney. They considered a periodic array of rectangular potential barriers, but we will simplify things further by taking the barriers to be delta functions. Although this is not a very realistic model, it turns out to have qualitative features similar to those found in electronic states in materials with periodic structure.

Consider the "comb" potential

$$V(x) = V_0\, a \sum_{n=-\infty}^{\infty} \delta(x - na). \tag{19.24}$$

We will assume that V_0 is positive; note that it has units of energy, since $\delta(x-na)$ has units of inverse length. To solve for the stationary states of $H = P^2/2m + V(x)$, we start by observing that when $x \neq na$ for integer n, the Hamiltonian in the position representation is just that of a free particle. We therefore know that we can write the wavefunctions within the domain $0 \leq x \leq a$ as linear combinations of e^{ikx} and e^{-ikx}, or as linear combinations of $\sin(kx)$ and $\cos(kx)$, where

$$E = \hbar^2 k^2/2m. \tag{19.25}$$

Our task is to do this while matching the wavefunctions through boundary conditions at $x = na$. Once we have specified q, we only need to do this at one lattice site, say the one at $x = 0$, and then the Bloch form of the wavefunction will extend it for us to the whole domain $-\infty < x < \infty$.

Actually, the analysis becomes somewhat simpler if we choose to write the wavefunction for $0 \leq x \leq a$ as the following linear combination:

$$\psi(x) \;=\; \sin(kx) + A\sin(k(x-a)), \qquad (0 \leq x \leq a), \tag{19.26}$$

where A is a constant to be determined. We choose the coefficient of $\sin(kx)$ to be unity, treating the normalization of the state as arbitrary. The other function $\sin(k(x-a))$ is chosen for convenience, since it is a linear combination of $\sin(kx)$ and $\cos(kx)$ that vanishes at $x = a$. From the Bloch form in (19.7), we have

$$u(x) = e^{-iqx}\left[\sin(kx) + A\sin(k(x-a))\right], \qquad (0 \leq x \leq a). \tag{19.27}$$

Now, from the periodicity of $u(x)$, we have $\psi(x) = e^{iqx}u(x) = e^{iqx}u(x+a)$ for all x, which we evaluate in the domain $-a \leq x \leq 0$ to find

$$\psi(x) \;=\; e^{-iqa}\left[\sin(k(x+a)) + A\sin(kx)\right], \qquad (-a \leq x \leq 0). \tag{19.28}$$

Comparing (19.26) and (19.28) at $x = 0$, we learn that $-A\sin(ka) = e^{-iqa}\sin(ka)$, or

$$A = -e^{-iqa}. \tag{19.29}$$

The special case that ka is an integer multiple of π, so that $\sin(ka) = 0$, is a nonsingular limit; see Exercise 19.1. Note that replacing $k \to -k$ does not provide an independent solution, because it just multiplies the wavefunction by -1, so we can take k to be positive without loss of generality. In contrast, taking $q \to -q$ does provide a distinct solution as long as $qa \neq 0, \pm\pi$, since it changes the relative phase between the wavefunction terms.

Having found the wavefunction form from consistency of the Bloch wave periodicity, it remains to relate k, and thus the energy $E = \hbar^2k^2/2m$, to q. We do this by applying the boundary condition for a delta function potential, using (6.47) with $x_0 = 0$. This gives

$$\psi'(0^+) - \psi'(0^-) = \frac{2w}{a}\psi(0), \tag{19.30}$$

where we have defined a dimensionless parameter characterizing the potential strength,

$$w = mV_0a^2/\hbar^2. \tag{19.31}$$

Now we use (19.26) and (19.28) to get

$$\psi(0) = e^{-iqa}\sin(ka), \tag{19.32}$$

$$\psi'(0^+) = k\left(1 - e^{-iqa}\cos(ka)\right), \tag{19.33}$$

$$\psi'(0^-) = ke^{-iqa}\left(\cos(ka) - e^{-iqa}\right). \tag{19.34}$$

Inserting these in (19.30), and simplifying, we obtain

$$\cos(qa) = \cos(ka) + w\sin(ka)/ka. \tag{19.35}$$

This implicitly relates the energy of the state to the modulation wavenumber q.

Equation (19.35) cannot be solved analytically (except in special limits, as in Exercise 19.1), but does allow a graphical solution. In Fig. 19.1, we plot the function

$$f(ka) = \cos(ka) + w\sin(ka)/ka \tag{19.36}$$

as a function of ka/π. To have a solution, it is necessary and sufficient that $-1 \leq f(ka) \leq 1$, since that is the allowed range for $\cos(qa)$. It follows that the solutions are limited to bands in ka, shown as the shaded ranges, with gaps between them

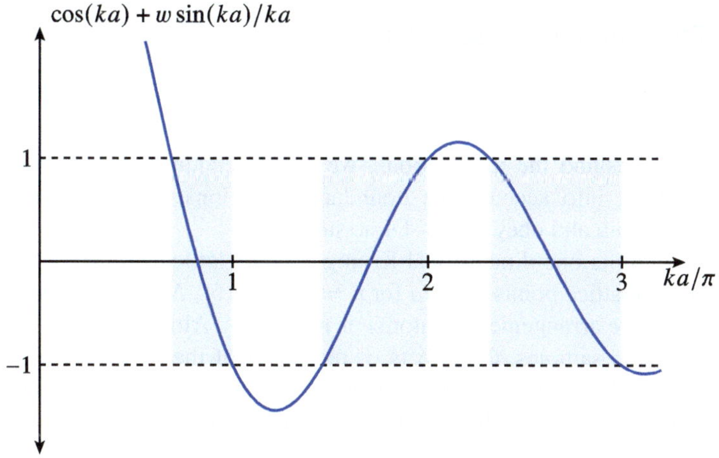

Fig. 19.1 Graphical solution of the delta-function Kronig–Penney model. The shaded ranges admit solutions for the equation $\cos(qa) = \cos(ka) + w \sin(ka)/ka$ (in this case with $w = 4$), where a is the lattice spacing, $E = \hbar^2 k^2/2m$ is the energy, and q is the Bloch-wave quasi-momentum. The upper dashed line corresponds to $q = 0$, and the lower dashed line corresponds to $qa = \pm\pi$

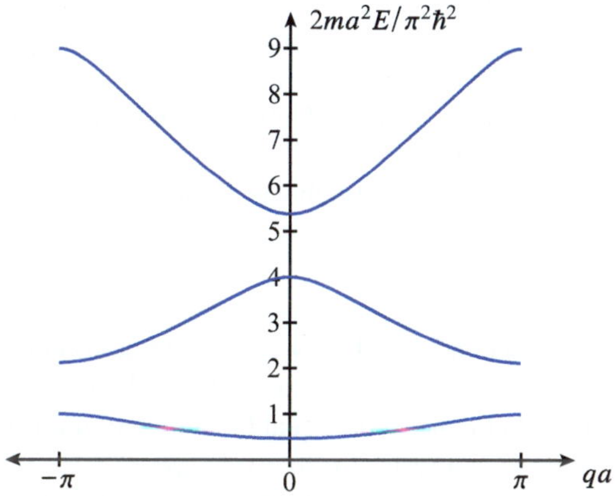

Fig. 19.2 Energies, in units of $\hbar^2\pi^2/2ma^2$, for the lowest three bands in the delta-function Kronig–Penney model (in this case with $w = 4$), as a function of qa. The vertical axis is also equal to $(ka/\pi)^2$

where no solution exists. The lowest allowed energy occurs for $q = 0$ and has ka equal to the smallest solution of the equation $\cos(ka) + w \sin(ka)/ka = 1$. The highest energy in the lowest band occurs for $|qa| = ka = \pi$, with $E = \hbar^2\pi^2/2ma^2$. Moving up in energy, the second band starts at $q = \pm\pi/a$ and ends at $q = 0$, while the third band starts at $q = 0$ and ends at $q = \pm\pi/a$, etc. The allowed energies for the lowest three bands are shown as functions of qa in Fig. 19.2.

19.3 Electronic Band Structure

In solid materials with crystalline structure, the electronic states have band structures that are qualitatively similar to those of the delta-function Kronig–Penney model. To understand the implications, we need to count the numbers of states available, taking into account the boundary conditions, and use the facts that electrons have spin and obey Fermi–Dirac statistics.

Consider a finite one-dimensional Kronig–Penney lattice with N primitive cells associated with lattice points $x = na$ for $n = 0, 1, 2, \ldots, N-1$. Each primitive cell contains the same arrangement of atoms or molecules. Although N is usually very large for realistic samples, the effects of the edges of the lattice will, in general, modify the Bloch wave structure. Numerically, this complication should have a minor effect, but to avoid dealing with it at all, we use the standard trick of imposing periodic boundary conditions. We treat the system as if it had an infinite length and an infinite number of primitive cells, but then impose

$$\psi(x) = \psi(x + Na) \tag{19.37}$$

for all x. This has the effect that the number of independent degrees of freedom corresponds to N primitive cells, lying on a circle rather than a line. This should not affect the local properties of the stationary states provided that the length of the sample Na is very large compared to a and the length scales set by the Hamiltonian, which are typically all microscopic, very roughly of order the Bohr radius for electronic states.

In terms of the Bloch wavefunction, the periodic boundary condition implies

$$u(x) = e^{iNqa}u(x + Na) = e^{iNqa}u(x). \tag{19.38}$$

In the second equality, we used the periodicity property $u(x + a) = u(x)$, which implies $u(x + Na) = u(x)$ as well. We now see that the allowed q are restricted to those that make Nqa equal to an integer multiple of 2π. Recall that for each energy band we can restrict q to $-\pi < qa \leq \pi$, or if we choose, $0 \leq q < 2\pi/a$. For our present purposes, the second choice is slightly more convenient, as it gives

$$q = \frac{2\pi n}{Na}, \qquad (n = 0, 1, 2, \ldots, N - 1). \tag{19.39}$$

Thus, the resulting number of allowed values of q is precisely N, the same as the number of primitive cells.

Similarly, for three-dimensional crystal potentials with N primitive cells, the number of stationary states in each band (not taking into account spin, yet) is also equal to N. To see this, consider a chunk of material with $N = N_1N_2N_3$ primitive cells, conveniently shaped in the form of a parallelepiped with macroscopic sides $N_1\boldsymbol{a}_1$ and $N_2\boldsymbol{a}_2$ and $N_3\boldsymbol{a}_3$, where the $\boldsymbol{a}_j$ are the primitive lattice vectors defined in

Sect. 19.1. To model this, it is convenient to start with an infinite lattice and impose periodic boundary conditions along each of the three lattice vectors $\boldsymbol{a}_1$, $\boldsymbol{a}_2$, and $\boldsymbol{a}_3$. That is, we take

$$\psi(\boldsymbol{r}) = \psi(\boldsymbol{r} + N_j \boldsymbol{a}_j), \qquad (j = 1, 2, 3). \tag{19.40}$$

Then, from (19.14), we must have $e^{i\boldsymbol{q}\cdot\boldsymbol{a}_j N_j} = 1$ for each of $j = 1, 2, 3$, so that

$$\boldsymbol{q} \cdot \boldsymbol{a}_j = 2\pi n_j / N_j \tag{19.41}$$

for integers n_j. Each band of states with nearly continuous (for large N) energies corresponds to one primitive reciprocal cell for $\boldsymbol{q}$, restricted to the range $0 \leq \boldsymbol{q}\cdot\boldsymbol{a}_j < 2\pi$ as in (19.22). So, we learn that

$$n_j = 0, 1, 2, \ldots, N_j - 1 \tag{19.42}$$

for each of $j = 1, 2, 3$. Thus, there are $N_1 N_2 N_3 = N$ solutions for $\boldsymbol{q}$, each corresponding to a distinct stationary state in the energy band.

So far we have neglected the fact that electrons have spin. Assuming approximate degeneracy with respect to the two orthogonal spin states for the electron, we conclude that for a chunk of material with N primitive cells, there are $2N$ states available in each energy band.[1] If the temperature is low, these states will fill up starting from the lowest energy, in accord with the Pauli exclusion principle.

Now suppose that each primitive cell comes with n_e electrons, so that the total number of electrons is $n_e N$. If n_e is an even number, it follows that each of the lowest $n_e/2$ energy bands will be completely filled at low temperatures, and all higher bands will be empty. Then if the energy gap between the highest-energy filled band (called the valence band) and the lowest-energy unfilled band (called the conduction band) is sufficiently large, it will be difficult for the electrons to respond to external electric fields by changing states, and the material will be an insulator. Some examples of this include table salt (NaCl) with a low-temperature band gap of 9.0 eV, and diamond (C) with a band gap of 5.5 eV. If n_e is even but the band gap is much smaller, then external electric fields, thermal excitations, and/or impurities allow states in the conduction band to be accessible, and the material is a semiconductor. Examples include silicon (Si) with a band gap of 1.17 eV, and germanium (Ge) with a band gap of 0.74 eV.

There are two simple prototypical cases that allow materials to be good conductors. First, if n_e is an odd number, then there will be a half-filled band. Examples of materials in this category include alkali metals (Li, Na, K, Rb, Cs) and noble metals (Cu, Ag, Au). Second, it may be that, taking into account the three-dimensional

---

[1] We will not consider it here, but a strong external magnetic field or a significant spin coupling that removes the spin degeneracy can split the band into two separate bands, each containing N states.

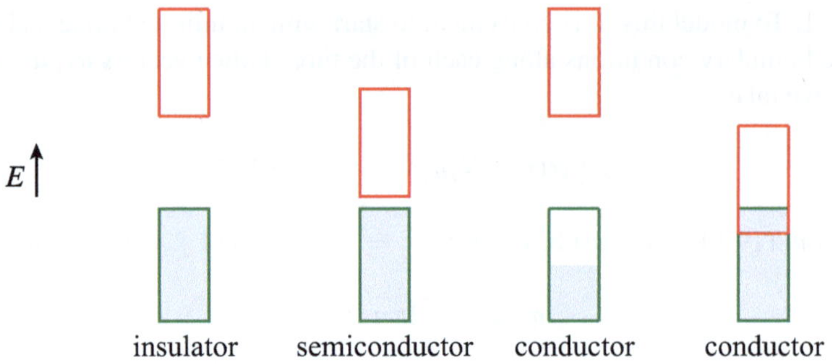

Fig. 19.3 Schematic filling of energy band structures at low temperature, for prototypical insulators, semiconductors, and conductors. Insulators have a large gap separating the filled valence band from the unfilled conduction band, inhibiting current flow. Semiconductors have a small gap between the valence and conduction bands. Conductors either have a half-filled band or bands that overlap in energy, allowing easy acceleration of charges through transitions between states with nearby energies

nature of the material, two bands that might otherwise be considered distinct, with one filled and the other unfilled, actually overlap in energy. Examples of elements in this category include Be, Mg, Ca, Sr, Ba from Group 2 of the periodic table, and Zn, Cd, Hg from Group 12. In both cases, the energy cost for the highest energy states to transition to nearby states is very low, explaining their good electrical conductivity.

These cases are illustrated in Fig. 19.3. Of course, real-world materials are more complicated than the Kronig–Penney model or simple variations of it. The most interesting and useful materials have electronic properties that do not fit into these prototypical categories neatly, or at all. These are covered in more specialized texts, and by active research.

Exercises

19.1 Consider the delta-function Kronig–Penney model with $w = mV_0a^2/\hbar^2$ as treated in Sect. 19.2.

(a) Carefully treat the special case that ka is an integer multiple of π. What happens to q and to the stationary state wavefunctions, and what is the physical interpretation?

(b) Working to first order in small w, show that the lowest energy band is

$$V_0 \leq E \leq \hbar^2\pi^2/2ma^2, \qquad \text{(small } w\text{)}. \tag{19.43}$$

What are the corresponding bounds on the energies for the second and third energy bands, to first order in w?

(c) In the limit of large w, show that the lowest energy band is

$$\frac{\hbar^2\pi^2}{2ma^2}\left(1-\frac{n}{w}\right) \le E \le \frac{\hbar^2\pi^2}{2ma^2}, \qquad \text{(large } w\text{)}, \tag{19.44}$$

where n is a certain positive integer that you will discover. What are the corresponding bounds on the energies for the second and third energy bands, to first order in $1/w$?

19.2 Consider the Kronig–Penney model with potential

$$V(x) = \begin{cases} 0 & \text{(for } 0 < x < a\text{)}, \\ V_0 & \text{(for } a < x < a+b\text{)}, \end{cases} \tag{19.45}$$

taken to be periodic in x with period $a+b$. Try a wavefunction solution for $E > V_0$ with

$$\psi(x) = \begin{cases} Ae^{ikx} + Be^{-ikx} & \text{(for } 0 \le x \le a\text{)}, \\ Ce^{ik'x} + De^{-ik'x} & \text{(for } a \le x \le a+b\text{)}, \end{cases} \tag{19.46}$$

where $k = \sqrt{2mE}/\hbar$ and $k' = \sqrt{2m(E-V_0)}/\hbar$. The Bloch wave form then gives, in the region just to the left of the origin,

$$\psi(x) = e^{-iq(a+b)}\left[Ce^{ik'(x+a+b)} + De^{-ik'(x+a+b)}\right] \qquad \text{(for } -b \le x \le 0\text{)}. \tag{19.47}$$

(a) Enforce the boundary conditions at $x = 0$ and at $x = a$ to show that the wavefunction will be a solution if

$$\begin{pmatrix} 1 & 1 & -e^{i(k'-q)(a+b)} & -e^{-i(k'+q)(a+b)} \\ k & -k & ? & ? \\ e^{ika} & e^{-ika} & ? & ? \\ ke^{ika} & -ke^{-ika} & ? & ? \end{pmatrix} \begin{pmatrix} A \\ B \\ C \\ D \end{pmatrix} = 0, \tag{19.48}$$

where you will fill in the missing entries denoted by question marks.

(b) For a non-trivial solution, the determinant of the matrix must vanish. Show that this gives

$$\cos(q(a+b)) = \cos(ka)\cos(k'b) - \frac{k^2+k'^2}{2kk'}\sin(ka)\sin(k'b), \tag{19.49}$$

providing the relationship between q and the energy $E = \hbar^2k^2/2m$.

Simple Molecules

20

20.1 Hierarchies of Scales and the Born-Oppenheimer Approximation

Molecules are bound states consisting of multiple atoms, or more fundamentally electrons and nuclei. Understanding the quantum energy levels and structure of molecules is a complicated problem requiring strategic thinking, as there are many degrees of freedom.

One way of simplifying the problem is to make use of the hierarchies of energy and distance scales caused by the large ratio (more than three orders of magnitude) of the nuclear masses to the electron masses. In a first approximation, the slowly moving nuclei can be treated as essentially at rest on the time scales relevant for the motion of electrons. The nuclear displacements from their equilibrium positions are small compared to the size of the molecule. One can therefore solve first for the electronic states while treating the nuclei as fixed. The nuclear wavefunctions are then treated separately, recognizing that the electrons will adjust themselves relatively quickly to changes in the nuclear positions. To determine the state describing the nuclear relative position degrees of freedom, one uses a potential that can be thought of as the result of averaging over the electronic positions, using the electronic wavefunction. This way of organizing the calculation in terms of a separation between light, fast degrees of freedom (the electrons) and heavy, slow degrees of freedom (the nuclei) is called the **Born-Oppenheimer approximation**, after the work of Max Born and J. Robert Oppenheimer in 1927. It is an example of a more general strategy called the **adiabatic approximation**, in which slow degrees of freedom are treated as nearly constant.

Let us start by making parametric estimates of the distance and energy scales involved. For fixed nuclear positions, dimensional analysis tells us that the characteristic size Δx_e of electronic states is comparable to the Bohr radius, since it is the only length scale that we can construct out of $\hbar$, the electron mass m, and the

S. P. Martin, *Quantum Mechanics*, Graduate Texts in Physics,
https://doi.org/10.1007/978-3-032-21119-4_20

electronic charge e. The corresponding electronic energy magnitudes E_e should be of order a Rydberg. So:

$$\Delta x_e \sim a_0 = \frac{\hbar^2}{e^2 m}, \qquad E_e \sim \frac{e^2}{2a_0} = \frac{e^4 m}{2\hbar^2} = \frac{\hbar^2}{2ma_0^2}. \tag{20.1}$$

The vibrational modes of the nuclei about their equilibrium positions can be modeled as effective harmonic oscillators with angular frequencies that we can call ω_N. To estimate ω_N crudely, we note that if we were to displace the position of a nucleus by a distance of order a_0, the change in energy should be of order a Rydberg. Using M to denote the nuclear mass, we therefore estimate the energy of the oscillator as $e^4 m/2\hbar^2 \sim M\omega_N^2 a_0^2/2$, which implies

$$\omega_N \sim \left(\frac{m}{M}\right)^{1/2} \frac{e^2}{\hbar a_0}. \tag{20.2}$$

The corresponding length and minimum energy scales associated with nuclear vibrations can therefore be estimated, using the harmonic oscillator results of (7.24) and (7.37), as

$$\Delta x_N \sim \sqrt{\frac{\hbar}{M\omega_N}} \sim \left(\frac{m}{M}\right)^{1/4} a_0, \qquad E_{\text{vib}} \sim \hbar\omega_N \sim \left(\frac{m}{M}\right)^{1/2} \frac{e^2}{a_0}. \tag{20.3}$$

Molecules also have rotational degrees of freedom, whose minimum energy scale can be estimated by noting that the largest moment of inertia should be of order $I \sim Ma_0^2$, and the smallest angular momentum difference is of order $L \sim \hbar$. Therefore, we estimate

$$E_{\text{rot}} \sim \frac{L^2}{2I} \sim \frac{\hbar^2}{Ma_0^2} \sim \frac{m}{M}\frac{e^2}{a_0}. \tag{20.4}$$

The hierarchy of nuclear and electron masses therefore leads to a hierarchy of electronic, vibrational, and rotational energy magnitudes, with very roughly

$$E_e : E_{\text{vib}} : E_{\text{rot}} = 1 : \left(\frac{m}{M}\right)^{1/2} : \frac{m}{M}. \tag{20.5}$$

Putting in the numbers, transitions between electronic states will result in emission of light that is typically in the UV or visible range, while transitions between nuclear vibrational modes are associated with the IR, and rotational modes with microwave radiation.

Consider a molecule consisting of N_e electrons and N_n nuclei. Let us work in position space, neglecting spin degrees of freedom for simplicity. The electrons have coordinates $\boldsymbol{r}_i$ with $i = 1, \ldots, N_e$, and the nuclei have coordinates $\boldsymbol{R}_I$ where

$I = 1, \ldots, N_n$, with masses M_I and integer charges Z_I. To realize the Born-Oppenheimer strategy, the Hamiltonian can be divided into parts that do and do not involve the nuclear masses:

$$H = H_e + H_N, \tag{20.6}$$

$$H_e = -\sum_i \frac{\hbar^2\nabla_i^2}{2m} + \sum_{i<j} \frac{e^2}{|\boldsymbol{r}_i - \boldsymbol{r}_j|} - \sum_i \sum_I \frac{Z_I e^2}{|\boldsymbol{r}_i - \boldsymbol{R}_I|} + \sum_{I<J} \frac{Z_I Z_J e^2}{|\boldsymbol{R}_I - \boldsymbol{R}_J|}, \tag{20.7}$$

$$H_N = -\sum_I \frac{\hbar^2\nabla_I^2}{2M_I}. \tag{20.8}$$

The first step in carrying out the Born-Oppenheimer approach is to find solutions for the electronic eigenvalue equation

$$H_e\, \phi_{n,\boldsymbol{R}}(\boldsymbol{r}_i) = \mathcal{E}_{n,\boldsymbol{R}}\, \phi_{n,\boldsymbol{R}}(\boldsymbol{r}_i), \tag{20.9}$$

in which the nuclear coordinates $\boldsymbol{R}_I$ are treated as fixed, and the labels n distinguish different bound eigenstates. The presence of the subscript $\boldsymbol{R}$ on the energy eigenvalues $\mathcal{E}_{n,\boldsymbol{R}}$ and the multi-electron wavefunctions $\phi_{n,\boldsymbol{R}}(\boldsymbol{r}_i)$ is to remind us that they depend on the choice of fixed $\boldsymbol{R}_I$. The wavefunctions also depend on all of the electronic coordinates $\boldsymbol{r}_i$. As noted above, since the nuclear masses are not involved in (20.9), the eigenvalues $\mathcal{E}_{n,\boldsymbol{R}}$ should be roughly of order a Rydberg, and the length scale over which the wavefunctions $\phi_{n,\boldsymbol{R}}(\boldsymbol{r}_i)$ vary appreciably should be comparable to the Bohr radius. Finding these solutions may be quite difficult, but with sufficient computing power they can be estimated using the variational principle. Note that the nucleus-nucleus repulsion term proportional to $Z_I Z_J$ in (20.7) is simply a constant in this part of the calculation.

Since $\phi_{n,\boldsymbol{R}}(\boldsymbol{r}_i)$ are the eigenstates of the Hermitian operator H_e, they can be chosen as an orthonormal basis for the electronic bound state part of the Hilbert space. This means that we can always write the full (multi-electron, multi-nucleus) wavefunction as

$$\Psi(\boldsymbol{r}_i, \boldsymbol{R}_I) = \sum_n \psi_n(\boldsymbol{R}_I)\phi_{n,\boldsymbol{R}}(\boldsymbol{r}_i), \tag{20.10}$$

for some coefficients $\psi_n(\boldsymbol{R}_I)$ that will serve as the wavefunctions for the nuclear position degrees of freedom. The next step is to look for eigenstates of this form for the full Hamiltonian H with eigenvalue E. Using (20.9) and (20.8) in (20.6), we have

$$\sum_n \Big(-\sum_I \frac{\hbar^2\nabla_I^2}{2M_I} + \mathcal{E}_{n,\boldsymbol{R}} - E \Big) \left[\psi_n(\boldsymbol{R}_I)\phi_{n,\boldsymbol{R}}(\boldsymbol{r}_i) \right] = 0. \tag{20.11}$$

Now using the product rule for Laplacians, $\nabla^2(fg) = f\nabla^2 g + g\nabla^2 f + 2(\nabla f)\cdot(\nabla g)$, this becomes

$$\sum_n \phi_{n,\boldsymbol{R}}(\boldsymbol{r}_i)\Big(-\sum_I \frac{\hbar^2\nabla_I^2}{2M_I} + \mathcal{E}_{n,\boldsymbol{R}} - E\Big)\psi_n(\boldsymbol{R}_I) =$$

$$\sum_n\sum_I \frac{\hbar^2}{M_I}\left[(\nabla_I\psi_n)\cdot(\nabla_I\phi_{n,\boldsymbol{R}}) + \frac{1}{2}\psi_n\nabla_I^2\phi_{n,\boldsymbol{R}}\right]. \tag{20.12}$$

The Born-Oppenheimer approximation entails neglecting the right-hand side of this equation (or treating it as a small perturbation), with the partial justification that the amplitude of nuclear vibrations is smaller than the distances between nuclei by a factor of order $(m/M)^{1/4}$, as noted in (20.3). To see how this works, recall that the wavefunction for an electronic state in a hydrogen atom, for example, involves a factor e^{-r/a_0}, where r is the separation between the nucleus and an electron. Therefore, considering the effect of changing a nuclear position slightly, we expect

$$\frac{|\nabla_I\phi_{n,\boldsymbol{R}}|}{\phi_{n,\boldsymbol{R}}} \sim \frac{1}{a_0}. \tag{20.13}$$

In comparison, we can approximate the wavefunction for each nucleus I by that of a harmonic oscillator with angular frequency $\omega_I = (m/M_I)^{1/2}e^2/\hbar a_0$ as in (20.2), so that ψ_n is proportional to $e^{-M_I\omega_I|\boldsymbol{R}_I-\boldsymbol{R}_{I0}|^2/2\hbar}$, where $\boldsymbol{R}_{I0}$ is the equilibrium position of the nucleus. We then estimate $|\nabla_I\psi_n|/\psi_n \sim |\boldsymbol{R}_I - \boldsymbol{R}_{I0}|M_I\omega_I/\hbar$. Inserting the estimates for ω_I and $\Delta x_N = |\boldsymbol{R}_I - \boldsymbol{R}_{I0}|$ from (20.2) and (20.3), we obtain

$$\frac{|\nabla_I\psi_n|}{\psi_n} \sim \left(\frac{M_I}{m}\right)^{1/4}\frac{1}{a_0}. \tag{20.14}$$

Thus, each transfer of a derivative ∇_I from the nuclear wavefunction ψ_n to the electronic wavefunction ϕ_n costs a relative suppression factor of $(m/M_I)^{1/4}$, in accord with the intuition that the heavy nuclei are much more sharply localized than the electrons.

Replacing the right side of (20.12) by 0, we realize the added benefit that since the wavefunctions $\phi_{n,\boldsymbol{R}}(\boldsymbol{r}_i)$ form an orthonormal basis, each term in the sum over n must vanish separately, and so we can write

$$\Big(-\sum_I \frac{\hbar^2\nabla_I^2}{2M_I} + \mathcal{E}_{n,\boldsymbol{R}} - E\Big)\psi_n(\boldsymbol{R}_I) = 0 \tag{20.15}$$

as the energy eigenstate equation for the nuclear excitations. For each electronic state labeled by n, there is a family of nuclear vibrational and rotational states and energies associated with the solutions to (20.15). In this Schrödinger-like equation, the $\mathcal{E}_{n,\boldsymbol{R}}$, which were obtained by solving for the electronic state energies, serve as the potential for the nuclear wavefunction.

To recapitulate, the Born-Oppenheimer approximation separates the molecular problem into solving (20.9) for the electronic states with fixed nuclei, and (20.15) for the nuclear degrees of freedom with the electronic states providing the potential. While it is tempting to say that we have treated the nuclear kinetic term H_N in (20.8) as a perturbation, this is not really the same sort of perturbation theory as in Chap. 15. As we have seen, the Born-Oppenheimer approach is an expansion in $(m/M)^{1/4}$, while standard perturbation theory would instead imply an expansion in powers of m/M. (The latter type of perturbative expansion does not work here, because the dynamics of the nuclear degrees of freedom cannot be well-approximated by an unperturbed Hamiltonian without any kinetic energy term.)

In the remaining sections of this chapter, we will discuss only a few selected aspects of molecular physics, restricted to simple cases. First, in Sect. 20.2, we will find an approximation for the solution of (20.9) for the electronic state of the minimal example of a molecule, the H_2^+ ion. In Sect. 20.3, we will then treat the vibrational nuclear state determination of (20.15) for more general diatomic molecules, based on a widely used and empirically successful potential. The general subject of quantized rigid body rotations, and the application to the rotational energy levels of molecules, are given in Sect. 20.4. Sect. 20.5 discusses the vibrational and rotational spectra for diatomic molecules, including numerical parameters for some common ones. Section 20.6 contains a brief account of the rotational and vibrational excitations for the water molecule H_2O.

20.2 The Simplest Molecule: the H_2^+ Ion

The H_2^+ ion is the simplest molecule, consisting of one electron and two protons. In the Born-Oppenheimer approximation, we first fix the positions of the protons at $\boldsymbol{R}_A$ and $\boldsymbol{R}_B$, with the distance between them denoted by

$$R = |\boldsymbol{R}_A - \boldsymbol{R}_B|, \tag{20.16}$$

and describe the electron's position with a vector $\boldsymbol{r}$ with corresponding Laplacian ∇^2. The Hamiltonian for the electron's state is, in the position representation,

$$H_e = -\frac{\hbar^2\nabla^2}{2m} + e^2\left(\frac{1}{R} - \frac{1}{|\boldsymbol{r}-\boldsymbol{R}_A|} - \frac{1}{|\boldsymbol{r}-\boldsymbol{R}_B|}\right), \tag{20.17}$$

with the corresponding geometry shown in Fig. 20.1.

Our first goal is to estimate the ground-state energy $\mathcal{E}_R$ and wavefunction $\phi_R(\boldsymbol{r})$, using a variational principle. The resulting $\mathcal{E}_R$ then serves as the effective potential energy for the nuclear separation dynamics, as we discussed in (20.15), the second

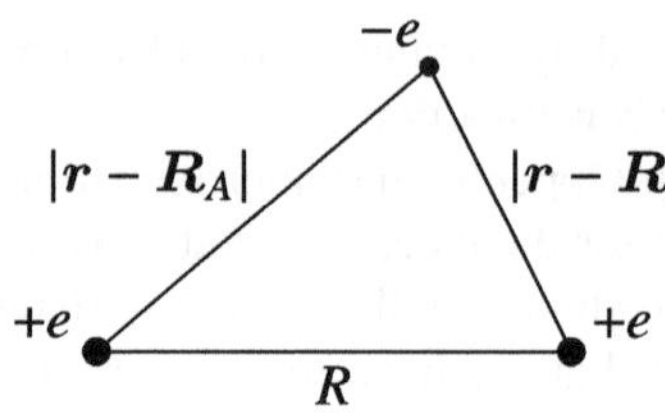

Fig. 20.1 The H_2^+ ion molecule consists of two protons with charges $+e$ at positions $\boldsymbol{R}_A$ and $\boldsymbol{R}_B$, and an electron with charge $-e$ at position $\boldsymbol{r}$. In the Born-Oppenheimer approximation, we first treat the proton positions $\boldsymbol{R}_A$ and $\boldsymbol{R}_B$ as fixed, to find the binding potential energy as a function of $R = |\boldsymbol{R}_A - \boldsymbol{R}_B|$

part of the Born-Oppenheimer approximation. From this, we will determine the binding energy and the nuclear separation of the H_2^+ ion.

To construct a trial state, let us start with unit-norm atomic orbital wavefunctions

$$\varphi_A(\boldsymbol{r}) = \sqrt{\frac{Z^3}{\pi a_0^3}} e^{-Z|\boldsymbol{r}-\boldsymbol{R}_A|/a_0}, \qquad \varphi_B(\boldsymbol{r}) = \sqrt{\frac{Z^3}{\pi a_0^3}} e^{-Z|\boldsymbol{r}-\boldsymbol{R}_B|/a_0}. \tag{20.18}$$

These have the same form as the ground state of a hydrogen atom centered at each of the two proton positions, except that we have included a variational parameter Z, which governs the size of the orbitals. Our trial wavefunction $\phi(\boldsymbol{r})$ will be a linear combination of φ_A and φ_B. Without loss of generality, we can take

$$\boldsymbol{R}_B = -\boldsymbol{R}_A = (R/2)\hat{z}, \tag{20.19}$$

so that the Hamiltonian has even parity, $[\Pi, H_e] = 0$. This ensures that the energy eigenstates can be chosen to have well-defined parity $= \pm 1$, so we try unnormalized variational states

$$|\phi_\pm\rangle = |\varphi_A\rangle \pm |\varphi_B\rangle, \tag{20.20}$$

with $\Pi|\phi_\pm\rangle = \pm|\phi_\pm\rangle$. From here on, the trial state and energy will be denoted just $|\phi\rangle$ and $\mathcal{E}_{Z,R}$ (without the subscript $\pm$) to avoid clutter, with the understanding that we are trying both parity possibilities simultaneously.

Applying the variational principle of Chap. 16, the ground-state energy estimate is

$$\mathcal{E}_{Z,R} = \frac{\langle\phi|H_e|\phi\rangle}{\langle\phi|\phi\rangle} = \frac{\langle\varphi_A|H_e|\varphi_A\rangle \pm \langle\varphi_B|H_e|\varphi_A\rangle}{1 \pm \langle\varphi_B|\varphi_A\rangle}. \tag{20.21}$$

In writing the last expression, we have used the facts that $\langle\varphi_A|\varphi_A\rangle = \langle\varphi_B|\varphi_B\rangle = 1$, and $\langle\varphi_A|\varphi_B\rangle = \langle\varphi_B|\varphi_A\rangle$, since the atomic orbital wavefunctions are normalized and real, and $\langle\varphi_A|H_e|\varphi_B\rangle = \langle\varphi_B|H_e|\varphi_A\rangle$, since H_e is Hermitian, and $\langle\varphi_B|H_e|\varphi_B\rangle =$

$\langle\varphi_A|H_e|\varphi_A\rangle$ by symmetry, since the orbitals have the same form up to a displacement by distance R.

The electronic state depends on two parameters: R (the distance between the protons, which is temporarily fixed as part of the Born-Oppenheimer approximation) and Z (the variational parameter). Before doing the detailed calculations, let us consider two extreme special cases of R in which the problem simplifies and we know what to expect. First, if R is very large compared to a_0, then the electron will essentially be forced to choose between the two protons, and the ground state can be thought of as a symmetrized tensor product between a hydrogen atom and a lonely distant proton. In that limit, we expect that Z will be very close to 1, and the eigenvalue of H_e will be almost exactly -1 Rydberg $= -e^2/2a_0$. In the opposite extreme case that R is very small compared to a_0, the situation is the same as if there were only a single nucleus with charge $+2e$, so we expect that Z will approach 2 in that limit. This will be accompanied by a diverging positive contribution to the H_e eigenvalue, due to the repulsive energy between the two protons (the last term in (20.17)).

For intermediate R, the minimum $\mathcal{E}_R$ should be achieved for $1 < Z < 2$. To minimize the energy, the preferred position for the electron is between the two protons, where it can benefit from its attraction to them both. However, for the odd-parity case, the wavefunction $\phi(\boldsymbol{r})$ vanishes at the midpoint $\boldsymbol{r} = 0$, and therefore it must be small everywhere nearby, by continuity. In the even-parity case the wavefunction is enhanced, due to constructive interference, in the region of lower potential between the protons. We therefore expect the ground state with strongest binding to have even parity. The essential reason for the bond is that the electron can be shared by the two nuclei to which it is attracted, lowering the potential energy.

To proceed, we need to find the matrix elements $\langle\varphi_B|\varphi_A\rangle$, $\langle\varphi_A|H_e|\varphi_A\rangle$, and $\langle\varphi_B|H_e|\varphi_A\rangle$ in (20.21). In each case, the evaluation in the position representation involves integrals with integrands that depend on $|\boldsymbol{r} - \boldsymbol{R}_A|$ and $|\boldsymbol{r} - \boldsymbol{R}_B|$, which at first might appear severely daunting. Fortunately, things simplify if we carry out the integrations using **elliptical coordinates** (μ, ν, ϕ), of which the last is the usual spherical coordinate angle measured around the z-axis, and the first two are defined by

$$|\boldsymbol{r} - \boldsymbol{R}_A| = (\mu + \nu)R/2, \qquad |\boldsymbol{r} - \boldsymbol{R}_B| = (\mu - \nu)R/2. \tag{20.22}$$

This is convenient because the integrands for the needed matrix elements are simple functions of μ and ν. Using $\boldsymbol{R}_B = -\boldsymbol{R}_A = (R/2)\hat{z}$ one finds the ranges $1 \le \mu < \infty$ and $-1 \le \nu \le 1$, and

$$r = \sqrt{\mu^2 + \nu^2 - 1}\, R/2, \qquad \cos\theta = \mu\nu/\sqrt{\mu^2 + \nu^2 - 1}, \tag{20.23}$$

from which we find, by computing the Jacobian determinant, the volume element

$$d^3\boldsymbol{r} = r^2 dr\, d(\cos\theta)\, d\phi = \frac{R^3}{8}(\mu^2 - \nu^2) d\mu\, d\nu\, d\phi. \tag{20.24}$$

It is also convenient to define a rescaled constant distance between the protons,

$$\rho = ZR/a_0. \tag{20.25}$$

With these variable changes, the matrix elements needed for (20.21) are expressed as

$$\langle\varphi_B|\varphi_A\rangle = \frac{\rho^3}{4}\int_1^\infty d\mu \int_{-1}^1 d\nu\ (\mu^2 - \nu^2)e^{-\rho\mu}, \tag{20.26}$$

$$\langle\varphi_A|H_e|\varphi_A\rangle = \frac{e^2}{2a_0}Z\rho^2\int_1^\infty d\mu \int_{-1}^1 d\nu\ f(\mu,\nu)e^{-\rho(\mu+\nu)}, \tag{20.27}$$

$$\langle\varphi_B|H_e|\varphi_A\rangle = \frac{e^2}{2a_0}Z\rho^2\int_1^\infty d\mu \int_{-1}^1 d\nu\ f(\mu,\nu)e^{-\rho\mu}, \tag{20.28}$$

with the last two involving a common polynomial factor

$$f(\mu,\nu) = \left(\frac{1}{2} - \frac{Z\rho}{4}\right)(\mu^2 - \nu^2) + (Z-2)\mu - Z\nu, \tag{20.29}$$

which comes from evaluating $H_e\varphi_A(\boldsymbol{r})$ in the position representation. Doing the integrations,

$$\langle\varphi_B|\varphi_A\rangle = \left(1 + \rho + \rho^2/3\right)e^{-\rho}, \tag{20.30}$$

$$\langle\varphi_A|H_e|\varphi_A\rangle = \frac{e^2}{2a_0}Z\left[Z - 2 + (2 + 2/\rho)e^{-2\rho}\right], \tag{20.31}$$

$$\langle\varphi_B|H_e|\varphi_A\rangle = \frac{e^2}{2a_0}Z\left[2/\rho - 2 - 10\rho/3 + Z(1 + \rho - \rho^2/3)\right]e^{-\rho}. \tag{20.32}$$

The resulting energy as a function of the variational parameter Z is therefore

$$\mathcal{E}_{Z,R} = \frac{e^2}{2a_0}\frac{Z\{Z - 2 + (2 + 2/\rho)e^{-2\rho} \pm [2/\rho - 2 - 10\rho/3 + Z(1 + \rho - \rho^2/3)]e^{-\rho}\}}{1 \pm (1 + \rho + \rho^2/3)e^{-\rho}}. \tag{20.33}$$

Now, for each R, one can minimize $\mathcal{E}_{Z,R}$ with respect to Z to obtain the energy $\mathcal{E}_R$, which according to (20.15) serves as the potential energy function in the Schrödinger equation for determining the nuclear separation and vibrational states.

The minimization to obtain Z for each R is best done by computer; note that there is a dependence on Z hidden in the definition of ρ in (20.25). The resulting

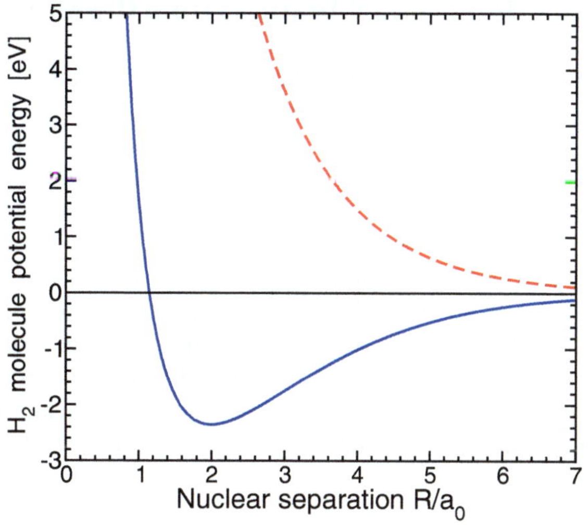

Fig. 20.2 The energy potential of the H_2^+ ion ground state, $\mathcal{E}_R + e^2/2a_0$, as a function of the nuclear separation R in units of the Bohr radius, as estimated by the variational principle for the even-parity binding (solid curve) and odd-parity anti-binding (dashed curve) solutions. For each R, the energy is obtained by minimizing (20.33) with respect to the variational parameter Z. The minimum is $\mathcal{E}_{R,\min} + e^2/2a_0 = -2.35$ eV, obtained for $R/a_0 = 2.003$ and $Z = 1.238$. The actual experimental energy of the H_2^+ ion is $E = -2.79$ eV

$\mathcal{E}_R + e^2/2a_0$ for each of the two parity choices is shown in Fig. 20.2. The odd-parity state has its minimum at $R = \infty$, and so is rejected, as an anti-binding solution. For the even-parity binding state, for very large R one finds $Z = 1$ and $\mathcal{E}_R + e^2/2a_0 = 0$ corresponding to dissociation of the molecule, and for very small R one finds $Z \to 2$ and a large positive energy from the proton-proton repulsion term, as expected. The minimum occurs for

$$R/a_0 = 2.003, \qquad Z = 1.238, \qquad \mathcal{E}_{R,\min} + e^2/2a_0 = -2.35 \text{ eV}. \tag{20.34}$$

This is about 16% smaller in magnitude than the experimental binding energy of the H_2^+ ion, $E = -2.79$ eV. As usual, the variational method gives an upper bound on the energy, and using a trial wavefunction with more parameters and a more general shape would get us even closer to the true value. However, our simple try is already sufficient to get a reasonable approximation with the correct features of the exact answer.

A similar approach can be used to find the binding energy and the nuclear separation potential for the neutral H_2 molecule with two electrons and two protons, which proceeds similarly but with slightly more involved integrals. This also holds for more complicated molecules with more complicated wavefunctions. Discussions can be found in more specialized books.

20.3 Diatomic Bonds and Vibrations from the Morse Potential

The example in the preceding section illustrates the calculation of the potential energy that determines the properties of the wavefunctions associated with the nuclear-position degrees of freedom, in the Born-Oppenheimer approximation. In principle, we can now use this sort of potential to find the expectation value of the nuclear separation R, and the excitation energies associated with nuclear vibrations and rotations. However, in practice the potential we found in Fig. 20.2 is only known approximately and numerically after minimizing with respect to the variational parameter Z for each R, even in the relatively easy case of H_2^+. It is useful to have a simpler analytic potential $V(R)$ with the same qualitative features, containing parameters that can either be estimated from the variational principle applied to the electronic wavefunction, or determined empirically by experiment, for general diatomic molecules. The appropriate qualitative features are that the potential should:

- diverge positively for $R \to 0$, due to the electrostatic repulsion of the nuclei,
- have a stable minimum for some R roughly comparable to the Bohr radius,
- approach a constant for $R \to \infty$, where the molecule separates into two atoms,
- have a large, but finite, number of vibrational states with nearly equal energy differences, in accord with experimental observations.

The last two points are consistent with the intuition that if we try to put too much energy into the molecule, we will break it into its atomic constituents. All of these qualitative features are realized by the approximate potential that we found for the H_2^+ ion in Fig. 20.2.

For general diatomic molecules, there is a simple three-parameter form, called the **Morse potential** after its invention by Philip M. Morse in 1929, which is both tractable for calculations and empirically successful. It is, using r from here on to denote the nuclear separation distance,

$$V(r) = V_0\left(1 - e^{-\beta(r-a)}\right)^2 . \tag{20.35}$$

Here V_0 is the depth of the potential, a is the position of its minimum where $V(r) = 0$, and β is the parameter governing the restoring force for small oscillations about the minimum. For large r the potential approaches V_0, which is therefore the classical approximation to the energy needed to break up the molecule. For small r, the potential approaches $V_0(e^{\beta a} - 1)^2$, which is positive, but not infinite. This slight deviation from the first of our desired features is of little practical consequence, because experience shows that $e^{\beta a} \gg 1$ for molecules, and thus the potential is so large at $r = 0$ that it does not matter exactly how large it is; the probability density for small r is tiny in any case.

For small deviations about the minimum position, the Morse potential is close to that of a harmonic oscillator,

$$V(r) \approx V_0\beta^2(r-a)^2. \tag{20.36}$$

If we denote the two nuclear masses as M_1 and M_2, then the reduced mass for the relative two-body problem (as discussed in Sect. 4.2) can be defined as

$$M = M_1M_2/(M_1 + M_2). \tag{20.37}$$

Then, by setting (20.36) equal to $\frac{1}{2}M\omega^2(r-a)^2$, we see that low-level vibrational excitations should be associated with a harmonic oscillator angular frequency

$$\omega = \beta\sqrt{2V_0/M}. \tag{20.38}$$

However, the potential also includes a deviation from harmonicity which will become increasingly important for higher energy vibrational states.

Let us now solve for the energy eigenvalues and wavefunctions of the Morse potential, taking the orbital angular momentum to be 0 so that the wavefunction depends only on r. (Rotational excitations have relatively small energies, suppressed by a factor $(m/M)^{1/2}$ as discussed in Sect. 20.1, and will be treated in the next section.) Writing $\psi(r) = U(r)/r$ as in Sect. 10.1, the time-independent Schrödinger equation becomes

$$\left(-\frac{\hbar^2}{2M}\frac{d^2}{dr^2} + V_0\left[1-e^{-\beta(r-a)}\right]^2\right)U = EU, \tag{20.39}$$

for energy eigenvalues E. We now define dimensionless quantities

$$x = \beta(r-a), \qquad \kappa = \sqrt{2MV_0}/\hbar\beta, \qquad \mathcal{E} = E/V_0, \tag{20.40}$$

which transform the eigenvalue equation into

$$\frac{d^2U}{dx^2} + \kappa^2\left[\mathcal{E} - \left(1-e^{-x}\right)^2\right]U = 0, \tag{20.41}$$

to be solved in the domain $-\beta a < x < \infty$. The boundary conditions are that $U = 0$ for $x = -\beta a$ (so that ψ is finite at $r = 0$) and at $x = \infty$ (so that we have a bound state). However, in order to find exact solutions, we cheat slightly by extending the domain to $-\infty < x < \infty$, and impose boundary conditions $U(-\infty) = U(\infty) = 0$, the same as for a one-dimensional bound state. As mentioned above, this modification makes no practical difference for $e^{\beta a} \gg 1$, and it will allow us to solve the problem exactly.

To get rid of the pesky exponential, we now make the further change of variable

$$z = \kappa e^{-x}, \qquad \frac{d}{dx} = -z\frac{d}{dz}, \tag{20.42}$$

so that the differential equation becomes

$$\frac{d^2U}{dz^2} + \frac{1}{z}\frac{dU}{dz} + \left[\frac{\kappa^2\mathcal{E}}{z^2} - \left(1 - \frac{\kappa}{z}\right)^2\right]U = 0. \tag{20.43}$$

The domain boundaries, where we require that U vanishes, are at $z = 0$ (corresponding to $r \to \infty$) and $z = \infty$ (corresponding, approximately, to $r = 0$).

The remaining process of finding the exact eigensolutions U and $\mathcal{E}$ is quite reminiscent of the methods we used to find the bound states of the one-dimensional harmonic oscillator in Sect. 7.2 and the isotropic three-dimensional harmonic oscillator in Sect. 10.5, so we will be brief. First, we note that for large z, the differential equation (20.43) becomes just $U'' \approx U$, with solutions proportional to $e^{\pm z}$. Choosing the negative exponent in order to have a normalizable bound state with good behavior at large z, we therefore write

$$U(z) = e^{-z}S(z), \tag{20.44}$$

where $S(z)$ must not grow exponentially at large z and must satisfy the differential equation

$$\frac{d^2S}{dz^2} + \left(\frac{1}{z} - 2\right)\frac{dS}{dz} + \left(\frac{\kappa^2(\mathcal{E}-1)}{z^2} + \frac{2\kappa - 1}{z}\right)S = 0. \tag{20.45}$$

For small z, trying a power law $S \approx c_0 z^p$ gives $p^2 = \kappa^2(1-\mathcal{E})$. The positive root for p must be the correct one, in order to satisfy the boundary condition $S = 0$ at $z = 0$, so

$$p = \kappa\sqrt{1-\mathcal{E}}, \tag{20.46}$$

which is consistent with the requirement that $\mathcal{E} = E/V_0 < 1$ in order to have a bound state. Therefore, let us try a power series solution

$$S(z) = z^p \sum_{n=0}^{\infty} c_n z^n. \tag{20.47}$$

Substituting this into (20.45), we find the recurrence relation

$$c_{n+1}/c_n = \frac{1 + 2n + 2p - 2\kappa}{(1+n+p)^2 - p^2}. \tag{20.48}$$

Now we can make the standard argument that the series must terminate so that $S(z)$ is a polynomial; otherwise $c_{n+1}/c_n \approx 2/n$ for large n, and we would find $S(z) \sim e^{2z}$. Therefore, the allowed energy eigenvalues $\mathcal{E}_n$ correspond to the vanishing of the numerator of (20.48) for some non-negative integer $n = 0, 1, 2, 3 \ldots$. Plugging (20.46) into $1 + 2n + 2p - 2\kappa = 0$ gives

$$\sqrt{1 - \mathcal{E}_n} = 1 - \frac{n + 1/2}{\kappa}. \tag{20.49}$$

Since the left side is positive, there can be only a finite number of bound state energy levels, with

$$n < \kappa - 1/2. \tag{20.50}$$

Solving (20.49) for $\mathcal{E}_n$ and multiplying by V_0, we finally obtain the remarkably simple result for the vibrational energy levels of the Morse potential,

$$E_n = \hbar\omega(n + 1/2) - \hbar\omega\chi(n + 1/2)^2, \tag{20.51}$$

where $\omega = \beta\sqrt{2V_0/M}$ was defined in (20.38), and the dimensionless number

$$\chi = \frac{1}{2\kappa} = \frac{\hbar\beta}{2\sqrt{2MV_0}} \tag{20.52}$$

is called the **anharmonicity parameter**. For molecules, χ is generally quite small. Thus, for small n so that the χ term can be neglected, the energy levels E_n are nearly equally spaced, with behavior close to that of a one-dimensional harmonic oscillator corresponding to vibrations of the nuclei along their common axis. The effect of the anharmonicity is to decrease the energy spacings as n increases toward its maximum allowed value, with E_n approaching V_0 from below. The total number of vibrational states is the largest integer less than $\kappa + 1/2 = (1 + 1/\chi)/2$. Data from fits to $\hbar\omega$ and χ for a few common molecules will be shown in Sect. 20.5, after we discuss the much smaller rotational contributions to the molecular energy, in the next section.

20.4 Rigid Bodies and Molecular Rotational States

In the preceding section, we considered molecular vibrations of diatomic molecules corresponding to nuclear oscillations along their common axis. We now turn to rotational excitations of molecules. This could be done by adding the centrifugal potential term $\hbar^2 l(l+1)/2Mr^2$ to the Morse potential Schrödinger equation (20.39), and then finding solutions with nonzero total angular momentum quantum number l. However, in keeping with the Born-Oppenheimer philosophy, we will instead follow the somewhat simpler path of treating the molecule as a rigid body with fixed electronic state and nuclear separation, and with moments of inertia of order

Table 20.1 Empirical values for vibrational ($\hbar\omega$ and $\hbar\omega\chi$) and rotational (B and α) energy parameters for some common diatomic molecules, in units of eV. Here D is the deuterium isotope of hydrogen. In the specialized literature, these energy parameters are usually given as equivalent photon wavenumbers with units cm^{-1}, obtained by dividing the numbers below by $2\pi\hbar c = 1.2398 \times 10^{-4}$ eV·cm. Also shown in the second column is the reduced nuclear mass $M = M_1 M_2/(M_1 + M_2)$ in units of the proton mass

molecule	M/m_p	$\hbar\omega$	$\hbar\omega\chi$	B	α
H_2	0.50	0.546	0.0151	0.00755	3.80×10^{-4}
D_2	1.00	0.387	0.00767	0.00378	1.34×10^{-4}
HCl	0.97	0.371	0.00655	0.00131	3.81×10^{-5}
N_2	7.00	0.293	0.00178	2.48×10^{-4}	2.13×10^{-6}
CO	6.86	0.269	0.00165	2.40×10^{-4}	2.17×10^{-6}
O_2	8.00	0.196	0.00148	1.79×10^{-4}	1.98×10^{-6}
Cl_2	17.7	0.0694	3.34×10^{-4}	3.03×10^{-5}	1.88×10^{-7}
NaCl	13.9	0.0452	2.20×10^{-4}	2.71×10^{-5}	2.02×10^{-7}

is always an asymmetry in the electronic state wavefunction, which gives rise to a permanent electric dipole moment pointing along the axis common to the nuclei. This electric dipole moment couples to the electromagnetic field. Such molecules are therefore **infrared active**, with efficient transitions between adjacent vibrational levels, $n \to n+1$ for absorption of a photon and $n \to n-1$ for emission. In a first approximation relevant for low-resolution experiments, one can neglect the anharmonicity of the vibrational modes and the rotational contributions, implying a single spectral line corresponding to photons with energies near $\hbar\omega$, which is usually in the infrared.

At higher resolution, it is necessary to take into account the change in angular momentum of the molecule. The spectrum depends not only on the anharmonicity χ but also on the fact that there is a selection rule $\Delta l = \pm 1$ for electric dipole transitions, similar to the one we will derive for single-electron transitions in atoms in Sect. 23.4. The absorbed or emitted photon energies can therefore be obtained by taking differences between the levels in (20.69), with the results (neglecting the small correction α for simplicity):

$$\Delta E_{n\to n+1}^{l\to l+1} = \hbar\omega(1 - 2(n+1)\chi) + 2(l+1)B, \tag{20.71}$$

$$\Delta E_{n\to n+1}^{l+1\to l} = \hbar\omega(1 - 2(n+1)\chi) - 2(l+1)B. \tag{20.72}$$

This shows that there is actually no photon absorption or emission line at $\hbar\omega_n$, where $\omega_n = \omega(1 - 2(n+1)\chi)$ is the angular frequency difference corrected for the anharmonicity. Instead, the spectrum for each n is as shown in Fig. 20.4, and features a series of lines equally spaced by $2B$ above $\hbar\omega_n$ for $l \to l+1$, and another series of lines equally spaced by $2B$ below $\hbar\omega_n$ for $l+1 \to l$, with a gap of $4B$ between them due to the selection rule that l must change by 1 unit. Measuring the energy

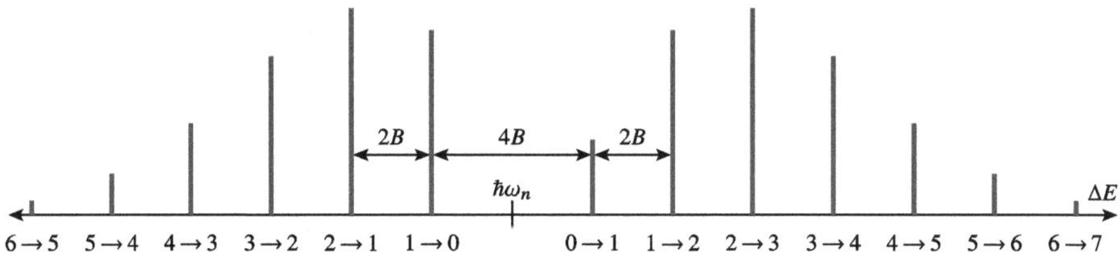

Fig. 20.4 Infrared vibrational-rotational absorption spectrum for a heteronuclear diatomic molecule undergoing transitions $n \to n+1$, with either $l \to l+1$ (right) or $l+1 \to l$ (left) as labeled. The central energy is at $\hbar\omega_n = \hbar\omega(1 - 2(n+1)\chi)$, where ω is the harmonic angular frequency and χ is the anharmonicity. However, the would-be line with $E_{\text{photon}} = \hbar\omega_n$ is absent, due to the selection rule $\Delta l = \pm 1$. The energy differences between lines are $2B$ and $4B$ as shown, with B the rotational energy parameter. The relative occupancies of initial states depend on temperature, and as shown here assume $k_B T = 10B$. Due to Boltzmann suppression of higher vibrational states, the most prominent absorption lines have $n = 0$, with $\omega_0 = \omega(1 - 2\chi)$ called the fundamental angular frequency. At low resolution, the rotational lines shown and the lines for different n coalesce into a single line with energy $E \approx \hbar\omega_0$

gaps reveals the moment of inertia of the molecule. The relative occupancies of the initial states with eigenvalues l are governed by the statistical factor

$$(2l+1)e^{-l(l+1)B/k_B T}. \tag{20.73}$$

For small l, this factor rises with l due to the rotational degeneracy factor $2l+1$, and for large l it is suppressed by the exponential Boltzmann factor. The relative initial-state occupancies shown in Fig. 20.4 correspond to $k_B T = 10B$. The number of prominent lines on either side of the central value $\hbar\omega_n$ increases with the temperature.

For **homonuclear** diatomic molecules (those with identical nuclei), there are two special considerations. First, according to the spin-statistics principle, the interchange of the two nuclei must multiply the state ket by a minus sign if they are fermions, or a plus sign if they are bosons. For a nucleus with spin s, this sign equals $(-1)^{2s}$. To match this, recall that the interchange of nuclei in the relative coordinates just coincides with a parity transformation (replacing each nuclear coordinate vector by minus itself), and therefore is associated with a factor of $(-1)^l$ for orbital angular momentum l, as in (8.179). Also, if the two spins s combine to form a total spin S, then the exchange of the two nuclear spin states results in a factor of $(-1)^{2s-S}$, as can be seen from (12.93), or from the symmetric and antisymmetric labels in (12.112). Therefore, combining the wavefunction and spin-state exchange factors, we must have

$$(-1)^{2s} = (-1)^l (-1)^{2s-S}. \tag{20.74}$$

It follows that l and S must both be even, or both odd. The molecular states with symmetric nuclear spins and $(-1)^{2s-S} = 1$ are referred to as "ortho-", while those with antisymmetric nuclear spin states and $(-1)^{2s-S} = -1$ are called "para-".[1]

Counting states using (12.112) for spin s, the total number of ortho- states with $S = 2s, 2s-2, \ldots$ is $(s+1)(2s+1)$, while the total number of para- states with $S = 2s-1, 2s-3, \ldots$ is $s(2s+1)$. To get this counting correct, one includes $2S+1$ states for each S; the grand total number of nuclear spin states is $(2s+1)^2$. For molecules in thermal equilibrium at typical lab temperatures, $k_B T = 0.0255$ eV is larger than the rotational energy differences, so each nuclear spin state has close to the same occupancy. Therefore, for diatomic molecules with identical nuclei, the ratio of the total number of para- states to the total number of ortho- states is

$$\frac{N_{\text{para}}}{N_{\text{ortho}}} = \frac{s}{s+1}. \tag{20.75}$$

This can be used to experimentally determine the spins of nuclei that form diatomic molecules. For example, for H_2 gas at lab temperature with $s = 1/2$, the abundance of ortho-H_2 with $S = 1$ and therefore odd l is about 3 times that of para-H_2 with $S = 0$ and even l. In contrast, for deuterium with $s = 1$, the abundance at lab temperature of ortho-D_2 with $S = 2$ or $S = 0$ and even l is about 2 times that of para-D_2 with $S = 1$ and odd l. For lower temperatures, the right side of (20.75) must be multiplied by a degeneracy-and-Boltzmann statistics factor

$$\frac{\sum_{l=1,3,5,\ldots}(2l+1)e^{-l(l+1)B/k_B T}}{\sum_{l=0,2,4,\ldots}(2l+1)e^{-l(l+1)B/k_B T}} \tag{20.76}$$

when $2s$ is even, or its reciprocal when $2s$ is odd. This factor is close to 1 at high temperatures, and decreases from more than 0.99 when $k_B T/B > 3$ (which includes typical lab temperature for all molecules) to 0 at lower temperatures. For example, at very low temperatures, so that most of the molecules are in the $l = 0$ level, H_2 is almost all para-, while D_2 is mostly ortho-. For nuclei with $s = 0$, like the main oxygen isotope, the para- states (which would require odd l) do not exist at all, so only even-l ortho- states occur.

The second special consideration for homonuclear diatomic molecules is that they cannot have a permanent electric dipole moment, since by symmetry there is no special direction along which it could point. Such molecules therefore have a much smaller coupling to electromagnetic fields, and are considered "inactive" in the infrared and microwave ranges.

However, the vibrational and rotational energy levels of homonuclear molecules can still be observed by inelastic scattering of photons, called the **Raman effect**.

[1] The terminology is similar to that for atomic helium as discussed in Sect. 18.3, in the sense that symmetric spin states are ortho- and antisymmetric spin states are para-. However, for atomic helium this refers to the electronic spins, while here the spins referred to are the nuclear ones.

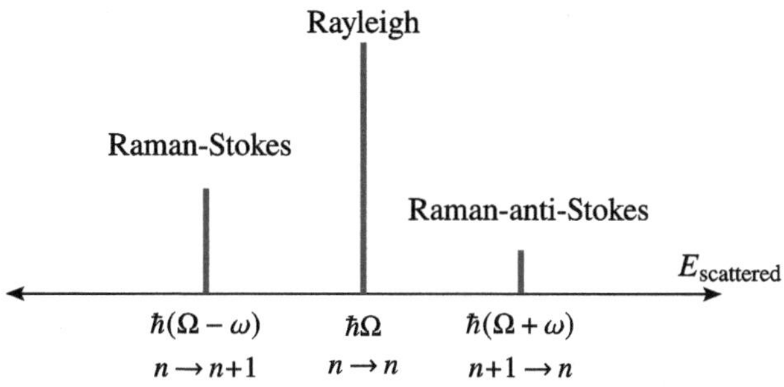

Fig. 20.5 The vibrational Raman effect spectrum of scattered photon energies for diatomic molecules. The incident photon has energy $\hbar\Omega$. If the molecule remains in the same state after the scattering, the scattered photon energy is also $\hbar\Omega$, giving the Rayleigh scattering line at center. The Stokes line results from the molecule transitioning to the next-higher energy vibrational state by $n \to n+1$, and the anti-Stokes line is due to the molecule moving to the next-lower vibrational state by $n+1 \to n$. The anti-Stokes line is less intense than the Stokes line, due to the relative Boltzmann suppression $e^{-\hbar\omega/k_BT}$ for the initial occupancy

In Raman scattering, a photon with known energy $\hbar\Omega$ impacts the molecule, sometimes changing its vibrational or rotational state. The molecule's change in energy is then seen as an opposite change in energy of the scattered photon. The Raman effect also occurs in heteronuclear molecules.

When the molecular state does not change, the final-state photon energy is also $\hbar\Omega$; this is called elastic Rayleigh scattering. In a first approximation in which anharmonicity and rotational excitations are neglected, if the molecule absorbs some energy by transitioning to the next higher vibrational state $n \to n+1$, the scattered photon will have energy $\hbar(\Omega - \omega)$, called Raman–Stokes scattering, and if the molecule gives up energy by transitioning to the next lower vibrational state by $n+1 \to n$, then the scattered photon will have energy $\hbar(\Omega + \omega)$, called Raman–anti-Stokes scattering. The main Rayleigh scattering line is therefore accompanied by two equally spaced lines, as shown in Fig. 20.5. Except at very high temperatures, the $n = 0$ state has the highest initial occupancy, and the anti-Stokes line is much less intense than the Stokes line for vibrational Raman scattering, because the initial states for anti-Stokes have higher energies and thus a relative Boltzmann suppression $e^{-\hbar\omega/k_BT}$.

The Raman effect can also result from inelastic scattering with transitions between states with different rotational energies. Here, it is important that there is a selection rule (which we will not attempt to derive) for Raman scattering, $\Delta l = 0$ or ± 2. Using the approximation $E_l = Bl(l+1)$, one therefore finds for the allowed scattered photon energies

$$E^{l\to l+2}_{\text{scattered}} = \hbar\Omega - (4l+6)B, \qquad \text{(Stokes)}, \tag{20.77}$$

$$E^{l+2\to l}_{\text{scattered}} = \hbar\Omega + (4l+6)B, \qquad \text{(anti-Stokes)}, \tag{20.78}$$

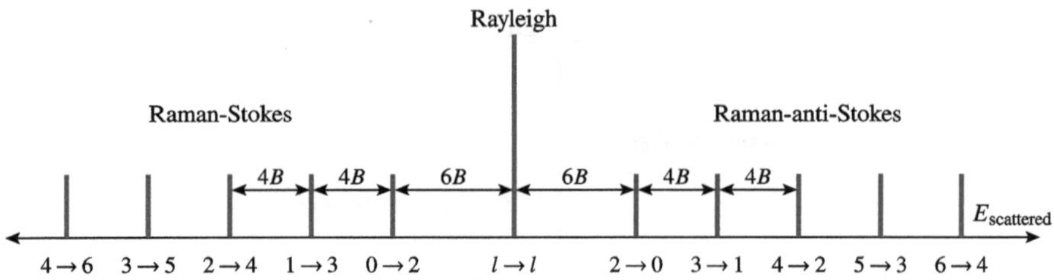

Fig. 20.6 The rotational Raman effect spectrum of scattered photon energies for diatomic molecules. The incident photon has energy $\hbar\Omega$. When the molecule remains in the same state, the scattered photon energy is also $\hbar\Omega$, giving the Rayleigh scattering line in the center. The Stokes lines on the left result from the molecule transitioning to a higher energy rotational state via $l \to l+2$ as labeled for various l, and the anti-Stokes lines on the right are due to the molecule transitioning to lower energy states by $l+2 \to l$. The energy separations between lines are $4B$ or $6B$ as shown, where B is the rotational energy parameter, inversely proportional to the moment of inertia. The relative intensities of the lines are not shown, and depend on the temperature and whether the molecule is homonuclear or heteronuclear

along with the Rayleigh line $E^{l\to l}_{\text{scattered}} = \hbar\Omega$. The positions of the spectral lines for the rotational Raman effect are shown in Fig. 20.6. There is an energy gap $6B$ on either side of the Rayleigh line to the nearest rotational Raman lines, and gaps $4B$ between different rotational Raman–Stokes and Raman–anti-Stokes lines. The relative intensities depend on the initial-state occupancies for l, which for homonuclear molecules are proportional to

$$(s+1)(2l+1)e^{-l(l+1)B/k_BT} \qquad \text{(homonuclear ortho- states)}, \tag{20.79}$$

$$s\,(2l+1)e^{-l(l+1)B/k_BT} \qquad \text{(homonuclear para- states)}, \tag{20.80}$$

alternating between odd and even l. For heteronuclear molecules, the initial-state occupancies are simply proportional to the factor already given in (20.73).

20.6 The Most Important Molecule: H_2O

Polyatomic molecules are obviously more complicated than diatomic ones, and mostly beyond our scope. However, given the importance of water to life and happiness on Earth, it seems worthwhile to give a brief phenomenological account of its vibrational and rotational quantum excitations.

The H_2O molecule has a bent (nonlinear) shape, with the geometry of the three nuclei shown in Fig. 20.7. The bond angle measured from the oxygen nucleus is famously about 104.5°. The electrons on the hydrogen atoms are attracted to the oxygen nucleus, leaving the hydrogen atoms with a net positive charge density and ensuring that the molecule has a permanent electric dipole moment in its ground state, pointing in the direction of the principal axis labeled $\hat{B}$. The reasons for this geometry and the nature of the bonding are discussed in more specialized books.

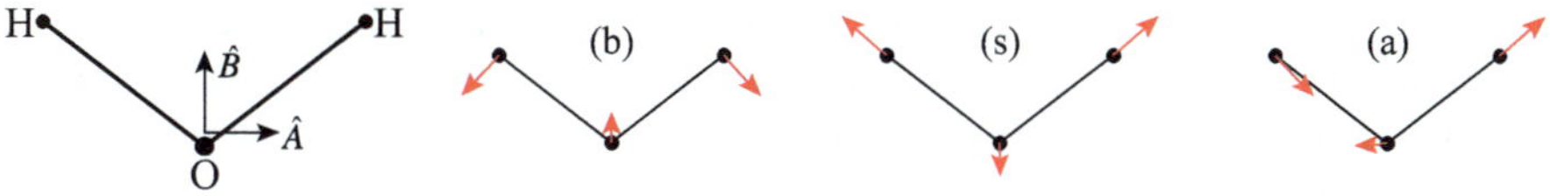

Fig. 20.7 The equilibrium geometry of the nuclei in the H_2O molecule is shown on the left. The rectangular coordinates of the nuclear positions, with respect to the center of mass, are O at (0, −0.6563) and H at (7.572, 5.209) and (−7.572, 5.209), in units of 10^{-11} meters. The angle between the OH bonds is 104.5°. The principal axes in which the moment-of-inertia tensor is diagonal are shown from the center of mass position as $\hat{A}$ and $\hat{B}$, with $\hat{C}$ out of the plane of this page. The three diagrams on the right illustrate the three vibrational modes: bending (b), symmetric stretching (s), and asymmetric stretching (a)

There are three[2] vibrational modes, called bending (b), symmetric stretching (s), and asymmetric stretching (a). The classical depictions of these modes are also shown in Fig. 20.7. The bending mode can be thought of as an oscillation of the bond angle about its equilibrium value. In the symmetric stretching mode, the two hydrogen nuclei (protons) move farther from and closer to the center of mass in unison, while in the asymmetric stretching mode one moves closer to the center of mass as the other moves farther from it. Considering each vibrational mode separately, for simplicity, one can parameterize the energies in terms of a Morse-potential-inspired spectrum,

$$E_n^{(m)} = \hbar\omega_m(n + 1/2) - \hbar\omega_m\chi_m(n + 1/2)^2, \qquad (m = b, s, a). \quad (20.81)$$

For each mode, ω_m is often called the harmonic angular frequency, and the slightly smaller difference in angular frequencies between the $n = 0$ and $n = 1$ energies, $\omega_m(1-2\chi_m)$, is referred to as the fundamental angular frequency. The experimental values in this parameterization are

$$\hbar\omega_b = 0.2046 \text{ eV}, \qquad \hbar\omega_b\chi_b = 0.00335 \text{ eV}, \quad (20.82)$$

$$\hbar\omega_s = 0.4754 \text{ eV}, \qquad \hbar\omega_s\chi_s = 0.0109 \text{ eV}, \quad (20.83)$$

$$\hbar\omega_a = 0.4892 \text{ eV}, \qquad \hbar\omega_a\chi_a = 0.0116 \text{ eV}. \quad (20.84)$$

These are all in the IR spectrum, as argued on general grounds for molecular vibrations in Sect. 20.1. The fundamental frequencies for the b, s, and a modes are, respectively, $\nu = \omega(1 - 2\chi)/2\pi = 4.785 \times 10^{13}$, 1.097×10^{14}, and 1.127×10^{14} Hz, with vacuum wavelengths 6.27, 2.73, and 2.66 microns. These modes are easily

[2] In general, a molecule has $3N$ degrees of freedom associated with the position coordinates of the N nuclei. Of these, 3 degrees of freedom correspond to translations of the molecule as a whole, and 3 more are associated with the rotations of the molecule if it is nonlinear, or 2 more if it is linear with $I_C = 0$. Therefore, in the approximation of harmonic oscillations about equilibrium, there are $3N - 6$ vibrational modes for a nonlinear molecule like H_2O, and $3N - 5$ vibrational modes for a linear molecule like CO_2 or any diatomic molecule.

excited by electromagnetic radiation due to the polar nature of the molecule, leading to strong IR absorption features. The anharmonicities help to effectively broaden the absorption spectrum, with an even stronger broadening effect from interactions between neighboring molecules in liquid water. Fortunately for us, water remains mostly transparent to the higher energy photons in the visible range. This is because the lowest electronic state excitations have much larger energies corresponding to UV photons, leaving a window in the liquid water absorption spectrum for frequencies from roughly $\nu = 3.75 \times 10^{14}$ to 1.5×10^{15} Hz, corresponding to the visible and the very near IR and UV.

At lower energies, rotational modes are important. Since H_2O is not a linear molecule, it behaves as a quantum asymmetric top for rotational excitations. Given the geometry in Fig. 20.7 and its caption, it is a short exercise in arithmetic to compute the moments of inertia for the principal axes $\hat{A}$ and $\hat{B}$ in the plane common to the nuclei, and $\hat{C}$ perpendicular to that plane, using oxygen's nuclear mass $M_O = 15.875 m_p$. This successfully predicts results within a few percent of the empirical rotational energy parameters for H_2O,

$$e_A = \hbar^2/2I_A = 0.003459 \text{ eV}, \qquad e_B = \hbar^2/2I_B = 0.001801 \text{ eV}, \quad (20.85)$$

$$e_C = \hbar^2/2I_C = 0.001152 \text{ eV}, \quad (20.86)$$

in terms of which the asymmetric top Hamiltonian is

$$H = e_A(\mathcal{L}_A/\hbar)^2 + e_B(\mathcal{L}_B/\hbar)^2 + e_C(\mathcal{L}_C/\hbar)^2. \quad (20.87)$$

As discussed in Sect. 20.4, there is no simple formula for the energies of an asymmetric top that works for all l, but for the $l = 1$ states the three energy eigenvalues are, from (20.63),

$$E_1^{(1)} = e_A + e_B = 0.00526 \text{ eV}, \qquad E_2^{(1)} = e_A + e_C = 0.00461 \text{ eV}, \quad (20.88)$$

$$E_3^{(1)} = e_B + e_C = 0.00295 \text{ eV}, \quad (20.89)$$

each with degeneracy 3. For the $l = 2$ states, the five energy eigenvalues are, numerically,

$$E_1^{(2)} = 0.00987 \text{ eV}, \qquad E_2^{(2)} = 0.01181 \text{ eV}, \qquad E_3^{(2)} = 0.01679 \text{ eV}, \quad (20.90)$$

$$E_4^{(2)} = 0.01694 \text{ eV}, \qquad E_5^{(2)} = 0.00870 \text{ eV}, \quad (20.91)$$

each with degeneracy 5. It is left to Exercise 20.4 to derive the general symbolic versions of these $l = 2$ energy eigenvalues in terms of the moments of inertia for a general asymmetric top.

Transitions between the preceding rotational states, and many others with higher l relevant for lab temperature $k_B T = 0.0255$ eV, correspond to photons with wavelengths in the far-infrared to near-microwave range. However, they are all much shorter than the 12.24 cm wavelength used by modern consumer microwave ovens to heat food. Microwave ovens efficiently heat liquid water, but do not operate by resonantly exciting rotational states (or the vibrational states described above, which have even higher energies). Instead, they produce time dependence in the superpositions of the rotational states, corresponding semi-classically to oscillatory changes in orientation rather than complete rotations, causing dissipation of energy into heat through interactions with neighboring molecules. Microwave ovens are much less efficient at heating ice, which has nearly locked-in molecular orientations.

Exercises

20.1 This problem is a one-dimensional toy-model analog of the H_2^+ molecule. Consider a light particle with mass m and two heavy particles each with mass M. The Hamiltonian is

$$H = \frac{P_1^2}{2m} + \frac{P_2^2}{2M} + \frac{P_3^2}{2M} + \frac{1}{2}k(X_1 - X_2)^2 + \frac{1}{2}k(X_1 - X_3)^2 - \frac{1}{2}K(X_2 - X_3)^2, \tag{20.92}$$

where k and K are both positive. Note that the interaction between the heavy particles is therefore repulsive, while the light particle feels an attractive force to each heavy particle.

(a) First we will solve the problem exactly. Define new coordinates $Q_1 = X_2 - X_3$ (heavy-particle separation) and $Q_2 = X_1 - (X_2 + X_3)/2$ (light particle position relative to the heavy-particle midpoint) and $Q_3 = [mX_1 + M(X_2 + X_3)]/(m+2M)$ (center of mass). Find their canonical momenta, and use them to rewrite the Hamiltonian in terms of two uncoupled harmonic oscillators and an effective free particle. We will ignore the latter, which is equivalent to restricting to zero momentum for the center of mass. Find the exact solutions for the energy eigenvalues, in the form $E_{n,n'} = \hbar\Omega(n + 1/2) + \hbar\Omega'(n' + 1/2)$, where $n, n' = 0, 1, 2, \ldots$, and you will solve for Ω and Ω' (with $\Omega > \Omega'$) in terms of the given quantities. It should be apparent from your results that stable states exist only if $K < k/2$. Compute the expectation value of $(X_2 - X_3)^2$ in each stationary state $|n, n'\rangle$.

(b) Now let us redo the problem with the Born–Oppenheimer approach appropriate for $m \ll M$. Treating $X_2 = x_2$ and $X_3 = x_3$ as fixed numbers, solve the Schrödinger equation for the light particle to find the energies

$$E_n(x_2, x_3) = \hbar\omega(n + 1/2) + c(x_2 - x_3)^2, \tag{20.93}$$

where $n = 0, 1, 2, \ldots$ and you will find ω and c in terms of the given quantities. Then, using $E_n(x_2, x_3)$ as the effective potential for the heavy particles, find the energy eigenvalues for the Hamiltonian

$$H_{\text{BO}} = \frac{P_2^2}{2M} + \frac{P_3^2}{2M} + E_n(X_2, X_3). \tag{20.94}$$

You should find energy eigenvalues $E_{n,n'} = \hbar\omega(n + 1/2) + \hbar\omega'(n' + 1/2)$, not including the free translation mode corresponding to the movement of the center of mass. Compute the expectation value of $(X_2 - X_3)^2$ in each stationary state.

(c) How do the Born–Oppenheimer approximation results ω and ω' compare to the exact results Ω and Ω' after an expansion in the small ratio m/M? How do the two results for $\langle n, n'|(X_2 - X_3)^2|n, n'\rangle$ compare? (You should find perfect agreement in two of the three comparisons; this is not typical of the Born–Oppenheimer approximation, but occurs in this particular example due to the special harmonic nature of the potential.)

20.2 The radial wavefunctions of the Morse potential discussed in Sect. 20.3 can be written exactly in terms of the associated Laguerre polynomials L_N^{α} defined in (10.121). Prove this, by using (20.49) and the associated Laguerre differential equation (10.122) to show that the differential equation (20.45) with $\mathcal{E}_n = 1 - (1 - (n + 1/2)/\kappa)^2$ is solved by $S_n(z) = z^q L_n^{2\kappa-2n-1}(2z)$, where q is a quantity that you will find in terms of κ and n.

20.3 Prove that the principal-axis angular momentum components of a rigid body, defined in (20.54), commute with the ordinary angular momentum components, as in (20.56). Also show that they obey the commutation relations in (20.57), with the notable minus signs.

20.4 In this problem, we will find the five $l = 2$ rotational energy levels, each with degeneracy 5, for a general asymmetric top.

(a) Start by writing the 5×5 matrix representations for generic angular momentum components J_x, J_y, and J_z, making use of (8.66) and (8.67). Then substitute $J_{x,y,z} \to \mathcal{L}_{A,B,C}$, and use them in (20.55) to find the 5×5 Hamiltonian matrix for $l = 2$. For convenience, write it in terms of $e_A = \hbar^2/2I_A$ and $e_B = \hbar^2/2I_B$ and $e_C = \hbar^2/2I_C$.

(b) Find the energy eigenvalues of the $l = 2$ Hamiltonian matrix obtained in part (a). This should involve three linear equations with solutions

$$E_1 = e_A + e_B + 4e_C, \quad E_2 = e_A + e_C + 4e_B, \quad E_3 = e_B + e_C + 4e_A, \tag{20.95}$$

and one quadratic equation with solutions

$$E_{4,5} = 2\left(e_A + e_B + e_C \pm \sqrt{e_A^2 + e_B^2 + e_C^2 - e_A e_B - e_A e_C - e_B e_C}\right). \tag{20.96}$$

Heisenberg and Interaction Representations 21

21.1 The Heisenberg Picture and Equations of Motion

Everything we have done so far uses a formulation that is sometimes called the **Schrödinger picture** of quantum mechanics. This just means that the state of the system evolves according to Postulate 6 of Sect. 3.1,

$$i\hbar \frac{d}{dt} |\psi(t)\rangle = H |\psi(t)\rangle . \tag{21.1}$$

However, there are other, equivalent, formulations of quantum mechanics, called "pictures", which are related to the Schrödinger picture by doing a time-dependent unitary transformation on the state of the system and a corresponding transformation to all operators. Done consistently, this cannot affect the predictions for physically measurable quantities, but the intermediate steps for solving problems will differ. A reason for considering other pictures is to make some calculations easier to do or understand.

Recall from the discussion in Sect. 3.4 that the solution to (21.1), starting from an initial state $|\psi(t_0)\rangle$, is

$$|\psi(t)\rangle = U(t, t_0) |\psi(t_0)\rangle , \tag{21.2}$$

where $U(t, t_0)$ is a unitary operator satisfying the differential equation

$$i\hbar \frac{d}{dt} U(t, t_0) = H(t) U(t, t_0). \tag{21.3}$$

If H has no explicit time dependence, then the solution to (21.3) is relatively easy, just $U(t, t_0) = e^{-i(t-t_0)H/\hbar}$. In that case, there is no particular advantage to the Heisenberg or interaction pictures described in what follows.

S. P. Martin, *Quantum Mechanics*, Graduate Texts in Physics,
https://doi.org/10.1007/978-3-032-21119-4_21

In the **Heisenberg picture**, all time dependence is transferred from the state of the system to the operators. To accomplish this, define

$$|\psi_H\rangle = U(t, t_0)^\dagger |\psi(t)\rangle , \tag{21.4}$$

$$A_H(t) = U(t, t_0)^\dagger A(t) U(t, t_0), \tag{21.5}$$

where $|\psi_H\rangle$ is the state ket in the Heisenberg picture, and $A_H(t)$ is the Heisenberg-picture operator corresponding to each Schrödinger-picture operator $A(t)$. Combining (21.2) and (21.4) gives

$$|\psi_H\rangle = |\psi(t_0)\rangle . \tag{21.6}$$

Thus, the Heisenberg-picture state $|\psi_H\rangle$ is independent of the time t; it only depends on the initial condition at t_0. Note that there are really infinitely many Heisenberg pictures, one for each choice of the reference initial time t_0. We include a subscript H to denote the state of the system or an operator as being in the Heisenberg picture. If an object is not written with the H subscript, it can be assumed to be in the Schrödinger picture. For convenience and simplicity, we will often write $U = U(t, t_0)$ in the rest of this section.

A possible point of confusion is that it is only the state ket of the system that we are defining differently in the Heisenberg and Schrödinger pictures via the unitary transformation in (21.4). Both the Schrödinger-picture state ket $|\psi(t)\rangle$ and the Heisenberg-picture state ket $|\psi_H\rangle$ can be expressed as (different) linear combinations of a set of *fixed* orthobasis kets, which are defined independently of the choice of picture. A convenient way to choose the fixed orthobasis kets is as time-independent eigenstates of some appropriate time-independent CSCO in the Schrödinger picture. Once so chosen, the orthobasis kets do not depend on t, by definition.

Inner products and matrix elements are the same in both pictures, since they are related by a unitary transformation. To see this, consider a matrix element of an arbitrary observable A between two possible states of the system $|\psi\rangle$ and $|\chi\rangle$ that both obey Schrödinger's equation in the Schrödinger picture. We have

$$\langle\chi(t)|A(t)|\psi(t)\rangle = \langle\chi(t)|UU^\dagger A(t)UU^\dagger|\psi(t)\rangle = \langle\chi_H|A_H(t)|\psi_H\rangle . \tag{21.7}$$

This shows the physical equivalence of the two pictures.

Even though many important observable operators (for example, $A =$ position, momentum, or spin) are time-independent in the Schrödinger picture, the corresponding Heisenberg-picture operators $A_H(t)$ will generally depend on t, and often in a complicated way, depending on the Hamiltonian. This is the price to be paid in the Heisenberg picture for the convenience that the state of the system has no t dependence.

However, there is an important special case in which $A_H(t)$ does not depend on t. Suppose that in the Schrödinger picture, H and A do not depend on t, and

$[A, H] = 0$. Then A commutes with $U(t, t_0)$ as well, since the latter is built out of H. In that case,

$$A_H = U^\dagger A U = U^\dagger U A = A, \tag{21.8}$$

and we say that A is a **constant of motion**, following classical physics terminology. Note that this corresponds to a conserved quantity, as defined in Sect. 5.1 in the Schrödinger picture. In particular, if H does not depend explicitly on t, then $H_H = H$, and the Schrödinger and Heisenberg-picture Hamiltonians are equal.

More generally, we can ask how $A_H(t)$ evolves in time. To find out, we calculate, using the product rule for derivatives,

$$\frac{dA_H}{dt} = \left[\frac{d}{dt}U(t, t_0)^\dagger\right] AU + U^\dagger A \left[\frac{d}{dt}U(t, t_0)\right] + U^\dagger \frac{\partial A}{\partial t} U. \tag{21.9}$$

Evaluating the derivatives in square brackets using the differential equation (21.3) gives

$$\frac{dA_H}{dt} = \frac{i}{\hbar} U^\dagger HAU - \frac{i}{\hbar} U^\dagger AHU + U^\dagger \frac{\partial A}{\partial t} U, \tag{21.10}$$

which can be rewritten as

$$\frac{dA_H}{dt} = \frac{i}{\hbar}[H_H, A_H] + \left(\frac{\partial A}{\partial t}\right)_H. \tag{21.11}$$

The meaning of the last term is that we take the derivative of A with respect to its explicit dependence on t in the Schrödinger picture, and then convert the result to the Heisenberg picture. Equation (21.11) is called the **Heisenberg equation of motion** for the operator A_H. Since the state of the system does not change in the Heisenberg picture in the absence of a measurement, this equation captures all physical effects of unitary time evolution. It is the quantum analog of the classical equation of motion.

The form of **equal-time commutation relations** does not change if one switches pictures. If one has a commutation relation

$$[A, B] = C, \tag{21.12}$$

in the Schrödinger picture, then

$$[A_H(t), B_H(t)] = [U^\dagger AU, U^\dagger BU] = U^\dagger[A, B]U = U^\dagger CU = C_H(t). \tag{21.13}$$

So, for example,

$$[X_H(t),\ P_H(t)] = [X, P] = i\hbar, \tag{21.14}$$

and if $\boldsymbol{J}$ is an angular momentum operator, then

$$[J_{Ha}(t),\ J_{Hb}(t)] = i\hbar\epsilon_{abc}J_{Hc}(t), \qquad (a, b, c = x, y, z). \tag{21.15}$$

It is crucial that we specified equal-time commutation relations, because in general the commutator $[A_H(t),\ B_H(t')]$ does not have a simple relationship to $[A, B]$ unless $t' = t$. For example,

$$[X_H(t),\ P_H(t')] \neq i\hbar \tag{21.16}$$

in general, unless $t' = t$.

To illustrate how the Heisenberg picture works in the simple case that the Hamiltonian does not explicitly depend on time, suppose that in the Schrödinger picture we have the usual Hamiltonian for a spinless particle of mass m moving in a potential,

$$H = \frac{P^2}{2m} + V(X), \tag{21.17}$$

where $V(x) = \sum_n v_n x^n$, with v_n a set of fixed real numbers. Then

$$P_H^2 = U^\dagger PUU^\dagger PU = U^\dagger P^2 U, \tag{21.18}$$

and similarly

$$(X_H)^n = U^\dagger X^n U. \tag{21.19}$$

Therefore, the Heisenberg-picture Hamiltonian is

$$H_H = U^\dagger HU = \frac{P_H^2}{2m} + V(X_H). \tag{21.20}$$

The Schrödinger picture operators X and P have no explicit time dependence,

$$\partial X/\partial t = 0, \qquad \partial P/\partial t = 0, \tag{21.21}$$

and so we obtain from (21.11), by using (21.14),

$$\frac{dX_H}{dt} = \frac{i}{\hbar}[H_H,\ X_H] = \frac{P_H}{m}, \tag{21.22}$$

$$\frac{dP_H}{dt} = \frac{i}{\hbar}[H_H,\ P_H] = -V'(X_H). \tag{21.23}$$

The Heisenberg equations of motion (21.22) and (21.23) have the same form as the classical equations of motion, but with the classical position and momentum promoted to the corresponding Heisenberg-picture operators. Using them, one can check that

$$\frac{d}{dt}H_H = 0, \tag{21.24}$$

which verifies that $H_H = H$ is indeed a constant of motion.

21.2 The Interaction Picture and Transition Amplitudes

Suppose that we have a Hamiltonian (in the Schrödinger picture) of the form

$$H(t) = H_0 + W(t), \tag{21.25}$$

where $W(t)$ may depend explicitly on time, but H_0 does not. Then one can define another picture called the **interaction picture**, also known as the **Dirac picture**.

The idea of the interaction picture is to transfer from the state ket to the operators only the part of the time dependence that H_0 is responsible for. This is done by defining, for the ket describing the Schrödinger picture state of the system $|\psi(t)\rangle$ and for all observables $A(t)$,

$$|\psi_I(t)\rangle = e^{itH_0/\hbar}\,|\psi(t)\rangle\,, \tag{21.26}$$

$$A_I(t) = e^{itH_0/\hbar}A(t)e^{-itH_0/\hbar}. \tag{21.27}$$

The subscript I is used to denote the interaction picture; when it is not present, the state or operator should be assumed to be in the Schrödinger picture. The interaction-picture states and operators coincide with their Schrödinger picture counterparts at time $t = 0$. (One could instead choose some other time to do the matching, but there is no particular advantage to doing so.) Note that $(H_0)_I = e^{itH_0/\hbar}H_0e^{-itH_0/\hbar} = H_0$, so a distinguishing subscript for H_0 is not needed. However, in general, $W_I(t) \neq W(t)$, so we must be careful to distinguish them from now on. Note that (21.26) and (21.27) can be considered a change of basis, since $e^{itH_0/\hbar}$ is a unitary operator.

Clearly, if $W(t) = 0$, then the interaction and Heisenberg pictures are the same, while if $H_0 = 0$, then the interaction and Schrödinger pictures are the same. More generally, the interaction picture is intermediate between the Schrödinger and Heisenberg pictures. At first, it might seem that the interaction picture combines the worst features of both, since both the state and ordinary observable operators evolve nontrivially in time. However, it turns out to be useful for an elegant formulation of time-dependent perturbation theory, when the effects of H_0 are large but easily calculable, and the effects of $W(t)$ can be treated as the perturbation.

By an argument exactly analogous to the derivation of (21.11), one can find the equation giving the time evolution of operators in the interaction picture,

$$\frac{dA_I}{dt} = \frac{i}{\hbar}[H_0, A_I] + \left(\frac{\partial A}{\partial t}\right)_I . \tag{21.28}$$

Here, the meaning of the last term is that one should take the derivative of A with respect to its explicit time dependence in the Schrödinger picture, and then convert the result to the interaction picture using (21.27).

Let us now define the unitary operator $\mathcal{U}_I(t, t_0)$ which gives the time evolution of $|\psi_I(t)\rangle$ in terms of its initial condition at an arbitrary time $t = t_0$, according to

$$|\psi_I(t)\rangle = \mathcal{U}_I(t, t_0)\,|\psi_I(t_0)\rangle . \tag{21.29}$$

To avoid confusion, it is important to note that $\mathcal{U}_I(t, t_0)$ is not the same as $U_I(t, t_0)$, which is defined to be the interaction-picture version of the Schrödinger picture unitary time-evolution operator $U(t, t_0)$. Directly from the definition (21.27), we have

$$U_I(t, t_0) \equiv e^{itH_0/\hbar}\, U(t, t_0)\, e^{-itH_0/\hbar} . \tag{21.30}$$

So, using (21.2), (21.26), and (21.29), these three unitary operators are related by

$$\mathcal{U}_I(t, t_0) = e^{itH_0/\hbar} U(t, t_0) e^{-it_0 H_0/\hbar} = U_I(t, t_0) e^{i(t-t_0)H_0/\hbar} . \tag{21.31}$$

In the special case that H has no explicit time dependence, we can now use the known solution $U(t, t_0) = e^{i(t_0-t)H/\hbar}$ from (3.26) to learn that

$$\mathcal{U}_I(t, t_0) = e^{itH_0/\hbar} e^{i(t_0-t)H/\hbar} e^{-it_0 H_0/\hbar} , \qquad \text{(for time-independent } H\text{)}. \tag{21.32}$$

Our goal now is to solve for $\mathcal{U}_I(t, t_0)$ more generally. To accomplish this, we calculate

$$\begin{aligned} i\hbar\frac{d}{dt}\,|\psi_I(t)\rangle &= i\hbar\frac{d}{dt}\left(e^{itH_0/\hbar}\,|\psi(t)\rangle\right) \\ &= -e^{itH_0/\hbar} H_0\,|\psi(t)\rangle + i\hbar e^{itH_0/\hbar}\frac{d}{dt}\,|\psi(t)\rangle . \end{aligned} \tag{21.33}$$

Using the Schrödinger equation $i\hbar\frac{d}{dt}\,|\psi\rangle = (H_0 + W)\,|\psi\rangle$, this becomes

$$i\hbar\frac{d}{dt}\,|\psi_I(t)\rangle = e^{itH_0/\hbar} W(t)\,|\psi\rangle , \tag{21.34}$$

which we can rewrite in a form involving only interaction-picture objects, as

$$i\hbar\frac{d}{dt}\,|\psi_I(t)\rangle \;=\; W_I(t)\,|\psi_I(t)\rangle\,. \tag{21.35}$$

Plugging the definition (21.29) into (21.35), we get the operator equation

$$i\hbar\frac{d}{dt}\mathcal{U}_I(t,t_0) \;=\; W_I(t)\,\mathcal{U}_I(t,t_0). \tag{21.36}$$

Integrating both sides gives

$$i\hbar\big[\mathcal{U}_I(t,t_0)-\mathcal{U}_I(t_0,t_0)\big] \;=\; \int_{t_0}^{t}dt'\,W_I(t')\,\mathcal{U}_I(t',t_0). \tag{21.37}$$

Rewriting this using $\mathcal{U}_I(t_0,t_0)=I$, we find the useful integral equation

$$\mathcal{U}_I(t,t_0) \;=\; I-\frac{i}{\hbar}\int_{t_0}^{t}dt'\,W_I(t')\,\mathcal{U}_I(t',t_0). \tag{21.38}$$

Taking the Hermitian adjoint of this, and using the unitarity relation $\mathcal{U}_I(t,t_0)^\dagger = \mathcal{U}_I(t,t_0)^{-1} = \mathcal{U}_I(t_0,t)$ and the Hermiticity of $W_I(t')$, one finds $\mathcal{U}_I(t_0,t) = I+\frac{i}{\hbar}\int_{t_0}^{t}dt'\,\mathcal{U}_I(t_0,t')\,W_I(t')$. Then, interchanging the roles of t and t_0 gives an alternative version of the integral equation,

$$\mathcal{U}_I(t,t_0) \;=\; I-\frac{i}{\hbar}\int_{t_0}^{t}dt'\,\mathcal{U}_I(t,t')\,W_I(t'). \tag{21.39}$$

Although (21.38) and (21.39) contain the same information, the latter will be the more useful form in the context of scattering problems, as we will see in Sect. 24.2.

Solving the integral equation (21.38) by iteration gives

$$\mathcal{U}_I(t,t_0) \;=\; I-\frac{i}{\hbar}\int_{t_0}^{t}dt'\,W_I(t')+\left(-\frac{i}{\hbar}\right)^2\int_{t_0}^{t}dt'\int_{t_0}^{t'}dt''\,W_I(t')W_I(t'')+\cdots, \tag{21.40}$$

or, rewriting more systematically,

$$\mathcal{U}_I(t,t_0) = \sum_{N=0}^{\infty}\left(-\frac{i}{\hbar}\right)^N\int_{t_0}^{t}dt_N\int_{t_0}^{t_N}dt_{N-1}\cdots \int_{t_0}^{t_2}dt_1\,W_I(t_N)W_I(t_{N-1})\cdots W_I(t_1). \tag{21.41}$$

Equation (21.41) is the **Dyson series** for the time evolution of the state in the interaction picture; compare to the similar structure of (3.54). This solution can be checked by plugging it directly into the differential equation (21.36). In general,

$W_I(t)$ may not commute with $W_I(t')$ when $t \neq t'$. Therefore, we are careful to put the operators in increasing time order, from right to left, as specified by the nested integrations with $t_0 < t_1 < t_2 < \cdots < t_{N-1} < t_N < t$. In practice, (21.41) is usually approximated by truncating it to the first few terms, treating W_I as small. A more compact (but less directly useful) way of writing (21.41) is

$$\mathcal{U}_I(t, t_0) = T \exp\left(-\frac{i}{\hbar}\int_{t_0}^{t} dt'\, W_I(t')\right), \tag{21.42}$$

where T is the time-ordering symbol introduced in the discussion surrounding (3.55), to which (21.42) can be compared. The symbol T is an instruction to re-order the operators following it, so that after expanding the exponential, $W_I(t_j)$ appears to the right of $W_I(t_k)$ whenever $t_j < t_k$. The time-ordering rule is that earlier operators act first (meaning closer to the ket when written left-to-right).

Given our solution for $\mathcal{U}_I(t, t_0)$ in (21.41), one can express the time dependence of the state back in the Schrödinger picture. Using (21.2) and (21.31),

$$|\psi(t)\rangle = U(t, t_0)\,|\psi(t_0)\rangle = e^{-itH_0/\hbar}\,\mathcal{U}_I(t, t_0)e^{it_0H_0/\hbar}\,|\psi(t_0)\rangle\,. \tag{21.43}$$

It is convenient to express this in terms of an orthobasis of eigenstates $|n\rangle$ of H_0 with eigenvalues $\mathcal{E}_n$. Using completeness twice, we have

$$|\psi(t)\rangle = \sum_m\sum_n e^{-i\mathcal{E}_m t/\hbar}e^{i\mathcal{E}_n t_0/\hbar}\,|m\rangle\,\langle m|\mathcal{U}_I(t, t_0)|n\rangle\,\langle n|\psi(t_0)\rangle\,. \tag{21.44}$$

Now, define the **transition amplitudes** for the eigenstates of H_0 as

$$a_{n\to m}(t, t_0) \equiv \langle m|\mathcal{U}_I(t, t_0)|n\rangle\,. \tag{21.45}$$

Then, given the initial-state coefficients

$$c_n(t_0) = \langle n|\psi(t_0)\rangle\,, \tag{21.46}$$

it follows that the Schrödinger picture state of the system at time t is

$$|\psi(t)\rangle = \sum_m c_m(t)\,|m\rangle\,, \tag{21.47}$$

where, from (21.44),

$$c_m(t) = \sum_n e^{i(\mathcal{E}_n t_0 - \mathcal{E}_m t)/\hbar}\,a_{n\to m}(t, t_0)\,c_n(t_0). \tag{21.48}$$

Note that $a_{n\to m}(t_0, t_0) = \langle m|n\rangle = \delta_{nm}$, so that (21.48) is consistent for $t = t_0$. The nontrivial part of the time evolution due to $W(t)$ is encoded in the transition amplitudes $a_{n\to m}(t, t_0)$, which in turn rely on our solution for $\mathcal{U}_I(t, t_0)$ in (21.41).

As a special case, suppose we are only interested in transitions between eigenstates of H_0, but with time evolution governed by the full Hamiltonian $H = H_0 + W(t)$. The probability to start at time t_0 in one of the H_0 eigenstates $|n\rangle$ and find the system at time t in another H_0 eigenstate $|m\rangle$ is

$$\mathcal{P}_{n\to m} = |a_{n\to m}(t, t_0)|^2. \tag{21.49}$$

More generally, the transition amplitudes $a_{n\to m}$ defined in (21.45) can be used to find the time evolution of an arbitrary state $|\psi(t)\rangle$, through (21.47) and (21.48). This is the basis of time-dependent perturbation theory discussed in the next chapter.

Exercises

21.1 Consider the one-dimensional harmonic oscillator for a particle of mass m and natural angular frequency ω.

(a) Use Theorem 2.7 to find the Heisenberg-picture operators $a_H(t)$ and $a_H^\dagger(t)$ in terms of the corresponding Schrödinger-picture operators a and $a^\dagger$, taking them to coincide at $t = t_0 = 0$. Use the results to find $X_H(t)$ and $P_H(t)$ in terms of a and $a^\dagger$, and in terms of X and P. Check that your results satisfy the differential equations from (21.11).
(b) Find the commutators $[X_H(t), P_H(t')]$ and $[X_H(t), X_H(t')]$ and $[P_H(t), P_H(t')]$.
(c) Find the correlation functions $\langle 0|X_H(t)X_H(t')|0\rangle$ and $\langle 0|P_H(t)P_H(t')|0\rangle$ in the ground state.

21.2 Consider a particle moving in one dimension under the influence of a time-dependent external force $f(t)$, so that the Hamiltonian is $H = P^2/2m - f(t)X$. Find the Heisenberg-picture operators $X_H(t)$ and $P_H(t)$ in terms of $f(t)$ and the Schrödinger-picture operators X and P, with boundary condition $X_H(t_0) = X$ and $P_H(t_0) = P$. [Hint: The function $g(t) = \int_{t_0}^{t} dt'\, f(t')$, sometimes known as the impulse, should play a prominent role in your answers.]

21.3 A particle of mass m has a Hamiltonian $H = H_0 + W$, where $H_0 = P^2/2m$ and $W = V(X)$ does not depend on time.

(a) Find the interaction-picture operators $X_I(t)$ and $P_I(t)$ in terms of the corresponding Schrödinger-picture operators, with which they coincide at $t = 0$.
(b) Find the commutators $[X_I(t), P_I(t')]$ and $[P_I(t), P_I(t')]$ and $[X_I(t), X_I(t')]$. In particular, show that the last one does not vanish except when $t' = t$.
(c) Repeat parts (a) and (b) with the roles of H_0 and W reversed, so that $H_0 = V(X)$ and $W = P^2/2m$.

Time-Dependent Perturbation Theory 22

22.1 The Short-Time and Sudden Approximations

As a warm-up for a more general treatment of time-dependent perturbation theory, suppose that the Hamiltonian for a system is

$$H(t) = \begin{cases} H_0 & (t < 0), \\ H_{\text{int}} & (0 < t < \delta), \\ H_0' & (t > \delta). \end{cases} \tag{22.1}$$

The starting and ending Hamiltonians are assumed to have no time dependence and to have known eigenstates and eigenvalues, $|n\rangle$ and $\mathcal{E}_n$ for H_0, and $|m'\rangle$ and $\mathcal{E}_{m'}$ for H_0'. In the interregnum $0 < t < \delta$, the Hamiltonian is H_{int}. Consider the **short-time approximation**, in which δ is so small that any explicit time dependence of H_{int} over that time interval is negligible. The idea is to take advantage of the smallness of δ to develop an approximation for the state of the system at late times, despite the possibly complicated form of H_{int}, even if we do not know its eigenstates and eigenvalues.

For each of $t \leq 0$ and $t \geq \delta$, we can use the Schrödinger equation to write the form of the state as

$$|\psi(t)\rangle = \sum_n b_n e^{-it\mathcal{E}_n/\hbar} |n\rangle \qquad (t \leq 0), \tag{22.2}$$

$$|\psi(t)\rangle = \sum_{m'} c_{m'} e^{-i(t-\delta)\mathcal{E}_{m'}/\hbar} |m'\rangle \qquad (t \geq \delta), \tag{22.3}$$

for some constants b_n and $c_{m'}$. For later convenience, a constant phase $e^{i\mathcal{E}_{m'}\delta/\hbar}$ has been included in (22.3), by choice of the definitions of $c_{m'}$. Typically, the b_n are known, and we would like to find the $c_{m'}$.

S. P. Martin, *Quantum Mechanics*, Graduate Texts in Physics,
https://doi.org/10.1007/978-3-032-21119-4_22

For $0 < t < \delta$, the Schrödinger equation is

$$i\hbar \frac{d}{dt} |\psi(t)\rangle = H_{\text{int}} |\psi(t)\rangle . \tag{22.4}$$

So, to first order in δ,

$$|\psi(\delta)\rangle = \left(I - i\frac{\delta}{\hbar} H_{\text{int}}\right) |\psi(0)\rangle . \tag{22.5}$$

Comparing this to (22.2) with $t = 0$, and (22.3) with $t = \delta$, we find

$$\sum_{m'} c_{m'} |m'\rangle = \left(I - i\frac{\delta}{\hbar} H_{\text{int}}\right) \sum_n b_n |n\rangle . \tag{22.6}$$

Now taking the inner product with another H_0' eigenstate $\langle k'|$, and then relabeling $k' \to m'$, we get the solution

$$c_{m'} = \sum_n b_n a_{n\to m'}, \tag{22.7}$$

where the transition amplitude is

$$a_{n\to m'} = \langle m'|n\rangle - i\frac{\delta}{\hbar}\langle m'|H_{\text{int}}|n\rangle. \tag{22.8}$$

Given the initial state, specified by the coefficients b_n of the H_0 eigenstates, the final state is determined by the coefficients $c_{m'}$ of the H_0' eigenstates.

One application of (22.8) is to the special case $H_0' = H_0$, so that $H_{\text{int}} - H_0$ acts for a short time as a perturbation to an otherwise constant Hamiltonian. Then $\langle m|n\rangle = \delta_{nm}$, and still to first order in δ,

$$a_{n\to m} = \delta_{nm} - i\frac{\delta}{\hbar} \langle m|H_{\text{int}}|n\rangle . \tag{22.9}$$

For example, if the initial state is an eigenstate $|n\rangle$ of H_0, so that only one of the b_n is nonzero, then the probability of finding the system in a different H_0 eigenstate $|m\rangle$ after $t > \delta$ is

$$\mathcal{P}_{n\to m} = \frac{\delta^2}{\hbar^2} |\langle m|H_{\text{int}}|n\rangle|^2, \qquad (m \neq n). \tag{22.10}$$

In the next section, we will see that this result can be obtained as a special case application of a general treatment that we will develop for time-dependent perturbation theory; see (22.26) and (22.27) with $W = H_{\text{int}} - H_0$ for $0 < t < \delta$.

The **sudden approximation** arises if we assume that δ is so small that $H_{\rm int}$ is irrelevant, but $H_0' \neq H_0$. Then, if the system was originally in an H_0 eigenstate $|n\rangle$ for $t < 0$, the probability of finding the system in an H_0' eigenstate $|m'\rangle$ for $t > 0$ is

$$\mathcal{P}_{n\to m'} = |c_{m'}|^2 = |\langle m'|n\rangle|^2. \tag{22.11}$$

This follows from the $\delta \to 0$ limit of (22.8).

As an example, suppose that a one-dimensional harmonic oscillator is in its ground state, when the angular frequency in the Hamiltonian suddenly doubles, due to some outside agent. Let us compute the probability that the oscillator will be found in the ground state of the new Hamiltonian. Working in the position representation, the normalized wavefunctions of the ground states are

$$\langle x|0\rangle_\omega = \left(\frac{m\omega}{\pi\hbar}\right)^{1/4} \exp(-m\omega x^2/2\hbar) \qquad \text{(before)}, \tag{22.12}$$

$$\langle x|0\rangle_{2\omega} = \left(\frac{2m\omega}{\pi\hbar}\right)^{1/4} \exp(-m\omega x^2/\hbar) \qquad \text{(after)}. \tag{22.13}$$

Then, in the sudden approximation, we compute

$$\begin{aligned} {}_{2\omega}\langle 0|0\rangle_\omega &= \int_{-\infty}^{\infty} dx\; {}_{2\omega}\langle 0|x\rangle \langle x|0\rangle_\omega = \sqrt{\frac{\sqrt{2}m\omega}{\pi\hbar}} \int_{-\infty}^{\infty} dx\, \exp(-3m\omega x^2/2\hbar) \\ &= 2^{3/4}/3^{1/2}, \end{aligned} \tag{22.14}$$

which results in a probability $\mathcal{P}_{0_\omega\to 0_{2\omega}} = |{}_{2\omega}\langle 0|0\rangle_\omega|^2 = 2\sqrt{2}/3 \approx 0.9428$.

As a more practical example, consider the effect of the decay of the tritium (^{3}H) nucleus on the atomic electron state. The tritium nucleus consists of 1 proton and 2 neutrons, so $Z = 1$. It undergoes a β decay to a helium isotope and an electron and an antineutrino,

$$^3\text{H} \to {}^3\text{He}^+ + e^- + \overline{\nu}_e. \tag{22.15}$$

The antineutrino has only weak interactions, and so can be ignored for the present purposes. The β-decay electron is emitted with average (maximum) kinetic energy 0.0057 MeV (0.0186 MeV), so that its average (maximum) speed can be computed to be $0.15c$ ($0.26c$). These are much larger than the typical semi-classical speeds of electrons in a $Z = 1$ atom, $v = \alpha c \approx c/137$. Therefore, it is a good approximation to also ignore the β-decay electron, as it leaves the scene very quickly.

The remaining $^3\text{He}^+$ ion consists of the nucleus with 2 protons and 1 neutron, so $Z = 2$, together with the left-over atomic electron from the ^{3}H atom. So, in the sudden approximation the probability of the atomic electron starting in the ground state of ^{3}H and ending up in the ground state of the $^3\text{He}^+$ ion is $\mathcal{P}_{1,0,0\to 1,0,0} = |a_{1,0,0\to 1,0,0}|^2$, where

$$a_{1,0,0\to 1,0,0} = \int d^3\boldsymbol{r}\,\left[\psi^{Z=2}_{1,0,0}(\boldsymbol{r})\right]^* \psi^{Z=1}_{1,0,0}(\boldsymbol{r}). \tag{22.16}$$

Recalling that $\psi^{Z}_{1,0,0}(\boldsymbol{r}) = \sqrt{\frac{Z^3}{\pi a_0^3}}\, e^{-Zr/a_0}$, we have

$$a_{1,0,0\to 1,0,0} = 4\pi \int_0^\infty dr\, r^2\, \frac{2^{3/2}}{\pi a_0^3} e^{-3r/a_0} = 2^{9/2}/3^3, \tag{22.17}$$

giving the probability for the electron to remain in the atomic ground state,

$$\mathcal{P}_{1,0,0\to 1,0,0} = 2^9/3^6 = 512/729 \approx 0.702. \tag{22.18}$$

22.2 Transition Amplitudes and Probabilities in Perturbation Theory

In Sect. 3.4, we learned how to solve for the time-dependence of quantum states. If the Hamiltonian does not depend on time, and one can solve its eigenvalue problem, then the time evolution can always be found by writing the unitary operator $U(t, t_0)$ in terms of the stationary states with appropriate phase factors that depend on the energy; see (3.24) and (3.30). However, if the Hamiltonian does depend on time, then as we saw in Sect. 21.2, the time evolution is given in terms of a Dyson series that typically cannot be obtained exactly in closed form. The approximation that follows from truncating the Dyson series in the interaction picture is known as time-dependent perturbation theory.

As in our discussion of the interaction picture, suppose that the full Hamiltonian is

$$H(t) = H_0 + W(t), \tag{22.19}$$

where H_0 does not depend explicitly on time, and let us assume that we know (at least approximately) the solutions to its eigenvalue problem,

$$H_0\,|n\rangle = \mathcal{E}_n\,|n\rangle\,. \tag{22.20}$$

Since H_0 is an observable, the kets $|n\rangle$ form a complete orthobasis. The remaining part of the full Hamiltonian, $W(t)$, contains the explicit time dependence. Then, given the state of the system at time t_0, the solution for the state at a later time t is found in terms of the transition amplitudes $a_{n\to m}$ defined by (21.45), which relies on (21.41), where

$$W_I(t) = e^{iH_0 t/\hbar}\, W(t)\, e^{-iH_0 t/\hbar} \tag{22.21}$$

is the perturbation part of the Hamiltonian, translated into the interaction picture.

The time-dependent perturbative expansion simply amounts to assuming that $W(t)$ is small, so that we can expand the transition amplitude in (21.45) as

$$a_{n\to m}(t, t_0) = a^{(0)}_{n\to m} + a^{(1)}_{n\to m} + a^{(2)}_{n\to m} + a^{(3)}_{n\to m} + \cdots, \tag{22.22}$$

where each term $a^{(N)}_{n\to m}$ contains N factors of W, following from the corresponding terms in the Dyson series (21.41). In practice, (22.22) is then truncated at some finite N, usually at $N = 1$ or $N = 2$. Note that the zeroth order contribution is just

$$a^{(0)}_{n\to m} = \langle m|n\rangle = \delta_{nm}, \tag{22.23}$$

as follows from the identity operator ($N = 0$) part of $\mathcal{U}_I$.

The first-order contribution from the Dyson series (21.41) is

$$a^{(1)}_{n\to m} = -\frac{i}{\hbar}\int_{t_0}^{t} dt_1\, \langle m|e^{iH_0 t_1/\hbar}\, W(t_1)\, e^{-iH_0 t_1/\hbar}|n\rangle. \tag{22.24}$$

Fortunately, each H_0 can be evaluated acting on a neighboring eigenket or eigenbra. So, defining

$$\omega_m = \mathcal{E}_m/\hbar, \qquad \omega_n = \mathcal{E}_n/\hbar, \qquad \text{and} \qquad \omega_{mn} = \omega_m - \omega_n, \tag{22.25}$$

we obtain

$$a^{(1)}_{n\to m} = -\frac{i}{\hbar}\int_{t_0}^{t} dt_1\, e^{i\omega_{mn}t_1}\, \langle m|\, W(t_1)\, |n\rangle\,. \tag{22.26}$$

This gives the time evolution of a general state from t_0 to t, at the leading nontrivial order in time-dependent perturbation theory, using (21.47)–(21.48) with $a_{n\to m} = \delta_{nm} + a^{(1)}_{n\to m}$.

Before moving on to consider the higher-order contributions, consider the special case that the system starts at time t_0 in an eigenstate $|n\rangle$ of H_0, and we then measure H_0 at time t. The transition probability to transfer from the stationary state $|n\rangle$ to a *different* eigenstate $|m\rangle$ of H_0, at the leading nontrivial order in the perturbation, is simply

$$\mathcal{P}_{n\to m} = |a^{(1)}_{n\to m}|^2, \qquad \text{(for } m \neq n\text{)}. \tag{22.27}$$

The validity of the approximation requires that we find $|a^{(1)}_{n\to m}| \leq 1$, so that the probability does not exceed 1. This is a necessary but not sufficient condition, because the sum of $|a^{(1)}_{n\to m}|^2$ for multiple final states $|m\rangle$ could still exceed 1.

And, even if that does not happen, the series expansion in (22.22) could still fail to converge when higher-order terms are included. The perturbation expansion is often, but not always, more reliable if $|a^{(1)}_{n\to m}| \ll 1$.

It is also instructive to consider what happens if we try to use the first-order approximation to directly compute the probability to remain in the initial stationary state $|n\rangle$. Since $a^{(0)}_{n\to n} = 1$, we get, instead of (22.27),

$$\mathcal{P}_{n\to n} = \left|1 + a^{(1)}_{n\to n} + a^{(2)}_{n\to n} + \cdots\right|^2. \tag{22.28}$$

Now, as part of the Hamiltonian, $W(t_1)$ must be Hermitian, so its expectation values are always real, and

$$a^{(1)}_{n\to n} = -\frac{i}{\hbar}\int_{t_0}^{t} dt_1 \; \langle n|W(t_1)|n\rangle \tag{22.29}$$

is a purely imaginary number. Therefore, although one might naively hope to neglect $a^{(2)}_{n\to n}$ as being of higher order, (22.28) would then give $1 + |a^{(1)}_{n\to n}|^2$, which is clearly unacceptable for a probability unless the first-order correction vanishes. To get a meaningful result from this direct calculation, we must work consistently to at least second order, by also including the contribution from $a^{(2)}_{n\to n}$ in the transition amplitude. Then the result is, dropping contributions of third order in $W(t)$,

$$\mathcal{P}_{n\to n} = 1 + |a^{(1)}_{n\to n}|^2 + 2\mathrm{Re}[a^{(2)}_{n\to n}]. \tag{22.30}$$

Alternatively, the probability to remain in the initial state $|n\rangle$ can be obtained indirectly, from the complementary probability,

$$\mathcal{P}_{n\to n} = 1 - \sum_{m\neq n} \mathcal{P}_{n\to m} = 1 - \sum_{m\neq n} |a^{(1)}_{n\to m}|^2, \tag{22.31}$$

which is also second order in the perturbation. Comparing these, we learn that

$$2\mathrm{Re}[a^{(2)}_{n\to n}] = -\sum_{m} |a^{(1)}_{n\to m}|^2, \tag{22.32}$$

where the sum over m now includes n.

Now consider the second-order contribution to the transition amplitude, obtained from (21.41) and (21.45), using (22.21),

$$a^{(2)}_{n\to m} = \left(-\frac{i}{\hbar}\right)^2 \int_{t_0}^{t} dt_2 \int_{t_0}^{t_2} dt_1 \; \langle m|e^{iH_0 t_2/\hbar} W(t_2) e^{iH_0(t_1-t_2)/\hbar} W(t_1) e^{-iH_0 t_1/\hbar}|n\rangle. \tag{22.33}$$

To evaluate this, we use completeness, by inserting a sum over all eigenstates of H_0,

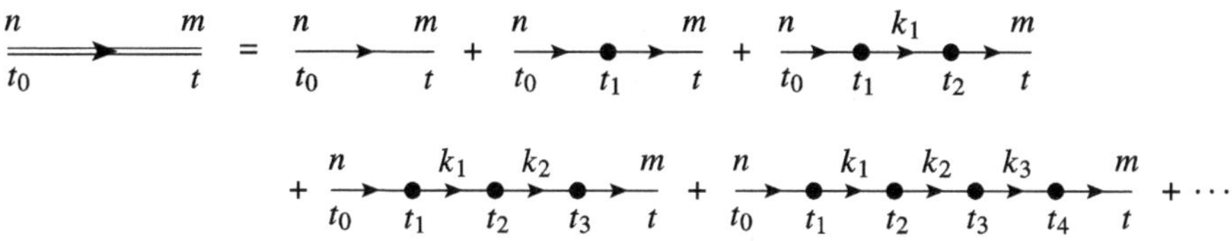

Fig. 22.1 A diagrammatic representation of the time-dependent perturbation theory expansion for the transition amplitude $a_{n\to m}(t, t_0)$, represented by the double line with arrow on the left. In the sum on the right side of the equality, each contribution with N vertices represents $a^{(N)}_{n\to m}(t, t_0)$. Each single line with arrow represents an H_0 eigenstate, as labeled. Each vertex labeled by a time t_j and incoming line k and outgoing line k' represents a factor of $-\frac{i}{\hbar}e^{i\omega_{k'k}t_j}\langle k'|W(t_j)|k\rangle$, with an integration $\int_{t_0}^{t_{j+1}} dt_j$. This is done for each of the intermediate times $t_1, \ldots, t_N$, with the identification $t_{N+1} = t$. All internal H_0 eigenstate labels k_j are summed over

$$I = \sum_{k_1} |k_1\rangle\langle k_1|\,, \tag{22.34}$$

immediately before $W(t_1)$. This puts every H_0 adjacent to one of its eigenstate bras or kets, allowing it to be replaced by $\mathcal{E}_n$, $\mathcal{E}_m$, or $\mathcal{E}_{k_1}$, with the result

$$a^{(2)}_{n\to m} = \left(-\frac{i}{\hbar}\right)^2 \int_{t_0}^{t} dt_2 \int_{t_0}^{t_2} dt_1 \sum_{k_1} e^{i\omega_{mk_1}t_2}\langle m|W(t_2)|k_1\rangle\, e^{i\omega_{k_1n}t_1}\langle k_1|W(t_1)|n\rangle. \tag{22.35}$$

It is left to Exercise 22.3 to perform the check of taking the special case $|m\rangle = |n\rangle$ and manipulating the result to verify that (22.32) indeed holds.

In the same way, the third-order contribution is found to be

$$a^{(3)}_{n\to m} = \left(-\frac{i}{\hbar}\right)^3 \int_{t_0}^{t} dt_3 \int_{t_0}^{t_3} dt_2 \int_{t_0}^{t_2} dt_1 \sum_{k_1}\sum_{k_2} e^{i\omega_{mk_2}t_3}\langle m|W(t_3)|k_2\rangle$$
$$e^{i\omega_{k_2k_1}t_2}\langle k_2|W(t_2)|k_1\rangle\, e^{i\omega_{k_1n}t_1}\langle k_1|W(t_1)|n\rangle\,. \tag{22.36}$$

A diagrammatic representation of the contributions to $a_{n\to m}$ is shown in Fig. 22.1. Each line with an arrow represents an eigenstate of H_0, as labeled. For each vertex with incoming state line k, outgoing state line k', and time label t_j, one writes a factor

$$-\frac{i}{\hbar}e^{i\omega_{k'k}t_j}\langle k'|W(t_j)|k\rangle. \tag{22.37}$$

Then one integrates $\int_{t_0}^{t_{j+1}} dt_j$ for each of the intermediate times $t_1, \ldots, t_N$ in the diagram for $a^{(N)}_{n\to m}$, with the identification $t_{N+1} = t$. All internal state labels k_j are summed over. This diagrammatic mnemonic allows us to quickly write down the formula for $a^{(N)}_{n\to m}$, corresponding to each sub-diagram starting with n and ending with m.

22.3 Applying First-Order Time-Dependent Perturbation Theory, and Fermi's Golden Rule

In this section and the next, we consider the application of time-dependent perturbation theory in the approximation of keeping only the first-order contribution to the transition amplitude, (22.26).

As a first example, consider a one-dimensional harmonic oscillator with the usual unperturbed Hamiltonian $H_0 = \frac{P^2}{2m} + \frac{1}{2}m\omega^2 X^2$, taken to be in its ground state $|0\rangle$ in the far past, at time $t = -\infty$. The particle has charge q, and a perturbation

$$W(t) = -qE_0 e^{-t^2/\tau^2} X \tag{22.38}$$

is applied, corresponding to a continuous turning on and then off of a uniform electric field E_0. The constant τ controls the time scale of the turn-on and turn-off. Our goal is to find the probability that the particle is found in a given excited state $|n\rangle$ when the perturbation has become negligible in the far future as $t \to \infty$.

Applying (22.26) with $\mathcal{E}_n = \hbar\omega(n + 1/2)$ and $t_0 \to -\infty$, the amplitude for the transition in the far future, when the perturbation has effectively turned off, is

$$a^{(1)}_{0\to n}(t = \infty,\, t_0 = -\infty) = -\frac{i}{\hbar}\int_{-\infty}^{\infty} dt_1\, e^{in\omega t_1}(-qE_0 e^{-t_1^2/\tau^2})\,\langle n|X|0\rangle\,. \tag{22.39}$$

The necessary matrix element is

$$\langle n|X|0\rangle = \sqrt{\frac{\hbar}{2m\omega}}\,\delta_{n1}, \tag{22.40}$$

so that within the first-order approximation, there is a selection rule that only the first excited state can be reached. The integral in (22.39) can then be done using the change of variables $u = t_1/\tau - i\omega\tau/2$, resulting in a transition probability

$$\mathcal{P}_{0\to 1} = |a^{(1)}_{0\to 1}|^2 = \frac{\pi q^2 E_0^2 \tau^2}{2m\hbar\omega} e^{-\omega^2\tau^2/2}, \tag{22.41}$$

while $a^{(1)}_{0\to n}$ and $\mathcal{P}_{0\to n}$ vanish for all $n > 1$. Since $\mathcal{P}$ cannot exceed 1, the approximation leading to (22.41) must fail if the applied field is too large (that is, if $q^2E_0^2\tau^2$ is too large compared to $m\hbar\omega e^{\omega^2\tau^2/2}$).

As a second example with many applications, we consider a generic unspecified H_0, and a perturbation that turns on at $t = 0$ and is constant thereafter,

$$W(t) = \begin{cases} 0 & (\text{for } t < 0), \\ V & (\text{for } t > 0). \end{cases} \tag{22.42}$$

To be as general as possible, let V be an arbitrary Hermitian operator, with the only restriction that it has no time dependence. Now consider the probability that the perturbation will induce a transition from an initial H_0 eigenstate $|i\rangle$ to a different final H_0 eigenstate $|f\rangle$. From (22.26), the amplitude for the transition is

$$a^{(1)}_{i\to f} = -\frac{i}{\hbar}\langle f|V|i\rangle \int_0^t dt_1\, e^{i\omega_{fi}t_1}, \tag{22.43}$$

where we have exploited the fact that the matrix element of V does not depend on time to pull it out of the integral, and defined $\omega_{fi} = (\mathcal{E}_f - \mathcal{E}_i)/\hbar$. Evaluating the integral, and squaring the magnitude of the result, we obtain the transition probability for $t > 0$,

$$\mathcal{P}_{i\to f}(t) = |a^{(1)}_{i\to f}|^2 = \frac{4\sin^2(t\omega_{fi}/2)}{\hbar^2\omega_{fi}^2}\,|\langle f|V|i\rangle|^2, \tag{22.44}$$

valid at first order in perturbation theory. For future convenience, we define the function

$$F(\mathcal{E}_f - \mathcal{E}_i,\, t) \equiv \frac{4}{(\mathcal{E}_f - \mathcal{E}_i)^2}\sin^2\left(\frac{t(\mathcal{E}_f - \mathcal{E}_i)}{2\hbar}\right), \tag{22.45}$$

so that

$$\mathcal{P}_{i\to f}(t) = F(\mathcal{E}_f - \mathcal{E}_i,\, t)\,|\langle f|V|i\rangle|^2. \tag{22.46}$$

The transitions at first order in perturbation theory are limited to those that satisfy

$$\langle f|V|i\rangle \neq 0, \tag{22.47}$$

an example of a matrix element selection rule.

When applying (22.46) in practice, it is important to distinguish the cases for which $|f\rangle$ is a discrete state with an isolated H_0 eigenvalue $\mathcal{E}_f$ (for example, a bound state of an electron in an atom) from those in which $|f\rangle$ is a state in a continuum (for example, a free-particle plane wave, or an unbound state of an electron ionized from an atom but still under the influence of the Coulomb potential of the nucleus).

First, suppose that $|f\rangle$ is discrete. In that case, (22.46) can be directly applied. For very small t, the transition probability grows quadratically, because for small t,

$$F(\mathcal{E}_f - \mathcal{E}_i, t) \approx t^2/\hbar^2, \tag{22.48}$$

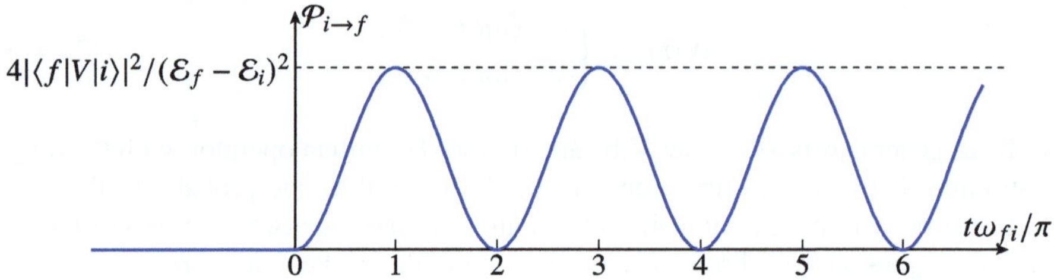

Fig. 22.2 The transition probability at first order in time-dependent perturbation theory, $\mathcal{P}_{i\to f} = |a^{(1)}_{i\to f}|^2$, as a function of the time t, for a constant perturbation V that turns on at time $t = 0$. For short times, the probability grows quadratically as $t^2|\langle f|V|i\rangle|^2/\hbar^2$, and for longer times oscillates between 0 and a maximum value $4|\langle f|V|i\rangle|^2/(\mathcal{E}_f - \mathcal{E}_i)^2$

which is independent of the unperturbed energies. Therefore, (22.46) becomes

$$\mathcal{P}_{i\to f} \approx \frac{t^2}{\hbar^2}|\langle f|V|i\rangle|^2 \qquad \text{(small } t\text{)}. \tag{22.49}$$

This agrees with the result we had already found in the short-time approximation in (22.10). Of course, (22.49) must fail for sufficiently large t, as the probability cannot exceed 1. For longer time scales, the transition probability for a given discrete final state oscillates between 0 and a maximum value $4|\langle f|V|i\rangle|^2/(\mathcal{E}_f - \mathcal{E}_i)^2$, as illustrated[1] in Fig. 22.2.

Because of the denominator, the final states $|f\rangle$ that have $\mathcal{E}_f \approx \mathcal{E}_i$ will tend to have the largest probabilities, provided that the matrix element $\langle f|V|i\rangle$ is not too small in magnitude. Also, the probability for a given final state $|f\rangle$ vanishes whenever $t = 2\pi n/\omega_{fi} = 2\pi\hbar n/(\mathcal{E}_f - \mathcal{E}_i)$ for integer n. These features are illustrated by the graph of $F(\mathcal{E}_f - \mathcal{E}_i, t)$ as a function of $\mathcal{E}_f - \mathcal{E}_i$, shown in Fig. 22.3. The final states $|f\rangle$ most likely to be reached at a given time t are those with energies that lie under the main peak around $\mathcal{E}_i$, which is bounded by

$$|\mathcal{E}_f - \mathcal{E}_i| < 2\pi\hbar/t. \tag{22.50}$$

The main peak gets narrower and taller for larger t, so that in the limit of long times, the most likely transitions are those for which the difference in unperturbed energies is constrained, increasingly strongly, by (22.50).

Now consider the case that $|f\rangle$ is not discrete, so that the unperturbed energy eigenvalues $\mathcal{E}_f$ and the matrix elements $\langle f|V|i\rangle$ each form a continuum. (For

[1] As a function of the time t, the probability oscillates with a similar appearance to the Rabi oscillations found for a magnetic spin resonance system in Sect. 9.2; compare to Fig. 9.2. However, the physical reason behind the oscillations is quite different, because in the present case the excitation is driven by a constant interaction rather than a harmonic driving frequency.

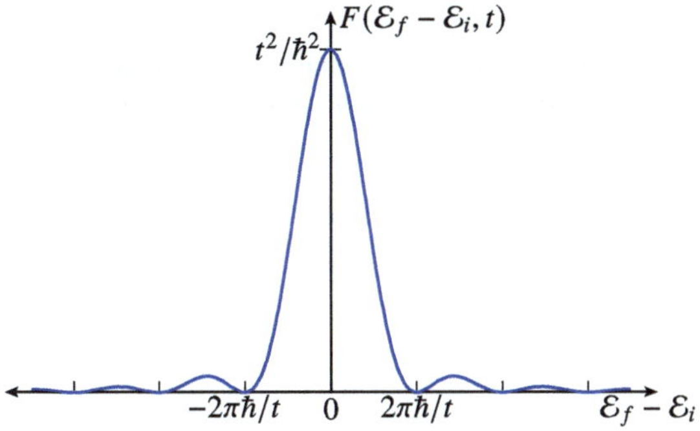

Fig. 22.3 The function $F(\mathcal{E}_f - \mathcal{E}_i, t)$ defined by (22.45) and appearing in the transition probability for first-order time-dependent perturbation theory in (22.46), as a function of $\mathcal{E}_f - \mathcal{E}_i$. The main peak height $t^2/\hbar^2$ grows quadratically with time, while its width shrinks proportionally to $1/t$. The total area under the curve is $2\pi t/\hbar$

example, the degeneracy labels could include the angular direction of a final-state momentum vector.) In that case, we write

$$|f\rangle = |\mathcal{E}_f, u_f\rangle, \tag{22.51}$$

where u_f represents the degeneracy label (or labels) for the unperturbed energy. We also define the density of states $\rho(\mathcal{E}_f, u_f)$ according to

$$\rho(\mathcal{E}_f, u_f)\, d\mathcal{E}_f\, du_f = \begin{pmatrix} \text{number of states between } \mathcal{E}_f \text{ and } \mathcal{E}_f + d\mathcal{E}_f \\ \text{and between } u_f \text{ and } u_f + du_f \end{pmatrix}, \tag{22.52}$$

where, for now, the degeneracy labels are also assumed to be continuous. The corresponding contribution to the transition probability is, from (22.46),

$$d\mathcal{P}_{i\to f} = d\mathcal{E}_f\, du_f\, \rho(\mathcal{E}_f, u_f)\, F(\mathcal{E}_f - \mathcal{E}_i, t)\, |\langle f|V|i\rangle|^2, \tag{22.53}$$

in terms of the function F defined in (22.45). In practice, (22.53) must be integrated over some ranges of $\mathcal{E}_f$ and u_f in order to give a numerical probability of transition to that range of states,

$$\Delta\mathcal{P} = \int d\mathcal{P}_{i\to f}. \tag{22.54}$$

If we want only the transition probability for some of the final states (for example, those with energy in some range, or momentum direction inside some cone), this can be enforced either by integrating only over the appropriate ranges of energy and degeneracy labels, or by putting a projection operator before V within the matrix element $\langle f|V|i\rangle$.

In the limit of large time t, we can take advantage of the fact that $F(\mathcal{E}_f - \mathcal{E}_i, t)$ as a function of $\mathcal{E}_f$ becomes increasingly sharply peaked near $\mathcal{E}_i$, and so approaches proportionality to a delta function. To understand the proportionality, we note that

$$\int_{-\infty}^{\infty} d\mathcal{E}_f\, F(\mathcal{E}_f - \mathcal{E}_i, t) = 2\pi t/\hbar, \tag{22.55}$$

so that for large t we can make the replacement

$$F(\mathcal{E}_f - \mathcal{E}_i, t) \to \frac{2\pi t}{\hbar}\delta(\mathcal{E}_f - \mathcal{E}_i). \tag{22.56}$$

This can be checked by taking $x = \mathcal{E}_f - \mathcal{E}_i$ and $\Delta = \hbar/t$ in (2.38). The transition probability (22.53) is then directly proportional to t,

$$d\mathcal{P}_{i\to f} = t\, dR_{i\to f}, \tag{22.57}$$

where the proportionality can therefore be interpreted as a transition rate per unit time,

$$dR_{i\to f} = \frac{2\pi}{\hbar} |\langle f|V|i\rangle|^2\, \delta(\mathcal{E}_f - \mathcal{E}_i)\, \rho(\mathcal{E}_f, u_f)\, d\mathcal{E}_f\, du_f. \tag{22.58}$$

The total transition rate to a group of states with $\mathcal{E}_f \approx \mathcal{E}_i$ is obtained by integration over $\mathcal{E}_f$, and integrating (or summing) over the degeneracy label(s) u_f. Doing the energy integration, and using the notation of (22.51), we get the transition rate

$$R = \frac{2\pi}{\hbar} \int du_f\, \rho(\mathcal{E}_f, u_f) \left|\langle \mathcal{E}_f, u_f|V|i\rangle\right|^2, \tag{22.59}$$

where it is now understood that only final states with $\mathcal{E}_f = \mathcal{E}_i$ are included, and again projection operators can be used inside the matrix element to restrict to particular classes of final states with some desired properties. Equation (22.59) [or its equivalent (22.58)] is one version of **Fermi's golden rule**. (It was actually originally obtained by Dirac, but Fermi popularized it.) If the energy degeneracy labels u_f are discrete rather than continuous, then one makes the replacement $\int du_f \to \sum_{u_f}$.

Although the derivation of the golden rule turned F into a delta function by assuming that t is large, we have to be careful if t is *too* large. This is simply because interpretation of (22.59) as a constant rate clearly requires that $t < 1/R$, so that the probability does not exceed 1. More generally, for times that are not short compared to $1/R$, the transition rate decreases as the initial state $|i\rangle$ becomes depleted. If we write $\mathcal{P}_{i\to i}(t) = e^{-R_{\text{tot}}t}$ for the probability of remaining in the initial state, including the effects of changing to *all* possible final states, then the transition rate to a particular class of final states approaches $Re^{-R_{\text{tot}}t}$ at late times.

The golden rule can also be applied as an approximation in the case of a group of many discrete but closely spaced final states with approximately the same matrix elements $\langle \mathcal{E}_f, u_f|V|i\rangle$. This works if the unperturbed energy differences $\delta\mathcal{E}_f$ between neighboring discrete final states are small enough that there are many states under the main peak in the function $F(\mathcal{E}_f - \mathcal{E}_i, t)$. This condition requires that

$$t\,\delta\mathcal{E}_f \ll 2\pi\hbar, \tag{22.60}$$

so that the elapsed time is short compared to the time scale set by the inverse energy spacing between adjacent states.

22.4 Harmonic Time-Dependent Perturbations

An even more common application of time-dependent perturbation theory is the case of a **harmonic time-dependent perturbation**, in which $W(t)$ is proportional to a sine or cosine function of time. For example, this includes the case of electromagnetic waves interacting with a charged particle. Suppose that the Hamiltonian perturbation turns on at time $t = 0$, so

$$W(t) = \begin{cases} 0 & (\text{for } t < 0), \\ Ve^{-i\omega t} + V^\dagger e^{i\omega t} & (\text{for } t > 0), \end{cases} \tag{22.61}$$

where V is an operator that does not depend on time, and ω is a driving frequency, not to be confused with $\omega_i = \mathcal{E}_i/\hbar$ and $\omega_f = \mathcal{E}_f/\hbar$ associated with the energies of the initial and final states. Note that $W(t)$ is Hermitian, but V need not be. This perturbation can also be written as $(V + V^\dagger)\cos(\omega t) + i(V^\dagger - V)\sin(\omega t)$ for $t > 0$, with $(V + V^\dagger)$ and $i(V^\dagger - V)$ both Hermitian operators.

Applying first-order time-dependent perturbation theory in the form of (22.26), the transition amplitude is

$$\begin{aligned} a^{(1),\omega}_{i\to f}(t) &= -\frac{i}{\hbar}\int_0^t dt' \left(e^{i(\omega_f-\omega_i-\omega)t'}\langle f|V|i\rangle + e^{i(\omega_f-\omega_i+\omega)t'}\langle f|V^\dagger|i\rangle\right) \\ &= \frac{1-e^{i(\omega_f-\omega_i-\omega)t}}{\hbar(\omega_f-\omega_i-\omega)}V_{fi} + \frac{1-e^{i(\omega_f-\omega_i+\omega)t}}{\hbar(\omega_f-\omega_i+\omega)}V^\dagger_{fi}, \end{aligned} \tag{22.62}$$

where we have used a shorthand notation for the time-independent matrix elements,

$$V_{fi} = \langle f|V|i\rangle, \qquad V^\dagger_{fi} = \langle f|V^\dagger|i\rangle. \tag{22.63}$$

Requiring that these matrix elements do not vanish provides a selection rule for first-order time-dependent perturbation theory.

The first term in (22.62) is resonant only if $\omega_f \approx \omega_i + \omega$, and the second term only if $\omega_f \approx \omega_i - \omega$. For $\omega \neq 0$, at most one of these conditions can be satisfied for any particular $\mathcal{E}_f$, so only one term will be important within the resonance approximation. Therefore, we can make the approximation of ignoring the cross terms in $|a_{i\to f}^{(1),\omega}(t)|^2$. Then the probability of transition between eigenstates of H_0 is

$$\mathcal{P}_{i\to f} = F(\mathcal{E}_f - \mathcal{E}_i - \hbar\omega,\, t)\,|V_{fi}|^2 + F(\mathcal{E}_f - \mathcal{E}_i + \hbar\omega,\, t)\,|V_{fi}^\dagger|^2, \quad (22.64)$$

where the same function F defined by (22.45) appears again. The application of (22.64) is slightly different depending on whether the final state is discrete or part of a continuum, although the final results will have a similar appearance.

First, consider the continuum case, denoting the final states with H_0 eigenvalue $\mathcal{E}_f$ and degeneracy label u_f as

$$|f\rangle = |\mathcal{E}_f, u_f\rangle. \quad (22.65)$$

For large t, the same argument that led to (22.58) now gives

$$d\mathcal{P}_{i\to f} = t\, dR_{i\to f}, \quad (22.66)$$

where the differential contribution to the transition rate is

$$dR_{i\to f} = \frac{2\pi}{\hbar}\Big(\delta(\mathcal{E}_f - \mathcal{E}_i - \hbar\omega)|V_{fi}|^2 + \delta(\mathcal{E}_f - \mathcal{E}_i + \hbar\omega)|V_{fi}^\dagger|^2\Big) \times \rho(\mathcal{E}_f, u_f)\, d\mathcal{E}_f\, du_f. \quad (22.67)$$

The effect of the first term can be referred to as **absorption**; since $\mathcal{E}_f = \mathcal{E}_i + \hbar\omega$, the Hamiltonian perturbation V causes the initial state to gain energy $\hbar\omega$ and jump up to the final state. The second term is **emission**, as the perturbation $V^\dagger$ causes the initial state to lose energy $\hbar\omega$ and jump down to the final state. Since we ignored the cross-terms, and the final state energies must be different in the two cases anyway, it makes sense to separate these, and write

$$dR_{i\to f}^{\text{abs}} = \frac{2\pi}{\hbar}|V_{fi}|^2\, \delta(\mathcal{E}_f - \mathcal{E}_i - \hbar\omega)\, \rho(\mathcal{E}_f, u_f)\, d\mathcal{E}_f\, du_f, \quad (22.68)$$

$$dR_{i\to f}^{\text{em}} = \frac{2\pi}{\hbar}|V_{fi}^\dagger|^2\, \delta(\mathcal{E}_f - \mathcal{E}_i + \hbar\omega)\, \rho(\mathcal{E}_f, u_f)\, d\mathcal{E}_f\, du_f. \quad (22.69)$$

After integration over $\mathcal{E}_f$,

$$dR_{i\to f}^{\text{abs}} = \frac{2\pi}{\hbar}\, |\langle \mathcal{E}_f, u_f|V|i\rangle|^2\, \rho(\mathcal{E}_f, u_f)\, du_f, \quad (22.70)$$

$$dR_{i\to f}^{\text{em}} = \frac{2\pi}{\hbar}\, |\langle \mathcal{E}_f, u_f|V^\dagger|i\rangle|^2\, \rho(\mathcal{E}_f, u_f)\, du_f, \quad (22.71)$$

where it is now understood that the energy conservation conditions

$$\mathcal{E}_f = \mathcal{E}_i + \hbar\omega, \qquad \text{(absorption)}, \tag{22.72}$$

$$\mathcal{E}_f = \mathcal{E}_i - \hbar\omega, \qquad \text{(emission)}, \tag{22.73}$$

are enforced, due to the delta functions. Equations (22.70) and (22.71), or their equivalents (22.68) and (22.69), are **Fermi's golden rule for harmonic perturbations**. If the energy degeneracy labels u_f are continuous, then they should be integrated over, and if they are discrete then the differential du_f is removed and they can be summed over.

If the transition from $|i\rangle$ to $|f\rangle$ by absorbing energy $\hbar\omega$ from V is allowed, then it follows that for the same ω one can also have the reverse process of $|f\rangle$ to $|i\rangle$ by emitting energy $\hbar\omega$ to $V^\dagger$; both the energy conservation and matrix element selection rules work. For both absorption and emission, we can obtain the differential rate for the opposite process $dR_{f\to i}$ by interchanging $\mathcal{E}_f \leftrightarrow \mathcal{E}_i$ and $|V_{fi}|^2 \leftrightarrow |V^\dagger_{if}|^2$. Note that the relevant matrix elements are simply related by

$$V^\dagger_{if} = \langle i|V^\dagger|f\rangle = (\langle f|V|i\rangle)^* = \left(V_{fi}\right)^*, \tag{22.74}$$

so $|V^\dagger_{if}|^2 = |V_{fi}|^2$. Let us now change the i and f labels to a more neutral notation a and b, since each can play the role of initial or final state. If we multiply (22.71) with $i = a$ and $f = b$ by $\rho(\mathcal{E}_a, u_a)du_a$, the result is the same as multiplying (22.70) with $i = b$ and $f = a$ by $\rho(\mathcal{E}_b, u_b)du_b$, after integrating over the same intervals in the degeneracy labels in each case. This shows that

$$\rho(\mathcal{E}_a, u_a)\,\frac{dR^{\text{em}}_{a\to b}}{du_b} = \rho(\mathcal{E}_b, u_b)\,\frac{dR^{\text{abs}}_{b\to a}}{du_a}. \tag{22.75}$$

This is known as the **detailed balance** relation between emission and absorption rates and the densities of states.

Let us now reconsider (22.64), this time assuming that the final state $|f\rangle$ is discrete. We now suppose that the perturbations occur as an incoherent distribution with a range of angular frequencies. By "incoherent", it is implied that interference contributions will be negligible, so that, to a good approximation, the probabilities, not the amplitudes, add for different driving angular frequencies ω. To see how this works, consider instead of (22.61), a perturbation that is a superposition of many such terms with infinitesimally spaced ω's with smoothly varying real amplitudes c_ω and, crucially, phases φ_ω that are assumed to be random, with no phase relation (no coherence) whatsoever between neighboring values of ω,

$$W(t) = \sum_\omega c_\omega\left(e^{i\varphi_\omega}Ve^{-i\omega t} + e^{-i\varphi_\omega}V^\dagger e^{i\omega t}\right) \qquad \text{(for } t > 0\text{)}. \tag{22.76}$$

After isolating the emission and absorption resonant pieces as before, the square of the transition amplitude $a^{(1)}_{i\to f}$ will contain terms of the form

$$|a^{(1)}_{i\to f}|^2 = \sum_{\omega}\sum_{\omega'} c_\omega c_{\omega'} e^{i(\varphi_\omega-\varphi_{\omega'})}\cdots, \tag{22.77}$$

where the $\cdots$ represents quantities that are slowly varying functions of ω and ω'. Because of the random phases and the slow variation of the summand, all terms with $\omega' \neq \omega$ now average to 0 in the limit of infinitesimal spacing of the sums, leaving behind a single sum of the form $\sum_\omega c_\omega^2\cdots$, coming from the subset of terms with $\omega' = \omega$, which add with constructive interference. This sum can now be converted to an integral

$$\sum_\omega c_\omega^2\cdots \to \int d\omega\, \rho(\omega)\cdots, \tag{22.78}$$

which means that the transition probabilities for continuous ω are assumed[2] to be weighted by a probability density function $\rho(\omega)$ that characterizes the perturbation, and should not be confused with a density of states. For example, if the perturbations take the form of electromagnetic radiation, one can relate $\rho(\omega)$ to the intensity spectrum of the incoherent light, as we will do in Sect. 23.2. In the presence of the integral over continuous ω, even though $\mathcal{E}_f$ is discrete, we can again replace

$$F(\mathcal{E}_f - \mathcal{E}_i \pm \hbar\omega,\, t) \to \frac{2\pi t}{\hbar}\delta(\mathcal{E}_f - \mathcal{E}_i \pm \hbar\omega), \tag{22.79}$$

making usc of the $\delta(x)$ formula (2.38), and so obtain the differential transition rates

$$dR^{\rm abs}_{i\to f} = \frac{2\pi}{\hbar}|V_{fi}|^2\,\delta(\mathcal{E}_f - \mathcal{E}_i - \hbar\omega)\,\rho(\omega)\,d\omega, \tag{22.80}$$

$$dR^{\rm em}_{i\to f} = \frac{2\pi}{\hbar}|V^\dagger_{fi}|^2\,\delta(\mathcal{E}_f - \mathcal{E}_i + \hbar\omega)\,\rho(\omega)\,d\omega. \tag{22.81}$$

After integrating over ω using $\int d\omega\, \delta(\mathcal{E}_f - \mathcal{E}_i \pm \hbar\omega) = 1/\hbar$, we have

$$R^{\rm abs}_{i\to f} = \frac{2\pi}{\hbar^2}\,|\langle f|V|i\rangle|^2\,\rho(\omega), \tag{22.82}$$

$$R^{\rm em}_{i\to f} = \frac{2\pi}{\hbar^2}\,|\langle f|V^\dagger|i\rangle|^2\,\rho(\omega), \tag{22.83}$$

[2] If the perturbation is coherent in frequency, meaning that the phases φ_ω vary slowly with ω, then one cannot neglect interference terms in the transition amplitude $a^{(1)}_{i\to f}$. We do not consider that case.

where now it is understood that ω is fixed in terms of $\mathcal{E}_f$ and $\mathcal{E}_i$ by (22.72) for absorption or (22.73) for emission. The energy degeneracies of the final state $|f\rangle$ are assumed here to be also discrete, and so are not indicated explicitly, but can be summed over.

Although (22.70)–(22.71) and (22.82)–(22.83) look similar, it bears emphasis that they apply in different circumstances. In (22.82)–(22.83), the final states are discrete, and the driving angular frequency ω is selected (by energy conservation) from an incoherent continuum of perturbations to the Hamiltonian with some assumed probability density function with respect to angular frequency, $\rho(\omega)$. In (22.70)–(22.71), it is the final state that is selected by energy conservation from among the continuous final states with density with respect to energy $\rho(\mathcal{E}_f, u_f)$, for a driving angular frequency that can be treated as a single monochromatic Hamiltonian perturbation. Both of these situations arise in the interaction of matter with electromagnetic waves, depending on whether the final electronic state is bound or unbound. This is the subject of the next chapter.

Exercises

22.1 At the end of Sect. 22.1, we found $\mathcal{P} = 512/729$ for the probability that a ^{3}H atom in its ground state would end up in the ground state of ^{3}He$^+$ after beta decay. Find the probability of ending up in each of the $n = 2$ or $n = 3$ levels instead. Check that the total probability for ending up in the $n = 1, 2$, and 3 levels does not exceed 1.

22.2 A particle of mass m moves freely in a one-dimensional box of length a, with the left edge at $x = 0$ and the right edge at $x = a$. At time $t = 0$, the particle is in the ground state, and the box is suddenly lengthened by expanding the right edge out to $x = b$.

(a) Find the probability that the particle ends up in each of the energy eigenstates of the final box, labeled by positive integers n.
(b) Use your results from part (a), and your knowledge of the laws of probability, to evaluate the following mathematical infinite sum

$$S(a, b) = \sum_{n=1}^{\infty} \frac{\sin^2(n\pi a/b)}{(n^2a^2 - b^2)^2}. \tag{22.84}$$

Quantum mechanics knows how to do nontrivial sums!
(c) As a check, for the special case $b = 2a$, evaluate the individual probabilities numerically for each integer up to $n = 10$, and their total with 6 significant digits. (Be careful for $n = 2$; the contribution is nonzero and finite.)

22.3 Show that the real part of the second-order transition amplitude for a state to itself, $a^{(2)}_{n\to n}$, as found in (22.35) with $|m\rangle = |n\rangle$, agrees with the sum rule (22.32).

22.4 A particle of mass m is bound in a one-dimensional harmonic oscillator of natural frequency ω. At $t = 0$, it is in the ground state $|0\rangle$, and a Hamiltonian perturbation is turned on, of the form $W(X, t) = \lambda X^2 e^{-t/T}$ for $t \geq 0$, where λ and T are constants.

(a) Use first-order time-dependent perturbation theory to find the probabilities that, after a very long time ($t \gg T$), the particle will have jumped to each of the excited states $|n\rangle$.
(b) Using your answers for part (a), infer the probability that after a long time the particle will be found in the ground state.
(c) Use second-order time-dependent perturbation theory to directly find the probability of finding the particle in the ground state. Check that your answer agrees with part (b).

22.5 A hydrogen atom is in its ground state at time $t = 0$, when a decaying electric field $\boldsymbol{E} = E_0\hat{z}e^{-t/T}$ is applied. Use first-order time-dependent perturbation theory to find the probability for the atom to be found at $t \gg T$ in each of the first excited states $[(n, l, m) = (2, 0, 0)$ and $(2, 1, 1)$ and $(2, 1, 0)$ and $(2, 1, -1)]$.

22.6 Consider the magnetic spin resonance problem discussed in Sect. 9.2.

(a) Treating $\widetilde{B}$ (and therefore Γ) as small, use first-order time-dependent perturbation theory to calculate the probability for transition from the state $|\downarrow\rangle$ to the state $|\uparrow\rangle$, as a function of t. How does this compare to the exact answer found in (9.41)?
(b) When ω is very close to ω_B, show that the perturbative answer you got in part (a) can be interpreted in terms of a rate $\approx N\gamma^2\widetilde{B}^2\delta(\omega - \omega_B)$, where N is a certain number that you will find. [Hint: use (2.38)] To apply this in practice, one would integrate over a spectrum of driving frequencies ω.

Absorption and Emission of Light 23

23.1 Electrons in the Presence of Electromagnetic Waves

In this chapter, we will be discussing the interactions of electrons with a classical electromagnetic wave. The Hamiltonian is obtained from the Coulomb gauge result in (4.55), with $q = -e$ and $\gamma = -g_e e/2m_e c \approx -e/m_e c$ for electrons,

$$H = H_0 + \frac{e}{m_e c}\boldsymbol{A} \cdot \boldsymbol{P} + \frac{e^2}{2m_e c^2}A^2 - e\Phi + \frac{e}{m_e c}\boldsymbol{S} \cdot \boldsymbol{B}, \tag{23.1}$$

where the electron's Hamiltonian in the absence of the electromagnetic wave is

$$H_0 = \frac{P^2}{2m_e} + U(\boldsymbol{R}). \tag{23.2}$$

The additional potential U includes effects not associated with the wave, for example the Coulomb attraction of the electron to a fixed heavy nucleus, or repulsion by other electrons. Let the wave have unit polarization vector $\hat{\varepsilon}$, and wavevector

$$\boldsymbol{k} = \hat{k}\omega/c, \tag{23.3}$$

so that the classical electric and magnetic fields are

$$\boldsymbol{E}(\boldsymbol{r}, t) = \hat{\varepsilon} E_0 \sin(\boldsymbol{k} \cdot \boldsymbol{r} - \omega t), \tag{23.4}$$

$$\boldsymbol{B}(\boldsymbol{r}, t) = (\hat{k} \times \hat{\varepsilon}) E_0 \sin(\boldsymbol{k} \cdot \boldsymbol{r} - \omega t), \tag{23.5}$$

which satisfy Maxwell's equations provided that $\hat{\varepsilon} \cdot \hat{k} = 0$. In Coulomb gauge, the potentials describing this classical wave are

S. P. Martin, *Quantum Mechanics*, Graduate Texts in Physics,
https://doi.org/10.1007/978-3-032-21119-4_23

$$\Phi(\boldsymbol{r}, t) = 0, \tag{23.6}$$

$$\boldsymbol{A}(\boldsymbol{r}, t) = -\hat{\varepsilon}\frac{c}{\omega}E_0 \cos(\boldsymbol{k} \cdot \boldsymbol{r} - \omega t), \tag{23.7}$$

as you can check using (4.28)–(4.29) and (4.54).

If the wave is not too intense, it is a good approximation to drop the nonlinear term proportional to A^2, so that our Hamiltonian simplifies to $H = H_0 + W$, where

$$W(t) = \frac{e}{m_e c}(\boldsymbol{A} \cdot \boldsymbol{P} + \boldsymbol{B} \cdot \boldsymbol{S}) \tag{23.8}$$

$$= \frac{eE_0}{m_e \omega}\left[-\cos(\boldsymbol{k} \cdot \boldsymbol{R} - \omega t)\,\hat{\varepsilon} \cdot \boldsymbol{P} + \sin(\boldsymbol{k} \cdot \boldsymbol{R} - \omega t)\,(\boldsymbol{k} \times \hat{\varepsilon}) \cdot \boldsymbol{S}\right] \tag{23.9}$$

This can be treated as a perturbation using the formalism of Sect. 22.4, specifically by writing

$$W(t) = V e^{-i\omega t} + V^{\dagger} e^{i\omega t}, \tag{23.10}$$

as in (22.61), with

$$V = -\frac{eE_0}{2m_e\omega} e^{i\boldsymbol{k}\cdot\boldsymbol{R}}\left[\hat{\varepsilon} \cdot \boldsymbol{P} + i(\boldsymbol{k} \times \hat{\varepsilon}) \cdot \boldsymbol{S}\right]. \tag{23.11}$$

To put this into a more compact form, we define a vector operator

$$\mathcal{P}_{\boldsymbol{k}} \equiv e^{i\boldsymbol{k}\cdot\boldsymbol{R}}\big(\boldsymbol{P} - i\boldsymbol{k} \times \boldsymbol{S}\big), \tag{23.12}$$

so that, making use of the identity $(\boldsymbol{k} \times \hat{\varepsilon}) \cdot \boldsymbol{S} = -\hat{\varepsilon} \cdot (\boldsymbol{k} \times \boldsymbol{S})$,

$$V = -\frac{eE_0}{2m_e\omega}\,\hat{\varepsilon} \cdot \mathcal{P}_{\boldsymbol{k}}. \tag{23.13}$$

Note that $\mathcal{P}_{\boldsymbol{k}}$ has the same units as momentum, but it is not Hermitian. Its matrix elements will be used in our discussion of absorption and emission of electromagnetic waves in the next few sections. The polarization vector $\hat{\varepsilon}$ can now be taken to be complex, for example to describe circularly polarized waves, but must satisfy $\hat{\varepsilon} \cdot \hat{k} = 0$ and $\hat{\varepsilon}^* \cdot \hat{\varepsilon} = 1$.

The Hamiltonian for a system of N_e electrons in an external classical electromagnetic wave can be similarly treated using an appropriate H_0 (which will include the repulsive potentials between each pair of electrons, their attractive potentials due to the heavy nucleus, and spin-orbit and spin-spin couplings) together with a harmonic perturbation

$$V = -\frac{eE_0}{2m_e\omega}\sum_{i=1}^{N_e} \hat{\varepsilon} \cdot \mathcal{P}_{\boldsymbol{k},i}. \tag{23.14}$$

Here, each $\mathcal{P}_{\boldsymbol{k},i}$ is defined in terms of the position, momentum, and spin operators $\boldsymbol{R}_i$, $\boldsymbol{P}_i$, and $\boldsymbol{S}_i$ for the individual electrons, as in (23.12). This is the linearized approximation of a Hamiltonian perturbation describing the interaction with an external electromagnetic wave.

23.2 Absorption of Electromagnetic Waves

Consider an electron described by some Hamiltonian H_0, which then encounters a classical electromagnetic wave. As described in the previous section, we can now apply time-dependent perturbation theory as discussed in Sect. 22.4, with

$$V = -\frac{eE_0}{2m_e\omega}\hat{\varepsilon}\cdot\mathcal{P}_{\boldsymbol{k}} \tag{23.15}$$

in (22.61), where E_0 and $\hat{\varepsilon}$ are the electric field amplitude and unit polarization vector of the wave, and the operator $\mathcal{P}_{\boldsymbol{k}}$ was defined in (23.12) in terms of the wavevector. We will compute results in terms of matrix elements of $\mathcal{P}_{\boldsymbol{k}}$, and then work out the simpler electric dipole and higher multipole approximations in Sects. 23.4 and 23.5, using an expansion in the angular frequency ω (or equivalently the wavenumber $k = \omega/c$).

Consider the rate for an initial H_0 eigenstate $|i\rangle$ to absorb energy from the wave and end up in a group of final H_0 eigenstates

$$|f\rangle = |\mathcal{E}_f, u_f\rangle. \tag{23.16}$$

For now, we take these to have continuous eigenvalues $\mathcal{E}_f$ and degeneracy labels u_f and density of states $\rho(\mathcal{E}_f, u_f)$ as defined in (22.52). Fermi's golden rule, (22.68), then gives the differential rate for absorption transitions,

$$dR^{\text{abs}}_{i\to f} = \frac{2\pi}{\hbar}\left(\frac{eE_0}{2m_e\omega}\right)^2 \left|\langle f|\hat{\varepsilon}\cdot\mathcal{P}_{\boldsymbol{k}}|i\rangle\right|^2 \rho(\mathcal{E}_f, u_f)\, du_f\, \delta(\mathcal{E}_f - \mathcal{E}_i - \hbar\omega)\, d\mathcal{E}_f. \tag{23.17}$$

The time-averaged flux of energy in the incident wave causing these transitions is

$$\text{energy flux in wave} = \frac{\text{energy}}{\text{(time)(area)}} = \frac{cE_0^2}{8\pi}. \tag{23.18}$$

This can be obtained by time-averaging either cu_{EM} in (4.26), or the magnitude of the Poynting vector $|\boldsymbol{S}|$ in (4.27). Since both (23.17) and (23.18) are proportional to E_0^2, it makes sense to define a measurable quantity proportional to their ratio, in which the incident wave intensity cancels out. The appropriate ratio is called the **absorption cross section**:

$$\sigma^{\text{abs}} = \frac{\text{absorbed energy/time}}{\text{energy flux in incident wave}}. \tag{23.19}$$

Note that $\sigma^{\rm abs}$ has units of area. The energy absorbed in each $i \to f$ transition is $\hbar\omega$, so

$$d\sigma^{\rm abs}_{i\to f} = \hbar\omega \, dR^{\rm abs}_{i\to f}/(cE_0^2/8\pi), \tag{23.20}$$

or, in terms of the fine-structure constant $\alpha = e^2/\hbar c$,

$$d\sigma^{\rm abs}_{i\to f} = \frac{4\pi^2\alpha\hbar}{m_e^2\omega} |\langle f|\hat{\varepsilon}\cdot\mathcal{P}_{\boldsymbol{k}}|i\rangle|^2 \rho(\mathcal{E}_f, u_f)\, du_f \, \delta(\mathcal{E}_f - \mathcal{E}_i - \hbar\omega)\, d\mathcal{E}_f. \tag{23.21}$$

If the final state energies $\mathcal{E}_f$ are continuous near $\mathcal{E}_i + \hbar\omega$, one can immediately integrate over them, with the result

$$d\sigma^{\rm abs}_{i\to f} = \frac{4\pi^2\alpha\hbar}{m_e^2\omega} |\langle f|\hat{\varepsilon}\cdot\mathcal{P}_{\boldsymbol{k}}|i\rangle|^2 \rho(\mathcal{E}_f, u_f)\, du_f, \tag{23.22}$$

where it is now implicitly required that $\mathcal{E}_f = \mathcal{E}_i + \hbar\omega$. Equation (23.22) also applies as an approximation if the relevant final states are discrete but very closely spaced. If some of the degeneracy labels u_f are discrete rather than continuous, then one should omit du_f for those labels and sum over them rather than integrating.

Now suppose instead that the final states selected by energy conservation have discrete H_0 eigenvalues $\mathcal{E}_f$. We will discuss this situation in three different ways, appropriate for different circumstances.

First, suppose that the electromagnetic fields are plane waves with fixed propagation direction $\hat{k}$ and polarization $\hat{\varepsilon}$, but with incoherent phases and a range of angular frequencies wide enough to cover the final state in question. Then, as discussed in Sect. 22.4, the incoherent phases mean that we can sum the probabilities (not the amplitudes) to obtain a transition rate. Using (23.15) in (22.80),

$$R^{\rm abs}_{i\to f} = \int_0^\infty d\omega\, \rho(\omega) \frac{2\pi}{\hbar} \delta(\mathcal{E}_f - \mathcal{E}_i - \hbar\omega) \left(\frac{eE_0}{2m_e\omega}\right)^2 |\langle f|\hat{\varepsilon}\cdot\mathcal{P}_{\boldsymbol{k}}|i\rangle|^2, \tag{23.23}$$

where $\rho(\omega)$ is the density per unit angular frequency of incoherent electromagnetic wave perturbations of the form (22.61). Let I be the total intensity (power per unit area) in the electromagnetic radiation, so that

$$dI = \frac{cE_0^2}{8\pi}\rho(\omega)d\omega. \tag{23.24}$$

We thus obtain, in terms of the intensity per unit angular frequency, $dI/d\omega$,

$$R^{\rm abs}_{i\to f} = \int_0^\infty d\omega \frac{dI}{d\omega} \delta(\mathcal{E}_f - \mathcal{E}_i - \hbar\omega) \frac{4\pi^2\alpha}{m_e^2\omega^2} |\langle f|\hat{\varepsilon}\cdot\mathcal{P}_{\boldsymbol{k}}|i\rangle|^2. \tag{23.25}$$

Doing the integration over ω, we find[1] an absorption rate

$$R_{i\to f}^{\mathrm{abs}} = \frac{4\pi^2\alpha}{m_e^2\hbar\omega^2}\frac{dI}{d\omega}\,|\,\langle f|\hat{\varepsilon}\cdot\mathcal{P}_{\boldsymbol{k}}|i\rangle\,|^2, \tag{23.26}$$

where $\omega = (\mathcal{E}_f - \mathcal{E}_i)/\hbar$ has now been implicitly fixed.

As a second scenario, suppose that the system is bathed in electromagnetic radiation that is not only phase-incoherent and broadband in ω, but also may be coming from all directions and with all possible polarization vectors. Although we have been discussing the electromagnetic field as a classical wave, in this case we choose to frame our discussion in terms of the numbers of photon quanta with energy $\hbar\omega$ and particular wavenumbers and polarizations. To do so, we take the system to be in a large cubic box of side L, with periodic boundary conditions giving allowed wavevectors

$$\boldsymbol{k} = \hat{k}\omega/c = \frac{2\pi}{L}(\hat{x}n_x + \hat{y}n_y + \hat{z}n_z) \tag{23.27}$$

where (n_x, n_y, n_z) are integers. In the large-volume continuum limit, a discrete sum over these wavevectors will correspond to the integration

$$\sum_{\boldsymbol{k}} \to \left(\frac{L}{2\pi}\right)^3\int d^3\boldsymbol{k} = \left(\frac{L}{2\pi c}\right)^3\int_0^\infty d\omega\,\omega^2\int d\Omega_{\boldsymbol{k}}. \tag{23.28}$$

Denote by $N_{\boldsymbol{k},\hat{\varepsilon}}$ the occupation number of photons in a mode with wavevector $\boldsymbol{k}$ and polarization vector $\hat{\varepsilon}$. Then the time-averaged energy associated with that mode is

$$\hbar\omega\,N_{\boldsymbol{k},\hat{\varepsilon}} = (\text{volume})\left(\frac{\text{energy}}{\text{volume}}\right) = L^3\left(E_0^{\boldsymbol{k},\hat{\varepsilon}}\right)^2/8\pi, \tag{23.29}$$

where $E_0^{\boldsymbol{k},\hat{\varepsilon}}$ is the corresponding classical electric field amplitude.

Going back to (22.64), and using the assumption of incoherence to sum the probabilities rather than amplitudes, the total probability for absorption is

$$\mathcal{P}_{i\to f}^{\mathrm{abs}} = \sum_{\boldsymbol{k}}\sum_{\hat{\varepsilon}}\left(\frac{eE_0^{\boldsymbol{k},\hat{\varepsilon}}}{2m_e\omega}\right)^2\,|\,\langle f|\hat{\varepsilon}\cdot\mathcal{P}_{\boldsymbol{k}}|i\rangle\,|^2\,F(\mathcal{E}_f - \mathcal{E}_i - \hbar\omega,\,t). \tag{23.30}$$

Now, using (23.28) and (23.29), this becomes

$$\mathcal{P}_{i\to f}^{\mathrm{abs}} = \int_0^\infty d\omega\int d\Omega_{\boldsymbol{k}}\sum_{\hat{\varepsilon}}\frac{e^2\hbar\omega}{4\pi^2m_e^2c^3}N_{\boldsymbol{k},\hat{\varepsilon}}\,|\,\langle f|\hat{\varepsilon}\cdot\mathcal{P}_{\boldsymbol{k}}|i\rangle\,|^2\,F(\mathcal{E}_f - \mathcal{E}_i - \hbar\omega,\,t). \tag{23.31}$$

[1] This could also be obtained directly from (22.82).

Note that the polarizations $\hat{\varepsilon}$ that are being summed over depend on the direction of $\boldsymbol{k}$, because of the transversality constraint $\hat{\varepsilon} \cdot \hat{k} = 0$ for electromagnetic waves. We now apply $F(\mathcal{E}_f - \mathcal{E}_i - \hbar\omega,\, t) = (2\pi t/\hbar)\delta(\mathcal{E}_f - \mathcal{E}_i - \hbar\omega)$ from (22.79), and do the integral over ω to obtain $\mathcal{P}^{\text{abs}}_{i\to f} = t\, R^{\text{abs}}_{i\to f}$, where the absorption rate is

$$R^{\text{abs}}_{i\to f} = \frac{\alpha\omega}{2\pi m_e^2 c^2} \int d\Omega_{\boldsymbol{k}} \sum_{\hat{\varepsilon}} N_{\boldsymbol{k},\hat{\varepsilon}} \left| \langle f|\hat{\varepsilon} \cdot \mathcal{P}_{\boldsymbol{k}}|i\rangle \right|^2, \tag{23.32}$$

in terms of the photon occupation numbers $N_{\boldsymbol{k},\hat{\varepsilon}}$ for modes with $\omega = c|\boldsymbol{k}| = (\mathcal{E}_f - \mathcal{E}_i)/\hbar$. As one might expect, the contribution to the rate from each mode is proportional to the number of photons available in that mode. Comparing (23.26) to (23.32), we see that

$$\frac{dI}{d\omega} \to \frac{\hbar\omega^3}{8\pi^3 c^2} \int d\Omega_{\boldsymbol{k}} \sum_{\hat{\varepsilon}} N_{\boldsymbol{k},\hat{\varepsilon}} \tag{23.33}$$

is the map between the effective intensity per unit angular frequency and the photon occupation numbers.

The third way of treating the absorption to a discrete final state is to compute a cross section, as we have already done for the case of absorption by a continuous group of final states, in (23.22). To do so, it is necessary to realize that there will always be at least a small effective broadening of the discrete energy level of the final state, for two reasons. First, $|f\rangle$ must be able to decay, at least back to the lower-energy initial state $|i\rangle$ if not others, leading to a finite lifetime. The second reason, which is often more important in practice, is that there are environmental effects on the H_0 system from perturbations due to nearby atoms with a thermal distribution of velocities, and from fluctuations in the electromagnetic fields.

If $\mathcal{E}_{f0}$ is a nominal discrete H_0 eigenvalue before the line-broadening is taken into account, then naively $\rho(\mathcal{E}_f, u_f) = \delta(\mathcal{E}_f - \mathcal{E}_{f0})$. The line-broadening effects modify this to a Breit–Wigner resonance, with a small energy width $\hbar\gamma$ that may need to be obtained empirically. Then, when using (23.21), the density of final states is parameterized as

$$\rho(\mathcal{E}_f, u_f) = f_{\text{BW}}(\mathcal{E}_f - \mathcal{E}_{f0}, \hbar\gamma) \equiv \frac{\hbar\gamma}{2\pi} \frac{1}{(\mathcal{E}_f - \mathcal{E}_{f0})^2 + \hbar^2\gamma^2/4}. \tag{23.34}$$

Here $\hbar\gamma$ is the width of the **Breit–Wigner lineshape** (named after Gregory Breit and Wigner, and also known as the **Lorentzian lineshape**) distribution for the energy $\mathcal{E}_f$, which is now continuous. Note that the normalization is

$$\int_{-\infty}^{\infty} d\mathcal{E}_f\, f_{\text{BW}}(\mathcal{E}_f - \mathcal{E}_{f0}, \hbar\gamma) = 1, \tag{23.35}$$

so that in the narrow-width limit, we recover

$$\lim_{\gamma\to 0} f_{\mathrm{BW}}(\mathcal{E}_f - \mathcal{E}_{f0}, \hbar\gamma) = \delta(\mathcal{E}_f - \mathcal{E}_{f0}), \tag{23.36}$$

as expected [compare to (2.36)]. The absorption cross section for a nominally discrete final state $|f\rangle$ becomes, after using (23.34) in (23.21) and integrating over $d\mathcal{E}_f$,

$$\sigma^{\mathrm{abs}}_{i\to f} = \frac{2\pi\alpha\hbar^2\gamma}{m_e^2\omega} \frac{\left|\langle f|\hat{\varepsilon}\cdot\mathcal{P}_k|i\rangle\right|^2}{(\mathcal{E}_{f0} - \mathcal{E}_i - \hbar\omega)^2 + \hbar^2\gamma^2/4}. \tag{23.37}$$

Here we have removed the du_f under the assumption that the degeneracy labels for the final state, such as angular momentum quantum numbers, are also discrete and can be summed over. This cross section is a strongly peaked function of ω, becoming sharper in the limit $\gamma \to 0$.

23.3 Induced and Spontaneous Emission of Light

We now turn to the case of transitions between electronic states involving emission of energy in the form of electromagnetic radiation. The presence of a classical electromagnetic wave with angular frequency ω will cause an initial H_0 eigenstate $|i\rangle$ to fall to a state $|f\rangle$ of lower energy, provided that the H_0 eigenvalues satisfy $\mathcal{E}_i - \mathcal{E}_f = \hbar\omega$. If $dI/d\omega$ is the intensity of incoherent electromagnetic waves per unit ω, then we can apply exactly the same reasoning we used to arrive at (23.26), except that we use (22.81) for emission rather than (22.80) for absorption. The result is the rate for **induced emission** (or **stimulated emission**),

$$R^{\mathrm{ind.\,em.}}_{i\to f} = \frac{4\pi^2\alpha}{m_e^2\hbar\omega^2}\frac{dI}{d\omega}\left|\langle f|\hat{\varepsilon}^*\cdot\mathcal{P}^\dagger_k|i\rangle\right|^2, \tag{23.38}$$

where now $\omega = (\mathcal{E}_i - \mathcal{E}_f)/\hbar$ is implicit. The complex conjugation of the polarization vector has no effect for linear polarizations, but makes a difference for circular polarization vectors [such as $\hat{\varepsilon} = (\hat{x} \pm i\hat{y})/\sqrt{2}$ if $\hat{k} = \hat{z}$].

Equation (23.38) assumes that the wavevector direction $\hat{k}$ is fixed. For the case of light incoming from all directions, the transition rate can be expressed in terms of the photon occupation numbers using (23.33), again just as we did for the absorption case in arriving at (23.32),

$$R^{\mathrm{ind.\,em.}}_{i\to f} = \frac{\alpha\omega}{2\pi m_e^2 c^2}\int d\Omega_k \sum_{\hat{\varepsilon}} N_{k,\hat{\varepsilon}}\left|\langle f|\hat{\varepsilon}^*\cdot\mathcal{P}^\dagger_k|i\rangle\right|^2. \tag{23.39}$$

The reason for referring to this as "induced" is that it is proportional to the intensity of incident light, or equivalently to the numbers of photons already present.

Despite the fact that our calculations so far do not show it, the transition from $|i\rangle$ to $|f\rangle$ with the emission of a photon can also occur even if there are no incident classical waves or photons present. That process is called **spontaneous emission**, and we will now show that it yields a result equal to (23.39) but with the factor $N_{\boldsymbol{k},\hat{\varepsilon}}$ replaced by 1. The phenomenon of spontaneous emission, in which an electronic state decays to a lower energy state by emitting a photon, is the quantum mechanical analog of the classical Larmor radiation of an accelerating charge. However, it can occur only if there is a quantum state with lower energy, while classical Larmor radiation would always occur, as discussed in Sect. 1.1.

To obtain the rate for spontaneous emission in the most direct way, the electromagnetic fields should be treated quantum mechanically, rather than classically as we have done. Recall that our method so far was to treat the potentials Φ and $\boldsymbol{A}$ and their derivatives $\boldsymbol{E}$ and $\boldsymbol{B}$ as classical quantities, which are then evaluated in terms of the position and momentum operators for the electron that witnesses them. In a more logical and complete approach, both the electronic and radiation degrees of freedom should be treated as a unified quantum system. Then Φ and $\boldsymbol{A}$ will act on a Hilbert space that describes not just the electronic state, but the photons as well. This allows a direct computation of spontaneous emission (and more complicated phenomena). In this section, instead of pursuing that direct method, we will derive the results for spontaneous emission using an indirect argument, based on inference from requiring the consistency of statistical mechanics, called the **Einstein A and B coefficients** method.

Consider the thought experiment of a cavity kept in thermal equilibrium at temperature T with the electromagnetic radiation inside of it. The walls of the cavity are made of a very large number of atoms that include two states $|a\rangle$ and $|b\rangle$ with energy difference

$$\mathcal{E}_b - \mathcal{E}_a = \hbar\omega. \tag{23.40}$$

The photon absorption process $a + \hbar\omega \to b$ and the emission process $b \to a + \hbar\omega$ are both occurring constantly to ensure the equilibrium. Let $N_{\boldsymbol{k},\hat{\varepsilon}}$ be the number of photons in the mode with wavevector $\boldsymbol{k}$ (with $k = \omega/c$) and polarization $\hat{\varepsilon}$.

In equilibrium, each $N_{\boldsymbol{k},\hat{\varepsilon}}$, which we will refer to as N for short in the following, is constant. Therefore, we must have

$$\left(\frac{dN}{dt}\right)_{\text{spon. em.}} + \left(\frac{dN}{dt}\right)_{\text{ind. em.}} + \left(\frac{dN}{dt}\right)_{\text{absorption}} = 0, \tag{23.41}$$

with contributions from spontaneous emission, induced emission, and absorption rates that are respectively of the forms

$$\left(\frac{dN}{dt}\right)_{\text{spon. em.}} = n_b A, \tag{23.42}$$

$$\left(\frac{dN}{dt}\right)_{\text{ind. em.}} = N n_b B_{ba}, \tag{23.43}$$

$$\left(\frac{dN}{dt}\right)_{\text{absorption}} = -Nn_a B_{ab}, \tag{23.44}$$

where it is crucial that A, B_{ab}, and B_{ba} can, in principle, be calculated in terms of the properties of the states $|a\rangle$ and $|b\rangle$ and the interaction Hamiltonian, and so cannot depend on the temperature. In writing (23.42)–(23.44), we have used the fact that the spontaneous and induced emission contributions to the rate for creation of photons must both be proportional to the number of atoms n_b in the state $|b\rangle$, while the absorption rate contribution to the rate for subtraction of photons is proportional to the number of atoms n_a in the state $|a\rangle$. Also, the induced emission and absorption rates are both proportional to the number of photons already present N, but by definition the spontaneous emission rate has no such factor. Our goal is to determine the ratios of the coefficients of proportionality, A, B_{ba}, and B_{ab}. These are in turn proportional to the transition rates $R^{\text{spon. em.}}_{b\to a}$, $R^{\text{ind. em.}}_{b\to a}$, and $R^{\text{abs}}_{a\to b}$, respectively, but with the corresponding factors of N extracted from the last two, since we have put them explicitly in (23.43) and (23.44).

The numbers n_a and n_b are unknown, but since the system is in thermal equilibrium, we can find their ratio. According to statistical mechanics, and our discussion in Sect. 3.5, the relative probability for a state with energy E is proportional to the Boltzmann factor $e^{-E/k_B T}$, where k_B is Boltzmann's constant. So,

$$n_a/n_b = e^{-(\mathcal{E}_a-\mathcal{E}_b)/k_B T} = e^{\hbar\omega/k_B T}. \tag{23.45}$$

Using (23.42)–(23.45) in (23.41), we can solve for the photon occupation number,

$$N = \frac{A}{B_{ab}e^{\hbar\omega/k_B T} - B_{ba}}. \tag{23.46}$$

Now, since $n\hbar\omega$ is the energy of the state with n photons, we also know that the average number of photons in that mode is equal to

$$N = \left(\sum_{n=0}^{\infty} n e^{-n\hbar\omega/k_B T}\right) \Big/ \sum_{n=0}^{\infty} e^{-n\hbar\omega/k_B T} = \frac{1}{e^{\hbar\omega/k_B T} - 1}, \tag{23.47}$$

where the last expression is obtained using the same method as described after (1.18). The key point is that the only way to reconcile the functional dependences of (23.46) and (23.47) on T is to require

$$A = B_{ab} = B_{ba}. \tag{23.48}$$

The equality $B_{ab} = B_{ba}$ is just a check of the detailed balance equality between $R^{\text{abs}}_{a\to b}$ and $R^{\text{ind. em.}}_{b\to a}$, which we had already found by direct computation in (23.32) and (23.39). The new requirement that $A = B_{ba}$ says that, because N was already factored out of B_{ba}, the rate we calculated for induced emission alone is related to the total emission rate by

$$\text{induced} \to \text{induced} + \text{spontaneous} \tag{23.49}$$

$$N_{k,\hat{\varepsilon}} \to N_{k,\hat{\varepsilon}} + 1. \tag{23.50}$$

Thus, the rate for spontaneous emission is obtained by simply removing the factor of $N_{k,\hat{\varepsilon}}$ from the formula (23.39) for induced emission,

$$R_{i\to f}^{\text{spon. em.}} = \frac{\alpha\omega}{2\pi m_e^2 c^2} \int d\Omega_k \sum_{\hat{\varepsilon}} \left| \langle f | \hat{\varepsilon}^* \cdot \mathcal{P}_k^\dagger | i \rangle \right|^2. \tag{23.51}$$

A more direct calculation that treats the electromagnetic fields quantum-mechanically gives the same result.

23.4 Electric Dipole Approximation

The case in which the wavelength of the electromagnetic wave is much larger than the spatial extent of the initial-state wavefunction is called the **electric dipole approximation**. To see why this might be a valid approximation, suppose that the electron is in a hydrogen-like atomic orbital state with principal quantum number n and atomic number Z. Then the minimum photon energy needed to ionize the state is $\hbar\omega = Z^2e^2/2a_0n^2$ from (11.16), while the characteristic spatial extent of the initial wavefunction is $\langle R \rangle = 3n^2a_0/2Z$, using (11.65) with $l = 0$. For larger l in the initial state, the characteristic size $\langle R \rangle$ is even smaller (for fixed n), according to (11.65). Comparing the atomic size scale to the inverse wavenumber of the light (with minimum ionizing energy),

$$k \langle R \rangle = \frac{\langle R \rangle}{(c/\omega)} \sim \frac{3\alpha Z}{4}. \tag{23.52}$$

This is smaller than 1, even when Z is large, since $\alpha = e^2/\hbar c \approx 1/137$. For transitions between bound states, the relevant wavelength is even longer because the change in energy is smaller. As an extreme case, the wavelength is macroscopic for transitions between states with the same principal quantum number, which have very nearly the same energy. This means that in transitions between bound states we can make the long-wavelength approximation

$$e^{i\boldsymbol{k}\cdot\boldsymbol{R}} \approx 1 \tag{23.53}$$

over the relevant spatial extent of the atom. In the same limit, one can also neglect the $\boldsymbol{k} \times \boldsymbol{S}$ contribution compared to the momentum operator $\boldsymbol{P}$ in $\mathcal{P}_k$ of (23.12). This is made plausible by the rough estimate

$$\frac{|\boldsymbol{k} \times \boldsymbol{S}|}{|\boldsymbol{P}|} \sim \frac{(\omega/c)\hbar}{\hbar/\langle R\rangle} \tag{23.54}$$

(using the uncertainty relation for position and momentum in the denominator), which is parametrically of the same order as (23.52), and so again suppressed by a factor αZ.

Therefore, the electric dipole (long-wavelength) approximation is defined by taking

$$\langle f|\mathcal{P}_{\boldsymbol{k}}|i\rangle \approx \langle f|\,\boldsymbol{P}\,|i\rangle \tag{23.55}$$

in the absorption and emission rates. Now we can use a trick, by noticing that if the only momentum dependence of H_0 is in the kinetic term, then

$$\left[H_0,\, \boldsymbol{R}\right] = \frac{1}{2m_e}\left[P^2,\, \boldsymbol{R}\right] = -i\frac{\hbar}{m_e}\boldsymbol{P}. \tag{23.56}$$

So, in the case of absorption,

$$\begin{aligned}\langle f|\,\hat{\varepsilon}\cdot\boldsymbol{P}\,|i\rangle &= \frac{im_e}{\hbar}\hat{\varepsilon}\cdot\langle f|[H_0,\boldsymbol{R}]|i\rangle = \frac{im_e}{\hbar}(\mathcal{E}_f-\mathcal{E}_i)\,\langle f|\hat{\varepsilon}\cdot\boldsymbol{R}\,|i\rangle\\ &= im_e\omega\,\langle f|\hat{\varepsilon}\cdot\boldsymbol{R}\,|i\rangle\,.\end{aligned} \tag{23.57}$$

For emission, the only difference is that $\hat{\varepsilon} \to \hat{\varepsilon}^*$ and $\omega \to -\omega$, so

$$\langle f|\hat{\varepsilon}^*\cdot\boldsymbol{P}\,|i\rangle = -im_e\omega\,\langle f|\hat{\varepsilon}^*\cdot\boldsymbol{R}\,|i\rangle\,. \tag{23.58}$$

For example, using (23.57) in (23.22), we have for the absorption cross section in the case of continuous final-state energies

$$d\sigma_{i\to f}^{\text{abs, dipole}} = 4\pi^2\alpha\hbar\omega\left|\,\langle f|\hat{\varepsilon}\cdot\boldsymbol{R}\,|i\rangle\,\right|^2\rho(\mathcal{E}_f,u_f)\,du_f. \tag{23.59}$$

Recalling that $\hat{\varepsilon}$ is the direction of the polarization of the absorbed electromagnetic wave, we see that the relevant matrix element in the electric dipole approximation is proportional to the position operator along the polarization direction. More generally, the electric dipole approximation amounts to making the substitutions

$$|\,\langle f|\hat{\varepsilon}\cdot\mathcal{P}_{\boldsymbol{k}}|i\rangle\,|^2 \to m_e^2\omega^2\,|\,\langle f|\hat{\varepsilon}\cdot\boldsymbol{R}\,|i\rangle\,|^2 \qquad \text{(electric dipole absorption)} \tag{23.60}$$

in the absorption rates for discrete final states with incoherent plane-wave light in (23.26), incoherent light from all directions in (23.32), and in the cross section for a line-broadened discrete final state in (23.37). Similarly, one makes the replacement

$$|\,\langle f|\hat{\varepsilon}^*\cdot\mathcal{P}_{\boldsymbol{k}}^\dagger\,|i\rangle\,|^2 \to m_e^2\omega^2\,|\,\langle f|\hat{\varepsilon}^*\cdot\boldsymbol{R}\,|i\rangle\,|^2 \qquad \text{(electric dipole emission)} \tag{23.61}$$

in the induced emission rates (23.38) for plane waves with a fixed direction of propagation and (23.39) for light from all directions in terms of the photon occupation numbers, and in (23.51) for spontaneous emission.

In the case of spontaneous emission in (23.51), the matrix element $\langle f|\hat{\varepsilon}^* \cdot \boldsymbol{R}|i\rangle$ in the electric dipole approximation does not depend directly on the wavevector magnitude, but does depend indirectly on the wavevector direction because of the requirement $\hat{\varepsilon}^* \cdot \hat{k} = 0$. Temporarily fixing $\hat{k} = \hat{z}$, the two independent polarization directions to be summed over can be taken along $\hat{x}$ and $\hat{y}$, so that

$$\sum_{\hat{\varepsilon}} \left| \langle f|\hat{\varepsilon}^* \cdot \boldsymbol{R}\,|i\rangle \right|^2 = \left| \langle f|X|i\rangle \right|^2 + \left| \langle f|Y|i\rangle \right|^2. \tag{23.62}$$

Then, integrating over all possible directions for $\hat{k}$ will effectively give an average of the two terms over three possible directions, so

$$\int d\Omega_{\boldsymbol{k}} \sum_{\hat{\varepsilon}} \left| \langle f|\hat{\varepsilon}^* \cdot \boldsymbol{R}\,|i\rangle \right|^2 = (4\pi)\frac{2}{3}\left(\left| \langle f|X|i\rangle \right|^2 + \left| \langle f|Y|i\rangle \right|^2 + \left| \langle f|Z|i\rangle \right|^2 \right) \tag{23.63}$$

$$= \frac{8\pi}{3} \langle f|\,\boldsymbol{R}\,|i\rangle \cdot \langle i|\,\boldsymbol{R}\,|f\rangle\,. \tag{23.64}$$

This reduces (23.51) to a simple formula for the rate for an initial state $|i\rangle$ to decay to a final state $|f\rangle$ by the spontaneous emission of a photon,

$$R_{i\to f}^{\text{spon. em.}} = \frac{4\alpha\omega^3}{3c^2} \left| \langle f|\,\boldsymbol{R}\,|i\rangle \right|^2. \tag{23.65}$$

Note that the right side involves

$$\left| \langle f|\,\boldsymbol{R}\,|i\rangle \right|^2 \equiv \langle f|\,\boldsymbol{R}\,|i\rangle \cdot \langle i|\,\boldsymbol{R}\,|f\rangle\,, \tag{23.66}$$

the square of the magnitude of a vector matrix element.

Suppose that there is no background electromagnetic radiation to induce emission or absorption, and let the number of atoms in the state $|i\rangle$ be $N_i(t)$ at time t. Then at time $t + \Delta t$,

$$N_i(t + \Delta t) = N_i(t) - \Delta t\, R_i N_i(t), \tag{23.67}$$

where the total decay rate of the initial state is

$$R_i = \sum_f R_{i\to f}^{\text{spon. em.}}. \tag{23.68}$$

Then

$$\frac{N_i(t+\Delta t)-N_i(t)}{\Delta t} = -R_i N_i(t), \tag{23.69}$$

which in the limit $\Delta t \to 0$ becomes the differential equation

$$\frac{dN_i}{dt} = -R_i N_i, \tag{23.70}$$

with the solution

$$N_i(t) = N_i(t_0)e^{-(t-t_0)R_i}. \tag{23.71}$$

Therefore $\tau_i = 1/R_i$ is the lifetime of the state $|i\rangle$, defined as the amount of time needed for the number of atoms in that state to decrease by a factor of $1/e \approx 0.367879$ in the absence of other perturbations. The half-life, defined as the amount of time required for the number to decrease by half, is related to this by $t_{1/2} = \tau_i \ln 2 \approx 0.693147/R_i$.

The preceding results imply matrix element selection rules that govern both absorption and emission in the electric dipole approximation when $|i\rangle$ and $|f\rangle$ are angular momentum eigenstates. These follow immediately from (13.54)–(13.57). First, (13.54) tells us that the change in the orbital angular momentum quantum number l must be exactly 1 unit, $\Delta l = l_f - l_i = \pm 1$. As a corollary, the parities of the initial and final states must be opposite, since the parity of a state with orbital angular momentum l is $(-1)^l$ from (8.179); this can also be seen directly from the parity selection rule, since the operator $\boldsymbol{R}$ has odd parity. Also, (13.55)–(13.57) tell us that the z-component of the orbital angular momentum changes by 1 unit or less: $\Delta m_l = m_{l,f} - m_{l,i} = 0, \pm 1$. As special cases, if the polarization of the wave is along $\hat{\varepsilon} = \hat{z}$, then (13.55) says $\Delta m_l = 0$, and if it is in the x, y plane, then (13.56)–(13.57) demand $\Delta m_l = \pm 1$. The spin is not affected by the position operator, so that $m_{s,f} = m_{s,i}$.

We therefore have the **electric dipole selection rules** for single-electron transitions,

$$\Delta l = \pm 1, \quad \Delta m_l = 0, \pm 1, \quad \Delta m_s = 0, \qquad \text{(electric dipole).} \tag{23.72}$$

These generalize to the case of multi-electron transitions (in which more than one electron changes orbitals in going from the initial to the final state), for which the electric dipole selection rules are

$$\pi_i \pi_f = -1, \quad \Delta J = 0, \pm 1, \quad \Delta m_J = 0, \pm 1, \quad J_i + J_f \neq 0, \quad \text{(electric dipole).} \tag{23.73}$$

The first three of these follow from requiring the non-vanishing of the matrix element of the parity-odd electric dipole operator $\hat{\varepsilon} \cdot \boldsymbol{R}$. The ΔJ rule follows from the fact that $\boldsymbol{R}$ is a vector operator, and therefore carries total angular momentum 1 in the sense explained in Chapter 13. In terms of addition of angular momentum, $J \otimes 1 = (J-1) \oplus J \oplus (J+1)$ for $J \geq 1$.

The last selection rule in (23.73) is completely general, even if we do not make the electric dipole approximation. It says just this: a total angular momentum $J = 0$ state can never go to another $J = 0$ state by absorbing or emitting a single photon. The reason is that a single photon can always be written as a linear combination of transverse circular polarizations, carrying intrinsic angular momentum component along the propagation direction with eigenvalue either $+\hbar$ or $-\hbar$, which cannot be matched by the initial and final electronic states if they both have no angular momentum. Therefore, we have a general rule

$$J_i + J_f \neq 0, \qquad \text{(all single-photon transitions).} \tag{23.74}$$

However, this rule (and all other selection rules listed here) can be violated if one considers emission or absorption of multiple photons, either as distinct sequential events involving intermediate states, or by going beyond first order in time-dependent perturbation theory or through perturbations caused by the environment of the atom. Multi-photon transition rates are therefore suppressed by additional powers of the fine-structure constant, and so can often be neglected when they compete with single-photon transitions.

In the approximation of the LS coupling scheme, the total spin, total orbital, and total angular momentum orthobasis quantum numbers L, S, J, m_J are good quantum numbers for the initial and final stationary states. Then, because the total spin does not appear at all in the electric dipole moment operator, one has the additional rules

$$\Delta S = 0, \quad \Delta L = 0, \pm 1, \quad L_i + L_f \neq 0, \quad \text{(electric dipole, } LS \text{ coupling).} \tag{23.75}$$

Recall, from the discussion in Sect. 18.4, that the LS coupling scheme holds to a good approximation for atoms with atomic number Z not too large.

As an example, consider the lifetimes of the $n = 2$ states of the hydrogen atom within the electric dipole approximation, neglecting fine-structure and hyperfine effects. The only kinematically allowed final state is the ground state with $n = 1$ and $l = 0$, so the $(n, l, m) = (2, 0, 0)$ state cannot decay by single photon emission in this approximation, since it would violate the $\Delta l = \pm 1$ electric dipole selection rule. For the $(2, 1, m_l)$ states, we can apply (23.65). The angular frequency of the emitted photon is determined from

$$\hbar\omega = E_{n=2} - E_{n=1} = \left(\frac{e^2}{2a_0}\right)\left(-\frac{1}{4} + 1\right) = 10.2 \text{ eV}, \tag{23.76}$$

which gives $\omega = 1.55 \times 10^{16}\,\mathrm{s}^{-1}$. The matrix elements $\langle f|\boldsymbol{R}|i\rangle$ were the subject of Exercise 11.4. The results can be summarized as

$$\langle 210|\boldsymbol{R}|100\rangle = \bar{a}\hat{z}, \tag{23.77}$$

$$\langle 211|\boldsymbol{R}|100\rangle = \bar{a}(-\hat{x} - i\hat{y})/\sqrt{2}, \tag{23.78}$$

$$\langle 21{-}1|\boldsymbol{R}|100\rangle = \bar{a}(\hat{x} - i\hat{y})/\sqrt{2}, \tag{23.79}$$

where $\bar{a} = 2^{15/2}a_0/3^5$. So, in all three cases, $|\langle f|\boldsymbol{R}|i\rangle|^2 = \bar{a}^2$. Plugging into (23.65) gives

$$R_{2p\to 1s} = \frac{131072}{177147}\frac{\alpha\omega^3 a_0^2}{c^2} = 6.25 \times 10^8\,\mathrm{s}^{-1}, \tag{23.80}$$

yielding a lifetime $\tau_{2p\to 1s} = 1/R_{2p\to 1s} = 1.60 \times 10^{-9}$ s.

23.5 Magnetic Dipole, Electric Quadrupole, and Higher Orders

Transitions that violate the electric dipole selection rules of (23.72) and (23.73) have rates that are typically suppressed by extra powers of $Z\alpha$, or by extra inverse powers of the wavelength of the photon emitted or absorbed. For this reason, they are often referred to as "forbidden". Traditionally, this terminology does not mean that they are literally forbidden, but rather that they take place at highly suppressed rates. By the same terminology, transitions that take place through the electric dipole operator are often called "allowed".

To go beyond the electric dipole approximation and its selection rules, consider the small k expansion of the operator $e^{i\boldsymbol{k}\cdot\boldsymbol{R}}\,(\boldsymbol{P} - i\boldsymbol{k} \times \boldsymbol{S})$ in (23.12). To make things a little simpler, let us temporarily choose the wavevector of the light as $\boldsymbol{k} = k\hat{z}$ and the polarization as $\hat{\varepsilon} = \hat{x}$, so

$$\hat{\varepsilon}\cdot\mathcal{P}_{\boldsymbol{k}} = e^{i\boldsymbol{k}\cdot\boldsymbol{R}}\,\hat{\varepsilon}\cdot(\boldsymbol{P} - i\boldsymbol{k} \times \boldsymbol{S}) = e^{ikZ}(P_x + ikS_y). \tag{23.81}$$

Now, expanding $e^{ikZ} \approx 1 + ikZ + \cdots$, and using $ZP_x = (ZP_x - XP_z)/2 + (ZP_x + XP_z)/2 = L_y/2 + (ZP_x + P_zX)/2$, we have

$$\hat{\varepsilon}\cdot\mathcal{P}_{\boldsymbol{k}} = P_x + i\frac{\omega}{2c}\left[L_y + 2S_y + (ZP_x + P_zX)\right] + \cdots . \tag{23.82}$$

Recasting this back into a form valid for general orthogonal $\hat{k}$ and $\hat{\varepsilon}$ gives

$$\hat{\varepsilon}\cdot\mathcal{P}_{\boldsymbol{k}} = \hat{\varepsilon}\cdot\boldsymbol{P} + i\frac{\omega}{2c}\left[(\hat{k}\times\hat{\varepsilon})\cdot(\boldsymbol{L}+2\boldsymbol{S}) + (\hat{k}\cdot\boldsymbol{R}\,\hat{\varepsilon}\cdot\boldsymbol{P} + \hat{k}\cdot\boldsymbol{P}\,\hat{\varepsilon}\cdot\boldsymbol{R})\right] + \cdots, \tag{23.83}$$

which is the result to next-to-leading order in the long-wavelength expansion. In the last term, the ordering of operators $\boldsymbol{R}$ and $\boldsymbol{P}$ does not matter; their commutator does not contribute because of the transversality condition $\hat{\varepsilon} \cdot \hat{k} = 0$.

Therefore, for the harmonic perturbation appearing in (22.61) and subsequent equations, we can write

$$V = V_{\mathrm{E1}} + V_{\mathrm{M1}} + V_{\mathrm{E2}} + \cdots , \tag{23.84}$$

where the separated contributions,

$$V_{\mathrm{E1}} = -\frac{eE_0}{2m_e\omega}\hat{\varepsilon} \cdot \boldsymbol{P}, \tag{23.85}$$

$$V_{\mathrm{M1}} = -i\frac{eE_0}{4m_e c}(\hat{k} \times \hat{\varepsilon}) \cdot (\boldsymbol{L} + 2\boldsymbol{S}), \tag{23.86}$$

$$V_{\mathrm{E2}} = -i\frac{eE_0}{4m_e c}(\hat{k} \cdot \boldsymbol{R}\, \hat{\varepsilon} \cdot \boldsymbol{P} + \hat{k} \cdot \boldsymbol{P}\, \hat{\varepsilon} \cdot \boldsymbol{R}) \tag{23.87}$$

are called electric dipole (E1), magnetic dipole (M1), and electric quadrupole (E2), respectively. Note that in the magnetic dipole operator, the vector $(\hat{k} \times \hat{\varepsilon})$ is the unit vector along the magnetic field direction in the wave; the magnetic field couples to the combination $\boldsymbol{L} + 2\boldsymbol{S}$.

Single-photon transitions that are dominantly magnetic dipole in character can now be evaluated from the general formulas in Sects. 23.2–23.3, by using

$$\left| \langle f | \hat{\varepsilon} \cdot \mathcal{P}_{\boldsymbol{k}} | i \rangle \right|^2 \to \frac{\omega^2}{4c^2} \left| \langle f | (\hat{k} \times \hat{\varepsilon}) \cdot (\boldsymbol{L} + 2\boldsymbol{S}) | i \rangle \right|^2 \qquad \text{(M1 absorption)}, \tag{23.88}$$

$$\left| \langle f | \hat{\varepsilon}^* \cdot \mathcal{P}_{\boldsymbol{k}}^\dagger | i \rangle \right|^2 \to \frac{\omega^2}{4c^2} \left| \langle f | (\hat{k} \times \hat{\varepsilon}^*) \cdot (\boldsymbol{L} + 2\boldsymbol{S}) | i \rangle \right|^2 \qquad \text{(M1 emission)}. \tag{23.89}$$

For the electric quadrupole term, we can do a trick similar to the one we used in the electric dipole case to get (23.60) and (23.61). Assuming that H_0 does not depend on the momentum operator except through the kinetic term $P^2/2m_e$, we find from (23.56) that

$$\left[H_0,\, \hat{k} \cdot \boldsymbol{R}\, \hat{\varepsilon} \cdot \boldsymbol{R}\right] = -i\frac{\hbar}{m_e}\left(\hat{k} \cdot \boldsymbol{R}\, \hat{\varepsilon} \cdot \boldsymbol{P} + \hat{k} \cdot \boldsymbol{P}\, \hat{\varepsilon} \cdot \boldsymbol{R}\right), \tag{23.90}$$

where we have made use of $\hat{\varepsilon} \cdot \hat{k} = 0$. Therefore,

$$V_{\mathrm{E2}} = \frac{eE_0}{4\hbar c}\left[H_0,\, \hat{k} \cdot \boldsymbol{R}\, \hat{\varepsilon} \cdot \boldsymbol{R}\right], \tag{23.91}$$

and in the matrix elements, one can evaluate H_0 on the adjacent ket $|i\rangle$ or bra $\langle f|$, giving a factor of $\mathcal{E}_f - \mathcal{E}_i = \hbar\omega$ for absorption and $-\hbar\omega$ for emission. It follows that in matrix elements for single-photon absorption, one can make the replacement[2]

$$V_{\text{E2}} \rightarrow \frac{eE_0\omega}{4c}\hat{k}\cdot\overleftrightarrow{Q}\cdot\hat{\varepsilon}, \tag{23.92}$$

where the symmetric, traceless rank-2 tensor quadrupole moment operator is

$$Q_{ab} = R_a R_b - \delta_{ab}R^2/3, \tag{23.93}$$

for $a, b = x, y, z$. This uses the fact that the δ_{ab} part of Q_{ab} does not contribute to V_{E2} because of $\hat{\varepsilon}\cdot\hat{k} = 0$. Within matrix elements for single-photon emission,

$$V_{\text{E2}}^{\dagger} \rightarrow -\frac{eE_0\omega}{4c}\hat{k}\cdot\overleftrightarrow{Q}\cdot\hat{\varepsilon}^* \tag{23.94}$$

similarly applies. Comparing (23.92) and (23.94) to (23.15), we see that for dominantly electric quadrupole transitions, the general formulas in Sects. 23.2–23.3 apply with

$$\left|\langle f|\,\hat{\varepsilon}\cdot\mathcal{P}_k\,|i\rangle\right|^2 \rightarrow \frac{m_e^2\omega^4}{4c^2}\left|\langle f|\,\hat{k}\cdot\overleftrightarrow{Q}\cdot\hat{\varepsilon}\,|i\rangle\right|^2 \qquad \text{(E2 absorption)}, \tag{23.95}$$

$$\left|\langle f|\,\hat{\varepsilon}^*\cdot\mathcal{P}_k^{\dagger}\,|i\rangle\right|^2 \rightarrow \frac{m_e^2\omega^4}{4c^2}\left|\langle f|\,\hat{k}\cdot\overleftrightarrow{Q}\cdot\hat{\varepsilon}^*\,|i\rangle\right|^2 \qquad \text{(E2 emission)}, \tag{23.96}$$

for single-photon absorption and emission, respectively.

Let us now work out the selection rules for M1 and E2 transitions implied by the preceding results. First, let us use (23.88) and (23.89) to find the requirements on the initial and final states in order for a magnetic dipole transition to occur. In transitions involving a single electron, neither $\boldsymbol{L}$ nor $\boldsymbol{S}$ changes the orbital quantum number l, and the spin quantum number s is of course fixed at $1/2$. However, the components of $\boldsymbol{L}$ are linear combinations of L_z, $L_\pm$, and the components of $\boldsymbol{S}$ are linear combinations of S_z, $S_\pm$, so they can change m_l and m_s by zero or one unit. Furthermore, magnetic dipole transitions cannot occur between states with different principal (radial) quantum numbers, because $\boldsymbol{L}$ and $\boldsymbol{S}$ do not operate on the radial wavefunctions and do not change l, so the matrix element calculation in the position representation will be proportional to

$$\int_0^\infty dr\, r^2 R_{nl}(r)R_{n'l}(r) = \delta_{nn'}. \tag{23.97}$$

[2] For two vectors $\boldsymbol{v}$ and $\boldsymbol{w}$ with components v_a and w_a, and a rank-2 tensor $\overleftrightarrow{T}$ with components T_{ab}, the notation $\boldsymbol{v}\cdot\overleftrightarrow{T}\cdot\boldsymbol{w}$ means $\sum_{a=x,y,z}\sum_{b=x,y,z} v_a T_{ab} w_b$. We have already encountered the quadrupole moment operator Q_{ab} in (13.29)–(13.32).

Therefore, the **magnetic dipole selection rules** are, for single-electron transitions,

$$\Delta n = 0, \qquad \Delta l = 0, \qquad \Delta m_l = 0, \pm 1, \qquad \text{(magnetic dipole).} \tag{23.98}$$

For multi-electron transitions, the selection rules from (23.88) and (23.89) are

$$\pi_i \pi_f = 1, \quad \Delta J = 0, \pm 1, \quad \Delta m_J = 0, \pm 1, \quad J_i + J_f \neq 0, \quad \text{(magnetic dipole).} \tag{23.99}$$

The parity selection rule follows from the fact that the operator $\boldsymbol{L} + 2\boldsymbol{S}$ has even parity, and the ΔJ and Δm_J selection rules follow from the fact that it is a vector, and therefore as explained in Chapter 13 it carries total angular momentum 1. The rule $J_i + J_f \neq 0$ is completely general, as we already noted in our discussion of the electric dipole case. In the LS coupling scheme approximation, where initial and final stationary states can be assigned total angular momentum basis quantum numbers L, S, J, m_J, there are additional rules

$$\Delta L = 0, \qquad \Delta S = 0 \qquad \text{(magnetic dipole, } LS \text{ coupling).} \tag{23.100}$$

Note that the electric dipole and magnetic dipole contributions cannot both be relevant for a given initial and final state. Magnetic dipole transitions never change the parity, while electric dipole transitions always do.

The 21 centimeter hyperfine transition line within the ground state of hydrogen, already discussed in Sects. 12.2 and 17.2, is a magnetic dipole transition, as it involves a spin flip but does not change the orbital angular momentum or principal quantum number, which are $l = 0$ and $n = 1$ in both the initial and final state. Since the resulting photon has an extremely long wavelength compared to the atomic size, the hyperfine transition occurs at a highly suppressed rate (compared to what one might expect if it were an electric dipole transition), because $k = \omega/c \ll 1/a_0$. The rate is calculated in Exercise 23.3.

Now consider the matrix element selection rules that apply to the electric quadrupole term. Depending on the choices for $\hat{k}$ and $\hat{\varepsilon}$, the matrix elements in (23.95) and (23.96) can involve linear combinations of the operators XY, XZ, YZ, $X^2 - Y^2$, and $X^2 + Y^2 - 2Z^2$, which are the components of the quadrupole tensor $\overleftrightarrow{Q}$. These operators all have parity $\pi_Q = +1$, so electric quadrupole transitions can only connect states with the same parity. Acting on single-electron orbital angular momentum eigenstates, these operators can change l by 0 or 2 units (but not 1 unit, as that would not satisfy the parity selection rule), and m_l by 0, 1, or 2 units. Unlike the magnetic dipole interaction, the electric quadrupole operator does act nontrivially on the radial wavefunction, so there is no restriction on changes in the principal quantum number. However, it does not change the spin state. Finally, it acts on $l = 0$ states to give $l = 2$ states, so it cannot connect two $l = 0$ states. So, we have the **electric quadrupole selection rules** for single-electron transitions,

$$\Delta l = 0, \pm 2, \quad \Delta m_l = 0, \pm 1, \pm 2, \quad \Delta m_s = 0,$$
$$l_i + l_f \neq 0, \quad \text{(electric quadrupole).} \tag{23.101}$$

For transitions involving multiple electrons, the selection rules are

$$\pi_i \pi_f = 1, \quad |\Delta J| \leq 2, \quad |\Delta m_J| \leq 2, \quad J_i + J_f > 1, \quad \text{(electric quadrupole),} \tag{23.102}$$

and in the LS coupling scheme approximation in which the initial and final multi-electron stationary states have good quantum numbers L, S, J, m_J, there are additional selection rules

$$\Delta S = 0, \quad |\Delta L| \leq 2, \quad L_i + L_f > 1 \quad \text{(electric quadrupole, } LS \text{ coupling).} \tag{23.103}$$

The general constraint that angular momentum can change by at most two units in electric quadrupole transitions follows from the Wigner–Eckart theorem 13.4 and the fact that the quadrupole operator is an irreducible tensor of order $j = 2$, as we noted in Sect. 13.1. The electric quadrupole transitions can occur only between states that have the same parity, so they can never occur if electric dipole transitions are allowed. If the initial and final states in question have $\Delta l = 0$ or $\Delta L = 0$, then the transition can be both magnetic dipole and electric quadrupole in character, but if $\Delta L = 2$ or $\Delta J = 2$ or it changes m_l or m_J by two units, then it will be electric quadrupole at leading order in the long-wavelength approximation.

If one continues the expansion in $\boldsymbol{k} \cdot \boldsymbol{R}$ of $\mathcal{P}_{\boldsymbol{k}}$, the resulting electric and magnetic multipole contributions are called Eq and Mq with $q = 1, 2, 3, \ldots$, where M2 is magnetic quadrupole, E3 is electric octupole, etc. The general selection rules for these contributions are

$$\pi_i \pi_f = (-1)^q, \qquad \text{(E}q\text{),} \tag{23.104}$$
$$\pi_i \pi_f = (-1)^{q-1}, \qquad \text{(M}q\text{),} \tag{23.105}$$
$$|\Delta J| \leq q \leq J_i + J_f, \qquad |\Delta m_J| \leq q, \qquad \text{(E}q\text{) and (M}q\text{).} \tag{23.106}$$

In the long wavelength (small k) limit, the leading Eq or Mq contributions for a transition between two states are those with the smallest q that satisfies these selection rules. For example, in a possible transition between an odd-parity state with $J = 2$ and an even-parity state with $J = 0$, the single-photon transition would be uniquely determined to be M2, magnetic quadrupole. For two states with the same parity and both with $J = 1$, the transition could be either M1 or E2. For two states both with $J = 0$, there is no allowed single-photon transition, as we have already discussed, and there will instead be a highly suppressed two-photon emission.

23.6 Photoelectric Effect for Atoms

In this section, we will compute the photoelectric-effect cross section for atoms, as an application of the results obtained in Sect. 23.2. Instead of considering electrons in a metal, we will take the initial electron to occupy a hydrogen-like atomic bound state $|i\rangle$. The light to be absorbed has angular frequency ω, polarization $\hat{\varepsilon}$, and wavevector $\boldsymbol{k} = \hat{k}\omega/c$. The final state $|f\rangle$ for the ejected electron is taken to be a continuum momentum eigenstate with eigenvalue $\boldsymbol{p}_f = \hbar\boldsymbol{K}$. As an approximation, we are neglecting the influence of the electric field of the nucleus on the final-state wavefunction. This approximation makes the problem more tractable, and works best in the limit that the final-state electron energy $\hbar^2 K^2/2m_e$ is large compared to the binding energy of the initial state. Since we are not assuming that the wavelength of the electromagnetic wave is large compared to the de Broglie wavelength $2\pi/K$ of the final state, we will *not* use the multipole expansion of the previous two sections in the following derivation.

To be specific, let the initial state of the electron have a hydrogen-like atomic ground-state wavefunction

$$\langle \boldsymbol{r}\,|i\rangle = \psi_{1,0,0}(\boldsymbol{r}) = \sqrt{\frac{Z^3}{\pi a_0^3}}\, e^{-Zr/a_0}, \tag{23.107}$$

which also applies approximately for $1s$ electrons in multi-electron atoms, with

$$\mathcal{E}_i = -Z^2e^2/2a_0. \tag{23.108}$$

The final-state electronic wavefunction is approximated by

$$\langle \boldsymbol{r}|f\rangle = \psi_{\boldsymbol{K}}(\boldsymbol{r}) = Ae^{i\boldsymbol{K}\cdot\boldsymbol{r}}, \tag{23.109}$$

where A is a normalization constant and the momentum and energy are

$$\boldsymbol{p}_f = \hbar\boldsymbol{K}, \qquad \text{and} \qquad \mathcal{E}_f = \frac{\hbar^2K^2}{2m_e}. \tag{23.110}$$

The delta function in (23.21) then requires that the photon energy is

$$\hbar\omega = \hbar ck = \frac{\hbar^2K^2}{2m_e} + \frac{Z^2e^2}{2a_0}. \tag{23.111}$$

This shows that there is a minimum possible value $\hbar\omega_{\min} = Z^2e^2/2a_0$ in order for the process to take place, achieved when the kinetic energy of the final-state electron is small. This is in agreement with the observation that the photoelectric effect requires a minimum incident photon energy quantum $\hbar\omega$, regardless of the intensity of the light, as noted in Sect. 1.3.

However, our calculation will work best in the opposite limit of large K, so that the plane-wave approximation of (23.109) is more nearly valid. In that limit, we have $\hbar^2 K^2/2m_e \gg Z^2 e^2/2a_0$, which can be rewritten as $K \gg Z/a_0$ by using $a_0 = \hbar^2/m_e e^2$. This also implies $\hbar\omega \gg Z^2 e^2/2a_0$, which can be rewritten as $\omega/c \gg Z^2\alpha/2a_0$. The velocity of the ejected electron is then

$$v_f = \frac{p_f}{m_e} = \frac{\hbar K}{m_e} \approx \frac{2\omega}{K}, \tag{23.112}$$

where we used (23.111) with the approximation that the binding energy is small compared to the photon energy. Therefore, since we will be treating the electron as nonrelativistic $v_f \ll c$, we need $\omega \ll cK$. So, the following results will be most reliable when

$$K \gg \omega/c \gg Z^2\alpha/2a_0, \qquad \text{and} \qquad K \gg Z/a_0. \tag{23.113}$$

The first condition allows us to neglect the spin contribution $\boldsymbol{k} \times \boldsymbol{S} = (\omega/c)\hat{k} \times \boldsymbol{S}$ in (23.12) compared to the $\boldsymbol{P}$ term.

In order to have a well-defined normalization factor A in (23.109), the final state is taken to be confined to a very large cubic box with side $L \gg a_0/Z$; as a check, we will show that the dependence on L cancels in the final result. Then the normalization condition is

$$1 = \int_{\text{box}} d^3\boldsymbol{r}\ |\psi_{\boldsymbol{K}}(\boldsymbol{r})|^2 = |A|^2 \int_{\text{box}} d^3\boldsymbol{r} = |A|^2 L^3, \tag{23.114}$$

so $A = L^{-3/2}$. Imposing periodic boundary conditions on the wavefunction implies that

$$\psi_{\boldsymbol{K}}(\boldsymbol{r}) = \psi_{\boldsymbol{K}}(\boldsymbol{r} + L[n_x\hat{x} + n_y\hat{y} + n_z\hat{z}]), \tag{23.115}$$

for integers n_x, n_y, n_z, so we obtain the allowed wavevector components

$$(K_x, K_y, K_z) = \frac{2\pi}{L}(n_x, n_y, n_z). \tag{23.116}$$

We will need the density of final states, $\rho(\mathcal{E}_f, \cos\theta_f, \phi_f)$, as defined by (22.52), where (θ_f, ϕ_f) are the spherical coordinate angles of the final-state electron's momentum vector $\boldsymbol{K}$, and serve as the degeneracy labels for the energy. For a volume of $\boldsymbol{K}$-space $dK_x dK_y dK_z$, the number of allowed final states is

$$\left(\frac{L}{2\pi}\right)^3 dK_x dK_y dK_z = \left(\frac{L}{2\pi}\right)^3 d^3\boldsymbol{K} = \left(\frac{L}{2\pi}\right)^3 K^2 dK\, d\phi_f\, d(\cos\theta_f). \tag{23.117}$$

Now, since

$$d\mathcal{E}_f = \frac{\hbar^2}{m_e} K\, dK, \tag{23.118}$$

we have, not counting the spin degree of freedom,

$$\begin{pmatrix} \text{number of states in the} \\ \text{range } d\mathcal{E}_f\, d\phi_f\, d(\cos\theta_f) \end{pmatrix} = \left(\frac{L}{2\pi}\right)^3 \frac{m_e}{\hbar^2} K\, d\mathcal{E}_f\, d\phi_f\, d(\cos\theta_f), \tag{23.119}$$

from which we obtain, by comparison with (22.52), the density of final electron states

$$\rho(\mathcal{E}_f, \cos\theta_f, \phi_f) = \left(\frac{L}{2\pi}\right)^3 \frac{m_e K}{\hbar^2} \tag{23.120}$$

in the plane-wave approximation.

We are assuming that the matrix element does not depend on spin, due to (23.113), so (23.12) becomes $\mathcal{P}_{\boldsymbol{k}} = e^{i\boldsymbol{k}\cdot\boldsymbol{R}}\hat{\varepsilon}\cdot\boldsymbol{P}$. Applying (23.22), we now obtain the differential cross section for absorption of the photon,

$$\frac{d\sigma}{d\Omega_f} = \frac{4\pi^2\alpha K}{m_e\hbar\omega}\left(\frac{L}{2\pi}\right)^3 \left|\langle f|e^{i\boldsymbol{k}\cdot\boldsymbol{R}}\hat{\varepsilon}\cdot\boldsymbol{P}\,|i\rangle\right|^2, \tag{23.121}$$

where $d\Omega_f = d(\cos\theta_f)\,d\phi_f$. It remains to evaluate the matrix element. Because of the transverse nature of electromagnetic waves $\hat{\varepsilon}\cdot\boldsymbol{k} = 0$, we have the commutator $\left[\boldsymbol{k}\cdot\boldsymbol{R},\, \hat{\varepsilon}\cdot\boldsymbol{P}\right] = 0$, so

$$\langle f|e^{i\boldsymbol{k}\cdot\boldsymbol{R}}\hat{\varepsilon}\cdot\boldsymbol{P}\,|i\rangle = \langle f|\hat{\varepsilon}\cdot\boldsymbol{P}e^{i\boldsymbol{k}\cdot\boldsymbol{R}}\,|i\rangle = \hbar\,\hat{\varepsilon}\cdot\boldsymbol{K}\,\langle f|e^{i\boldsymbol{k}\cdot\boldsymbol{R}}|i\rangle\,, \tag{23.122}$$

where we used $\boldsymbol{p}_f = \hbar\boldsymbol{K}$. The last matrix element is evaluated in the position representation as

$$\begin{aligned}\langle f|e^{i\boldsymbol{k}\cdot\boldsymbol{R}}|i\rangle &= \int d^3\boldsymbol{r}\left(\frac{1}{\sqrt{L^3}}e^{-i\boldsymbol{K}\cdot\boldsymbol{r}}\right)e^{i\boldsymbol{k}\cdot\boldsymbol{r}}\left(\sqrt{\frac{Z^3}{\pi a_0^3}}e^{-Zr/a_0}\right) \\ &= \sqrt{\frac{Z^3}{\pi a_0^3 L^3}}\,I(q),\end{aligned} \tag{23.123}$$

where we have defined the integral

$$I(q) = \int d^3\boldsymbol{r}\, e^{-Zr/a_0}e^{i\boldsymbol{q}\cdot\boldsymbol{r}} \tag{23.124}$$

with

$$q \equiv k - K, \tag{23.125}$$

so that $\hbar q$ is the momentum transferred from the photon to the atom.

Due to the spherical symmetry of the initial state wavefunction, we see that $I(q)$ only depends on the magnitude of q, not its direction. So, for the purpose of evaluating $I(q)$, we can temporarily choose $q = q\hat{z}$, so that $q \cdot r = qr\cos\theta$, giving

$$I(q) = \int_0^\infty dr\, r^2 \int_{-1}^{1} d(\cos\theta) \int_0^{2\pi} d\phi\; e^{-r(Z/a_0 - iq\cos\theta)}. \tag{23.126}$$

The ϕ integration trivially gives a factor of 2π. Doing the $\cos\theta$ integration next, and the r integration last,

$$\begin{aligned} I(q) &= -\frac{2\pi i}{q} \int_0^\infty dr\, r \left(e^{-r(Z/a_0 - iq)} - e^{-r(Z/a_0 + iq)} \right) \\ &= \frac{8\pi a_0^3}{Z^3(1 + q^2 a_0^2/Z^2)^2}. \end{aligned} \tag{23.127}$$

Putting together (23.121)–(23.123) and (23.127) we arrive at

$$\frac{d\sigma}{d\Omega_f} = \frac{32\alpha\hbar K}{m_e\omega} \left|\hat{\varepsilon} \cdot K\right|^2 \frac{a_0^3}{Z^3(1 + q^2 a_0^2/Z^2)^4}, \tag{23.128}$$

in which the dependence on the size of the box L (from the density of states, and from the matrix element) has canceled, as promised. Note that the differential cross section is largest when the electron is ejected in the direction of the polarization of the electric field of the wave.

It is now convenient to choose a coordinate system in which (without loss of generality, since the initial electron wavefunction is spherically symmetric) the incident photon momentum is in the $\hat{z}$ direction, and has polarization along the $\hat{x}$ direction, so

$$k = \hat{z}k = \hat{z}\omega/c, \qquad \hat{\varepsilon} = \hat{x}. \tag{23.129}$$

Writing, from here on, $\theta = \theta_f$ and $\phi = \phi_f$ for the spherical coordinate direction angles of the ejected electron's momentum, with $K = \hat{r}K$, we have

$$\hat{\varepsilon} \cdot K = K\sin\theta\cos\phi, \tag{23.130}$$

$$q^2 = (k - K)^2 = K^2 + (\omega/c)^2 - 2K(\omega/c)\cos\theta, \tag{23.131}$$

so that (23.128) becomes

$$\frac{d\sigma}{d\Omega} = \frac{32\alpha\hbar}{m_e\omega}\frac{a_0^3K^3}{Z^3(N_1 - N_2\cos\theta)^4}\sin^2\theta\cos^2\phi, \tag{23.132}$$

in which we have introduced dimensionless quantities

$$N_1 = 1 + a_0^2(K^2 + \omega^2/c^2)/Z^2, \tag{23.133}$$

$$N_2 = 2a_0^2K\omega/cZ^2. \tag{23.134}$$

To find the total cross section, we use the integrals

$$\int_0^{2\pi} d\phi\, \cos^2\phi = \pi, \tag{23.135}$$

$$\int_{-1}^{1} d(\cos\theta)\frac{\sin^2\theta}{(N_1 - N_2\cos\theta)^4} = \frac{4}{3(N_1^2 - N_2^2)^2}, \tag{23.136}$$

to obtain

$$\sigma = \int d\Omega\frac{d\sigma}{d\Omega} = \frac{128\pi\alpha\hbar}{3m_e\omega}\frac{a_0^3K^3}{Z^3(N_1^2 - N_2^2)^2}. \tag{23.137}$$

If the light is polarized along $\hat{y}$ rather than $\hat{x}$, then $\cos^2\phi$ should be replaced by $\sin^2\phi$ in the differential cross section (23.132). If the light is unpolarized we should instead make the replacement based on the average of the two transverse polarizations, $\cos^2\phi \to (\cos^2\phi + \sin^2\phi)/2 = 1/2$, which gives, after integrating over ϕ,

$$\frac{d\sigma}{d(\cos\theta)} = \frac{32\pi\alpha\hbar}{m_e\omega}\frac{a_0^3K^3}{Z^3(N_1 - N_2\cos\theta)^4}\sin^2\theta. \tag{23.138}$$

The total cross section is independent of the polarization, because the angle ϕ is integrated over.

Now consider the high-energy limit $K \gg \omega/c$ and $K \gg Z/a_0$ of (23.113), which we showed was appropriate for our plane-wave approximation for the final-state electron. Expanding to next-to-leading order in large K, we have

$$\frac{1}{(N_1 - N_2\cos\theta)^4} = \left(\frac{Z}{Ka_0}\right)^8\left(1 + \frac{8\omega}{cK}\cos\theta + \cdots\right), \tag{23.139}$$

so that (23.132) and (23.137) become

$$\frac{d\sigma}{d\Omega} = \frac{32\alpha\hbar}{m_e\omega}\left(\frac{Z}{Ka_0}\right)^5\left(1 + \frac{8\omega}{cK}\cos\theta\right)\sin^2\theta\cos^2\phi, \tag{23.140}$$

$$\sigma = \frac{128\pi\alpha\hbar}{3m_e\omega}\left(\frac{Z}{Ka_0}\right)^5. \tag{23.141}$$

This can be rewritten to eliminate K in favor of ω using the approximate energy conservation condition $K = \sqrt{2m_e\omega/\hbar}$, which follows from (23.111) in the limit of small binding energy compared to the photon energy. The result for the total cross section is

$$\sigma = \frac{256\pi}{3}\alpha\left(\frac{a_0}{Z}\right)^2\left(\frac{|\mathcal{E}_i|}{\hbar\omega}\right)^{7/2}, \tag{23.142}$$

where a_0/Z is the characteristic length scale of the initial state, and $|\mathcal{E}_i| = Z^2e^2/2a_0$ is the binding energy, and $\hbar\omega$ is the incident photon energy.

The preceding results apply not just to single-electron wavefunctions for hydrogen-like ions, but to the ejection of electrons from the K-shell[3] of heavier atoms. There are two K-shell electrons (one for each spin), so the cross section we obtained should be multiplied by 2 to find the total K-shell ejection cross section for multi-electron atoms. The most important qualitative features are that the cross section grows with increasing Z like Z^5, and falls sharply with increasing incident photon energy like $1/(\hbar\omega)^{7/2}$. The ejected electrons are preferentially emitted along the photon's electric field polarization, which is perpendicular to the photon's propagation direction, but with a slight preference for the forward direction ($\cos\theta > 0$) as a subleading effect in the high-energy expansion in (23.140).

In the opposite limit of small K, (23.132) and (23.137) appear to show a threshold behavior for the cross section proportional to K^3. However, in that regime, the calculation we did is not to be trusted, because the plane-wave approximation for the final state is quite bad.

Exercises

23.1 In this problem you will find the lifetime of the $|n, l, m\rangle = |3, 2, 2\rangle$ state of an isolated hydrogen atom, due to spontaneous emission of a single photon. Ignore fine, hyperfine, and Lamb shift splitting effects, and use the electric dipole approximation.

(a) According to the electric dipole selection rules, what state or states could be the final state? What are the angular frequency ω, the wavelength λ, and the color of the emitted light? Note that $\lambda \gg a_0$, so that the electric dipole approximation is indeed valid.

[3] The traditional jargon "K-shell" simply means the innermost shell with principal quantum number $n = 1$, called $1s$ in Sect. 18.4. The $2s$, $2p$ electrons are called L-shell, and $3s$, $3p$ and $3d$ are called M-shell, etc. In each shell with principal quantum number n, there can be up to $2\sum_{l=0}^{n-1}(2l+1) = 2n^2$ electrons.

(b) Use (23.65) to evaluate the emission rate R, and compute $\tau = 1/R$ to find the lifetime, both symbolically in terms of quantities like α, c, $\hbar$, a_0, and ω, and then numerically in seconds.
(c) What E2 and/or M1 spontaneous emission processes are available to the $|3, 2, 2\rangle$ state? Give the possible final states and rough parametric estimates of their contribution to the lifetime.

23.2 Consider an electron of mass m_e trapped in a three-dimensional isotropic harmonic oscillator potential with angular frequency Ω. The electron is initially in the first excited state with $l = 1$, $m = 0$, and spontaneously emits a photon to decay to the ground state.

(a) How is the angular frequency ω of the emitted photon related to the angular frequency Ω of the harmonic oscillator potential?
(b) Compute the lifetime τ of the first excited state.
(c) How does the average radiated power $\hbar\omega/\tau$ by the first excited state compare to the classical prediction of Larmor's formula?

23.3 Calculate the rate for the spontaneous decay of each triplet state to the singlet state in the hydrogen atom $n = 1$ level, corresponding to the 21 cm line hyperfine transition as discussed near (12.38)-(12.40) and (17.48). The three triplet states have the same lifetime for spontaneous decay. What is it, in years?
[Hints: use the selection rules to convince yourself that this is a magnetic dipole (M1) transition. Therefore, make use of (23.51) with (23.89). Both states involved in the transition have the same principal quantum number $n = 1$, and $l = 0$, so only the electron spin $\boldsymbol{S}$ in the matrix element contributes, and you do not need to worry about the radial or angular wavefunctions or the operator $\boldsymbol{L}$ at all. In terms of the electron and proton spins, the relevant states are

$$\text{triplet states:} \quad |\uparrow_e\uparrow_p\rangle, \qquad |\downarrow_e\downarrow_p\rangle, \qquad \left(|\uparrow_e\downarrow_p\rangle + |\downarrow_e\uparrow_p\rangle\right)/\sqrt{2}, \tag{23.143}$$

$$\text{singlet state:} \quad \left(|\uparrow_e\downarrow_p\rangle - |\downarrow_e\uparrow_p\rangle\right)/\sqrt{2}, \tag{23.144}$$

but the proton spin operator $\boldsymbol{S}_p$ also does not enter into the matrix element in our approximation. Note that for each emitted photon direction $\hat{k}$, there are two orthogonal polarization vectors $\hat{\varepsilon}$, leading to a factor of 2. Furthermore, $\hat{n} = \hat{k} \times \hat{\varepsilon}$ is always a unit vector, so instead of integrating over the $\hat{k}$ direction with $d\Omega_{\hat{k}}$, you can integrate $|\langle f|\hat{n} \cdot \boldsymbol{S}_e|i\rangle|^2$ over the $\hat{n}$ direction with $d\Omega_{\hat{n}}$. You should show that the rate is

$$R = \frac{\alpha\hbar^2\omega^3}{3m_e^2c^4}. \tag{23.145}$$

Recall that the angular frequency ω of the emitted photon is determined by the energy difference in (12.40).]

Scattering in Three Dimensions

24

24.1 Cross Sections and Scattering Amplitudes

Consider the problem of scattering of free particles from a localized target in three dimensions. Suppose that the incoming particles are in a monoenergetic beam, all moving in the same direction, which we will take to be $\hat{z}$ by a choice of coordinate system. If the beam is uniform over some area large compared to the size of the target, it can be described by a flux

$$n = \frac{\text{number of incident particles}}{\text{(area)(time)}}. \tag{24.1}$$

Here, the area is measured in the xy plane, perpendicular to the beam propagation direction. The experiment then counts (or infers, perhaps from measurements of energy deposited in a detector) the number of particles scattered into an infinitesimal differential of solid angle

$$d\Omega = d\phi\, d(\cos\theta). \tag{24.2}$$

Let the number of scattered particles as a function of direction be described by

$$ds = \frac{\text{number of scattered particles in } d\Omega}{\text{time}}. \tag{24.3}$$

Clearly, ds must be proportional to n, and to $d\Omega$. We therefore define the proportionality factor to be the differential **scattering cross section**,

$$\frac{d\sigma}{d\Omega} = \frac{ds}{n\, d\Omega}. \tag{24.4}$$

S. P. Martin, *Quantum Mechanics*, Graduate Texts in Physics,
https://doi.org/10.1007/978-3-032-21119-4_24

Note that this has units of area, like the absorption cross section for light discussed in Chap. 23. The total cross section is then defined as

$$\sigma = \int d\Omega \frac{d\sigma}{d\Omega} = \int_0^{2\pi} d\phi \int_{-1}^{1} d(\cos\theta) \frac{d\sigma}{d\Omega}, \tag{24.5}$$

also with units of area. Intuitively, the cross section is the effective area of the target for the type of scattering under consideration.

It is often true that the differential cross section has **azimuthal symmetry**, which means that it does not depend on ϕ. This occurs if the target and the properties of the scattering process are invariant under rotations about the initial beam propagation direction. In that case, one can give the differential cross section in terms of

$$\frac{d\sigma}{d(\cos\theta)} = \int_0^{2\pi} d\phi \frac{d\sigma}{d\Omega} = 2\pi \frac{d\sigma}{d\Omega}. \tag{24.6}$$

In particular, this will occur if the target is spherically symmetric and the incident particles are either spinless or have randomly oriented spins.

For classical scattering problems, the cross section can be computed in terms of specific trajectories followed by the incident particles. It is useful to define the **impact parameter** b as the distance between a particular initial particle trajectory line (when it is still very far from the target) and the line parallel to it through a suitably chosen central point in the target, which we take to be the origin of our coordinate system.

Consider first a classical[1] example, **hard-sphere scattering**. Particles are fired at a target ball (of radius R, and heavy enough that it does not move when hit). Gravitational effects are neglected, so that the incident particles move in straight lines. For a particular scattering event, the impact parameter b is the distance between the initial particle trajectory line and the line parallel to it through the center of the target ball, as shown in Fig. 24.1.To find the differential cross section, the general strategy is to relate b to the scattering angle θ. Using elementary geometry, the angles α and θ are related to each other and the impact parameter by $b = R\sin\alpha$ and $\theta = \pi - 2\alpha$, which implies

$$b = R\sin(\pi/2 - \theta/2) = R\cos(\theta/2). \tag{24.7}$$

Assuming a uniform flux n of incident particles moving in the $\hat{z}$ direction, the number of them per unit time with impact parameter between b and $b + db$ is equal to the product of n with the area of the annulus in the xy plane with inner radius b and outer radius $b + db$,

[1] We will later obtain the result for the corresponding quantum mechanical problem, in Sect. 24.5.

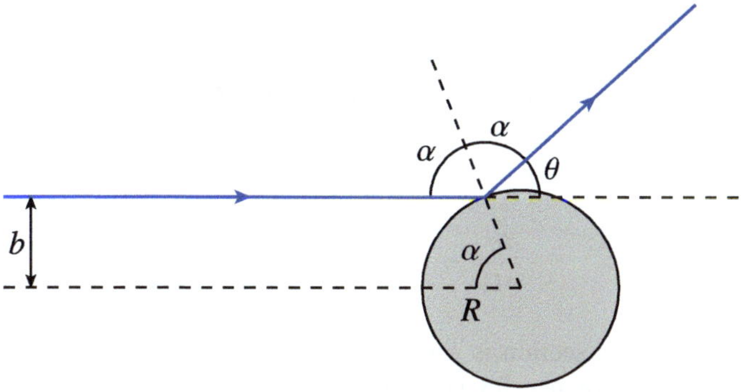

Fig. 24.1 A classical trajectory for scattering from a hard sphere of radius R, with impact parameter b and scattering angle θ measured from the forward direction $\hat{z}$ (the direction pointing to the right). The geometry gives the relation $b = R\cos(\theta/2)$

$$n(2\pi b\, db) = n2\pi R\cos(\theta/2)\, R\, d(\cos(\theta/2)) = \frac{1}{2}n\pi R^2\, d(\cos\theta). \quad (24.8)$$

This means that the rate of scattered particles in solid angle $d\Omega$ in (24.3) is

$$ds = \frac{1}{2}n\pi R^2 d(\cos\theta)\frac{d\phi}{2\pi} = \frac{1}{4}nR^2 d\Omega. \quad (24.9)$$

We therefore obtain, from the definition (24.4), the differential cross section

$$\frac{d\sigma}{d\Omega} = R^2/4. \quad (24.10)$$

Because of the azimuthal symmetry, this is equivalently

$$\frac{d\sigma}{d(\cos\theta)} = \pi R^2/2, \quad (24.11)$$

which integrates to give the total cross section

$$\sigma = \int_{-1}^{1} d(\cos\theta)\frac{d\sigma}{d(\cos\theta)} = \pi R^2. \quad (24.12)$$

This example has two rather special features. First, the differential cross section is isotropic; it does not depend on θ. Second, the total cross section is simply equal to the geometrical cross-sectional area of the target; the scattering takes place if and only if the impact parameter does not exceed the radius R.

Another classical (and classic) example is Rutherford scattering of a light spinless particle with charge $Z_1 e$ (for example, an α particle with $Z_1 = 2$) and initial energy E due to a potential

$$V(r) = Z_1 Z_2 e^2/r, \tag{24.13}$$

caused, for example, by a heavy nucleus containing Z_2 protons. From the classical hyperbolic trajectories, one can show (again by finding θ in terms of the impact parameter distance defined when the scattering particle is very far away) that

$$\frac{d\sigma}{d\Omega} = \frac{Z_1^2 Z_2^2 e^4}{16E^2 \sin^4(\theta/2)}. \tag{24.14}$$

This differential cross section is certainly not constant, and in fact blows up with a non-integrable divergence for small θ. To get a finite value for the total cross section, one must modify the question slightly and count only those events in which the particle is scattered by more than some minimal cut-off angle $\theta_{\rm min}$. (In real-world experiments, this is justified by the finite resolution of detectors, and by the difficulties encountered if one tries to put a detector too close to the beam.) Then

$$\begin{aligned}\sigma(\theta > \theta_{\rm min}) &= \int_0^{2\pi} d\phi \int_{-1}^{\cos\theta_{\rm min}} d(\cos\theta) \frac{Z_1^2 Z_2^2 e^4}{16E^2 \sin^4(\theta/2)} \\ &= \frac{\pi Z_1^2 Z_2^2 e^4}{4E^2} \left(\frac{1+\cos\theta_{\rm min}}{1-\cos\theta_{\rm min}}\right).\end{aligned} \tag{24.15}$$

The divergence is recovered in the limit $\theta_{\rm min} \to 0$. The interpretation is that in scattering from a $1/r$ potential, the initial charged particle is always scattered at least a little, with sufficient strength that the unrestricted total cross section is infinite; we say that the Coulomb potential has infinite range.

Most scattering problems are intermediate between the extremes of hard-sphere scattering (with a sharp cut-off limited range of interaction) and Coulomb scattering (with an infinite range). Even if scattering occurs for every impact parameter or scattering angle, the integrated total cross section σ can still be finite, if the differential cross section rises less quickly than $1/(1 - \cos\theta)$ for small θ. If so, then the potential giving rise to it is said to have finite range.

Enough classical warm-ups; we now turn to the quantum theory of scattering. Consider a beam of particles with fixed momentum $\boldsymbol{p} = \hbar\boldsymbol{k}$ that encounter a target in the form of a potential $V(\boldsymbol{r})$ that is significant only within a finite volume near the origin. If the potential were completely absent, the beam could be described by a plane wavefunction

$$\psi(\boldsymbol{r}) = \langle \boldsymbol{r}|\boldsymbol{k}\rangle = \frac{1}{(2\pi)^{3/2}} e^{i\boldsymbol{k}\cdot\boldsymbol{r}} = \frac{1}{(2\pi)^{3/2}} e^{ikz}. \tag{24.16}$$

In writing the last equality, we chose our coordinate system so that the initial beam momentum is along the z direction, so

$$\boldsymbol{k} = \hat{z}k. \tag{24.17}$$

The wavefunction (24.16) cannot be normalized to unity, so we have chosen instead to normalize it so that it satisfies Dirac orthonormality and completeness relations

$$\langle \boldsymbol{k}'|\boldsymbol{k}\rangle = \delta^{(3)}(\boldsymbol{k}' - \boldsymbol{k}), \qquad \int d^3\boldsymbol{k}\ |\boldsymbol{k}\rangle\langle\boldsymbol{k}| = I. \tag{24.18}$$

The probability of finding a particle per unit volume is then

$$\frac{\text{probability}}{\text{volume}} = |\psi(\boldsymbol{r},t)|^2 = |e^{-iEt/\hbar}\psi(\boldsymbol{r})|^2 = \frac{1}{(2\pi)^3}, \tag{24.19}$$

a constant in both time and position. This normalization is rather arbitrary, but will cancel in the differential cross section. The total flux of particles (probability for a particle in the incident beam to pass through a unit area, per unit time) described by the wavefunction is therefore

$$n = \frac{\text{probability}}{\text{(area)(time)}} = \left(\frac{\text{probability}}{\text{volume}}\right)\text{(speed)} = \frac{1}{(2\pi)^3}\frac{\hbar k}{m}. \tag{24.20}$$

Let us next anticipate the form of a stationary-state wavefunction for large r, taking into account the presence of the scattering potential:

$$\psi(\boldsymbol{r}) = \frac{1}{(2\pi)^{3/2}}\left(e^{ikz} + \frac{e^{ikr}}{r}f_k(\theta,\phi)\right), \qquad (r \to \infty). \tag{24.21}$$

We will justify this form before the end of Sect. 24.2. For now, we just note that the first term represents the incident wave, while the quantity

$$f_k(\theta,\phi) \tag{24.22}$$

characterizes the outgoing flux of probability density, and is called the **scattering amplitude**. It multiplies a factor e^{ikr}, which corresponds to a wave traveling away from the origin, and a factor $1/r$ which indicates that the probability amplitude varies inversely with distance. This gives a probability density that falls off like $1/r^2$, which is the requisite behavior so that the probability to detect a particle scattered within a given solid angle $d\Omega$, subtended by detector area $r^2 d\Omega$, is independent of the distance to the detector r, when r is large. Due to energy conservation and the assumption that the scattering potential vanishes at large distances, the magnitude of the wavevector is the same k for both incoming and outgoing waves. The scattering amplitude $f_k(\theta,\phi)$ has units of length and depends only on the magnitude k and the spherical coordinate angles, interpreted as those of the propagation direction of a scattered particle. In cases with azimuthal symmetry, we will write the scattering amplitude as $f_k(\theta)$.

A small detector at (θ,ϕ) covering a solid angle $d\Omega$, at a large distance r from the target, will see a total scattering probability per unit time

$$ds = \left(\frac{\text{probability}}{\text{volume}}\right)(\text{area})(\text{speed}) = \left|\frac{1}{(2\pi)^{3/2}} f_k(\theta, \phi) \frac{e^{ikr}}{r}\right|^2 \left(r^2 d\Omega\right)\left(\frac{\hbar k}{m}\right)$$
$$= \frac{|f_k(\theta, \phi)|^2}{(2\pi)^3} \frac{\hbar k}{m} d\Omega. \tag{24.23}$$

Here we have neglected the interference with the incident part of the wavefunction proportional to e^{ikz}, with the justification that in the real world the incoming particle beam has only a finite extent that will not overlap with sufficiently distant detectors measuring the scattered particles. (To summarize the assumptions made: the beam size perpendicular to its direction is large compared to the target, but small compared to the distance to the detector.) Using (24.20) and (24.23) in (24.4), we obtain the important result

$$\frac{d\sigma}{d\Omega} = |f_k(\theta, \phi)|^2. \tag{24.24}$$

The differential cross section is equal to the squared magnitude of the scattering amplitude.

24.2 The Scattering Matrix (S-Matrix) Operator and the Lippmann–Schwinger Equation

A general scattering problem involves an initial state, denoted $|a\rangle$, of distant free particles in a superposition of plane-wave momentum eigenstates. The symbol "a" is a shorthand for all of the data about the identity and momentum and spin quantum numbers of the incoming particles. Interactions then occur in a localized scattering region, described by a wavefunction that might be quite complicated. Finally, one observes particles moving far away from the scattering region as a superposition of plane waves, in a state denoted $|b\rangle$.

If the incoming and outgoing particles were pure plane waves, then they would fill all space at all times. In realistic scattering events, the states are actually wave packets, consisting of superpositions of plane-wave states with some spread of momenta, like the Gaussian wave packets of Sect. 6.2. Instead of dealing directly with wave packets, the problem can be finessed by taking our scattering potential to be localized in time as well as position. We imagine that at some times in both the far past and the far future the non-kinetic Hamiltonian terms (which are otherwise constant) are slowly turned off. Thus, particles in our wave packet that are far away are free in those two extremes, with a purely kinetic-energy Hamiltonian H_0. It is convenient to make use of the interaction picture of Sect. 21.2, so that the plane-wave states have no time dependence at all at those times in the far past and future.

Suppose that the state is a simple plane-wave ket $|a\rangle$ in the interaction picture at some very early time, which we can write (somewhat loosely for now) as $t = -\infty$. As we learned in Sect. 21.2, the interaction-picture state at later times t will be

$$\mathcal{U}_I(t, -\infty)\,|a\rangle\,, \tag{24.25}$$

where $\mathcal{U}_I$ is the unitary time-evolution operator. This motivates a formal definition of the **scattering matrix** or **S-matrix** operator as its very late-time limit,

$$S = \mathcal{U}_I(+\infty, -\infty), \tag{24.26}$$

where $+\infty$ represents a large positive time. It follows that the probability amplitude to find the system in the far future in another plane-wave state $|b\rangle$ is the S-matrix element

$$\langle b|S|a\rangle\,. \tag{24.27}$$

According to this formal definition, the S-matrix operator is unitary, which is a way of saying that probability is conserved. However, below we will modify the formal definition to treat more carefully the $t = \pm\infty$ behavior. As we will see, this modification maintains the unitarity of S, in a somewhat subtle way.

Since $\mathcal{U}_I$ is built out of the interaction (non-kinetic) Hamiltonian, the S-matrix is invariant under any symmetries that the system might have. Notably, S must commute with the angular momentum operator if the interaction Hamiltonian is rotationally invariant. This means that S will be diagonal in a suitably chosen basis of eigenstates of those symmetries. The S-matrix also encodes further information about the interaction Hamiltonian, including its characteristic length and energy scales, and any bound states or interesting substructure that it may have. Understanding the S-matrix through scattering is therefore a useful probe of physics at short-distance scales that might otherwise be difficult to access directly.

One way to evaluate the S-matrix elements in (24.27) is to work at a fixed intermediate time, which we can choose to be $t = 0$. To do this, let us define two scattering states $|\psi_a^+\rangle$ and $|\psi_b^-\rangle$ according to

$$|\psi_a^+\rangle = \mathcal{U}_I(0, -\infty)\,|a\rangle\,, \tag{24.28}$$

$$\langle\psi_b^-| = \langle b|\,\mathcal{U}_I(\infty, 0). \tag{24.29}$$

In words, $|\psi_a^+\rangle$ is the (generally quite complicated) result at time $t = 0$ of evolving the plane-wave state $|a\rangle$ from the distant past $t = -\infty$, in the interaction picture. The interpretation of $|\psi_b^-\rangle$ is perhaps somewhat less intuitive: it is the complicated state at time $t = 0$ that is obtained by evolving the simple plane-wave state $|b\rangle$ backwards in time from the far future $t = \infty$. This can be seen by using $\mathcal{U}_I(\infty, 0)^\dagger = \mathcal{U}_I(0, \infty)$ in the adjoint of (24.29), which implies

$$|\psi_b^-\rangle = \mathcal{U}_I(0, \infty)\,|b\rangle\,. \tag{24.30}$$

The point of making these definitions is that we can now write the S-matrix elements between plane-wave states as the inner product

$$\langle b|S|a\rangle = \langle\psi_b^-|\psi_a^+\rangle\,, \tag{24.31}$$

evaluated at the fixed time $t = 0$.

Suppose that the interaction Hamiltonian has no explicit time-dependence, so that energy is conserved. Then initial and final states connected by nonzero S-matrix elements must have the same energy $E_a = E_b$, and we can write

$$\langle b|S|a\rangle = \langle b|a\rangle - 2\pi i\delta(E_a - E_b)\,\langle b|T|a\rangle\,, \tag{24.32}$$

which defines an operator T, called the **transition operator**. It is the part of the S-matrix associated with genuinely nontrivial scattering. The factor of $-2\pi i$ is for future convenience, as we will see soon when we identify T more explicitly.

The preceding discussion has been very general, as we have not specified whether the free particles in the initial and final configurations are governed by nonrelativistic or relativistic kinematics, or whether their numbers and identities change in the scattering process. Quantum field theory is the natural framework for addressing the numerous difficult issues that can arise, including changes in particle number, infrared divergences associated with massless particles, and ultraviolet divergences associated with very high energies of intermediate states. In the rest of this book, we will be content to specialize to the simpler case that the scattering interaction is a potential $V(\boldsymbol{R})$ that depends on a single coordinate vector operator, and not on time. This $\boldsymbol{R}$ could either describe a single particle that appears in both the initial and final states and interacts with an external potential, or it could be the relative coordinate between two particles that interact with each other through $V(\boldsymbol{R})$. In both cases, we assume that the scattering is elastic, so that particles do not change into particles of another type, and that the kinematics conserves the nonrelativistic energy, so that the Hamiltonian has the time-independent form

$$H = H_0 + V(\boldsymbol{R}), \qquad H_0 = P^2/2m. \tag{24.33}$$

The potential is assumed to be short-range, which in practice usually means weaker than a Coulomb potential at large distances, so that $rV(\boldsymbol{r}) \to 0$ for $r \to \infty$. From here on, we will replace the labels a and b by wavevectors $\boldsymbol{k}$ and $\boldsymbol{k}'$ associated with momentum eigenvalues $\hbar\boldsymbol{k}$ and $\hbar\boldsymbol{k}'$, with

$$H_0\,|\boldsymbol{k}\rangle = E\,|\boldsymbol{k}\rangle\,, \qquad E = \hbar^2k^2/2m. \tag{24.34}$$

Spin labels, if any, are suppressed, so that the S-matrix elements are denoted $\langle\boldsymbol{k}'|S|\boldsymbol{k}\rangle$.

With these assumptions, in the interaction picture we have

$$V_I(t') = e^{it'H_0/\hbar}\,V e^{-it'H_0/\hbar}, \tag{24.35}$$

and recall from (21.32) that the unitary time-evolution operator for *finite* times is

$$\mathcal{U}_I(t,t') = e^{itH_0/\hbar}e^{-i(t-t')H/\hbar}e^{-it'H_0/\hbar}. \tag{24.36}$$

However, the latter formula has problematic limits for t or $t' \to \pm\infty$, due to its oscillatory nature. To get around this, we define modified versions of $\mathcal{U}_I(0,-\infty)$ and $\mathcal{U}_I(0,+\infty)$, by inserting exponential damping factors $e^{\epsilon t'/\hbar}$ and $e^{-\epsilon t'/\hbar}$ in the integral equation (21.39),

$$\Omega_+ \equiv I - \frac{i}{\hbar}\int_{-\infty}^{0} dt'\, e^{\epsilon t'/\hbar}\,\mathcal{U}_I(0,t')\,V_I(t'), \quad [\text{modified } \mathcal{U}_I(0,-\infty)], \tag{24.37}$$

$$\Omega_- \equiv I + \frac{i}{\hbar}\int_{0}^{\infty} dt'\, e^{-\epsilon t'/\hbar}\,\mathcal{U}_I(0,t')\,V_I(t'), \quad [\text{modified } \mathcal{U}_I(0,\infty)], \tag{24.38}$$

with positive infinitesimal ϵ in both cases, to turn off the interactions in the far past and in the far future, respectively. Inserting (24.35) and (24.36) with $t = 0$ into (24.37) and (24.38), we find

$$\Omega_+ = I - \frac{i}{\hbar}\int_{-\infty}^{0} dt' e^{it'(H-i\epsilon)/\hbar}\,V\,e^{-it'H_0/\hbar}, \tag{24.39}$$

$$\Omega_- = I + \frac{i}{\hbar}\int_{0}^{\infty} dt' e^{it'(H+i\epsilon)/\hbar}\,V\,e^{-it'H_0/\hbar}. \tag{24.40}$$

These are called **Møller operators**, after Christian Møller.

Now, acting with $\Omega_\pm$ on the initial plane-wave state $|\boldsymbol{k}\rangle$, and taking advantage of $e^{-it'H_0/\hbar}\,|\boldsymbol{k}\rangle = e^{-it'E/\hbar}\,|\boldsymbol{k}\rangle$, we find

$$|\psi_{\boldsymbol{k}}^{+}\rangle \equiv \Omega_+\,|\boldsymbol{k}\rangle = \left(I - \frac{i}{\hbar}\int_{-\infty}^{0} dt' e^{it'(H-E-i\epsilon)/\hbar}\,V\right)|\boldsymbol{k}\rangle\,, \tag{24.41}$$

$$|\psi_{\boldsymbol{k}}^{-}\rangle \equiv \Omega_-\,|\boldsymbol{k}\rangle = \left(I + \frac{i}{\hbar}\int_{0}^{\infty} dt' e^{it'(H-E+i\epsilon)/\hbar}\,V\right)|\boldsymbol{k}\rangle\,. \tag{24.42}$$

Doing the t' integrals, we get

$$|\psi_{\boldsymbol{k}}^{\pm}\rangle = \left(I + \frac{1}{E - H \pm i\epsilon}V\right)|\boldsymbol{k}\rangle\,. \tag{24.43}$$

The states $|\psi_{\boldsymbol{k}}^{\pm}\rangle$ are seen to differ in how they avoid the possible singularity in the formal limit $H \to E$, through opposite $\pm\, i\epsilon$ prescriptions in the denominator.

Equation (24.43) contains an inverse operator involving the full Hamiltonian, which is not at all convenient for practical evaluation. It is therefore useful to rearrange it, by multiplying through by $E - H \pm i\epsilon$ and then using $H = H_0 + V$, to write

$$(E - H_0 \pm i\epsilon)|\psi_{\boldsymbol{k}}^{\pm}\rangle = (E - H_0 \pm i\epsilon)\,|\boldsymbol{k}\rangle + V|\psi_{\boldsymbol{k}}^{\pm}\rangle. \tag{24.44}$$

Finally we multiply by $(E - H_0 \pm i\epsilon)^{-1}$ to arrive at

$$|\psi_{\boldsymbol{k}}^{\pm}\rangle = |\boldsymbol{k}\rangle + \frac{1}{E - H_0 \pm i\epsilon} V|\psi_{\boldsymbol{k}}^{\pm}\rangle. \tag{24.45}$$

This is called the **Lippmann–Schwinger equation**, after Bernard Lippmann and Julian Schwinger. It will play an essential role in the following.

Another consequence of (24.44) is obtained by using $H_0\,|\boldsymbol{k}\rangle = E\,|\boldsymbol{k}\rangle$ and $H_0 = H - V$ again, to find $(H - E)|\psi_{\boldsymbol{k}}^{\pm}\rangle = \pm i\epsilon(|\psi_{\boldsymbol{k}}^{\pm}\rangle - |\boldsymbol{k}\rangle)$. In the limit $\epsilon \to 0$, this reduces to

$$H|\psi_{\boldsymbol{k}}^{\pm}\rangle = E|\psi_{\boldsymbol{k}}^{\pm}\rangle. \tag{24.46}$$

Thus we have the nice result that $|\psi_{\boldsymbol{k}}^{+}\rangle$ and $|\psi_{\boldsymbol{k}}^{-}\rangle$ are both stationary states of the full Hamiltonian H, each with exactly the same energy eigenvalue $E = \hbar^2 k^2/2m$ as the H_0 eigenvalue of the corresponding plane-wave state. As we will see, $|\psi_{\boldsymbol{k}}^{+}\rangle$ in particular has a wavefunction with the large-r form that we anticipated in (24.21).

Another remarkable result is that the scattering states have the same orthonormality properties as the corresponding plane-wave states,

$$\langle\psi_{\boldsymbol{k}'}^{+}|\psi_{\boldsymbol{k}}^{+}\rangle = \langle\psi_{\boldsymbol{k}'}^{-}|\psi_{\boldsymbol{k}}^{-}\rangle = \langle\boldsymbol{k}'|\boldsymbol{k}\rangle = \delta^{(3)}(\boldsymbol{k}' - \boldsymbol{k}), \tag{24.47}$$

where the last equality reflects the normalization assumed in (24.18). To prove (24.47), start by considering the matrix element $\langle\psi_{\boldsymbol{k}'}^{+}|V|\psi_{\boldsymbol{k}}^{+}\rangle$, which can be rewritten in two ways, using the Lippmann–Schwinger equation for $|\psi_{\boldsymbol{k}}^{+}\rangle$, and its adjoint for $\langle\psi_{\boldsymbol{k}'}^{+}|$,

$$\langle\psi_{\boldsymbol{k}'}^{+}|V|\psi_{\boldsymbol{k}}^{+}\rangle = \langle\psi_{\boldsymbol{k}'}^{+}|\,V\Big(|\boldsymbol{k}\rangle + \frac{1}{E - H_0 + i\epsilon} V|\psi_{\boldsymbol{k}}^{+}\rangle\Big), \tag{24.48}$$

$$\langle\psi_{\boldsymbol{k}'}^{+}|V|\psi_{\boldsymbol{k}}^{+}\rangle = \Big(\langle\boldsymbol{k}'| + \langle\psi_{\boldsymbol{k}'}^{+}|\,V\frac{1}{E' - H_0 - i\epsilon}\Big)V|\psi_{\boldsymbol{k}}^{+}\rangle, \tag{24.49}$$

where $E = \hbar^2 k^2/2m$ and $E' = \hbar^2 k'^2/2m$. Taking the difference of these two equations, we find

$$\langle\psi_{\boldsymbol{k}'}^{+}|V|\boldsymbol{k}\rangle - \langle\boldsymbol{k}'|V|\psi_{\boldsymbol{k}}^{+}\rangle + \langle\psi_{\boldsymbol{k}'}^{+}|\,V\left(\frac{1}{E - H_0 + i\epsilon} - \frac{1}{E' - H_0 - i\epsilon}\right)V|\psi_{\boldsymbol{k}}^{+}\rangle = 0. \tag{24.50}$$

The terms in parentheses can be rewritten using

$$\frac{1}{E - H_0 + i\epsilon} - \frac{1}{E' - H_0 - i\epsilon} = \frac{E' - E - i\epsilon}{(E' - H_0 - i\epsilon)(E - H_0 + i\epsilon)}, \quad (24.51)$$

where we have changed $-2i\epsilon$ to $-i\epsilon$ in the numerator, since ϵ is infinitesimal and its only purpose is to avoid singular denominators. Now we use this in (24.50), and divide by $E' - E - i\epsilon$. The result can be expressed, using $E\,|\boldsymbol{k}\rangle = H_0\,|\boldsymbol{k}\rangle$ in the first term and $E'\,\langle \boldsymbol{k}'| = \langle \boldsymbol{k}'|\,H_0$ in the second term, as

$$\begin{aligned}\langle \psi^{+}_{\boldsymbol{k}'}|\,V \frac{1}{E' - H_0 - i\epsilon}\,|\boldsymbol{k}\rangle + \langle \boldsymbol{k}'|\,\frac{1}{E - H_0 + i\epsilon} V\,|\psi^{+}_{\boldsymbol{k}}\rangle +\\ \langle \psi^{+}_{\boldsymbol{k}'}|\,V \frac{1}{E' - H_0 - i\epsilon}\,\frac{1}{E - H_0 + i\epsilon} V\,|\psi^{+}_{\boldsymbol{k}}\rangle &= 0. \quad (24.52)\end{aligned}$$

Again applying the Lippmann–Schwinger equation and its adjoint, this becomes

$$\left(\langle \psi^{+}_{\boldsymbol{k}'}| - \langle \boldsymbol{k}'|\right)|\boldsymbol{k}\rangle + \langle \boldsymbol{k}'|\left(|\psi^{+}_{\boldsymbol{k}}\rangle - |\boldsymbol{k}\rangle\right) + \left(\langle \psi^{+}_{\boldsymbol{k}'}| - \langle \boldsymbol{k}'|\right)\left(|\psi^{+}_{\boldsymbol{k}}\rangle - |\boldsymbol{k}\rangle\right) = 0, \quad (24.53)$$

which finally simplifies to $\langle \psi^{+}_{\boldsymbol{k}'}|\psi^{+}_{\boldsymbol{k}}\rangle = \langle \boldsymbol{k}'|\boldsymbol{k}\rangle$. The proof of $\langle \psi^{-}_{\boldsymbol{k}'}|\psi^{-}_{\boldsymbol{k}}\rangle = \langle \boldsymbol{k}'|\boldsymbol{k}\rangle$ is identical, except for $i\epsilon \to -i\epsilon$ everywhere.

Equation (24.47) implies, using the definitions in (24.41) and (24.42), that the operators Ω_+ and Ω_- preserve the norms of states, as they obey

$$\Omega_+^\dagger \Omega_+ = \Omega_-^\dagger \Omega_- = I. \quad (24.54)$$

It may therefore come as a surprise to learn that these operators are usually *not* unitary. In a finite-dimensional space, an operator A that satisfies $A^\dagger A = I$ is invertible, with $A^{-1} = A^\dagger$, and also satisfies $AA^\dagger = I$, and so is unitary. However, in an infinite-dimensional Hilbert space, one can have $A^\dagger A = I$ even if A is not invertible, in which case $AA^\dagger \neq I$. Such an operator that is not unitary, but preserves inner products, is called an **isometric operator**. In our present situation, the modification by inserting the convergence factors $e^{\pm\epsilon t'/\hbar}$ causes $\Omega_\pm$ to be isometric but not necessarily unitary, unlike the unitary operators $\mathcal{U}_I(t, t')$ for finite times t, t'.

To better understand this, note that applying completeness of the plane-wave states $|\boldsymbol{k}\rangle$ to (24.41) and (24.42) yields

$$\Omega_\pm = \int d^3\boldsymbol{k}\,|\psi^{\pm}_{\boldsymbol{k}}\rangle\langle \boldsymbol{k}|, \qquad \Omega_\pm^\dagger = \int d^3\boldsymbol{k}\,|\boldsymbol{k}\rangle\langle \psi^{\pm}_{\boldsymbol{k}}|. \quad (24.55)$$

Now, as a check of (24.54),

$$\begin{aligned}\Omega_\pm^\dagger \Omega_\pm &= \int d^3\boldsymbol{k} \int d^3\boldsymbol{q}\,|\boldsymbol{k}\rangle\langle \psi^{\pm}_{\boldsymbol{k}}|\psi^{\pm}_{\boldsymbol{q}}\rangle\langle \boldsymbol{q}| = \int d^3\boldsymbol{k} \int d^3\boldsymbol{q}\,\delta^{(3)}(\boldsymbol{k} - \boldsymbol{q})|\boldsymbol{k}\rangle\langle \boldsymbol{q}|\\ &= \int d^3\boldsymbol{k}\,|\boldsymbol{k}\rangle\langle \boldsymbol{k}| = I. \quad (24.56)\end{aligned}$$

However, while the scattering states $|\psi_{\boldsymbol{k}}^{+}\rangle$ satisfy Dirac orthonormality, they are not necessarily a complete orthobasis, because we know that they have positive energy, while the full Hamiltonian H including the potential V may have bound states $|\chi_n\rangle$ with negative energy. Therefore, completeness for them reads

$$\int d^3\boldsymbol{k}|\psi_{\boldsymbol{k}}^{+}\rangle\langle\psi_{\boldsymbol{k}}^{+}| + \sum_n |\chi_n\rangle\langle\chi_n| = I, \tag{24.57}$$

and similarly for $|\psi_{\boldsymbol{k}}^{-}\rangle$. It follows from (24.55) and (24.57) that

$$\Omega_+\Omega_+^\dagger = \Omega_-\Omega_-^\dagger = I - \sum_n |\chi_n\rangle\langle\chi_n|\,. \tag{24.58}$$

So, the Møller operators are unitary (and invertible) only if the potential V has no bound states.

The S-matrix operator (and its adjoint) is now defined by

$$S = \Omega_-^\dagger\Omega_+, \qquad S^\dagger = \Omega_+^\dagger\Omega_-. \tag{24.59}$$

The unitarity of S defined in this way may not be immediately obvious, but it is true. To prove it, note first that since the scattering states $|\psi_{\boldsymbol{k}}^{\pm}\rangle$ and the bound states $|\chi_n\rangle$ are eigenstates of the same Hermitian operator H with different eigenvalues, Theorem 2.13 guarantees us that they are orthogonal. It follows immediately from (24.55) that

$$\Omega_\pm^\dagger\,|\chi_n\rangle = 0, \qquad \langle\chi_n|\,\Omega_\pm = 0. \tag{24.60}$$

So, we can now compute

$$S^\dagger S = \Omega_+^\dagger\Omega_-\Omega_-^\dagger\Omega_+ = \Omega_+^\dagger\Big(I - \sum_n |\chi_n\rangle\,\langle\chi_n|\Big)\Omega_+ = \Omega_+^\dagger\Omega_+ = I, \tag{24.61}$$

and similarly

$$SS^\dagger = \Omega_-^\dagger\Omega_+\Omega_+^\dagger\Omega_- = \Omega_-^\dagger\Big(I - \sum_n |\chi_n\rangle\,\langle\chi_n|\Big)\Omega_- = \Omega_-^\dagger\Omega_- = I. \tag{24.62}$$

Thus we have the crucial result that the unitarity of the S-matrix survives the exponential time-damping modification for infinitesimal ϵ, even though the Møller operators that it is built out of are not unitary.

Let us investigate the S-matrix element between two arbitrary plane-wave states,

$$\langle \boldsymbol{k}'|S|\boldsymbol{k}\rangle = \langle \psi^-_{\boldsymbol{k}'}|\psi^+_{\boldsymbol{k}}\rangle = \langle \psi^+_{\boldsymbol{k}'}|\psi^+_{\boldsymbol{k}}\rangle + \big(\langle \psi^-_{\boldsymbol{k}'}| - \langle \psi^+_{\boldsymbol{k}'}|\big)|\psi^+_{\boldsymbol{k}}\rangle \tag{24.63}$$

$$= \delta^{(3)}(\boldsymbol{k}-\boldsymbol{k}') + \langle \boldsymbol{k}'|\, V\Big(\frac{1}{E'-H+i\epsilon} - \frac{1}{E'-H-i\epsilon}\Big)|\psi^+_{\boldsymbol{k}}\rangle, \tag{24.64}$$

where we used (24.47) and the adjoint of (24.43) with $\boldsymbol{k}\to\boldsymbol{k}'$ and $E' = \hbar^2 k'^2/2m$ to get the last equality. Now, thanks to (24.46), we can replace H by $E = \hbar^2 k^2/2m$, to get

$$\langle \boldsymbol{k}'|S|\boldsymbol{k}\rangle = \delta^{(3)}(\boldsymbol{k}-\boldsymbol{k}') - 2i\frac{\epsilon}{(E-E')^2+\epsilon^2}\,\langle \boldsymbol{k}'|V|\psi^+_{\boldsymbol{k}}\rangle\,. \tag{24.65}$$

It might be tempting to suppose that the last term should vanish in the limit $\epsilon\to 0$, but special care is needed when $E = E'$. Recalling, from (2.36), that

$$\lim_{\epsilon\to 0}\frac{\epsilon}{x^2+\epsilon^2} = \pi\,\delta(x), \tag{24.66}$$

we actually find

$$\langle \boldsymbol{k}'|S|\boldsymbol{k}\rangle = \delta^{(3)}(\boldsymbol{k}-\boldsymbol{k}') - 2\pi i\delta(E-E')\,\langle \boldsymbol{k}'|V|\psi^+_{\boldsymbol{k}}\rangle\,. \tag{24.67}$$

Comparing with (24.32), we see that the transition operator T must satisfy

$$T\,|\boldsymbol{k}\rangle = V|\psi^+_{\boldsymbol{k}}\rangle, \tag{24.68}$$

so that

$$\langle \boldsymbol{k}'|S|\boldsymbol{k}\rangle = \delta^{(3)}(\boldsymbol{k}-\boldsymbol{k}') - 2\pi i\delta(E-E')\,\langle \boldsymbol{k}'|T|\boldsymbol{k}\rangle\,. \tag{24.69}$$

Since the plane-wave kets $|\boldsymbol{k}\rangle$ form an orthobasis, by superposition (24.68) provides a complete definition of T acting on any state.

Our next goal is to relate the matrix elements of T to the scattering amplitude $f_k(\theta,\phi)$ introduced in (24.21)–(24.24) of the previous section. To do this, we will work with the Lippmann–Schwinger equation (24.45) in the position representation. From here on, we drop the subscript $\boldsymbol{k}$ for the scattering state kets and wavefunctions to reduce clutter, so that $|\psi^\pm\rangle$ are understood to be the scattering states related to $|\boldsymbol{k}\rangle$ by time evolution from $t=\mp\infty$ to $t=0$, with wavefunctions $\psi^\pm(\boldsymbol{r}) = \langle \boldsymbol{r}|\psi^\pm\rangle$. The wavefunction form of the Lippmann–Schwinger equation is then

$$\langle \boldsymbol{r}|\psi^\pm\rangle = \langle \boldsymbol{r}|\boldsymbol{k}\rangle + \int d^3r'\,\langle \boldsymbol{r}|\frac{1}{E-H_0\pm i\epsilon}|\boldsymbol{r}'\rangle\,\langle \boldsymbol{r}'|V|\psi^\pm\rangle\,, \tag{24.70}$$

where we have used completeness with respect to position eigenstates $|\boldsymbol{r}'\rangle$. The last term in (24.70) involves a matrix element that we write as

$$\langle \boldsymbol{r}|\frac{1}{E - H_0 \pm i\epsilon}|\boldsymbol{r}'\rangle = \frac{2m}{\hbar^2} G_\pm(\boldsymbol{r}, \boldsymbol{r}'), \tag{24.71}$$

which defines $G_\pm(\boldsymbol{r}, \boldsymbol{r}')$ as a function independent of the scattering potential V.

To evaluate the function $G_\pm(\boldsymbol{r}, \boldsymbol{r}')$, we will now use completeness again, this time with respect to plane-wave states with momenta $\hbar\boldsymbol{q}$, normalized just as in (24.16), so

$$\langle \boldsymbol{r}|\boldsymbol{q}\rangle = \frac{1}{(2\pi)^{3/2}} e^{i\boldsymbol{q}\cdot\boldsymbol{r}}, \tag{24.72}$$

with orthonormality and completeness relations $\langle \boldsymbol{q}'|\boldsymbol{q}\rangle = \delta^{(3)}(\boldsymbol{q}' - \boldsymbol{q})$ and $\int d^3\boldsymbol{q}\ |\boldsymbol{q}\rangle\langle\boldsymbol{q}| = I$. Thus (24.71) becomes

$$G_\pm(\boldsymbol{r}, \boldsymbol{r}') = \frac{\hbar^2}{2m} \int d^3\boldsymbol{q}\ \langle \boldsymbol{r}|\frac{1}{E - H_0 \pm i\epsilon}|\boldsymbol{q}\rangle\ \langle \boldsymbol{q}|\boldsymbol{r}'\rangle\,. \tag{24.73}$$

The reason for doing this is that now the operator H_0 can be evaluated as the number $\hbar^2 q^2/2m$ acting on $|\boldsymbol{q}\rangle$, so that

$$\langle \boldsymbol{r}|\frac{1}{E - H_0 \pm i\epsilon}|\boldsymbol{q}\rangle = \frac{1}{E - \hbar^2 q^2/2m \pm i\epsilon}\ \langle \boldsymbol{r}|\boldsymbol{q}\rangle\,. \tag{24.74}$$

Remembering that $E = \hbar^2 k^2/2m$ is the energy of the incident beam particles, we find an integral expression for the function defined by (24.71),

$$G_\pm(\boldsymbol{r}, \boldsymbol{r}') = \int \frac{d^3\boldsymbol{q}}{(2\pi)^3} \frac{e^{i\boldsymbol{q}\cdot(\boldsymbol{r}-\boldsymbol{r}')}}{k^2 - q^2 \pm i\epsilon}. \tag{24.75}$$

Before doing this integral, as an aside we observe that $G_\pm(\boldsymbol{r}, \boldsymbol{r}')$ is also the solution to a differential equation. To see this, note that

$$\nabla^2 G_\pm(\boldsymbol{r}, \boldsymbol{r}') = \int \frac{d^3\boldsymbol{q}}{(2\pi)^3} \frac{(-q^2)\, e^{i\boldsymbol{q}\cdot(\boldsymbol{r}-\boldsymbol{r}')}}{k^2 - q^2 \pm i\epsilon}, \tag{24.76}$$

where the Laplacian was taken with respect to the coordinate $\boldsymbol{r}$, taking advantage of the simple exponential dependence. Therefore,

$$(\nabla^2 + k^2) G_\pm(\boldsymbol{r}, \boldsymbol{r}') = \int \frac{d^3\boldsymbol{q}}{(2\pi)^3} e^{i\boldsymbol{q}\cdot(\boldsymbol{r}-\boldsymbol{r}')} = \delta^{(3)}(\boldsymbol{r} - \boldsymbol{r}'). \tag{24.77}$$

One says that $G_{\pm}(\boldsymbol{r}, \boldsymbol{r}')$ is a **Green function** (named after mathematician George Green) for the Helmholtz differential operator $\nabla^2 + k^2$, with a delta-function source at $\boldsymbol{r}'$.

To evaluate the Green function, go back to (24.75) and put the integral $\int d^3\boldsymbol{q}$ into spherical coordinates,

$$G_{\pm}(\boldsymbol{r}, \boldsymbol{r}') = \frac{1}{(2\pi)^3} \int_0^{\infty} dq\, q^2 \int_0^{2\pi} d\phi \int_{-1}^{1} d(\cos\theta) \frac{e^{iq|\boldsymbol{r}-\boldsymbol{r}'|\cos\theta}}{k^2 - q^2 \pm i\epsilon}, \tag{24.78}$$

where the angles θ and ϕ of $\boldsymbol{q}$ are measured from the direction $\boldsymbol{r} - \boldsymbol{r}'$. Now do the angular integrals to get

$$G_{\pm}(\boldsymbol{r},\, \boldsymbol{r}') = \frac{i}{4\pi^2|\boldsymbol{r} - \boldsymbol{r}'|} \int_0^{\infty} dq\, \frac{q}{q^2 - k^2 \mp i\epsilon} \left(e^{iq|\boldsymbol{r}-\boldsymbol{r}'|} - e^{-iq|\boldsymbol{r}-\boldsymbol{r}'|}\right). \tag{24.79}$$

Let $x = |\boldsymbol{r} - \boldsymbol{r}'|$, and call the integral $I_{\pm}(x, k)$. It can be evaluated (using the method of complex variable contour integration, for example) as $I_{\pm}(x, k) = i\pi e^{\pm ikx}$. Therefore,

$$G_{\pm}(\boldsymbol{r},\, \boldsymbol{r}') = -\frac{e^{\pm ik|\boldsymbol{r}-\boldsymbol{r}'|}}{4\pi\,|\boldsymbol{r} - \boldsymbol{r}'|}. \tag{24.80}$$

Plugging this result into the Lippmann–Schwinger equation (24.70), using (24.71), gives

$$\langle \boldsymbol{r}|\psi^{\pm}\rangle = \langle \boldsymbol{r}|\boldsymbol{k}\rangle - \frac{2m}{\hbar^2}\int d^3\boldsymbol{r}'\, \frac{e^{\pm ik|\boldsymbol{r}-\boldsymbol{r}'|}}{4\pi\,|\boldsymbol{r} - \boldsymbol{r}'|}\, \langle \boldsymbol{r}'|\, V\, |\psi^{\pm}\rangle\,. \tag{24.81}$$

Assuming that the incoming particles have momentum in the $\hat{z}$ direction, so that $\boldsymbol{k} = k\hat{z}$, and using $\langle \boldsymbol{r}'|\, V\, |\psi^{\pm}\rangle = V(\boldsymbol{r}')\,\langle \boldsymbol{r}'|\psi^{\pm}\rangle$, we find the Lippmann–Schwinger equation in the position representation

$$\psi^{\pm}(\boldsymbol{r}) = \frac{e^{ikz}}{(2\pi)^{3/2}} - \frac{2m}{\hbar^2}\int d^3\boldsymbol{r}'\, \frac{e^{\pm ik|\boldsymbol{r}-\boldsymbol{r}'|}}{4\pi\,|\boldsymbol{r} - \boldsymbol{r}'|}\, V(\boldsymbol{r}')\,\psi^{\pm}(\boldsymbol{r}'). \tag{24.82}$$

This splits the scattering wavefunction $\psi^{\pm}(\boldsymbol{r})$ into a part describing the incident particle beam and an integral describing the effect of scattering by the potential.

We now specialize to points $\boldsymbol{r}$ that are very far away from the points $\boldsymbol{r}'$ in the neighborhood of the origin where $V(\boldsymbol{r}')$ is significant. We do this by expanding for $r' \ll r$, so that

$$|\boldsymbol{r} - \boldsymbol{r}'| = r\sqrt{1 - 2\hat{r}\cdot\boldsymbol{r}'/r + r'^2/r^2} = r - \hat{r}\cdot\boldsymbol{r}' + O(1/r). \tag{24.83}$$

Let us define a wavevector $\boldsymbol{k}'$, with the same magnitude as the incoming wavevector $\boldsymbol{k} = k\hat{z}$, but directed away from the scattering region (in the radial direction),

$$\boldsymbol{k}' = k\hat{r}. \tag{24.84}$$

Then (24.83) implies

$$e^{\pm ik|\boldsymbol{r}-\boldsymbol{r}'|} \approx e^{\pm ikr} e^{\mp i\boldsymbol{k}'\cdot\boldsymbol{r}'}. \tag{24.85}$$

and (24.82) becomes, for large r,

$$\psi^{\pm}(\boldsymbol{r}) = \frac{e^{ikz}}{(2\pi)^{3/2}} - \frac{m}{2\pi\hbar^2}\frac{e^{\pm ikr}}{r}\int d^3\boldsymbol{r}'\, e^{\mp i\boldsymbol{k}'\cdot\boldsymbol{r}'}V(\boldsymbol{r}')\,\psi^{\pm}(\boldsymbol{r}'). \tag{24.86}$$

We now see that the factor e^{+ikr} in $\psi^{+}(\boldsymbol{r})$ corresponds to a scattered wave moving *away* from the target potential. The factor e^{-ikr} in $\psi^{-}(\boldsymbol{r})$ corresponds to the unphysical situation of an incoming radial wave converging on a target from all directions and cleverly matching itself onto a plane wave. Therefore we concentrate on

$$\psi^{+}(\boldsymbol{r}) = \frac{1}{(2\pi)^{3/2}}\left(e^{ikz} + \frac{e^{ikr}}{r} f_k(\theta,\phi)\right), \qquad (r \to \infty), \tag{24.87}$$

where we have now identified an expression for the scattering amplitude,

$$f_k(\theta,\phi) = -\frac{m}{2\pi\hbar^2}(2\pi)^{3/2}\int d^3\boldsymbol{r}'\, e^{-i\boldsymbol{k}'\cdot\boldsymbol{r}'}V(\boldsymbol{r}')\,\psi^{+}(\boldsymbol{r}'), \tag{24.88}$$

with the angles (θ,ϕ) defined to be the spherical coordinate angles of $\boldsymbol{k}'$ when the coordinate system has been chosen so that $\boldsymbol{k} = k\hat{z}$. Equation (24.87) indeed has the form promised in (24.21). However, (24.87) and (24.88) do not yet constitute a solution, because (24.87) writes ψ^{+} in terms of $f_k(\theta,\phi)$, which in turn depends on ψ^{+}. Our remaining task in this section is to see how to solve these equations for $f_k(\theta,\phi)$ by eliminating ψ^{+}.

Since the ket $|\boldsymbol{k}'\rangle$ is the state with position wavefunction $e^{i\boldsymbol{k}'\cdot\boldsymbol{r}}/(2\pi)^{3/2}$, an equivalent way of writing (24.88) is

$$f_k(\theta,\phi) = -\frac{4\pi^2 m}{\hbar^2}\langle\boldsymbol{k}'|V|\psi^{+}\rangle. \tag{24.89}$$

Now, using the definition of the transition operator in (24.68), this becomes a matrix element between completely known (plane-wave) states,

$$f_k(\theta, \phi) = -\frac{4\pi^2 m}{\hbar^2} \langle \boldsymbol{k}'|T|\boldsymbol{k}\rangle . \tag{24.90}$$

The problem of finding the scattering amplitude has therefore been reduced to solving for the matrix elements of T.

From the Lippmann–Schwinger equation (24.45) with the + sign selected, and again using the definition of the transition operator in (24.68), we find

$$\begin{aligned} T\,|\boldsymbol{k}\rangle &= V\left(|\boldsymbol{k}\rangle + \frac{1}{E - H_0 + i\epsilon} V|\psi^+\rangle\right) \\ &= \left(V + V\frac{1}{E - H_0 + i\epsilon} T\right)|\boldsymbol{k}\rangle . \end{aligned} \tag{24.91}$$

Since this is true for every plane-wave ket $|\boldsymbol{k}\rangle$, we have an operator relation between T, V, H_0, and the incident particle energy E,

$$T = V + V\frac{1}{E - H_0 + i\epsilon} T. \tag{24.92}$$

This can be formally solved for T, with the result

$$T = \left(I - V\frac{1}{E - H_0 + i\epsilon}\right)^{-1} V. \tag{24.93}$$

Together with (24.90), we can finally say that we have a solution for the scattering amplitude. However, practical evaluation of it is still a nontrivial task that typically requires approximation. Different approximation schemes amount to practical methods for evaluating the formal inverse operator on the right side of (24.93).

24.3 The Optical Theorem

In this short section, we derive an important consequence of probability conservation:

Theorem 24.1 (Optical Theorem) *The total integrated scattering cross section is related to the imaginary part of the forward scattering amplitude*[2] $f_k(\theta = 0)$ *by*

$$\sigma = \frac{4\pi}{k} Im\,[f_k(0)] . \tag{24.94}$$

[2] Note that we write $f_k(0)$ for $\theta = 0$ even though we are *not* assuming azimuthal symmetry, because the coordinate ϕ is both meaningless and irrelevant when $\theta = 0$.

Proof We start with

$$\mathrm{Im}\,[f_k(0)] = -\frac{4\pi^2 m}{\hbar^2}\,\mathrm{Im}\left(\langle \boldsymbol{k}|V|\psi^+\rangle\right), \tag{24.95}$$

which is obtained from (24.89) by setting $\langle \boldsymbol{k}'| = \langle \boldsymbol{k}|$ for forward scattering, and then taking the imaginary part. To evaluate the right side, we can apply the Hermitian adjoint of the Lippmann–Schwinger equation (24.45) with the + sign selected,

$$\langle \boldsymbol{k}| = \langle \psi^+| - \langle \psi^+|\,V\frac{1}{E - H_0 - i\epsilon}, \tag{24.96}$$

to obtain

$$\mathrm{Im}\left(\langle \boldsymbol{k}|V|\psi^+\rangle\right) = \mathrm{Im}\left(\langle \psi^+|V|\psi^+\rangle\right) - \mathrm{Im}\left(\langle \psi^+|V\frac{1}{E - H_0 - i\epsilon}V|\psi^+\rangle\right). \tag{24.97}$$

The first term on the right side is 0, because V is Hermitian, and therefore always has real expectation values. In contrast, the second term does not vanish, because the operator $1/(E - H_0 - i\epsilon)$ is not Hermitian, due to the $i\epsilon$. Using the definition of the transition operator in (24.68),

$$\begin{aligned}
\mathrm{Im}\left(\langle \boldsymbol{k}|V|\psi^+\rangle\right) &= -\mathrm{Im}\left(\langle \boldsymbol{k}|T^\dagger\frac{1}{E - H_0 - i\epsilon}T|\boldsymbol{k}\rangle\right) && (24.98)\\
&= -\frac{1}{2i}\langle \boldsymbol{k}|T^\dagger\left(\frac{1}{E - H_0 - i\epsilon} - \frac{1}{E - H_0 + i\epsilon}\right)T|\boldsymbol{k}\rangle && (24.99)\\
&= -\langle \boldsymbol{k}|T^\dagger\frac{\epsilon}{(E - H_0)^2 + \epsilon^2}T|\boldsymbol{k}\rangle\,. && (24.100)
\end{aligned}$$

Inserting a completeness relation $\int d^3\boldsymbol{k}'|\boldsymbol{k}'\rangle\langle \boldsymbol{k}'| = I$ just before the T operator, this becomes

$$\mathrm{Im}\left(\langle \boldsymbol{k}|V|\psi^+\rangle\right) = -\int d^3\boldsymbol{k}'\,|\langle \boldsymbol{k}'|T|\boldsymbol{k}\rangle|^2\,\pi\delta(E - E'), \tag{24.101}$$

where we made use of $\lim_{\epsilon\to 0}\epsilon/(x^2 + \epsilon^2) = \pi\delta(x)$ from (2.36). Now, recalling that the angles of $\boldsymbol{k}'$ with respect to the fixed $\boldsymbol{k}$ are (θ, ϕ) with solid angle differential $d\Omega$, the integration measure $\int d^3\boldsymbol{k}'$ separates into $\int d\Omega \int_0^\infty dk'\,k'^2$. Using $\delta(E - E') = (m/\hbar^2 k)\delta(k - k')$ to do the radial integration over k', (24.101) turns into

$$\mathrm{Im}\left(\langle \boldsymbol{k}|V|\psi^+\rangle\right) = -\frac{\pi m k}{\hbar^2}\int d\Omega\,|\langle \boldsymbol{k}'|T|\boldsymbol{k}\rangle|^2. \tag{24.102}$$

Putting this into (24.95) gives the imaginary part of the forward scattering amplitude in terms of the matrix elements of the transition operator,

$$\mathrm{Im}\,[f_k(0)] = \frac{4\pi^3 m^2 k}{\hbar^4} \int d\Omega\, |\langle \boldsymbol{k}'|T|\boldsymbol{k}\rangle|^2. \tag{24.103}$$

Now we use (24.90) and then (24.24) to write

$$|\langle \boldsymbol{k}'|T|\boldsymbol{k}\rangle|^2 = \frac{\hbar^4}{16\pi^4 m^2}|f_k(\theta,\phi)|^2 = \frac{\hbar^4}{16\pi^4 m^2}\frac{d\sigma}{d\Omega}. \tag{24.104}$$

Inserting this in (24.103) and doing the angular integration gives

$$\mathrm{Im}\,[f_k(0)] = \frac{k}{4\pi}\sigma, \tag{24.105}$$

which concludes the proof. □

The optical theorem is a consequence of conservation of probability. Heuristically, it holds because the total cross section σ is an effective area quantifying the probability that a particle in the incident beam will be removed by scattering. This reduction in the intensity of the beam must come from destructive interference with the part of the scattering wavefunction in the forward direction, which turns out to be proportional to the imaginary part of $f_k(0)$. The factor of $1/k$ is necessary because of dimensional analysis, as σ has units of area, while f_k and $1/k$ each have units of length.

The optical theorem is useful for at least two reasons. First, it is sometimes easier to obtain $\mathrm{Im}[f_k(0)]$ than it is to find the total cross section directly. Second, it can provide a check of the consistency of a calculation. In particular, it tells us that $\mathrm{Im}[f_k(0)]$ must be positive. However, as a warning, it can appear to fail if one makes approximations to the full scattering amplitude. We will see an example of this in the next section.

24.4 Born Approximation

The **Born approximation** is a type of perturbative expansion applied to scattering problems, in which the wavefunction within the target region is assumed to be only mildly altered by the potential. In the first-order (or "leading") Born approximation, for the purposes of (24.88) we simply take the wavefunction inside the target to be the same as the incident plane wave,

$$\psi^+(\boldsymbol{r}') \approx \frac{1}{(2\pi)^{3/2}} e^{ikz'}. \tag{24.106}$$

Recall that the wavevectors of the incident and scattered wavefunctions are $\boldsymbol{k} = k\hat{z}$ and $\boldsymbol{k}' = k\hat{r}$ respectively, so that $kz' = \boldsymbol{k} \cdot \boldsymbol{r}'$. Equation (24.88) then reduces to an integral that can be evaluated for any given potential,

$$f_k(\theta, \phi) \approx -\frac{m}{2\pi\hbar^2} \int d^3\boldsymbol{r}'\, e^{i(\boldsymbol{k}-\boldsymbol{k}')\cdot\boldsymbol{r}'}\, V(\boldsymbol{r}'). \tag{24.107}$$

In this first-order Born approximation, the sign of the potential does not make any difference, since it enters linearly in the scattering amplitude, which then gets squared to give $d\sigma/d\Omega = |f_k(\theta, \phi)|^2$. In particular, within the leading Born approximation, an attractive potential gives the same differential cross section as a repulsive potential with the same magnitude.

Before proceeding to use it, let us first address the question of the validity of the Born approximation. Intuitively, this relies on the potential not being too strong. Consider the wavefunction form of the Lippmann–Schwinger equation valid inside the target region, as it appeared in (24.82) before we took the large r limit. To make a rough estimate, let us call a the characteristic range of the potential, meaning that $V(\boldsymbol{r})$ is significant only for $r < a$. Then we can require that in the right side of (24.82) the correction term is much smaller in magnitude than the plane-wave term,

$$\frac{2m}{\hbar^2} \left| \int d^3\boldsymbol{r}'\, \frac{e^{ik|\boldsymbol{r}-\boldsymbol{r}'|}}{4\pi|\boldsymbol{r}-\boldsymbol{r}'|} V(\boldsymbol{r}')\, e^{ikz'} \right| \ll 1, \tag{24.108}$$

for $r < a$, where we have presumptively replaced the wavefunction ψ^+ by its putative approximation (24.106). To make things easier, we can just consider $\boldsymbol{r}$ close to the origin, where the correction might be expected to be largest, or at least not much smaller than at other points. Then a rough condition for validity of the Born approximation is

$$\frac{m}{2\pi\hbar^2} \left| \int d^3\boldsymbol{r}'\, \frac{1}{r'} e^{ik(r'+z')} V(\boldsymbol{r}') \right| \ll 1. \tag{24.109}$$

Now, we make the further approximation of taking $k \approx 0$ and replacing the potential by its constant average value $\overline{V}$ for $r' < a$, and 0 for $r' > a$. The rest of the integral is then

$$\int d^3\boldsymbol{r}'\, \frac{1}{r'} = 4\pi \int_0^a dr'\, r' = 2\pi a^2. \tag{24.110}$$

Thus our estimate for the validity condition of the Born approximation is

$$\frac{ma^2}{\hbar^2} |\overline{V}| \ll 1, \tag{24.111}$$

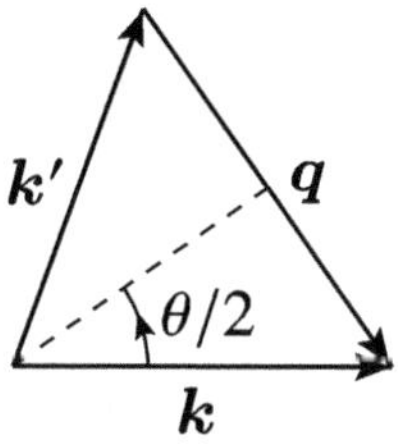

Fig. 24.2 Geometry of vectors involved in elastic scattering from a potential. The initial and final wavevectors $\boldsymbol{k}$ and $\boldsymbol{k}'$ have equal length k. The vector $\boldsymbol{q} = \boldsymbol{k} - \boldsymbol{k}'$, where $\hbar\boldsymbol{q}$ is the momentum transferred from the initial particle to the scattering potential, has length $q = 2k\sin(\theta/2)$, where θ is the angle between $\boldsymbol{k}$ and $\boldsymbol{k}'$

where a and $\overline{V}$ are the characteristic range and strength of the potential. This could have been guessed merely on dimensional grounds, at least in the low-energy limit of small k. However, this estimate for the range of validity of the Born approximation can often be too conservative, particularly for larger k, because the oscillation of the integrand can produce efficient cancellation in the left side of (24.108), not accounted for in our rough estimate. In that case, dimensional analysis suggests that the weaker condition $ma|\overline{V}|/\hbar^2 k \ll 1$ could suffice.

It is also an amusing fact that the optical theorem completely fails in the leading Born approximation. This is immediately obvious from the fact that (24.107) is real when $\boldsymbol{k}' = \boldsymbol{k}$, so that $\mathrm{Im}[f_k(0)] = 0$, but the total cross section does not vanish. Going beyond the leading order Born approximation, as we will do before the end of this section, is necessary to restore the (approximate) validity of the optical theorem.

Let us now consider examples of the leading Born approximation. An important special case is that of a spherically symmetric potential, so that the scattering amplitude has azimuthal symmetry (no dependence on ϕ). Define

$$\boldsymbol{q} = \boldsymbol{k} - \boldsymbol{k}', \tag{24.112}$$

so that $\hbar\boldsymbol{q}$ is the momentum transferred from the initial particle to the scattering potential. (See Fig. 24.2.) Then $f_k(\theta)$ only depends on k and θ through the combination

$$q = |\boldsymbol{k} - \boldsymbol{k}'| = 2k\sin(\theta/2). \tag{24.113}$$

Indeed, (24.107) becomes

$$f_k(\theta) = -\frac{m}{2\pi\hbar^2}\int d^3\boldsymbol{r}'\; V(r')\, e^{iqr'\cos\theta'}, \tag{24.114}$$

where we have taken advantage of the spherical symmetry to choose our $\boldsymbol{r}'$ coordinate system in the integral so that $\boldsymbol{q}$ is in the $\theta' = 0$ direction. It follows that

$$
\begin{aligned}
f_k(\theta) &= -\frac{m}{2\pi\hbar^2}\int_0^\infty dr' r'^2 \int_0^{2\pi} d\phi' \int_{-1}^{1} d(\cos\theta')\, V(r')\, e^{iqr'\cos\theta'} \\
&= \frac{im}{\hbar^2 q}\int_0^\infty dr' r' V(r')\big(e^{iqr'} - e^{-iqr'}\big),
\end{aligned}
\tag{24.115}
$$

which we can rewrite, without the distracting primes, as

$$
f_k(\theta) = -\frac{2m}{\hbar^2 q}\int_0^\infty dr\, r\, V(r)\sin(qr). \tag{24.116}
$$

This is the first-order Born approximation result for the scattering amplitude with an arbitrary spherically symmetric potential. In this case, $f_k(\theta)$ is always real. Note that the large-distance contribution to the integral converges if $V(r)$ decreases faster than $1/r$ for large r.

In the low-energy approximation $q \to 0$, one can use $\sin(qr) \approx qr$ to find that

$$
f_k(\theta) = -\frac{2m}{\hbar^2}\int_0^\infty dr\, r^2 V(r), \qquad \text{(small } q \text{ limit)}, \tag{24.117}
$$

which is independent of q, and therefore independent of θ. In the opposite limit of very large q, (24.116) gives $f_k(\theta) \to 0$ because of the q in the denominator together with the cancellation due to the very rapid oscillation of $\sin(qr)$ in the integrand.

As an example, consider the **Yukawa potential**, named after Hideki Yukawa, who used it in 1935 to model the strong interactions between nucleons due to the exchange of mesons. It is

$$
V(r) = -\frac{g}{r}e^{-r/a}, \tag{24.118}
$$

where g and a are constants with units of [(energy)(length)] and [length], respectively. For $r \ll a$, this has the same form as a Coulomb potential. However, for large $r \gg a$, the potential approaches 0 exponentially faster than the Coulomb potential does. For that reason it is sometimes also known as the **screened Coulomb potential**, and a can be thought of as the range of the potential. Applying (24.116) gives

$$
f_k(\theta) = \frac{2mg}{\hbar^2 q}\int_0^\infty dr\, e^{-r/a}\sin(qr) = \frac{2mg}{\hbar^2}\frac{1}{q^2 + 1/a^2}. \tag{24.119}
$$

Now we can use $q^2 = 4k^2\sin^2(\theta/2) = 2k^2(1 - \cos\theta)$, to find

$$
\frac{d\sigma}{d\Omega} = \left|f_k(\theta)\right|^2 = \left(\frac{2mg}{\hbar^2}\right)^2 \frac{1}{[2k^2(1-\cos\theta) + 1/a^2]^2} \tag{24.120}
$$

in the Born approximation. If we take the long-range limit $a \to \infty$ and let $-g = Z_1 Z_2 e^2$, then

$$\frac{d\sigma}{d\Omega} = \frac{Z_1^2 Z_2^2 e^4 m^2}{4\hbar^4 k^4 \sin^4(\theta/2)} = \frac{Z_1^2 Z_2^2 e^4}{16E^2 \sin^4(\theta/2)}, \tag{24.121}$$

where in the last expression we used $E = \hbar^2 k^2/2m$. This is the same differential scattering cross section as we found for the corresponding classical Rutherford scattering problem in Sect. 24.1.

In the limit of low-energy scattering from the Yukawa potential, the rough condition for validity of the Born approximation is obtained by taking $\overline{V} = g/a$ in (24.111), so

$$|g| \ll \hbar^2/ma. \tag{24.122}$$

This can be compared to the condition necessary for a bound state to exist, which can be shown (for example, applying the variational method with a carefully chosen trial wavefunction) to be approximately $g > 0.8399\,\hbar^2/ma$. The general lesson is that if the potential allows a bound state, then it is too strong for the Born approximation to be valid for low E. However, it can be valid for larger E, such that $k \gg 1/a$.

Returning to the general scattering problem, let us consider how to extend the Born approximation beyond leading order. We can start with the expression (24.93) for the transition operator, and use the expansion $(I-X)^{-1} = I+X+X^2+X^3+\cdots$. The result is

$$T = V + V\frac{1}{E - H_0 + i\epsilon}V + V\frac{1}{E - H_0 + i\epsilon}V\frac{1}{E - H_0 + i\epsilon}V + \cdots. \tag{24.123}$$

The ordering of operators matters, because H_0 and V do not commute unless the latter is a trivial (constant) potential. Recalling, from (24.90), that

$$f_{\mathbf{k}}(\theta, \phi) = -\frac{4\pi^2 m}{\hbar^2}\langle \mathbf{k}'|T|\mathbf{k}\rangle, \tag{24.124}$$

we obtain an expansion in powers of the potential,

$$f_{\mathbf{k}}(\theta, \phi) = f_k^{(1)}(\theta, \phi) + f_k^{(2)}(\theta, \phi) + f_k^{(3)}(\theta, \phi) + \cdots, \tag{24.125}$$

where $f_k^{(N)}(\theta, \phi)$ has N factors of the potential V. In particular,

$$f_k^{(1)}(\theta, \phi) = -\frac{4\pi^2 m}{\hbar^2}\langle \mathbf{k}'|V|\mathbf{k}\rangle \tag{24.126}$$

is the leading Born approximation that we have already studied, which just amounts to setting $T = V$ in (24.124). The higher-order contributions in the Born expansion are now seen to be

$$f_k^{(2)}(\theta, \phi) = -\frac{4\pi^2 m}{\hbar^2} \langle \boldsymbol{k}' | V \frac{1}{E - H_0 + i\epsilon} V | \boldsymbol{k} \rangle , \tag{24.127}$$

$$f_k^{(3)}(\theta, \phi) = -\frac{4\pi^2 m}{\hbar^2} \langle \boldsymbol{k}' | V \frac{1}{E - H_0 + i\epsilon} V \frac{1}{E - H_0 + i\epsilon} V | \boldsymbol{k} \rangle , \tag{24.128}$$

etc.

We can now work out these contributions in the wavefunction representation, in terms of the Green function, by using completeness with respect to position and (24.72) and (24.71) for the plane wave and Green function. The results are

$$f_k^{(1)}(\theta, \phi) = -\frac{1}{4\pi}\left(\frac{2m}{\hbar^2}\right) \int d^3\boldsymbol{r}_1 \, e^{i\boldsymbol{k}\cdot\boldsymbol{r}_1} V(\boldsymbol{r}_1) e^{-i\boldsymbol{k}'\cdot\boldsymbol{r}_1}, \tag{24.129}$$

$$f_k^{(2)}(\theta, \phi) = -\frac{1}{4\pi}\left(\frac{2m}{\hbar^2}\right)^2 \int d^3\boldsymbol{r}_1 \int d^3\boldsymbol{r}_2 \, e^{i\boldsymbol{k}\cdot\boldsymbol{r}_1} V(\boldsymbol{r}_1) G_+(\boldsymbol{r}_1, \boldsymbol{r}_2) V(\boldsymbol{r}_2) e^{-i\boldsymbol{k}'\cdot\boldsymbol{r}_2}, \tag{24.130}$$

$$f_k^{(3)}(\theta, \phi) = -\frac{1}{4\pi}\left(\frac{2m}{\hbar^2}\right)^3 \int d^3\boldsymbol{r}_1 \int d^3\boldsymbol{r}_2 \int d^3\boldsymbol{r}_3 \, e^{i\boldsymbol{k}\cdot\boldsymbol{r}_1} V(\boldsymbol{r}_1) G_+(\boldsymbol{r}_1, \boldsymbol{r}_2) V(\boldsymbol{r}_2) G_+(\boldsymbol{r}_2, \boldsymbol{r}_3) V(\boldsymbol{r}_3) e^{-i\boldsymbol{k}'\cdot\boldsymbol{r}_3}. \tag{24.131}$$

A diagrammatic version of these contributions is shown in Fig. 24.3. The formula for $f_k^{(N)}(\theta, \phi)$ contains N vertices where the scattering potential acts. At each scattering vertex, labeled $n = 1, \ldots, N$, there is an integration

$$\text{vertex} \leftrightarrow \frac{2m}{\hbar^2} \int d^3\boldsymbol{r}_n \, V(\boldsymbol{r}_n), \tag{24.132}$$

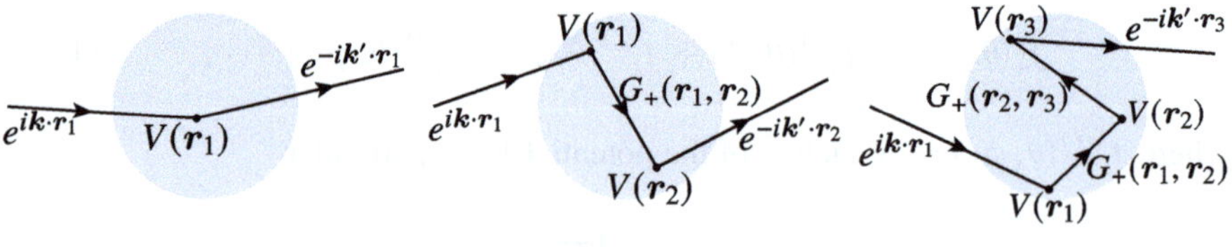

Fig. 24.3 Diagrammatic representation of the first three terms in the Born approximation expansion for the scattering amplitude, $f_k^{(1)}$, $f_k^{(2)}$, and $f_k^{(3)}$ from (24.129)–(24.131). Each of the vertex positions $\boldsymbol{r}_n$ is integrated over, with an associated factor of $2mV(\boldsymbol{r}_n)/\hbar^2$

and we assign factors

$$\text{internal lines} \leftrightarrow G_+(\boldsymbol{r}_n, \boldsymbol{r}_{n+1}) \tag{24.133}$$

to the lines between consecutive vertices $\boldsymbol{r}_n$ and $\boldsymbol{r}_{n+1}$. There are also two factors

$$\text{external lines} \leftrightarrow e^{i\boldsymbol{k}\cdot\boldsymbol{r}_1} e^{-i\boldsymbol{k}'\cdot\boldsymbol{r}_N} \tag{24.134}$$

associated with the initial and final momenta, and a single factor of $-1/4\pi$.

An intuitive interpretation for the Born expansion is that the particle is described by plane waves except for interactions with the potential. Between potential interactions, it propagates according to the Green function of the Helmholtz operator, and the initial and final wavefunctions are free plane waves. All possible numbers of interactions are summed over, and then all possible interaction positions are integrated over, to give the total scattering amplitude.

24.5 Spherical Potential Scattering and the Partial Wave Expansion

Consider scattering from a spherically symmetric potential $V(\boldsymbol{r}) = V(r)$. As usual, we take the initial particle wavevector to be $\boldsymbol{k} = \hat{z}k$, so that we have azimuthal symmetry and the scattering amplitude can only depend on θ, not ϕ. This implies that we can write a **partial wave expansion** in the orbital angular momentum quantum number l,

$$f_k(\theta) = \sum_{l=0}^{\infty} (2l+1) P_l(\cos\theta)\, f_l(k). \tag{24.135}$$

The $P_l(\cos\theta)$ are the Legendre polynomials, and the factor of $2l+1$ is a convention. This equation defines the quantities $f_l(k)$, called **partial wave amplitudes**, that appear on the right side. They can equivalently be viewed as functions $f_l(E)$ of the energy $E = \hbar^2 k^2/2m$.

To better understand the partial wave expansion, recall the expansion of a plane wave in the absence of scattering, worked out at the end of Sect. 10.2,

$$e^{ikz} = e^{ikr\cos\theta} = \sum_{l=0}^{\infty} (2l+1) P_l(\cos\theta)\, i^l j_l(kr). \tag{24.136}$$

The $j_l(kr)$ are spherical Bessel functions. For large r, we find from (10.27) that

$$i^l j_l(kr) \approx \frac{e^{ikr} - e^{-ikr+i\pi l}}{2ikr}, \qquad (r \gg 1/k). \tag{24.137}$$

Plugging (24.135)–(24.137) into the scattering wavefunction of (24.87), and collecting like terms, we find

$$\psi^+(\boldsymbol{r}) = \frac{1}{(2\pi)^{3/2}2ik}\sum_{l=0}^{\infty}(2l+1)P_l(\cos\theta)\left(S_l\frac{e^{ikr}}{r}+(-1)^{l+1}\frac{e^{-ikr}}{r}\right), \quad (24.138)$$

where we have defined

$$S_l = 1 + i2kf_l(k), \quad (24.139)$$

and the large distance limit $r \to \infty$ is now implicit. The terms proportional to e^{ikr} and e^{-ikr} are outgoing and incoming spherical waves, respectively; compare to the discussion around (10.34)–(10.36). They are called the **partial waves** of angular momentum l. Note that the outgoing spherical waves depend on the potential $V(r)$ through the coefficients S_l. In contrast, the incoming spherical waves do not depend on the S_l and therefore do not depend on $V(r)$; they are just a rewriting of the incoming beam.

The quantities S_l have a simple interpretation: they are the S-matrix elements in an angular momentum basis. To see this, consider general spherical coordinate angles $\Omega_{\boldsymbol{k}} = (\theta_{\boldsymbol{k}}, \phi_{\boldsymbol{k}})$ and $\Omega_{\boldsymbol{k}'} = (\theta_{\boldsymbol{k}'}, \phi_{\boldsymbol{k}'})$ for the directions of $\boldsymbol{k}$ and $\boldsymbol{k}'$, respectively. The angle between these two vectors is the θ appearing in the preceding equations. Therefore, the spherical harmonics addition formula (8.164) allows us to rewrite the scattering amplitude in (24.135) as

$$f_{\boldsymbol{k}}(\theta) = 4\pi\sum_{l=0}^{\infty}\sum_{m=-l}^{l} Y_l^m(\Omega_{\boldsymbol{k}'})^*\, Y_l^m(\Omega_{\boldsymbol{k}})\,\frac{S_l-1}{i2k}. \quad (24.140)$$

So, (24.90) tells us that the transition operator matrix elements are

$$\langle\boldsymbol{k}'|T|\boldsymbol{k}\rangle = \frac{i\hbar^2}{2\pi mk}\sum_{l,m} Y_l^m(\Omega_{\boldsymbol{k}'})^*\, Y_l^m(\Omega_{\boldsymbol{k}})\,(S_l-1), \quad (24.141)$$

and so the S-matrix elements are, from (24.69),

$$\langle\boldsymbol{k}'|S|\boldsymbol{k}\rangle = \delta^{(3)}(\boldsymbol{k}-\boldsymbol{k}') + \frac{\hbar^2}{mk}\delta(E-E')\sum_{l,m} Y_l^m(\Omega_{\boldsymbol{k}'})^*\, Y_l^m(\Omega_{\boldsymbol{k}})\,(S_l-1). \quad (24.142)$$

This can be simplified using $\delta(E-E') = (m/\hbar^2 k)\delta(k-k')$, and the definition

$$\delta^{(3)}(\boldsymbol{k}-\boldsymbol{k}') = \frac{1}{k^2}\delta(k-k')\delta(\phi_{\boldsymbol{k}}-\phi_{\boldsymbol{k}'})\delta(\cos\theta_{\boldsymbol{k}}-\cos\theta_{\boldsymbol{k}'}), \quad (24.143)$$

and the spherical harmonic completeness formula (8.125), reproduced here for convenience, $\delta(\phi - \phi')\delta(\cos\theta - \cos\theta') = \sum_{l=0}^{\infty}\sum_{m=-l}^{l} Y_l^m(\theta', \phi')^* \, Y_l^m(\theta, \phi)$. The result is

$$\langle \boldsymbol{k}'|S|\boldsymbol{k}\rangle = \frac{1}{k^2}\delta(k - k')\sum_{l=0}^{\infty}\sum_{m=-l}^{l} Y_l^m(\Omega_{\boldsymbol{k}'})^* \, Y_l^m(\Omega_{\boldsymbol{k}}) \, S_l. \tag{24.144}$$

Now consider a basis of free-particle angular momentum eigenstates $|k, l, m\rangle$, of the type discussed in Sect. 10.2. These states are related to the plane-wave states by

$$|k, l, m\rangle = \int d\Omega_{\boldsymbol{k}} \; Y_l^m(\Omega_{\boldsymbol{k}})^* \; |\boldsymbol{k}\rangle \,, \tag{24.145}$$

with the inverse relation

$$|\boldsymbol{k}\rangle = \sum_{l=0}^{\infty}\sum_{m=-l}^{l} Y_l^m(\Omega_{\boldsymbol{k}}) \; |k, l, m\rangle \,, \tag{24.146}$$

and satisfy orthonormality and completeness relations

$$\langle k', l', m'|k, l, m\rangle = \frac{1}{k^2}\delta(k - k') \, \delta_{ll'} \, \delta_{mm'}, \tag{24.147}$$

$$\int_0^{\infty} dk \, k^2 \sum_{l=0}^{\infty}\sum_{m=-l}^{l} |k, l, m\rangle\langle k, l, m| = I. \tag{24.148}$$

It is a short exercise to show that the content of (24.144) can now be expressed as

$$\langle k', l', m'|S|k, l, m\rangle = \frac{1}{k^2}\delta(k - k') \, \delta_{ll'} \, \delta_{mm'} \, S_l, \tag{24.149}$$

or, in yet another way, as

$$S = \int_0^{\infty} dk \, k^2 \sum_{l=0}^{\infty}\sum_{m=-l}^{l} S_l \, |k, l, m\rangle\langle k, l, m| \,. \tag{24.150}$$

The fact that the S-matrix is diagonal in the basis of orbital angular momentum eigenstates is due to the assumed rotational symmetry of the potential.

The unitarity of the S-matrix now implies

$$
\begin{aligned}
I &= S^{\dagger} S \\
&= \int_0^{\infty} dk'\, k'^2 \sum_{l',m'} \int_0^{\infty} dk\, k^2 \sum_{l,m} S^*_{l'} S_l \,|k', l', m'\rangle\, \langle k', l', m'|k, l, m\rangle\, \langle k, l, m| \\
&= \int_0^{\infty} dk\, k^2 \sum_{l,m} |S_l|^2\, |k, l, m\rangle\langle k, l, m|
\end{aligned} \tag{24.151}
$$

where the last equality made use of the orthonormality relation (24.147) to eliminate the k' integral and the l' and m' sums. Comparison with (24.148) establishes that $|S_l|^2 = 1$ for each l. This crucial fact is called **partial wave unitarity**, and it is a way of restating the assumption that the potential is only scattering the particles, not creating or destroying them. It means that we can write, for the complex factors S_l in (24.138),

$$
S_l = 1 + i2kf_l(k) = e^{i2\delta_l}, \tag{24.152}
$$

which defines a set of real numbers δ_l called the **phase shifts** of the angular momentum l partial waves. (The factor of 2 in the exponent is a traditional convention.) Equation (24.152) exposes that the effect of the scattering is just to give each outgoing partial wave of orbital angular momentum l an extra phase factor $e^{i2\delta_l}$ relative to the corresponding incoming partial wave.

The general scattering problem for a spherically symmetric potential thus reduces to finding the phase shifts δ_l appearing in the wavefunction for large r obtained by rewriting (24.138),

$$
\psi^+(\mathbf{r}) = \frac{1}{(2\pi)^{3/2}} \sum_{l=0}^{\infty} (2l+1) P_l(\cos\theta) \frac{e^{i2\delta_l} e^{ikr} - e^{-i(kr - l\pi)}}{2ikr}, \qquad (r \to \infty). \tag{24.153}
$$

In the limit that the potential $V(r)$ vanishes, $\delta_l = 0$, and more generally the δ_l parameterize the effect of $V(r)$ on each l partial wave. We will show before the end of this section that it is a good approximation to use only the lowest few partial waves $l = 0, 1, 2, \ldots$, except in the case of very high-energy scattering.

Some useful equivalent ways of writing the relation between the phase shifts and the partial wave amplitudes $f_l(k)$ are, from (24.152),

$$
f_l(k) = \frac{e^{i2\delta_l} - 1}{2ik} = \frac{e^{i\delta_l} \sin\delta_l}{k} = \frac{1}{k(\cot\delta_l - i)}. \tag{24.154}
$$

Using the next-to-last expression of (24.154) in (24.135), we have the scattering amplitude in terms of the phase shifts,

$$f_k(\theta) = \frac{1}{k}\sum_{l=0}^{\infty}(2l+1)e^{i\delta_l}\sin\delta_l\, P_l(\cos\theta). \tag{24.155}$$

This can be used to compute the total cross section,

$$\sigma = 2\pi\int_{-1}^{1} d(\cos\theta)|f_k(\theta)|^2 \tag{24.156}$$

$$= \frac{2\pi}{k^2}\sum_{l=0}^{\infty}\sum_{l'=0}^{\infty}(2l+1)(2l'+1)e^{i(\delta_l-\delta_{l'})}\sin(\delta_l)\sin(\delta_{l'}) \int_{-1}^{1} d(\cos\theta)P_{l'}(\cos\theta)P_l(\cos\theta). \tag{24.157}$$

To simplify this, apply the Legendre polynomial orthonormality relation (8.139), reproduced here for convenience: $\int_{-1}^{1} du\, P_{l'}(u)P_l(u) = \delta_{ll'}2/(2l+1)$. We arrive at

$$\sigma = \frac{4\pi}{k^2}\sum_{l=0}^{\infty}(2l+1)\sin^2\delta_l. \tag{24.158}$$

Note that even though the differential cross section $d\sigma/d\Omega = |f_k(\theta)|^2$ certainly has interferences between partial wave amplitudes with different l, there is no such interference in the total cross section. As a check, or an alternative derivation, the optical theorem tells us that

$$\sigma = \frac{4\pi}{k}\mathrm{Im}\big[f_k(\theta=0)\big] = \frac{4\pi}{k}\left(\frac{1}{k}\sum_{l=0}^{\infty}(2l+1)P_l(1)\sin(\delta_l)\mathrm{Im}[e^{i\delta_l}]\right), \tag{24.159}$$

which reproduces (24.158) after using $\mathrm{Im}[e^{i\delta_l}] = \sin\delta_l$ and the Legendre polynomial normalization condition $P_l(1) = 1$.

An important consequence of (24.154), as encoded in (24.158), follows if we write

$$\sigma = \sum_{l=0}^{\infty}\sigma_l, \tag{24.160}$$

where σ_l is the contribution from orbital angular momentum l. Then $\sin^2\delta_l \le 1$ implies

$$\sigma_l \le \frac{4\pi}{k^2}(2l+1). \tag{24.161}$$

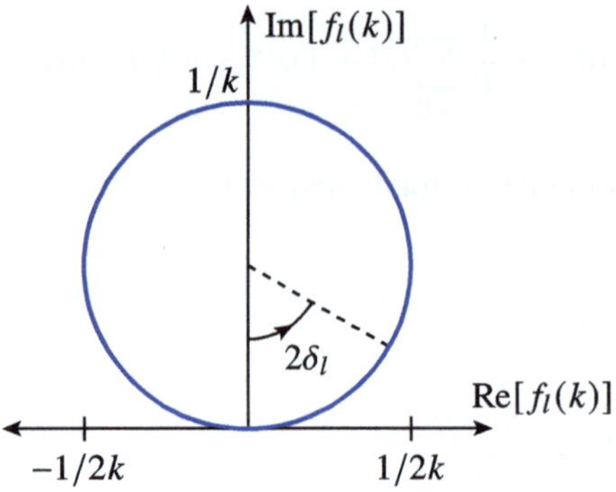

Fig. 24.4 Partial wave unitarity illustrated: the possible values of the partial wave amplitude $f_l(k) = (e^{i2\delta_l} - 1)/2ik$ lie on a circle with radius $1/2k$ in the upper-half complex plane, and are related to the phase shift angle δ_l as shown. The Born approximation limit is realized near the bottom of the circle, for small $\sin\delta_l$, which gives small and nearly real $f_l(k)$. The partial wave unitarity bound (24.161) is saturated for $f_l(k) = i/k$, at the top of the circle, with $\sin^2\delta_l = 1$, so $\delta_l = (n + 1/2)\pi$ for integer n

This is known as the **partial wave unitarity bound**. Equality is achieved only if $\sin^2\delta_l = 1$, which implies $\delta_l = (n + 1/2)\pi$ for integer n. In that case, σ_l is said to saturate the partial wave unitarity bound. Remarkably, this bound on the contribution σ_l for each partial wave applies no matter how big the scattering potential is. It becomes stronger with increasing energy, as it can be rewritten

$$\sigma_l \leq \frac{2\pi\hbar^2}{mE}(2l + 1), \tag{24.162}$$

and it should be viewed as a requirement imposed by the conservation of probability.

Partial wave unitarity also restricts the values that $f_l(k)$ can take to a circle of radius $1/2k$ in the complex plane, as can be seen from the first equality of (24.154) and illustrated in Fig. 24.4. If δ_l is small, then $f_l(k)$ is also small in magnitude, and is almost purely real. This is realized in the leading Born approximation limit. In the opposite case that $|f_l(k)|$ is maximal, so that the partial wave unitarity bound (24.161) is saturated, then $f_l(k)$ is at the top of the circle and is purely imaginary, equal to i/k.

Consider potentials that vanish completely[3] outside of a finite range, so that $V(r) = 0$ for $r > R$. Then the exact (not just the large r limit) wavefunction for a stationary scattering state can be written as, taking into account the azimuthal symmetry,

[3] This assumption is false for many interesting potentials. However, the results we are about to obtain are often a good approximation, provided that the potential $V(r)$ is suitably weak for $r > R$.

$$\psi^+(\mathbf{r}) = \frac{1}{(2\pi)^{3/2}} \sum_{l=0}^{\infty} (2l+1) P_l(\cos\theta) F_l(r), \tag{24.163}$$

where the radial wavefunction for the angular momentum l partial wave is

$$F_l(r) = \begin{cases} A_l(r) & (\text{for } r \leq R), \\ B_l(r) & (\text{for } r \geq R). \end{cases} \tag{24.164}$$

The exact form of the function $A_l(r)$ may be quite difficult to obtain, since it depends on the scattering potential. However, because we are taking the potential to vanish for $r \geq R$, the function $B_l(r)$ must be a superposition of the free-particle solutions that we found in Sect. 10.2,

$$B_l(r) = c_l h_l^{(1)}(kr) + d_l h_l^{(2)}(kr), \tag{24.165}$$

where c_l and d_l are constant coefficients, and $h_l^{(1)}$ and $h_l^{(2)}$ are the spherical Hankel functions

$$h_l^{(1)}(kr) = j_l(kr) + i n_l(kr) \sim -i\, e^{i(kr - l\pi/2)}/kr, \tag{24.166}$$

$$h_l^{(2)}(kr) = j_l(kr) - i n_l(kr) \sim i\, e^{-i(kr - l\pi/2)}/kr, \tag{24.167}$$

where the large r asymptotic forms from (10.34) are shown. Matching to our scattering wavefunction for large r, (24.153), we can solve for the constants c_l and d_l in terms of the phase shifts δ_l. The result is

$$B_l(r) = \frac{1}{2} e^{il\pi/2} \left[e^{i2\delta_l} h_l^{(1)}(kr) + h_l^{(2)}(kr) \right], \tag{24.168}$$

or equivalently,

$$B_l(r) = e^{il\pi/2} e^{i\delta_l} \left[\cos(\delta_l)\, j_l(kr) - \sin(\delta_l)\, n_l(kr) \right]. \tag{24.169}$$

To obtain scattering state solutions, one can now adopt the following strategy. First, the difficult part: solve the time-independent Schrödinger differential equation for $A_l(r)$ in the region of nonzero potential, $r \leq R$. Because this is a second-order differential equation in r, in general there will be two linearly independent solutions with arbitrary coefficients. To fix the coefficients, match these solutions to (24.169) at $r = R$, using continuity of the wavefunction,

$$A_l(R) = B_l(R), \tag{24.170}$$

and, if the potential is finite at $r = R$, continuity of the derivative

$$A_l'(R) = B_l'(R). \tag{24.171}$$

Since the $B_l(R)$ are given in (24.169) in terms of known functions, one can solve for the δ_l. To write the solution as compactly as possible, it is useful to define the quantities

$$\alpha_l \equiv A_l'(R)/A_l(R), \tag{24.172}$$

in which the overall normalization cancels. Then, since (24.170) and (24.171) imply

$$\alpha_l = B_l'(R)/B_l(R), \tag{24.173}$$

we obtain the general solution for the phase shift using (24.169),

$$\tan\delta_l = \frac{\alpha_l j_l(kR) - kj_l'(kR)}{\alpha_l n_l(kR) - kn_l'(kR)}. \tag{24.174}$$

Summarizing: once the α_l have been found using (24.172), the phase shifts can be calculated from (24.174), and then used in (24.155) to obtain $f_k(\theta)$, which in turn gives the differential cross section $|\,f_k(\theta)|^2$.

At leading order in an expansion in small kR, (10.24) and (10.25) for the spherical Bessel and Neumann functions can be used to find

$$\tan\delta_l = \frac{(kR)^{2l+1}(l - R\alpha_l)}{(2l-1)!!\,(2l+1)!!\,(l+1+R\alpha_l)}. \tag{24.175}$$

For $l = 0, 1, 2, 3$, and 4, the denominator factor $(2l-1)!!\,(2l+1)!!$ is respectively $1, 3, 45, 1575$, and 99,225, so the coefficient of $(kR)^{2l+1}$ gets smaller rapidly as l increases. For low and intermediate energies (meaning as long as kR is not very large compared to 1), it is therefore justified to keep only the first few partial wave contributions, as we had promised to show earlier.

Let us now specialize to the low energy limit, $kR \ll 1$, where it is justified to neglect all but the s-wave phase shift δ_0 contribution, which behaves like $\tan\delta_0 \propto k$ and provides an isotropic differential cross section. Expanding the $l = 0$ case of (24.174) to the next-to-leading order in k, the result can be parameterized as

$$k\cot\delta_0 = -\frac{1}{a} + \frac{1}{2}r_0 k^2 + \cdots, \tag{24.176}$$

which defines length scales a and r_0, called the **scattering length** and the **effective range** respectively. In terms of R and $\alpha_0 = A_0'(R)/A_0(R)$, one finds

$$a = \frac{\alpha_0 R^2}{1+\alpha_0 R}, \tag{24.177}$$

and the determination of r_0 is left to Exercise 24.8. The scattering length a can be either positive or negative depending on the sign of α_0, and can be much larger in magnitude than the length scale R built into the potential, as we will see. In terms of these parameters, we can evaluate the low-energy, and therefore s-wave, cross section from (24.158),

$$\sigma = \frac{4\pi}{k^2}\frac{1}{1+\cot^2\delta_0} = \frac{4\pi a^2}{1+(1-r_0/a)a^2k^2}, \tag{24.178}$$

where terms of higher order than k^2 have been consistently neglected in the denominator. In the extreme low-energy approximation, σ has a simple dependence on the scattering length,

$$\sigma \approx 4\pi a^2. \tag{24.179}$$

In principle, a, r_0, and higher-order contributions to (24.176) can be obtained experimentally from the energy dependence and interference effects in the low-energy limit.

24.6 Bound States, Resonances, and Poles in Scattering Amplitudes

The partial wave scattering amplitudes and cross sections, viewed as functions of the energy E analytically continued to the complex domain, have an analytic structure that conveys information about the potential. As we will explain in this section, a pole (divergence) in the partial wave scattering amplitude corresponds either to a true bound state, or to a **resonance**, also known as a **quasi-bound state**, of the potential $V(r)$.

First, consider the case that the potential has one or more discrete bound states. We assume that the additive constant part of the potential has been fixed so that it vanishes at large distances, $V(\infty) = 0$, so that the condition for a bound state is an isolated eigenvalue $E_{\text{bound}} < 0$ for the time-independent Schrödinger equation. To uncover the connection between such states and the scattering amplitude, consider the unbound scattering solutions for the wavefunction $\psi_E(\boldsymbol{r})$ with positive $E = \hbar^2k^2/2m$. Here k is the wavenumber far from the origin as discussed in Sects. 24.1 and 24.2, with the asymptotic form of the wavefunction for large r as given in (24.153). Suppose we try the trick of making the replacement $k \to i\kappa$ in that wavefunction, where κ is now taken to be a real number. The solution so obtained, if it exists, will have energy $E_{\text{bound}} = -\hbar^2\kappa^2/2m < 0$ and therefore will be a bound state.

The only obstacle to constructing bound state wavefunction solutions in this way is that they will almost always be hopelessly non-normalizable, and therefore unphysical. To see this, note that the asymptotic form for large r given in (24.153) will be, after $k \to i\kappa$,

$$\psi(\mathbf{r}) = \frac{1}{(2\pi)^{3/2}} \sum_{l=0}^{\infty} (2l+1) P_l(\cos\theta) \frac{-S_l e^{-\kappa r} + (-1)^l e^{\kappa r}}{2\kappa r}, \qquad (r \to \infty). (24.180)$$

The term proportional to $e^{-\kappa r}$ is sensible, but the term $e^{\kappa r}$ evidently gives a probability density that diverges exponentially for large r. It appears that this wavefunction would describe a physically absurd state in which the probability to find the particle in any finite region is vanishingly small compared to the probability to find it farther away from the origin.

However, this solution can be salvaged for special, isolated values of the energy. Since we are considering a spherically symmetric potential, we can look for eigenstates of the total orbital angular momentum. Taking a term of fixed l in (24.180), and renormalizing the wavefunction by multiplying by a constant factor $-2(2\pi)^{3/2}/(2l+1)S_l$, we have

$$\psi_l(\mathbf{r}) = P_l(\cos\theta) \left(e^{-\kappa r} + (-1)^{l+1} \frac{e^{\kappa r}}{S_l} \right) / \kappa r, \qquad (r \to \infty). \qquad (24.181)$$

In order to re-interpret this wavefunction as a sensible (normalizable) energy and angular momentum eigenstate wavefunction at large r, it is necessary to eliminate the offending second term by requiring

$$S_l \to \infty. \qquad (24.182)$$

We conclude that a true bound state for the potential $V(r)$ with negative energy E_{bound} must correspond to a pole in the S-matrix element $S_l = e^{i2\delta_l}$, and thus in the scattering amplitude $f_l(k) = (e^{i2\delta_l} - 1)/(2ik)$, when viewed as a function of energy analytically continued to $E = E_{\text{bound}} < 0$. Due to the factor $P_l(\cos\theta)$, the orbital angular momentum quantum number of the bound state is l. So, one can relate the bound state energies, and even their angular momenta, to the poles in the partial-wave scattering amplitudes.

Since scattering solutions always have positive energy, and the bound states always have negative energy, one never actually hits the bound state pole directly in scattering experiments. Indeed, this is clear from the fact that for positive E, partial wave unitarity tells us that $e^{i2\delta_l}$ has magnitude 1. Nevertheless, the presence of the bound states can be inferred from the functional dependence of $e^{i2\delta_l}$ (and thus the cross section) on E, particularly if $-E_{\text{bound}}$ is small.

As an example, consider low-energy (small k) scattering as discussed in Eqs. (24.176)–(24.179) of the previous section. Taking $k \to i\kappa$ in (24.176), we see from (24.154) that the s-wave scattering amplitude can have a pole at the solution to

$$\kappa = \frac{1}{a} + \frac{1}{2} r_0 \kappa^2, \qquad (24.183)$$

with κ positive to ensure that the bound-state wavefunction is well-behaved at $r \to \infty$. Thus, if the scattering length a is found to be positive and not too small, one can predict a weakly bound state with $\kappa \approx 1/a$ and energy close to

$$E_{\text{bound}} \approx -\frac{\hbar^2}{2ma^2}. \tag{24.184}$$

Including the effect of r_0, this is more precisely

$$E_{\text{bound}} \approx -\frac{\hbar^2\kappa^2}{2m}, \tag{24.185}$$

where κ is the smaller solution to the quadratic equation (24.183), approximately

$$\kappa \approx \frac{1}{a} + \frac{r_0}{2a^2}. \tag{24.186}$$

The low-energy s-wave cross section (24.178) in the presence of a weakly bound state with energy $E_{\text{bound}} < 0$ can now be written in the convenient form

$$\sigma_0 \approx \frac{4\pi a^2}{1 - E/E_{\text{bound}}}, \tag{24.187}$$

as dictated by the $E = 0$ limit found in (24.179), together with the requirement of a pole at $E = E_{\text{bound}}$.

Returning to the case of general (not necessarily small) k, the bound-state energies are not the only possible poles in the partial wave amplitude. To see this, let us look for peaks in the cross section. The partial wave cross section from (24.158) can be rewritten as

$$\sigma_l = \frac{4\pi}{k^2}\frac{2l+1}{1+\cot^2\delta_l}, \tag{24.188}$$

where $\cot^2\delta_l$ is often quite large, as for example in the Born approximation limit. However, if it happens for some E that $\cot\delta_l \approx 0$, then σ_l will peak at that energy, saturating the partial wave unitarity bound. Suppose that $E = E_{\text{res}}$ is a resonant energy that makes $\cot\delta_l = 0$. For energies that are close to E_{res}, we can expand $\cot\delta_l$ in $E - E_{\text{res}}$, so that

$$\cot(\delta_l) = \cot(\delta_l)\Big|_{E=E_{\text{res}}} + (E - E_{\text{res}})\frac{d}{dE}\cot(\delta_l)\Big|_{E=E_{\text{res}}} + O\Big((E - E_{\text{res}})^2\Big). \tag{24.189}$$

The first term on the right vanishes by assumption, and we define the quantity Γ, with units of energy, according to

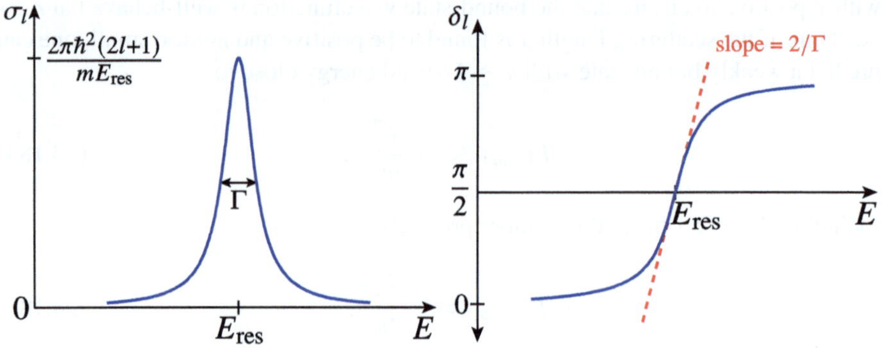

Fig. 24.5 Behavior of the partial-wave cross section σ_l and the phase shift δ_l as a function of the scattering energy E near a resonance pole $E_{res} - i\Gamma/2$. The partial-wave cross section is given by the Breit–Wigner lineshape shown on the left. The phase shift δ_l increases by π near each resonance, as shown on the right

$$\frac{2}{\Gamma} = -\frac{d}{dE}\cot(\delta_l)\bigg|_{E=E_{res}}, \tag{24.190}$$

so that, near the resonant energy,

$$\cot(\delta_l) \approx -2(E - E_{res})/\Gamma. \tag{24.191}$$

Using this in (24.188) gives, for E close to E_{res},

$$\sigma_l \approx \frac{4\pi(2l+1)}{k^2}\frac{\Gamma^2/4}{(E - E_{res})^2 + \Gamma^2/4}. \tag{24.192}$$

Excluding the energy dependence of the $1/k^2 = \hbar^2/2mE$ factor, this is the Breit–Wigner, or Lorentzian, lineshape with full width at half maximum (FWHM) Γ for the quasi-bound state resonance with angular momentum l and energy E_{res}. It is depicted in Fig. 24.5.

Near $E = E_{res}$, (24.191) tells us that the phase shift behaves like

$$\delta_l \approx \pi\,(n + 1/2) + \arctan\left(\frac{E - E_{res}}{\Gamma/2}\right), \tag{24.193}$$

for some integer n. This behavior is also sketched in Fig. 24.5, showing that the phase shift rises through $\pi(n + 1/2)$ near a resonant energy. These resonances correspond to poles in the partial wave scattering amplitude as a function of complex energy, at $E = E_{res} - i\Gamma/2$, because

$$f_l(k) = \frac{1}{k\,(\cot\delta_l - i)} \approx -\frac{\Gamma/2k}{E - E_{res} + i\Gamma/2}, \tag{24.194}$$

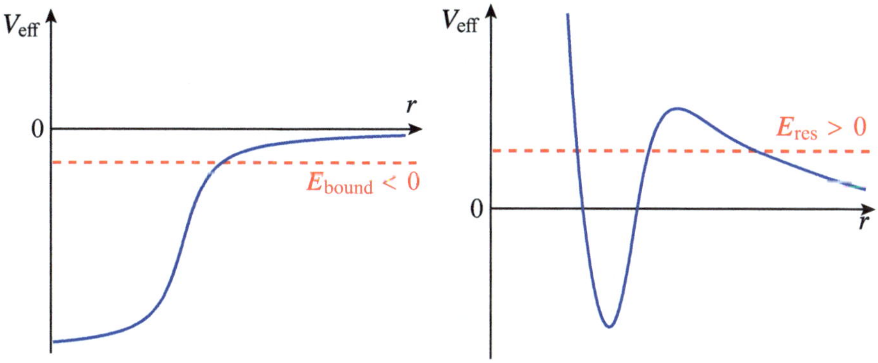

Fig. 24.6 Sketches of example effective potentials $V_{\text{eff}}(r)$ and energies for a true bound state with $l = 0$ (left) and a quasi-bound state resonance (right), which will cause the scattering amplitude and the cross section to have poles at $E = E_{\text{bound}} < 0$ or at $E = E_{\text{res}} - i\Gamma/2$, respectively. In the latter case, the local minimum of the effective potential could be either negative (as shown) or positive, and could arise from the combination of the repulsive centrifugal core $\hbar^2 l(l+1)/2mr^2$ and a potential $V(r)$ that is attractive at intermediate distances

which follows from (24.154). Recall from the discussion at the end of Sect. 5.6 that a complex energy $E_{\text{res}} - i\Gamma/2$ is the tell-tale signature of an unstable state. Therefore, the resonance can be identified with an unstable quasi-bound state with lifetime $\hbar/\Gamma$.

Figure 24.6 shows plausible potential shapes that could give rise to poles in the scattering amplitude from a true bound state (left panel) or a resonant quasi-bound state (right panel). We saw at the end of Sect. 10.4 that in three dimensions, the existence of a true bound state with energy $E_{\text{bound}} < 0$ is not guaranteed unless the potential well is sufficiently deep. Unlike a true bound state, a quasi-bound state resonant energy E_{res} is positive. One way that such a quasi-bound state can arise is from the interplay between an attractive central potential $V(r)$ and the orbital angular momentum contribution to the effective potential for the radial wavefunction in the Schrödinger equation (recall the discussion in Sect. 10.1),

$$V_{\text{eff}}(r) = V(r) + \frac{\hbar^2 l(l+1)}{2mr^2}. \tag{24.195}$$

For $l \neq 0$, the centrifugal term is repulsive and grows stronger at small r. The potential shown in the right panel of Fig. 24.6 has a finite barrier height between its minimum and the asymptotic region at very large distances where it vanishes. Any corresponding state peaked inside the effective well with positive energy is not a true bound state; the wavefunction will always leak outside by tunneling. This is the reason for the terminology "quasi-bound state". Since a quasi-bound state resonant pole occurs only for complex E, it again can never be hit directly in scattering experiments, but can be inferred from the complex analytic singularity structure of

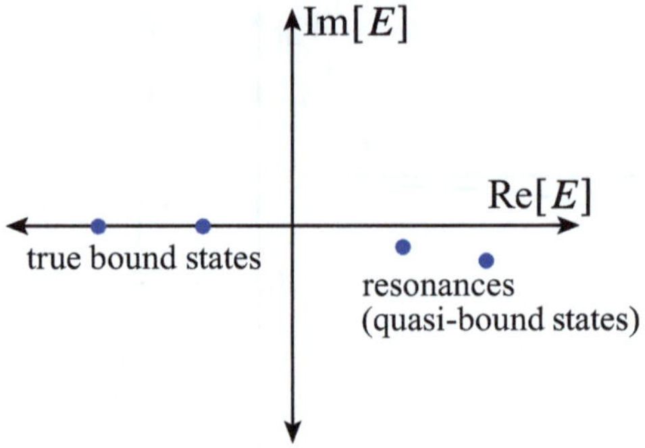

Fig. 24.7 Positions of poles of the partial wave amplitudes and S-matrix elements $e^{i2\delta_l}$, in the complex energy plane. Resonances have poles in the analytic continuation to complex E below the real axis, at $E = E_{\text{res}} - i\Gamma/2$, corresponding to quasi-bound states with lifetime $\hbar/\Gamma$. True bound states have poles on the negative real E axis, at $E = E_{\text{bound}} < 0$. Physical scattering energies are only on the positive real E axis. Therefore, these poles are never directly accessed in scattering, where $e^{i2\delta_l}$ is always finite with unit magnitude, but their influence can be seen in the dependence of the cross section on energy

the scattering amplitude functions. The smaller Γ is, the closer one can approach the resonance pole. However, $e^{i2\delta_l}$ always has unit magnitude for physical (real positive) values of the scattering energy E.

A schematic map of the two kinds of poles in the complex E plane for $e^{i2\delta_l}$, and therefore for $f_l(k)$, is shown in Fig. 24.7. They consist of true bound states on the negative real E axis and quasi-bound states below the positive real E axis.

24.7 Examples of Scattering from Spherical Potentials

Consider the scattering of a particle from a hard (impenetrable) sphere, so that the potential is

$$V(r) = \begin{cases} \infty & (\text{for } r \leq R), \\ 0 & (\text{for } r \geq R). \end{cases} \tag{24.196}$$

This is the quantum-mechanical version of the classical scattering of a particle from a hard sphere, discussed in Sect. 24.1. In this particular case, it is not necessary to use (24.171)–(24.174), because the solution for $r \leq R$ is trivial,

$$\psi^+(\boldsymbol{r}) = 0 \qquad (r \leq R), \tag{24.197}$$

corresponding to $A_l(r) = 0$. Meanwhile, for $r \geq R$, we write

$$\psi^+(\mathbf{r}) = \frac{1}{(2\pi)^{3/2}} \sum_{l=0}^{\infty} (2l+1) P_l(\cos\theta)\, B_l(r), \tag{24.198}$$

with $B_l(r)$ given by (24.169). To match to the vanishing solution at $r = R$, we need $B_l(R) = 0$, so

$$\tan\delta_l = j_l(kR)/n_l(kR), \qquad (l = 0, 1, 2, \ldots) \tag{24.199}$$

for each of the s-wave, p-wave, d-wave, ... partial wave phase shifts. This is especially simple for the s-wave,

$$\tan\delta_0 = [\sin(kR)/kR]\,/\,[-\cos(kR)/kR] = -\tan(kR), \tag{24.200}$$

with the solution

$$\delta_0 = -kR. \tag{24.201}$$

Therefore, for $l = 0$,

$$\begin{aligned} B_0(r) \propto \cos\delta_0 \frac{\sin(kr)}{kr} + \sin\delta_0 \frac{\cos(kr)}{kr} &= \frac{1}{kr}\sin(kr + \delta_0) \\ &= \frac{1}{kr}\sin(k(r - R)). \end{aligned} \tag{24.202}$$

As illustrated in Fig. 24.8, the outgoing s-wave is just shifted out by a distance R, compared to the free-particle solution.

More generally, consider all δ_l in the low-energy limit $kR \ll 1$ in which the wavenumber is much smaller than the inverse size of the hard sphere. In that case,

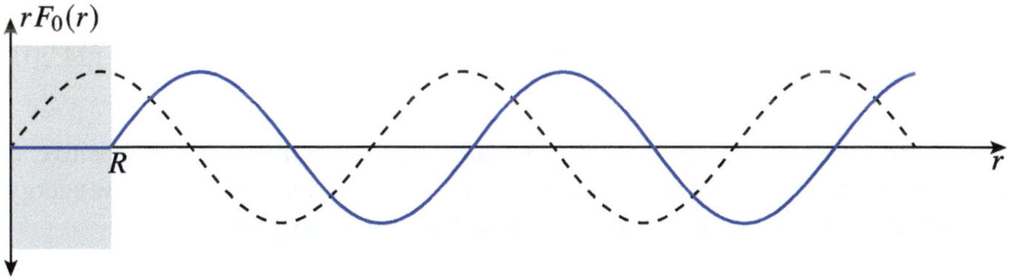

Fig. 24.8 The solid line shows the s-wave radial wavefunction $rF_0(r)$ as a function of r, for scattering from a hard sphere with $V = \infty$ in the shaded region $r < R$. The dashed line shows what $rF_0(r)$ would be if the potential were absent, corresponding to free particle propagation. The hard-sphere potential pushes the wavefunction out by a distance R, giving a phase shift $\delta_0 = -kR$

$$j_l(kR) \approx \frac{(kR)^l}{(2l+1)!!}, \qquad n_l(kR) \approx -\frac{(2l-1)!!}{(kR)^{l+1}}, \tag{24.203}$$

where, for $l = 0$, one should recall the definition $(-1)!! = 1$. So, from (24.199),

$$\tan\delta_l = -\frac{(kR)^{2l+1}}{(2l-1)!!\,(2l+1)!!}, \tag{24.204}$$

which quickly becomes very small as l is increased. [Note that this result for $\tan\delta_l$ could also be obtained simply as the $\alpha_l \to \infty$ limit of (24.175).]

In the extreme low energy limit $kR \ll 1$, only the s-wave contributes, resulting in

$$\frac{d\sigma}{d\Omega} \approx |f_0(k)|^2 = \frac{\sin^2\delta_0}{k^2} = \frac{\sin^2(kR)}{k^2} \approx R^2, \qquad (kR \ll 1). \tag{24.205}$$

Since this is isotropic (constant in both ϕ and $\cos\theta$), the total cross section is

$$\sigma = \int d\Omega\,\frac{d\sigma}{d\Omega} = 4\pi R^2, \qquad (kR \ll 1), \tag{24.206}$$

which is 4 times the classical result $\sigma_{\text{classical}} = \pi R^2$ that we found in (24.12).

In the opposite limit, many δ_l can contribute for very high scattering energy. For $kR \gg 1$, one can show that $\sigma \approx 2\pi R^2$, which is still twice the classical result. This is related to the fact that since $V(r)$ is discontinuous at $r = R$, no matter how big E is, the wavefunction varies sharply on length scales $\ll 1/k$. To get $\sigma = \pi R^2 = \sigma_{\text{classical}}$, it is necessary to take a smoothed potential (a "softened hard sphere"), in which the potential decreases to $V = 0$ over a distance scale much smaller than R but much larger than $1/k$.

As another example, consider a finite well or barrier, with potential

$$V(r) = \begin{cases} V_0 & (\text{for } r \le R), \\ 0 & (\text{for } r \ge R), \end{cases} \tag{24.207}$$

where the constant V_0 can be either negative (an attractive well) or positive (a repulsive core potential). Let us only consider s-wave scattering, valid at low energy. Outside the potential's range, the radial wavefunction is proportional to

$$F_0(r) = e^{i\delta_0}\left[\cos\delta_0\, j_0(kr) - \sin\delta_0\, n_0(kr)\right] = e^{i\delta_0}\frac{\sin(kr+\delta_0)}{kr}, \qquad (r \ge R), \tag{24.208}$$

with the normalization chosen arbitrarily, and

$$\frac{\hbar^2 k^2}{2m} = E. \tag{24.209}$$

Inside the potential's range, assuming $E > V_0$, the radial wavefunction is

$$F_0(r) = C j_0(k'r) = C \sin(k'r)/k'r, \qquad (r \le R), \tag{24.210}$$

with a relative normalization constant C, and

$$\frac{\hbar^2 k'^2}{2m} = E - V_0. \tag{24.211}$$

The solution $n_0(k'r)$ is rejected here, because it is not normalizable at $r = 0$, as discussed in Sect. 10.2. Now we match the solutions at $r = R$, to obtain an equation that determines C,

$$C \sin(k'R)/k'R = e^{i\delta_0} \sin(kR + \delta_0)/kR. \tag{24.212}$$

We also match $d\psi^+/dr$ at $r = R$, by specializing the convenient general results of (24.172) and (24.174). After some algebra,

$$\tan\delta_0 = \frac{(k/k')\sin(k'R)\cos(kR) - \cos(k'R)\sin(kR)}{(k/k')\sin(k'R)\sin(kR) + \cos(k'R)\cos(kR)}. \tag{24.213}$$

The preceding assumed $E > V_0$. If instead $E < V_0$, then k' will be imaginary, so that (24.212) and (24.213) are valid after making the replacements $\sin(k'R)/k' \to \sinh(\kappa R)/\kappa$ and $\cos(k'R) \to \cosh(\kappa R)$, where $\hbar^2\kappa^2/2m = V_0 - E$.

Graphs of the $l = 0$ wavefunction normalized by a factor of the radial coordinate are shown in Fig. 24.9 for the attractive case $V_0 < 0$ so that $k' > k$, and the repulsive case $V_0 > 0$ so that $k' < k$, respectively. In the attractive case, δ_0 is positive, and the potential "pulls in" the $l = 0$ wave by a shift δ_0/k in the position of the nodes. In contrast, for the repulsive case, δ_0 is negative, and the potential can be thought of as "pushing out" the $l = 0$ wave by an amount $|\delta_0|/k$. Things are a little more complicated for the partial waves with $l > 0$, and for potentials that are not piecewise constant. However, the essential qualitative feature remains that outgoing partial waves are pulled in by attractive potentials (because the Schrödinger equation dictates that the wavefunction oscillates faster in the region where the potential energy is smaller) and pushed out by repulsive potentials (because the wavefunction oscillates more slowly, or not at all, in regions with larger potential energy).

The phase shift δ_0 is small in magnitude for $E \gg |V_0|$. In the case of an attractive potential ($V_0 < 0$), the phase shift increases as E decreases until it reaches $\delta_0 = \pi/2$, where the s-wave cross section is as large as it can possibly be, for a given energy,

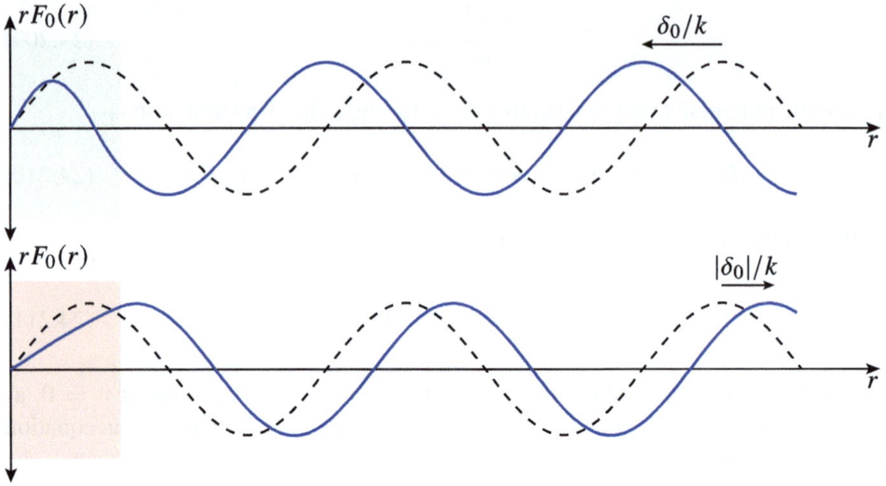

Fig. 24.9 The solid lines show examples of the s-wave radial wavefunction $rF_0(r)$ as a function of r, for scattering from a spherical-well potential in the shaded region, if the potential is attractive (top) or repulsive (bottom). The dashed line shows the corresponding result if the potential is absent. The potential pulls in (for the attractive case, $\delta_0 > 0$) or pushes out (for the repulsive case, $\delta_0 < 0$) the wavefunction by a distance $|\delta_0|/k$

$$\sigma_{l=0}\Big|_{\max} = \frac{4\pi}{k^2}\sin^2(\pi/2) = \frac{2\pi\hbar^2}{mE}. \tag{24.214}$$

In this case, the scattering has saturated the partial wave unitarity bound. Decreasing E even more, eventually one may reach $\delta_0 = \pi$, for which

$$\sigma_{l=0}\Big|_{\min} = \frac{4\pi}{k^2}\sin^2(\pi) = 0. \tag{24.215}$$

This is the Ramsauer–Townsend effect; there is no scattering even though $V(r)$ is non-negligible and attractive, because the de Broglie wavelength happens to match the characteristic length scale of the potential. We have seen this type of behavior already in a one-dimensional scattering example, at the end of Sect. 6.6, and in the real world it was first observed as suppressions in the cross sections for electron scattering from inert gas atoms Ar, Kr, and Xe.

In the low-energy limit, one can find the scattering length and effective range defined in (24.176), by expanding (24.213) for small k. The results for $V_0 < 0$ are

$$a = R - \tan(k_0 R)/k_0, \tag{24.216}$$

$$r_0 = R - \frac{R^3}{3a^2} - \frac{1}{ak_0^2}, \tag{24.217}$$

where we have defined

$$k_0 = \sqrt{-2mV_0}/\hbar. \tag{24.218}$$

If $k_0 R$ happens to be slightly larger than $\pi/2$, then the scattering length can be large, $a \gg R$. This corresponds to the existence of an s-wave bound state with small binding energy $E_{\text{bound}} \approx -\hbar^2/2ma^2$, and then $\sigma \approx 4\pi a^2/(1 - E/E_{\text{bound}})$ as discussed in the previous section.

24.8 Neutron-Proton Scattering and the Deuteron

The low-energy scattering of neutrons and protons provides a practical illustration of some of the ideas discussed in Sects. 24.5–24.7. The masses of the proton and neutron are respectively $m_p = 938.272\ \text{MeV}/c^2$ and $m_n = 939.565\ \text{MeV}/c^2$, with a reduced mass in the two-body problem (see Sect. 4.2),

$$\mu = \frac{m_n m_p}{m_n + m_p} \approx 469.459\ \text{MeV}/c^2. \tag{24.219}$$

The potential between nucleons arises from the rather complicated strong nuclear force, but the spectrum of bound states for two nucleons is very simple. There are no pp or nn bound states, and there is only one bound state for the neutron and the proton. This is the deuteron (d), a state with spin 1 and bound-state energy

$$E^d_{\text{bound}} = -2.2246\ \text{MeV}, \tag{24.220}$$

which is determined by measuring the energy of the photon (γ) emitted when a proton captures a neutron, $n + p \to d + \gamma$. The binding energy 2.2246 MeV is rather small in comparison to other nuclear binding energies per nucleon. Since the deuteron is weakly bound, there should be a pole in the neutron-proton scattering cross section when extrapolated to negative energy.

Let us now understand the angular momentum and parity quantum numbers of the deuteron, as a prerequisite to figuring out its role in neutron-proton scattering. Since the full Hamiltonian must be invariant under rotations, the energy eigenstates are also eigenstates of the square of the total angular momentum operator $\boldsymbol{J}$, obtained by combining the spins of the two constituents $\boldsymbol{S}_p$ and $\boldsymbol{S}_n$ and their total orbital angular momentum $\boldsymbol{L}$ in the center-of-mass frame. The statement that the deuteron has spin 1 means that the operator J^2 has eigenvalue $\hbar^2 J(J+1)$, where $J = 1$. The neutron and proton each have spin 1/2, so the possible total constituent spin combinations for the deuteron are $S = 0$ and $S = 1$. The $J = 1$ spin of the deuteron is then obtained by combining this with the total orbital angular momentum quantum number L in the center-of-mass frame, which can therefore only be $L = 0$, 1, or 2. Now, the magnetic dipole moment of the deuteron is known experimentally to be $\mu_d = 0.85744\mu_N$, which is very close to the sum of the dipole moments of the neutron and proton, $\mu_p + \mu_n = 2.79285\mu_N - 1.91304\mu_N = 0.87981\mu_N$. This indicates that the deuteron magnetic moment is very nearly realized when the spins are aligned in the $S = 1$ state, without much contribution from the orbital motion of the charged proton. Therefore, the deuteron must have predominantly $L = 0$. However, there are also two clear experimental indications that it cannot be a pure $L = 0$ eigenstate. First, there is the small but significant discrepancy in the magnetic moment sum noted above. Second, the deuteron is found experimentally to have a nonzero electric quadrupole moment, which is inconsistent with the perfect spherical symmetry of a pure $L = 0$ eigenstate. The strong interaction Hamiltonian responsible for binding the deuteron is known (from its more fundamental formulation, quantum chromodynamics, or QCD) to commute with parity. Therefore, (8.182) applies, and it must be possible to assign the deuteron a definite parity eigenvalue $\eta_d = (-1)^L \eta_p \eta_n$. Here η_p and η_n are the intrinsic parities of the proton and neutron, which are conventionally taken to be $\eta_p = \eta_n = 1$. To have a definite parity eigenvalue $\eta_d = \pm 1$, the deuteron must be a superposition of states with either all L even, or all L odd. Since we already found that the state is mostly $L = 0$, this implies $\eta_d = +1$ and rules out the possibility of any $L = 1$ component. Therefore, the deuteron must be a linear combination of $L = 0$ and $L = 2$. In order to give $J = 1$, it follows that the combination of constituent spins can only be pure $S = 1$. In spectroscopic notation, the deuteron state is predominantly ${}^{2S+1}L_J = {}^3S_1$, but with a few percent ${}^3D_1$ component.

For our present purposes, the most important feature of the previous paragraph is that the deuteron is a pure $S = 1$ combination. The absence of a bound state in the $S = 0$ combination shows that the strong nuclear force is spin-dependent, which means that we should also keep track of the $S = 0$ and $S = 1$ combinations separately in scattering. The data for low-energy scattering of neutrons and protons in the $S = 0$ and $S = 1$ combinations yield scattering lengths and effective ranges, as defined in (24.176), of

$$a^{S=1} = 5.42 \text{ fm}, \qquad r_0^{S=1} = 1.75 \text{ fm}, \tag{24.221}$$

$$a^{S=0} = -23.7 \text{ fm}, \qquad r_0^{S=0} = 2.7 \text{ fm}, \tag{24.222}$$

in units of 1 fm $= 10^{-15}$ meters. The positive scattering length in the $S = 1$ channel supports the hypothesis of a bound state. Solving (24.183) gives a prediction for the bound state of

$$\kappa = 0.231 \text{ fm}^{-1}, \tag{24.223}$$

which then yields the estimate

$$E_{\text{bound}} = -\frac{\hbar^2 \kappa^2}{2\mu} = -2.22 \text{ MeV}, \tag{24.224}$$

in good agreement[4] with the experimental deuteron binding energy quoted in (24.220). Now, using the facts that the $E = 0$ cross section for s-wave scattering is $4\pi a^2$, and that it must have a pole at $E = E^d_{\text{bound}}$, we have as in (24.187),

$$\sigma^{S=1} = \frac{4\pi \left(a^{S=1}\right)^2}{1 - E/E^d_{\text{bound}}} \approx \frac{3.69\ \text{b}}{1 + E/2.2246\ \text{MeV}}, \tag{24.225}$$

where we have used the traditional nuclear physics unit of cross section

$$1\ \text{b} = 100\ \text{fm}^2 = 10^{-28}\ \text{m}^2, \tag{24.226}$$

called[5] a **barn**.

In the $S = 0$ channel for neutron-proton scattering, the fact that the scattering length $a^{S=0}$ is negative confirms that there is no bound state. Still, we can write the cross section in a form similar to (24.225), by using (24.178),

$$\sigma^{S=0} = \frac{4\pi \left(a^{S=0}\right)^2}{1 + (ka^{S=0})^2(1 - r_0^{S=0}/a^{S=0})} \approx \frac{70.6\ \text{b}}{1 + E/(0.066\ \text{MeV})}. \tag{24.227}$$

However, it should be emphasized that there really is no neutron-proton bound state with energy near -0.066 MeV. Such a pole in the scattering cross section at negative real E, but with a negative scattering length a and therefore not associated with an actual bound state, is sometimes called a **virtual bound state**. It can be interpreted as an indication that if the $S = 0$ neutron-proton potential were just slightly more attractive, then the scattering length would be positive and a bound state would exist. Such virtual bound state poles should not be confused with the quasi-bound state resonances with positive energy discussed in Sect. 24.6.

The initial neutron and proton spins are often random and unmeasured, so that the four spin states (1 for $S = 0$ and 3 for $S = 1$) are equally likely. Averaging over them, the prediction for the total neutron-proton cross section in the $E \to 0$ limit is

$$\sigma = \frac{1}{4}\sigma^{S=0} + \frac{3}{4}\sigma^{S=1} = 20.4\ \text{b}, \tag{24.228}$$

in good agreement with experimental observation.

---

[4] Neglecting $r_0^{S=1}$ here would give much worse agreement. However, it must be admitted that the excellence of the agreement as presented here is partly accidental. Since (24.176) was truncated beyond order k^2, rather than solving the quadratic equation (24.183) for κ, one could just as consistently approximate it by (24.186), which would have given instead $\kappa = 0.214\ \text{fm}^{-1}$ and thus $E_{\text{bound}} = -1.91$ MeV.

[5] The origin of the unit name refers to the fact that, by nuclear physics standards, this is not a small area, jokingly considered as easy to hit as "the broad side of a barn" in idiomatic North American English. Collider experiments at the present high-energy frontier often discuss cross sections in units of picobarns or femtobarns.

The data in (24.221) and (24.222) can also be used to make a crude model for the neutron-proton potential in the form of a spherical well as in (24.207), by solving (24.216)–(24.218) numerically for R and $k_0 = \sqrt{-2\mu V_0}/\hbar$. The results are

$$R^{S=1} = 2.07 \text{ fm}, \qquad k_0^{S=1} = 0.91 \text{ fm}^{-1}, \qquad V_0^{S=1} = -34 \text{ MeV}, \tag{24.229}$$

$$R^{S=0} = 2.59 \text{ fm}, \qquad k_0^{S=0} = 0.58 \text{ fm}^{-1}, \qquad V_0^{S=0} = -14 \text{ MeV}. \tag{24.230}$$

From the analysis of Sect. 10.4, the condition for such a spherical-well potential to have n bound states with $l = 0$ can be written as [see (10.84), and recall that V_0 and a there are $-V_0$ and R here]

$$k_0 R/\pi > n - 1/2. \tag{24.231}$$

Putting in the numbers, the $S = 1$ potential model predicts exactly one bound state, and the $S = 0$ potential model barely misses having a bound state, in accord with the observed facts. The deuteron is surprisingly weakly bound, in the sense that the binding energy (2.2246 MeV) is more than an order of magnitude smaller than the depth of the potential (34 MeV).

24.9 Scattering of Identical Particles

So far, we have neglected the possibility that the scattering particles might be identical. To remedy this, first consider the scattering of two identical spin-0 bosons (for example, α particles, also known as ^{4}He nuclei). The total wavefunction will then be of the form

$$\Psi(\boldsymbol{r}_1, \boldsymbol{r}_2) = e^{i\boldsymbol{p}\cdot(\boldsymbol{r}_1+\boldsymbol{r}_2)/2\hbar}\psi(\boldsymbol{r}), \tag{24.232}$$

where, following (4.14) and (4.15), $\boldsymbol{p}$ is the total vector momentum of the two particles, $(\boldsymbol{r}_1 + \boldsymbol{r}_2)/2$ is the center-of-mass position, and

$$\boldsymbol{r} = \boldsymbol{r}_1 - \boldsymbol{r}_2 \tag{24.233}$$

is the relative position vector. By choosing the center-of-mass frame, we can take $\boldsymbol{p} = 0$. Because the particles are spinless bosons, this wavefunction must be symmetric under $\boldsymbol{r}_1 \leftrightarrow \boldsymbol{r}_2$, so the relative-position wavefunction must obey

$$\psi(\boldsymbol{r}) = \psi(-\boldsymbol{r}), \tag{24.234}$$

in other words, it must have even parity. Therefore, instead of a scattering wavefunction proportional to $e^{ikz} + f_k(\theta)e^{ikr}/r$, we must have (dropping the overall normalization in this section)

$$\psi(\boldsymbol{r}) = e^{ikz} + e^{-ikz} + [f_k(\theta) + f_k(\pi - \theta)] \frac{e^{ikr}}{r} \tag{24.235}$$

for the scattering wavefunction in the center-of-mass frame. In this way, both incoming particles, and both outgoing particles, are described on an equal footing. The differential cross section is therefore of the form

$$\begin{aligned} \frac{d\sigma}{d\Omega} &= |f_k(\theta) + f_k(\pi - \theta)|^2 \\ &= |f_k(\theta)|^2 + |f_k(\pi - \theta)|^2 + 2\mathrm{Re}\left[f_k^*(\theta) f_k(\pi - \theta)\right]. \end{aligned} \tag{24.236}$$

This incorporates the fact that if one outgoing particle is detected at angle θ, then by momentum conservation the other must be at angle $\pi - \theta$. The intrinsic indistinguishability of the particles guarantees that the differential cross sections at those two angles are equal.

As a consequence, there must be constructive interference in the amplitudes for identical spinless bosons for right-angle ($\theta = \pi/2$) scattering,

$$\left.\frac{d\sigma}{d\Omega}\right|_{\theta=\pi/2} = 4|f_k(\pi/2)|^2. \tag{24.237}$$

Also, only even-l partial waves can contribute in the identical boson scattering case, because

$$\begin{aligned} f_k(\theta) + f_k(\pi - \theta) &= \sum_{l=0}^{\infty} (2l + 1) \left[P_l(\cos\theta) + P_l(\cos(\pi - \theta))\right] f_l(k) \\ &= 2 \sum_{\text{even } l} (2l + 1) P_l(\cos\theta) f_l(k), \end{aligned} \tag{24.238}$$

where the terms with odd l have canceled because $P_l(\cos(\pi - \theta)) = (-1)^l P_l(\cos\theta)$.

Now consider the case of scattering of electrons, or other identical spin-1/2 fermions. This can be divided into two cases, depending on the total spin quantum number S.

First, suppose that $S = 0$. Since that is an antisymmetric spin state, the spatial wavefunction must be symmetric under exchange of the labels 1, 2, so $\psi(\boldsymbol{r}) = \psi(-\boldsymbol{r})$, and

$$\frac{d\sigma}{d\Omega}^{S=0} = |f_k(\theta) + f_k(\pi - \theta)|^2, \tag{24.239}$$

and so all of the remarks just made for scattering of identical bosons go through as before.

If instead the total spin state is the symmetric $S = 1$ combination, then the spatial wavefunction must be antisymmetric under exchange of the labels 1, 2, so

$\psi(\boldsymbol{r}) = -\psi(-\boldsymbol{r})$, with odd parity. Therefore, the scattering wavefunction must be proportional to

$$\psi(\boldsymbol{r}) = e^{ikz} - e^{-ikz} + [f_k(\theta) - f_k(\pi - \theta)]\,\frac{e^{ikr}}{r}, \tag{24.240}$$

which implies

$$\frac{d\sigma}{d\Omega}^{S=1} = |f_k(\theta)|^2 + |f_k(\pi - \theta)|^2 - 2\mathrm{Re}\left[f_k^*(\theta) f_k(\pi - \theta)\right]. \tag{24.241}$$

It follows that for identical fermions in the symmetric spin state there is perfect destructive interference for right-angle scattering,

$$\left.\frac{d\sigma}{d\Omega}^{S=1}\right|_{\theta=\pi/2} = 0. \tag{24.242}$$

Furthermore, only partial waves with odd l contribute.

In many cases, the fermion spins are random and unmeasured. If so, then four spin states (1 for $S = 0$ and 3 for $S = 1$) are equally likely, so the observed unpolarized cross section for identical fermions is the weighted average:

$$\left.\frac{d\sigma}{d\Omega}\right|_{\text{unpolarized}} = \frac{1}{4}\frac{d\sigma}{d\Omega}^{S=0} + \frac{3}{4}\frac{d\sigma}{d\Omega}^{S=1} \tag{24.243}$$

$$= |f_k(\theta)|^2 + |f_k(\pi - \theta)|^2 - \mathrm{Re}\left[f_k^*(\theta) f_k(\pi - \theta)\right], \tag{24.244}$$

where the absence of a factor of 2 in front of the interference term is not a typographical error. For right-angle scattering, the result is partial destructive interference.

In all of the cases of scattering of identical particles, it is important to note that when integrating the differential cross section to get the total cross section, one must include a factor of $1/2$ to avoid double counting. This is because in the center-of-mass frame a state with one particle at (θ, ϕ) also has an identical particle at $(\pi - \theta, \phi + \pi)$. Therefore, the state specified by the presence of a particle at (θ, ϕ) and the state specified by $(\pi - \theta, \phi + \pi)$ are actually the same state, and should not be counted twice. So, one has

$$\sigma = \frac{1}{2}\int d\Omega\,\frac{d\sigma}{d\Omega} \qquad \text{(identical particles)} \tag{24.245}$$

for both the identical boson and identical fermion cases. Note that this factor of $1/2$ for identical particles comes in at the level of the total cross section, not the differential cross section.

Exercises

24.1 Particles of mass m and incident momentum $\boldsymbol{p} = \hbar k \hat{z}$ scatter from a potential

$$V(r) = \begin{cases} V_0 & (\text{for } r < a) \\ 0 & (\text{for } r > a), \end{cases} \tag{24.246}$$

where V_0 and a are constants. Apply the first-order Born approximation to:

(a) Find the differential scattering cross section for small $|V_0|$. Write your answer in terms of $q = 2k\sin(\theta/2)$. Check that your answer has units of area.
(b) Show that in the limit of small ka, the differential cross section found in part (a) is constant with respect to the scattering angle, and show that the total cross section is $\sigma = nV_0^2 a^6$ where n is a quantity that you will find. (Check the units.)
(c) Working now to the next-to-leading order in an expansion in small ka, show that the differential cross section has the form $d\sigma/d\Omega = b + c\cos(\theta)$, and find the quantities b and c.

24.2 Particles of mass m and momentum $\boldsymbol{p} = \hbar k \hat{z}$ scatter from a potential $V(r) = V_0 e^{-r/a}$. Use the leading Born approximation to find the differential cross section $d\sigma/d\Omega$, and integrate it to find the total cross section.

24.3 Consider the scattering of a spinless point particle with mass m, momentum $\boldsymbol{p} = \hbar k \hat{z}$, and charge $-e$, due to a point charge Ze fixed at the origin, which is screened at large distances by a uniform thin spherical shell of opposite charge at $r = R$, so that

$$V(r) = \begin{cases} -Ze^2(1/r - 1/R) & (\text{for } r < R), \\ 0 & (\text{for } r > R). \end{cases} \tag{24.247}$$

Use the first-order Born approximation for scattering to:

(a) Find the differential cross section.
(b) Find the differential cross section and the total cross section in the low energy limit, keeping the first two nonzero contributions in the expansion in small $E = \hbar^2 k^2/2m$.
(c) In the high-energy limit, find an asymptotic expression for the differential cross section at a fixed angle θ, keeping the first two nonzero contributions in an expansion in large kR.
(d) Again in the high-energy limit, find the total cross section at leading order in the expansion in large kR. (Note that this can *not* be obtained from the leading term in the fixed-angle high-energy expansion of the previous part. What goes wrong?)

24.4 Suppose that the differential cross section for particles of momentum $\hbar \boldsymbol{k}$ to scatter from a potential $V_0(\boldsymbol{r})$ centered at the origin is known to be $d\sigma_0/d\Omega$ in the leading Born approximation. Now consider scattering from N such potentials centered at positions $\boldsymbol{a}_n$, so that the total potential is $V(\boldsymbol{r}) = \sum_{n=1}^{N} V_0(\boldsymbol{r} - \boldsymbol{a}_n)$. Show that in the leading Born approximation the differential cross section is

$$\frac{d\sigma}{d\Omega} = \left| \sum_{n=1}^{N} \exp\left(i(\boldsymbol{k} - \boldsymbol{k}') \cdot \boldsymbol{a}_n\right) \right|^2 \frac{d\sigma_0}{d\Omega}. \tag{24.248}$$

24.5 A beam of electrons with momentum $\boldsymbol{p} = \hbar k \hat{z}$ scatters from the classical potential $V(r)$ produced by a hydrogen atom in its ground state, which (from Gauss' law) obeys

$$\frac{dV}{dr} = \frac{q_e^2}{r^2}\left(1 - \int_0^r dr' 4\pi r'^2 |\psi_{1,0,0}(r')|^2\right). \tag{24.249}$$

Here the electron charge is q_e, to distinguish it from the base of the natural logarithms.

(a) Obtain dV/dr from the above equation by doing the integral, and deduce that $V(r) = -q_e^2(1/r + A/a_0)e^{-Br/a_0}$, where A and B are certain constants that you will determine. [Hint: take the derivative of this form of $V(r)$ and match it to the dV/dr that you found.]
(b) Find the differential cross section in the Born approximation, ignoring spin. Show that it can be written as $d\sigma/d\Omega = n_1(N_2 + a_0^2 q^2)^2/(N_3 + a_0^2 q^2)^4$, where n_1 is a quantity that you will find, and N_2 and N_3 are certain positive integers, and $q = 2k \sin(\theta/2)$.
(c) In the low energy limit, what is $d\sigma/d\Omega$? How does it depend on θ?
(d) In the high energy limit, show that $d\sigma/d\Omega$ approaches the Rutherford differential cross section. (Can you explain why, even though the atom is electrically neutral?)
(e) Find the total cross section. Does it diverge, as for Rutherford scattering?

24.6 For scattering of particles with mass m and momentum $\boldsymbol{p} = \hbar k \hat{z}$ from a spherically symmetric potential $V(r)$, show that in the leading Born approximation the phase shifts in the partial wave expansion are

$$\delta_l = N \frac{mk}{\hbar^2} \int_0^\infty dr\, r^2 V(r) \left(j_l(kr)\right)^2, \tag{24.250}$$

where N is a certain integer that you will discover.

24.7 The differential cross section of a beam of spinless particles of mass m with fixed low kinetic energy E on a target of spherically symmetric atoms is found to be

$$\frac{d\sigma}{d(\cos\theta)} = \frac{\pi\hbar^2}{mE}[c_0 + c_1\cos(\theta) + c_2\cos^2(\theta)], \tag{24.251}$$

where c_0, c_1, and c_2 depend on E but not on θ. Assume that the target is very small compared to the experimental apparatus, that multiple scatterings within the target are not an issue, and that the results should be interpreted as a combination of s-wave and p-wave scattering.

(a) Find real equations relating the phase shifts δ_0 and δ_1 to the measured quantities c_0, c_1, c_2.
(b) Use partial wave unitarity to find theoretical upper bounds on c_0 and c_2.
(c) What are the theoretical upper and lower bounds on c_1? [Hint: both upper and lower bounds are realized by phase shifts satisfying $\delta_0 = \pi n_0/6$ and $\delta_1 = \pi n_1/6$ where n_0 and n_1 are positive integers less than 6. You may use this fact without proving it.]

24.8 The low-energy (s-wave) limit of scattering from a potential that vanishes outside of some range R is treated at the end of Sect. 24.5. By expanding (24.174) with $l = 0$, verify (24.176) with the scattering length a given in (24.177), and show that the effective range is $r_0 = nR\left(1 - 1/\alpha_0 R + 1/(\alpha_0 R)^2\right)$, where n is a certain positive rational number that you will discover.

Subsystems: Entanglement, Evolution, and Generalized Measurements

25

25.1 Open and Bipartite Systems and Entanglement

An **open system** is a quantum system that is part of a larger one. The rest of the larger system may be referred to as the **environment**, and we often want to study the dynamics of the open system in such a way that its interactions with the environment are simplified or averaged over in some way. Typically, the dynamics of the environment are not completely accessible to us, or are too complicated to understand in full detail, but will still have important impacts on the open system of primary interest.

More generally, we wish to consider **bipartite quantum systems**, defined to be those whose Hilbert space is the tensor product of two Hilbert spaces for subsystems a and b, so that $\mathcal{H}_{ab} = \mathcal{H}_a \otimes \mathcal{H}_b$. Often, subsystem a is the open system whose behavior we want to understand, and b is its environment (perhaps the whole rest of the universe). Another interpretation is that we might consider b to contain a quantum measurement apparatus that works through its correlations with the object of measurement, subsystem a. Yet another interpretation is that a and b might be associated with observers Alice and Bob who are in different locations, and therefore each have access to only part of a quantum state. Understanding the quantum correlations that arise between two subsystems through entanglement gives insights into the dynamical origins of measurements, the emergence of macroscopic classical-like behavior in quantum mechanics, and the properties of information in quantum mechanics.

To describe the quantum physics of an open system, it is natural to use the density matrix operator description introduced in Sect. 3.5. To understand why, take the simple example of the entangled state that we introduced in (14.1) and used in our discussion of the EPR problem, which we rewrite here in the notation

$$|\psi\rangle = \frac{1}{\sqrt{2}}\Big(|\!\uparrow\rangle_a \otimes |\!\downarrow\rangle_b - |\!\downarrow\rangle_a \otimes |\!\uparrow\rangle_b\Big). \tag{25.1}$$

S. P. Martin, *Quantum Mechanics*, Graduate Texts in Physics,
https://doi.org/10.1007/978-3-032-21119-4_25

Consider how Alice will describe the situation when she can observe only system a, assuming no communication with Bob. No matter what spin components of her particle Alice measures, if the experiments are repeated many times, the results are perfectly described as a random ensemble with equal probability $1/2$ to be a spin-up or spin-down state. Considered as an open system, Alice can therefore describe her particle's spin a using a density operator on $\mathcal{H}_a$ alone:

$$\rho_a = \frac{1}{2}\Big(|\uparrow\rangle\langle\uparrow| + |\downarrow\rangle\langle\downarrow|\Big). \tag{25.2}$$

Even if Bob is busily making measurements of spin b, as long as Alice does not (yet) know the results, she will still describe her measurements in terms of ρ_a. This density operator is that of a completely random ensemble, and contains no information. In this way, the density operator formalizes the absence of instantaneous communication to Alice due to Bob's local interactions with spin b.

In contrast, suppose the combined state is instead a non-entangled ket, say

$$|\psi'\rangle = |\uparrow\rangle_a \otimes |\downarrow\rangle_b \tag{25.3}$$

for example. In this case, for the purposes of Alice's experiments, subsystem a is always spin-up, so the density operator is

$$\rho_a' = |\uparrow\rangle\langle\uparrow|, \tag{25.4}$$

corresponding to a pure ensemble.

As these examples suggest, if a combined system ab is in a product (non-entangled) pure state, then the subsystem a can always be described by a pure ensemble, but if the combined system is in an entangled state, then a considered by itself is described by a mixed state. We will now proceed to understand this in the general case.

Let us denote orthobases for the two subsystem Hilbert spaces by $|\varphi_i\rangle_a$ with $i = 1, \ldots, d_a$ and $|\chi_j\rangle_b$ with $j = 1, \ldots, d_b$. Then the $d_a d_b$ kets $|\varphi_i\rangle \otimes |\chi_j\rangle$ form an orthobasis for $\mathcal{H}_{ab}$. (Here, and often in the following, we suppress the subscripts a and b on the subsystem kets when the meaning should be clear.)

Now, suppose Alice is measuring an observable A that acts only on $\mathcal{H}_a$ and has eigenvalues (the possible measurement outcomes) α. These outcomes are associated with projection operators

$$P_\alpha \otimes I, \tag{25.5}$$

where P_α acts on subsystem a, and the identity operator acts on subsystem b. To be as general as possible, suppose that we have prepared a mixed ensemble of states of the system ab, which is therefore described by a density operator ρ_{ab}. Using (3.67), the probability of getting the measurement result α for A is

$$\mathcal{P}(\alpha) = \mathrm{Tr}[\rho_{ab}(P_\alpha \otimes I)]$$
$$= \sum_i \sum_j \Big(\langle\varphi_i| \otimes \langle\chi_j| \Big) \rho_{ab} \, (P_\alpha \otimes I) \Big(|\varphi_i\rangle \otimes |\chi_j\rangle \Big). \tag{25.6}$$

To put this into a nicer form, define the **reduced density operator** for subsystem a,

$$\rho_a = \mathrm{Tr}_b[\rho_{ab}] = \sum_j \langle\chi_j|\rho_{ab}|\chi_j\rangle. \tag{25.7}$$

Here we have taken the trace only over $\mathcal{H}_b$, called a **partial trace**. The notation is a bit tricky, since the objects $\langle\chi_j|\rho_{ab}|\chi_j\rangle$ are matrix elements with respect to the Hilbert space $\mathcal{H}_b$, but are still operators on $\mathcal{H}_a$. In terms of the reduced density operator, (25.6) becomes simply

$$\mathcal{P}(\alpha) = \sum_i \langle\varphi_i|\rho_a P_\alpha|\varphi_i\rangle = \mathrm{Tr}_a[\rho_a P_\alpha]. \tag{25.8}$$

Thus, ρ_a as defined by (25.7) contains all of the information that is available to Alice[1] by only observing subsystem a. It follows from (25.8) that the average result of measurements of A is

$$\overline{A} = \mathrm{Tr}_a[\rho_a A]. \tag{25.9}$$

These results are in accord with (3.67) and (3.68).

Recall that in general the trace is basis independent, so $\rho_a = \mathrm{Tr}_b(\rho_{ab})$ does not depend on the choice of basis for $\mathcal{H}_b$. It is also a short exercise to check that ρ_a obeys the three crucial properties required of any density operator. First, it is a Hermitian operator on $\mathcal{H}_a$. Second, it has unit trace, $\mathrm{Tr}_a(\rho_a) = 1$. Third, it is a **positive operator**. In general, a Hermitian operator A is said to be positive[2] if $\langle\psi|A|\psi\rangle \geq 0$ for any ket $|\psi\rangle$. More generally, if we wish to describe an open system a without any direct reference to a particular environment system, it is essential to require that it be described by a density operator ρ_a satisfying these three properties.

Consider the special case of a pure state for the combined system. The general form of such a state is

$$|\psi\rangle = \sum_i \sum_j c_{ij} \, |\varphi_i\rangle \otimes |\chi_j\rangle, \tag{25.10}$$

---

[1] Conversely, Bob, who makes observations only on subsystem b, will describe his experiences using $\rho_b = \mathrm{Tr}_a[\rho_{ab}] = \sum_i \langle\varphi_i|\rho_{ab}|\varphi_i\rangle$. Of course, the observables he can measure do not include A.

[2] A better terminology would be "non-negative", but it is hard to fight tradition.

density operator ρ'_a of (25.4). But let us generalize the first example slightly, by replacing the minus sign in (25.1) by an arbitrary relative phase $e^{i\theta}$, so that

$$|\psi\rangle = \frac{1}{\sqrt{2}}\Big(|\uparrow\rangle_a \otimes |\downarrow\rangle_b + e^{i\theta}\,|\downarrow\rangle_a \otimes |\uparrow\rangle_b\Big). \tag{25.15}$$

To an observer who is capable of making measurements on the combined system ab, this relative phase is certainly physically relevant; for example, you can check that the observable $\Omega = \sigma_x \otimes \sigma_x$ has expectation value $\langle\psi|\Omega|\psi\rangle = \cos\theta$. Now, the density operator for the combined system is

$$\begin{aligned}\rho_{ab} &= |\psi\rangle\,\langle\psi| \\ &= \frac{1}{2}\Big[\Big(|\uparrow\rangle\,\langle\uparrow|\Big)_a \otimes \big(|\downarrow\rangle\,\langle\downarrow|\big)_b + \Big(|\downarrow\rangle\,\langle\downarrow|\Big)_a \otimes \big(|\uparrow\rangle\,\langle\uparrow|\big)_b \\ &\quad + e^{i\theta}\big(|\downarrow\rangle\,\langle\uparrow|\big)_a \otimes \big(|\uparrow\rangle\,\langle\downarrow|\big)_b + e^{-i\theta}\big(|\uparrow\rangle\,\langle\downarrow|\big)_a \otimes \big(|\downarrow\rangle\,\langle\uparrow|\big)_b\Big].\end{aligned} \tag{25.16}$$

Taking the partial trace over the b spin using $\mathrm{Tr}_b(\rho_{ab}) = \langle\uparrow|_b\,\rho_{ab}\,|\uparrow\rangle_b + \langle\downarrow|_b\,\rho_{ab}\,|\downarrow\rangle_b$, the last two terms in (25.16) do not contribute, and we find

$$\rho_a = \frac{1}{2}\Big(|\uparrow\rangle\,\langle\uparrow| + |\downarrow\rangle\,\langle\downarrow|\Big). \tag{25.17}$$

The relative phase information present in the combined system state as the factor $e^{i\theta}$ is therefore completely inaccessible to Alice alone.[3] Recall, from the discussion surrounding (3.61) and (3.62), that this loss of phase information is a characteristic feature of mixed ensembles.

25.2 Dynamical Maps and Evolution of Open Systems

Consider an open system described at a given time by a density operator ρ. (Going forward, we refer to the subsystem a of a bipartite system ab as simply "the open system", and often leave off the subscript a for its states and operators.) What is the density operator ρ' that describes the system at some unspecified later time? For the special case of a closed system, the answer was given in (3.79), which says that $\rho' = U\rho U^\dagger$ for some unitary operator U. However, this is not true for open systems. In general, we can write

$$\rho \to \rho' = \mathcal{E}(\rho), \tag{25.18}$$

[3] To see this without using the density operator language, but in a morally equivalent way, note that the most general combined Hilbert space Hermitian operator that acts nontrivially only on Alice's spin is $A = \big(r_0 I + r_1\sigma_x + r_2\sigma_y + r_3\sigma_z\big) \otimes I$, where $r_{0,1,2,3}$ are arbitrary real numbers. It is a short exercise to compute $\langle\psi|A|\psi\rangle$ and check that it does not depend on θ.

where $\mathcal{E}$ is a map taking operators to operators, sometimes called a **superoperator**. We will refer to $\mathcal{E}$ as a **quantum dynamical map**, but it has many names; other sources refer to it as a **quantum channel**, or a **quantum operation**. Our goal is to understand the possible forms allowed by quantum mechanics for dynamical maps of open systems.

One way to investigate the issue is to introduce a specific environment subsystem with orthobasis $|\chi_n\rangle_b$ where $n = 1, \ldots, d_b$. Let us further assume that the initial condition for the combined system is a product density operator

$$\rho_{ab} = \rho \otimes (|0\rangle \langle 0|), \tag{25.19}$$

where $|0\rangle$ is some reference state of subsystem b. We now let the combined system evolve in time, just as in (3.79), which says that the final density operator is $U\rho_{ab}U^\dagger$ for some unitary operator U that acts on the combined system. The reduced density operator for the open system after the evolution is obtained by taking the partial trace over the environment subsystem b,

$$\mathcal{E}(\rho) = \mathrm{Tr}_b(U\rho_{ab}U^\dagger) = \sum_n \langle \chi_n|U|0\rangle\, \rho\langle 0|U^\dagger|\chi_n\rangle. \tag{25.20}$$

The objects $\langle \chi_n|U|0\rangle$ and $\langle 0|U^\dagger|\chi_n\rangle$ are matrix elements of the environment subsystem b, but are still operators acting on the open system a. They reflect the fact that the unitary evolution of the combined state causes the environment state to transition from $|0\rangle$ to a superposition of basis states, correlated with the evolution of the open system. Giving them the names

$$K_n = \langle \chi_n|U|0\rangle\,, \qquad K_n^\dagger = \langle 0|U^\dagger|\chi_n\rangle \qquad (n = 1, \ldots, d_b), \tag{25.21}$$

we have the important result

$$\mathcal{E}(\rho) = \sum_n K_n \rho K_n^\dagger, \tag{25.22}$$

where it is easy to check from (25.21), using completeness, that

$$\sum_n K_n^\dagger K_n = I, \tag{25.23}$$

the identity operator on the open system. In general, equations of the form (25.22) and (25.23) constitute an **operator-sum representation** or **Kraus representation** for the dynamical map $\mathcal{E}$, and the K_n are called **Kraus operators**, after Karl Kraus.

If there is only a single nonzero Kraus operator, then (25.23) tells us that it is a unitary operator. In that special case, we say that the dynamical map is unitary, as it has the same form as the unitary time evolution of a closed system in (3.79), namely $\rho' = U\rho U^\dagger$.

Dynamical maps defined by operator-sum representations, as in (25.22) and (25.23), can be sequentially combined. This means that if $\mathcal{E}_1$ and $\mathcal{E}_2$ are dynamical maps with operator-sum representations, then so is

$$\mathcal{E}_2 \circ \mathcal{E}_1 (\rho) = \mathcal{E}_2(\mathcal{E}_1(\rho)). \tag{25.24}$$

The proof is left to Exercise 25.2. A dynamical map $\mathcal{E}$ may have an inverse dynamical map $\mathcal{E}^{-1}$, so that $\mathcal{E}^{-1}(\mathcal{E}(\rho)) = \rho$. However, if $\mathcal{E}^{-1}$ is required to also have an operator-sum representation obeying (25.22) and (25.23), this happens if and only if the map is unitary. The most interesting dynamical maps for open systems are non-unitary and thus non-invertible.

As a simple example of a non-unitary dynamical map for an open system, consider a model for the decay of an atom in an excited state $|1\rangle_a$ to its ground state $|0\rangle_a$ by emitting a photon. These two states are the orthobasis for the open system. The environment orthobasis consists of states with and without a photon from the decay, denoted $|1\rangle_b$ and $|0\rangle_b$ respectively. The combined system thus has an orthobasis of four states $|n\rangle_a \otimes |m\rangle_b$ with $n, m = 0, 1$. The atomic state can be either excited or not, but we assume that there is no photon present initially. This means that the combined system always starts in a superposition of the two orthobasis states

$$|0\rangle_a \otimes |0\rangle_b \qquad \text{and} \qquad |1\rangle_a \otimes |0\rangle_b\,. \tag{25.25}$$

We now define a unitary map U to describe the evolution of the combined system,

$$U\big(\,|0\rangle \otimes |0\rangle\,\big) = |0\rangle \otimes |0\rangle\,, \tag{25.26}$$

$$U\big(\,|1\rangle \otimes |0\rangle\,\big) = \sqrt{1-p}\,|1\rangle \otimes |0\rangle + \sqrt{p}\,|0\rangle \otimes |1\rangle\,, \tag{25.27}$$

where p is a real number between 0 and 1. The interpretation of the first equation is that nothing happens if the atom starts in its ground state. The second equation says that if it starts in the excited state, then with probability p it will decay to the ground state while adding the photon to the environment state, and with probability $1 - p$ everything will remain as it is. The operator U preserves inner products, in the sense that the states in (25.26) and (25.27) are orthogonal and each has norm 1. This is an important requirement for our assertion that U is a unitary operator on the complete Hilbert space.[4]

[4] Actually, we have only bothered to specify how U acts on two of the four orthobasis states for the combined system, because that is all we will directly need, but you can easily complete it by constructing $U(\,|0\rangle \otimes |1\rangle\,)$ and $U(\,|1\rangle \otimes |1\rangle\,)$ such that U is unitary. More generally, it is true that if a linear operator preserves all inner products on a subspace, it can always be extended to a unitary operator on the full Hilbert space. The formal statement and a constructive proof of this assertion are postponed to Theorem 25.4 near the end of this section.

Following the construction of (25.21), we see that there are two Kraus operators in this example, one for each of our two environment orthobasis states. We can use that equation as a recipe to construct their actions on the orthobasis of the open system,

$$K_0 \, |0\rangle_a = \langle 0|_b \, U \left(|0\rangle \otimes |0\rangle \right) = |0\rangle_a \,, \tag{25.28}$$

$$K_0 \, |1\rangle_a = \langle 0|_b \, U \left(|1\rangle \otimes |0\rangle \right) = \sqrt{1-p} \, |1\rangle_a \,, \tag{25.29}$$

$$K_1 \, |0\rangle_a = \langle 1|_b \, U \left(|0\rangle \otimes |0\rangle \right) = 0, \tag{25.30}$$

$$K_1 \, |1\rangle_a = \langle 1|_b \, U \left(|1\rangle \otimes |0\rangle \right) = \sqrt{p} \, |0\rangle_a \,, \tag{25.31}$$

or, in matrix representation form,

$$K_0 = \begin{pmatrix} 1 & 0 \\ 0 & \sqrt{1-p} \end{pmatrix}, \qquad K_1 = \begin{pmatrix} 0 & \sqrt{p} \\ 0 & 0 \end{pmatrix}. \tag{25.32}$$

These indeed satisfy the completeness requirement $\sum_n K_n^\dagger K_n = I$. Now, given an initial density matrix of the open system (the atom),

$$\rho = \begin{pmatrix} \rho_{00} & \rho_{01} \\ \rho_{10} & \rho_{11} \end{pmatrix}, \tag{25.33}$$

the dynamical map resulting from the Kraus operator-sum representation of (25.22) is

$$\mathcal{E}(\rho) = K_0 \rho K_0^\dagger + K_1 \rho K_1^\dagger = \begin{pmatrix} \rho_{00} + p\rho_{11} & \sqrt{1-p}\,\rho_{01} \\ \sqrt{1-p}\,\rho_{10} & (1-p)\rho_{11} \end{pmatrix}. \tag{25.34}$$

So far, we have not associated the dynamical map with a specific time interval. To remedy this, we can take (25.34) to correspond to an infinitesimal time interval Δt, with an excited state atomic decay probability $p = \Gamma \Delta t$. We then apply the map N times. In doing so, we take the initial state of the environment to be the no-photon state $|0\rangle_b$ each time, on the grounds that any photons from decays in earlier iterations of the dynamical map will have fled the scene anyway, and therefore can and should be ignored. The result is

$$\rho \to \begin{pmatrix} \rho_{00} + (1-(1-p)^N)\rho_{11} & (\sqrt{1-p})^N \, \rho_{01} \\ (\sqrt{1-p})^N \, \rho_{10} & (1-p)^N \rho_{11} \end{pmatrix}. \tag{25.35}$$

Now taking the limit $N \to \infty$, with $t = N\Delta t$ held fixed, and using the defining property of the exponential function,

$$\lim_{N\to\infty}\left(1-\frac{\Gamma t}{N}\right)^N = e^{-\Gamma t}, \tag{25.36}$$

we find, after making use of the trace condition $\rho_{00} = 1 - \rho_{11}$,

$$\rho(t) = \begin{pmatrix} 1 - e^{-\Gamma t}\rho_{11} & e^{-\Gamma t/2}\rho_{01} \\ e^{-\Gamma t/2}\rho_{10} & e^{-\Gamma t}\rho_{11} \end{pmatrix}. \tag{25.37}$$

The interpretation of the diagonal entries is simply that a population of excited states decays to the ground state with a probability $1 - e^{-\Gamma t}$, which approaches 1 at late times. The off-diagonal entries of the density operator matrix also decay, with a lifetime twice as large. This dynamical map is not unitary, and no matter what mixed state we started with, it always evolves into the pure ground state at late times. In particular, the entropy for this open system decreases to 0, unlike what we found for a unitary evolution of a closed system in (3.80).

In the construction of (25.21), the number of Kraus operators was d_b, the same as the dimension of the environment subsystem Hilbert space. This is not necessarily convenient, especially if the environment is a very large and complicated system. However, the operator-sum representation is not unique, and different versions for the same map $\mathcal{E}(\rho)$ can have different numbers of Kraus operators. One can define new Kraus operators by

$$\widetilde{K}_\mu = \sum_n v_{\mu n} K_n, \tag{25.38}$$

after which the same map can be rewritten as $\mathcal{E}(\rho) = \sum_\mu \widetilde{K}_\mu \rho \widetilde{K}_\mu^\dagger$ with $\sum_\mu \widetilde{K}_\mu^\dagger \widetilde{K}_\mu = I$, provided that $\sum_\mu v_{\mu n'}^* v_{\mu n} = \delta_{nn'}$. This condition is equivalent to saying that the (non-square, in general) matrix of elements $v_{\mu n}$ can be extended to a unitary matrix. Fortunately, for any given dynamical map $\mathcal{E}$, we will soon show that there is always an operator-sum representation with no more than d^2 Kraus operators, where d is the dimension of the open system Hilbert space, regardless of the (possibly enormous) dimension d_b of the environment Hilbert space.

We now inquire about more general dynamical maps, beyond the type we defined in (25.20). One obvious possibility is to take the initial condition for the combined system to involve a general density operator for the environment, rather than a pure state $|0\rangle$, so that initially[5]

$$\rho_{ab} = \rho \otimes \rho_b, \tag{25.39}$$

[5] The initial condition of (25.39) is still far from general, as it is a product of density operators for the subsystems, with no initial correlations between system and environment. It is an understatement to say that the general case of an entangled density operator for the initial condition is more subtle. It will not be addressed here.

and then $\mathcal{E}(\rho) = \mathrm{Tr}_b[U(\rho \otimes \rho_b)U^\dagger]$ for some unitary joint-state operator U. It is left to Exercise 25.3 to show that in this case the map $\mathcal{E}(\rho)$ can be written in the same operator-sum form, but with d_b^2 Kraus operators instead of only d_b, although as noted in the previous paragraph it is often possible to reduce this number.

Given this, it is a plausible guess that dynamical maps can always be written in the operator-sum form of (25.22) and (25.23). As we will now show, this is indeed true, provided we assume some reasonable conditions motivated by the requirement that $\mathcal{E}(\rho)$ must be a legitimate density operator for the open system. These conditions are:

- $\mathcal{E}$ is a linear map, satisfying $\mathcal{E}(c_1 A_1 + c_2 A_2) = c_1\mathcal{E}(A_1) + c_2\mathcal{E}(A_2)$ for complex numbers $c_{1,2}$ and operators $A_{1,2}$. A motivation to impose this condition is that any collection of mixed ensembles can be viewed as a single mixed ensemble, so one must have for any density operators ρ_k with population ratios p_k,

$$\mathcal{E}\Big(\sum_k p_k \rho_k\Big) = \sum_k p_k \mathcal{E}(\rho_k). \tag{25.40}$$

- The map $\mathcal{E}$ preserves traces, so that $\mathrm{Tr}[\mathcal{E}(A)] = \mathrm{Tr}(A)$. When $A = \rho$ and $\mathrm{Tr}(\rho) = 1$, this is equivalent to the conservation of probability.
- The map $\mathcal{E}$ is a completely positive map.

The last condition requires a terminology explanation. A superoperator $\mathcal{E}$ is said to be a **positive map** if $\mathcal{E}(A)$ is a positive operator whenever A is a positive operator. A positive map is said to be **completely positive** if *every* extension of it to include an environment system, on which it acts trivially, is also positive. In other words, a positive map $\mathcal{E}(A)$ for operators A on $\mathcal{H}_a$ is completely positive if, for every extended system $\mathcal{H}_a \otimes \mathcal{H}_b$, the map $\mathcal{E}_{\mathrm{ext}}$ defined on product operators by

$$\mathcal{E}_{\mathrm{ext}}(A \otimes B) = \mathcal{E}(A) \otimes B, \tag{25.41}$$

and extended by linearity to all operators on $\mathcal{H}_a \otimes \mathcal{H}_b$, is also a positive map. A motivation for requiring dynamical maps to be completely positive is that if $\mathcal{E}(\rho)$ is a valid quantum evolution, then it ought to remain one if we add an arbitrary environment to which it is very weakly coupled. Naively, this might seem trivial, but it is not, because there do exist linear trace-preserving maps that are positive but not completely positive. A minimal example will be given at the end of this section.

It is a short exercise to confirm that a dynamical map with the operator-sum form of (25.22) and (25.23) satisfies the three conditions above. Conversely, we can now make, and prove, a strong statement about the general form of dynamical maps on open systems:

Theorem 25.3 (Operator-Sum Representation of Dynamical Maps) *If a map $\rho \to \rho' = \mathcal{E}(\rho)$ for density operators on an open system Hilbert space of dimension* *d is linear, trace-preserving, and completely positive, then it has an operator-sum representation*

$$\mathcal{E}(\rho) = \sum_{n=1}^{N} K_n \rho K_n^\dagger, \qquad \sum_{n=1}^{N} K_n^\dagger K_n = I, \tag{25.42}$$

where it is always possible to choose N to be d^2 or less.

Proof For the open system Hilbert space $\mathcal{H}_a$, choose an orthobasis $|\varphi_i\rangle$ with $i = 1, \ldots, d$. Now introduce[6] an auxiliary system Hilbert space $\mathcal{H}_b$ with the same dimension d, so that we can choose an orthobasis $|\chi_i\rangle$ for it, again with $i = 1, \ldots, d$. This allows us to assign to each ket $|\varphi_i\rangle$ a uniquely associated $\langle\chi_i|$, a fact that will be important soon. Now consider the following normalized state defined on the combined system with Hilbert space $\mathcal{H}_a \otimes \mathcal{H}_b$:

$$|\Psi\rangle = \frac{1}{\sqrt{d}} \sum_{i=1}^{d} |\varphi_i\rangle \otimes |\chi_i\rangle . \tag{25.43}$$

[As an aside, this is a maximally entangled ket, with Schmidt number d and entanglement entropy $\ln(d)$.] Consider the extension of $\mathcal{E}$ to a map $\mathcal{E}_{\text{ext}}$ for operators on the combined Hilbert space, defined in general as in (25.41), so that

$$\mathcal{E}_{\text{ext}}\left(|\Psi\rangle\langle\Psi|\right) = \frac{1}{d} \sum_{i=1}^{d} \sum_{j=1}^{d} \mathcal{E}(|\varphi_i\rangle\langle\varphi_j|) \otimes (|\chi_i\rangle\langle\chi_j|). \tag{25.44}$$

Because $\mathcal{E}$ is completely positive by assumption, $\mathcal{E}_{\text{ext}}$ is a positive map, by definition. Therefore, since $|\Psi\rangle\langle\Psi|$ is a valid density operator on $\mathcal{H}_a \otimes \mathcal{H}_b$, it follows that so is $\mathcal{E}_{\text{ext}}\left(|\Psi\rangle\langle\Psi|\right)$, since it satisfies the conditions of being a positive Hermitian operator with trace 1. As a density operator, it must be possible to express it in terms of an orthobasis of its eigenkets $|\Phi_n\rangle$, as in (3.70),

$$\mathcal{E}_{\text{ext}}\left(|\Psi\rangle\langle\Psi|\right) = \sum_{n=1}^{d^2} p_n |\Phi_n\rangle\langle\Phi_n| . \tag{25.45}$$

Here we have used the fact that the number of orthobasis eigenkets $|\Phi_n\rangle$ is d^2, because that is the dimension of $\mathcal{H}_a \otimes \mathcal{H}_b$. Equating (25.44) and (25.45), and then acting from the left with $\langle\chi_i|$ and the right with $|\chi_j\rangle$ and multiplying by d, we obtain

[6] It is important that this auxiliary system b is introduced only as a calculation trick, and is not the physical environment system, which we will not need to make any reference to in this proof.

$$\mathcal{E}(|\varphi_i\rangle\langle\varphi_j|) = d\sum_{n=1}^{d^2} p_n \langle\chi_i|\Phi_n\rangle\langle\Phi_n|\chi_j\rangle. \tag{25.46}$$

In the last expression, $\langle\chi_i|\Phi_n\rangle$ is obtained by taking an inner product in $\mathcal{H}_b$, but it is a ket in the open system Hilbert space $\mathcal{H}_a$. We now make the crucial observation that, because there is a unique association of $|\varphi_i\rangle$ with $\langle\chi_i|$, we can define a linear operator K_n on $\mathcal{H}_a$ by specifying its action on each of the orthobasis kets,

$$K_n|\varphi_i\rangle = \sqrt{d\,p_n}\,\langle\chi_i|\Phi_n\rangle\,. \tag{25.47}$$

This allows us to rewrite (25.46) as

$$\mathcal{E}(|\varphi_i\rangle\langle\varphi_j|) = \sum_{n=1}^{d^2} K_n|\varphi_i\rangle\langle\varphi_j|K_n^\dagger. \tag{25.48}$$

Since $\mathcal{E}$ is linear by assumption, and any density operator can be expressed as a linear combination $\rho = \sum_{i,j}\rho_{ij}|\varphi_i\rangle\langle\varphi_j|$, we obtain the result for $\mathcal{E}(\rho)$ in the first part of (25.42). Finally, we compute from (25.48) that

$$\begin{aligned}
\mathrm{Tr}\big[\mathcal{E}(|\varphi_i\rangle\langle\varphi_j|)\big] &= \sum_k\sum_n \langle\varphi_k|K_n|\varphi_i\rangle\langle\varphi_j|K_n^\dagger|\varphi_k\rangle \\
&= \sum_n\langle\varphi_j|K_n^\dagger\Big(\sum_k|\varphi_k\rangle\langle\varphi_k|\Big)K_n|\varphi_i\rangle \\
&= \langle\varphi_j|\Big(\sum_n K_n^\dagger K_n\Big)|\varphi_i\rangle.
\end{aligned} \tag{25.49}$$

Now, using the assumption that $\mathcal{E}$ is trace-preserving, this must equal $\mathrm{Tr}(|\varphi_i\rangle\langle\varphi_j|) = \delta_{ij}$ for all i, j, which shows that $\sum_n K_n^\dagger K_n$ is the identity operator, the second part of (25.42). □

We conclude this section by tying up two loose ends. First, in the footnote following (25.27), it was claimed that any linear operator that preserves all inner products on a subspace can be extended to a unitary operator on the whole Hilbert space. More formally:

Theorem 25.4 (Construction of a Unitary Operator from Its Action on a Subspace) *Let $\mathcal{H}$ be a finite-dimensional Hilbert space with a subspace $\mathcal{H}_s$. Suppose that U_s is a linear operator that acts on kets in the subspace $\mathcal{H}_s$, mapping them to kets in the full space $\mathcal{H}$ while preserving inner products. This means that for all kets $|v\rangle$ and $|w\rangle$ in $\mathcal{H}_s$, the kets $U_s|v\rangle$ and $U_s|w\rangle$ are elements of $\mathcal{H}$, and*

$$\langle w|U_s^\dagger U_s|v\rangle = \langle w|v\rangle\,. \tag{25.50}$$

Then U_s can be extended to a unitary operator U defined on all of $\mathcal{H}$. In other words, one can find an operator U such that $U|v\rangle = U_s|v\rangle$ for all $|v\rangle$ in $\mathcal{H}_s$, and $U^\dagger U = UU^\dagger = I$.

Proof We will construct a unitary operator U that satisfies the theorem. Let the dimensions of $\mathcal{H}$ and $\mathcal{H}_s$ be d and d_s, respectively. Choose an orthobasis $\{|\phi_j\rangle\}$ for $\mathcal{H}_s$, with $j = 1, \ldots, d_s$. Now, let $\mathcal{H}^\perp$ be the subspace of kets in $\mathcal{H}$ that are perpendicular to all kets in $\mathcal{H}_s$. Then $\mathcal{H}^\perp$ is a Hilbert space, and one can use the Gram-Schmidt procedure to find an orthobasis $\{|\overline{\phi}_k\rangle\}$ for it, with $k = 1, \ldots, d - d_s$. Note that $\{|\phi_j\rangle, |\overline{\phi}_k\rangle\}$ is an orthobasis for $\mathcal{H}$. Next, note that from (25.50), $U_s|\phi_j\rangle$ form a set of d_s orthonormal kets in $\mathcal{H}$. The set of kets perpendicular to all of them is a Hilbert space with dimension $d - d_s$, and one can therefore find an orthobasis $\{|\widehat{\phi}_k\rangle\}$ for it, again with $k = 1, \ldots, d - d_s$. Now define

$$U = \sum_{j=1}^{d_s} U_s|\phi_j\rangle\langle\phi_j| + \sum_{k=1}^{d-d_s} |\widehat{\phi}_k\rangle\langle\overline{\phi}_k|. \tag{25.51}$$

Since $\langle\overline{\phi}_k|\phi_i\rangle = 0$ and $\langle\phi_j|\phi_i\rangle = \delta_{ij}$, it is clear that $U|\phi_i\rangle = U_s|\phi_i\rangle$ for all orthobasis elements of $\mathcal{H}_s$, as required. Finally, we check that U is indeed unitary:

$$\begin{aligned} U^\dagger U &= \Big(\sum_{j'} |\phi_{j'}\rangle\langle\phi_{j'}|U_s^\dagger + \sum_{k'} |\overline{\phi}_{k'}\rangle\langle\widehat{\phi}_{k'}|\Big)\Big(\sum_j U_s|\phi_j\rangle\langle\phi_j| + \sum_k |\widehat{\phi}_k\rangle\langle\overline{\phi}_k|\Big) \\ &= \sum_{j,j'} |\phi_{j'}\rangle\langle\phi_{j'}|\phi_j\rangle\langle\phi_j| + \sum_{k,k'} |\overline{\phi}_{k'}\rangle\langle\widehat{\phi}_{k'}|\widehat{\phi}_k\rangle\langle\overline{\phi}_k| \\ &= \sum_j |\phi_j\rangle\langle\phi_j| + \sum_k |\overline{\phi}_k\rangle\langle\overline{\phi}_k| \\ &= I, \end{aligned} \tag{25.52}$$

where we have used (25.50) and the orthonormality and completeness properties of $|\phi_j\rangle$, $|\overline{\phi}_k\rangle$, and $|\widehat{\phi}_k\rangle$ on their respective subspaces. Since $\mathcal{H}$ is finite dimensional, $U^\dagger U = I$ implies $UU^\dagger = I$ as well, and U is unitary. □

The other loose end is that after (25.41) we promised to exhibit a linear, trace-preserving, positive map that is not completely positive. A standard example is the transpose map, $\mathcal{E}(\rho) = \rho^T$, which is defined to act on subsystem a by taking the transpose of the density matrix in some orthobasis. This map is linear. It is also trace-preserving and positive because ρ and ρ^T have the same eigenvalues, and the trace is the sum of the eigenvalues while a positive operator is one with all non-negative eigenvalues. However, its extension $\mathcal{E}_{\text{ext}}$ as defined by (25.41) is not a positive map. To see this in a minimal case, consider two 2-dimensional subsystems a and b, with orthobases $|0\rangle_a$, $|1\rangle_a$ and $|0\rangle_b$, $|1\rangle_b$ respectively. Let us use a notation

$|jk\rangle = |j\rangle_a \otimes |k\rangle_b$ for $j, k = 0, 1$. Then the map $\mathcal{E}_{\text{ext}}(\rho)$ acts to exchange the subsystem a labels only, according to

$$\mathcal{E}_{\text{ext}}\big(\,|jk\rangle\,\langle nm|\,\big) = |nk\rangle\,\langle jm|\,. \tag{25.53}$$

Now consider the combined system density operator

$$\rho_{ab} = \frac{1}{2}\big(\,|00\rangle\,\langle 00| + |00\rangle\,\langle 11| + |11\rangle\,\langle 00| + |11\rangle\,\langle 11|\,\big). \tag{25.54}$$

This is the density operator for the pure state $(|00\rangle + |11\rangle)/\sqrt{2}$, so it is clearly a positive operator. Applying (25.53) to it gives

$$\mathcal{E}_{\text{ext}}(\rho_{ab}) = \frac{1}{2}\big(\,|00\rangle\,\langle 00| + |10\rangle\,\langle 01| + |01\rangle\,\langle 10| + |11\rangle\,\langle 11|\,\big). \tag{25.55}$$

Now consider the ket $|\psi\rangle = |01\rangle - |10\rangle$. One finds

$$\langle\psi|\mathcal{E}_{\text{ext}}(\rho_{ab})|\psi\rangle = -1, \tag{25.56}$$

so $\mathcal{E}_{\text{ext}}(\rho_{ab})$ is not a positive operator, and by definition $\mathcal{E}(\rho)$ is not a completely positive map.

25.3 Generalized Measurements

The postulates of quantum mechanics presented in Sect. 3.1 give rules for observables and measurements in closed systems. The allowed results of a measurement are the eigenvalues α of a Hermitian operator, and the probabilities and post-measurement state of the system are dictated by Hermitian projection operators P_α constructed out of the corresponding eigenstates. That type of measurement, used up to now in this book, is called a **projective measurement**.

For open systems, a more general type of measurement process is available. As a model, the open system a is coupled to an environment subsystem b, which we can call the "meter". To conduct a measurement, the meter is initially prepared in a specific reference state. We then allow a unitary transformation[7] to take place, causing the combined system state to become entangled. In a sense, this is already a kind of measurement on the open system, since there are now correlations between it and the meter, and even if the open system was originally in a pure state, it is now described by the density operator of a mixed state. To expose this, a projective measurement is then performed on the meter subsystem. The combination of the unitary entanglement followed by the projective measurement on the meter is

[7] The unitary transformation of the combined system is typically associated with some time evolution, but we need not be specific about the time interval or the Hamiltonian.

called a **generalized measurement** of the open system. It will give us probabilities associated with the possible measurement outcomes, and for each outcome will also tell us the post-measurement density operator of the open system.

To be specific, let the quantum meter subsystem have dimension $d_b = N$, with an orthobasis of states $|\chi_n\rangle$ where $n = 1, 2, \ldots, N$. To carry out the measurement process, we require that the meter system is always prepared initially in the same reference state $|0\rangle$ (not necessarily one of the $|\chi_n\rangle$). Then, for any open system state $|\psi\rangle$, define the entangling unitary transformation by

$$U\big(|\psi\rangle \otimes |0\rangle\big) = \sum_n \big(M_n |\psi\rangle \otimes |\chi_n\rangle\big), \tag{25.57}$$

where the operators M_n are called **measurement operators** on the open system. The M_n can be recognized as nothing other than the Kraus operators for a dynamical map. In fact, in the following we will essentially be revisiting the discussion surrounding (25.19)–(25.23), just with K_n replaced by M_n and in a cosmetically different way.

In order for U to be unitary, it must preserve inner products between any two states $|\psi\rangle \otimes |0\rangle$ and $|\phi\rangle \otimes |0\rangle$. This implies $\sum_k \sum_n \langle\phi|M_n^\dagger M_k|\psi\rangle \langle\chi_n|\chi_k\rangle = \langle\phi|\psi\rangle$. Due to the orthonormality of the meter states $\langle\chi_n|\chi_k\rangle = \delta_{nk}$, this simplifies to $\langle\phi| \sum_n M_n^\dagger M_n |\psi\rangle = \langle\phi|\psi\rangle$. Since this is supposed to hold for any $|\phi\rangle$ and $|\psi\rangle$, it is necessary that

$$\sum_n M_n^\dagger M_n = I, \tag{25.58}$$

as in (25.23). Theorem 25.4 guarantees that (25.58) is also sufficient; having defined U to preserve all inner products when acting on the subspace of states of the form $|\psi\rangle \otimes |0\rangle$, it can be extended to a unitary operator on the whole combined Hilbert space.

Now suppose, to be general, that we have a mixed ensemble of open system states, so that we start with a combined density operator of the form

$$\rho_{ab} = \rho \otimes |0\rangle \langle 0| \,. \tag{25.59}$$

Applying the unitary transformation U, the new density operator of the combined system is

$$\rho'_{ab} = U\rho_{ab}U^\dagger = \sum_{k,m} M_k \rho M_m^\dagger \otimes |\chi_k\rangle \langle\chi_m| \,. \tag{25.60}$$

To complete the measurement process, we assume that an observer, with access to the meter only, makes a projective measurement to collapse the meter to one of the states $|\chi_n\rangle$, using mutually orthogonal projection operators that act nontrivially only on the meter subsystem,

$$P_n = I \otimes |\chi_n\rangle \langle\chi_n| . \tag{25.61}$$

Specifically, the Hermitian observable for this projective measurement is

$$Q = \sum_{n=1}^{N} n P_n, \tag{25.62}$$

which is a spectral decomposition in terms of the P_n. The outcome of the measurement is thus one of the Q eigenvalues, $n = 1, \ldots, N$. Applying (3.67), the probability of getting each outcome n is

$$\mathcal{P}(n) = \mathrm{Tr}\big(\rho'_{ab} P_n\big) = \sum_{k,m} \mathrm{Tr}_a\big(M_k \rho M_m^\dagger\big)\mathrm{Tr}_b\big(|\chi_k\rangle \langle\chi_m|\chi_n\rangle \langle\chi_n|\big), \tag{25.63}$$

which simplifies, using $\mathrm{Tr}_b\big(|\chi_k\rangle \langle\chi_m|\chi_n\rangle \langle\chi_n|\big) = \delta_{mn} \langle\chi_n|\chi_k\rangle = \delta_{mn}\delta_{kn}$, to

$$\mathcal{P}(n) = \mathrm{Tr}\big(\rho M_n^\dagger M_n\big). \tag{25.64}$$

Also, if the measurement process outcome is n, then the post-measurement combined system density operator, obtained by applying the general formula (3.85), will be

$$\rho''_{ab} = \frac{P_n \rho'_{ab} P_n}{\mathcal{P}(n)} = \frac{M_n \rho M_n^\dagger \otimes |\chi_n\rangle \langle\chi_n|}{\mathrm{Tr}\big(\rho M_n^\dagger M_n\big)}. \tag{25.65}$$

Taking the partial trace over the meter subsystem, using $\mathrm{Tr}_b\big(|\chi_n\rangle \langle\chi_n|\big) = \langle\chi_n|\chi_n\rangle = 1$, gives

$$\rho_n = \frac{M_n \rho M_n^\dagger}{\mathcal{P}(n)} = \frac{M_n \rho M_n^\dagger}{\mathrm{Tr}\big(\rho M_n^\dagger M_n\big)}. \tag{25.66}$$

for the density operator of the open subsystem after obtaining the outcome n.

In the special case that the open system starts in a pure state $|\psi\rangle$,

$$\mathcal{P}(n) = \langle\psi|M_n^\dagger M_n|\psi\rangle, \qquad |\psi_n\rangle = \frac{M_n |\psi\rangle}{\sqrt{\mathcal{P}(n)}}. \tag{25.67}$$

Note that in this special case the post-measurement state $|\psi_n\rangle$ is also a pure one, due to the collapse of the meter state.

To recapitulate and slightly augment the preceding discussion, a generalized measurement on an open system can be defined by choosing *any* set of measurement operators M_n that obey the completeness relation (25.58). The probability of getting the outcome n from the measurement is (25.64), and the open system density operator after the measurement outcome n will be (25.66). These rules can be taken as definitions of a generalized measurement that do not refer to a specific environment, although we had our meter subsystem in mind while deriving them. As consistency checks, note first that the sum of all of the measurement result probabilities is

$$\sum_n \mathcal{P}(n) = \sum_n \mathrm{Tr}\big(\rho M_n^\dagger M_n\big) = \mathrm{Tr}(\rho) = 1. \tag{25.68}$$

This shows that the completeness relation, $\sum_n M_n^\dagger M_n = I$, plays the important role of ensuring the conservation of probability. Second, (25.66) implies that the open system density operator after the measurement obeys $\mathrm{Tr}(\rho_n) = 1$ and $\rho_n = \rho_n^\dagger$, and that ρ_n is a positive operator (with non-negative eigenvalues), provided that the initial ρ satisfied these properties.

There is a close relationship between generalized measurements and dynamical maps on open systems. In this connection, the generalized measurement operators M_n can simply be identified with the Kraus operators K_n of a dynamical map. To see this more generally (without reference to the derivations above in terms of specific environment models), suppose we start with an open system density operator ρ, and make a generalized measurement with operators M_n, but do not know the result. Then the open system will be described by a density operator for a mixed state obtained by summing over the outcomes using (25.64) and (25.66),

$$\rho' = \sum_n \mathcal{P}(n)\rho_n = \sum_n M_n \rho M_n^\dagger, \tag{25.69}$$

which is nothing but the dynamical map definition of (25.42) with $K_n = M_n$. Thus, any dynamical evolution of an open system can be viewed as a generalized measurement of it by the environment, if the result of the generalized measurement is unknown or disregarded. Conversely, any generalized measurement can be viewed as a dynamical map followed by an ordinary projective measurement, but with the key feature that the state collapse is banished from the open system being measured, and instead occurs in the meter system.

Projective measurements are special cases of generalized measurements, as the latter's name implies. They arise if one imposes the further conditions that the measurement operators M_n are Hermitian, and are orthogonal projection operators in the sense that for any n and m,

$$M_n M_m = \delta_{nm} M_n \qquad \text{(projective measurements)}. \tag{25.70}$$

This implies that $M_n^\dagger M_n = M_n$ can play the role of the orthogonal Hermitian projection operators P_α in Postulates 4 and 5 of Sect. 3.1, and their equivalents for density operators, (3.67) and (3.85).

Generalized measurements can do interesting things that projective measurements restricted to the open system cannot. First, recall that Postulate 3 tells us that the outcome of doing a projective measurement is always one of the eigenvalues α of a Hermitian operator. In particular, the number of distinct possible outcomes cannot exceed the dimension of the state space. In contrast, the outcomes for generalized measurements are not necessarily associated with the eigenvalues of a Hermitian operator, and the number N of distinct possible outcomes can be arbitrarily large. Also, because the M_n are not necessarily orthogonal projection operators, if one does a generalized measurement on an open system and finds the result n, one is not guaranteed to get the same result if one immediately repeats the same measurement. For the same reason, the post-measurement states need not be orthogonal to each other for different outcomes of a generalized measurement, again unlike the case for projective measurements. Finally, while a projective measurement cannot decrease the entropy when the outcome is unknown or disregarded [as we showed in the discussion leading to (3.91)], a generalized measurement can do so, as we will see by example soon.

In some applications, we may care only about the probabilities, and not what the density operator is after the measurement. In that case, we can define a measurement by specifying Hermitian operators[8] E_n with the completeness requirement

$$\sum_n E_n = I, \tag{25.71}$$

and the positivity requirement that $\langle\psi|E_n|\psi\rangle \geq 0$ for each n and all $|\psi\rangle$. With these conditions, one can show that it is always possible to find operators M_n such that $E_n = M_n^\dagger M_n$. One can therefore copy (25.64) by assigning outcome probabilities

$$\mathcal{P}(n) = \mathrm{Tr}\big(\rho E_n\big). \tag{25.72}$$

In this context, a choice of operators E_n is often called a **positive operator-valued measure**, or **POVM**. For any choice of POVM, there are actually many inequivalent generalized measurements, because there are many different operators M_n such that $E_n = M_n^\dagger M_n$. In particular, if a POVM $\{E_n\}$ is obtained as $E_n = M_n^\dagger M_n$ from a set $\{M_n\}$, then clearly it can also be obtained as $E_n = M_n'^\dagger M_n'$ from the set $\{M_n'\} = \{V_n M_n\}$, where the V_n are arbitrary unitary operators chosen independently for each n. Choosing a POVM $\{E_n\}$ fully determines the outcome probabilities, but it certainly does not determine the open system post-measurement density operator, because (25.66) does not just involve the E_n. To completely define the generalized measurement, one must choose the specific operators M_n.

Suppose we want to measure a system to detect the presence of a photon. As a very simplified model, consider an open system with two orthobasis states $|0\rangle$ (no photon) and $|1\rangle$ (photon is present). One way to proceed is to define a projective measurement using

[8] The notation E_n for the operators $M_n^\dagger M_n$ is a standard tradition, and they have nothing to do with energy.

$$P_{\text{photon}} = |1\rangle \langle 1| , \tag{25.73}$$

which has eigenvalues $\alpha = 0$ and 1, with eigenstates $|0\rangle$ and $|1\rangle$. For an input density matrix

$$\rho = \begin{pmatrix} \rho_{00} & \rho_{01} \\ \rho_{10} & \rho_{11} \end{pmatrix} , \tag{25.74}$$

the probability of obtaining the result $\alpha = 1$ ("yes, the photon is present") will be $\mathcal{P}(1) = \rho_{11}$. After that result, the system will definitely be in the pure state $|1\rangle$; according to the standard rules of projective measurements, detecting the photon ensures that any future measurement will also detect it. But this shows that our simple model is *too* simple. In the real world, the process of observing a photon typically destroys the photon by converting it and collecting its energy in a detector. A more realistic description of the measurement should have the property that if the photon was detected, the post-measurement state should be $|0\rangle$.

To do better while still maintaining simplicity, we can invent a generalized measurement; let us call it a **destructive measurement** of the photon, by defining two measurement operators

$$M_0 = |0\rangle \langle 0| , \qquad M_1 = |0\rangle \langle 1| . \tag{25.75}$$

The corresponding POVM operators are then

$$E_0 = M_0^\dagger M_0 = |0\rangle \langle 0| , \qquad E_1 = M_1^\dagger M_1 = |1\rangle \langle 1| , \tag{25.76}$$

so that the completeness condition $\sum_n M_n^\dagger M_n = I$ is indeed satisfied. Because $M_1^\dagger M_1$ is equal to P_{photon}, the probability for the outcome $n = 1$ is exactly the same, $\mathcal{P}(1) = \rho_{11}$. The difference is entirely in the post-measurement state for this outcome. Applying (25.66) gives, for $\rho_{11} \neq 0$,

$$\rho_1 = \frac{M_1 \rho M_1^\dagger}{\mathcal{P}(1)} = \frac{1}{\rho_{11}} \begin{pmatrix} 0 & 1 \\ 0 & 0 \end{pmatrix} \begin{pmatrix} \rho_{00} & \rho_{01} \\ \rho_{10} & \rho_{11} \end{pmatrix} \begin{pmatrix} 0 & 0 \\ 1 & 0 \end{pmatrix} = \begin{pmatrix} 1 & 0 \\ 0 & 0 \end{pmatrix} . \tag{25.77}$$

This describes the pure state $|0\rangle$ with no photon, as expected, since if it was there we destroyed it by measuring it. If we try to measure it again, we will not find it. Similarly, for $\rho_{00} \neq 0$, the results for the outcome $n = 0$ are a probability $\mathcal{P}(0) = \rho_{00}$ and a post-measurement density operator

$$\rho_0 = \frac{M_0 \rho M_0^\dagger}{\mathcal{P}(0)} = \frac{1}{\rho_{00}} \begin{pmatrix} 1 & 0 \\ 0 & 0 \end{pmatrix} \begin{pmatrix} \rho_{00} & \rho_{01} \\ \rho_{10} & \rho_{11} \end{pmatrix} \begin{pmatrix} 1 & 0 \\ 0 & 0 \end{pmatrix} = \begin{pmatrix} 1 & 0 \\ 0 & 0 \end{pmatrix} . \tag{25.78}$$

Regardless of the outcome of the generalized measurement, the post-measurement state of the open system is the pure state with no photon, so the entropy of the open system has been reduced to 0.

As POVMs, both the projective measurement and the destructive measurement of the photon have the same operators $E_0 = |0\rangle\langle 0|$ and $E_1 = |1\rangle\langle 1|$, which in this example happen to be projection operators. The only difference between them is that for the projective measurement we chose $M_1 = |1\rangle\langle 1|$, while for the destructive measurement we chose $M_1 = |0\rangle\langle 1|$, which is not a projection operator. Both choices gave the same probabilities, but the latter choice gave a better model for the impact of the environment (detector) on the photon. It is left to Exercise 25.4 to consider the case $M_1 = \sqrt{p}\,|0\rangle\langle 1| + \sqrt{1-p}\,|1\rangle\langle 1|$, which interpolates between the two choices for the detector impact on an initial-state photon while keeping the same POVM operator E_1.

We close this section with another example of the useful application of a generalized measurement: the problem of distinguishing two states. Suppose that Alice is presented with a spin state which is known to be one of two possibilities, $|\psi_1\rangle$ or $|\psi_2\rangle$. Can she do a measurement to determine which it is?

If the two possible states are orthogonal, for example $|\psi_1\rangle = |\uparrow\rangle$ and $|\psi_2\rangle = |\downarrow\rangle$, then the task is trivial. Alice can simply measure the projection operator observable $P_\uparrow = |\uparrow\rangle\langle\uparrow|$, which has eigenvalues 0 and 1. Depending on which result she gets, she knows which state she was given, with certainty.

However, if the two possible states are not orthogonal, then absolute certainty is not possible. Consider, for example, the problem of distinguishing

$$|\psi_1\rangle = |\uparrow\rangle\,, \qquad |\psi_2\rangle = \frac{1}{\sqrt{2}}\big(|\uparrow\rangle + |\downarrow\rangle\big). \tag{25.79}$$

Alice could again make a projective measurement with the observable $P_\uparrow = |\uparrow\rangle\langle\uparrow|$. If the state was $|\psi_1\rangle$, then she will certainly get the result 1, but then she cannot know which state it was, because it could have been $|\psi_2\rangle$. On the other hand, if the state was $|\psi_2\rangle$, then she will have probability $1/2$ to get each of the results 0 and 1 for $P_\uparrow$. If, and only if, the result was 0, then she will know with certainty that the state was not $|\psi_1\rangle$, so it must have been $|\psi_2\rangle$. Therefore, given an ensemble of experiments in which she is given the two states randomly with equal probability, Alice will know with certainty which state it was in only a fraction $1/4$ of the trials.

Alice is clever and knows about generalized measurements, so she devises one with the following three POVM operators in matrix form in the $|\uparrow\rangle$, $|\downarrow\rangle$ basis,

$$E_1 = \frac{a}{2}\begin{pmatrix} 1 & -1 \\ -1 & 1 \end{pmatrix}, \qquad E_2 = a\begin{pmatrix} 0 & 0 \\ 0 & 1 \end{pmatrix},$$

$$E_3 = \begin{pmatrix} 1 - a/2 & a/2 \\ a/2 & 1 - 3a/2 \end{pmatrix}, \tag{25.80}$$

where a is a real number to guarantee Hermiticity. Note that these operators are designed to satisfy $\sum_n E_n = I$. The positivity requirement limits the range of a, because requiring E_1 and E_2 to have non-negative eigenvalues and to be distinct operators implies $a > 0$, and requiring E_3 to have non-negative eigenvalues implies $a \leq 2 - \sqrt{2}$. Now if Alice is given $|\psi_1\rangle$, you can use $\mathcal{P}(n) = \langle\psi_1|E_n|\psi_1\rangle$ to calculate that the measurement outcome probabilities are

$$\mathcal{P}(1) = a/2, \qquad \mathcal{P}(2) = 0, \qquad \mathcal{P}(3) = 1 - a/2, \tag{25.81}$$

while if she is given $|\psi_2\rangle$, then the outcome probabilities are

$$\mathcal{P}(1) = 0, \qquad \mathcal{P}(2) = a/2, \qquad \mathcal{P}(3) = 1 - a/2. \tag{25.82}$$

So, if Alice measures the outcome $n = 1$, she will know with certainty that the state was $|\psi_1\rangle$, and if she measures $n = 2$, she will know that the state was definitely $|\psi_2\rangle$. Alice will never wrongly identify the state, but if instead the outcome is $n = 3$, then she gains no information. To maximize her discrimination ability, Alice should choose a to be the maximum allowed, which results in a probability $a/2 = 1 - 1/\sqrt{2} \approx 0.293$ that she will be able to identify the state with absolute certainty, no matter which it is. In exchange for this opportunity to sometimes identify either state with certainty, the state will remain unknown if the outcome is $n = 3$, with probability $1 - a/2 = 1/\sqrt{2} \approx 0.707$.

25.4 Summary of Rules of Quantum Mechanics for Open Systems

In the preceding sections of this chapter, we derived rules for open quantum systems based on models of interactions with an environment. An alternative is to treat these rules for open systems as fundamental principles, made without reference to an environment, and comparable to the postulates that we listed in Sect. 3.1. For convenient reference and review, they are collected in this short section, for the case of a Hilbert space with finite dimension.

First, the state of an open system is specified, not by a ket, but by a density operator ρ, which must be Hermitian, unit-trace, and positive. In other words, it must satisfy $\rho^\dagger = \rho$ and $\mathrm{Tr}(\rho) = 1$ and $\langle\psi|\rho|\psi\rangle \geq 0$ for all kets $|\psi\rangle$ in the Hilbert space on which it acts. Second, a generalized measurement corresponds to a collection of some number N (which can be arbitrarily large) of operators M_n that satisfy

$$\sum_{n=1}^{N} M_n^\dagger M_n = I. \tag{25.83}$$

In other words, the operators $E_n = M_n^\dagger M_n$ form a positive operator-valued measure. Third, the possible outcomes of a generalized measurement are associated with the labels $n = 1, \ldots, N$ for the measurement operators. Fourth, the probability of obtaining the result n for a generalized measurement is

$$\mathcal{P}(n) = \mathrm{Tr}\left(\rho M_n^\dagger M_n\right) = \mathrm{Tr}\left(\rho E_n\right). \tag{25.84}$$

Fifth, the state after the generalized measurement outcome n is given by the density operator

$$\rho_n = \frac{M_n \rho M_n^\dagger}{\mathcal{P}(n)}. \tag{25.85}$$

Sixth, the evolution of the open-system state is given by a trace-preserving completely positive dynamical map, which can be written in the operator-sum representation

$$\rho \to \rho' = \sum_n K_n \rho K_n^\dagger, \tag{25.86}$$

for some set of Kraus operators satisfying

$$\sum_n K_n^\dagger K_n = I. \tag{25.87}$$

This evolution is unitary only in the special case that there is only one Kraus operator.

Some sources give postulates allowing generalized measurements even for closed systems, which might be viewed as more permissive than the postulates we presented in Sect. 3.1. However, since any system other than the whole universe is actually an open system, in practice this does not really allow anything new. This follows because we showed that generalized measurements can be derived from unitary time evolution and a projective measurement after adding an environment system to the open system being measured.

The sixth rule for open systems might be considered too vague or incomplete, since it contains no statement about how the Kraus operators are constructed or how they are related to the elapsed time of the evolution. This is because, in general, the time evolution of the density operator of an open system is governed by a differential equation only approximately, if at all. This is the subject of the next section.

25.5 Local Time Evolution Approximation and the Lindblad Equation

How do open systems evolve in time? For a closed system, we learned in (3.77) that the density operator obeys a simple linear first-order differential equation, $i\hbar\dot{\rho} = [H, \rho]$, which was derived by applying the Schrödinger equation to the states in a mixed ensemble. It would be nice to have a similar differential equation for the density operator of an open system interacting with an environment, but it turns out that this is possible only in an approximation. Intuitively, the obstacle is that interactions transfer energy and information to the environment, from which their effects can return after a non-infinitesimal time delay. Even if the combined system has unitary time evolution, the effective dynamics of the open subsystem will not be local in time, and so cannot be expressed using a differential equation in t with a finite number of derivatives. Instead, $\dot{\rho}(t)$ can depend not just on $\rho(t)$, but on $\rho(t')$ for all earlier times $t' < t$.

Fortunately, in many cases the dynamics of an open system are still *approximately* local in time; this is sometimes called the **Markovian approximation**. This will occur if the environment tends to "forget" what has happened on time scales longer than some characteristic value τ_{env}, for example by thermal averaging or dissipation processes. This is particularly plausible if the environment is very large, or if parts of it with direct coupling to the open system are a small subset of the whole. For example, the open system may emit light or vibrational quanta, which might be unlikely to return to it within any particular time interval. The fact that some of the energy flowing into the environment may be effectively lost forever also leads to damping of the open system on some time scale τ_{damping}. Also relevant are time scales of order $\hbar/\Delta E$, where the ΔE are energy eigenvalue differences for the open system treated in isolation, which would describe unitary time evolution of the open system if the environment were absent. Let us denote the smallest of those time scales by τ_{unitary}. If τ_{env} is very short, then we can hope to describe the averaged dynamics, for an interval dt that is much larger than τ_{env} but much smaller than τ_{damping} and τ_{unitary}, by

$$\rho(t + dt) \approx \rho(t) + dt\, \mathcal{L}(\rho(t)), \tag{25.88}$$

where

$$\mathcal{L}(\rho) = d\rho/dt \tag{25.89}$$

is called the **Liouvillian**. It is a superoperator map from the open system density operator $\rho(t)$ to its approximate time derivative. Our next goal is to understand its form.

The density operator $\rho(t + dt)$ can be obtained from $\rho(t)$ by a dynamical map. As we learned in (25.42), if the open system Hilbert space has dimension d, the dynamical map can always be expressed in an operator-sum form with at most d^2

Kraus operators,

$$\rho(t+dt) = \sum_{n=0}^{d^2-1} K_n\, \rho(t)\, K_n^\dagger, \tag{25.90}$$

where, for future convenience, we have chosen to start the index n from 0. To conserve probability, they must obey

$$\sum_{n=0}^{d^2-1} K_n^\dagger K_n = I. \tag{25.91}$$

The assumed approximation of locality in time, (25.88), implies that $\rho(t+dt)$ must be linear in dt, and since it is quadratic in the Kraus operators, the latter must have an expansion in $\sqrt{dt}$. Therefore, we can write for each n,

$$K_n = K_n^{(0)} + \sqrt{dt}\, K_n^{(1)} + dt\, K_n^{(2)}, \tag{25.92}$$

where we are consistently dropping contributions of order $(dt)^{3/2}$ and higher. The (perhaps unexpected) appearance of $\sqrt{dt}$ in this expansion occurs because dt here is not strictly infinitesimal.

Now consider the case $dt = 0$, which tells us that

$$\sum_{n=0}^{d^2-1} K_n^{(0)} \rho\, K_n^{(0)\dagger} = \rho \tag{25.93}$$

for all ρ. Choosing an orthobasis $|\varphi_i\rangle$ for the open system, and looking at the specific case $\rho = |\varphi_i\rangle\langle\varphi_i|$, then applying $\langle\varphi_j|$ from the left and $|\varphi_j\rangle$ from the right, we learn that

$$\sum_{n=0}^{d^2-1} \left|\langle\varphi_j|K_n^{(0)}|\varphi_i\rangle\right|^2 = \delta_{ij} \qquad (i, j = 1, \ldots, d). \tag{25.94}$$

The left side is a sum of non-negative numbers. This is only possible if $\langle\varphi_j|K_n^{(0)}|\varphi_i\rangle = 0$ for each n when $i \neq j$, so the matrix representations of all $K_n^{(0)}$ are diagonal. Since the orthobasis choice was arbitrary, and the only operators that are diagonal in every orthobasis are proportional to the identity, we must have $K_n^{(0)} = k_n I$ for each n, for some complex numbers k_n, which must then satisfy $\sum_n |k_n|^2 = 1$. Now, by doing a unitary rotation among the Kraus operators of the form (25.38), we can arrange that one of the k_n (let us choose $n = 0$) is equal to 1, and the others vanish.

Having selected $K_0^{(0)} = I$, and $K_n^{(0)} = 0$ for all other n, next consider the parts of (25.90) and (25.91) proportional to $\sqrt{dt}$, which give, respectively,

$$K_0^{(1)}\rho + \rho K_0^{(1)\dagger} = 0, \qquad K_0^{(1)} + K_0^{(1)\dagger} = 0. \tag{25.95}$$

It follows that $K_0^{(1)}$ is anti-Hermitian and $[K_0^{(1)}, \rho] = 0$ for every ρ, which together imply that $K_0^{(1)}$ must be a purely imaginary multiple of the identity operator, $K_0^{(1)} = iaI$ for some real a. The dynamical map contribution that depends on a comes from

$$\begin{aligned} K_0 \rho K_0^\dagger &= (I + iaI\sqrt{dt} + \ldots)\,\rho\,(I - iaI\sqrt{dt} + \ldots) \\ &= (1 + a^2 dt + \ldots)\rho, \end{aligned} \tag{25.96}$$

which shows that the effect of a can always be absorbed into a redefinition of $K_0^{(2)}$. Therefore, without loss of generality, we can take $a = 0$, so $K_0^{(1)} = 0$. Also, since any linear operator can be expressed as a sum of Hermitian and anti-Hermitian operators, we write $K_0^{(2)} = G - \frac{i}{\hbar}H$, which defines Hermitian operators G and H. So, we have

$$K_0 = I + dt\Big(G - \frac{i}{\hbar}H\Big), \tag{25.97}$$

$$K_n = \sqrt{dt}\,L_n, \qquad (n = 1, \ldots, d^2 - 1). \tag{25.98}$$

In the last expression we have renamed $K_n^{(1)} = L_n$ to agree with a traditional notation, and dropped $K_n^{(2)}$ since it now contributes to (25.90) and (25.91) only at order $(dt)^{3/2}$, because of $K_n^{(0)} = 0$ for $n > 0$.

Plugging (25.97) and (25.98) into (25.91) gives

$$I = \sum_{n=0}^{d^2-1} K_n^\dagger K_n = I + dt\left(2G + \sum_{n=1}^{d^2-1} L_n^\dagger L_n\right), \tag{25.99}$$

so that we can solve for

$$G = -\frac{1}{2}\sum_{n=1}^{d^2-1} L_n^\dagger L_n. \tag{25.100}$$

Finally, using (25.97), (25.98), and (25.100) in (25.90), we arrive at

$$\frac{d\rho}{dt} = \mathcal{L}(\rho) = -\frac{i}{\hbar}[H, \rho] + \sum_{n=1}^{d^2-1}\left(L_n \rho L_n^\dagger - \tfrac{1}{2}L_n^\dagger L_n \rho - \tfrac{1}{2}\rho L_n^\dagger L_n\right). \tag{25.101}$$

This is known as the **Lindblad equation** after Göran Lindblad; a slightly different form of it was obtained earlier by Vittorio Gorini, Andrzej Kossakowski, and E.C. George Sudarshan. It follows that the Liouvillian $\mathcal{L}(\rho)$ is linear in ρ, and always obeys $\text{Tr}[\mathcal{L}(\rho)] = 0$, so that $\text{Tr}[\rho(t)] = \text{Tr}[\rho(0)] = 1$.

Equation (25.101) is the (Markovian approximation) generalization to open quantum systems of the unitary time evolution for closed systems, (3.77). The operators L_n are known as **Lindblad operators**, or **jump operators** because the terms $L_n \rho L_n^\dagger$ are responsible for transitions ("jumps") in the open system that occur solely due to the influence of the environment. The Lindblad operator terms often produce damping due to a net energy flow from the open system to the environment, but on a much shorter time scale they can also cause an irreversible loss of coherence among the open system states. These effects will be explained and explored in more detail, with examples, in Chap. 26.

We can interpret H in (25.101) as an effective Hamiltonian for the open system, although it need not be the same as the Hamiltonian for the limit in which interactions with the environment are removed. In fact, H is not even uniquely determined by the previous discussion. As an exercise, you can check that the Liouvillian $\mathcal{L}(\rho)$ is unchanged under a transformation

$$L_n \rightarrow L_n + c_n I, \tag{25.102}$$

$$H \rightarrow H + \frac{i}{2}\hbar \sum_{n=1}^{d^2-1} (c_n L_n^\dagger - c_n^* L_n), \tag{25.103}$$

where the c_n are arbitrary complex numbers. Note that this maintains the Hermiticity of H. This freedom can be used to choose a representation such that each Lindblad operator is traceless, $\text{Tr}(L_n) = 0$. There is a further freedom to do unitary rotations of the Lindblad operators among each other without changing $\mathcal{L}(\rho)$, as $L_n \rightarrow \sum_m v_{nm} L_m$, where v_{nm} are the elements of a unitary $(d^2 - 1) \times (d^2 - 1)$ matrix. In many cases, some of the Lindblad operators are 0, so the sums over n in the preceding can be taken over some smaller number than $d^2 - 1$.

Exercises

25.1 Consider the density matrix $\rho_a = \text{Tr}_b(\rho_{ab})$ for subsystem a obtained by tracing over the complementary subsystem b, as in Sect. 25.1.

(a) Show that ρ_a is Hermitian, has trace 1, and is a positive operator.
(b) Show that if there is a unitary time evolution that acts only on subsystem b, then ρ_a is unchanged.
(c) Show that if an arbitrary projective measurement is made on subsystem b, but the results are unknown, then ρ_a is unchanged.

25.2 Prove that if $\mathcal{E}_1$ and $\mathcal{E}_2$ are dynamical maps, then so is $\mathcal{E}_2 \circ \mathcal{E}_1(\rho) = \mathcal{E}_2(\mathcal{E}_1(\rho))$. Do this by giving the Kraus operator representation of $\mathcal{E}_2 \circ \mathcal{E}_1$ in terms of the Kraus operators for $\mathcal{E}_1$ and $\mathcal{E}_2$, and verifying that they satisfy (25.23).

25.3 For an arbitrary environment density operator ρ_b and unitary joint-state operator U, consider the dynamical map $\mathcal{E}(\rho) = \mathrm{Tr}_b[U(\rho \otimes \rho_b)U^\dagger]$. Show that $\mathcal{E}(\rho)$ can be written in operator-sum form with at most d_b^2 Kraus operators, where d_b is the dimension of the environment Hilbert space.

25.4 An open system has two orthobasis kets $|0\rangle$ and $|1\rangle$ representing the absence and presence of a photon respectively, as in the discussion involving (25.73)–(25.78). Consider a generalized measurement defined by operators

$$M_0 = |0\rangle\langle 0|, \qquad M_1 = \sqrt{p}\,|0\rangle\langle 1| + \sqrt{1-p}\,|1\rangle\langle 1|, \tag{25.104}$$

which allows a photon to have a nonzero chance to survive the generalized measurement.

(a) Check that the corresponding POVM operators are $E_0 = |0\rangle\langle 0|$ and $E_1 = |1\rangle\langle 1|$.
(b) For a general initial density matrix $\begin{pmatrix} \rho_{00} & \rho_{01} \\ \rho_{10} & \rho_{11} \end{pmatrix}$, find the probabilities of the outcomes $n = 0$ and $n = 1$, and the post-measurement density operator and its entropy in each case.

25.5 Show that the Liouvillian $\mathcal{L}(\rho)$ defined by (25.101) is not changed under the transformation of Lindblad operators in (25.102) and (25.103).

25.6 Show that for time evolution governed by the Lindblad equation (25.101), the von Neumann entropy cannot decrease if the Lindblad operators are all Hermitian.

26 Decoherence

26.1 Emergence of Classical Behavior

Quantum mechanics is most commonly applied to phenomena that, by the standards of everyday experience, involve very small amounts of time, distance, energy, and angular momentum. These processes include scattering and transitions involving electrons, photons, atoms, molecules, nuclei, and elementary particles. In contrast, our everyday experiences with larger objects are described well using classical physics. Some aspects of this are not too hard to understand. For example, the tiny value of $\hbar$ when expressed in macroscopic measurement units immediately implies that in our daily lives we can usually ignore the position-momentum uncertainty principle, and the discrete quantization of angular momentum.

A more subtle feature is that superposition and interference never seem to play any role in our efforts to understand and describe our macroscopic experiences, despite being absolutely essential and ubiquitous features of quantum mechanics. When trying to hit, catch, or throw in baseball or cricket, it is not necessary to take into account the possibility that the ball might be in a superposition of states that suffer visible interference. While the smallness of $\hbar$ again plays a role, there is still something to be explained: why is it so hard to prepare a baseball in a state where interference might become an obvious issue? Even the best pitchers in baseball and bowlers in cricket, whatever their other formidable skills in velocity, placement, and deception, never seem to achieve interference nodes or distinct wavepacket peaks separated on a scale of, say, centimeters. The explanation for this is **decoherence**, which can be defined as the quantum processes, due to interaction with the environment, by which certain non-classical-like coherent superposition states of macroscopic objects become impossible, for all practical purposes.

A famous example, known even in popular culture, is the thought experiment of Schrödinger's cat. A cat is confined to a box which also includes a radioactive nucleus which can decay in an infinitesimal time interval Δt with probability $\Gamma \Delta t$. If the decay occurs, an apparatus within the box kills the poor cat. As an idealization,

S. P. Martin, *Quantum Mechanics*, Graduate Texts in Physics,
https://doi.org/10.1007/978-3-032-21119-4_26

the cat, the nucleus, and the lethal apparatus are all treated as a single quantum system undergoing unitary time evolution. After some time comparable to $1/\Gamma$, which could be hours or days for a suitably chosen nucleus, it might be supposed that the system will therefore be in a coherent superposition of states in which the cat has significant probability amplitudes to be both alive and dead, like

$$e^{-\Gamma t/2} \left|\text{alive}\right\rangle + \sqrt{1 - e^{-\Gamma t}} \left|\text{dead}\right\rangle . \tag{26.1}$$

This supposedly continues until we open the box and observe the contents, presumably only then collapsing the state to one in which the cat is definitely alive or dead.

Schrödinger presented this idealized picture as contrary to both common sense and experience. Real cats seem to always be in either a state $\left|\text{alive}\right\rangle$ or $\left|\text{dead}\right\rangle$, and never in a superposition. However, general superpositions of states are perfectly valid and commonly encountered in quantum mechanics for electrons, and even for larger composite objects like nuclei and molecules. What makes the cat different? There are two obvious and related features to take into account. First, the cat is composed of a much larger number of constituent particles, something like 10^{28} total electrons, protons, and neutrons, if it is well-fed. Surely this makes a difference. Second, contrary to the idealized assumption, the cat (and any apparatus potent enough to kill it) cannot be isolated in reality. Instead, the cat-apparatus-nucleus quantum system has an enormous number of interactions with the walls of the box, and even the photons in the cosmic microwave background radiation. These environmental interactions might be viewed as effectively conducting continuous measurements on the state of the cat's health, collapsing the state into one in which the cat is definitely dead or alive. While the influence of the environment is generally accepted to be crucial, the exact mechanism by which decoherence works is not obvious. Furthermore, we ought to be able to treat the walls of the box, and any photons within it, as also parts of the quantum system undergoing unitary time evolution, although this might be prohibitively difficult to analyze.

Despite many remaining questions, a general picture of decoherence, pioneered by H. Dieter Zeh starting in the early 1970s and continuing with Wojciech Zurek a decade later, can be summarized qualitatively as follows. The dynamics of interacting subsystems entangles them, in such a way that the density operator of any particular open system evolves towards a mixed ensemble (not a superposition!) of special states, called **robust states** or **classical-like states** or **pointer states**. For macroscopic open systems, the entanglement involves an enormous number of degrees of freedom and is very rapid, even if the other effects of the environment are not so dramatic. The robust states are the ones that correspond to everyday experience. Other states, including almost all superpositions of the robust states, have such extremely low probabilities to endure after a very short time that they may as well not exist. The interactions of each open system with its environment select the robust states for possible survival.

In terms of Schrödinger's thought experiment, the Hamiltonian couplings of the cat and its environment are supposed to pick out the states $|\text{alive}\rangle$ and $|\text{dead}\rangle$ as robust states. Instead of (26.1), the density operator will evolve almost instantaneously towards the time-dependent diagonal form

$$\rho(t) = e^{-\Gamma t} |\text{alive}\rangle \langle\text{alive}| + (1 - e^{-\Gamma t}) |\text{dead}\rangle \langle\text{dead}| \tag{26.2}$$

The missing off-diagonal terms in the density operator, proportional in this case to $|\text{alive}\rangle \langle\text{dead}|$ and $|\text{dead}\rangle \langle\text{alive}|$, are called **coherences**, while the diagonal entries are called **populations**. The key point is that the coherences are absent here because their coefficients are presumed to have a time dependence like $e^{-\gamma_{\text{deco}} t}$ or perhaps $e^{-\gamma_{\text{deco}}^2 t^2}$, with an incredibly small decoherence time $1/\gamma_{\text{deco}}$, many orders of magnitude shorter than humans (or even cats) can resolve. In particular, $1/\gamma_{\text{deco}}$ can easily be much shorter than the time scale $1/\Gamma$ associated with the nuclear decay and the cat's demise. The absence of the coherences after a very short time asserts that, even before we open the box, for all practical purposes the cat is either dead or alive, rather than a coherent superposition. If we rewrite this density operator in any other orthobasis, say,

$$\frac{1}{\sqrt{2}}\big(|\text{alive}\rangle + |\text{dead}\rangle\big), \qquad \frac{1}{\sqrt{2}}\big(|\text{alive}\rangle - |\text{dead}\rangle\big), \tag{26.3}$$

it is not diagonal; evidently, these states are not robust. Another way of saying the same thing is that for times much larger than $1/\gamma_{\text{deco}}$, the system is described almost perfectly as in (26.2) by a mixed ensemble of only states in which the cat is either alive or dead, and not by the coherent superposition pure state in (26.1).

Underpinning all of this is the idea that if we can calculate $1/\gamma_{\text{deco}}$, it will be shown to be tiny because $\hbar$ is small in everyday units. However, calculating the details of this in terms of cats is too complicated (and cat-killing experiments are unlikely to be funded anyway), so we will resort to some much simpler models to illustrate how decoherence should work.

To explain in general terms how the coherences in a density operator could be reduced, consider an open system a coupled to an environment b in such a way that the unitary time evolution operator for the combined system has the special form

$$U = \sum_n \big(|\varphi_n\rangle \langle\varphi_n|\big) \otimes V_n, \tag{26.4}$$

where V_n are unitary operators acting on the environment, and the states $|\varphi_n\rangle$ form an orthobasis, and will turn out to be the robust states of the open system in this example. Suppose that initially the density operator for the combined system has the form

$$\rho_{\text{com}} = \rho \otimes (|0\rangle \langle 0|), \tag{26.5}$$

where $|0\rangle$ is some normalized reference environment state, perhaps its ground state. Writing $\rho_{nm} = \langle\varphi_n|\rho|\varphi_m\rangle$ for the initial density operator matrix elements of the open system, the combined system density operator after the evolution is then

$$\rho'_{\text{com}} = U\rho_{\text{com}}U^\dagger = \sum_{n,m} \rho_{nm}\Big(|\varphi_n\rangle\langle\varphi_m|\Big) \otimes \Big(V_n|0\rangle\langle 0|V_m^\dagger\Big). \qquad (26.6)$$

To find the density operator ρ' for the open system after the evolution, we trace ρ'_{com} over the environment subsystem using $\text{Tr}_b\big(V_n|0\rangle\langle 0|V_m^\dagger\big) = \langle 0|V_m^\dagger V_n|0\rangle$. Defining new unit-norm environment states $|\chi_n\rangle = V_n|0\rangle$, we arrive at

$$\rho' = \sum_{n,m} \langle\chi_m|\chi_n\rangle\,\rho_{nm}\,|\varphi_n\rangle\langle\varphi_m|\,, \qquad (26.7)$$

so that, for each n and m,

$$\rho'_{nm} = \langle\chi_m|\chi_n\rangle\,\rho_{nm}. \qquad (26.8)$$

Now, the unitarity of V_n ensures that $\langle\chi_n|\chi_n\rangle = 1$, so the diagonal elements ρ_{nn} of the open system density matrix are unaffected by the evolution. However, for the off-diagonal entries, the Cauchy–Schwarz inequality (2.21) implies that the matrix elements $\langle\chi_m|\chi_n\rangle$ have magnitude less than or equal to 1, with equality only in the special case that $V_n|0\rangle$ and $V_m|0\rangle$ are proportional. Thus, the coherences will be reduced in magnitude. In this example, the special set of states $|\varphi_n\rangle$ that interact with the environment in the special form of (26.4) are the ones whose populations remain intact, while their coherences are reduced.

The previous discussion leaves unanswered the question of why the coherences should decay exponentially (or faster) with time, rather than some more mild reduction. This is indeed the key question to explain decoherence for any particular system. For a toy model example to address this, consider an open system whose density operator evolves according to the Lindblad equation (25.101), but with the very simple assumption that there is only one nonzero Lindblad operator, which is proportional to the open system Hamiltonian itself, $L = \sqrt{\alpha}H$, where α is a constant with units of $(\text{energy})^{-2}(\text{time})^{-1}$. We therefore have

$$\frac{d\rho}{dt} = -\frac{i}{\hbar}[H,\rho] + \alpha\Big(H\rho H - \tfrac{1}{2}H^2\rho - \tfrac{1}{2}\rho H^2\Big). \qquad (26.9)$$

This is easy to evaluate in an orthobasis of energy eigenstates satisfying $H|\varphi_n\rangle = E_n|\varphi_n\rangle$, with the result that the density operator matrix elements $\rho_{nm} = \langle\varphi_n|\rho|\varphi_m\rangle$ satisfy

$$\frac{d}{dt}\rho_{nm} = \Big[-\frac{i}{\hbar}(E_n - E_m) - \tfrac{1}{2}\alpha(E_n - E_m)^2\Big]\rho_{nm}. \qquad (26.10)$$

This is an especially easy differential equation, with the solution

$$\rho_{nm}(t) = \exp\left(\left[-\frac{i}{\hbar}(E_n - E_m) - \frac{1}{2}\alpha(E_n - E_m)^2\right]t\right)\rho_{nm}(0). \tag{26.11}$$

The solution shows that the diagonal elements (populations) with $n = m$ are constant in time. The off-diagonal elements (coherences) have a time-dependent phase, but more importantly their magnitudes suffer an exponential decay with mean lifetime $2/\alpha(E_n - E_m)^2$. In this simple example, the robust classical-like states that survive decoherence at late times are the energy eigenstates, but we will soon see that this is not a general feature.

26.2 Loss of Coherence for a Damped Harmonic Oscillator

In Sect. 7.4, we observed that the energy eigenstates of the harmonic oscillator behave nothing like our macroscopic experience, as their expectation values for position and momentum do not change in time. This motivated us to introduce the quasi-classical, or coherent, states, which behave as much as possible like classical oscillators. These are indeed the robust states called for by the decoherence idea. Note that in this case, the robust states are not energy eigenstates. They also do not form an orthobasis, since no two coherent states are orthogonal.

However, noting the mere existence of such states only solves half of the problem of classical behavior; the other half is to understand why they are inevitable for macroscopic masses and displacements. What would happen if we were somehow able to prepare a superposition of such states with a macroscopic position gap? How quickly does decoherence occur? Why are non-classical-like states essentially impossible to obtain? In this section we will show firstly that coherent states robustly survive the introduction of damping by an environment, and secondly that these states will decohere from each other at an incredibly fast rate.

Let us take a simple model for the damped harmonic oscillator as an open system, based on the Lindblad equation (25.101). To incorporate damping, we choose a single nonzero Lindblad operator

$$L = \sqrt{\gamma}a, \tag{26.12}$$

where γ is a positive real parameter with units of 1/time, and a is the lowering operator of the oscillator. As we will confirm soon, this Lindblad operator evolves the open system by continuously lowering the energy, which is transferred to the environment. The environment therefore is effectively providing a frictional force. For the open system Hamiltonian in the Lindblad equation, we take $H = \hbar\omega(a^\dagger a + 1/2)$. So, the equation governing the time evolution of the density operator of the oscillator is

$$\frac{d\rho}{dt} = -i\omega\big[a^\dagger a, \rho\big] + \gamma\Big(a\rho a^\dagger - \tfrac{1}{2}a^\dagger a\rho - \tfrac{1}{2}\rho a^\dagger a\Big). \tag{26.13}$$

Our first goal is to show that there are pure coherent state solutions $|\alpha(t)\rangle$ to this equation. Referring to the formula for coherent states, (7.110), we therefore try

$$\rho = |\alpha\rangle\langle\alpha| = e^{-|\alpha|^2}\sum_{n=0}^{\infty}\sum_{m=0}^{\infty}\frac{\alpha^n}{\sqrt{n!}}\frac{\alpha^{*m}}{\sqrt{m!}}|n\rangle\langle m|, \tag{26.14}$$

where α is a function of time to be determined. It is convenient to proceed by defining the matrix elements

$$\rho_{nm} = \langle n|\rho|m\rangle = e^{-|\alpha|^2}\frac{\alpha^n}{\sqrt{n!}}\frac{\alpha^{*m}}{\sqrt{m!}}, \tag{26.15}$$

so that

$$\rho = \sum_{n,m}\rho_{nm}|n\rangle\langle m|. \tag{26.16}$$

One can now compute the various pieces needed in (26.13),

$$\frac{d\rho}{dt} = \sum_{n,m}\left(\frac{n}{\alpha}\frac{d\alpha}{dt} + \frac{m}{\alpha^*}\frac{d\alpha^*}{dt} - \frac{d}{dt}|\alpha|^2\right)\rho_{nm}|n\rangle\langle m|, \tag{26.17}$$

$$a^\dagger a\rho = \sum_{n,m} n\rho_{nm}|n\rangle\langle m|, \tag{26.18}$$

$$\rho a^\dagger a = \sum_{n,m} m\rho_{nm}|n\rangle\langle m|, \tag{26.19}$$

$$a\rho a^\dagger = \sum_{n,m}\sqrt{n}\sqrt{m}\rho_{nm}|n-1\rangle\langle m-1| = |\alpha|^2\sum_{n,m}\rho_{nm}|n\rangle\langle m|, \tag{26.20}$$

where the last equality was obtained by using $\sqrt{n}\sqrt{m}\rho_{nm} = |\alpha|^2\rho_{n-1,m-1}$ and then relabeling the summation indices according to $n-1 \to n$ and $m-1 \to m$. Plugging (26.17)–(26.20) into (26.13), and collecting like terms, one finds for each n and m

$$\left[n\left(\frac{d}{dt}\ln\alpha + i\omega + \gamma/2\right) + m\left(\frac{d}{dt}\ln\alpha^* - i\omega + \gamma/2\right)\right.$$
$$\left.-\frac{d}{dt}|\alpha|^2 - \gamma|\alpha|^2\right]\rho_{nm} = 0. \tag{26.21}$$

These equations have the neat common solution

$$\alpha(t) = \alpha_0 \exp\left[-(i\omega + \gamma/2)t\right], \tag{26.22}$$

where α_0 is an arbitrary complex number that serves as the initial condition. Thus, a pure coherent state remains a pure coherent state, with a simple time dependence including an exponential damping. Note that with $\gamma = 0$, we recover the time evolution of the coherent state found in (7.131). Recalling from (7.135) that $|\alpha|$ is proportional to the position amplitude of the coherent state oscillation, we see that the parameter γ is associated with a damping factor $e^{-\gamma t/2}$ in the amplitude, and therefore $e^{-\gamma t}$ in the energy.

To address the question of decoherence, suppose we somehow manage to prepare a pure state that is a superposition of two coherent states,

$$|\psi\rangle = \frac{1}{\sqrt{N}}\left(|\alpha_0\rangle + |\beta_0\rangle\right). \tag{26.23}$$

(Recall that coherent states are never orthogonal, so imposing $\langle\psi|\psi\rangle = 1$ means that $N = 2 + 2\mathrm{Re}\left(\langle\beta_0|\alpha_0\rangle\right)$ is not equal to 2.) The initial condition for the density operator is then

$$\rho(0) = |\psi\rangle\langle\psi| = \frac{1}{N}\left(|\alpha_0\rangle\langle\alpha_0| + |\beta_0\rangle\langle\beta_0| + |\alpha_0\rangle\langle\beta_0| + |\beta_0\rangle\langle\alpha_0|\right). \tag{26.24}$$

We have already seen how the first two terms in this density operator evolve, so we can make an educated guess

$$\rho(t) = \frac{1}{N}\left(|\alpha\rangle\langle\alpha| + |\beta\rangle\langle\beta| + F(t)\,|\alpha\rangle\langle\beta| + F(t)^*\,|\beta\rangle\langle\alpha|\right), \tag{26.25}$$

where, as in (26.22),

$$\alpha = \alpha_0 \exp\left[-(i\omega + \gamma/2)t\right], \qquad \beta = \beta_0 \exp\left[-(i\omega + \gamma/2)t\right], \tag{26.26}$$

and the other time dependence is in the decoherence factor $F(t)$, which remains to be found.

Plugging the guess of (26.25) into the Lindblad equation (26.13), and using the same sort of strategy as in (26.15)–(26.21), one finds success provided that

$$\frac{d}{dt}F(t) = \gamma\left(-\frac{1}{2}|\alpha|^2 - \frac{1}{2}|\beta|^2 + \alpha\beta^*\right)F(t), \tag{26.27}$$

which can be rewritten as

$$\frac{d}{dt}\ln(F) = \left(-\frac{1}{2}|\alpha_0|^2 - \frac{1}{2}|\beta_0|^2 + \alpha_0\beta_0^*\right)\gamma\, e^{-\gamma t}. \tag{26.28}$$

Integrating both sides with respect to t, then exponentiating, and imposing the initial condition $F(0) = 1$, we find after rewriting the α_0, β_0 part slightly,

$$F(t) = \exp\left[\left(-\frac{1}{2}|\alpha_0 - \beta_0|^2 + i\,\mathrm{Im}(\alpha_0\beta_0^*)\right)\left(1 - e^{-\gamma t}\right)\right]. \tag{26.29}$$

When α_0 and β_0 are macroscopic and different, the magnitude of this decoherence factor quickly approaches a very tiny value, essentially eliminating the off-diagonal terms in $\rho(t)$. Expanding in the short time limit $\gamma t \ll 1$ gives

$$F(t) \approx e^{i\Omega t}e^{-\gamma_{\text{deco}}t}, \tag{26.30}$$

where $\Omega = \gamma\,\mathrm{Im}(\alpha_0\beta_0^*)$ is a mostly uninteresting phase, and

$$\gamma_{\text{deco}} = \frac{1}{2}\gamma|\alpha_0 - \beta_0|^2 \tag{26.31}$$

is the much more interesting decoherence rate.

Equation (26.31) shows that the environmental energy damping γ is accompanied by a much faster exponential damping of the coherences between classical-like states. Recall, from (7.135), that α_0 is related to the position amplitude by $\alpha_0 = \sqrt{\frac{m\omega}{2\hbar}}x_0$. So, writing

$$|\alpha_0 - \beta_0|^2 = \frac{m\omega}{2\hbar}(\Delta x)^2, \tag{26.32}$$

we have

$$\frac{\gamma_{\text{deco}}}{\gamma} = \frac{m\omega(\Delta x)^2}{4\hbar}. \tag{26.33}$$

This shows that the very short decoherence time $1/\gamma_{\text{deco}}$ for macroscopic states is indeed due to the smallness of $\hbar$. Let us put in some numbers to get an idea. Take the same undergraduate lab oscillator discussed in Sect. 7.4, with $m = 0.2$ kg and $\omega = 10$ radians/second, and consider the decoherence of two coherent states with peaks $\Delta x = 1$ centimeter apart. If the energy damping rate is a generous $1/\gamma = 1000$ seconds, corresponding to a quite low-friction environment, then the time scale needed for a loss of coherence is

$$1/\gamma_{\text{deco}} \approx 2 \times 10^{-27}\ \text{s}, \tag{26.34}$$

which is a tiny fraction of the time needed for light to traverse a proton's diameter.

One should certainly question whether the above estimate is accurate for a realistic case, since the physical processes leading to damping operate on much longer time scales than the $1/\gamma_{\text{deco}}$ that we found. Real-world macroscopic oscillators have internal structures with all kinds of vibrational modes, and there are numerous environmental factors that contribute to damping in different ways not captured in our simple model. However, the key conceptual result stands: decoherence implies a ridiculously short time for a macroscopic system to become almost perfectly described by a mixed ensemble of robust coherent states, and never a superposition of such states subject to interference.

26.3 Measurement and Decoherence: Phase Damping of a Spin

The process of **phase damping** of a spin gives another instructive simple example of decoherence, without appealing to the Lindblad equation approximation.

Consider an open system a consisting of a single spin with orthobasis $|\uparrow\rangle$ and $|\downarrow\rangle$, coupled to an environment b with three orthobasis states labeled $|0\rangle$, $|1\rangle$, and $|2\rangle$. The environment starts in the reference state $|0\rangle$, and it subsequently evolves according to a dynamical map defined by a unitary operator on the combined system, with[1]

$$U\big(|\uparrow\rangle \otimes |0\rangle\big) = \sqrt{1-p}\,|\uparrow\rangle \otimes |0\rangle + \sqrt{p}\,|\uparrow\rangle \otimes |1\rangle\,, \tag{26.35}$$

$$U\big(|\downarrow\rangle \otimes |0\rangle\big) = \sqrt{1-p}\,|\downarrow\rangle \otimes |0\rangle + \sqrt{p}\,|\downarrow\rangle \otimes |2\rangle\,, \tag{26.36}$$

where $0 < p < 1$. This map has the property that it does not change the spin at all, but instead has some probability amplitude to move the environment to a different state that is correlated with the spin. If the spin is up, then the environment may be changed to the state $|1\rangle$, but if the spin is down, then it may be changed to the state $|2\rangle$ instead. In this way, the dynamical map can be thought of as making a sort of imperfect measurement on the open system, by entangling it with the environment. An observer who projectively measures the post-map environment state will have completed a generalized measurement, as discussed in Sect. 25.3. If the observer finds it to be $|1\rangle$, the open system must be in the spin-up state, and if it is $|2\rangle$, it is spin-down. If found to be $|0\rangle$, the spin will remain unknown.

The dynamical map on the open system induced by U can be written in operator-sum form in terms of three Kraus operators K_0, K_1, and K_2, one for each of the environment orthobasis states $|0\rangle$, $|1\rangle$, and $|2\rangle$. To construct them, we compute their actions on the two open system orthobasis kets using (25.21),

$$K_0\,|\uparrow\rangle = \langle 0|\; U\big(|\uparrow\rangle \otimes |0\rangle\big) = \sqrt{1-p}\,|\uparrow\rangle\,, \tag{26.37}$$

$$K_0\,|\downarrow\rangle = \langle 0|\; U\big(|\downarrow\rangle \otimes |0\rangle\big) = \sqrt{1-p}\,|\downarrow\rangle\,, \tag{26.38}$$

$$K_1\,|\uparrow\rangle = \langle 1|\; U\big(|\uparrow\rangle \otimes |0\rangle\big) = \sqrt{p}\,|\uparrow\rangle\,, \tag{26.39}$$

[1] Equations (26.35)–(26.36) only explicitly define U acting on 2 of the 6 orthobasis states for the combined system, but Theorem 25.4 assures us that, since it preserves inner products, it can be completed to a unitary operator on the whole combined Hilbert space. We will do this soon.

$$K_1 |{\downarrow}\rangle = \langle 1| \, U \left(|{\downarrow}\rangle \otimes |0\rangle \right) = 0, \tag{26.40}$$

$$K_2 |{\uparrow}\rangle = \langle 2| \, U \left(|{\uparrow}\rangle \otimes |0\rangle \right) = 0, \tag{26.41}$$

$$K_2 |{\downarrow}\rangle = \langle 2| \, U \left(|{\downarrow}\rangle \otimes |0\rangle \right) = \sqrt{p} \, |{\downarrow}\rangle , \tag{26.42}$$

from which we learn

$$K_0 = \sqrt{1-p} \, I, \qquad K_1 = \sqrt{p} \, |{\uparrow}\rangle \langle{\uparrow}| , \qquad K_2 = \sqrt{p} \, |{\downarrow}\rangle \langle{\downarrow}| . \tag{26.43}$$

These satisfy $K_0^\dagger K_0 + K_1^\dagger K_1 + K_2^\dagger K_2 = I$ as required, and the dynamical map is $\mathcal{E}(\rho) = K_0 \rho K_0^\dagger + K_1 \rho K_1^\dagger + K_2 \rho K_2^\dagger$. Given an initial density matrix for the spin,

$$\rho = \begin{pmatrix} \rho_{\uparrow\uparrow} & \rho_{\uparrow\downarrow} \\ \rho_{\downarrow\uparrow} & \rho_{\downarrow\downarrow} \end{pmatrix}, \tag{26.44}$$

the result of the dynamical map is found to be

$$\mathcal{E}(\rho) = \begin{pmatrix} \rho_{\uparrow\uparrow} & (1-p)\rho_{\uparrow\downarrow} \\ (1-p)\rho_{\downarrow\uparrow} & \rho_{\downarrow\downarrow} \end{pmatrix}. \tag{26.45}$$

In words, the spin populations (the diagonal entries) are unchanged, while the spin coherences in this basis are reduced by a factor $1 - p$.

So far, we have not associated the dynamical map with a specific elapsed time interval. This can be remedied by treating the dynamical map as corresponding to an infinitesimal time Δt, and then repeating the process a large number of times N to build up the result for a finite time $t = N\Delta t$. However, there are two ways to do this, which lead to quite different outcomes. For the first way, which we will call Version 1, we can suppose that each iteration starts from the combined system state left over from the previous iteration. Since the environment is left in a different state than it started, this means that each dynamical map is actually a slightly different one than its predecessor. The second way, Version 2, proceeds by assuming that, after each iteration, the environment state is restored to $|0\rangle$, perhaps by some relaxation process not captured in our simple model. For example, the states $|1\rangle$ and $|2\rangle$ might quickly turn back into $|0\rangle$ by emitting a photon that always leaves the scene without further effect. In that case, we just use the same map (26.45) over and over.[2] Let us see what these two versions predict.

To implement Version 1, the simplest way is to go back to the unitary map U defined in (26.35)–(26.36), and complete it to the whole combined Hilbert space.

[2] Note that the treatment of the decaying atomic state example surrounding (25.35)–(25.37) corresponds to Version 2 here, for the good reason explained there that the outgoing photon should not be relevant to the atom. This would be revealed in a more detailed treatment that took into account the photon kinematics.

Working in a matrix representation with the orthobasis $|\uparrow\rangle \otimes |0\rangle$, $|\uparrow\rangle \otimes |1\rangle$, $|\downarrow\rangle \otimes |0\rangle$, $|\downarrow\rangle \otimes |2\rangle$, $|\uparrow\rangle \otimes |2\rangle$, $|\downarrow\rangle \otimes |1\rangle$, in that order, you can check that the map is implemented by the unitary matrix

$$U = \begin{pmatrix} \sqrt{1-p} & -\sqrt{p} & 0 & 0 & 0 & 0 \\ \sqrt{p} & \sqrt{1-p} & 0 & 0 & 0 & 0 \\ 0 & 0 & \sqrt{1-p} & -\sqrt{p} & 0 & 0 \\ 0 & 0 & \sqrt{p} & \sqrt{1-p} & 0 & 0 \\ 0 & 0 & 0 & 0 & 1 & 0 \\ 0 & 0 & 0 & 0 & 0 & 1 \end{pmatrix}. \tag{26.46}$$

To realize this as a continuous unitary evolution in the form of an exponential $U = \exp(-iHt/\hbar)$, we invent a Hermitian Hamiltonian

$$H = \hbar\omega \begin{pmatrix} 0 & -i & 0 & 0 & 0 & 0 \\ i & 0 & 0 & 0 & 0 & 0 \\ 0 & 0 & 0 & -i & 0 & 0 \\ 0 & 0 & i & 0 & 0 & 0 \\ 0 & 0 & 0 & 0 & 0 & 0 \\ 0 & 0 & 0 & 0 & 0 & 0 \end{pmatrix}, \tag{26.47}$$

which enables us to identify

$$\sqrt{p} = \sin(\omega t), \qquad \sqrt{1-p} = \cos(\omega t). \tag{26.48}$$

This means that (26.45) becomes

$$\mathcal{E}(\rho) = \begin{pmatrix} \rho_{\uparrow\uparrow} & \cos^2(\omega t)\,\rho_{\uparrow\downarrow} \\ \cos^2(\omega t)\,\rho_{\downarrow\uparrow} & \rho_{\downarrow\downarrow} \end{pmatrix}, \qquad \text{(Version 1)}, \tag{26.49}$$

where $\cos^2(\omega t)$ has replaced $1 - p$ as the parameterization of the dynamical map. This is the result if the environment evolves continuously without being reset.

To implement Version 2, we take $p = \Gamma\Delta t = \Gamma t/N$ in the infinitesimal map, and then use $\lim_{N\to\infty}(1 - \Gamma t/N)^N = e^{-\Gamma t}$ to write the N-fold iteration of (26.45) as

$$\mathcal{E}(\rho) = \begin{pmatrix} \rho_{\uparrow\uparrow} & e^{-\Gamma t}\,\rho_{\uparrow\downarrow} \\ e^{-\Gamma t}\,\rho_{\downarrow\uparrow} & \rho_{\downarrow\downarrow} \end{pmatrix}, \qquad \text{(Version 2)}. \tag{26.50}$$

This is the result if the environment is continually being restored to the reference state $|0\rangle$.

Both versions lead to decoherence in the preferred basis of robust states $|\uparrow\rangle$ and $|\downarrow\rangle$, but in different ways. In Version 1, the decoherence is quadratic in t for short times, with $1-p \approx 1-\omega^2 t^2$, and becomes complete at time $t = \pi/2\omega$. However, the density operator is periodic, and the coherences are restored at regular time intervals. In Version 2, the decoherence is linear in t for short times, but is monotonic, irreversible, and complete for long times. This irreversibility is due to the fact that the information flowing to the environment is eliminated by the resetting of the environment state, so that any coherence present in the initial spin state will be lost forever. In both versions, after the dynamical map has taken place, the completion of a generalized measurement by projectively measuring the environment may provide some information about the open system state. The probability that the spin state will be successfully determined from the environment measurement is equal to p, and so is maximized if $p = 1$, corresponding to complete decoherence. Both versions are only simplified models for how an environment will cause decoherence of a spin state, but in a realistic scenario there will almost always be some irreversible component to the loss of coherence. In the rapidly growing enterprise of quantum information, one often seeks practical ways to minimize this.

26.4 Decoherence for a Spin in an Exactly Solvable Model

In the example of decoherence of a damped harmonic oscillator in Sect. 26.2, we used the Lindblad equation to model the interaction with the environment. However, the Lindblad equation is an approximation, so one might have residual doubts about the results obtained. In the case of phase damping of a spin in Sect. 26.3, we studied two versions. In the first version, an exact unitary time evolution of the combined system resulted in complete decoherence, but this was temporary and the coherences were restored at regular intervals. The second version resulted in permanent decoherence at late times, but at the cost of appealing to a non-unitary reset of the environment state.

In this section, we will study a slightly more complicated model, proposed by Wojciech Zurek in 1982, for decoherence of a spin, which can be solved exactly and leads to effectively complete decoherence at late times in a preferred basis of robust states, despite having quasi-periodic time-dependence. Since we can solve it exactly, we will not have to use the Markovian approximation of local time dependence for the open system, or appeal to non-unitary time evolution in the environment.

The open system of interest will consist of a single spin with orthobasis $|\uparrow\rangle$, $|\downarrow\rangle$, which will turn out to be the robust states. It interacts with an environment consisting of a large number N of spins, each with orthobasis $|\uparrow\rangle_k$, $|\downarrow\rangle_k$ for $k = 1, \ldots, N$. The Hamiltonian is taken to be

$$H = \sigma_z \otimes H_{\text{env}}, \tag{26.51}$$

where σ_z acts on the open-system spin, and the environment part of the Hamiltonian is

$$H_{\text{env}} = \frac{\hbar}{2}\sum_{k=1}^{N}\omega_k\sigma_z^{(k)}, \tag{26.52}$$

where $\sigma_z^{(k)}$ acts on the environment spin k, with coupling constants ω_k that have units of angular frequency. Since H is diagonal acting on each spin in our chosen orthobasis, it will be simple to exponentiate.

We start at time $t = 0$ in a product density operator for the combined system,

$$\rho_{\text{com}}(0) = \rho(0) \otimes |0\rangle\langle 0|, \tag{26.53}$$

where the initial environment state is itself a product state,

$$|0\rangle = \bigotimes_{k=1}^{N}\big(\alpha_k|\uparrow\rangle_k + \beta_k|\downarrow\rangle_k\big), \tag{26.54}$$

with $|\alpha_k|^2 + |\beta_k|^2 = 1$ for each k, and we write the initial open system density operator as

$$\rho(0) = \rho_{\uparrow\uparrow}|\uparrow\rangle\langle\uparrow| + \rho_{\uparrow\downarrow}|\uparrow\rangle\langle\downarrow| + \rho_{\downarrow\uparrow}|\downarrow\rangle\langle\uparrow| + \rho_{\downarrow\downarrow}|\downarrow\rangle\langle\downarrow|. \tag{26.55}$$

Now, using $U = e^{-iHt/\hbar}$, we obtain the time evolutions

$$U\big(|\uparrow\rangle\otimes|0\rangle\big) = |\uparrow\rangle\otimes|\chi_+\rangle, \qquad U\big(|\downarrow\rangle\otimes|0\rangle\big) = |\downarrow\rangle\otimes|\chi_-\rangle, \tag{26.56}$$

where the unit-norm environment states correlated with the open system spins are

$$|\chi_\pm\rangle = \bigotimes_{k=1}^{N}\big(\alpha_k e^{\mp i\omega_k t/2}|\uparrow\rangle_k + \beta_k e^{\pm i\omega_k t/2}|\downarrow\rangle_k\big). \tag{26.57}$$

The combined system density operator at time t is therefore

$$\begin{aligned}\rho_{\text{com}}(t) &= U\rho_{\text{com}}(0)U^\dagger \\ &= \rho_{\uparrow\uparrow}|\uparrow\rangle\langle\uparrow|\otimes|\chi_+\rangle\langle\chi_+| + \rho_{\uparrow\downarrow}|\uparrow\rangle\langle\downarrow|\otimes|\chi_+\rangle\langle\chi_-| + \\ &\quad \rho_{\downarrow\uparrow}|\downarrow\rangle\langle\uparrow|\otimes|\chi_-\rangle\langle\chi_+| + \rho_{\downarrow\downarrow}|\downarrow\rangle\langle\downarrow|\otimes|\chi_-\rangle\langle\chi_-|. \end{aligned} \tag{26.58}$$

To obtain the reduced density operator for the open system, we need to trace ρ_{com} over all of the environment spins. This can be done easily by noticing that

$$\text{Tr}\big(|\chi_+\rangle\langle\chi_+|\big) = \langle\chi_+|\chi_+\rangle = 1, \tag{26.59}$$

$$\text{Tr}\big(|\chi_-\rangle\langle\chi_-|\big) = \langle\chi_-|\chi_-\rangle = 1, \tag{26.60}$$

$$\text{Tr}\big(|\chi_+\rangle\langle\chi_-|\big) = \langle\chi_-|\chi_+\rangle = F(t), \tag{26.61}$$

where

$$F(t) = \prod_{k=1}^{N} \left(|\alpha_k|^2 e^{-i\omega_k t} + |\beta_k|^2 e^{i\omega_k t}\right), \tag{26.62}$$

which has initial value $F(0) = 1$. The resulting reduced density operator for the open system, in the matrix representation in the $|\uparrow\rangle, |\downarrow\rangle$ basis, is

$$\rho(t) = \begin{pmatrix} \rho_{\uparrow\uparrow} & F(t)\rho_{\uparrow\downarrow} \\ F(t)^*\rho_{\downarrow\uparrow} & \rho_{\downarrow\downarrow} \end{pmatrix}. \tag{26.63}$$

We therefore recognize that $F(t)$ is the decoherence factor responsible for suppressing the coherences of the robust states $|\uparrow\rangle$ and $|\downarrow\rangle$. Using the normalization condition $|\alpha_k|^2 + |\beta_k|^2 = 1$, you can show from (26.62) that

$$|F(t)| = \prod_{k=1}^{N} \sqrt{1 - 4|\alpha_k\beta_k|^2 \sin^2(\omega_k t)}, \tag{26.64}$$

which is a product of N factors, each between 0 and 1. If the $\alpha_k\beta_k$ are generic, with frequencies ω_k that are not all commensurate, then $|F(t)| < 1$ for all $t > 0$. However, because it is a product of periodic functions, it is quasi-periodic for any finite N, which means that it will return arbitrarily close to 1 for some sufficiently large t. This property is called **Poincaré recurrence**.[3]

However, the Poincaré recurrence time t required to have $|F(t)| > 1 - \epsilon$ for any fixed ϵ grows extremely quickly with N. So, for a sufficiently large environment with slightly different ω_k values, the recurrence time will be many orders of magnitude longer than the age of the universe, and therefore has no practical significance. For large N, we expect that the suppression of coherences is otherwise extremely efficient. An analysis by Zurek, F.M. Cucchietti, and J.P. Paz showed that for large N and ω_k distributed randomly with a finite standard deviation, for almost all t one will have Gaussian decoherence,

$$|F(t)| \approx e^{-\gamma_{\text{deco}}^2 t^2}, \qquad \gamma_{\text{deco}}^2 = 2\sum_{k=1}^{N} |\alpha_k|^2|\beta_k|^2\omega_k^2. \tag{26.65}$$

[3] The recurrence property was originally pointed out as a general feature of Hamiltonian classical physics by Henri Poincaré in 1890. For any dynamical system that preserves phase-space volumes and has only bounded orbits, almost every trajectory will return arbitrarily close to its initial conditions, with a recurrence time that is finite. However, as in our present case, the recurrence time is often irrelevantly large, especially if the number of degrees of freedom is more than a few. For example, if you opened a bottle of classical gas molecules inside a sealed room, the molecules will in principle eventually all return to the open bottle, for almost all initial conditions. But, obviously, it is pointless to try to observe this phenomenon. For every practical purpose, it cannot happen, because the recurrence time scale is much too large.

Indeed, this has the same expansion for small t as (26.64). We conclude that the form of the environmental interactions selects $|\uparrow\rangle$ and $|\downarrow\rangle$ as robust states, and for large N ensures that at late times the open system will be described by a mixed ensemble of them, rather than a coherent superposition.

Exercises

26.1. A spin-1/2 system obeys the Lindblad equation (25.101) with $H = \omega S_z$ and a single jump operator $L = \sqrt{\gamma}\, S_z/\hbar$, where ω and γ are constants with units 1/time. This could be interpreted as a coupling of the spin to a magnetic field pointing in the $\hat{z}$ direction, which also acts as an environment that monitors the spin.

(a) Find the differential equations for the four elements of the 2×2 density matrix, and solve for them in terms of their initial conditions. What are the robust states that always survive decoherence at late times?
(b) Suppose that the system starts in the pure state $(2\,|\uparrow\rangle + |\downarrow\rangle)/\sqrt{5}$. Write the density matrix in the $|\uparrow\rangle, |\downarrow\rangle$ basis as a function of t. Use it to find $\langle S_z\rangle$ and $\langle S_x\rangle$ as functions of time.
(c) Express the density matrix solution from the previous part in the diagonal form $\rho(t) = p_1\,|\psi_1\rangle\langle\psi_1| + p_2\,|\psi_2\rangle\langle\psi_2|$, identifying the eigenvalues p_1, p_2, and eigenvectors $|\psi_1\rangle$, $|\psi_2\rangle$.

26.2. A (very) simplified decoherence model for quantum Brownian motion of a particle with mass m moving in one dimension uses a single Lindblad operator $L = \sqrt{2\gamma}\, X/\lambda_{\text{th}}$, where X is the position operator, γ is a coupling rate (with units of 1/time) characterizing the rate of monitoring of the position by the environment, and $\lambda_{\text{th}} = \hbar/\sqrt{2mk_BT}$ is (up to a factor that depends on convention) the **thermal de Broglie wavelength**. Solve the Lindblad equation (25.101), ignoring the H term for extra simplicity, in position space in terms of $\rho(x, x', t) = \langle x|\rho(t)|x'\rangle$, to show that the exponential decay of spatial coherences is

$$\rho(x, x', t) = e^{-\gamma_{\text{deco}} t}\rho(x, x', 0), \tag{26.66}$$

where the ratio of the decoherence rate for density matrix separations $|x - x'|$ to the monitoring rate is

$$\frac{\gamma_{\text{deco}}}{\gamma} = \frac{(x - x')^2}{\lambda_{\text{th}}^2} = \frac{2mk_BT(x - x')^2}{\hbar^2}. \tag{26.67}$$

The diagonal part of the density matrix with $x = x'$ is unaffected. What is $\gamma_{\text{deco}}/\gamma$ for $x - x' = 10^{-9}$ meters, for a nanoparticle of mass $m = 10^{-18}$ grams at room temperature?

Invitation to Quantum Information

27

27.1 Qubits and the Bloch Ball

In this chapter, we will briefly introduce some of the basic concepts of quantum information theory. This is an exciting and rapidly developing field, with the potential to revolutionize how computations are done and what they can accomplish. However, the practical realization of quantum computations faces significant challenges. Although some simple quantum computers have been built, it is not clear at present which of many types of physical implementations will scale in a useful way. Accordingly, our scope will be limited to theoretical considerations, which represent an idealized limit for what might be attainable in the real world. We will avoid some of the specialized terminology and tools, including quantum circuit diagrams, which can be found in references that treat the subject in more depth and detail.

In classical information theory, the basic unit of information is a bit, which can be either 0 or 1, and can be thought of as the answer to a single yes/no question. To encode or process information, one can use a register, which is an ordered collection of N bits. Since each bit has two possible values, there are 2^N possible classical states for the register. The binary representation of the register is an integer n, in the range from 0 to $2^N - 1$. Thus, we can write

$$n = (01011010 \cdots 1) \tag{27.1}$$

to express the fact that the classical bit register can be used to store or transmit the information corresponding to n. Classical computations consist of operations on the register to change its value in some desired way.

The analogous basic unit of quantum information is a quantum bit, or **qubit**, a system with an orthobasis consisting of two states. Physically, a qubit could be the spin state of an electron or nucleus, the polarization state of a photon or electromagnetic field mode in a cavity, a pair of energy levels for a trapped ion or a neutral atom, a spin or charge state of a semiconductor quantum dot, or the state

S. P. Martin, *Quantum Mechanics*, Graduate Texts in Physics,
https://doi.org/10.1007/978-3-032-21119-4_27

of a superconducting circuit. In the quantum information context it is usual to label the orthobasis states as $|0\rangle$ and $|1\rangle$, rather than $|\uparrow\rangle$ and $|\downarrow\rangle$ in the case of a spin-1/2 system. The most general pure state of a qubit is

$$|\psi\rangle = a|0\rangle + b|1\rangle, \tag{27.2}$$

where a and b are two complex numbers. We can always rescale $|\psi\rangle$ to have unit norm, so we can impose the restriction $|a|^2 + |b|^2 = 1$, eliminating one real degree of freedom. The global phase is also not physical, eliminating a second real degree of freedom. Therefore, the physical degrees of freedom in a qubit amount to two real numbers, or one complex number.

However, measurements on a quantum state can only extract partial information about it. For a single qubit, any projective measurement can have at most two possible results, since that is the dimension of the state space. For example, if the observable is σ_z, the measurement will project onto the basis vectors $|0\rangle$ and $|1\rangle$. For a general qubit state $a|0\rangle + b|1\rangle$, the outcome of this measurement is not deterministic, with probabilities $|a|^2$ and $|b|^2$ for the two possible outcomes. Furthermore, the act of measurement collapses the qubit to one of $|0\rangle$ or $|1\rangle$, permanently erasing some of the information it contained, and more measurements will reveal nothing more about what it was initially. To do better, it is necessary to devise measurements that act on multiple qubits in a carefully chosen way, but even this can only determine some of the information contained in the quantum state. Much of quantum information theory amounts to extracting a desired subset of information from multiple entangled qubits.

A qubit will certainly interact with its environment. Sometimes this is what we want, as in controlled operations such as quantum computations, or measurements to reveal partial information about its state. However, interactions with the environment can also be a curse that we want to suppress as much as possible. The environment can induce decoherence of the qubit, corrupting it in a way analogous to noisy errors in classical computations. Treating the qubit as an open system by tracing over its environment means that it should be described by a density matrix, as explained in Sect. 25.1.

The density matrix for a qubit in the $|0\rangle, |1\rangle$ basis is a 2×2 Hermitian matrix with trace 1, for which the most general form is

$$\rho = \frac{1}{2}\begin{pmatrix} 1+b_3 & b_1 - ib_2 \\ b_1 + ib_2 & 1 - b_3 \end{pmatrix}, \tag{27.3}$$

for some real numbers b_1, b_2, and b_3. By constructing the vector $\boldsymbol{b} = b_1\hat{x} + b_2\hat{y} + b_3\hat{z}$, we can rewrite the density matrix as

$$\rho = \frac{1}{2}(I + \boldsymbol{b} \cdot \boldsymbol{\sigma}), \tag{27.4}$$

where $\boldsymbol{\sigma} = \sigma_x \hat{x} + \sigma_y \hat{y} + \sigma_z \hat{z}$ is the vector of Pauli matrices. The eigenvalues of ρ are

$$p_\pm = \frac{1}{2}\left(1 \pm \sqrt{b_1^2 + b_2^2 + b_3^2}\right) = \frac{1}{2}\left(1 \pm |\boldsymbol{b}|\right). \tag{27.5}$$

As we learned in Sect. 3.5, p_+ and p_- can be interpreted as mixed ensemble probabilities for the orthobasis states consisting of the eigenvectors of ρ. As probabilities, they must be in the range from 0 to 1 inclusive, so $\boldsymbol{b}$ must be a vector with length $|\boldsymbol{b}| \leq 1$.

The vector $\boldsymbol{b}$ that specifies the density matrix of a qubit is called its **Bloch vector**, and it lies in the **Bloch ball** of radius 1, both named for Felix Bloch. In the case that $\boldsymbol{b}$ is a unit vector, lying anywhere on the spherical surface of the Bloch ball, the qubit is in a pure state. A complete specification of such a pure state with $p_+ = 1$ and $p_- = 0$ requires the two spherical coordinate direction angles (θ_b, ϕ_b) of the Bloch vector, corresponding to the two real degrees of freedom mentioned above. The pure state is the eigenstate of $\boldsymbol{b} \cdot \boldsymbol{\sigma}$ with eigenvalue 1,

$$|\psi\rangle = \cos(\theta_b/2)\,|0\rangle + e^{i\phi_b} \sin(\theta_b/2)\,|1\rangle\,. \tag{27.6}$$

To specify a mixed state of the qubit requires not just the direction but the length $|\boldsymbol{b}|$. The unique density matrix for the completely random ensemble has $\boldsymbol{b} = 0$, the point at the center of the Bloch ball, with $p_+ = p_- = 1/2$.

The information content in a qubit can be measured by the von Neumann entropy

$$\sigma = -p_+ \log(p_+) - p_- \log(p_-), \tag{27.7}$$

as defined in (3.72), which we revisited as entanglement entropy in the context of bipartite systems in (25.13). In (27.7), we have normalized σ by using base-2 logarithms. If the qubit's density matrix is that of a completely random ensemble, so that we have no information about its state, then the entropy is $-2(1/2)\log(1/2) = 1$. At the other extreme, if the qubit's state is perfectly known, then $p_+ = 1$ and $p_- = 0$, and the entropy vanishes. Figure 27.1 shows how σ depends on p_+, illustrating that the entropy decreases monotonically with more information content.

27.2 Quantum Registers and Parallelism

A **register** of N qubits is an arbitrary superposition of tensor product states of the form

$$|\psi_1\rangle \otimes |\psi_2\rangle \otimes \cdots \otimes |\psi_N\rangle = |\psi_1 \psi_2 \cdots \psi_N\rangle\,. \tag{27.8}$$

The register has a canonical orthobasis of unentangled states of the form, for example,

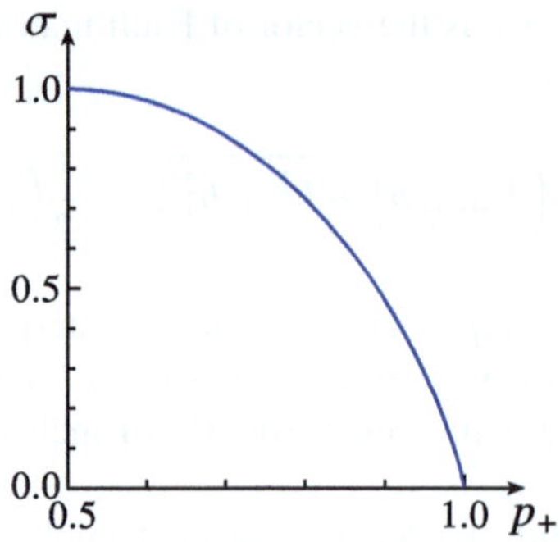

Fig. 27.1 The von Neumann entropy $\sigma = -p_+ \log(p_+) - p_- \log(p_-)$ for the density matrix of a qubit, as a function of the larger eigenvalue p_+. The normalization of σ is fixed here by using base-2 logarithms. The entropy is maximum for the completely random ensemble with $p_+ = 1/2$ and minimum for pure states with $p_+ = 1$, decreasing monotonically as the information content increases

$$|n\rangle = |01011010 \cdots 1\rangle , \tag{27.9}$$

where n is a convenient integer label for the binary representation. This means that the complete information about the state of the register can be specified in terms of 2^N complex numbers c_n, as

$$|\Psi\rangle = \sum_{n=0}^{2^N-1} c_n |n\rangle . \tag{27.10}$$

The normalization of $|\Psi\rangle$ has no physical significance, and can be set to 1 by requiring $\sum_n |c_n|^2 = 1$, eliminating one real degree of freedom. The global phase of $|\Psi\rangle$ is also not observable, eliminating another real degree of freedom (say, the phase of one of the nonzero c_n). This leaves $2^N - 1$ complex numbers to completely specify the state of the register.

There are two general types of things that we can do with a register of qubits. First, we can apply a unitary transformation to it, which is equivalent to evolving it in time with some appropriate Hamiltonian, as we learned in Sect. 3.4. However, in theoretical quantum computation studies, one often does not bother to specify either the elapsed time or the Hamiltonian. Instead, one simply assumes that *any* unitary operation on the qubit register is allowed, and can be realized somehow. The corresponding transformation is called a **quantum gate**, by analogy with logic gates in ordinary computation. The simplest gates operate on only one or two qubits at a time, and more complicated gates can be constructed out of simpler ones. Since unitary transformations are invertible, the information contained in the register can be recovered and is not changed. The second type of thing that one can do to a register is to conduct measurements on it to find out partial information about its state, for example to read a message or to learn something about the result of a computation. Measurements collapse the state, so some of the information that had been present in the register will necessarily be destroyed.

A first clue as to the power of quantum information over classical information is the enormous difference in information content. As we have just seen, a quantum register of N qubits stores $2^N - 1$ continuous complex numbers worth of information, while the classical N bit register only stores an integer in the range from 0 to $2^N - 1$, equivalent to the responses to N yes/no questions. The possibility of entanglement of the qubits in the register is a big part of the quantum advantage, as most of the states with more than one nonzero coefficient c_n are entangled states. Entanglement is not just a puzzling phenomenon to be understood, but the key resource behind much of quantum information and computing capabilities.

To see the benefit of entanglement, consider a function f, which maps an integer n in the range from 0 to $2^N - 1$ to another integer $f(n)$ in the range from 0 to $2^M - 1$. The numbers n and $f(n)$ can each be given a binary representation in an N-bit and M-bit register, respectively. If we want to evaluate all of the values of the function f using classical processes, we would need to do the computation 2^N times, one for each possible value of n. Quantum computation provides a powerful alternative. Consider a tensor product of an N-qubit register with an M-qubit register, and the map U_f defined by

$$U_f\Big(\,|n\rangle \otimes |m\rangle\,\Big) = |n\rangle \otimes |m + f(n)\rangle\,, \tag{27.11}$$

where the sum $m + f(n)$ is taken modulo 2^M. Since the states labeled $|n\rangle \otimes |m\rangle$ form an orthobasis, by linearity this completely specifies the map for any state of the combined system. It is also a very short exercise to show that this map is unitary.

Now we apply U_f to a particular useful choice. For each qubit, define the orthobasis[1] of eigenstates of σ_x,

$$|+\rangle = \frac{1}{\sqrt{2}}\big(\,|0\rangle + |1\rangle\,\big), \qquad |-\rangle = \frac{1}{\sqrt{2}}\big(\,|0\rangle - |1\rangle\,\big), \tag{27.12}$$

and construct from $|+\rangle$ the N-qubit register state $|{+}{+}\cdots{+}\rangle = |+\rangle \otimes |+\rangle \otimes \cdots \otimes |+\rangle$. By expanding the tensor product, we find

$$|{+}{+}\cdots{+}\rangle = \frac{1}{\sqrt{2^N}} \sum_{n=0}^{2^N-1} |n\rangle\,. \tag{27.13}$$

[1] Some jargon: in quantum information and computing, the unitary transformation between the $\{|0\rangle\,, |1\rangle\}$ and $\{|+\rangle\,, |-\rangle\}$ orthobases is called the **Hadamard transformation** or **Hadamard gate**, and is usually denoted H (not to be confused with a Hamiltonian). It obeys $H = H^{-1} = H^\dagger = \frac{1}{\sqrt{2}}\begin{pmatrix} 1 & 1 \\ 1 & -1 \end{pmatrix}$ in either basis.

This is a very useful state, as it is an equal-weight superposition of all 2^N canonical basis states of the N-qubit register. By linearity, the map defined by (27.11) acts according to

$$U_f\Big(\,|{+}{+}\cdots{+}\rangle \otimes |m\rangle\,\Big) = \frac{1}{\sqrt{2^N}}\sum_{n=0}^{2^N-1} |n\rangle \otimes |m + f(n)\rangle\,. \qquad (27.14)$$

For each m, this single state is a superposition containing information about all of the results $f(n)$ for every n. It does this by entangling the information for $f(n)$ with the information for n. The point is that, in principle, we can use a single operation to perform the computational task simultaneously for all of the different n, which is 2^N computations at once. This is not possible in classical computations, except by duplicating the hardware or by extending the computation time, which quickly becomes impractical for increasing N. This beautiful aspect of quantum information is called **parallelism**.

However, for parallelism to be useful it is not enough to perform the calculation; it is also necessary to find a way to extract information from the state. Although quantum registers can store and process vastly more information than classical registers, only a subset of the information can be extracted through measurement. For example, one might gain information about $f(n)$ for some particular n by conducting a measurement that projects onto the input register state $|n\rangle$. But this collapses the state, permanently erasing all of the information about $f(n)$ for other values of n. Fortunately, some global information about the function (that is, information that depends on the values of $f(n)$ for multiple different inputs n) can be extracted by making further use of superposition and devising measurements that are cleverer than just projecting onto $|n\rangle$.

To illustrate how this might work in a simple example, consider the following problem in quantum computation, due to Ethan Bernstein and Umesh Vazirani as a variation on earlier problems proposed and solved by David Deutsch and Richard Jozsa. To make things simple, suppose that the result for the function $f(n)$ is either 0 or 1, so that the second register in the discussion above has $M = 1$. Furthermore, suppose the function $f(n)$ coded into the unitary operation U_f is not known, but is asserted to be of the form

$$f(n) = a \cdot n \pmod{2}, \qquad (27.15)$$

for some unknown integer a in the same range as n, from 0 to $2^N - 1$. The meaning of the dot notation is

$$a \cdot n \equiv \sum_{k=1}^{N} a_k n_k, \qquad (27.16)$$

where $(a_1a_2\ldots a_N)$ and $(n_1n_2\ldots n_N)$ are the binary representations of the integers a and n, with each a_k and n_k equal to 0 or 1. Our assignment is to determine what a is. In classical information theory, we would need N evaluations of $f(n)$ in order to determine the N bits a_k. However, using quantum mechanics, the problem can be solved using only one operation of U_f on a cleverly chosen input state, followed by one measurement.

To see how, let us first remark on a useful change of basis. The original canonical basis of N-qubit register states, based on $|0\rangle$ and $|1\rangle$ for the individual qubits, is

$$|n\rangle = |n_1n_2\cdots n_N\rangle\,, \qquad (\text{each } n_k = 0 \text{ or } 1). \tag{27.17}$$

Now, using the orthobasis $|+\rangle$, $|-\rangle$ for each qubit, we can construct another canonical basis for the N-qubit states,

$$|\tilde{s}\rangle = |\tilde{s}_1\tilde{s}_2\cdots\tilde{s}_N\rangle\,, \qquad (\text{each } \tilde{s}_k = + \text{ or } -). \tag{27.18}$$

Expanding the implicit tensor product gives the relation between the two orthobases,

$$|\tilde{s}\rangle = |\tilde{s}_1\tilde{s}_2\cdots\tilde{s}_N\rangle = \frac{1}{\sqrt{2^N}}\sum_{n=0}^{2^N-1}(-1)^{s\cdot n}\,|n\rangle\,, \tag{27.19}$$

where the dot notation means

$$s\cdot n \equiv \sum_{k=1}^{N} s_k n_k\,, \qquad \text{where } s_k = \begin{cases} 0 & \text{if } \tilde{s}_k = +, \\ 1 & \text{if } \tilde{s}_k = -. \end{cases} \tag{27.20}$$

Note that (27.13) was a special case of this, in which all $\tilde{s}_k = +$, so that $s\cdot n = 0$ for all n. Conversely, by taking the inner product $\langle n|\tilde{s}\rangle = \langle\tilde{s}|n\rangle$, we find

$$|n\rangle = \frac{1}{\sqrt{2^N}}\sum_{\tilde{s}_1,\ldots,\tilde{s}_N=\pm}(-1)^{s\cdot n}\,|\tilde{s}\rangle\,. \tag{27.21}$$

To solve the Bernstein–Vazirani problem, we start with the state $|{+}{+}\cdots{+}\rangle\otimes|-\rangle$ and act on it with the unitary operator U_f defined as in (27.11). The result, using (27.14), is

$$U_f\Big(|{+}{+}\cdots{+}\rangle\otimes|-\rangle\Big) = \frac{1}{\sqrt{2^{N+1}}}\sum_{n=0}^{2^N-1}|n\rangle\otimes\Big(|f(n)\rangle - |f(n)+1\rangle\Big). \tag{27.22}$$

Recall that the addition in the last ket label is taken mod 2. It follows that $|f(n)\rangle - |f(n)+1\rangle = (-1)^{f(n)}\big(|0\rangle - |1\rangle\big) = (-1)^{a\cdot n}\big(|0\rangle - |1\rangle\big)$, and therefore

$$U_f\Big(|{+}{+}\cdots{+}\rangle \otimes |-\rangle \Big) = \frac{1}{\sqrt{2^N}} \sum_{n=0}^{2^N-1} (-1)^{a\cdot n} |n\rangle \otimes |-\rangle \,. \tag{27.23}$$

Using (27.19), this can be rewritten as simply

$$U_f\Big(|{+}{+}\cdots{+}\rangle \otimes |-\rangle \Big) = |\tilde{a}_1\tilde{a}_2\cdots\tilde{a}_N\rangle \otimes |-\rangle \,, \tag{27.24}$$

with

$$\tilde{a}_k = \begin{cases} + & \text{if } a_k = 0, \\ - & \text{if } a_k = 1. \end{cases} \tag{27.25}$$

Now we conduct a measurement by projecting the state onto the $|\tilde{s}\rangle$ basis, using the observable

$$Q = \Bigg(\sum_{\tilde{s}_1,\ldots,\tilde{s}_N=\pm} s\,|\tilde{s}\rangle\,\langle\tilde{s}| \Bigg) \otimes I, \tag{27.26}$$

where s is the integer whose binary representation is $(s_1s_2\cdots s_N)$. This solves the problem, since exactly one of the $|\tilde{s}\rangle$ is $|\tilde{a}_1\tilde{a}_2\cdots\tilde{a}_N\rangle$. With probability 1, the result of the measurement of Q on the state $U_f\,(|{+}{+}\cdots{+}\rangle \otimes |-\rangle)$ will be the number a.

As promised, quantum information methods solve the Bernstein–Vazirani problem with only one unitary transformation and one measurement, a task that would take N evaluations of $f(n)$ with classical computation methods. This is not necessarily very impressive, as doing 1 computation with $N+1$ qubits is not obviously easier (with present technology) than doing N classical computations on N bits. However, it shows the general principle that quantum parallelism provides advantages. Although it is beyond our scope to show here, there are some computation problems that are believed to be not feasible in practical terms using classical methods, but become easy (in principle) by exploiting quantum parallelism in more sophisticated ways that make use of entanglement and superposition.

The most famous example is Peter Shor's 1994 breakthrough quantum computing algorithm for factoring an integer n as a product of primes. The reason for its fame is that it cracks the most popular modern public-key encryption schemes, with profound implications for human events far beyond mere scientific curiosity. In these schemes, anyone can encrypt a message using a public key that depends on some very large integer n, but to decrypt and read the message one must know the prime factorization of n, which is not publicly provided. Thus, the security of the code relies on the widely held, but unproved, belief that if n has more than a few hundred digits, prime factorization is a prohibitively hard problem for classical computing,[2] with a computation time that

[2] At least, this is true for *publicly known* classical algorithms.

grows faster than any polynomial in the number of bits of n. The running time for the best known algorithm (the general number field sieve) scales for large n approximately like $\exp([(64/9)(\ln n)(\ln\ln n)^2]^{1/3})$. In contrast, Shor's factoring algorithm on a quantum computer should have a running time that scales like only $(\ln n)^2(\ln\ln n)(\ln\ln\ln n)$. The quantum algorithm is not deterministic, and gives the answer only with high probability rather than unit probability, but checking a candidate factorization is trivial in terms of computation time, as it requires only one multiplication. As of this writing, the immediate implications of Shor's factoring algorithm are only academic, since implementation of it for large n on a quantum computer is not feasible yet,[3] for mundane practical reasons. However, those with important secrets to keep should not bet on this remaining true forever.

Shor's factoring algorithm, and algorithms for many other classically hard problems, make use of the **quantum Fourier transform**, a unitary map defined by

$$U_{\mathrm{FT}}\,|n\rangle = \frac{1}{\sqrt{2^N}}\sum_{k=0}^{2^N-1}\exp\left(2\pi ink/2^N\right)|k\rangle\,. \tag{27.27}$$

This provides a change of basis that is useful for parallelizing calculations and extracting relevant information through measurement. It is crucial that it is possible to implement the quantum Fourier transform as a subroutine acting on qubit registers. The factoring problem is a special case of a more general type of problem called the hidden subgroup problem, which contains many other interesting and useful examples that can be attacked in similar ways by quantum computing. An example of a completely different type of problem is Grover's search algorithm, discovered by Lov Grover in 1996, which uses quantum parallelism to give a quadratic speed improvement over classical search methods. It is also worth mentioning that quantum computation may be the natural way to attack difficult quantum mechanics problems.

27.3 Sending Information Using Entanglement: Teleportation and Dense Coding

Transmission of information using classical bits is a well-exploited feature of modern life, underlying much familiar technology. Electromagnetic waves can be modulated in various ways to encode the 0s and 1s in a register. In quantum information, the problem is more subtle, since even a single qubit contains a continuum of information.

Let us establish some notation before continuing. First, the Pauli matrices acting on qubits will occur often, and position operators will not appear at all, so

[3] At least, this is true for *publicly known* quantum computing hardware.

throughout the rest of this chapter we will follow a common practice in the quantum information literature by writing

$$X \equiv \sigma_x, \qquad Y \equiv \sigma_y, \qquad Z \equiv \sigma_z, \tag{27.28}$$

so that $Z\,|0\rangle = |0\rangle$ and $Z\,|1\rangle = -\,|1\rangle$, and $X\,|0\rangle = |1\rangle$ and $X\,|1\rangle = |0\rangle$, and $Y\,|0\rangle = i\,|1\rangle$ and $Y\,|1\rangle = -i\,|0\rangle$. When there is more than one qubit, a subscript on X, Y, or Z may be used to indicate which qubit is being acted on. The simplest cases will involve entanglement of two qubits, like the ones familiar to Alice and Bob from the Einstein–Podolsky–Rosen and Bohm problem in Sect. 14.1. Let us define four maximally entangled states of a two-qubit system,

$$|\phi^{\pm}\rangle = \frac{1}{\sqrt{2}}\big(\,|00\rangle \pm |11\rangle\,\big), \tag{27.29}$$

$$|\psi^{\pm}\rangle = \frac{1}{\sqrt{2}}\big(\,|01\rangle \pm |10\rangle\,\big). \tag{27.30}$$

These are called **Bell states** or **EPRB states**. In many, but not all, cases of interest, the two qubits are spatially separated, after which they cannot be directly accessed simultaneously by a single observer. If we identify $|0\rangle$ and $|1\rangle$ with spin-up and spin-down respectively, then the state featured in the EPRB discussion in Sect. 14.1 was $|\psi^-\rangle$ in the present notation, but the problem could have been equally well formulated in terms of any one of the other Bell states instead. Note that $|\phi^+\rangle$, $|\phi^-\rangle$, $|\psi^+\rangle$, $|\psi^-\rangle$ form a complete orthobasis for the two-qubit system, and the four unentangled states $|00\rangle$, $|01\rangle$, $|10\rangle$, and $|11\rangle$ form another orthobasis.

There is a very useful measurement that projects any two-qubit state onto the four Bell states. The corresponding observable may be defined explicitly as

$$Q = |\phi^+\rangle\langle\phi^+| + 2|\phi^-\rangle\langle\phi^-| + 3|\psi^+\rangle\langle\psi^+| + 4|\psi^-\rangle\langle\psi^-|, \tag{27.31}$$

with possible outcomes equal to the four eigenvalues 1, 2, 3, and 4. This specific choice of eigenvalues is rather arbitrary, but the important point is that the result of the measurement of Q can be expressed using 2 classical bits. This is an example of a **collective measurement**, as it acts nontrivially on more than one qubit.

Now consider the following problem: Alice has a qubit state

$$|\gamma\rangle_c = c_0\,|0\rangle + c_1\,|1\rangle\,, \tag{27.32}$$

which she would like to transmit to Bob in a distant location, without sending the physical qubit. Can this be accomplished by transmitting only classical bits? Naively, it might seem impossible, since the qubit contains two continuous real numbers worth of information, while the classical bits to be transmitted are only a discrete set. Indeed, it will not do for Alice to make any kind of measurement on the qubit $|\gamma\rangle_c$ alone. For example, measuring Z by projecting onto the basis $|0\rangle$ and

$|1\rangle$ will not allow reconstruction of $|\gamma\rangle_c$, and even worse, it will collapse the state, making further measurements useless. However, it can be done, provided that Alice and Bob already share a known Bell state, which was prepared for this occasion, and upon which they have not yet conducted any measurements. It is not important which entangled Bell state they share, but to be concrete let us suppose it is $|\phi^+\rangle_{ab}$. Alice can act on the first qubit of this entangled pair either by measuring it or by unitarily evolving it, but she is not able to do anything with the second qubit. The converse is true for Bob in the obvious way.

To accomplish the task, Alice makes a collective measurement on the two qubits that she has access to, labeled a (the one entangled with Bob's qubit b) and c (the one to be sent). Specifically, Alice measures the observable Q_{ac} from (27.31), projecting onto the four Bell orthobasis states $|\phi^+\rangle_{ac}$, $|\phi^-\rangle_{ac}$, $|\psi^+\rangle_{ac}$, and $|\psi^-\rangle_{ac}$. (This is equivalent to the simultaneous measurement of the two commuting observables $Z_a Z_c$ and $X_a X_c$, as you will show in Exercise 27.3.) The result of Alice's measurement will be one of four possibilities, corresponding to these four orthogonal states. She then sends this information, equivalent to only two classical bits, to Bob by some ordinary communication channel. Bob then applies a unitary operator U_b to his qubit of the shared Bell state, following this rule:

$$U_b = I, \text{ if Alice's measurement result corresponded to } |\phi^+\rangle_{ac}, \tag{27.33}$$

$$U_b = Z, \text{ if Alice's measurement result corresponded to } |\phi^-\rangle_{ac}, \tag{27.34}$$

$$U_b = X, \text{ if Alice's measurement result corresponded to } |\psi^+\rangle_{ac}, \tag{27.35}$$

$$U_b = -iY, \text{ if Alice's measurement result corresponded to } |\psi^-\rangle_{ac}. \tag{27.36}$$

As we are about to show, this will put Bob's qubit into the state $(c_0\,|0\rangle + c_1\,|1\rangle)_b$, an exact copy of the state that Alice wanted to transfer. This is called **quantum teleportation** of the qubit from Alice to Bob.

To see why the procedure works, first note that we can write the initial joint state of all three qubits as

$$|\phi^+\rangle_{ab} \otimes |\gamma\rangle_c = \frac{1}{\sqrt{2}}\big(\,|00\rangle + |11\rangle\,\big)_{ab} \otimes \big(c_0\,|0\rangle + c_1\,|1\rangle\,\big)_c \tag{27.37}$$

$$= \frac{1}{\sqrt{2}}\big(c_0\,|000\rangle + c_0\,|110\rangle + c_1\,|001\rangle + c_1\,|111\rangle\,\big)_{abc}. \tag{27.38}$$

Now the key trick: you can check by a routine expansion that this is also equal to

$$\frac{1}{2}\,|\gamma\rangle_b \otimes |\phi^+\rangle_{ac} + \frac{1}{2}\big(Z\,|\gamma\rangle\,\big)_b \otimes |\phi^-\rangle_{ac} + \frac{1}{2}\big(X\,|\gamma\rangle\,\big)_b \otimes |\psi^+\rangle_{ac}$$
$$+\frac{1}{2}\big(iY\,|\gamma\rangle\,\big)_b \otimes |\psi^-\rangle_{ac}. \tag{27.39}$$

When Alice makes her measurement of Q_{ac}, the state is projected onto one of these four terms, and her ordinary 2-bit message to Bob tells him which one. Bob's application of the appropriate unitary operator selected from the list in (27.33)–(27.36) then turns qubit b into the state $|\gamma\rangle_b$, due to the Pauli-matrix relations $X^2 = Y^2 = Z^2 = I$.

The teleportation of the qubit has several interesting features to keep in mind. First, it is only possible because of entanglement; using an unentangled qubit pair in place of $|\phi^+\rangle_{ab}$ would certainly not work. Second, Alice's original qubit $|\gamma\rangle_c$ is necessarily destroyed by the process of teleporting it to Bob, in the sense that its state is collapsed by Alice's measurement on it. This means that the qubit has been transferred, but not cloned, which turns out to be impossible on general grounds as we will show in Sect. 27.4. Third, neither Alice nor Bob has learned anything about the teleported state during the process. No matter what the teleported qubit was, Alice's measurement has equal probabilities $1/4$ for each of the four outcomes, and the measurement result only gives information about the joint state of the qubits a and c, not c alone. For his part, Bob has only applied a unitary transformation, not measured anything. Finally, the teleportation of the state is certainly not instantaneous communication, as the transfer speed is limited by that of the two classical bits sent from Alice to Bob.

Teleporting a whole register of N qubits (including their possible entanglements with each other) is a straightforward generalization, as can be understood using the same ideas, with some proliferation of indices. Alice and Bob will need to start with N prepared Bell pairs,

$$|\Psi\rangle_{ab} = |\phi^+\rangle_{a_1 b_1} \otimes \cdots \otimes |\phi^+\rangle_{a_N b_N}, \tag{27.40}$$

which can be thought of as an entangled state living in the tensor product of an N-qubit register controlled by Alice and an N-qubit register controlled by Bob. Let us write the register to be teleported from Alice to Bob as

$$|\Gamma\rangle_c = \sum_{\{\gamma_k\}} c_{\gamma_1 \cdots \gamma_N} |\gamma_1 \cdots \gamma_N\rangle, \tag{27.41}$$

where each γ_k is summed over 0 and 1. The initial joint state is thus a $3N$-qubit register,

$$|\Psi\rangle_{ab} \otimes |\Gamma\rangle_c = \sum_{\{\gamma_k\}} c_{\gamma_1 \cdots \gamma_N} \Big(|\phi^+\rangle_{a_1 b_1} \otimes |\gamma_1\rangle_{c_1} \Big) \otimes \cdots \otimes \Big(|\phi^+\rangle_{a_N b_N} \otimes |\gamma_N\rangle_{c_N} \Big). \tag{27.42}$$

Now each of the N factors in parentheses can be re-expressed, using exactly the same trick as for a single qubit, as

$$|\phi^+\rangle_{a_k b_k} \otimes |\gamma_k\rangle_{c_k} = \frac{1}{2}|\gamma_k\rangle_{b_k} \otimes |\phi^+\rangle_{a_k c_k} + \frac{1}{2}\big(Z\,|\gamma_k\rangle\big)_{b_k} \otimes |\phi^-\rangle_{a_k c_k}$$
$$+\frac{1}{2}\big(X\,|\gamma_k\rangle\big)_{b_k} \otimes |\psi^+\rangle_{a_k c_k} + \frac{1}{2}\big(iY\,|\gamma_k\rangle\big)_{b_k} \otimes |\psi^-\rangle_{a_k c_k} \quad (27.43)$$

For each $k = 1, \ldots, N$, Alice makes a projective measurement $Q_{a_k c_k}$ on the four Bell states for the two qubits a_k, c_k, and sends the results to Bob as $2N$ classical bits of information. Bob then applies an appropriate unitary transformation

$$U_b = U_{b_1} \otimes U_{b_2} \otimes \cdots \otimes U_{b_N} \quad (27.44)$$

on his register, with each U_{b_k} chosen just as in (27.33)–(27.36). The result will be that the N-qubit register Bob controls will be left in the state $|\Gamma\rangle_b$.

Now we turn to a question that may at first seem unrelated, but is actually just the inverse of quantum teleportation. Suppose that Alice needs to send a message composed of classical bits to Bob. She decides to encode the information using qubits,[4] since she knows that they are denser in information than classical bits. How much classical bit information can Alice send to Bob, encoded in a register of N qubits?

First consider what happens if Alice only sends a single unentangled qubit. Clearly, she can use it to encode a single classical bit, for example by preparing the qubit in one of the two states $|0\rangle$ and $|1\rangle$, which will then be measured by Bob using the observable Z. A little thought reveals that this is also the best that can be done. No matter how Alice prepares the qubit, any projective measurement that Bob can do will have one of at most 2 results, since that is the dimension of the one-qubit state space. Importantly, Bob's measurement will also collapse the qubit's state into a completely known one, making all further measurements useless. So only one classical bit can be sent with a single unentangled qubit.

But now suppose that Alice and Bob share a known Bell state, prepared earlier, say $|\phi^+\rangle$, with each having access to one qubit of the pair. To send her message, Alice does one of the 4 unitary transformations $U = I$, Z, X, or iY on the qubit a that she controls, causing the Bell state to change in one of the following ways:

$$I_a|\phi^+\rangle = |\phi^+\rangle, \quad (27.45)$$
$$Z_a|\phi^+\rangle = |\phi^-\rangle, \quad (27.46)$$
$$X_a|\phi^+\rangle = |\psi^+\rangle, \quad (27.47)$$
$$iY_a|\phi^+\rangle = |\psi^-\rangle. \quad (27.48)$$

[4] Physical delivery of qubits over a distance is a challenge, especially because one must avoid decoherence due to environmental interactions. Photonic polarization qubits can be transmitted with lasers or over optical fibers, but direct transportation of spins is more difficult. We consider the idealized case.

Alice sends the qubit a to Bob, who now controls both qubits. Bob can measure the observable Q from (27.31), projecting onto the Bell state basis and giving the result 1, 2, 3, or 4 depending on the choice Alice made. Thus, Alice has successfully transmitted two classical bits to Bob, by sending only one qubit. The transfer of two classical bits by sending one entangled qubit is known as **superdense coding**, or slightly more modestly as **dense coding**. Entanglement effectively doubles the information capacity of the communication.

One might object that dense coding requires the involvement of two qubits to encode and send two classical bits. However, it is also true that Alice never had direct access to Bob's qubit. In fact, the Bell pair could have been created and separated to the two parties long before Alice decided what message to send. Bob could even have prepared the Bell state himself, then sent Alice's share to her. Then, by modifying the one qubit and sending it back, Alice accomplished the mission (perhaps without even knowing for sure that Bob's share of the Bell pair existed).

Another benefit to the dense coding protocol, likely more important than information density, is security. Alice and Bob have an evil nemesis, Eve, who wants to eavesdrop on their conversations. If Alice sent classical bits, then Eve might be able to read them by listening in. If a single unentangled qubit is sent instead, then Eve could still read it by measuring the observable Z, provided that she can intercept the qubit and knows that it is encoded in the $|0\rangle$, $|1\rangle$ basis. Eve could then send the qubit on its way to Bob, without Alice or Bob ever knowing. But, in the dense coding method, any measurement done by Eve provides her no information, because Eve only has access to the half of the entangled pair that was sent by Alice. A formal way of expressing this is that the density matrix for the qubit sent by Alice, obtained by tracing over the qubit held by Bob, is $\rho_a = I/2$, corresponding to a completely random ensemble.

A new pristine entangled qubit pair is needed for each pair of classical bits transmitted, as they are "used up" by the measurement collapse needed by Bob to extract the message. Therefore, by sending a register of N qubits, Alice can transmit $2N$ classical bits of information extractable by Bob. This is the converse of the result found above that Alice can send $2N$ classical bits to teleport a general N-qubit register.

We have seen that in teleportation, sending two classical bits is enough to transmit one qubit, while in dense coding sending one qubit is enough to transmit two classical bits. In that sense, they are opposites or inverses. However, this does not mean that a qubit is equivalent to two classical bits; the statements refer to the optimal extraction of information aided by entanglement, rather than equivalence of information contained.

27.4 Copying Quantum Information: the No-Cloning Theorem and Redundant Encoding

Suppose we have a system in a state $|\psi\rangle$, which we would like to copy without changing the original. Let us try to do this by producing the copy on another system of the same type, but starting in some state $|\beta\rangle$ that is not of interest and will be overwritten. In other words, we might like to devise a method of starting with the tensor product state $|\psi\rangle \otimes |\beta\rangle$ and ending up with $|\psi\rangle \otimes |\psi\rangle$, no matter what $|\psi\rangle$ is. If this were possible, it would be very useful for quantum information applications. For example, we could make many exact copies of a quantum state, and then measure them in many different ways, including with non-commuting operators, to determine the exact form of $|\psi\rangle$. However, it cannot be done.

To see why, it is easiest to take the simple case that $|\psi\rangle$ is a single qubit. Suppose that we have a device or method, realized as a unitary operator, whose effect is the transformation

$$|0\rangle \otimes |\beta\rangle \rightarrow |0\rangle \otimes |0\rangle \,, \tag{27.49}$$

$$|1\rangle \otimes |\beta\rangle \rightarrow |1\rangle \otimes |1\rangle \,. \tag{27.50}$$

Then, acting on a general qubit $|\psi\rangle = a\,|0\rangle + b\,|1\rangle$, by linearity we must have

$$\big(a\,|0\rangle + b\,|1\rangle\big) \otimes |\beta\rangle \rightarrow a\,|0\rangle \otimes |0\rangle + b\,|1\rangle \otimes |1\rangle \,. \tag{27.51}$$

The point is simply that this is not the same state as

$$|\psi\rangle \otimes |\psi\rangle = a^2\,|0\rangle \otimes |0\rangle + ab\,|0\rangle \otimes |1\rangle + ab\,|1\rangle \otimes |0\rangle + b^2\,|1\rangle \otimes |1\rangle \,, \tag{27.52}$$

except in the very special case that $ab = 0$, which implies $a = 0$ or $b = 0$. A straightforward generalization of this proof gives the following no-go result:

Theorem 27.1 (No Cloning) *It is not possible to produce an identical copy of a general state $|\psi\rangle$ without modifying the original. In other words, given a system consisting of a tensor product of two subsystems of the same type, there does not exist a linear operator that takes $|\psi\rangle \otimes |\beta\rangle \rightarrow |\psi\rangle \otimes |\psi\rangle$ for every state $|\psi\rangle$.*

This is consistent with the example of teleportation in Sect. 27.3, in which Alice had to destroy her qubit by measurement collapse in the process of sending it to Bob. More generally, the impossibility of cloning a state avoids paradoxes involving instantaneous communication. Consider an entangled Bell state $|\phi^+\rangle_{ab}$ shared by Alice and Bob. Alice wants to send a message to Bob, either yes or no. To say "yes", she measures Z_a, and to say "no", she instead measures X_a. (The actual result of the measurement is not relevant to the argument.) Now, suppose Bob could clone his qubit. He makes a large number of copies, and then measures Z_b on half of them and X_b on the other half. If Alice's message was "yes", Bob's measurements of Z_b will all have the same result, and his measurements of X_b will appear random. Conversely, if Alice's message was "no", then Bob's measurements of Z_b will appear random and his measurements of X_b will all have the same result. By cloning his qubit a sufficiently large number of times, Bob could learn Alice's message immediately, with arbitrarily high confidence. But this is just fiction; in reality Bob has access to only one copy of his qubit, and no instantaneous communication can occur.

Although cloning of states is impossible, it is possible to do something slightly different, and still very useful, called **redundant encoding**. Consider the unitary transformation of (27.51), which we can express in a more compact way by leaving out the irrelevant qubit $|\beta\rangle$, and using a double arrow to denote an encoding, as

$$a\,|0\rangle + b\,|1\rangle \;\Rightarrow\; a\,|00\rangle + b\,|11\rangle\,. \tag{27.53}$$

Although this perfectly valid operation does not clone the original qubit, what we have managed to do is to encode its information in a two-qubit register, in which the second qubit is redundant. By iterating this procedure, it is possible to encode a qubit $|\psi\rangle = a\,|0\rangle + b\,|1\rangle$ in a register with as many redundant physical qubits as we wish,

$$a\,|0\rangle + b\,|1\rangle \;\Rightarrow\; a\,|000\cdots 0\rangle + b\,|111\cdots 1\rangle\,. \tag{27.54}$$

The right-hand side is sometimes known as a **logical qubit** or a **synthetic qubit**, to distinguish it from the physical qubits that it is made of. Its information content is exactly the same as in the original qubit: one complex degree of freedom. Redundant encoding does not violate the no-cloning theorem, because the redundant qubits that form the synthetic qubit are fully entangled with each other. The key difference is that a cloned copy of the original qubit would have been an independent copy, not entangled with the original.

Redundant encoding can be done in usefully different ways, depending on the choice of basis. For example, we might choose to use for the redundant physical qubits the basis

$$|+\rangle = \frac{1}{\sqrt{2}}\big(\,|0\rangle + |1\rangle\,\big), \qquad |-\rangle = \frac{1}{\sqrt{2}}\big(\,|0\rangle - |1\rangle\,\big), \tag{27.55}$$

which are eigenstates of X with eigenvalues 1 and -1, respectively. Then one can construct a redundant encoding of a logical qubit that takes the form

$$a\,|0\rangle + b\,|1\rangle \;\Rightarrow\; a\,|{+}{+}{+}\ldots{+}\rangle \;+\; b\,|{-}{-}{-}\ldots{-}\rangle\,. \tag{27.56}$$

One can also nest this encoding procedure, by bundling synthetic qubits into bigger synthetic qubits. As we will see in Sect. 27.6, these synthetic qubit tricks are very useful for quantum error correction.

27.5 Keeping Secrets with Quantum Private Key Generation

As noted at the end of Sect. 27.2, the security of many public-key encryption schemes is threatened by the prospect of quantum computers solving problems that were previously believed to be too hard, based on classical computation algorithms. Shor's quantum factorization algorithm is the most famous example of this, but there are others. One way to improve public-key codes is to devise decryption problems for which the solution would take too long even for the best possible quantum computers, and some proposals have been made along these lines. However, a different approach is to use private keys. In this section, we will show how quantum information enables secure generation of private keys and their distribution over large distances.

The one-time private key encryption method is unbreakable, even classically. Suppose that Alice and Bob each have a copy of a prepared and secret register of random bits, $\kappa = (\kappa_1\kappa_2\cdots\kappa_N)$. This is the private key. Alice has a message that she wants to send to Bob, consisting of bits $\alpha = (\alpha_1\alpha_2\cdots\alpha_N)$. To encrypt the message, she adds the two classical registers bitwise to get $\epsilon = \alpha + \kappa = (\epsilon_1\epsilon_2\cdots\epsilon_N)$, where $\epsilon_k = \alpha_k + \kappa_k \pmod 2$ for each $k = 1, \ldots, N$. Alice then sends the encrypted message ϵ to Bob, who can read the message by bitwise subtracting mod 2 (or adding, which is the same thing) the key, $\alpha = \epsilon - \kappa$. An eavesdropper Eve who intercepts the message will gain absolutely no information without knowing the private key, because if the key κ consists of completely random bits, so does the encrypted message ϵ.

However, this absolute security requires that one cannot reuse the key or any part of it. Otherwise, it is conceivable that a clever Eve might be able to discern patterns that allow the key to be discovered. If Alice and Bob are at different locations, they will need a new random private key bit for each bit of message to be sent. What if they run out of their supply of private key bits? Any classical communication of a new private key gives Eve an opportunity to intercept it. Fortunately for Alice and Bob, there are quantum methods for private key generation and distribution, which can work even when they are spatially separated, and allow for any eavesdropping by Eve to be detected.

It should be emphasized that the overriding concern here is security, not speed or efficiency of communication. Also, only the private key generation and distribution, not the encryption or decryption, relies on the principles of quantum mechanics.

One way of generating and distributing private keys is the **EPR protocol**. Alice and Bob are far away, but share a large collection of Bell pairs, all in the same state

$$|\phi^+\rangle = \frac{1}{\sqrt{2}}\Big(|00\rangle + |11\rangle\Big) = \frac{1}{\sqrt{2}}\Big(|{+}{+}\rangle + |{-}{-}\rangle\Big). \tag{27.57}$$

When they need to generate a private key, Alice and Bob each measure X or Z, chosen randomly and independently, sequentially for each qubit that they control, getting one of two results in each case. They then communicate to each other the sequence of measurements that they made, but not the results of those measurements. There is no need to hide this communication, which is therefore considered public. In about half of the cases, they will have chosen to make different measurements, and their results will be uncorrelated; those are discarded. In the other cases, their measurements were either both X, or both Z. In those cases, Alice and Bob will always get the same result, and those will define the bits of their private key. (Alice and Bob agree that $X = 1$ or $Z = 1$ stand for a classical bit 0, and $X = -1$ or $Z = -1$ for a classical bit 1.) For each sequence of N Bell states that they process in this way, they can therefore expect to generate a private key of length approximately $N/2$, with random bits known only to them. This private key can then be used to encrypt and decrypt a message composed of ordinary classical bits.

Alice and Bob should worry that Eve might be trying to learn their private key. Eve could do this by intercepting and modifying the Bell pairs before they are used. For example, perhaps Eve is actually the vendor that sold the supposed Bell pairs to Alice and Bob, and she substituted states of her own choosing in place of $|\phi^+\rangle$. Fortunately for Alice and Bob, there is an easy way to detect this. After generating their private key, they tell each other publicly what their results are for a random selection of the private key bits, which of course will not be used for actual secret messages. If those bits fail to match, they will know that the two-qubit state provided to them was not actually $|\phi^+\rangle$, exposing Eve's fraud. By sacrificing a sufficient fraction of their private key bits, they can be reasonably sure that the states were not tampered with, and feel safe in using the other bits. This relies on the fact that $|\phi^+\rangle$ is, up to an overall phase, the unique two-qubit state that will always give the same result to Alice and Bob regardless of whether it was X or Z that they both measured. This also explains why they had to go to the trouble of randomly measuring X and Z, and then discarding the results when their choices did not match. If Eve knew in advance that they were both going to measure Z, then she could substitute the state $|00\rangle$ or $|11\rangle$ in place of the state $|\phi^+\rangle$, ensuring that the private key bit for that case would be 0 or 1, at Eve's discretion, and that Alice and Bob would not see anything amiss.

The use of Bell pairs in the preceding might give the impression that entanglement is a necessary ingredient for quantum private key distribution, but this is not true. To see this, consider the following variation on the scheme. Alice creates the Bell pairs $|\phi^+\rangle_{ab}$ herself, measures X or Z randomly on qubit a, and then sends qubit b to Bob. Then Bob measures either X or Z on qubit b, and the rest of the protocol is the same. This is formally equivalent to a protocol in which qubit a never existed at all, and Alice simply prepares qubit b randomly in one of the four states $|0\rangle, |1\rangle$ (the Z basis) or $|+\rangle, |-\rangle$ (the X basis). She then sends this qubit to Bob, who measures either X or Z. Alice and Bob announce to each other over a public channel their choices of the X or Z basis, and the rest of the protocol is the same. This version, which uses only a one-way sending of qubits and no entanglement, is called the **BB84 protocol**, after its 1984 proposal by Charles Bennett and Gilles Brassard. This has the clear advantage that Alice and Bob do not need a large supply of nonlocally entangled Bell states.

In the BB84 scheme, it is again crucial that Alice and Bob use both the X and Z bases randomly. If Eve can intercept the qubits that Alice sends, and knows in advance that a qubit is definitely an eigenstate of X or of Z, she will be able to measure it in the corresponding eigenstate basis and send on an exact copy to Bob. Without this knowledge, the no-cloning theorem prevents Eve from making a duplicate and keeping one copy for herself, to be measured after Alice and Bob have announced whether they both chose X or Z. Any measurements that Eve makes will collapse the qubits, unless Eve happens to guess whether they are eigenstates of X or Z. So, Alice and Bob will again discover any eavesdropping by sacrificing a portion of their private key bits and noting that sometimes they do not match.

In the real world, the qubits are typically encoded in photon polarizations, and sent either through optical fibers or through empty space. There are of course practical limitations, including noise, decoherence, and photon loss, which become more acute for long-distance communications. Also, the detection of Eve's possible shenanigans is only probabilistic, and one must account for the fact that in some cases the check bits will fail to match because of non-malicious faults. The need for tradeoffs and optimizations has given rise to more sophisticated protocols. The relative simplicity of one-way non-entangled photonic qubit transmission, and the market demand for privacy in communication, have already led to the availability of commercial implementations of quantum key generation and distribution.

27.6 Fighting Decoherence and Other Faults with Error Correction

An important issue in both classical and quantum information is dealing with errors that may change a register in undesirable ways. In the case of classical bits, errors can be introduced by stray electromagnetic fields, mechanical defects, thermal fluctuations, radioactivity, or cosmic rays. There are well-developed methods for limiting such errors, and detecting and correcting them when they occur, so that the engineering problem for classical bits is essentially a solved one. For qubits, the same sort of errors can occur, but there is the additional and especially pernicious problem of decoherence due to interactions with the environment. As we have seen in Chap. 26, decoherence is very hard to avoid completely, especially in systems with even small couplings to macroscopic degrees of freedom. Correction of decoherence errors is therefore a necessity in order to build practical large-scale quantum computers. In this section, we will give a brief account of error-correction strategies for qubits.

For the case of a single classical bit, a simple strategy for error correction is redundant encoding followed by majority-rule correction. For example, one might redundantly encode each logical bit as three physical bits, according to the rule

$$0 \Rightarrow (000), \qquad\qquad 1 \Rightarrow (111). \tag{27.58}$$

Then, if only one physical bit is modified by an error, the logical bit 0 might change to (100) or (010) or (001), while the logical bit 1 might change to (011) or (101) or (110). Such an error is easily diagnosed by the fact that the three physical bits are not all the same. The error is then corrected by changing the different bit to agree with the majority. This strategy fails if an error affects two or three physical bits at the same time, but this is hopefully rare, especially if the error correction is applied repeatedly after short time intervals. An obvious way to improve the reliability is to use more than three physical bits for each logical bit, but there are more sophisticated and efficient schemes for classical error-correcting codes.

Error correction of qubits is more difficult for several reasons. First, the no-cloning theorem declares that it is not possible to just make three independent identical copies of the qubit as in the classical case; therefore we will have to use redundant encoding as described in Sect. 27.4. Second, a measurement on a single physical qubit will inevitably collapse the state, erasing the information it contains. So, we will have to use cleverly chosen collective measurements, which act on more than one physical qubit at a time in such a way as to not destroy the information in the logical qubit. Third, there is only one type of error for a classical bit; it can flip, changing from 0 to 1 or 1 to 0. In contrast, qubits have multiple different error syndromes, because the general qubit state is an arbitrary linear combination of the basis states, including relative phases that have physical meaning.

To expand on the last point, let us consider several different error types. The most obvious counterpart to the classical bit error for a qubit is the basis bit-flip error, which changes the qubit basis state $|0\rangle$ to $|1\rangle$ and vice versa, so that the general physical qubit state changes according to

$$a\,|0\rangle + b\,|1\rangle \;\rightarrow\; a\,|1\rangle + b\,|0\rangle \qquad \text{(bit-flip error).} \tag{27.59}$$

Another type of error is a phase flip, which changes the relative phase between the kets $|0\rangle$ and $|1\rangle$ by a factor of -1, so that

$$a\,|0\rangle + b\,|1\rangle \;\rightarrow\; a\,|0\rangle - b\,|1\rangle \qquad \text{(phase-flip error).} \tag{27.60}$$

The error could also be a combination of a phase flip and a bit flip,

$$a\,|0\rangle + b\,|1\rangle \;\rightarrow\; a\,|1\rangle - b\,|0\rangle \qquad \text{(phase-and-bit-flip error).} \tag{27.61}$$

These three error types are realized by unitary operators, respectively,

$$U_{\text{err}} = X, \qquad \text{(bit-flip error),} \tag{27.62}$$

$$U_{\text{err}} = Z, \qquad \text{(phase-flip error),} \tag{27.63}$$

$$U_{\text{err}} = -iY, \qquad \text{(phase-and-bit-flip error).} \tag{27.64}$$

These are "large" errors, as they square to the identity (up to a global phase). In general, a qubit error could be an arbitrary unitary transformation. Since the identity matrix and the Pauli matrices together form a basis (under addition) for all 2×2 matrices, one can always write (again up to an irrelevant global phase factor),

$$U_{\text{err}} = \frac{I + i(\epsilon_X X + \epsilon_Y Y + \epsilon_Z Z)}{\sqrt{1 + \epsilon_X^2 + \epsilon_Y^2 + \epsilon_Z^2}}, \tag{27.65}$$

where the unitarity condition $U_{\text{err}}^{\dagger} U_{\text{err}} = I$ requires that ϵ_X, ϵ_Y, and ϵ_Z are real numbers, which may or may not be small. The three special cases of the bit flip, the phase flip, and the combined phase-and-bit flip are recovered in the limits of large ϵ_X, large ϵ_Z, and large ϵ_Y, respectively (once again, up to a global phase). Also included in this general parameterization are "small" errors that are continuously connected to the no-error case, but can build up over time. Ideally, a quantum error-correction method should be able to fix any error of this type, even if we have no way of knowing which error occurred. We will see that this is indeed possible.

The strategy for quantum error correction can be summarized as follows. First, each logical qubit is redundantly encoded as a register of physical qubits. After some time, an error may have been introduced in one of the physical qubits, but we don't know which error or which qubit, if any. Therefore, one makes measurements on the register to diagnose which error syndrome has occurred, but this must be done in a careful way so that the state is unaffected if there was actually no error. The joint result of the measurements will be one of a discrete finite set. If an error is detected, the measurements also project the register state onto a corresponding discrete set of known error-syndrome states. Finally, one applies an appropriate unitary operator, determined by the result of the diagnostic measurements, to correct the register back to its pre-error state. This error-correction process should be repeated at regular time intervals, short enough so that the probability of multiple errors within each interval is small.

For simplicity, the following discussion is framed in terms of pure-state qubits affected by an error that takes the form of a unitary transformation U_{err}, as above. However, it is very important that this class of algorithms also corrects decoherence errors that evolve the pure state of the qubit to a mixed state, through entanglement with the environment. The reason is that the diagnostic measurement step automatically projects the density matrix of the register onto a pure error-syndrome state, unentangled with the environment. The ensuing correction is a unitary operator on the register alone, so that the final restored state is always a pure one.

As a warm-up example, let us first consider the simplified case that errors are restricted to the bit-flip type on a single physical qubit. Inspired by the majority-rule scheme for classical bits, we use the triply redundant encoding rule

$$a\,|0\rangle + b\,|1\rangle \Rightarrow a\,|000\rangle + b\,|111\rangle\,. \tag{27.66}$$

A bit-flip error on one of the three physical qubits could change this state to any one of the following error-syndrome states:

Table 27.1 Error correction for single-qubit bit-flips, using the 3-physical-qubit encoding of one logical qubit. Simultaneous measurement of the compatible observables A_{12} and A_{13} diagnoses the error syndrome and projects onto the corresponding error syndrome state. The unitary operator U_{recovery} then corrects the error

Error syndrome	A_{12}	A_{13}	U_{recovery}
No error	1	1	I
Bit flip on qubit 1	-1	-1	X_1
Bit flip on qubit 2	-1	1	X_2
Bit flip on qubit 3	1	-1	X_3

$$a\,|100\rangle + b\,|011\rangle \qquad \text{(bit-flip error on qubit 1)}, \tag{27.67}$$

$$a\,|010\rangle + b\,|101\rangle \qquad \text{(bit-flip error on qubit 2)}, \tag{27.68}$$

$$a\,|001\rangle + b\,|110\rangle \qquad \text{(bit-flip error on qubit 3)}. \tag{27.69}$$

These states are illegal, according to the original encoding, and so must be corrected. Now, to diagnose the error, it would be bad to simply measure the three qubits one at a time, the strategy that worked in the classical case. If we measure Z_1 for the first qubit, that would give results 1 and -1 with probabilities $|a|^2$ and $|b|^2$ or vice versa, but this does not distinguish with certainty which qubit may have suffered an error. Even worse, the information stored will have been destroyed by the measurement collapse, which projects onto only states with definite $Z_1 = 1$ or -1, rather than the superposition in the original state. Measurement of a single physical qubit would irrevocably destroy the logical qubit we are trying to correct.

Instead, the correct strategy is to use only collective measurements, which act on more than one physical qubit. Specifically, we measure the compatible operators

$$A_{12} = Z_1 Z_2, \qquad \text{and} \qquad A_{13} = Z_1 Z_3, \tag{27.70}$$

which compare pairs of qubits without measuring them individually. (The notation means the same thing as $A_{12} = Z \otimes Z \otimes I$ and $A_{13} = Z \otimes I \otimes Z$.) Each of the four states appearing in (27.66)–(27.69) is always an eigenstate of both A_{12} and A_{13}, no matter what a and b are. This means that measuring one or both of them does not change the state, and there is no loss of information from collapse. Furthermore, the four possible results for the measurement of (A_{12}, A_{13}) correspond to the four possible error syndromes (including the trivial syndrome of no error), telling us which error occurred and how to correct it. If we measure $(A_{12}, A_{13}) = (-1, -1)$, the error was a bit-flip of qubit 1, as in (27.67). We can therefore correct the error by applying the unitary operator X_1. If instead $(A_{12}, A_{13}) = (-1, 1)$, the error was a bit-flip of qubit 2, and we should correct the error by applying X_2. If $(A_{12}, A_{13}) = (1, -1)$, then the error was a bit-flip of qubit 3, and we should apply the unitary operator X_3. Finally, the case $(A_{12}, A_{13}) = (1, 1)$ corresponds to no error, so no unitary correction is applied. This error-correction algorithm is summarized in Table 27.1.

The algorithm actually corrects a more general class of errors on a single physical qubit, namely those obtained with $\epsilon_Y = \epsilon_Z = 0$ and arbitrary ϵ_X in (27.65). For example, suppose that the error hits the first physical qubit, according to the unitary transformation

$$a|000\rangle + b|111\rangle \rightarrow \frac{1}{\sqrt{1+\epsilon_X^2}}\left[a|000\rangle + b|111\rangle + i\epsilon_X\left(a|100\rangle + b|011\rangle\right)\right], \tag{27.71}$$

where ϵ_X is a real number whose magnitude may or may not be small. (Note that $\epsilon_X = 0$ corresponds to no error, while $\epsilon_X \rightarrow \infty$ is the large bit-flip error case.) Now when the measurement of (A_{12}, A_{13}) takes place, there are two possible results. With probability $1/(1+\epsilon_X^2)$, the measurement will give the result $(1, 1)$ and collapse the state to $a|000\rangle + b|111\rangle$. In this case, no unitary recovery transformation is applied. With probability $\epsilon_X^2/(1+\epsilon_X^2)$, the measurement will give the result $(-1, -1)$, diagnosing the error syndrome as a bit-flip on qubit 1, and collapsing the state to $a|100\rangle + b|011\rangle$. In that case, the algorithm calls for applying $U_{\text{recovery}} = X_1$, changing the state back to $a|000\rangle + b|111\rangle$. So, in both cases, the final result for the state is the correct pre-error state. Even though there is nontrivial measurement collapse in this case, it is actually beneficial, because it forces the state into one of the discrete error syndromes that we know how to fix. The measurement allows us to correctly repair the state using one of three operations, every time, no matter what the continuous parameter ϵ_X was.

However, the preceding example does not protect at all against phase-flip errors. As a second warm-up, let us consider an algorithm that corrects all single-qubit phase-flip errors (but does not handle bit-flip errors). The strategy is actually very similar to the previous case, based on the observation that a phase-flip in the basis $|0\rangle, |1\rangle$ is equivalent to a bit-flip in the basis $|+\rangle, |-\rangle$, and vice versa, where $|\pm\rangle = (|0\rangle \pm |1\rangle)/\sqrt{2}$ are the eigenstates of X with eigenvalues ± 1. Consider the following 3-qubit redundant encoding:

$$a|0\rangle + b|1\rangle \Rightarrow a|{+}{+}{+}\rangle + b|{-}{-}{-}\rangle. \tag{27.72}$$

A phase-flip error amounts to changing $|+\rangle \leftrightarrow |-\rangle$ on one of the qubits, so the possible error-syndrome states are

$$a|{-}{+}{+}\rangle + b|{+}{-}{-}\rangle \qquad \text{(phase-flip error on qubit 1)}, \tag{27.73}$$

$$a|{+}{-}{+}\rangle + b|{-}{+}{-}\rangle \qquad \text{(phase-flip error on qubit 2)}, \tag{27.74}$$

$$a|{+}{+}{-}\rangle + b|{-}{-}{+}\rangle \qquad \text{(phase-flip error on qubit 3)}. \tag{27.75}$$

It is straightforward to check that measurement of the compatible operators

$$B_{12} = X_1X_2, \qquad \text{and} \qquad B_{13} = X_1X_3 \tag{27.76}$$

will uniquely diagnose the phase-flip error syndrome, with $(B_{12}, B_{13}) = (-1, -1)$, $(-1, 1)$, $(1, -1)$, or $(1, 1)$ for, respectively, a phase-flip of qubit 1, 2, 3, or no error. This also works for the case of any U_{err} in (27.65) with arbitrary ϵ_Z, as long as $\epsilon_X = \epsilon_Y = 0$. In general, the measurement projects onto one of the phase-flip error-syndrome states. The error is then corrected by applying the unitary operator Z_1, Z_2, Z_3, or I, respectively.

In 1995, Shor showed how to generalize these ideas into an algorithm that corrects *any* single-qubit error. One starts with the 9-qubit redundant encoding:

$$|0\rangle \Rightarrow \frac{1}{\sqrt{2^3}}\big(|000\rangle + |111\rangle\big) \otimes \big(|000\rangle + |111\rangle\big) \otimes \big(|000\rangle + |111\rangle\big), \tag{27.77}$$

$$|1\rangle \Rightarrow \frac{1}{\sqrt{2^3}}\big(|000\rangle - |111\rangle\big) \otimes \big(|000\rangle - |111\rangle\big) \otimes \big(|000\rangle - |111\rangle\big). \tag{27.78}$$

This encodes each logical qubit as three bundles of three physical qubits each. It can be viewed as a nested encoding, first triply redundantly in the $|+\rangle$, $|-\rangle$ basis, and then triply redundantly in the $|0\rangle$, $|1\rangle$ basis.

If a phase-flip error occurs in any single physical qubit, it will change the relative sign of the two terms within one of the three bundles. The resulting state depends only on which bundle contained the phase-flip, not which of the three qubits within the bundle suffered the error. Therefore, phase-flip errors can be completely diagnosed by simultaneously measuring the two compatible operators

$$B_{123456} = X_1X_2X_3X_4X_5X_6, \tag{27.79}$$

$$B_{123789} = X_1X_2X_3X_7X_8X_9, \tag{27.80}$$

each with eigenvalues ± 1. If, and only if, B_{123456} is found to be -1, then the phase-flip error must have been in either the first or the second bundle, and similarly if B_{123789} is found to be -1, then the error was in either the first or the third bundle. There are thus four possible error syndromes for the combined measurement of the two operators. After the measurement projects the state onto one of them, the phase-flip error can be corrected by applying a unitary transformation. For example, if it is determined that the phase-flip error occurred on the second bundle, one can undo the error by applying $U^{\text{phase}}_{\text{recovery}} = Z_4$ (although using Z_5 or Z_6 or the symmetric combination $Z_4Z_5Z_6$ instead would work just as well). The same encoding also allows one to detect and correct any bit-flip error. By measuring each of the pairs

$$A_{12} = Z_1Z_2, \qquad A_{13} = Z_1Z_3, \qquad A_{45} = Z_4Z_5, \tag{27.81}$$

$$A_{46} = Z_4Z_6, \qquad A_{78} = Z_7Z_8, \qquad A_{79} = Z_7Z_9, \tag{27.82}$$

one can detect which of the 9 qubits was bit-flipped, and apply the appropriate unitary recovery operator. All eight of these observables commute, and measurement collapse has no effect on the state in the special case that there was no error. The

Table 27.2 Error correction for the Shor 9-physical-qubit encoding of one logical qubit based on measurement of the 8 compatible observables B_{123456}, B_{123789}, A_{12}, A_{13}, A_{45}, A_{46}, A_{78}, and A_{79}. Measurement of B_{123456} and B_{123789} diagnoses any single-qubit phase-flip error and projects onto the corresponding error-syndrome state, which can then be corrected by applying the corresponding unitary operator $U^{\text{phase}}_{\text{recovery}}$. Measurement of A_{12} and A_{13} diagnoses any bit-flip error of qubit 1, 2, or 3, which is then corrected by the corresponding $U^{123}_{\text{recovery}}$, and similarly for bit-flip errors on qubits 4, 5, 6, 7, 8, or 9. This corrects any error on any single physical qubit, restoring the 9-qubit register to its pre-error state

Error syndrome	B_{123456}	B_{123789}	$U^{\text{phase}}_{\text{recovery}}$
No phase-flip error	1	1	I
Phase flip on qubit 1, 2, or 3	−1	−1	Z_1
Phase flip on qubit 4, 5, or 6	−1	1	Z_4
Phase flip on qubit 7, 8, or 9	1	−1	Z_7
Error syndrome	A_{12}	A_{13}	$U^{123}_{\text{recovery}}$
No bit-flip error on qubit 1, 2, or 3	1	1	I
Bit flip on qubit 1	−1	−1	X_1
Bit flip on qubit 2	−1	1	X_2
Bit flip on qubit 3	1	−1	X_3
Error syndrome	A_{45}	A_{46}	$U^{456}_{\text{recovery}}$
No bit-flip error on qubit 4, 5, or 6	1	1	I
Bit flip on qubit 4	−1	−1	X_4
Bit flip on qubit 5	−1	1	X_5
Bit flip on qubit 6	1	−1	X_6
Error syndrome	A_{78}	A_{79}	$U^{789}_{\text{recovery}}$
No bit-flip error on qubit 7, 8, or 9	1	1	I
Bit flip on qubit 7	−1	−1	X_7
Bit flip on qubit 8	−1	1	X_8
Bit flip on qubit 9	1	−1	X_9

error correction algorithm is summarized in Table 27.2. As you will check explicitly by doing Exercise 27.5, the algorithm corrects every possible single-qubit error of the general form given in (27.65). The essential reason for this is that the general error transformation in (27.65) is always a linear combination (superposition) of I, X, Z, and Y, which means no change, a bit flip, a phase flip, or a combination of phase and bit flips.

This error correction strategy relies on the assumption that only one physical qubit was affected. This may be a reasonable (but not guaranteed) assumption if errors are rare on the time scale between applications of the algorithm. However, if two qubits both had bit-flip errors, the algorithm would wrongly flip the qubit that did not change, instead of the two that did. Similarly, if two bundles had phase-flip errors, the wrong bundle would have its phase flipped.

Although Shor's encoding and error-correction algorithm is relatively simple to understand, it turns out not to be the most efficient. Later developments showed that it is possible to do the same thing with synthetic qubits containing only 5 (instead of 9) physical qubits. There are also more sophisticated methods that can, with high probability, correct errors that occur in more than one physical qubit simultaneously.

Exercises

27.1 The decoherence of a pure state qubit to a mixed state can be described by a dynamical map with Kraus operators K_n, according to $\rho \to \sum_n K_n \rho K_n^\dagger$, as discussed in Sect. 25.2. Consider the set of points on the unit-sphere surface of the Bloch ball, corresponding to all pure states of the qubit. What surface results from applying each of the following maps?

(a) the **depolarizing channel** map defined by $K_0 = \sqrt{1-3p/4}\,I$ and $K_1 = \sqrt{p/4}\,X$ and $K_2 = \sqrt{p/4}\,Y$ and $K_3 = \sqrt{p/4}\,Z$. [You should find that the effect of the map is to shrink the Bloch vector according to $\boldsymbol{b} \to (1-p)\boldsymbol{b}$. Thus, pure states on the unit-sphere surface of the Bloch ball are mapped to mixed states on a sphere of radius $1-p$.]
(b) the **bit-flip decoherence** map $K_0 = \sqrt{1-p}\,I$ and $K_1 = \sqrt{p}\,X$.
(c) the **phase-flip decoherence** map $K_0 = \sqrt{1-p}\,I$ and $K_1 = \sqrt{p}\,Z$.
(d) the **bit-and-phase-flip decoherence** map $K_0 = \sqrt{1-p}\,I$ and $K_1 = \sqrt{p}\,Y$.

27.2 Show that the map U_f defined by (27.11) is unitary.

27.3 Consider the operators $A = Z_1 Z_2$ and $B = X_1 X_2$ acting on a system of two qubits labeled 1 and 2. Show that $[A, B] = 0$, and that their common eigenstates are the states $|\phi^\pm\rangle$ and $|\psi^\pm\rangle$ in (27.29) and (27.30). Find the corresponding eigenvalues. Relate the eigenvalues of (A, B) to those of the operator Q in (27.31).

27.4 Consider the teleportation of a qubit $|\gamma\rangle_c$ from Alice to Bob as discussed in Sect. 27.3, but suppose that the initial shared Bell pair is in the state $|\psi^-\rangle_{ab}$, instead of $|\phi^+\rangle_{ab}$. What rules will Bob follow to receive the teleported state, replacing (27.33)–(27.36)?

27.5 For Shor's 9-qubit encoding of a logical qubit $a\,|0\rangle + b\,|1\rangle$ using (27.77) and (27.78), consider the most general unitary error transformation of the form in (27.65) applied to the first physical qubit. For the simultaneous measurement of the operators B_{123456}, B_{123789}, A_{12}, and A_{13}, give the possible outcomes and their probabilities, in terms of ϵ_X, ϵ_Y, and ϵ_Z. Give the corresponding post-measurement states, and check that the algorithm summarized in Table 27.2 always restores the qubit to the pre-error state. (Since there is nothing special about the first physical qubit, this generalizes to a unitary error transformation on any single qubit.)

28 Relativistic Quantum Mechanics

28.1 Special Relativity, Four-Vectors, and Lorentz Transformations

Einstein's special relativity treats the four dimensions of spacetime symmetrically, allowing them to mix under coordinate transformations. This presents a challenge for the quantum mechanics of a single particle as formulated above, where the three spatial coordinates are operators, but time is treated completely differently, as a mere parameter. In this chapter we will discuss how quantum mechanical wave equations can be made consistent with the symmetries of special relativity, leading to the Klein–Gordon equation for spin-0 particles and the Dirac equation for spin-1/2 particles.

Classically, for each event in spacetime with spatial coordinates (x, y, z) and time coordinate t, we can assign a four-vector coordinate,

$$x^\mu = (x^0, x^1, x^2, x^3) = (ct, x, y, z). \tag{28.1}$$

Greek indices $\mu, \nu, \rho, \ldots$ run over the values $0, 1, 2, 3$, and c is the speed of light in vacuum. It is a guiding principle of special relativity that the laws of physics should take the same form in any **inertial reference frame**, which means that the coordinates of a free classical particle change at a constant rate. It is often useful to change our coordinate system from one inertial reference frame to another, according to

$$x^\mu \to x'^\mu = L^\mu{}_\nu x^\nu. \tag{28.2}$$

Here $L^\mu{}_\nu$ is a constant 4×4 real matrix satisfying an orthogonality constraint to be found soon. The repeated (and therefore implicitly summed) index ν is said to be "contracted". Such a change of coordinates specified by $L^\mu{}_\nu$ is called a **Lorentz**

S. P. Martin, *Quantum Mechanics*, Graduate Texts in Physics,
https://doi.org/10.1007/978-3-032-21119-4_28

transformation. By convention, contracted indices must have opposite heights, one raised and one lowered.

As a matter of terminology, x^μ (with a raised index) is an example of a **contravariant** four-vector. By definition, all contravariant four-vectors transform in the same way,

$$a^\mu \to a'^\mu = L^\mu{}_\nu a^\nu. \tag{28.3}$$

Another contravariant four-vector is the **four-momentum** consisting of the energy E and spatial momentum $\boldsymbol{p}$ of a particle,

$$p^\mu = (E/c,\ \boldsymbol{p}) = (E/c,\ p_x, p_y, p_z). \tag{28.4}$$

In the rest frame of a particle of mass m, its four-momentum is $p^\mu = (mc, 0, 0, 0)$.

A simple example of a Lorentz transformation is the rotation about the z-axis by a fixed angle α. In the new coordinate system, $ct' = ct$, $x' = x\cos\alpha - y\sin\alpha$, $y' = x\sin\alpha + y\cos\alpha$, and $z' = z$. As another example, called a **boost**, we could go to a frame moving with respect to the original frame with velocity v along the $\hat{z}$ direction, with the origins of the two frames coinciding at time $t = t' = 0$. Then,

$$ct' = \gamma(ct - vz/c), \qquad x' = x, \qquad y' = y, \qquad z' = \gamma(z - vt), \tag{28.5}$$

where

$$\gamma = 1/\sqrt{1 - v^2/c^2}. \tag{28.6}$$

It follows that the Lorentz transformation for a boost v along the $\hat{z}$ direction is

$$L^\mu{}_\nu = \begin{pmatrix} \gamma & 0 & 0 & -\gamma v/c \\ 0 & 1 & 0 & 0 \\ 0 & 0 & 1 & 0 \\ -\gamma v/c & 0 & 0 & \gamma \end{pmatrix}, \tag{28.7}$$

and the inverse Lorentz transformation is

$$a^\mu = (L^{-1})^\mu{}_\nu a'^\nu, \qquad (L^{-1})^\mu{}_\nu = \begin{pmatrix} \gamma & 0 & 0 & \gamma v/c \\ 0 & 1 & 0 & 0 \\ 0 & 0 & 1 & 0 \\ \gamma v/c & 0 & 0 & \gamma \end{pmatrix}. \tag{28.8}$$

As one might expect, the inverse of a boost simply replaces the boost velocity $v\hat{z}$ by $-v\hat{z}$.

For any two events one can define the **spacetime interval** or **invariant interval**, which is independent of the choice of Lorentz frame and therefore tells us how far apart the two events are in a coordinate-independent sense. Consider two events occurring at x^μ and $x^\mu + a^\mu$, where a^μ is some four-vector displacement. The spacetime interval is

$$ds^2 = -(a^0)^2 + (a^1)^2 + (a^2)^2 + (a^3)^2 = g_{\mu\nu} a^\mu a^\nu, \tag{28.9}$$

where, in special relativity, the **metric tensor** is constant and diagonal,

$$g_{\mu\nu} = \begin{pmatrix} -1 & 0 & 0 & 0 \\ 0 & 1 & 0 & 0 \\ 0 & 0 & 1 & 0 \\ 0 & 0 & 0 & 1 \end{pmatrix}. \tag{28.10}$$

If $ds^2 > 0$, the quantity $\sqrt{ds^2}$ is called the proper distance between the events, and they are said to be spacelike separated. If $ds^2 < 0$, they have a timelike separation, and the quantity $\sqrt{-ds^2}/c$ is called the proper time interval. If $ds^2 = 0$, which is the case if a photon can intersect both events, the interval is said to be lightlike.

The **inverse metric** tensor $g^{\mu\nu}$ is defined so that

$$g^{\mu\nu} g_{\nu\rho} = \delta^\mu_\rho, \tag{28.11}$$

where $\delta^\mu_\nu = 1$ if $\mu = \nu$, and otherwise $= 0$. In special relativity, the matrix forms of $g^{\mu\nu}$ and $g_{\mu\nu}$ have the same numerical entries, and do not depend on the inertial reference frame.

The metric tensor allows us to define **covariant** four-vectors by lowering an index,

$$x_\mu = g_{\mu\nu} x^\nu = (-ct,\ x,\ y,\ z), \tag{28.12}$$

$$p_\mu = g_{\mu\nu} p^\nu = (-E/c,\ p_x,\ p_y,\ p_z), \tag{28.13}$$

which amounts to flipping the sign of the time-like component. More generally, any index can be lowered or raised using the metric tensor or its inverse:

$$a_\mu = g_{\mu\nu} a^\nu, \qquad a^\mu = g^{\mu\nu} a_\nu. \tag{28.14}$$

It follows that covariant four-vectors transform as

$$a'_\mu = L_\mu{}^\nu a_\nu, \tag{28.15}$$

where we define

$$L_\mu{}^\nu = g_{\mu\rho} L^\rho{}_\sigma g^{\sigma\nu}. \tag{28.16}$$

Note the positions of the indices; $L_\mu{}^\nu$ and $L^\nu{}_\mu$ are not the same thing. If a^μ and b^μ are contravariant four-vectors, then

$$g_{\mu\nu} a^\mu b^\nu = g^{\mu\nu} a_\mu b_\nu = a_\mu b^\mu = a^\mu b_\mu. \tag{28.17}$$

Since the four-vector indices have been all contracted, this quantity should be Lorentz-invariant, meaning that it is independent of the choice of inertial frame coordinates. This implies $g_{\mu\nu} a'^\mu b'^\nu = g_{\mu\nu} a^\mu b^\nu$, so that

$$g_{\mu\nu} L^\mu{}_\rho L^\nu{}_\sigma a^\rho b^\sigma = g_{\rho\sigma} a^\rho b^\sigma. \tag{28.18}$$

Since a^ρ and b^σ are arbitrary, it must be true that

$$g_{\mu\nu} L^\mu{}_\rho L^\nu{}_\sigma = g_{\rho\sigma}. \tag{28.19}$$

This is the orthogonality constraint that a Lorentz transformation matrix must satisfy.

Let us now consider some particular Lorentz transformations. To begin, we note that (28.19) can be expressed in matrix form as $L^T g L = g$, where T denotes the ordinary matrix transpose, so that $(L^T)_\rho{}^\mu = L^\mu{}_\rho$ is distinct from $L_\rho{}^\mu$ as defined by (28.16). This in turn implies $\det(L) = \pm 1$ because of the general facts $\det(AB) = \det A \det B$ and $\det A^T = \det A$. An **improper Lorentz transformation** is one with $\det(L) = -1$. An example is

$$L^\mu{}_\nu = \begin{pmatrix} -1 & 0 & 0 & 0 \\ 0 & 1 & 0 & 0 \\ 0 & 0 & 1 & 0 \\ 0 & 0 & 0 & 1 \end{pmatrix}. \tag{28.20}$$

This just flips the sign of the time coordinate:

$$x'^0 = -x^0 \qquad x'^1 = x^1 \qquad x'^2 = x^2 \qquad x'^3 = x^3. \tag{28.21}$$

Another improper Lorentz transformation is parity or space inversion,

$$L^\mu{}_\nu = \begin{pmatrix} 1 & 0 & 0 & 0 \\ 0 & -1 & 0 & 0 \\ 0 & 0 & -1 & 0 \\ 0 & 0 & 0 & -1 \end{pmatrix}, \tag{28.22}$$

so that

$$x'^0 = x^0, \qquad x'^1 = -x^1, \qquad x'^2 = -x^2, \qquad x'^3 = -x^3. \tag{28.23}$$

It was once thought that the laws of physics must be invariant under these transformations. Although this might seem intuitively reasonable, it was shown experimentally in the 1950s that parity is violated in the weak interactions, specifically in the weak decays of the ^{60}Co nucleus and π^+ and K^+ mesons. Likewise, experiments in the 1960s on the decays of K^0 mesons provided evidence that time-reversal is not a symmetry of the Hamiltonian governing the weak interactions.

However, all experiments up to now are consistent with invariance of the laws of physics under the subset of Lorentz transformations that are continuously connected to the identity; these are known as **proper orthochronous Lorentz transformations** and have $\det(L) = +1$ and $L^0{}_0 \geq 1$. They can all be built up out of infinitesimal Lorentz transformations

$$L^\mu{}_\nu = \delta^\mu_\nu + \omega^\mu{}_\nu + O(\omega^2). \tag{28.24}$$

For these, (28.19) reads $g_{\mu\nu}(\delta^\mu_\rho + \omega^\mu{}_\rho + \cdots)(\delta^\nu_\sigma + \omega^\nu{}_\sigma + \cdots) = g_{\rho\sigma}$. This can be simplified using the index lowering and raising convention (28.14), to give $g_{\rho\sigma} + \omega_{\sigma\rho} + \omega_{\rho\sigma} + \cdots = g_{\rho\sigma}$. Therefore,

$$\omega_{\sigma\rho} = -\omega_{\rho\sigma} \tag{28.25}$$

is an antisymmetric 4×4 matrix, with 6 independent entries. These nonzero elements correspond to 3 rotations (ρ, σ = 1,2 or 1,3 or 2,3) and 3 boosts (ρ, σ = 0,1 or 0,2 or 0,3). Any Lorentz transformation can be built up out of repeated infinitesimal boosts and rotations, perhaps combined with time-reversal and/or space inversion. Lorentz transformations obey the mathematical properties of a group, known as the **Lorentz group**. The subset of Lorentz transformations that can be built out of repeated infinitesimal boosts and rotations forms a smaller group, the restricted (or proper orthochronous) Lorentz group.

A **Lorentz tensor** is a multi-component object that carries a fixed number of spacetime vector indices, and transforms appropriately when one goes to a new reference frame. The objects $g^{\mu\nu}$ and $g_{\mu\nu}$ and δ^μ_ν are examples of constant tensors. Contravariant and covariant four-vectors are also tensors. In general, the defining characteristic of a tensor function $T^{\mu_1\mu_2\cdots}_{\nu_1\nu_2\cdots}(x)$ is that under a change of reference frame, it transforms so that in the primed coordinate system, the corresponding tensor T' is:

$$T'^{\mu_1\mu_2\cdots}_{\nu_1\nu_2\cdots}(x') = L^{\mu_1}{}_{\rho_1} L^{\mu_2}{}_{\rho_2} \cdots L_{\nu_1}{}^{\sigma_1} L_{\nu_2}{}^{\sigma_2} \cdots T^{\rho_1\rho_2\cdots}_{\sigma_1\sigma_2\cdots}(x). \tag{28.26}$$

A Lorentz scalar function is the special case of a Lorentz tensor that carries no vector indices, and is therefore independent of the choice of inertial reference frame.

For example, if p^μ and q^μ are the contravariant four-momenta of any particles, then $p^\mu q_\mu$ is a scalar, and must not depend on which inertial reference frame it is computed or measured in. In particular, this implies that a particle with mass m satisfies the **on-shell condition**

$$p^\mu p_\mu = -E^2/c^2 + \boldsymbol{p}^2 = -m^2c^2, \tag{28.27}$$

where the right side is obtained by evaluation in the rest frame with $p^\mu = (mc, 0, 0, 0)$, and the sign corresponds to our convention $g_{\mu\nu} = \text{diag}(-1,\ 1,\ 1,\ 1)$. Note that in this chapter only, we will use $\boldsymbol{a}^2 \equiv \boldsymbol{a} \cdot \boldsymbol{a}$ for a three-vector $\boldsymbol{a}$, to avoid confusion with $a^2 \equiv a^\mu a_\mu$ for a four-vector.

Spacetime derivatives of Lorentz scalar functions are four-vectors. Suppose that $f(x^\mu) = f(t, \boldsymbol{r})$ is a scalar function of x^μ. Then

$$\partial_\mu f \equiv \frac{\partial f}{\partial x^\mu} = \left(\frac{1}{c}\frac{\partial f}{\partial t},\ \nabla f\right) \tag{28.28}$$

satisfies the Lorentz transformation rule of a covariant four-vector, (28.15). By raising the index, one obtains a contravariant four-vector function

$$\partial^\mu f = g^{\mu\nu}\partial_\nu f = \left(-\frac{1}{c}\frac{\partial f}{\partial t},\ \nabla f\right). \tag{28.29}$$

One can obtain another scalar function by acting twice with the 4-dimensional derivative operator on f, contracting the indices,

$$\partial^\mu \partial_\mu f = -\frac{1}{c^2}\frac{\partial^2 f}{\partial t^2} + \nabla^2 f. \tag{28.30}$$

The differential operator $\partial^\mu \partial_\mu$ is thus a four-dimensional generalization of the Laplacian.

28.2 Klein–Gordon Wave Equation

As a first attempt at relativistic quantum mechanics, let us consider the simple case of a free spinless particle of mass m, whose wavefunction $\Phi(x^\mu)$ satisfies the Schrödinger equation

$$H\Phi = i\hbar\frac{\partial \Phi}{\partial t}. \tag{28.31}$$

To construct a Hamiltonian consistent with the symmetry of invariance under Lorentz transformations, we note that for a free particle, there should be an orthobasis of states Φ that are also eigenstates of $\boldsymbol{P} = -i\hbar\nabla$ with eigenvalues $\boldsymbol{p}$, the three-momentum of the particle. Now, using the special relativistic form $E = \sqrt{m^2c^4 + c^2\boldsymbol{p}^2}$, one can try

$$H = \sqrt{m^2c^4 + c^2\boldsymbol{P}^2}, \tag{28.32}$$

where $\boldsymbol{P}^2 = -\hbar^2\nabla^2$ acting on wavefunctions. This has a series expansion,

$$H = mc^2 - \frac{\hbar^2}{2m}\nabla^2 - \frac{\hbar^4}{8m^3c^2}(\nabla^2)^2 + \cdots . \tag{28.33}$$

If we keep only the first two terms, then we recover the standard nonrelativistic quantum mechanics of a free particle, since the constant rest-energy contribution mc^2 contributes only an unobservable global phase to the time evolution. However, in the fully relativistic case in which the series is not truncated, the presence of an infinite number of terms with arbitrarily large numbers of derivatives is quite problematic, including apparently nonlocal effects. Furthermore, the Schrödinger equation in this case is not manifestly relativistic, since it involves only a single time derivative but an infinite number of spatial derivatives.

Instead, one can consider $H^2\Phi = (m^2c^4 + c^2\boldsymbol{P}^2)\Phi$, avoiding the square root. Setting this equal to $(i\hbar\partial/\partial t)^2$ acting on Φ, in order to be consistent with the Schrödinger equation, we find

$$\left(-\hbar^2\frac{\partial^2}{\partial t^2} + c^2\hbar^2\nabla^2 - m^2c^4\right)\Phi = 0. \tag{28.34}$$

Using (28.30), this can be written in the manifestly Lorentz-invariant form

$$\left(\hbar^2\partial^\mu\partial_\mu - m^2c^2\right)\Phi = 0. \tag{28.35}$$

This is called the **Klein–Gordon** wave equation after the 1926 proposals of Oskar Klein and Walter Gordon.

It is easy to guess the solutions of the Klein–Gordon equation. If we try

$$\Phi(x^\mu) = \Phi_0 e^{ip_\mu x^\mu/\hbar}, \tag{28.36}$$

where Φ_0 is a normalization constant and $p^\mu = (p^0, \boldsymbol{p})$ is a constant four-vector, then $\partial_\mu\Phi = i(p_\mu/\hbar)\Phi$, and so

$$\partial^\mu\partial_\mu\Phi = -(p^\mu p_\mu/\hbar^2)\Phi. \tag{28.37}$$

Therefore, we obtain a basis of solutions by imposing $p^\mu p_\mu = -m^2c^2$. Any Φ of this form is an eigenstate of H^2 and $\boldsymbol{P}$, with energy $E = cp^0$ and three-momentum eigenvalue $\boldsymbol{p}$, provided that they satisfy the on-shell condition $E^2 = m^2c^4 + \boldsymbol{p}^2c^2$.

However, taking the Klein–Gordon equation as fundamental raises a serious issue. For every positive-energy solution $p^\mu = (E/c, \boldsymbol{p})$ with $E = \sqrt{m^2c^4 + c^2\boldsymbol{p}^2} > 0$, there is a corresponding negative-energy solution $p^\mu = (-E/c, \boldsymbol{p})$. By increasing $|\boldsymbol{p}|$, one can have arbitrarily large $|E|$, so the negative energies are not bounded from below. This would lead to the release of an infinite amount of energy as negative-energy particles interact, acquiring larger and larger three-momenta without bound.

28.3 Dirac Equation

In 1928, Dirac suggested an alternative, based on the observation that the problem with the Klein–Gordon equation seemed to be that it is quadratic in H and in $\partial/\partial t$, which is responsible for the sign ambiguity of E. Dirac could also have been motivated by the fact that the electron has spin; since it has two intrinsic degrees of freedom, trying to describe it with a single-component wavefunction $\Phi(x^\mu)$ is doomed to fail. Instead, Dirac proposed to write a relativistic Schrödinger equation, linear in $\partial/\partial t$ and with H linear in spatial derivatives, for a multi-component wavefunction $\Psi_i(x^\mu)$, where the index $i = 1, 2, \ldots, n$ runs over the components. These indices will be suppressed from here on, as we write Ψ as a column vector when its explicit form is needed.

Dirac reasoned that since the Schrödinger wave equation is linear in $\partial/\partial t$, and since relativity places t on the same footing as x, y, z, it should also be linear in derivatives of the spatial coordinates. Therefore, the equation ought to take the form

$$i\hbar\frac{\partial\Psi}{\partial t} = H\Psi = (c\boldsymbol{\alpha}\cdot\boldsymbol{P} + mc^2\beta)\Psi, \tag{28.38}$$

where $\alpha_x, \alpha_y, \alpha_z$, and β are dimensionless $n\times n$ matrices. Equation (28.38) is known as the Dirac equation, and Ψ is called a **Dirac spinor**.

To determine what $\boldsymbol{\alpha}$ and β should be, consider $\partial^2\Psi/\partial t^2$, which can be evaluated in two ways. First, the on-shell condition for E^2 gives, just as in the Klein–Gordon case,

$$-\hbar^2\frac{\partial^2\Psi}{\partial t^2} = (-c^2\hbar^2\nabla^2 + m^2c^4)\Psi. \tag{28.39}$$

On the other hand, expressing H^2 in terms of the right-hand side of (28.38), we find

$$-\hbar^2\frac{\partial^2\Psi}{\partial t^2} = \left[-c^2\hbar^2\alpha_a\alpha_b\frac{\partial^2}{\partial x^a\partial x^b}\right.$$
$$\left.-i\hbar mc^3(\alpha_a\beta + \beta\alpha_a)\frac{\partial}{\partial x^a} + m^2c^4\beta^2\right]\Psi, \tag{28.40}$$

where a and b are summed over x, y, z. Since partial derivatives commute,

$$\alpha_a\alpha_b\frac{\partial^2}{\partial x^a\partial x^b} = \frac{1}{2}(\alpha_a\alpha_b + \alpha_b\alpha_a)\frac{\partial^2}{\partial x^a\partial x^b}. \tag{28.41}$$

So, in order for (28.39) and (28.40) to agree, we must have

$$\beta^2 = I, \tag{28.42}$$

$$\alpha_a \beta + \beta \alpha_a = 0, \tag{28.43}$$

$$\alpha_a \alpha_b + \alpha_b \alpha_a = 2\delta_{ab} I. \tag{28.44}$$

It is also necessary that $\boldsymbol{\alpha}$ and β are Hermitian matrices, in order for H to be Hermitian.

For charged particles with nonzero mass, a solution for $\boldsymbol{\alpha}$ and β turns out to require at least $n = 4$ spinor indices.[1] This may be somewhat surprising, since naively one only needs $n = 2$ to describe a spin-1/2 particle like the electron. As we will see, the Dirac equation necessarily describes negative energy electrons in a way similar to the Klein–Gordon equation, accounting for the doubling. It is convenient to write the 4×4 matrices in 2×2 blocks. The necessary conditions are satisfied by

$$\boldsymbol{\alpha} = \begin{pmatrix} 0 & \boldsymbol{\sigma} \\ \boldsymbol{\sigma} & 0 \end{pmatrix}, \qquad \beta = \begin{pmatrix} I & 0 \\ 0 & -I \end{pmatrix}, \tag{28.45}$$

where $\boldsymbol{\sigma}$ are the Pauli matrices, I is the 2×2 identity, and "0" means a 2×2 block of 0's.

The Dirac equation automatically incorporates intrinsic angular momentum, or spin. To see this, we note that the Hamiltonian for a free particle should have the symmetry of rotations generated by a total angular momentum operator $\boldsymbol{J}$, so that $[\boldsymbol{J}, H] = 0$. Let us first consider how orbital angular momentum fits in. Since $\boldsymbol{L} = \boldsymbol{R} \times \boldsymbol{P}$, its components can be written as $L_a = \epsilon_{abc} R_b P_c$. Now we find the commutator of L_a with the Dirac Hamiltonian H. The $mc^2\beta$ term has no coordinate dependence, so it does not contribute to the commutator. Therefore, $[L_a, H] = c\alpha_b[L_a, P_b] = c\alpha_b\epsilon_{adc}[R_d, P_b]P_c = i\hbar c\alpha_b\epsilon_{adc}\delta_{bd}P_c = i\hbar c\epsilon_{abc}\alpha_b P_c$, so that

$$[\boldsymbol{L}, H] = i\hbar c \boldsymbol{\alpha} \times \boldsymbol{P}. \tag{28.46}$$

This does not vanish, so there must be another contribution to $\boldsymbol{J}$. By inspired guesswork, we try defining the spin operator

$$\boldsymbol{S} = \frac{\hbar}{2} \begin{pmatrix} \boldsymbol{\sigma} & 0 \\ 0 & \boldsymbol{\sigma} \end{pmatrix}. \tag{28.47}$$

[1] The fact that Dirac spinors and Lorentz vectors both have 4 components is coincidental, and one must not confuse the two types of 4-dimensional spaces. If we lived in D-dimensional spacetime, Dirac spinors would have $n = 2^{D/2}$ components for even D, or $n = 2^{(D-1)/2}$ for odd D.

To check this guess, note that the nonzero commutator of $\boldsymbol{S}$ with H is due entirely to the matrix nature of both,

$$\begin{aligned}[S_a, H] &= cP_b[S_a, \alpha_b] = \frac{\hbar}{2}cP_b \begin{pmatrix} 0 & [\sigma_a, \sigma_b] \\ [\sigma_a, \sigma_b] & 0 \end{pmatrix} \\ &= i\hbar c\epsilon_{abc}P_b\alpha_c, \end{aligned} \tag{28.48}$$

where we have made use of $[\sigma_a, \sigma_b] = 2i\epsilon_{abc}\sigma_c$. This gives

$$[\boldsymbol{S}, H] = -i\hbar c\boldsymbol{\alpha} \times \boldsymbol{P} = -[\boldsymbol{L}, H]. \tag{28.49}$$

Thus $[\boldsymbol{J}, H] = 0$, so that the Dirac Hamiltonian is indeed invariant under rotations generated by the total angular momentum operator $\boldsymbol{J} = \boldsymbol{L} + \boldsymbol{S}$.

The Dirac Hamiltonian also respects parity, but the way this works is slightly tricky. We can start by defining a unitary and Hermitian "coordinate parity" operator $\Pi_c = \Pi_c^{-1}$ to obey

$$\Pi_c \boldsymbol{R} \Pi_c = -\boldsymbol{R}, \tag{28.50}$$

from which it follows that

$$\Pi_c \boldsymbol{P} \Pi_c = -\boldsymbol{P} \tag{28.51}$$

as well. However, it is then easy to see that $\Pi_c H \Pi_c \neq H$, because the $\boldsymbol{\alpha} \cdot \boldsymbol{P}$ term flips sign, but the β term does not. Therefore, we define a spinor-matrix factor of the parity operator,

$$\Pi_m = \Pi_m^{-1} = \beta, \tag{28.52}$$

so that $\Pi_m \boldsymbol{\alpha} \Pi_m = -\boldsymbol{\alpha}$ and $\Pi_m \beta \Pi_m = \beta$. Then the complete parity operator defined by

$$\Pi = \Pi_c \Pi_m = \begin{pmatrix} \Pi_c & 0 \\ 0 & -\Pi_c \end{pmatrix} \tag{28.53}$$

satisfies $\Pi = \Pi^{-1}$ and $\Pi H \Pi = H$, and the Dirac Hamiltonian commutes with parity.

The Dirac equation (28.38) can be rewritten in a nicer way by multiplying it on the left by the matrix β, and defining gamma matrices γ^μ, carrying a raised Lorentz index, by

$$\gamma^0 = \beta, \qquad \gamma^a = \beta\alpha_a, \qquad (a = 1, 2, 3). \tag{28.54}$$

Since $\beta^2 = I$, the Dirac equation can be expressed as

$$\left[i\hbar\left(\gamma^0\frac{\partial}{\partial x^0} + \gamma^1\frac{\partial}{\partial x^1} + \gamma^2\frac{\partial}{\partial x^2} + \gamma^3\frac{\partial}{\partial x^3}\right) - mc\right]\Psi = 0, \qquad (28.55)$$

or, even more nicely, as

$$(i\hbar\gamma^\mu\partial_\mu - mc)\Psi = 0. \qquad (28.56)$$

The γ^μ matrices defined above are explicitly given by

$$\gamma^0 = \begin{pmatrix} I & 0 \\ 0 & -I \end{pmatrix}, \qquad \gamma^a = \begin{pmatrix} 0 & \sigma^a \\ -\sigma^a & 0 \end{pmatrix}, \qquad (a = 1, 2, 3). \qquad (28.57)$$

However, this representation is not unique. To see this, suppose U is any constant unitary 4×4 matrix satisfying $U^\dagger U = UU^\dagger = I$. Then the Dirac equation implies

$$U(i\hbar\gamma^\mu\partial_\mu - mc)U^\dagger U\Psi = 0, \qquad (28.58)$$

from which it follows that, writing $\gamma'^\mu = U\gamma^\mu U^\dagger$, and $\Psi' = U\Psi$,

$$(i\hbar\gamma'^\mu\partial_\mu - mc)\Psi' = 0. \qquad (28.59)$$

So, the new γ'^μ matrices together with the new spinor Ψ' are just as good as the old pair γ^μ, Ψ; there are an infinite number of different, equally valid choices. The choice in (28.45) and (28.57) is called the Dirac representation. Another choice, popular in treatments of high-energy particle physics and quantum field theory, is the chiral representation (see Exercise 28.1).

Some useful properties of the γ^μ matrices, independent of the representation choice, are

$$\gamma^{0\dagger} = \gamma^0, \qquad (28.60)$$

$$\gamma^{a\dagger} = -\gamma^a, \qquad (a = 1, 2, 3), \qquad (28.61)$$

$$\gamma^0\gamma^{\mu\dagger}\gamma^0 = \gamma^\mu, \qquad (28.62)$$

$$\mathrm{Tr}(\gamma^\mu\gamma^\nu) = -4g^{\mu\nu}, \qquad (28.63)$$

$$\gamma^\mu\gamma_\mu = -4, \qquad (28.64)$$

$$\gamma^\mu\gamma^\nu + \gamma^\nu\gamma^\mu = \{\gamma^\mu, \gamma^\nu\} = -2g^{\mu\nu}. \qquad (28.65)$$

On the right sides of each of (28.64) and (28.65), there is an implicit 4×4 unit matrix.

How does a Dirac spinor $\Psi(x)$ transform under a Lorentz transformation? It carries no vector index, so it is not a tensor. On the other hand, the fact that the Hamiltonian "mixes up" the four components of $\Psi(x)$ is a clue that it does not transform like an ordinary scalar function either. Instead, we might expect that the primed frame spinor is given by

$$\Psi'(x') = \Lambda\Psi(x), \tag{28.66}$$

where Λ is some 4×4 matrix that depends on the Lorentz transformation matrix $L^\mu{}_\nu$. It is left to Exercise 28.2 to show that this is indeed true, and that Lorentz invariance of the Dirac equation implies, for an infinitesimal Lorentz transformation $L^\mu{}_\nu = \delta^\mu_\nu + \omega^\mu{}_\nu$,

$$\Lambda = I - \omega_{\mu\nu}[\gamma^\mu, \gamma^\nu]/8 + \cdots . \tag{28.67}$$

To obtain the result for a non-infinitesimal proper Lorentz transformation, we apply the same infinitesimal transformation N times, and take $N \to \infty$. Letting $\Omega_{\mu\nu} = N\omega_{\mu\nu}$, so that $\omega^\mu{}_\nu = \Omega^\mu{}_\nu/N$ by raising the first index, we obtain

$$L^\mu{}_\nu = \lim_{N\to\infty} \left(\delta^\mu_\nu + \Omega^\mu{}_\nu/N\right)^N = [\exp(\Omega)]^\mu{}_\nu \tag{28.68}$$

by using $\lim_{N\to\infty}(1 + x/N)^N = e^x$. For the Dirac spinor, one finds similarly

$$\begin{aligned} \Psi'(x') &= \lim_{N\to\infty} \left(1 - \Omega_{\mu\nu}[\gamma^\mu, \gamma^\nu]/8N\right)^N \Psi(x) \\ &= \exp\left(-\Omega_{\mu\nu}[\gamma^\mu, \gamma^\nu]/8\right)\Psi(x), \end{aligned} \tag{28.69}$$

where $\Omega_{\mu\nu} = g_{\mu\rho}\Omega^\rho{}_\nu$. So, the Lorentz transformation matrix for Dirac spinors as defined in (28.66), corresponding to the $L^\mu{}_\nu$ that appears in (28.68), is

$$\Lambda = \exp\left(-\Omega_{\mu\nu}[\gamma^\mu, \gamma^\nu]/8\right). \tag{28.70}$$

We conclude this section by constructing plane-wave solutions to the free-particle Dirac equation. Consider first the special case of the rest frame of the particle, so that the $\boldsymbol{\alpha} \cdot \boldsymbol{P}$ term vanishes. The Dirac equation then has the diagonal form

$$i\hbar\frac{\partial\Psi}{\partial t} = mc^2\beta\Psi. \tag{28.71}$$

Defining 2-component constant spinors

$$\chi_\uparrow = \begin{pmatrix}1\\0\end{pmatrix}, \qquad \chi_\downarrow = \begin{pmatrix}0\\1\end{pmatrix}, \tag{28.72}$$

it is easy to check that a basis of four solutions is

$$\Psi_s^+ = e^{-imc^2t/\hbar}\begin{pmatrix}\chi_s\\0\end{pmatrix}, \qquad \Psi_s^- = e^{imc^2t/\hbar}\begin{pmatrix}0\\\chi_s\end{pmatrix}, \tag{28.73}$$

with positive and negative energy eigenvalues $E = mc^2$ and $E = -mc^2$ for the superscript labels $+$ and $-$, and S_z eigenvalues $\hbar/2$ and $-\hbar/2$ for the subscript labels $s = \uparrow$ and $\downarrow$. These spin assignments agree with the spin operator that we found in (28.47).

Dirac spinor wavefunctions for arbitrary nonzero three-momenta can now be constructed from these rest-frame solutions, by applying the Lorentz transformation for boosts and/or rotations using (28.70). Alternatively, one can find the eigenspinors of the Dirac Hamiltonian directly. Let us do this for the special case of three-momentum p along the $\hat{z}$ direction. Factoring out the plane-wave time and position dependence, we try

$$\Psi(\boldsymbol{r}, t) = e^{i(pz-Et)/\hbar}u, \tag{28.74}$$

where u is a constant spinor. Plugging this into the Dirac equation [either (28.38), or its equivalent (28.56)] gives

$$\begin{pmatrix}(E-mc^2)I & -pc\sigma_z\\ -pc\sigma_z & (E+mc^2)I\end{pmatrix} u = 0, \tag{28.75}$$

which has two energy eigenvalue solutions $E = \pm\sqrt{m^2c^4+p^2c^2}$, each with degeneracy 2. Up to normalization, the corresponding orthogonal basis spinors are, for $E = +\sqrt{m^2c^4+p^2c^2}$,

$$u_\uparrow^+ = \begin{pmatrix}\sqrt{m^2c^4+p^2c^2}+mc^2\\0\\pc\\0\end{pmatrix}, \qquad u_\downarrow^+ = \begin{pmatrix}0\\\sqrt{m^2c^4+p^2c^2}+mc^2\\0\\-pc\end{pmatrix} \tag{28.76}$$

and for $E = -\sqrt{m^2c^4+p^2c^2}$,

$$u_\uparrow^- = \begin{pmatrix}-pc\\0\\\sqrt{m^2c^4+p^2c^2}+mc^2\\0\end{pmatrix}, \qquad u_\downarrow^- = \begin{pmatrix}0\\pc\\0\\\sqrt{m^2c^4+p^2c^2}+mc^2\end{pmatrix} \tag{28.77}$$

These are again eigenstates of S_z, with eigenvalues $\hbar/2$ for $\uparrow$ and $-\hbar/2$ for $\downarrow$.

Like the Klein–Gordon equation, the Dirac equation has the embarrassment of solutions with negative energies with arbitrarily large magnitudes, even though it is linear in both H and $\partial/\partial t$. Undaunted, Dirac proposed to get around the problem of negative energy states by making use of the fact that spin-1/2 particles are fermions. Recall that the Pauli exclusion principle dictates that two fermions cannot occupy the same quantum state. This motivated Dirac to suggest that all of the negative energy states are normally occupied. This prevents electrons with positive energy from making disastrous transitions to the $E < 0$ states. The infinite number of filled $E < 0$ states is called the **Dirac sea**.

If one of the states in the Dirac sea becomes unoccupied, it leaves behind a **hole**. Since a hole is the absence of an $E < 0$ state, it behaves as a particle with energy $-E > 0$. An electron has charge $-e$, so the hole corresponding to its absence effectively has the opposite charge, $+e$. Since both electrons and holes obey $p^\mu p_\mu = -m^2c^2$, they have the same mass. Dirac's proposal thus predicts the existence of "anti-electrons" or positrons, with positive energy and positive charge. Because no other candidate was known at the time, he briefly entertained the incorrect suggestion that the anti-electron was the proton, until the positron was discovered in 1932 in cosmic ray experiments.

Despite this impressive and revolutionary success, the modern resolution of the problem of negative-energy states—for both the Klein–Gordon and Dirac cases—is to abandon the whole approach of defining fundamental operators for positions and momenta. Instead, relativistic particles are described by quantum field theories, in which the spatial position coordinates $\boldsymbol{r}$ join the time t as *parameters*, not operators. In quantum field theories, there are still objects $\Phi(x^\mu)$ for spinless particles and $\Psi(x^\mu)$ for spin-1/2 particles, but they are operators ("quantum fields") rather than wavefunctions. This automatically avoids states with negative energies of arbitrarily large magnitude, and introduces antiparticles without appealing to Dirac's rather *ad hoc* sea.[2] Nevertheless, the Dirac equation formalism and Dirac spinors remain essential for setting up the kinematics of quantum field theories for spin-1/2 particles. Much more could be said about the Dirac wave equation and its solutions, but we mostly leave this to texts that specialize in relativistic quantum field theory. In the remainder of this chapter, we will only discuss two important successes of the Dirac equation, related to the coupling of the electron to electromagnetic fields.

[2] In any case, Dirac's sea trick cannot work for the Klein–Gordon equation, or any other relativistic theory of bosons, since they do not have a Pauli exclusion principle. Unlike Dirac, we are now privileged to know of a scalar particle that is quite possibly fundamental, the Higgs boson. There are also spin-1 bosons that are plausibly fundamental based on all presently available evidence (the photon, the W and Z particles, and the gluon of the Standard Model).

28.4 Electromagnetic Fields and the Dirac Equation

In this section, we will work out how the Dirac equation incorporates the effects of background electromagnetic fields, and then recover the nonrelativistic limit.

As Einstein famously showed in 1905, Maxwell's electromagnetism is already a relativistic theory. To illustrate this, we start by revisiting some of the key equations of Sect. 4.3, putting them into a manifestly Lorentz-covariant form. The classical electromagnetic potentials are unified into a contravariant four-vector,

$$A^\mu = (\Phi, \boldsymbol{A}). \tag{28.78}$$

The gauge transformation $\boldsymbol{A} \to \boldsymbol{A} + \nabla\Lambda$ with $\Phi \to \Phi - \frac{1}{c}\frac{\partial \Lambda}{\partial t}$ is then realized as

$$A^\mu \to A^\mu + \partial^\mu \Lambda \tag{28.79}$$

for an arbitrary scalar function Λ. The relations between the potentials and the fields, $\boldsymbol{E} = -\nabla\Phi - \frac{1}{c}\frac{\partial \boldsymbol{A}}{\partial t}$ and $\boldsymbol{B} = \nabla \times \boldsymbol{A}$, can be expressed as

$$\begin{pmatrix} 0 & -E_x & -E_y & -E_z \\ E_x & 0 & B_z & -B_y \\ E_y & -B_z & 0 & B_x \\ E_z & B_y & -B_x & 0 \end{pmatrix} = F_{\mu\nu} = \partial_\mu A_\nu - \partial_\nu A_\mu, \tag{28.80}$$

which defines the antisymmetric field strength tensor $F_{\mu\nu}$. The classical densities of charge and current can also be unified into a contravariant four-vector, as

$$j^\mu = (c\rho, \boldsymbol{j}). \tag{28.81}$$

The sourced Maxwell equations (Gauss' law $\nabla \cdot \boldsymbol{E} = 4\pi\rho$ and Ampère's law $\nabla \times \boldsymbol{B} = \frac{1}{c}\frac{\partial \boldsymbol{E}}{\partial t} + \frac{4\pi}{c}\boldsymbol{j}$) are then given by

$$\partial_\nu F^{\mu\nu} = \frac{4\pi}{c} j^\mu, \tag{28.82}$$

or equivalently,

$$\partial_\nu \partial^\mu A^\nu - \partial_\nu \partial^\nu A^\mu = \frac{4\pi}{c} j^\mu. \tag{28.83}$$

The remaining Maxwell equations ($\nabla \cdot \boldsymbol{B} = 0$ and Faraday's law $\nabla \times \boldsymbol{E} = -\frac{1}{c}\frac{\partial \boldsymbol{B}}{\partial t}$) are equivalent to the identity

$$\partial_\rho F_{\mu\nu} + \partial_\mu F_{\nu\rho} + \partial_\nu F_{\rho\mu} = 0, \tag{28.84}$$

for $\mu, \nu, \rho = 0, 1, 2, 3$. Note that this equation is automatically true by construction in terms of A_μ, as can be seen using the definition $F_{\mu\nu} = \partial_\mu A_\nu - \partial_\nu A_\mu$.

Also, because $F^{\mu\nu}$ is antisymmetric and partial derivatives commute, we have $\partial_\mu \partial_\nu F^{\mu\nu} = 0$. It follows, by comparing with ∂_μ acting on (28.82), that

$$\partial_\mu j^\mu = 0, \tag{28.85}$$

which expresses the local conservation of charge, as it can be rewritten $\nabla \cdot \boldsymbol{j} + \partial\rho/\partial t = 0$.

To see how to incorporate electromagnetic effects into the Dirac equation, we recall from (4.36) that the space and time derivatives in the Schrödinger equation are modified, in the presence of $\boldsymbol{A}$ and Φ, according to

$$\boldsymbol{P} \to \boldsymbol{P} - \frac{q}{c}\boldsymbol{A}, \qquad i\hbar\frac{\partial}{\partial t} \to i\hbar\frac{\partial}{\partial t} - q\Phi. \tag{28.86}$$

Applying this rule to (28.38), we arrive at the Dirac Hamiltonian with electromagnetic potentials included,

$$H = c\boldsymbol{\alpha}\cdot\left(\boldsymbol{P} - \frac{q}{c}\boldsymbol{A}\right) + mc^2\beta + q\Phi. \tag{28.87}$$

Also, since $A_\mu = (-\Phi, \boldsymbol{A})$ and $\partial_\mu = ((1/c)\partial/\partial t, \nabla)$ and $\boldsymbol{P}$ is represented by $-i\hbar\nabla$, the two parts of (28.86) can be unified neatly into a Lorentz-covariant rule

$$i\hbar\partial_\mu \to i\hbar\partial_\mu + \frac{q}{c}A_\mu. \tag{28.88}$$

Applying this to (28.56), we arrive at the covariant form of the Dirac wave equation,

$$\left[\gamma^\mu\left(i\hbar\partial_\mu + \frac{q}{c}A_\mu\right) - mc\right]\Psi = 0, \tag{28.89}$$

for a particle of mass m and charge q in the presence of electromagnetic potentials.

We now wish to recover the nonrelativistic limit of the Dirac equation; this is a useful thing to do because it fixes the g-factor for the electron to be $g_e = 2$ (before corrections coming from virtual photons of the quantized electromagnetic field), and provides the fine-structure terms in the Hamiltonian for the hydrogen atom. To accomplish this, it is useful to write the 4-component Dirac spinor in terms of two 2-component spinors φ and χ, and factor out the largest part of the time dependence for positive-energy states,

$$\Psi = e^{-imc^2t/\hbar}\begin{pmatrix}\varphi\\ \chi\end{pmatrix}. \tag{28.90}$$

The Schrödinger equation for the Hamiltonian in (28.87) then becomes

$$i\hbar \frac{\partial \varphi}{\partial t} = c\boldsymbol{\Pi}\cdot\boldsymbol{\sigma}\chi + q\Phi\varphi, \tag{28.91}$$

$$i\hbar \frac{\partial \chi}{\partial t} = c\boldsymbol{\Pi}\cdot\boldsymbol{\sigma}\varphi + (q\Phi - 2mc^2)\chi, \tag{28.92}$$

where $\boldsymbol{\Pi} = \boldsymbol{P} - (q/c)\boldsymbol{A}$. Now, in the nonrelativistic ($|\boldsymbol{p}| \ll mc$) limit, we see from (28.76) that the positive-energy solutions are mostly φ, and from (28.77) that the negative-energy solutions are mostly χ. Since our goal is to take the nonrelativistic limit for the positive-energy solutions, our strategy is to eliminate the small components χ in favor of the large components φ. Accordingly, we rewrite (28.92) as

$$\chi = \frac{1}{2mc}\Omega\varphi + \frac{1}{2mc^2}\left(-i\hbar\frac{\partial}{\partial t} + q\Phi\right)\chi, \tag{28.93}$$

where we have defined, for convenience, the Hermitian operator

$$\Omega = \boldsymbol{\Pi}\cdot\boldsymbol{\sigma}. \tag{28.94}$$

By iteratively eliminating χ in the limit of large c on the right side of (28.93),

$$\chi = \frac{1}{2mc}\left[I + \frac{1}{2mc^2}\left(-i\hbar\frac{\partial}{\partial t} + q\Phi\right) + \cdots\right]\Omega\varphi. \tag{28.95}$$

Here, and in the following, $\partial/\partial t$ is understood to act on everything to its right, including both φ and the vector potential $\boldsymbol{A}$ hidden in Ω. Putting (28.95) into (28.91) gives an uncoupled differential equation for φ,

$$i\hbar\frac{\partial\varphi}{\partial t} = \left[\frac{1}{2m}\Omega^2 + q\Phi + \frac{1}{4m^2c^2}\Omega\left(-i\hbar\frac{\partial}{\partial t} + q\Phi\right)\Omega + \cdots\right]\varphi, \tag{28.96}$$

where terms with 4 or more powers of c in the denominator have been dropped.

It might be tempting to interpret (28.96) as the wave equation for the positive-energy solutions, but this would be wrong for a subtle reason: unitary time evolution does not preserve the norm of φ. To see this, note that the squared norm of the full Dirac spinor obeys

$$\begin{aligned} 1 &= \int d^3\boldsymbol{r}\, \Psi^\dagger\Psi = \int d^3\boldsymbol{r}\, (\varphi^\dagger\varphi + \chi^\dagger\chi) \\ &= \int d^3\boldsymbol{r}\left(\varphi^\dagger\varphi + \varphi^\dagger\frac{\Omega^2}{4m^2c^2}\varphi + \cdots\right), \end{aligned} \tag{28.97}$$

where in the last expression we have used the leading approximation for χ from (28.95). This shows that the two-component spinor ψ defined by

$$\psi = \left(I + \frac{\Omega^2}{8m^2c^2} + \cdots\right)\varphi \tag{28.98}$$

will have constant norm, and therefore unitary time evolution. Thus, ψ defined in this way (and not φ) is the correctly normalized wavefunction describing the nonrelativistic limit of a spin-1/2 particle with leading relativistic corrections. Plugging the inverse relation

$$\varphi = \left(I - \frac{\Omega^2}{8m^2c^2} + \cdots\right)\psi \tag{28.99}$$

into (28.96), and consistently dropping terms of order $1/c^4$, after a little algebra the result can be written

$$i\hbar\frac{\partial\psi}{\partial t} = \left(\frac{1}{2m}\Omega^2 + q\Phi + \Delta H + \cdots\right)\psi, \tag{28.100}$$

where the leading-order relativistic correction is

$$\Delta H = \frac{1}{4m^2c^2}\left(\Omega^2\left(-i\hbar\frac{\partial}{\partial t} + q\Phi\right) - \frac{1}{2}\left[\Omega, \left[\Omega, -i\hbar\frac{\partial}{\partial t} + q\Phi\right]\right]\right). \tag{28.101}$$

To put the nonrelativistic part into a more familiar form, we evaluate

$$\Omega^2 = (\boldsymbol{\Pi}\cdot\boldsymbol{\sigma})^2 = \boldsymbol{\Pi}^2 + i\boldsymbol{\sigma}\cdot(\boldsymbol{\Pi}\times\boldsymbol{\Pi}) = \left(\boldsymbol{P} - \frac{q}{c}\boldsymbol{A}\right)^2 - \frac{\hbar q}{c}\boldsymbol{\sigma}\cdot(\nabla\times\boldsymbol{A}), \tag{28.102}$$

where the second equality uses the Pauli matrix product identity $(\boldsymbol{a}\cdot\boldsymbol{\sigma})(\boldsymbol{b}\cdot\boldsymbol{\sigma}) = \boldsymbol{a}\cdot\boldsymbol{b} + i(\boldsymbol{a}\times\boldsymbol{b})\cdot\boldsymbol{\sigma}$, and the last term uses the position-representation differential form $\boldsymbol{P} = -i\hbar\nabla$. Now, using $\nabla\times\boldsymbol{A} = \boldsymbol{B}$, and $\boldsymbol{\sigma} = 2\boldsymbol{S}/\hbar$, we arrive at

$$i\hbar\frac{\partial\psi}{\partial t} = \left(\frac{1}{2m}\left(\boldsymbol{P} - \frac{q}{c}\boldsymbol{A}\right)^2 + q\Phi - \frac{q}{mc}\boldsymbol{S}\cdot\boldsymbol{B} + \Delta H + \ldots\right)\psi. \tag{28.103}$$

This agrees with the form of (4.48), provided that the gyromagnetic ratio is $\gamma = q/mc$. For the electron with $q = -e$, comparison with (4.39) gives

$$\gamma_e = -\frac{g_e e}{2mc} = -\frac{e}{mc}, \tag{28.104}$$

and we have learned that the gyromagnetic ratio with $g_e = 2$ is successfully predicted by the Dirac equation. The same holds for the muon and the tau lepton (just with different masses), as they are also fundamental fermions. However, the Dirac equation cannot give the correct gyromagnetic ratios for composite fermions like the proton and the neutron, because of large contributions from their internal structure.

Returning to the leading-order relativistic correction in (28.101), the first term simplifies because, from (28.96) with (28.99), $(-i\hbar\frac{\partial}{\partial t} + q\Phi)\psi = -\frac{\Omega^2}{2m}\psi$, up to higher-order terms in $1/c$. So we obtain

$$\Delta H = -\frac{1}{8m^3c^2}\Omega^4 - \frac{1}{8m^2c^2}\left[\Omega, \left[\Omega, -i\hbar\frac{\partial}{\partial t} + q\Phi\right]\right]. \tag{28.105}$$

From here on, we consistently drop terms in ΔH with $-(q/c)\boldsymbol{A}$. So, at leading order in $1/c$, we have $\Omega^4 = (\boldsymbol{P}^2)^2$. Also, we evaluate

$$\left[\Omega, -i\hbar\frac{\partial}{\partial t} + q\Phi\right] = -i\hbar q\boldsymbol{\sigma}\cdot\nabla\Phi = i\hbar q\boldsymbol{\sigma}\cdot\boldsymbol{E}, \tag{28.106}$$

and then,

$$\begin{aligned}\left[\Omega, \left[\Omega, -i\hbar\frac{\partial}{\partial t} + q\Phi\right]\right] &= i\hbar q\left(\boldsymbol{\sigma}\cdot\boldsymbol{P}\,\boldsymbol{\sigma}\cdot\boldsymbol{E} - \boldsymbol{\sigma}\cdot\boldsymbol{E}\,\boldsymbol{\sigma}\cdot\boldsymbol{P}\right)\\ &= \hbar^2 q\nabla\cdot\boldsymbol{E} + 2\hbar q\boldsymbol{\sigma}\cdot(\boldsymbol{E}\times\boldsymbol{P}) + i\hbar^2 q\boldsymbol{\sigma}\cdot(\nabla\times\boldsymbol{E}),\end{aligned} \tag{28.107}$$

after using the Pauli matrix product identity again. Dropping the last term because $\nabla\times\boldsymbol{E} = -\frac{1}{c}\frac{\partial\boldsymbol{B}}{\partial t}$ is higher order in the $1/c$ expansion, we finally get

$$\Delta H = -\frac{1}{8m^3c^2}(\boldsymbol{P}^2)^2 - \frac{q}{2m^2c^2}\boldsymbol{S}\cdot(\boldsymbol{E}\times\boldsymbol{P}) - \frac{\hbar^2 q}{8m^2c^2}\nabla\cdot\boldsymbol{E}. \tag{28.108}$$

These three terms are, respectively, the relativistic kinematic, spin-orbit, and Darwin terms that appeared in the fine-structure Hamiltonian for an electron in a central electrostatic potential, (17.3)–(17.5). In making this comparison, one uses $q = -e$, $m = m_e$, $\nabla\cdot\boldsymbol{E} = -\nabla^2\Phi$, and $\boldsymbol{E} = -\frac{1}{R}\frac{d\Phi}{dR}\boldsymbol{R}$ for a spherically symmetric potential, followed by $\boldsymbol{R}\times\boldsymbol{P} = \boldsymbol{L}$. In particular, the Dirac equation automatically gives the Thomas precession factor in the spin-orbit Hamiltonian, and the correct normalization of the Darwin term.

28.5 Dirac Equation Solutions for the Hydrogen Atom

In the previous section, we derived the leading relativistic corrections to the Hamiltonian for a spin-1/2 charged particle, at leading order in $1/c^2$. These can in turn be used to obtain the fine-structure energy shifts for the hydrogen atom, as we have already done in Sect. 17.1 using first-order perturbation theory. We will now show that these effects can also be obtained directly from an exact[3] solution of the Dirac equation, without relying on perturbation theory and to all orders in $1/c$. As a bonus, we will find the relativistic electron wavefunction.

Our task is to solve the equation

$$H\Psi(\boldsymbol{r}) = E\Psi(\boldsymbol{r}) \tag{28.109}$$

with the Dirac Hamiltonian in the position representation,

$$H = -i\hbar c\boldsymbol{\alpha}\cdot\nabla + mc^2\beta - Ze^2/r, \tag{28.110}$$

which arises from $q = -e$ and the four-vector potential $A^\mu = (Ze/r, 0, 0, 0)$ for a nucleus of charge Ze at rest. Now, (28.46) implies that L_z and L^2 do not commute with H, so we cannot hope to find energy eigenstates that are also eigenstates of orbital angular momentum with fixed l. However, it follows from (28.49) that $\boldsymbol{J} = \boldsymbol{L} + \boldsymbol{S}$ does commute with H, and the parity operator Π defined in (28.53) does as well. Therefore, an appropriate CSCO for our problem is H, J^2, J_z, and Π.

We will use spherical coordinates (r, θ, ϕ), and begin by constructing two-component spinor eigenstates of J^2 and J_z with fixed L^2 eigenvalue $\hbar^2 l(l+1)$, called **spinor harmonics**. Then we will form the four-component Dirac spinor energy eigenstates by combining two spinor harmonics with different l, but the same J^2 and J_z eigenvalues $\hbar^2 j(j+1)$ and $\hbar m$.

Fortunately, the problem of adding orbital and spin-1/2 angular momentum was already solved in Sect. 12.3. For a given j and m, we can construct two distinct spinor harmonics, one with $l = j - 1/2$ and another with $l = j + 1/2$. From (12.56) and (12.59), we see that they are respectively

$$\phi^+_{jm} = \frac{1}{\sqrt{2l+1}}\begin{pmatrix}\sqrt{l+1/2+m}\,Y_l^{m-1/2}\\ \sqrt{l+1/2-m}\,Y_l^{m+1/2}\end{pmatrix}, \qquad (l = j - 1/2), \tag{28.111}$$

$$\phi^-_{jm} = \frac{1}{\sqrt{2l+1}}\begin{pmatrix}\sqrt{l+1/2-m}\,Y_l^{m-1/2}\\ -\sqrt{l+1/2+m}\,Y_l^{m+1/2}\end{pmatrix}, \qquad (l = j + 1/2), \tag{28.112}$$

where in each case the upper and lower components have S_z eigenvalues $\hbar/2$ and $-\hbar/2$, and we have chosen their overall phases for future convenience. Note that m is always half-integer, and $Y_l^{m+1/2}(\theta, \phi)$ and $Y_l^{m-1/2}(\theta, \phi)$ are the usual spherical harmonic functions. For a given j, the spinor harmonics ϕ^+_{jm} and ϕ^-_{jm} have l differing by one unit, so they have opposite coordinate parity Π_c eigenvalues $(-1)^l$.

[3] However, the results below are not fully exact, because the Dirac equation for the electron is itself an approximation. It does not include the effects of nuclear spin, nuclear structure, and virtual photon effects from quantization of the electromagnetic field. Thus, the hyperfine splitting and the Lamb shift are not part of the exact solution to the Dirac equation, and are neglected in this section.

Because the potential has spherical symmetry and we will be using spherical coordinates, it is useful to consider the operator

$$\boldsymbol{\sigma}\cdot\hat{r} = \begin{pmatrix} \cos\theta & e^{-i\phi}\sin\theta \\ e^{i\phi}\sin\theta & -\cos\theta \end{pmatrix}. \tag{28.113}$$

Remarkably, this acts on the spinor harmonics by turning them into each other,

$$\phi^{+}_{jm} = \boldsymbol{\sigma}\cdot\hat{r}\,\phi^{-}_{jm}, \qquad \phi^{-}_{jm} = \boldsymbol{\sigma}\cdot\hat{r}\,\phi^{+}_{jm}. \tag{28.114}$$

To prove it, note that as an operator, $\boldsymbol{\sigma}\cdot\hat{r}$ commutes with J^2 and J_z and has odd parity with respect to the operator Π_c defined in (28.50). This implies that $\boldsymbol{\sigma}\cdot\hat{r}\,\phi^{-}_{jm}$ must have the same j and m and opposite parity as ϕ^{-}_{jm}. The unique spinor harmonic with this property is ϕ^{+}_{jm}, so it must be true that $\boldsymbol{\sigma}\cdot\hat{r}\,\phi^{-}_{jm} = C\phi^{+}_{jm}$ for some constant of proportionality C. To evaluate C, we can consider the special case $\theta = 0$, so that $\hat{r} = \hat{z}$ and $\boldsymbol{\sigma}\cdot\hat{z} = \begin{pmatrix} 1 & 0 \\ 0 & -1 \end{pmatrix}$. Now, using $Y_l^{m'}(0,\phi) = \delta_{m'0}\sqrt{\frac{2l+1}{4\pi}}$ as found in (8.146), we have

$$\phi^{+}_{jm} = \sqrt{\frac{2j+1}{8\pi}}\begin{pmatrix} \delta_{m,\frac{1}{2}} \\ \delta_{m,-\frac{1}{2}} \end{pmatrix}, \qquad \phi^{-}_{jm} = \sqrt{\frac{2j+1}{8\pi}}\begin{pmatrix} \delta_{m,\frac{1}{2}} \\ -\delta_{m,-\frac{1}{2}} \end{pmatrix} \qquad (\text{for } \theta = 0). \tag{28.115}$$

This shows that $C = 1$, establishing the first part of (28.114). The identity

$$(\boldsymbol{\sigma}\cdot\hat{r})^2 = I \tag{28.116}$$

then immediately implies the second part of (28.114) from the first.

We seek eigenstates of H that are also eigenstates of J^2, J_z, and Π. Given the spinor harmonic definitions of (28.111) and (28.112), and the form of the parity operator $\Pi = \Pi_c\beta$ defined in (28.53), we see that the four-component Dirac spinors

$$\Psi^{\pm}_{jm}(r,\theta,\phi) = \frac{1}{r}\begin{pmatrix} G^{\pm}_{jm}(r)\,\phi^{\pm}_{jm} \\ -iF^{\pm}_{jm}(r)\,\phi^{\mp}_{jm} \end{pmatrix} \tag{28.117}$$

have J^2, J_z, and Π eigenvalues $\hbar^2 j(j+1)$, $\hbar m$, and $(-1)^{j\mp 1/2}$, respectively. The radial wavefunctions $G^{\pm}_{jm}(r)$ and $F^{\pm}_{jm}(r)$ are to be determined by solving $H\Psi = E\Psi$, and the factors of $1/r$ and $-i$ are introduced for later convenience.

To evaluate the first term in $H\Psi$ using the Dirac Hamiltonian from (28.110), we need

$$\boldsymbol{\alpha}\cdot\nabla = \begin{pmatrix} 0 & \boldsymbol{\sigma}\cdot\nabla \\ \boldsymbol{\sigma}\cdot\nabla & 0 \end{pmatrix} \tag{28.118}$$

acting on our spinor harmonics multiplied by radial functions $F^{\pm}_{jm}(r)$ and $G^{\pm}_{jm}(r)$. We therefore calculate, for a generic function $f(r)$:

$$\begin{aligned} \boldsymbol{\sigma}\cdot\nabla\left(f(r)\,\phi^{\pm}_{jm}\right) &= \boldsymbol{\sigma}\cdot\hat{r}\,\boldsymbol{\sigma}\cdot\hat{r}\,\boldsymbol{\sigma}\cdot\nabla\, f(r)\,\phi^{\pm}_{jm} \\ &= \boldsymbol{\sigma}\cdot\hat{r}\left(\hat{r}\cdot\nabla + \frac{i}{r}\boldsymbol{\sigma}\cdot(\boldsymbol{r}\times\nabla)\right) f(r)\,\phi^{\pm}_{jm} \\ &= \boldsymbol{\sigma}\cdot\hat{r}\left(\frac{\partial}{\partial r} - \frac{1}{r}\frac{2}{\hbar^2}\boldsymbol{S}\cdot\boldsymbol{L}\right) f(r)\,\phi^{\pm}_{jm}. \end{aligned} \tag{28.119}$$

The first equality used $(\boldsymbol{\sigma}\cdot\hat{r})^2 = I$, the second employed the Pauli matrix product identity $(\boldsymbol{a}\cdot\boldsymbol{\sigma})(\boldsymbol{b}\cdot\boldsymbol{\sigma}) = \boldsymbol{a}\cdot\boldsymbol{b} + i(\boldsymbol{a}\times\boldsymbol{b})\cdot\boldsymbol{\sigma}$, and the third used $\boldsymbol{S} = \hbar\boldsymbol{\sigma}/2$ and the position representation identity $\boldsymbol{L} = \boldsymbol{R}\times\boldsymbol{P} = \boldsymbol{r}\times(-i\hbar\nabla)$. Now we use $2\boldsymbol{S}\cdot\boldsymbol{L} = J^2 - L^2 - S^2 = \hbar^2[j(j+1) - l(l+1) - 3/4]$, which can in turn be simplified using $l = j \mp 1/2$. Finally, using (28.114), the result can be written as

$$\boldsymbol{\sigma}\cdot\nabla\left(f(r)\,\phi^{\pm}_{jm}\right) = \left(\frac{df}{dr} + (1+\kappa^{\pm})\frac{f}{r}\right)\phi^{\mp}_{jm}, \tag{28.120}$$

where, for convenience, we have defined

$$\kappa^{\pm} = \mp(j+1/2). \tag{28.121}$$

Now plugging (28.117) into the Dirac equation (28.109) with the Hamiltonian (28.110), and using (28.120), gives the system of equations

$$\left(E - mc^2 + \frac{Ze^2}{r}\right)G + \hbar c\left(\frac{d}{dr} - \frac{\kappa}{r}\right)F = 0, \tag{28.122}$$

$$\left(E + mc^2 + \frac{Ze^2}{r}\right)F - \hbar c\left(\frac{d}{dr} + \frac{\kappa}{r}\right)G = 0, \tag{28.123}$$

where each of G, F, and κ has the same implicit superscript $\pm$, suppressed along with the subscripts jm from here on. For further simplicity, define dimensionless quantities

$$\mathcal{E} = E/mc^2, \qquad s = mcr/\hbar, \tag{28.124}$$

so that our first-order coupled differential equations become

$$\left(\frac{d}{ds} - \frac{\kappa}{s}\right) F - \left(1 - \mathcal{E} - \frac{Z\alpha}{s}\right) G = 0, \tag{28.125}$$

$$\left(\frac{d}{ds} + \frac{\kappa}{s}\right) G - \left(1 + \mathcal{E} + \frac{Z\alpha}{s}\right) F = 0. \tag{28.126}$$

Note that for a bound state, $E < mc^2$, so $\mathcal{E} < 1$.

We now examine the large-distance behaviors of G and F. For large s, the terms proportional to κ and $Z\alpha$ can be neglected, so we get $F' = (1 - \mathcal{E})G$ and $G' = (1 + \mathcal{E})F$, which imply $G'' = (1 - \mathcal{E}^2)G$ and $F'' = (1 - \mathcal{E}^2)F$. The solutions at large s therefore behave like

$$G, F \sim e^{-\sqrt{1-\mathcal{E}^2}s} \qquad \text{(for large } s\text{)}. \tag{28.127}$$

Conversely, for small s, one finds $F' = (\kappa F - Z\alpha G)/s$ and $G' = (-\kappa G + Z\alpha F)/s$, which can be disentangled to give

$$\left(s\frac{d}{ds}\right)^2 G = \gamma^2 G, \qquad \left(s\frac{d}{ds}\right)^2 F = \gamma^2 F, \tag{28.128}$$

with

$$\gamma = \sqrt{\kappa^2 - Z^2\alpha^2}. \tag{28.129}$$

This is solved by a simple power-law behavior

$$G, F \sim s^\gamma, \qquad \text{(for small } s\text{)}, \tag{28.130}$$

where the possibility $s^{-\gamma}$ is rejected so that the wavefunction has normalizable behavior near $s = 0$.

We now factor out these large- and small-distance behaviors by writing

$$G(s) = s^\gamma e^{-\sqrt{1-\mathcal{E}^2}s} g(s), \qquad F(s) = s^\gamma e^{-\sqrt{1-\mathcal{E}^2}s} f(s). \tag{28.131}$$

This defines new functions $g(s)$ and $f(s)$, which should be finite and nonzero at $s = 0$, and should not grow exponentially for large s. Plugging these into (28.125) and (28.126) gives

$$\left(\frac{d}{ds} - \sqrt{1 - \mathcal{E}^2} + \frac{\gamma - \kappa}{s}\right) f - \left(1 - \mathcal{E} - \frac{Z\alpha}{s}\right) g = 0, \tag{28.132}$$

$$\left(\frac{d}{ds} - \sqrt{1 - \mathcal{E}^2} + \frac{\gamma + \kappa}{s}\right) g - \left(1 + \mathcal{E} + \frac{Z\alpha}{s}\right) f = 0. \tag{28.133}$$

Since $g(s)$ and $f(s)$ are finite and nonzero at $s = 0$, we try series solutions

$$g(s) = \sum_{q=0}^{\infty} a_q s^q, \qquad f(s) = \sum_{q=0}^{\infty} b_q s^q. \tag{28.134}$$

Putting this into (28.132) and (28.133), and requiring that the coefficient of each power s^{q-1} vanishes, gives the coupled recurrence relations

$$(1-\mathcal{E})a_{q-1} + \sqrt{1-\mathcal{E}^2}\, b_{q-1} - Z\alpha a_q - (q+\gamma-\kappa)b_q = 0, \tag{28.135}$$

$$(1+\mathcal{E})b_{q-1} + \sqrt{1-\mathcal{E}^2}\, a_{q-1} + Z\alpha b_q - (q+\gamma+\kappa)a_q = 0. \tag{28.136}$$

Consider first the case $q = 0$. Since we are assuming $a_{-1} = b_{-1} = 0$, we learn that

$$\frac{b_0}{a_0} = \frac{Z\alpha}{\kappa - \gamma} = \frac{\kappa + \gamma}{Z\alpha}. \tag{28.137}$$

Note that the last equality is a check on γ in (28.129).

For general q, multiply (28.135) by $\sqrt{1+\mathcal{E}}$ and (28.136) by $\sqrt{1-\mathcal{E}}$, and then take the difference. The coefficients of a_{q-1} and b_{q-1} both cancel, and we find that the coefficients of the two series obey the proportionality

$$\frac{b_q}{a_q} = \frac{(q+\gamma+\kappa)\sqrt{1-\mathcal{E}} - Z\alpha\sqrt{1+\mathcal{E}}}{(q+\gamma-\kappa)\sqrt{1+\mathcal{E}} + Z\alpha\sqrt{1-\mathcal{E}}}. \tag{28.138}$$

Using these in (28.135) and (28.136), it follows that for large q,

$$\frac{a_q}{a_{q-1}} \approx \frac{b_q}{b_{q-1}} \approx \frac{2\sqrt{1-\mathcal{E}^2}}{q}. \tag{28.139}$$

We can therefore see that both series must terminate. Otherwise, we would find $g, f \sim e^{2\sqrt{1-\mathcal{E}^2}s}$, and thus $G, F \sim e^{\sqrt{1-\mathcal{E}^2}s}$, and the wavefunctions would grow exponentially at large distances, failing to be normalizable.

Therefore, we must have $a_{n'+1} = 0$ for some non-negative integer n', and then (28.138) implies that $b_{n'+1} = 0$ as well. Then (28.135) and (28.136) with $q = n'+1$ tell us

$$\frac{b_{n'}}{a_{n'}} = -\sqrt{\frac{1-\mathcal{E}}{1+\mathcal{E}}}. \tag{28.140}$$

Requiring that this ratio is equal to (28.138) with $q = n'$, we find the condition for a normalizable bound state,

$$Z\alpha\mathcal{E} = \sqrt{1-\mathcal{E}^2}\,(n'+\gamma). \tag{28.141}$$

This has the positive-energy solution

$$\mathcal{E} = \left[1+\left(\frac{Z\alpha}{n'+\gamma}\right)^2\right]^{-1/2}, \tag{28.142}$$

or, restoring E, j, and $Z\alpha$ using (28.124), (28.129), and then (28.121),

$$E = mc^2\left[1+\left(\frac{Z\alpha}{n'+\sqrt{(j+1/2)^2-Z^2\alpha^2}}\right)^2\right]^{-1/2}. \tag{28.143}$$

Now let

$$n = n'+j+1/2, \tag{28.144}$$

which is a positive integer, since n' is non-negative and $j+1/2$ is a positive integer. Then we can rewrite the bound-state energy eigenvalues as

$$E = mc^2\left[1+\frac{Z^2\alpha^2}{(n+\delta_j)^2}\right]^{-1/2}, \tag{28.145}$$

where

$$\delta_j \equiv \sqrt{(j+1/2)^2-Z^2\alpha^2}-(j+1/2). \tag{28.146}$$

Expanding in small $Z\alpha$,

$$E = mc^2\left[1-\frac{Z^2\alpha^2}{2n^2}+\frac{Z^4\alpha^4}{2n^3}\left(\frac{3}{4n}-\frac{1}{j+1/2}\right)+\cdots\right]. \tag{28.147}$$

The first term is the electron's rest energy, and the second term is the usual nonrelativistic energy eigenvalue $-Z^2e^2/2a_0n^2$, so we recognize n as the principal quantum number. The third term is the total fine-structure correction to the energy that we found (with $Z=1$) in (17.26).

Finally, we should carefully enumerate the valid solutions that we have found. For each j and m, there are two solutions with the same E, with wavefunctions $\Psi^{\pm}_{jm}$ in (28.117), but with one exception. The exception is that if $n'=0$, then we have from (28.137) and (28.140) that $b_0/a_0 = -\sqrt{(1-\mathcal{E})/(1+\mathcal{E})} = (\kappa+\gamma)/Z\alpha$, but this is only possible if κ is negative, because γ and $Z\alpha$ are positive. This means that in the special case $n'=0$ only $\kappa^+ = -(j+1/2)$ is allowed, and so the would-be solution Ψ^{-}_{jm} in (28.117) is not valid. Therefore, for each principal quantum number $n = 1, 2, 3, \ldots$, there are two bound-state solutions $\Psi^{\pm}_{jm}$ with $j = 1/2,\ 3/2,\ \ldots,\ n - 3/2$, and only one bound-state solution Ψ^{+}_{jm} with $j = n - 1/2$. Each of these has further degeneracy $2j + 1$ from the m quantum number. All of this is in accord with the results found at the end of Sect. 17.1 from the perturbative approach to the fine structure of the hydrogen atom.

Exercises

28.1. The chiral representation of gamma matrices for the Dirac equation is

$$\gamma^0 = \begin{pmatrix} 0 & I \\ I & 0 \end{pmatrix}, \qquad \gamma^a = \begin{pmatrix} 0 & \sigma^a \\ -\sigma^a & 0 \end{pmatrix}, \qquad (a = 1, 2, 3). \tag{28.148}$$

Find a unitary matrix U such that $\gamma^{\mu}_{\text{chiral}} = U\gamma^{\mu}_{\text{Dirac}}U^{\dagger}$, where $\gamma^{\mu}_{\text{Dirac}}$ are the matrices in (28.57). (Note that the spatial gamma matrices are the same in the two representations, so what is needed is a unitary transformation that affects only γ^0.)

28.2. In this problem, we will check the Lorentz invariance of the Dirac equation, and in the process determine the Lorentz transformation rule for Dirac spinors. Suppose that two coordinate systems are related by a Lorentz transformation $x'^{\mu} = L^{\mu}{}_{\nu}x^{\nu}$. The Dirac wavefunction $\Psi'(x')$ in the primed frame should be related to that in the unprimed frame by $\Psi'(x') = \Lambda\Psi(x)$ where Λ is a 4×4 matrix. Now, the Dirac equation in the unprimed frame is $(i\hbar\gamma^{\mu}\frac{\partial}{\partial x^{\mu}} - mc)\Psi(x) = 0$, and in the primed frame it is $(i\hbar\gamma^{\mu}\frac{\partial}{\partial x'^{\mu}} - mc)\Psi'(x') = 0$.

(a) Show that these equations are consistent provided that $\Lambda^{-1}\gamma^{\rho}L_{\rho}{}^{\mu}\Lambda = \gamma^{\mu}$, or equivalently $\Lambda^{-1}\gamma^{\mu}\Lambda = L^{\mu}{}_{\nu}\gamma^{\nu}$.
(b) Now suppose that $L^{\mu}{}_{\nu} = \delta^{\mu}_{\nu} + \omega^{\mu}{}_{\nu}$ with $\omega^{\mu}{}_{\nu}$ infinitesimal, where $\omega_{\mu\nu} = -\omega_{\nu\mu}$ is antisymmetric. Prove that the equation found in part (a) is satisfied if $\Lambda = 1 - \omega_{\mu\nu}[\gamma^{\mu}, \gamma^{\nu}]/8$.

28.3. Consider a Dirac spinor wavefunction Ψ satisfying the Dirac equation (28.89) with electromagnetic potentials. Derive the local conservation of probability,

$$\frac{\partial\rho}{\partial t} = -\nabla\cdot\boldsymbol{J}, \tag{28.149}$$

where $\rho = \Psi^{\dagger}\Psi$ is the probability density and you will need to identify the probability current $\boldsymbol{J}$. You may be surprised by the difference from the nonrelativistic case treated in Sect. 5.6.

28.4. What is the next order ($Z^6\alpha^6$) contribution to the hydrogen-like atom stationary state energies in (28.147)? For $Z = 1$, evaluate it numerically in eV for the $n = 1$ ground state ($j = 1/2$) and for the $n = 2$ states with $j = 1/2$ and $j = 3/2$, and compare these corrections to the hyperfine and Lamb shift contributions.

29 Feynman Path Integral Approach

29.1 Propagators

Consider a quantum system with a complete set of commuting observables Q_a, with eigenvalues q_a, where a is an index running from 1 to n. For example, for a single spinless particle moving in three-dimensional space, we have $n = 3$, and the Q_a with $a = 1, 2, 3$ could be chosen to be the components of the position coordinate vector $\boldsymbol{R}$, with eigenvalues $\boldsymbol{r}$. Alternatively, we could choose the Q_a to be the components of the momentum vector $\boldsymbol{P}$, with eigenvalues $\boldsymbol{p}$. For simplicity, we will often suppress the index a, and let the complete set of CSCO eigenvalues for a basis state be represented simply as the generic symbol q. We do this to make room for other subscript labels that will distinguish the values of the q's at various times. Given the state ket $|\psi(t)\rangle$ of the system, we can then write a wavefunction

$$\psi(q, t) = \langle q|\psi(t)\rangle, \tag{29.1}$$

where $|q\rangle$ are the CSCO eigenstates in which each of the observables Q_a is known to be equal to the corresponding eigenvalue q_a.

Recall, from Sect. 3.4, that if the state of the system at time t_0 was $|\psi(t_0)\rangle$, then the state at time t is

$$|\psi(t)\rangle = U(t, t_0)|\psi(t_0)\rangle, \tag{29.2}$$

where $U(t, t_0)$ is the unitary time evolution operator. Using completeness of the Q eigenstates at some initial time t_0, the wavefunction at any later time t is

$$\psi(q, t) = \int dq_0 \; \langle q|U(t, t_0)|q_0\rangle\langle q_0|\psi(t_0)\rangle. \tag{29.3}$$

S. P. Martin, *Quantum Mechanics*, Graduate Texts in Physics,
https://doi.org/10.1007/978-3-032-21119-4_29

It is now useful to define the **propagator** (or **kernel**, or **transition amplitude**) as the function

$$U(q,t;\, q_0,t_0) \equiv \langle q|U(t,t_0)|q_0\rangle, \tag{29.4}$$

so that

$$\psi(q,t) = \int dq_0\; U(q,t;\, q_0,t_0)\, \psi(q_0,t_0). \tag{29.5}$$

The utility of the propagator is that, given the wavefunction $\psi(q_0,t_0)$ at an initial time t_0, the wavefunction at any other time t is obtained by doing this integration over q_0.

The propagator is also the solution to a differential equation, which follows from (3.25),

$$i\hbar\frac{\partial}{\partial t}U(q,t;\, q_0,t_0) = i\hbar\langle q|\frac{\partial}{\partial t}U(t,t_0)|q_0\rangle = \langle q|H(t)U(t,t_0)|q_0\rangle. \tag{29.6}$$

Consider the example of a spinless particle moving in three dimensions in a potential, so that the Hamiltonian is $H = P^2/2m + V(\boldsymbol{R},t)$. Interpreting q as the position $\boldsymbol{r}$, (29.5) becomes

$$\psi(\boldsymbol{r},t) = \int d^3\boldsymbol{r}_0\, U(\boldsymbol{r},t;\, \boldsymbol{r}_0,t_0)\, \psi(\boldsymbol{r}_0,t_0), \tag{29.7}$$

where (29.6) tells us that

$$\left[i\hbar\frac{\partial}{\partial t} + \frac{\hbar^2\nabla^2}{2m} - V(\boldsymbol{r},t)\right]U(\boldsymbol{r},t;\, \boldsymbol{r}_0,t_0) = 0. \tag{29.8}$$

Since this is a first-order differential equation in t, in principle it can be solved given the boundary condition at $t = t_0$, which is $U(\boldsymbol{r},t_0;\, \boldsymbol{r}_0,t_0) = \delta^{(3)}(\boldsymbol{r}-\boldsymbol{r}_0)$. In practice, this might be quite difficult, depending on the choice of potential.

Suppose the Hamiltonian does not depend explicitly on time, so that $U(t,t_0) = e^{-i(t-t_0)H/\hbar}$, from (3.26). If we know all of the H eigenstates $|n\rangle$, with energies E_n, then using completeness we get the spectral decomposition

$$\begin{aligned} U(q,t;\, q_0,t_0) &= \sum_n e^{-i(t-t_0)E_n/\hbar}\langle q|n\rangle\langle n|q_0\rangle \\ &= \sum_n e^{-i(t-t_0)E_n/\hbar}\psi_n(q)\psi_n(q_0)^*, \end{aligned} \tag{29.9}$$

where $\psi_n(q) = \langle q|n\rangle$ are the wavefunctions of the orthonormal Hamiltonian eigenstates with energies E_n. Thus, the propagator can be evaluated as a sum over products of energy eigenstate wavefunctions, weighted by phase factors that vary linearly with time.

As a simple example, consider the propagator for a free particle of mass m, moving in one dimension, with the position x playing the role of q in the preceding. We have already done the work for this case, in Sect. 6.2, and the result of (6.29) can be rewritten as

$$U(x,t;\, x_0,t_0) = \left[\frac{m}{2\pi i\hbar(t-t_0)}\right]^{1/2} \exp\left[\frac{im(x-x_0)^2}{2\hbar(t-t_0)}\right]. \tag{29.10}$$

Repeating the procedure for a spinless free particle moving in three dimensions gives

$$U(\boldsymbol{r},t;\, \boldsymbol{r}_0,t_0) = \left[\frac{m}{2\pi i\hbar(t-t_0)}\right]^{3/2} \exp\left[\frac{im|\boldsymbol{r}-\boldsymbol{r}_0|^2}{2\hbar(t-t_0)}\right] \tag{29.11}$$

for the position-representation propagator.

For a somewhat more involved example, consider the one-dimensional harmonic oscillator, with the familiar Hamiltonian $H = P^2/2m + m\omega^2X^2/2$. One way of writing the propagator is to use the spectral decomposition as in (29.9),

$$U(x,t;x_0,t_0) = \sum_{n=0}^{\infty} \psi_n(x)\psi_n(x_0)^* e^{-i\omega(t-t_0)(n+1/2)}, \tag{29.12}$$

where $\psi_n(x)$ are the stationary state wavefunctions of (7.53), involving Hermite polynomials. It may therefore come as a surprise that the propagator can be written in a nice form that does not involve Hermite polynomials or an infinite sum,

$$U(x,t;\, x_0,t_0) = \left[\frac{m\omega}{2\pi i\hbar\sin(\omega(t-t_0))}\right]^{1/2} \exp\left(iS_{\rm cl}[x,t;\,x_0,t_0]/\hbar\right), \tag{29.13}$$

where, for general initial and final positions and times,

$$S_{\rm cl}[x_f,t_f;x_i,t_i] = \frac{m\omega}{2\sin(\omega(t_f-t_i))}\left\{(x_f^2+x_i^2)\cos(\omega(t_f-t_i)) - 2x_ix_f\right\}. \tag{29.14}$$

Remarkably, this is the action for the classical trajectory that starts at x_i at time t_i, and ends at x_f at time t_f. Recall that, in general, the **action** is defined by

$$S = \int_{t_i}^{t_f} dt\, L(x,\dot{x}), \tag{29.15}$$

where $L(x, \dot{x})$ is the Lagrangian. In the present case, $L = \frac{1}{2}m\dot{x}^2 - \frac{1}{2}m\omega^2x^2$, and you can check that (29.14) follows from

$$x_{\text{cl}}(t) = \frac{x_f \sin(\omega(t - t_i)) + x_i \sin(\omega(t_f - t))}{\sin(\omega(t_f - t_i))}, \tag{29.16}$$

which is the classical solution to the equation of motion $\ddot{x} = -\omega^2 x$ and the initial and final boundary conditions.

The quantum harmonic oscillator propagator claimed in (29.13) is somewhat nontrivial to derive from scratch, but once written down, it is not too hard to verify. This consists of checking that it obeys the first-order (in time) differential equation

$$\left(i\hbar\frac{\partial}{\partial t} + \frac{\hbar^2}{2m}\frac{\partial^2}{\partial x^2} - \frac{1}{2}m\omega^2x^2\right) U(x, t;\, x_0, t_0) = 0, \tag{29.17}$$

as in (29.8), and that it satisfies the correct boundary condition

$$U(x, t_0;\, x_0, t_0) = \delta(x - x_0), \tag{29.18}$$

which follows from taking the $t \to t_0$ limit of (29.13), with the help of the delta function representation (2.35).

The propagator for the oscillator in (29.13) has several features worthy of note. First, for the period $2\pi/\omega$ of the classical oscillator, the propagator is anti-periodic. This can be seen most clearly from the spectral decomposition form (29.12), which shows that for $t \to t + 2\pi/\omega$, the propagator acquires a phase factor -1 due to the zero-point energy (the $1/2$ in $n+1/2$ in the exponent). Second, one recovers the free particle propagator of (29.10) by taking the limit $\omega \to 0$. Finally, it is intriguing that the only dependence of the quantum propagator on the starting and ending positions x and x_0 comes through the $e^{iS/\hbar}$ factor for the classical trajectory that connects them. This is a suggestive clue for the developments in the next section.

29.2 Summing Over Paths

An alternative way of computing the propagator for a quantum system uses the **Feynman sum over paths** or **sum over histories**, developed by Richard P. Feynman in his 1942 PhD thesis. Let us write, very schematically for now,

$$U(q_f, t_f;\, q_i, t_i) = \sum_{\text{paths}} (\text{something}). \tag{29.19}$$

In this section, we denote the initial and final CSCO eigenvalue variables and times by (q_i, t_i) and (q_f, t_f) respectively, in order to distinguish them from intermediate times t and positions q in the following discussion. The "paths" here are trajectories

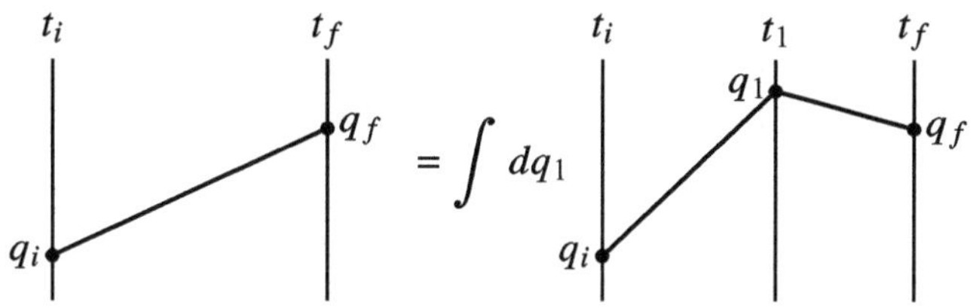

Fig. 29.1 Diagrammatic representation of the composition law (29.21). Each line between dots represents a propagator. The propagator $U(q_f, t_f; q_i, t_i)$ is equal to the integral (or sum, if the allowed values are discrete) over all possible q_1 of the product $U(q_f, t_f; q_1, t_1)U(q_1, t_1; q_i, t_i)$, for any choice of intermediate time t_1

$q(t)$ starting from q_i at time t_i and ending with q_f at time t_f. In the case of a spinless particle in three dimensions, a path could be a trajectory given by a function $\boldsymbol{r}(t)$. After choosing the CSCO eigenvalue variables, these paths are arbitrary, in the sense that *all* possible trajectories (not just those satisfying the classical equations of motion!) are included as long as they obey the initial and final boundary conditions.

The task before us now is to make (29.19), including its mysterious summand, more precise. To motivate this form, first note that

$$\begin{aligned} U(q_f, t_f;\, q_i, t_i) &= \langle q_f | U(t_f, t_1) U(t_1, t_i) | q_i \rangle \\ &= \int dq_1\, \langle q_f | U(t_f, t_1) | q_1 \rangle\langle q_1 | U(t_1, t_i) | q_i \rangle, \end{aligned} \tag{29.20}$$

where we have chosen an arbitrary time t_1 between t_i and t_f, split up the unitary time evolution operator accordingly using (3.53), and then used completeness of the eigenstates $|q_1\rangle$. This can be rewritten as a composition rule for propagators,

$$U(q_f, t_f;\, q_i, t_i) = \int dq_1\, U(q_f, t_f;\, q_1, t_1)\, U(q_1, t_1;\, q_i, t_i). \tag{29.21}$$

In words: to propagate from time t_i to time t_f, we integrate over all of the ways to propagate from t_i to t_1, and then from t_1 to t_f, as indicated in Fig. 29.1.

There is an intuitive analogy, **Feynman's disappearing screen**, for the preceding mathematical derivation of (29.21) and Fig. 29.1. Consider particles whose vertical positions are described by q. The particles are emitted at a source point q_i, and detected at a point q_f, which can be varied by moving the detector. The probability of detection is proportional to the squared magnitude of the complex propagator amplitude. Now if we place a large impenetrable screen between the source and the detector, then the amplitude will certainly vanish. But suppose that we then drill a small hole in the screen at a single point q_1. In that case, the only nonzero contribution corresponds to the particle first moving from q_i to q_1, and then from q_1 to q_f, so the propagator amplitude is the product of two separate amplitudes. If we drill a second hole at a different q_1, then the total amplitude will be the sum of two

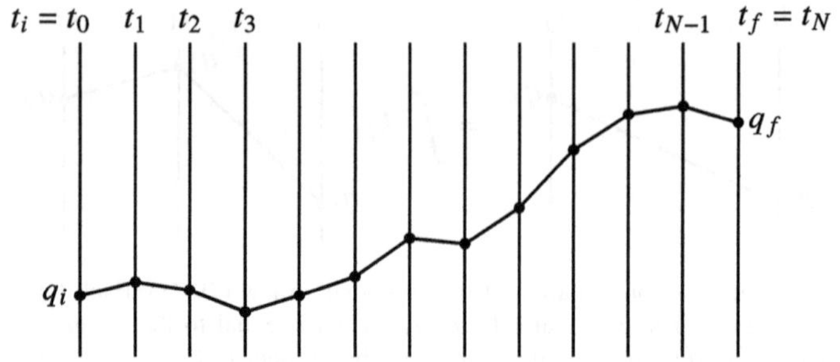

Fig. 29.2 One possible trajectory, or path, $q(t)$ with initial boundary condition q_i at time t_i and final boundary condition q_f at time t_f. Time increases moving to the right, and is divided into equal slices, with variables $q_1, q_2, \ldots, q_{N-1}$ at intermediate times $t_1, t_2, \ldots, t_{N-1}$. Each short line between dots represents a propagator. The path integral is obtained by taking the limit of $N \to \infty$ time steps with infinitesimal time intervals $\epsilon = t_j - t_{j-1} = (t_f - t_i)/N$, and integrating over all $q_1, q_2, \ldots, q_{N-1}$

such product amplitudes, which interfere to give a diffraction pattern of maxima and minima as we vary q_f. Drilling more holes in the screen will just add more product amplitudes to the superposition, corresponding to more intermediate points q_1. In the limit that we have drilled through *every* point on the screen, it will have completely disappeared. So the amplitude for no screen at all must be proportional to the superposition of all amplitudes for all possible points q_1, as on the right side of (29.21) and its pictorial version in Fig. 29.1.[1]

We now extend the same idea to subdivide the time interval from t_i to t_f into N time steps instead of just two, with intermediate times $t_1, t_2, \ldots, t_{N-1}$, as shown in Fig. 29.2. At each time step, we integrate (or sum, if the allowed values are discrete) over all possible q_j for $j = 1, 2, \ldots, N-1$, using completeness of the states $|q_j\rangle$ at each step. The result is

$$U(q_f, t_f;\, q_i, t_i) = \int dq_1 \int dq_2 \cdots \int dq_{N-1}\, U(q_f, t_f;\, q_{N-1}, t_{N-1})$$
$$U(q_{N-1}, t_{N-1};\, q_{N-2}, t_{N-2}) \cdots U(q_2, t_2;\, q_1, t_1) U(q_1, t_1;\, q_i, t_i). \tag{29.22}$$

Although this is not the ultimate realization of the sum over paths idea, it already illustrates the basic concept that we can find the propagator by adding up all possible ways to get from q_i at time t_i to q_f at time t_f.

Feynman's sum over paths can be thought of as the continuum limit of the discretized picture shown in Fig. 29.2. He proposed that the propagator

[1] A slight imperfection in the simple pictorial version of the analogy is that the line containing q_1 in Fig. 29.1 represents a fixed time, while the disappearing screen instead occupies a fixed spatial position.

$U(q_f, t_f;\, q_i, t_i)$ can be computed as a sum, over all possible space-time paths, of a phase factor determined by the classical action for the path. The paths to be summed over are functions $q(t)$ defined for $t_i \leq t \leq t_f$ and constrained to obey the boundary conditions

$$q(t_i) = q_i, \qquad q(t_f) = q_f. \tag{29.23}$$

The action for a given path is

$$S[q(t)] = \int_{t_i}^{t_f} dt\, L(q,\, \dot{q},\, t), \tag{29.24}$$

where $L(q,\, \dot{q},\, t)$ is the classical Lagrangian for the system, often the kinetic energy minus the potential energy. For example, for a spinless nonrelativistic particle in three dimensions, the position vector $\boldsymbol{r}(t)$ plays the role of $q(t)$, and the Lagrangian is

$$L(\boldsymbol{r},\, \dot{\boldsymbol{r}},\, t) = \frac{1}{2}m\dot{\boldsymbol{r}}^2 - V(\boldsymbol{r}, t). \tag{29.25}$$

The action is an example of a **functional**, an object that takes as its input a function [here, the path $q(t)$] and returns a number. Feynman proposed that (29.19) should be

$$U(q_f, t_f;\, q_i, t_i) \propto \sum_{\text{paths } q(t)} e^{iS[q(t)]/\hbar}. \tag{29.26}$$

This fulfilled an inspiring (but less concrete) suggestion made earlier by Dirac.

Because the number of paths is continuously infinite, it makes more sense to think of integrating over all paths rather than summing them. This suggests the arguably better names **Feynman path integral** or **functional integral**, and the notational change

$$\sum_{\text{paths } q(t)} e^{iS[q(t)]/\hbar} \to \int d[q(t)]\, e^{iS[q(t)]/\hbar}. \tag{29.27}$$

Just as ordinary definite integration takes a function and returns a number, a functional integral takes a functional (in this case, the action) and returns a number. The constant of proportionality implied in (29.26) can then be absorbed into the normalization for the functional differential, denoted by $d[q(t)]$.

In summary, the following can now be taken as an alternative to the Schrödinger equation Postulate 6 of Sect. 3.1:

Postulate 6′: Time Evolution as a Sum Over Paths For a system with eigenvalues $\boldsymbol{q}$ for a complete set of commuting observables, the propagator is

$$U(\boldsymbol{q}_f, t_f; \boldsymbol{q}_i, t_i) = \int d[\boldsymbol{q}(t)] \exp\left(iS[\boldsymbol{q}(t)]/\hbar\right), \tag{29.28}$$

where $S[\boldsymbol{q}(t)] = \int_{t_i}^{t_f} dt\, L(\boldsymbol{q}, \dot{\boldsymbol{q}}, t)$ is the action functional, and the functional integral is over all paths satisfying $\boldsymbol{q}(t_i) = \boldsymbol{q}_i$ and $\boldsymbol{q}(t_f) = \boldsymbol{q}_f$. Equivalently, the wavefunction satisfies

$$\psi(\boldsymbol{q}_f, t_f) = \int d[\boldsymbol{q}(t)] \exp\left(iS[\boldsymbol{q}(t)]/\hbar\right) \psi(\boldsymbol{q}(t_i), t_i), \tag{29.29}$$

but now the functional integral is over all paths satisfying $\boldsymbol{q}(t_f) = \boldsymbol{q}_f$, but with no constraint on $\boldsymbol{q}(t_i)$, instead integrating over those initial values as well, as in (29.5).

Note that while the Schrödinger version of Postulate 6 requires us to specify the Hamiltonian, the Feynman version, Postulate 6′, instead requires that the Lagrangian is the object to be specified to determine the dynamics of the system.

Although path integration has proved to be a very useful tool, especially in quantum field theory, it is not as straightforward to define as the Schrödinger differential equation. Physicists often simply ignore the associated problems of mathematical rigor. One way to define the functional integral, inspired by the composition rule for the propagator as in (29.22) and Fig. 29.2, is to discretize the path into steps, integrate over the values of q on each step, and then take the limit in which the number of steps is infinitely large.

As a specific example, consider the case of a spinless particle moving in a potential $V(\boldsymbol{r}, t)$ in three dimensions, with the coordinates $\boldsymbol{r}$ playing the role of $\boldsymbol{q}$. To compute the path integral, we partition the time interval (t_i, t_f) into N equal steps of duration

$$\epsilon = (t_f - t_i)/N, \tag{29.30}$$

so that

$$t_j = t_i + j\epsilon, \qquad (j = 0, 1, 2, \ldots, N), \tag{29.31}$$

with $t_i = t_0$ and $t_f = t_N$. We then define the discretized version of a path from $\boldsymbol{r}_i$ to $\boldsymbol{r}_f$ by specifying the positions $\boldsymbol{r}_j$ at each t_j for $j = 1, \ldots, N-1$, with $\boldsymbol{r}_0 = \boldsymbol{r}_i$ and $\boldsymbol{r}_N = \boldsymbol{r}_f$. The velocity of the particle in the interval between times t_{j-1} and t_j is $\boldsymbol{v} = (\boldsymbol{r}_j - \boldsymbol{r}_{j-1})/\epsilon$, so the kinetic energy is $\frac{1}{2}mv^2 = \frac{1}{2}m(|\boldsymbol{r}_j - \boldsymbol{r}_{j-1}|/\epsilon)^2$. Therefore, the discretized version of the action for the path is

$$S[\text{path}] = S(\boldsymbol{r}_0, \boldsymbol{r}_1, \ldots, \boldsymbol{r}_{N-1}, \boldsymbol{r}_N) = \epsilon \sum_{j=1}^{N} \left[\frac{m|\boldsymbol{r}_j - \boldsymbol{r}_{j-1}|^2}{2\epsilon^2} - V(\boldsymbol{r}_j, t_j) \right], \tag{29.32}$$

with $\boldsymbol{r}_0 = \boldsymbol{r}_i$ and $\boldsymbol{r}_N = \boldsymbol{r}_f$. In the limit $\epsilon \to 0$ with $N = (t_f - t_i)/\epsilon \to \infty$, this will give the action for the continuous path. We now write (29.29) as

$$\psi(\boldsymbol{r}_f, t_f) = \lim_{N\to\infty} \int \frac{d^3\boldsymbol{r}_0}{a^3} \int \frac{d^3\boldsymbol{r}_1}{a^3} \int \frac{d^3\boldsymbol{r}_2}{a^3} \cdots \int \frac{d^3\boldsymbol{r}_{N-1}}{a^3} \exp\big[i S[\text{path}]/\hbar\big]\, \psi(\boldsymbol{r}_0, t_0), \tag{29.33}$$

where a is a normalization factor, with units of length, to be found soon. So, to connect with the general form of (29.5), we remove the integration over $\boldsymbol{r}_0$ and the initial wavefunction $\psi(\boldsymbol{r}_0, t_0)$ to get

$$U(\boldsymbol{r}_f, t_f;\, \boldsymbol{r}_i, t_i) = \lim_{N\to\infty} \frac{1}{a^3} \int \frac{d^3\boldsymbol{r}_1}{a^3} \int \frac{d^3\boldsymbol{r}_2}{a^3} \cdots \int \frac{d^3\boldsymbol{r}_{N-1}}{a^3} \exp\big[i S[\text{path}]/\hbar\big]. \tag{29.34}$$

Here, it is important that there is one more factor of $1/a^3$ than the remaining number of integrations, in accord with the fact that the propagator has dimension $1/[\text{length}]^3$. Equation (29.34) is one possible meaning of the notation

$$U(\boldsymbol{r}_f, t_f;\, \boldsymbol{r}_i, t_i) = \int d[\boldsymbol{r}(t)]\, \exp\left(i S[\boldsymbol{r}(t)]/\hbar\right) \tag{29.35}$$

for the path integral.

Let us now show that the preceding proposal is indeed equivalent to the Schrödinger equation in the case of a spinless nonrelativistic particle. It is sufficient to consider only time evolution over a single infinitesimal time interval, since longer time intervals can be obtained using the composition law (29.21). For simplicity, we write $t_i = t_0 = t$, and after one infinitesimal time step, $t_f = t + \epsilon$. Then, using the $N = 1$ version of (29.33) with $t_0 = t$, the wavefunction at the final time in terms of the initial wavefunction is

$$\psi(\boldsymbol{r},\, t+\epsilon) = \int \frac{d^3\boldsymbol{r}_0}{a^3} e^{i S(\boldsymbol{r}, \boldsymbol{r}_0)/\hbar} \psi(\boldsymbol{r}_0, t). \tag{29.36}$$

Expanding the left side in small ϵ gives

$$\text{LS of (29.36)} = \psi(\boldsymbol{r},\, t+\epsilon) = \psi(\boldsymbol{r}, t) + \epsilon \frac{\partial}{\partial t} \psi(\boldsymbol{r}, t) + \cdots. \tag{29.37}$$

The right side of (29.36) is

$$\text{RS of (29.36)} = \int \frac{d^3\boldsymbol{r}_0}{a^3} \exp\left(\frac{i}{\hbar}\left[\frac{m|\boldsymbol{r}-\boldsymbol{r}_0|^2}{2\epsilon} - \epsilon V(\boldsymbol{r}, t+\epsilon)\right]\right) \psi(\boldsymbol{r}_0, t). \tag{29.38}$$

or, expanding the exponential part involving $V(\boldsymbol{r}, t)$ to first order in ϵ, and then defining a new integration variable $\boldsymbol{u} = \boldsymbol{r}_0 - \boldsymbol{r}$,

$$\text{RS of (29.36)} = \left[1 - i\frac{\epsilon}{\hbar}V(\boldsymbol{r}, t)\right] \int \frac{d^3\boldsymbol{u}}{a^3}\, e^{imu^2/2\hbar\epsilon}\, \psi(\boldsymbol{r} + \boldsymbol{u}, t). \quad (29.39)$$

Now, in the limit of small ϵ, there will be almost complete cancellation from the rapidly varying phase inside the integral. The only region that contributes as $\epsilon \to 0$ is where $u^2 \approx 0$, so that the phase factor $e^{imu^2/2\hbar\epsilon}$ is close to 1. Therefore, we can use the expansion for small u,

$$\psi(\boldsymbol{r} + \boldsymbol{u}, t) = \psi(\boldsymbol{r}, t) + \boldsymbol{u} \cdot \nabla\psi(\boldsymbol{r}, t) + \frac{1}{2}(\boldsymbol{u} \cdot \nabla)^2\psi(\boldsymbol{r}, t) + \cdots. \quad (29.40)$$

Inside the integration over $\boldsymbol{u}$, this can be simplified because the parts linear in u_x, u_y, or u_z will cancel. Thus, the term linear in $\boldsymbol{u}$ does not contribute at all, and using rectangular coordinates and the spherical symmetry of the integrand, you can check that inside the integral the last term can be replaced according to $(\boldsymbol{u} \cdot \nabla)^2 \to \frac{1}{3}u^2\nabla^2$. So, we have

$$\int d^3\boldsymbol{u}\, e^{imu^2/2\hbar\epsilon}\, \psi(\boldsymbol{r} + \boldsymbol{u}, t) = \psi(\boldsymbol{r}, t) \int d^3\boldsymbol{u}\, e^{imu^2/2\hbar\epsilon} + \frac{1}{6}\nabla^2\psi(\boldsymbol{r}, t) \int d^3\boldsymbol{u}\, u^2 e^{imu^2/2\hbar\epsilon} + \cdots, \quad (29.41)$$

in which the neglected terms have higher powers of u^2. Now we compute:

$$\int d^3\boldsymbol{u}\, e^{imu^2/2\hbar\epsilon} = 4\pi \int_0^\infty du\, u^2\, e^{imu^2/2\hbar\epsilon} = \left(\frac{2\pi i\hbar\epsilon}{m}\right)^{3/2}, \quad (29.42)$$

$$\int d^3\boldsymbol{u}\, u^2\, e^{imu^2/2\hbar\epsilon} = 4\pi \int_0^\infty du\, u^4\, e^{imu^2/2\hbar\epsilon} = i\frac{3\hbar\epsilon}{m}\left(\frac{2\pi i\hbar\epsilon}{m}\right)^{3/2}. \quad (29.43)$$

Note that each power of u^2 outside of the exponent leads to another power of ϵ, justifying our expansion in small u; effectively, we have $u \sim \sqrt{\epsilon}$. Plugging (29.42) and (29.43) into (29.39) gives

$$\text{RS of (29.36)} = \frac{1}{a^3}\left(\frac{2\pi i\hbar\epsilon}{m}\right)^{3/2}\left[1 + i\frac{\epsilon}{\hbar}\left(\frac{\hbar^2\nabla^2}{2m} - V(\boldsymbol{r}, t)\right) + O(\epsilon^2)\right]\psi(\boldsymbol{r}, t). \quad (29.44)$$

Now compare (29.37) and (29.44). Matching at leading order in ϵ just informs us that the normalization factor must be

$$a = \left(\frac{2\pi i\hbar\epsilon}{m}\right)^{1/2}. \tag{29.45}$$

Then, matching the terms at first order in ϵ gives

$$i\hbar\frac{\partial}{\partial t}\psi(\boldsymbol{r},t) = \left[-\frac{\hbar^2\nabla^2}{2m} + V(\boldsymbol{r},t)\right]\psi(\boldsymbol{r},t), \tag{29.46}$$

which is the Schrödinger equation in the position-wavefunction representation. This completes the demonstration that the Feynman sum over paths formulation is equivalent to the Schrödinger equation time evolution, for a spinless nonrelativistic particle.

29.3 Evaluation of the Path Integral for the Harmonic Oscillator

In the previous section, we demonstrated the equivalence of the path integral formulation and the Schrödinger equation, but this is not the same as actually computing a useful result directly in terms of the path integral. We will now show how to arrive at the propagator for the one-dimensional harmonic oscillator, already given in (29.13), in the path integral approach.

Consider the propagator as a path integral, starting at time $t = 0$ and ending at $t = T$,

$$U(x_f, T; x_i, 0) = \int d[x(t)]\, e^{iS[x(t)]/\hbar}. \tag{29.47}$$

In the case of the harmonic oscillator, we have a great advantage: we know the unique classical trajectory $x_{\rm cl}(t)$ for fixed endpoints, given by (29.16) with $t_f = T$ and $t_i = 0$. Accordingly, it will help us greatly to do a change of variables

$$x(t) = x_{\rm cl}(t) + y(t), \tag{29.48}$$

where the new shifted coordinate $y(t)$ can be interpreted as the quantum fluctuation, satisfying the boundary conditions

$$y(0) = y(T) = 0. \tag{29.49}$$

Now we can evaluate the action functional, with the result

$$S[x(t)] = S[x_{\rm cl}(t)] + \frac{m}{2}\int_0^T dt\,\left(\dot{y}^2 - \omega^2 y^2\right), \tag{29.50}$$

where the cross-terms involving both $x_{\rm cl}$ and y have conveniently canceled. To see this cancellation, note that those cross-terms are

$$m \int_0^T dt \left(\dot{x}_{\rm cl}\dot{y} - \omega^2 x_{\rm cl} y\right) = -m \int_0^T dt \left(\ddot{x}_{\rm cl} + \omega^2 x_{\rm cl}\right) y, \tag{29.51}$$

where we have integrated by parts, making use of the vanishing of the boundary terms from (29.49). The last expression vanishes because $x_{\rm cl}$ satisfies the classical equation of motion.

Now, since $S[x_{\rm cl}(t)]$ is a constant with respect to the new functional integration variable $y(t)$, we can pull it out of the path integral, to get

$$U(x_f, T; x_i, 0) = I\, e^{iS[x_{\rm cl}(t)]/\hbar}, \tag{29.52}$$

where it remains to evaluate the functional integral over $y(t)$,

$$I = \int d[y(t)] \exp\left[\frac{im}{2\hbar}\int_0^T dt \left(\dot{y}^2 - \omega^2 y^2\right)\right]. \tag{29.53}$$

We have already made good progress, since we have successfully obtained the correct $e^{iS[x_{\rm cl}(t)]/\hbar}$ factor in (29.13). Indeed, from its definition, the remaining factor I clearly has no dependence on x_f or x_i. Our remaining task is to show that, in accord with (29.13),

$$I = \left(\frac{m\omega}{2\pi i\hbar \sin(\omega T)}\right)^{1/2}. \tag{29.54}$$

We will do the path integral to get I in two different ways.

First, we use the discretization of the same type used in (29.34),

$$I = \lim_{N\to\infty} \frac{1}{a}\left(\prod_{j=1}^{N-1}\int_{-\infty}^{\infty}\frac{dy_j}{a}\right) \exp\left(\frac{im}{2\hbar\epsilon}\sum_{k=1}^{N}\left[(y_k - y_{k-1})^2 - \epsilon^2\omega^2 y_k^2\right]\right) \tag{29.55}$$

where $\epsilon = T/N$, and we have already obtained the normalization a in (29.45). Let $z_j = y_j/a$, and assemble $z_1, z_2, \ldots, z_{N-1}$ into an $(N-1)$-dimensional vector z. This gives

$$I = \lim_{N\to\infty} \frac{1}{a}\int d^{N-1}z \, \exp\left(-\pi z^T B_N z\right), \tag{29.56}$$

in which B_N is an $(N-1)\times(N-1)$ matrix,

$$B_N = \begin{pmatrix} 2-\delta & -1 & 0 & 0 & \cdots & 0 \\ -1 & 2-\delta & -1 & 0 & \cdots & 0 \\ 0 & -1 & 2-\delta & -1 & \cdots & 0 \\ 0 & 0 & -1 & 2-\delta & \cdots & 0 \\ \vdots & \vdots & \vdots & \vdots & \ddots & -1 \\ 0 & 0 & 0 & 0 & -1 & 2-\delta \end{pmatrix}, \tag{29.57}$$

where for convenience we have defined

$$\delta = \epsilon^2\omega^2 = \omega^2 T^2/N^2. \tag{29.58}$$

Now, since B_N is a real symmetric matrix, we can do a rotation in the $(N-1)$-dimensional space to new coordinates $\boldsymbol{u} = O\boldsymbol{z}$, where O is an orthogonal matrix, in such a way that the rotated matrix $\tilde{B}_N = OB_NO^T$ is diagonal. Since a real symmetric matrix is just a special kind of Hermitian matrix, and an orthogonal matrix is just a special kind of unitary matrix, Theorems 2.16 and 2.18 tell us that the diagonal entries of $\tilde{B}_N$ are the same as the eigenvalues of B_N. Calling those eigenvalues λ_j,

$$I = \lim_{N\to\infty} \frac{1}{a} \prod_{j=1}^{N-1} \int_{-\infty}^{\infty} du_j \, e^{-\pi\lambda_j u_j^2}. \tag{29.59}$$

Now each of the u_j integrals can be done separately, using $\int_{-\infty}^{\infty} du e^{-\pi\lambda u^2} = \lambda^{-1/2}$, giving

$$I = \lim_{N\to\infty} \frac{1}{a} \prod_{j=1}^{N-1} (\lambda_j)^{-1/2} = \lim_{N\to\infty} \frac{1}{a} (\det B_N)^{-1/2} \tag{29.60}$$

where we have used the fact from matrix algebra that the determinant of B_N is equal to the product of its eigenvalues.

Let us now find the determinant of the $(N-1)\times(N-1)$ matrix B_N, temporarily taking δ to be general, even though in our case it depends on N through (29.58). We have

$$\det B_2 = 2 - \delta, \tag{29.61}$$

$$\det B_3 = (2-\delta)^2 - 1, \tag{29.62}$$

and for larger N we can evaluate $\det B_N$ in terms of its minors for the two nonzero entries in the first row. Conveniently, these involve determinants of versions of the same matrix with dimension smaller by 1 and 2. This gives the recurrence relation

$$\det B_N = (2-\delta)\det B_{N-1} - \det B_{N-2}, \qquad (N = 4, 5, 6, \ldots). \quad (29.63)$$

It is possible to solve this recurrence relation systematically, but we will take the shortcut of writing down the answer and checking that it works. To do so, define numbers $\beta_{1,2}$ by

$$\beta_1 + \beta_2 = 2 - \delta, \qquad \beta_1 \beta_2 = 1. \quad (29.64)$$

Then we claim that

$$\det B_N = \frac{\beta_1^N - \beta_2^N}{\beta_1 - \beta_2}. \quad (29.65)$$

This is easily checked to work for $N = 2$ and 3, and then for the recurrence relation (29.63), by direct substitution.

We are now ready to take the large N limit. First, solving (29.64), we find that

$$\beta_{1,2} = 1 - \frac{\delta}{2} \pm i\sqrt{\delta - \delta^2/4} = 1 \pm i\omega T/N + \ldots, \quad (29.66)$$

where the ellipses represent terms of higher order in $1/N$. Therefore, for large N,

$$\begin{aligned}\det B_N &= \frac{(1+i\omega T/N)^N - (1-i\omega T/N)^N}{2i\omega T/N} = \frac{e^{i\omega T} - e^{-i\omega T}}{2i\omega T/N} \\ &= \frac{N\sin(\omega T)}{\omega T},\end{aligned} \quad (29.67)$$

where the definition of the exponential function, $e^x = \lim_{N\to\infty}(1 + x/N)^N$, was used to get the second equality. Using this in (29.60) along with $a = (2\pi i\hbar T/mN)^{1/2}$ from (29.45), we find

$$I = \lim_{N\to\infty} \left(\frac{mN}{2\pi i\hbar T}\right)^{1/2} \left(\frac{\omega T}{N\sin(\omega T)}\right)^{1/2} = \left(\frac{m\omega}{2\pi i\hbar\sin(\omega T)}\right)^{1/2}, \quad (29.68)$$

which agrees with the result found in (29.54).

Greedy for more insight, let us now consider a second way to calculate I, using a different realization of the path integral. Instead of enumerating the possible paths by discretizing in time, we write the quantum fluctuation as a general Fourier series expansion of terms that satisfy the boundary conditions $y(0) = y(T) = 0$,

$$y(t) = \sum_{n=1}^{\infty} y_n \sin(n\pi t/T). \quad (29.69)$$

We then interpret the path integral in (29.53) as an integration over all of the Fourier coefficients y_n,

$$I = \frac{1}{c}\left(\prod_{n=1}^{\infty}\int_{-\infty}^{\infty}\frac{dy_n}{b_n}\right)\exp\left[\frac{im}{2\hbar}\int_0^T dt\,\left(\dot{y}^2 - \omega^2 y^2\right)\right], \tag{29.70}$$

which introduces normalization factors c and b_n with dimensions of length. Now the action integral over t in the exponent can be easily computed using $\int_0^T dt\,\sin^2(n\pi t/T) = T/2$, yielding

$$I = \frac{1}{c}\left(\prod_{n=1}^{\infty}\int_{-\infty}^{\infty}\frac{dy_n}{b_n}\right)\exp\left[-\sum_{n=1}^{\infty}\frac{imT}{4\hbar}\left(\omega^2 - n^2\pi^2/T^2\right)y_n^2\right]. \tag{29.71}$$

This nicely separates into one-dimensional integrals over the y_n, as

$$I = \frac{1}{c}\prod_{n=1}^{\infty}\left(\int_{-\infty}^{\infty}\frac{dy_n}{b_n}\exp\left[-\frac{imT}{4\hbar}\left(\omega^2 - n^2\pi^2/T^2\right)y_n^2\right]\right). \tag{29.72}$$

Doing the individual Gaussian integrals,

$$I = \frac{1}{c}\prod_{n=1}^{\infty}\frac{1}{b_n}\left(\frac{4\pi i\hbar}{mT(n^2\pi^2/T^2 - \omega^2)}\right)^{1/2}. \tag{29.73}$$

To make further progress, we make use of a famous formula due to Euler,

$$\prod_{n=1}^{\infty}\left(1 - \frac{x^2}{n^2\pi^2}\right) = \frac{\sin(x)}{x}, \tag{29.74}$$

with $x = \omega T$ in our present case, to obtain

$$I = \frac{1}{c}\left(\frac{\omega T}{\sin(\omega T)}\right)^{1/2}\prod_{n=1}^{\infty}\frac{1}{b_n}\left(\frac{4i\hbar T}{n^2\pi m}\right)^{1/2}. \tag{29.75}$$

Comparing with (29.54), we see that we have correctly obtained the ω dependence, and that the whole formula for I is correct provided that the normalization factors are

$$b_n = \left(\frac{4i\hbar T}{n^2\pi m}\right)^{1/2}, \qquad c = \left(\frac{2\pi i\hbar T}{m}\right)^{1/2}. \tag{29.76}$$

The advantage of this Fourier expansion method for the harmonic oscillator path integral is that the separation into individual integrals was automatic.

The free-particle propagator can be recovered from a careful evaluation of the $\omega \to 0$ limit. In the preceding derivations, we were able to make use of a very special property shared by the free particle and the harmonic oscillator, namely that the action is quadratic in the configuration variable(s) and its time derivatives. This led to Gaussian integrations in either (29.59) or (29.72), which we were therefore able to perform analytically. Another simple application is given in the next section. For other problems in nonrelativistic quantum mechanics, the path integral method is only rarely efficient compared to methods based on the Schrödinger equation. However, the path integral approach is still extremely valuable, for in addition to the beautiful conceptual understanding it brings, it turns out to provide a practically useful way of systematically organizing certain approximation methods. In quantum field theories, this includes both Feynman rules for perturbative expansions about quadratic Lagrangians, and stochastic methods for non-perturbative calculations including lattice methods—topics beyond our scope here.

29.4 The Ehrenberg–Siday–Aharonov–Bohm Effect

The classical trajectory of a charged particle in an electromagnetic field is determined uniquely by its initial conditions and the electromagnetic fields $\boldsymbol{E}$ and $\boldsymbol{B}$ that it encounters locally along its path. In quantum mechanics, things are different. The Hamiltonian in the Schrödinger equation, and the Lagrangian in the path integral approach, depend on the gauge-dependent potentials Φ and $\boldsymbol{A}$. These potentials encode global information that is gauge-invariant but not accessible to an observer on a classical trajectory making only local measurements of $\boldsymbol{E}$ and $\boldsymbol{B}$. As a result, there can be observable consequences of electromagnetic fields even if the particle is absolutely forbidden to enter the region where $\boldsymbol{E}$ and $\boldsymbol{B}$ are nonzero!

The simplest example of this purely quantum phenomenon is most commonly known as the Aharonov–Bohm effect, after a 1959 paper by Yakir Aharonov and David Bohm, although it was actually proposed and analyzed a decade earlier by Werner Ehrenberg and Raymond Siday. Consider a spinless charged particle with coordinates $\boldsymbol{r}(t)$, subject to a static but non-uniform magnetic field $\boldsymbol{B} = \nabla \times \boldsymbol{A}$. Specializing (4.31) to $\Phi = 0$, the Lagrangian is

$$L = \frac{1}{2}m\left(\frac{d\boldsymbol{r}}{dt}\right)^2 + \frac{q}{c}\frac{d\boldsymbol{r}}{dt} \cdot \boldsymbol{A}(\boldsymbol{r}). \tag{29.77}$$

The path integral expression for the propagator,

$$U(\boldsymbol{r}_f, t_f; \boldsymbol{r}_i, t_i) = \int d[\boldsymbol{r}(t)] \exp\left(\frac{i}{\hbar} S_{\text{path}}[\boldsymbol{r}(t)]\right), \tag{29.78}$$

therefore naturally divides into two parts, according to

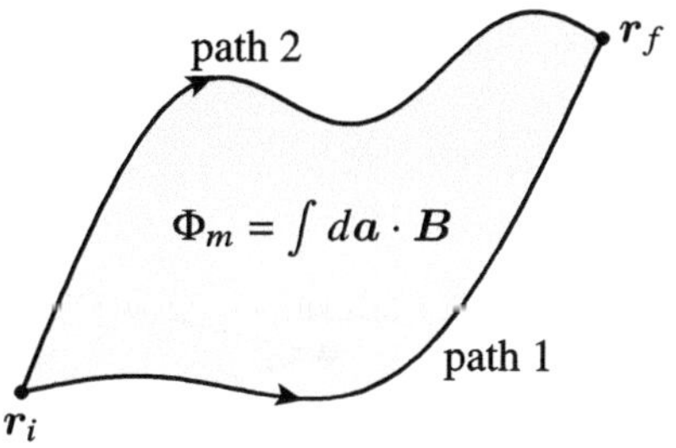

Fig. 29.3 The dependence of the difference in actions for two alternative paths for a charged spinless particle on a static magnetic field is proportional to the magnetic flux Φ_m through an area bounded by the two paths

$$S_{\text{path}} = S_{\text{path}}^{A=0} + \Delta S_{\text{path}}, \tag{29.79}$$

where the first term is the free-particle action, and the remaining part can be rewritten as a line integral along the path specified from $\boldsymbol{r}_i$ to $\boldsymbol{r}_f$ by $\boldsymbol{r}(t)$,

$$\Delta S_{\text{path}} = \frac{q}{c}\int_{t_i}^{t_f} dt\,\frac{d\boldsymbol{r}}{dt}\cdot \boldsymbol{A}(\boldsymbol{r}) = \frac{q}{c}\int_{\substack{\boldsymbol{r}_i\\ \text{path}}}^{\boldsymbol{r}_f} d\boldsymbol{r}\cdot \boldsymbol{A}(\boldsymbol{r}). \tag{29.80}$$

Thus each path contributes with a phase factor that depends on the line integral of $\boldsymbol{A}$ along the path.

When evaluating the probability of detecting the particle at $\boldsymbol{r}_f$ at time t_f, we need the relative phases between different paths. Consider any two paths contributing to the propagator, call them "path 1" and "path 2". (See Fig. 29.3.) Then the dependence on $\boldsymbol{A}$ of the relative phase between their contributions to the propagator is

$$\Delta\varphi \equiv (\Delta S_{\text{path 1}} - \Delta S_{\text{path 2}})/\hbar = \frac{q}{\hbar c}\left(\int_{\substack{\boldsymbol{r}_i\\ \text{path 1}}}^{\boldsymbol{r}_f} d\boldsymbol{r}\cdot \boldsymbol{A}(\boldsymbol{r}) - \int_{\substack{\boldsymbol{r}_i\\ \text{path 2}}}^{\boldsymbol{r}_f} d\boldsymbol{r}\cdot \boldsymbol{A}(\boldsymbol{r})\right). \tag{29.81}$$

Using the minus sign to reverse the direction of the line integral over path 2, we can combine the two paths into a single closed path P, so

$$\Delta\varphi = \frac{q}{\hbar c}\oint_P d\boldsymbol{r}\cdot \boldsymbol{A}. \tag{29.82}$$

Then, applying Stokes' Theorem gives the key result

$$\Delta\varphi = \frac{q}{\hbar c}\int_{\text{area}} d\boldsymbol{a}\cdot(\nabla\times \boldsymbol{A}) = \frac{q}{\hbar c}\int_{\text{area}} d\boldsymbol{a}\cdot \boldsymbol{B} = \frac{q\Phi_m}{\hbar c}. \tag{29.83}$$

Here Φ_m is the magnetic field flux through a surface area bounded by the two paths, with the area element direction $d\boldsymbol{a}$ related to the path P direction by the right-hand

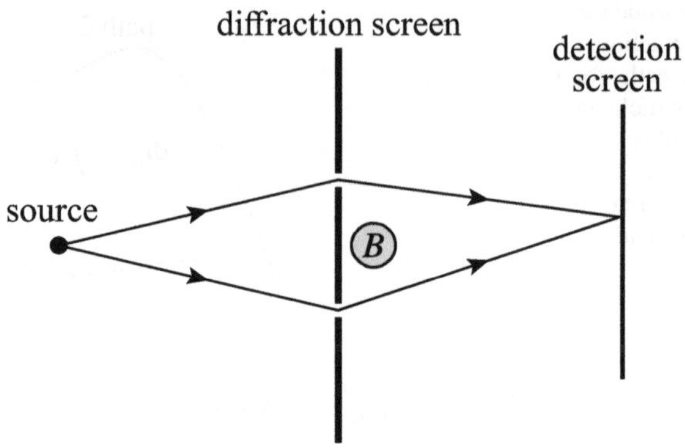

Fig. 29.4 The Ehrenberg–Siday–Aharonov–Bohm effect for an interference experiment for electrons. The shaded circle represents the cross-section of a solenoid, carrying a magnetic field pointing out of the plane. The presence of the magnetic field in the solenoid affects the interference pattern visible on the detection screen, even though the electrons are absolutely forbidden by shielding to enter the region with nonzero $\boldsymbol{B}$

rule. As usual when applying Stokes' Theorem, the surface is not unique; any such area bounding the two paths must[2] give the same result.

A two-slit interference version of the **Ehrenberg–Siday–Aharonov–Bohm effect** is depicted in Fig. 29.4. A source of electrons is separated from a flat detection screen by a diffraction screen with two small holes. Behind the diffraction screen is a solenoid (or perhaps a permanent magnet), which carries in its interior a magnetic field $\boldsymbol{B}$ pointing out of the plane of the page. The solenoid is tightly wound and long enough so that the magnetic field $\boldsymbol{B}$ outside of it can be neglected, and is shielded so that electrons cannot enter it. Therefore, classically the presence of the magnetic field inside the solenoid can have no effect on the electrons.

In quantum-mechanical reality, it is important that the vector potential outside of the solenoid is necessarily nonzero. For example, in terms of cylindrical coordinates (r, ϕ, z) coaxial with a very long solenoid of radius R, the magnetic field has the form

$$\boldsymbol{B} = \begin{cases} B\hat{z} & (\text{for } r < R), \\ 0 & (\text{for } r > R), \end{cases} \tag{29.84}$$

and the associated vector potential can be chosen to be

$$\boldsymbol{A} = \begin{cases} (Br/2)\hat{\phi} & (\text{for } r \leq R), \\ (BR^2/2r)\hat{\phi} & (\text{for } r \geq R). \end{cases} \tag{29.85}$$

[2] This relies on the standard assumption that magnetic monopoles do not exist, so that the flux through any closed surface is 0. For the possibility that magnetic monopoles do exist, see Sect. 29.5.

The resulting magnetic flux for any closed path encircling the solenoid once is $\Phi_m = \pi R^2 B$. Accordingly, the phase difference, due to the existence of the magnetic field, for electrons going on any path below the solenoid compared to those going on any path above the solenoid is, from (29.83) with $q = -e$,

$$\Delta\varphi = 2\pi\,\Phi_m/\Phi_0, \tag{29.86}$$

where we have defined the relevant quantum unit of magnetic flux,[3]

$$\Phi_0 \equiv \frac{2\pi\hbar c}{e} = 4.13567 \times 10^{-7}\,\text{gauss}\cdot\text{cm}^2 = 4.13567 \times 10^{-15}\,\text{Tesla}\cdot\text{m}^2. \tag{29.87}$$

To observe the effect, start with the solenoid turned off, and consider a reference point at the center of the detection screen where the interference is maximally constructive, with paths from both holes contributing equally to the peak intensity I_0. Now when the magnetic field is turned on, the intensity at that point will be changed to

$$\left|\frac{1}{2}e^{i\Delta\varphi} + \frac{1}{2}\right|^2 I_0 = \frac{1}{4}\left|e^{-2\pi i\Phi_m/\Phi_0} + 1\right|^2 I_0, \tag{29.88}$$

where the two terms correspond to the contributions from all paths 1 through the bottom hole and all paths 2 through the top hole, respectively. So, if the magnetic field strength B is adjusted so that $\Phi_m = (n + 1/2)\Phi_0$ where n is any integer, the interference will become completely destructive, and the intensity will vanish at the reference point. If $\Phi_m = n\Phi_0$, the interference will be maximally constructive again. This effect has been confirmed in many increasingly sophisticated experiments.

One can interpret the effect as illustrating that the vector potential $\boldsymbol{A}$ contains genuine physical information that is not apparent through local interaction of the charge with the field $\boldsymbol{B}$. This is the meaning of $\Delta\varphi$ in (29.82). However, the effect can also be described entirely in terms of the nontrivial magnetic flux, Φ_m, in the last expression in (29.83), which depends only on the gauge-invariant magnetic field $\boldsymbol{B}$. This dependence of interference on the flux is a topological one, in the sense that it occurs because the allowed paths of the charged particle cannot be continuously deformed into each other without crossing into the region with nonzero $\boldsymbol{B}$. In the path integral approach, we understand this in terms of the particle's ability (in fact, obligation) to coherently sample phases from all available paths, even though none of those paths actually encounters the nonzero $\boldsymbol{B}$ field.

[3] This same quantity also appeared in (9.88), in the discussion of Landau levels. In superconductivity applications, one often encounters a magnetic flux quantum that is half as big, corresponding to a Cooper pair charge of magnitude $2e$.

29.5 Dirac Quantization Condition for Magnetic Monopoles

The Ehrenberg–Siday–Aharonov–Bohm effect is an experimentally well-established phenomenon. The derivation of it can also be applied to learn something nontrivial about a more speculative hypothesis, the possibility of magnetic monopoles. Dirac showed that the extension of Maxwell's electrodynamics to include magnetic monopoles is only consistent with quantum mechanics if both magnetic and electric charges are quantized. There are several different ways to derive Dirac's result. In this section we will follow a non-historical path, first doing it using (29.82), and then again using the closely related version of (29.83).

The standard assumption that magnetic monopoles do not exist is one of the four Maxwell equations, $\nabla \cdot \boldsymbol{B} = 0$. To include magnetic monopoles in the theory, one generalizes this to

$$\nabla \cdot \boldsymbol{B} = 4\pi \rho_m, \tag{29.89}$$

where ρ_m is the density of magnetic charge. A point monopole with magnetic charge q_m, fixed at the origin, would then give rise to a magnetic field

$$\boldsymbol{B} = q_m \hat{r}/r^2 \tag{29.90}$$

expressed in spherical coordinates and cgs units. Equations (29.89) and (29.90) are the hypothetical analogs of, respectively, Gauss' law $\nabla \cdot \boldsymbol{E} = 4\pi\rho$ and $\boldsymbol{E} = q\hat{r}/r^2$, for ordinary electric charge density ρ and a point electric charge q. This similarity between electric and magnetic charges is one reason to take magnetic monopoles seriously as a hypothesis. Another is that they can be shown to be necessarily present in some extensions of the Standard Model of particle physics that may be motivated for other reasons, notably Grand Unified Theories. Although magnetic monopoles have not been observed despite extensive searches, it is possible that this is because they are extremely rare in the present universe. For example, they could be too heavy to produce with collision energies available today, and the cosmological inflation that is believed (for other reasons) to have occurred in the early universe would have diluted the density of them left over from the Big Bang to the extent that it is extremely unlikely for us to encounter one.

Suppose that there is a magnetic monopole q_m at rest at the origin, with $\boldsymbol{B}$ as given in (29.90). Consider the quantum mechanics of an ordinary particle with charge q, which is assumed to be restricted to a circle of radius R in the xy plane with its center at the origin, as shown in the left panel of Fig. 29.5. The propagator for the particle to start at some point $\boldsymbol{r}_i$ at time t_i and end up at $\boldsymbol{r}_f$ at time t_f can be written as a sum over contributions from paths that wind around the circle in different ways. Thus we can write

$$U(\boldsymbol{r}_f, t_f; \boldsymbol{r}_i, t_i) = \int_{\text{paths 1}} d[\boldsymbol{r}(t)] \exp\left(\frac{i}{\hbar} S[\boldsymbol{r}(t)]\right)$$

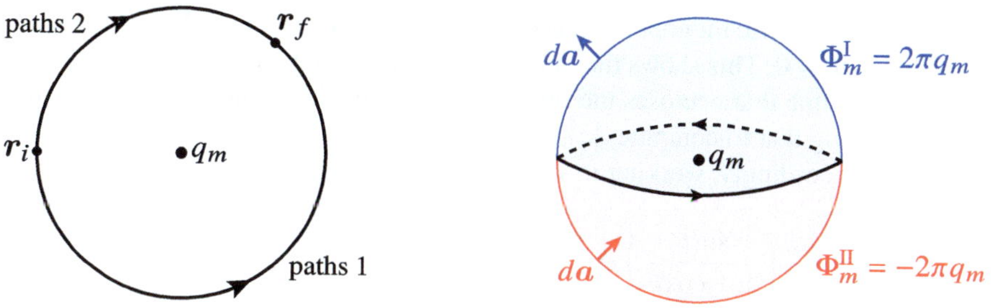

Fig. 29.5 A monopole of magnetic charge q_m is fixed at the origin, and a particle with electric charge q is restricted to a circle of fixed radius around it. As shown in the left panel, there are contributions to the propagator path integral from trajectories with net counterclockwise direction (paths 1) and net clockwise direction (paths 2). The presence of the monopole affects the relative phase between these classes of paths through $\Delta\varphi$, which can be obtained from (29.82) as a line integral of the vector potential around the closed circle. Alternatively, it can be obtained using (29.83) from the magnetic flux through either the upper hemisphere or the lower hemisphere, as depicted in the right panel

$$+ \int_{\text{paths 2}} d[\boldsymbol{r}(t)] \exp\left(\frac{i}{\hbar} S[\boldsymbol{r}(t)]\right). \tag{29.91}$$

The "paths 1" contribution is from paths that have a net counterclockwise direction in the figure, while "paths 2" have a net clockwise direction; of course, these paths $\boldsymbol{r}(t)$ need not be monotonic in direction around the circle. (We neglect contributions from paths that wind around the circle more than once. This simplification is not essential to the following argument.) The propagator can now be written in the form

$$U(\boldsymbol{r}_f, t_f; \boldsymbol{r}_i, t_i) = e^{i\alpha}\left(e^{i\Delta\varphi} c_1 + c_2\right), \tag{29.92}$$

where c_1 and c_2 are the contributions from paths 1 and paths 2 that would be present if the magnetic monopole were absent, $e^{i\alpha}$ is a common phase factor due to the magnetic monopole, and $e^{i\Delta\varphi}$ is the relative phase between the two classes of paths due to the magnetic monopole. If the particle was known to be at $\boldsymbol{r}_i$ at time t_i, then the probability that it will be found at $\boldsymbol{r}_f$ at time t_f is

$$\mathcal{P} = \left|e^{i\Delta\varphi} c_1 + c_2\right|^2. \tag{29.93}$$

Since the initial and final points are arbitrary, and other potentials felt by the charge q are not specified, c_1 and c_2 can be considered arbitrary complex numbers. Now $\Delta\varphi$ can be evaluated in two distinct ways, either of which leads to the Dirac quantization condition.

First, one can evaluate $\Delta\varphi$ using (29.82), the line integral of $\boldsymbol{A}$ around the full circle in the xy plane at radius R and in the $+\hat{\phi}$ direction. However, there is a problem associated with the vector potential, because the equation $\nabla \times \boldsymbol{A} = \boldsymbol{B}$

assumes no magnetic monopoles, since the vector calculus identity $\nabla \cdot (\nabla \times \boldsymbol{A}) = 0$ implies $\nabla \cdot \boldsymbol{B} = 0$. This shows that it is impossible to write down a single consistent $\boldsymbol{A}$ in any volume that encloses the origin. Fortunately, this difficulty can be evaded by recognizing that to compute $\Delta\varphi$, we only need $\boldsymbol{A}$ on the path arcs at $r = R$ with $\theta = \pi/2$. Accordingly, working in spherical coordinates, we can take either

$$\boldsymbol{A}_{\mathrm{I}} = \frac{q_m}{r}\left(\frac{\sin\theta}{1+\cos\theta}\right)\hat{\phi}, \qquad \text{or} \qquad \boldsymbol{A}_{\mathrm{II}} = -\frac{q_m}{r}\left(\frac{\sin\theta}{1-\cos\theta}\right)\hat{\phi}. \tag{29.94}$$

These both give $\nabla \times \boldsymbol{A} = \boldsymbol{B} = q_m\hat{r}/r^2$, except at the origin, and on the negative z axis for $\boldsymbol{A}_{\mathrm{I}}$ and on the positive z axis for $\boldsymbol{A}_{\mathrm{II}}$, where the respective denominators vanish. Since we only need $r = R$ and $\theta = \pi/2$ to evaluate the line integral, $\boldsymbol{A}_{\mathrm{I}}$ and $\boldsymbol{A}_{\mathrm{II}}$ are both equally valid for our purposes. However, they give two different results for $\Delta\varphi$, with opposite signs:

$$\Delta\varphi_{\mathrm{I}} = -\Delta\varphi_{\mathrm{II}} = \frac{q}{\hbar c}\int_0^{2\pi} R d\phi\, \frac{q_m}{R} = \frac{2\pi q q_m}{\hbar c}. \tag{29.95}$$

To avoid an unacceptable inconsistency in the physical prediction for the probability $|e^{i\Delta\varphi}c_1 + c_2|^2$ following from the two equally valid choices for $\boldsymbol{A}$, since c_1 and c_2 are arbitrary complex numbers, we must have $e^{i\Delta\varphi_{\mathrm{I}}} = e^{i\Delta\varphi_{\mathrm{II}}}$. This requirement is equivalent to $4\pi q q_m/\hbar c = 2\pi n$ for some integer n, or simply

$$\frac{2qq_m}{\hbar c} = n, \qquad (n = \text{integer}). \tag{29.96}$$

This is the **Dirac quantization condition for magnetic monopole charges.**

The second (alternative) way to evaluate $\Delta\varphi$ avoids any mention of the vector potential by using the second equality of (29.83). We have

$$\Delta\varphi = \frac{q}{\hbar c}\int_{\text{area}} d\boldsymbol{a} \cdot \boldsymbol{B} = \frac{qq_m}{\hbar c}\int_{\text{area}} d\boldsymbol{a} \cdot \frac{\hat{r}}{r^2} = \frac{q\Phi_m}{\hbar c}. \tag{29.97}$$

Here we must confront the issue that there are two equally good choices of area bounded by the circle in the xy plane, one above the origin and one below. For simplicity of computation, it is convenient to take these to be the upper and lower hemispheres of radius R, labeled I and II in the second panel of Fig. 29.5. The directions of their area elements point outward and inward, respectively, as dictated by the orientations of their circular paths and the right-hand rule in Stokes' Theorem. Therefore, the magnetic fluxes through them have opposite signs, and we have the two equally valid choices

$$\Phi_m^{\mathrm{I}} = -\Phi_m^{\mathrm{II}} = (2\pi R^2)(q_m/R^2) = 2\pi q_m. \tag{29.98}$$

Using these two choices in (29.97) immediately gives exactly the same ambiguity as in (29.95), with the same resolution: the Dirac quantization condition (29.96) must hold.

Since we can pick q to be the proton charge e or the electron charge $-e$, we learn from Dirac's result (29.96) that magnetic charges q_m must be quantized in integer multiples of a fundamental unit

$$q_m^{\text{Dirac}} = \frac{\hbar c}{2e} = \frac{e}{2\alpha}, \tag{29.99}$$

where $\alpha = e^2/\hbar c \approx 1/137$ is the dimensionless fine-structure constant. (Note that in the cgs system, magnetic and electric charges have the same units.) Conversely, if even one magnetic monopole with charge $q_m^{\text{Dirac}} = \hbar c/2e$ exists anywhere in the universe, then all electric charges must be integer multiples of e. Thus magnetic monopoles, if they exist, would give a compelling reason for the observed quantization of electric charge.

There is a subtlety here, because quarks and antiquarks exist with charges that are integer multiples of $e/3$. Therefore, one might naively think that the fundamental quantum of magnetic charge should actually be 3 times q_m^{Dirac}. However, quarks are confined inside hadrons by quantum chromodynamics (QCD), also known as the strong nuclear force. The QCD gauge potentials also contribute to the path integral, and it turns out (although showing it in detail is well beyond our scope here) that one can still have magnetic monopoles with charge q_m^{Dirac}, provided that they also carry magnetic-QCD charges, and so are confined. Thus, Dirac's q_m^{Dirac} is the true quantum of magnetic charge, even though he did not know about quarks or QCD when he derived it.

29.6 Classical Limit from the Sum Over Paths

In this section, we will show how the equations of motion for classical physics follow from an appropriate limit of quantum mechanics in the path integral formulation.

Consider a path $q(t)$ subject to the boundary conditions $q(t_i) = q_i$ and $q(t_f) = q_f$, and suppose we make a small change $\delta q(t)$ to it,

$$q(t) \to q(t) + \delta q(t), \tag{29.100}$$

as illustrated in Fig. 29.6. This results in a change in the action, which we write as

$$S[q(t)] \to S[q(t)] + \delta S. \tag{29.101}$$

Because we must maintain the boundary conditions at $t = t_i$ and $t = t_f$ when computing the propagator, we must require that

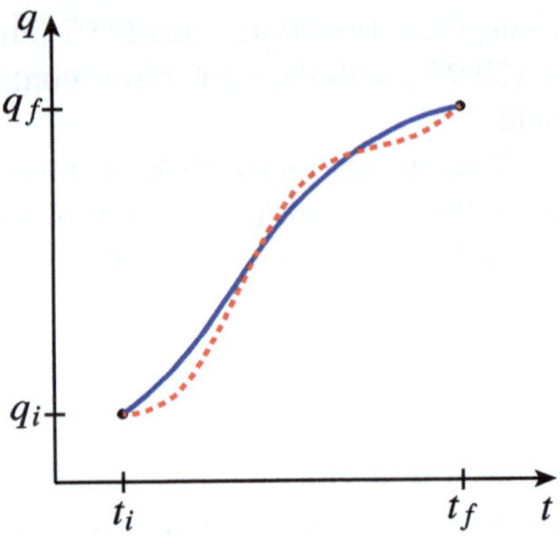

Fig. 29.6 A path $q(t)$ (solid line), and a nearby path $q(t) + \delta q(t)$ (dashed line), subject to the boundary conditions $q(t_i) = q_i$ and $q(t_f) = q_f$ and $\delta q(t_i) = \delta q(t_f) = 0$. The path is a stationary-action path if the change in the action δS vanishes for every infinitesimal change in path $\delta q(t)$

$$\delta q(t_i) = \delta q(t_f) = 0. \tag{29.102}$$

In general, if δS is nonzero, then the integrand of the path integral will acquire an additional phase factor $e^{i\delta S/\hbar}$ compared to the result from the original path. If we take the classical limit $\hbar \to 0$, this additional phase factor varies rapidly as δS changes, leading to very efficient cancellation between the contributions for nearby paths.

Therefore, in the classical limit, the propagator is dominated by **stationary-action paths**, defined to be those for which $\delta S = 0$ for every $\delta q(t)$ that satisfies the boundary conditions (29.102). The point is that those are the only paths that do not suffer efficient cancellations from the neighboring paths, with cancellation becoming perfectly efficient in the $\hbar \to 0$ limit. In quantum mechanics, the path integral samples all paths, but in the classical limit $\hbar \to 0$, only stationary-action paths contribute non-negligibly.

Suppose we are given a Lagrangian $L(q, \dot{q}, t)$ for a system, so that

$$S[q(t)] = \int_{t_i}^{t_f} dt\, L(q, \dot{q}, t). \tag{29.103}$$

The change in the action due to the change in path $\delta q(t)$ can be written as

$$\delta S = \int_{t_i}^{t_f} dt \left[\delta q(t) \frac{\partial L}{\partial q} + \delta \dot{q}(t) \frac{\partial L}{\partial \dot{q}} \right], \tag{29.104}$$

or, using $\delta \dot{q}(t) = \frac{d}{dt} \delta q(t)$,

$$\delta S = \int_{t_i}^{t_f} dt \left[\delta q(t) \frac{\partial L}{\partial q} + \frac{d}{dt} \delta q(t) \frac{\partial L}{\partial \dot{q}} \right]. \tag{29.105}$$

Integrating the second term by parts, we have

$$\delta S = \int_{t_i}^{t_f} dt\, \delta q(t) \left[\frac{\partial L}{\partial q} - \frac{d}{dt}\left(\frac{\partial L}{\partial \dot{q}}\right)\right] + \left(\delta q(t)\, \frac{\partial L}{\partial \dot{q}}\right)\Bigg|_{t_i}^{t_f}. \qquad (29.106)$$

The last term is 0 because of the boundary conditions (29.102). In order to have $\delta S = 0$ for all small variations $\delta q(t)$, we conclude that the integrand in (29.106) must vanish,

$$\frac{\partial L}{\partial q} - \frac{d}{dt}\left(\frac{\partial L}{\partial \dot{q}}\right) = 0. \qquad (29.107)$$

This is the Euler-Lagrange equation of motion familiar from classical mechanics. More generally, if there are multiple variables q_a for $a = 1, \ldots, n$, with a Lagrangian $L(q_a, \dot{q}_a, t)$, then by the same argument the classical limit implies

$$\frac{\partial L}{\partial q_a} - \frac{d}{dt}\left(\frac{\partial L}{\partial \dot{q}_a}\right) = 0 \qquad (29.108)$$

for each a. Thus, we have derived the classical equations of motion from the $\hbar \to 0$ limit of Feynman's path integral formulation of quantum mechanics.

In the case of a single spinless particle moving in three dimensions, we could of course take our CSCO eigenvalues to be $(q_1, q_2, q_3) \to \boldsymbol{r}$. However, in more general situations, we often need to describe processes that change the number of particles. Such processes include decays of one particle into two or more particles, like the decay of a Higgs boson into a quark–antiquark pair, or two gluons, or two photons. They also include the joining of two particles to make other particles, for example the annihilation of an electron–positron pair to make a Z boson, which then turns into a quark–antiquark pair. In such cases, the way we have been treating quantum mechanics is hopelessly clumsy, because even the number of coordinates needed to describe the positions or momenta of the particles changes depending on the state.

Furthermore, special relativity teaches us that time and space are on the same footing, while in most of this book we have been treating spatial coordinates as operators and using the time coordinate as a parameter, not an operator. This is particularly embarrassing since the symmetries of special relativity mix space and time coordinates.

The simultaneous resolution of these issues is that, instead of associating the position coordinates $\boldsymbol{r}$ with operators, they are taken to be labels, just like the time t. In other words, there are different operators, called **quantum fields**,

$$\Phi_n(\boldsymbol{r}, t) \qquad (29.109)$$

for each set of labels $(\boldsymbol{r}, t)$. The indices n are also labels, associated with the type of particle, the spin components, and other quantum numbers. The field operators can act on a vacuum state $|0\rangle$ (which is typically the lowest-energy state and describes empty space) to give other states that have any number of particles and antiparticles.

In the path integral formulation, the classical action functional looks schematically like

$$S[\phi_n(\boldsymbol{r},t)] = \int_{t_i}^{t_f} dt \int d^3\boldsymbol{r}\ \mathcal{L}(\phi_n(\boldsymbol{r},t), \partial_\mu \phi_n(\boldsymbol{r},t)), \tag{29.110}$$

where $\mathcal{L}$ is the classical Lagrangian density, $\phi_n(\boldsymbol{r},t)$ is the classical field corresponding to the quantum field $\Phi_n(\boldsymbol{r},t)$, and ∂_μ unifies the time and space derivatives as in (28.28).

The resulting type of theory, which also obeys the other postulates of Sect. 3.1, is called a **quantum field theory**. Relativistic quantum field theories have the important property that energies are bounded from below, which is not the case in relativistic one-particle quantum mechanics, as we saw in Chap. 28. In quantum field theory, the Schrödinger equation formulation is valid, but often less useful than the equivalent path integral formulation, which provides elegant methods for both conceptual understanding and practical calculations, using both perturbative and non-perturbative methods. But that's another story…

Exercises

29.1. Consider a particle of mass m inside a one-dimensional box $0 < x < L$.

(a) Use the spectral decomposition form (29.9) of the propagator to show that

$$U_{\text{box}}(x,t;x_0,0) = \frac{2}{L}\sum_{n=1}^{\infty} \sin(n\pi x/L)\sin(n\pi x_0/L)e^{-i\hbar\pi^2 n^2 t/2mL^2}. \tag{29.111}$$

(b) The Poisson resummation formula states that for any function $f(u)$,

$$\sum_{n=-\infty}^{\infty} f(n) = \sum_{\nu=-\infty}^{\infty} \int_{-\infty}^{\infty} du\ f(u)e^{2\pi i\nu u}. \tag{29.112}$$

Use it to show that the propagator for the particle in a box can be rewritten as

$$U_{\text{box}}(x,t;x_0,0) = \sum_{\nu=-\infty}^{\infty} \left[U(x,t;\ x_0+2\nu L,0) - U(x,t;\ -x_0+2\nu L,0)\right], \tag{29.113}$$

where U is the free-particle propagator of (29.10). Physically, the sum includes amplitudes for all possible paths in which the otherwise-free particle "bounces" off the walls. The terms with minus signs are necessary so that $U_{\text{box}}(0,t;x_0,0) = U_{\text{box}}(L,t;x_0,0) = 0$.

29.2. In this problem, we will generalize the derivation of the equivalence between the path integral and the Schrödinger equation, as given in (29.36)–(29.46), to the case of a spinless particle of charge q in the presence of electromagnetic potentials $\boldsymbol{A}$ and Φ. We start with the Lagrangian $L(\boldsymbol{r}, \dot{\boldsymbol{r}}, t) = \frac{1}{2}m\dot{\boldsymbol{r}}^2 + \frac{q}{c}\dot{\boldsymbol{r}}\cdot\boldsymbol{A}(\boldsymbol{r}, t) - q\Phi(\boldsymbol{r}, t)$. Consider the infinitesimal propagation from $(\boldsymbol{r}_0, t)$ to $(\boldsymbol{r}, t+\epsilon)$,

$$\psi(\boldsymbol{r}, t+\epsilon) = \int \frac{d^3\boldsymbol{r}_0}{a^3}\, e^{iS(\boldsymbol{r},\boldsymbol{r}_0)/\hbar}\, \psi(\boldsymbol{r}_0, t), \tag{29.114}$$

with the discretized action

$$S(\boldsymbol{r}, \boldsymbol{r}_0) = \epsilon\left(\frac{m}{2}\frac{|\boldsymbol{r}-\boldsymbol{r}_0|^2}{\epsilon^2} + \frac{q}{c}\frac{(\boldsymbol{r}-\boldsymbol{r}_0)}{\epsilon}\cdot\boldsymbol{A}(\boldsymbol{r}_m, t_m) - q\Phi(\boldsymbol{r}_m, t_m)\right), \tag{29.115}$$

where we used $\dot{\boldsymbol{r}} = (\boldsymbol{r}-\boldsymbol{r}_0)/\epsilon$ and the midpoint prescription $\boldsymbol{r}_m \equiv (\boldsymbol{r}_0+\boldsymbol{r})/2$ and $t_m \equiv t+\epsilon/2$. As in the text, change the integration variable to $\boldsymbol{u} = \boldsymbol{r}_0 - \boldsymbol{r}$, and expand appropriately in small ϵ and $\boldsymbol{u}$, to obtain the Schrödinger equation

$$i\hbar\frac{\partial}{\partial t}\psi(\boldsymbol{r}, t) = \left[\frac{1}{2m}\left(-i\hbar\nabla - \frac{q}{c}\boldsymbol{A}(\boldsymbol{r}, t)\right)^2 + q\Phi(\boldsymbol{r}, t)\right]\psi(\boldsymbol{r}, t). \tag{29.116}$$

You should find that the midpoint prescription for $\boldsymbol{r}_m$ is important for getting the correct ordering of terms involving $\boldsymbol{A}\cdot\nabla$ and $\nabla\cdot\boldsymbol{A}$, but is otherwise unimportant in the sense that one could have just used t instead of t_m everywhere, and $\boldsymbol{r}$ instead of $\boldsymbol{r}_m$ in Φ, when expanding the final result to first order in ϵ.

29.3. Consider the propagator for the harmonic oscillator, in (29.13) and (29.14), with $T = t - t_0$. Compare it to the general spectral decomposition form, as given in (29.9), to deduce the known result $E_n = \hbar\omega(n+1/2)$ for $n = 0, 1, 2, \ldots$, and check the known wavefunctions for $n = 0$, 1, and 2. [Hint: write $\cos(\omega T) = (1/z+z)/2$ and $\sin(\omega T) = (1/z-z)/2i$, where $z = e^{-i\omega T}$. Then do a formal expansion treating z as if it were small, even though it of course has $|z| = 1$.]

Index

S. P. Martin, *Quantum Mechanics*, Graduate Texts in Physics,
https://doi.org/10.1007/978-3-032-21119-4

T

MIX
Papier aus verantwortungsvollen Quellen
Paper from responsible sources
FSC® C105338

If you have any concerns about our products,
you can contact us on
ProductSafety@springernature.com

In case Publisher is established outside the EU,
the EU authorized representative is:
Springer Nature Customer Service Center GmbH
Europaplatz 3, 69115 Heidelberg, Germany

Printed by Libri Plureos GmbH
in Hamburg, Germany